Peter Lees
Department of Veterinary Basic Sciences
The Royal Veterinary College (University of London)
Hawkshead Campus
Hawkshead Lane
North Mymms
Hatfield, Herts AL9 7TA
United Kingdom

Antibiotic Policies: Theory and Practice

Antibiotic Policies: Theory and Practice

Edited by

Ian M. Gould

Aberdeen Royal Infirmary
Foresterhill, UK

and

Jos W. M. van der Meer

University Hospital Nijmegen
Nijmegen, The Netherlands

Kluwer Academic / Plenum Publishers

New York, Boston, Dordrecht, London, Moscow

Library of Congress Cataloging-in-Publication Data

Antibiotic policies: theory and practice / edited by Ian M. Gould and Jos W.M. van der Meer.
   p. ; cm.
   Includes bibliographical references and index.
   ISBN 0-306-48500-1
     1. Antibiotics—Government policy. 2. Drugs—Prescribing—Government policy.
     3. Drug resistance in microorganisms—Prevention—Government policy.
     [DNLM: 1. Anti-Bacterial Agents—administration & dosage. 2. Drug Resistance,
     Bacterial. 3. Drug and Narcotic Control. QV 350 A6264 2004] I. Gould, Ian M.
     II. Meer, J. W. M. van der.

RM267.A495 2004
362.17′82—dc22

                                                         2004042124

ISBN 0-306-48500-1

© 2005 Kluwer Academic / Plenum Publishers, New York
233 Spring Street, New York, New York 10013

http//www.wkap.nl/
10 9 8 7 6 5 4 3 2 1

A C.I.P. record for this book is available from the Library of Congress.

Permissions for books published in Europe: *permissions@wkap.nl*
Permissions for books published in the United States of America: *permissions@wkap.com*

Printed in the United States of America

Foreword

Bacterial resistance to antibiotics and the critical need to curtail this global public health problem have reached wide awareness among healthcare providers, hospital administrators, public health officials, and even the public. The professional and lay literature continues to report the consequence of antibiotic misuse and the rise in strains resistant to antibiotics. When trying to understand drug resistance and develop ways to control the mounting public health problem, there are many factors to consider, which have both health and ecologic considerations. This book addresses that challenge.

The major thrust of the authors is to define the problem and offer global and multidisciplinary approaches to resolve it as drug resistance confronts individuals, hospitals and communities. What do we need to know to help clinicians and public health officials correct the situation? What theories should be entertained and what policies and practice can be engaged to produce a change? These are some of the questions asked and discussed from a variety of viewpoints. Using theory, an understanding of practice, and the factors involved in creating and perpetuating the problem, the authors suggest policies and guidelines which target reversing drug resistance.

Antibiotic use, both in terms of quantity and length of application, is the major contributor to the selection and propagation of resistant strains. While some antibiotic resistance may emerge with each treatment, the steady rise in resistance frequency in an environment generally implies prolonged use and/or use at a concentration which is too low to effect cure, but high enough to encourage growth of resistant strains. Besides antibiotic use, other factors, including epidemiology, gene transfer, and even other drugs, influence the resistance phenomenon. Understanding and quantifying the relationship between drug use and resistance represent an on-going quest and are critical, not only for finding ways to reduce resistance, but also for improving quality of care in the treatment of infectious diseases.

The composition and breadth of information provided make this volume different from other books which deal more with the phenomenon of this infectious disease issue and less with its resolution. In a single volume, one learns about the extent of the problem in different geographic settings, the factors involved, and how this information can help develop guidelines and policies which, when put into practice, can lead to change.

One prevalent theme is the need to address drug resistance locally, nationally, and internationally. The importance of international cooperation and compliance is stressed. Such a public health endeavor would benefit the resident peoples as well as the whole world, since bacteria and other microorganisms travel so easily from country to country. The problem certainly necessitates altering physician and patient behaviors as well as establishing guidelines which can be accepted by the providers. Such changes will only succeed if the healthcare provider and/or system sees the need for change and seeks to implement it. Emphasis is placed on local monitoring and surveillance of drug use and susceptibility data, which help to pinpoint conditions geographically and define what means can effectively improve the situation. There is a potential role for pharmacists in overseeing appropriate drug prescription and usage, but this role is not generally appreciated. The inclusion of the pharmacist in efforts to improve antibiotic use and curtail resistance is just one of many suggestions in the authors' multidisciplinary approach.

Infection prevention and its impact on the resistance problem is examined in both the hospital and community environments. The appropriate use of disinfection and antisepsis needs to be appreciated better in both environments. Misuse of surface antibacterials, such as overuse in healthy homes, could affect the efficacy of the products needed to protect vulnerable patients and potentially increase the risk of antibiotic resistance. If used appropriately, antibacterials should control spread of infections, leading to fewer antibiotic prescriptions and consequently, decreased resistance.

Costs of resistance have been estimated in billions of US dollars, but financial resources to effect change and save money are limited. So, despite concerted interest, the means to obtain necessary data and to develop policy for change are lacking. Inroads into providing these sources of revenue are critically needed.

This book is a comprehensive, fact-filled, and welcomed addition to the library of public health officials, healthcare providers and research scientists. While it focuses chiefly on bacterial pathogens, it also addresses the resistance problem as it relates to viruses and fungi. The chapters are written by noted experts in the field who bring forth both their own medical and scientific experience as well as wide knowledge of the field. Together, the contributors produce a well-described composite of the resistance problem and the potential for its resolution.

Stuart B. Levy, M.D.
Director
Professor of Molecular Biology and Microbiology
Professor of Medicine
Tufts University School of Medicine
President, Alliance for Prudent Use of Antibiotics

Preface

Antibiotics have been a tremendous success story for over 50 years but this very success has led to major problems with antibiotic resistance. There is increasing interest in policies and guidelines to reduce antibiotic use and improve quality of prescribing and consequently, patient outcome. Much of this activity is concentrated in hospitals where antibiotic resistance problems are greatest due to the intensity of antibiotic use complicated by cross infection. Greater volumes of antibiotics are consumed in the community in patients and animals and key issues there, are also addressed.

Issues regarding the interaction between resistance and consumption are explored in detail such as control of resistance by modifying and/or reducing consumption, analysis of associations between resistance and consumption, developments in surveillance of resistance and consumption and statistical models.

Modern concepts of evidence medicine are explored as they apply to interventions to reduce antibiotic resistance and improve antibiotic use. The latest evidence about how to design, implement, and evaluate policies and guidelines is described. As well as dealing with the theory, several chapters will give practical advice as to how to implement the latest ideas in antibiotic stewardship.

Issues critical and unique to developing countries are explored as well.

Contents

Foreword   v

Preface   ix

1. Antibiotic Policies—A Historical Perspective   1
   IAN PHILLIPS

2. Guideline Implementation: It is Not Impossible   15
   PETER A. GROSS

3. UK Guidelines: Methodology and Standards of Care   23
   DILIP NATHWANI

4. Pneumonia Guidelines in Practice   37
   GAVIN BARLOW

5. Collecting, Converting, and Making Sense of Hospital
   Antimicrobial Consumption Data   67
   STEPHANIE NATSCH

6. How Do Measurements of Antibiotic Consumption
   Relate to Antibiotic Resistance?   75
   ROGER L. WHITE

7. Quantitative Measurement of Antibiotic Use   105
   FIONA M. MACKENZIE and IAN M. GOULD

8. Benchmarking   119
   HENRIK WESTH

9. Experiences with Antimicrobial Utilisation
   Surveillance and Benchmarking   133
   CATHERINE M. DOLLMAN

10. Interventions to Optimise Antibiotic Prescribing in
    Hospitals: The UK Approach   .   159
    ERWIN M. BROWN

11. Improving Prescribing in Surgical Prophylaxis   185
    JOS W. M. VAN DER MEER and MARJO VAN KASTEREN

12. Audits for Monitoring the Quality of Antimicrobial Prescriptions   197
    INGE C. GYSSENS

13. Multidisciplinary Antimicrobial Management Teams and the
    Role of the Pharmacist in Management of Infection   227
    KAREN KNOX, W. LAWSON, and A. HOLMES

14. Antibiotic Policy—Slovenian Experiences   251
    MILAN ČIŽMAN and BOJANA BEOVIĆ

15. Intensive Care Unit   261
    HAKAN HANBERGER, DOMINIQUE L. MONNET, and
    LENNART E. NILSSON

16. The Real Cost of MRSA   281
    STEPHANIE J. DANCER

17. Antifungal Agents: Resistance and Rational Use   311
    FRANK C. ODDS

18. Strategies for the Rational Use of Antivirals   331
    SHEILA M. L. WAUGH and WILLIAM F. CARMAN

19. Disinfection Policies in Hospitals and the Community   351
    EMINE ALP and ANDREAS VOSS

20. The Evolution of Antibiotic Resistance within Patients   367
    IAN R. BOOTH

21. Impact of Pharmacodynamics on Dosing Schedules:
    Optimizing Efficacy, Reducing Resistance, and
    Detection of Emergence of Resistance   387
    JOHAN W. MOUTON

22. Types of Surveillance Data and Meaningful Indicators
    for Reporting Antimicrobial Resistance   409
    HERVÉ M. RICHET

23. Data Mining to Discover Emerging Patterns of
Antimicrobic Resistance 421
J. A. POUPARD, R. C. GAGNON, and M. J. STANHOPE

24. Applications of Time-series Analysis to Antibiotic
Resistance and Consumption Data 447
JOSÉ-MARÍA LÓPEZ-LOZANO, DOMINIQUE L. MONNET,
PILAR CAMPILLOS ALONSO, ALBERTO CABRERA QUINTERO,
NIEVES GONZALO JIMÉNEZ, ALBERTO YAGÜE MUÑOZ,
CLAUDIA THOMAS, ARIELLE BEYAERT, MARK STEVENSON, and
THOMAS V. RILEY

25. Biocide Use and Antibiotic Resistance 465
JEAN-YVES MAILLARD

26. Interventions to Improve Antibiotic Prescribing in the Community 491
SANDRA L. ARNOLD

27. Education of Patients and Professionals 531
CHRISTINE BOND

28. The Influence of National Policies on Antibiotic Prescribing 545
MOYSSIS LELEKIS AND PANOS GARGALIANOS

29. Antibiotic Use in the Community 567
SIGVARD MÖLSTAD and OTTO CARS

30. Antibiotic Use in the Community: The French Experience 583
AGNÈS SOMMET and DIDIER GUILLEMOT

31. Antibiotic Policies in Developing Countries 593
ANÍBAL SOSA

32. Antimicrobial Resistance and its Containment in
Developing Countries 617
DENIS K. BYARUGABA

33. Antibiotic Use in Animals—Policies and Control
Measures Around Europe 649
PASCAL SANDERS

34. The Pharmaceutical Company Approach to Antibiotic Policies 673
ANTHONY R. WHITE

35. Antibiotic Use—Ecological Issues and Required Actions 701
IAN M. GOULD

Author Index 717

Subject Index 733

Chapter 1

Antibiotic Policies—A Historical Perspective

Ian Phillips
Calle Cabello 7, Málaga 29012, Spain

1. ORIGINS OF ANTIBIOTIC POLICIES

1.1. The pioneers

In 1945, at the end of World War II, Professor Sir Alexander Fleming (as he had then become) was asked by Butterworth, the medical publishers, to write a book on penicillin. He refused, on the grounds that he was too busy and that as a laboratory worker, he would have been unable to place penicillin therapy in a proper perspective in relation to other forms of medical and surgical treatment. However, he did agree to edit a book—and produced a volume (Fleming, 1946a) which is now something of a rarity—bringing together contributions from many of those who had gained experience of the compound during its early years of scarcity. Furthermore, he wrote an introductory chapter in which he overcame his modesty and set out some general rules for penicillin treatment (Fleming, 1946b). First, he wrote, it should be used only for the treatment of those infections caused by penicillin-sensitive microbes. His list of these runs: staphylococci, *Streptococcus pyogenes, Streptococcus viridans*, some anaerobic streptococci, pneumococcus, gonococcus, meningococcus, and so on. He also points out the importance of acquired resistance even though there was little of it at the time. However, despite the many casualties on this list, the principle remains sound. Second, the antibiotic must be given by an appropriate route, in adequate dosage, for an appropriate period of time—and despite uncertainties, the principles again remain true. *En passant*,

Antibiotic Policies: Theory and Practice. Edited by Gould and van der Meer
Kluwer Academic / Plenum Publishers, New York, 2005

he made a plea that doctors should resist patients' and press demands for penicillin to be used for a variety of inappropriate purposes, and listed cancer, tuberculosis, rheumatoid arthritis, psoriasis, and almost all the virus diseases among the many conditions that he had personally been asked to treat. Patients and the media may now be better informed, but irrational demands continue. He also mentions the problem of toxicity, dismissing it promptly in the context of penicillin, although L. P. Garrod gives a more balanced account in an other chapter (Garrod, 1946). Finally, Fleming discusses the assay of penicillin in blood, cerebrospinal fluid, urine, pus, and sputum and so can even be considered to have indicated the importance of pharmacokinetics (Fleming, 1946c). It seems to me that in his book, albeit in the somewhat discursive manner characteristic of the times, Fleming had defined rational therapy in terms of the infection to be treated, the causative organism and its *in vitro* antibiotic susceptibility and the importance of pharmacology including toxicity.

One more aspect of Fleming's book deserves attention in the context of the development of the concepts of antibiotic policies. In a chapter on the prophylactic use of penicillin, Porritt and Mitchell (1946) discuss the comparative merits of sulfonamides and penicillin for the prevention of infection in war wounds. The clinical trial conducted to clarify the issue would not nowadays commend itself to those who regulate most of the day-to-day relevance out of such things, but it did lead to the abandonment of sulfonamides and their replacement by penicillin, a policy that I found still in operation 40 years later! Was this indeed the first antibiotic policy?

1.2.   Early developments

In a chapter that I wrote in 1979 (Phillips, 1979), I argued that an antibiotic policy assumes that antimicrobial therapy will be rational for the individual patient—"that the antibiotic chosen is likely to cure or prevent infection; that the pathogen is sensitive to it in vitro; that the risk of side effects is minimised; and that pharmacological and pharmaceutical properties are appropriate." Many guides based on these considerations were produced in the 1970s (Bint and Reeves, 1978; Geddes, 1977; Wise, 1977). A policy is something superimposed on such rational use, taking into account the risk of development of resistance, cost, simplicity, and the personal preferences of the prescribing clinician (Figure 1). It depends on pragmatic consensus, but even that should not prevent a clinician ignoring it in what he believes to be the best interest of an individual patient.

This point of view represents the teaching of Fleming and the early pioneers as digested and passed on by my own teachers at St Thomas' Hospital, notably Professor Ronald Hare (who wrote his own version of the discovery of penicillin, based on a ring-side seat at St Mary's Hospital in the 1920s) and

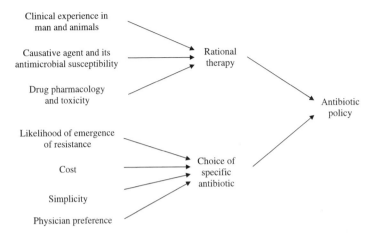

Figure 1. Components of an antibiotic policy (based on Phillips, 1979).

Dr Mark Ridley (who sadly died at a young age). Their views were in turn much influenced by the work of Dr Mary Barber, initially in Hare's department at St Thomas' and later at The Hammersmith Hospital, and co-author with L. P. Garrod of an influential textbook "*Antibiotic and Chemotherapy*," first published in 1963 (Barber and Garrod, 1963).

There seems little doubt that the first civilian antibiotic policies came into being with the emergence of the "Hospital Staphylococcus" in the late 1950s (Garrod and O'Grady, 1971; Phillips, 1979). At St Thomas' Hospital a Sepsis Committee was set up in 1959 "to review hospital infection and to suggest methods for its prevention" (Phillips, 1979). One of the major approaches to control, related to the use of antibiotics and the recommendations were based to a large degree on the specific experience of Mary Barber and her colleagues (Barber and Burston, 1955; Barber and Dutton, 1958; Barber and Garrod, 1963; Barber and Rozwadowska-Dowzenko, 1948; Barber *et al.*, 1958). For example, she recommended the use of erythromycin only in combination, as did Lowbury (Lowbury, 1957), and for a number of years our clinicians prescribed it only with a full dose of novobiocin. It is of interest that none of these clinicians complained about the toxicity of novobiocin and whether their patients did must be a matter of conjecture! The policy for the treatment of *Staphylococcus aureus* infection in operation from 1960 until 1967 (Phillips and Cooke, 1982) involved the use of penicillin for the 30–35% of hospital-isolates still susceptible to penicillin, and erythromycin plus novobiocin for the remainder unless there was resistance to either of them, in which case methicillin or later cloxacillin was to be used. Thus in the years 1959–60 we moved from a policy of free use of any antistaphylococcal agent, to restriction (of erythromycin and then methicillin/cloxacillin) to use of combinations (erythromycin plus novobiocin). Prof. (later Sir) Robert Williams and his colleagues, in their influential book *"Hospital Infection, Causes and Prevention"* (Williams *et al.*, 1960) listed restriction, diversification, rotation, and combination as the measures available to those who devise anti-biotic policies, and we had in some measure used all four. By 1967, it had become clear that, after its first appearance in the early 1960s, methicillin resis-tance, despite its early recognition, had not become a problem (Cookson and Phillips, 1988; Jepsen, 1986), and so erythromycin and novobiocin, and, for good measure, fusidic acid were restricted and methicillin, cloxacillin or, later, flucloxacillin were made freely available.

Did policies of the kind introduced in my own hospital overcome the prob-lem of staphylococcal resistance? We and others, in many parts of the world, thought so (Barber *et al.*, 1958, 1960; Goodier and Parry, 1959; Hinton and Orr, 1957; Kirby and Ahern, 1953; Lepper *et al.*, 1954; Lowbury, 1955; Phillips, 1979; Phillips and Cooke, 1982; Ridley *et al.*, 1970; Shooter, 1957, 1981; Wallmark and Finland, 1961)! Rosendal and her colleagues in Denmark, attributed the changes to a diminished use of streptomycin and tetracycline for the treatment of staphylococcal infection (Rosendal *et al.*, 1977). At its worst, the "Hospital Staphylococcus" was resistant to penicillin, streptomycin, tetra-cycline, chloramphenicol, erythromycin, novobiocin, and neomycin, and if fusidic acid was used, it often became resistant to that too. It was never resis-tant to methicillin, although other less common and less multiresistant strains were (Figure 2).

The early restriction of erythromycin was accompanied by a fall in resist-ance rates from 18% to 4%, returning to 20–25% only after 3-years use of erythromycin and novobiocin, an example, we thought, of the delaying effect on the emergence of resistance due to antibiotic combination. Restriction of methi-cillin/cloxacillin had, again we believed, resulted in our having less than 2% (and usually much less) of isolates resistant to these drugs. However, it was then made clear to me by the Director of the National Staphylococcal Reference Laboratory, Dr M. T. Parker, that similar events had been occurring nationally even in hospitals that had no antibiotic policies (Parker, 1971). The subsequent demise of the "Hospital Staphylococcus" from the mid-1960s onwards (Ayliffe *et al.*, 1979; Bulger and Sherris, 1968; Gransden *et al.*, 1982; Shooter, 1981), in the context of a relaxation of our policies reinforced the conclusion that much of what we observed had more to do with the natural waning of an epidemic than to our infection control and antibiotic policy interventions. Prof. Mouton and his colleagues were not alone in drawing attention to paradoxes (Mouton *et al.*, 1976). The unwisdom of attributing past events to uncontrolled and assumed causes was an early lesson which continues to be ignored (Phillips, 1998a, b)! This is not to say that some resistance was not driven by antibiotic use. Another personal experience relates to chloramphenicol-resistant staphylococci which

	Year																			
	1958	59	60	61	62	63	64	65	66	67	68	69	70	71	72	73	74	75	76	77
P	80	62	75	63	72	63	72	75	62	73	75	72	78	74	79	82	81	79	85	83
PST	20	32	35	25	31	32	18	24	14	9	5	7	7	8	5	4	3	0.4		
M								0.1	0.2	0.5	1	2	1	1	1	2	1	0.4	0.4	0.05
E	18	4	4	2	20	24	15	17	4	3	4	9	3	3	2.5	5	4.5	5	6	7
L														0	0.3	0.3	1	0.5	0.4	0.5
F													0	2	2.5	2	2	2.6	3.4	4
Policy	1	2	3							4				5			6			

	Year																			
	1978	79	80	81	82	83	84	85	86	87	88	89	90	91	92	93	94	95	96	97
P	85	85	84	86	76	63	92	84	88	81	50		83	92	90	89	93	95	84	87
M	0	1.2	0.9	0.9	3	0	21	8	3	0	0		0	6	3	2	10	39	49	42
E	4	6	7	6.4	8	5	25	16	6	3	5		10	10	15	7	13	45	46	47
G	2	1.8	1	2	8	0	17	4	6	3	0		4	4	5	4	20	26	13	10
C			0	0	3	0	0	0	0	0	0		6	6	10	2	5	44	43	44
Policy							7						8							

Figure 2. Trends in antibiotic resistance of *Staphylococcus aureus* isolated from inpatients in St Thomas' Hospital 1958–97 and antibiotic policy changes.

Notes: Policy: (1) free use; (2) erythromycin restricted; (3) penicillin, erythromycin plus novobiocin, and methicillin/cloxacillin; (4) penicillin and methicillin/cloxacillin, erythromycin, novobiocin, and fusidic acid restricted; (5) relaxation with increasing use of clindamycin and fusidic acid; (6) decreasing use of clindamycin because of pseudomembranous colitis; (7) increased use of vancomycin because of MRSA; and (8) further increased use of vancomycin due to successive epidemics of MRSA.

Antibiotic resistance (each figure represents the % resistant during the year): P, penicillin; PST, penicillin, streptomycin, and tetracycline combined (The Hospital Staphyloccus); M, methicillin (MRSA methicillin/multiply-resistant *S. aureus*); E, erythromycin; L, lincomycin; F, fusidic acid; G, gentamicin; C, ciprofloxacin.

Sources: Based on Phillips, 1980 (1958–77), Gransden *et al.*, 1982 (1978–81), Phillips and King, 1979 hospital bacteraemia isolates only, unpublished (1982–1997).

were isolated only in the wards of a particular surgeon who regularly used the drug (with no untoward effects on his patients) and which disappeared immediately upon the surgeon's retirement.

1.3.   Extension of policies

The work of Dr Maxwell Finland and his colleagues in Boston (Finland, 1970; Finland and Jones, 1956; Finland *et al.*, 1959) and work in specialised units such as the Birmingham Accident Hospital (Lowbury, 1955) extended interest to the development of resistance among Gram-negative organisms. One of the best known examples of control of resistant organisms by total antibiotic restriction—klebsiella infection in a neurosurgical unit—was reported by Price and Sleigh (1970). This expansion of interest in organisms other than *S. aureus*, together with the introduction of new classes of antibiotics, at ever increasing costs, in the 1950s and 1960s, led to a further justification for the introduction of antibiotic policies based on the understandable confusion of clinicians in the face of so many similar or to them apparently identical drugs. As early as 1955, Prof. L. P. Garrod concluded that "the choice is now so wide, and the indications are so complex, that few clinicians can keep fully abreast of knowledge about them" (Garrod, 1955).

These developments had already led to the creation of yet another type of committee, often called the Antibiotics Committee, which, at St Thomas', was set up in 1960 to "consider, continually review, recommend and give information on antibiotic policy in the hospital" (Phillips and Cooke, 1982). It continued to do this for 30 years, taking all of the considerations so far mentioned into account, including costs, until its work was taken over by a more general Use of Drugs committee, an example of a general Formulary Committee. I believe that it is fair to say that antibiotics were not high on the agenda of such committees since the amount expended on them was relatively little in comparison with drugs used to control chronic ill health, both physical and mental: in 1980, antibiotics accounted for only 13.4% of hospital drug costs (Cooke *et al.*, 1983). Whether for that or other reasons, the amount of effort given to controlling antibiotic use declined, and there was a hiatus before the renewed efforts described elsewhere in this volume, largely related to the increasing prevalence of resistance and an increased perception that it was a problem.

There were sceptics in relation to the ethics of antibiotic policies throughout. Selkon (1980) agreed with many that general policies involving restriction were an intolerable affront to clinical freedom, led to suboptimal treatment for individual patients, and had "rapidly lost credibility." He further argued that they complicated antimicrobial chemotherapy. Finally he described the policy in Newcastle where clinicians reserved the right to prescribe whatever antibiotics they thought appropriate, with the pharmacist informing the microbiologists of potentially harmful prescribing, and the microbiologists intervening on an individual basis. Perhaps the more *laissez faire* attitude that we developed in the 1970s (Figure 2) when there were few problems, was not very different from this.

1.4.   Policies for all

The original antibiotic policies, arising as they did from problems of control of infection with resistant bacteria in hospitals, were strictly for application within the individual hospitals that produced them. They were often amplified by policies specific to units within hospitals. The Burns unit in the Birmingham Accident Hospital was one of the pioneers (Lowbury, 1955; Lowbury *et al.*, 1957). One of our first restricted policies was developed for use within our Renal Unit: a specific example from that policy was the use of vancomycin administered in weekly dosage to anephric patients with staphylococcal infections around indwelling shunts for renal dialysis (Eykyn *et al.*, 1970), together with minimal use of the first-generation cephalosporin, cephaloridine, and of gentamicin for other severe infections (Phillips,1981). Interestingly, we had no vancomycin-resistant enterococcal infections at the time of their first description in a neighbouring teaching hospital and for many years thereafter. Later we produced a policy for our Urology Unit (Casewell *et al.*, 1981).

A final addition to our Antibiotic Committee was a general practitioner, when it became clear that resistance was becoming a problem in the community—for example, that *Escherichia coli* was becoming resistant at a rate of 1% per annum, to what had been considered first line agents for urinary tract infection (Phillips *et al.*, 1990). It also had become generally recognised that most antibiotics were used in the community, and, furthermore, were not particularly rationally used (Cooke *et al.*, 1985).

The apparent success of particular policies led to demands for their dissemination to other hospitals and even to NHS Regions. Ridley produced his *"Pocket Guide to Antimicrobial Chemotherapy"* based on the St Thomas' guide and policy in 1971 (Ridley, 1971), and the microbiologists of the South East Thames Regional Health Authority produced their "Guide to the use of antimicrobial drugs" in 1977 (SETRHA, 1997). However, many felt that a lack of local ownership of such guides, and the need to compromise, made them little more than the guides they purported to be and not true policies. Lowbury summed up the problem when he pointed out in 1975, that "in Dudley Road Hospital (Birmingham) a severe wound infection could be treated with kanamycin (whereas) in the Burns Unit of Birmingham Accident Hospital, kanamycin would be an incorrect choice," arguing for purely local policies (Lowbury, 1957). Nevertheless, at the other extreme, some countries, such as Czechoslovakia (Modr, 1978), developed and applied national policies.

1.5.   A fundamental approach

In the 1960s, there was some discussion of the possibilities for alternative approaches to the control of resistance. Pollock (1960) suggested that as well as reducing selection pressures, we might try to prevent mutation and recombination. There have been other calls for reversal of resistance by genetic means, although I have expressed reservations in relation to potential methods involving genetic engineering (Phillips, 1998a, b). I hope that I shall be proved wrong!

I have also called for fundamental rethinking about the possibility of controlled trials of antibiotic policies. Too often are we satisfied by our attributions of epidemiological trends in resistance to arbitrarily chosen—even if apparently logical—trends in antibiotic usage (Phillips, 1998a). Perhaps a few more charts relating increases in antibiotic resistance to sales of bananas and the like, might be helpful.

2.   PRACTICAL APPLICATION OF POLICIES

2.1.   Surveillance of antibiotic usage and resistance

Surveillance of resistance grew out of such studies as those carried out by Finland and his colleagues at the Boston City Hospital (Kislak *et al.*, 1964; McGowan and Finland, 1974a; Wallmark and Finland, 1961), followed up by specific surveys of antibiotic usage and resistance in many parts of the world (Kayser, 1978; Lawson and MacDonald, 1977; Moss *et al.*, 1981; Mouton *et al.*, 1976; Sheckler and Bennett, 1970; Swindell *et al.*, 1983). O'Brien and his colleagues reported specifically international comparisons (O'Brien *et al.*, 1978).

Having been appointed Infection Control Officer at St Thomas' Hospital in 1963, I was responsible for the collection of statistics on wound infection, visiting all of the hospital wards weekly as well as carrying out myself the phage typing of all hospital *S. aureus* isolates. My main finding was that although the prevalence of hospital staphylococcal infection did not decline, the prevalence of multiply resistant *S. aureus* certainly did! Whether this "result" fulfilled the expectations of the Sepsis Committee must be doubtful. However, another of the essential features of a policy had been put in place—the collection of statistics on the prevalence of resistance (Figure 2), continued from 1958 until my retirement in 1996 (Gransden *et al.*, 1982; Phillips, 1980; Phillips and King, unpublished; Ridley *et al.*, 1970). Something else was learned from this exercise, that the collection of statistics has absolutely no effect to the good on the prevalence of resistance, and this probably led to the addition of an Infection Control Nurse to a nascent Infection Control Team, as part of a national development.

For a time during the 1970s we too collected statistics on antibiotic usage and attempted to relate them to the prevalence of resistance (Gransden *et al.*, 1982). This turned out to be a period of respite when epidemigenic strains of *S. aureus* virtually disappeared from our wards (Gransden *et al.*, 1982) and there appeared to be little correlation between usage and resistance, and so we abandoned the collection—just before epidemic methicillin-resistant strains returned, apparently to stay (Figure 2).

2.2.   Ensuring compliance

The Antibiotics Committee learned a great deal about the development and application of antibiotic policies and their audit (Phillips and Cooke, 1982).

Mary Barber, a formidable lady, showed us the importance of authority and leadership! We further recognised the importance of the inclusion on the committee of influential physicians and surgeons on the staff of the hospital. They were very important in ensuring support from the staff in general, who were always consulted formally in full committee, on changes in policy, especially in relation to new drugs. It was this general support that allowed the hospital Pharmacist to stock only those drugs that were approved by the Antibiotics Committee, although it was always made clear that clinicians had the right to ignore policy if they felt that it jeopardised their patient. In fact, they seldom did this, preferring to discuss their problems with microbiologists, who, incidentally had to be available when the problems arose and not at their leisure. It is of interest that two physicians always voted against acceptance of policies since they were held to limit their clinical freedom (see Selkon, 1980), but in effect always followed recommendations. Incidentally, our surgeons commonly made no objections but were more likely to try to evade recommendations! A single psychiatrist complained when our policy substituted amoxycillin for ampicillin, but did not persist! It remains important in setting up policies to minimise what may be seen as unethical interference with a clinician's choice of drug, something that seems sometimes to be ignored in the international policies that are currently the vogue.

It will be clear that we recognised the importance of enforcing our policies. First we made available to all hospital medical staff, information on rational antibiotic use in the context of our policies, and this became a small, easily portable booklet in 1966. Many others produced similar guidelines or commented on their usefulness (Bint and Reeves, 1978; Geddes, 1977; Williams 1984; Wise, 1977). This emphasises the advice on the importance of education from colleagues such as Harold Neu in the United States (Neu and Howrey, 1975). We constantly reviewed the sources of information and education in the 1980s, including the important contribution of the pharmaceutical industry, whose representatives were encouraged to speak to members of our Antibiotics Committee before approaching other doctors (Cooke *et al.*, 1980, 1985). Second, pharmacists were asked to draw major aberrations of prescribing to the attention of the laboratory doctors who would then intercede with their colleagues, a process that might have been simplified by the availability of data from computerised prescribing, which we constantly expected but never attained. The development of ward pharmacy helped this goal. Third, the microbiology laboratory was authorised to practice selective testing and reporting, a practice advocated by others (Gould, 1960; Grüneberg, 1980). For example, of the eight agents that might be tested against staphylococci only three might be routinely reported, for example. This practice was helped by the increasing sophistication of microbiology laboratory computing (Phillips, 1978) but it was later compromised by the development of a hospital market place, in which clinicians had their own budgets and their own priorities for spending them. For some reason they never discovered the full extent of our activities on their behalf! Finally, with their strong infectious diseases tradition, physicians in the United States emphasised the importance of involving such experts in prescribing (Kunin *et al.*, 1973; McGowan and Finland, 1974b, 1976) whereas in the United Kingdom we used our medically qualified clinical microbiologists for the same purpose (Gransden *et al.*, 1990; Grüneberg, 1980).

2.3.   Measuring compliance with policies

Despite the great efforts put into the enforcement of antibiotic policies, there was little formal indication that prescribing physicians actually followed them. Audit of policy application developed from audits of the rationality of use, for example, in the United States (Kunin, 1977a, b), where such audits became one of the bases of hospital accreditation (Counts, 1977). Other examples have already been mentioned in relation to audits of rational use.

Having appointed a pharmacist with expertise in information science specifically for the purpose, we conducted a number of prevalence surveys to assess the degree to which our policies were followed. We found that the antibiotics allowed by the policy were those actually in use. For example, in 1980, in a 1-day prevalence survey involving 120 patients receiving systemic antibiotics, we found 135 prescriptions for freely available drugs, 38 for restricted drugs (requiring the approval of a senior clinician), and only 5 for strictly restricted drugs (requiring discussion with microbiologists). However, in many cases, the drugs, especially those that were freely available, were given at the wrong time and for the wrong duration (this applied particularly to prophylaxis) and in inappropriate dosage. Findings in 1983 were similar (Cooke *et al.*, 1983, 1985; Phillips and Cooke, 1985). We concluded that more education was needed—"knowing when to use an antibiotic is as important as knowing which antibiotic to choose."

3.   CONCLUSIONS

Antibiotic policies had their origins in the experiences of those who introduced penicillin into clinical practice. At first, they were little more than guides to rational therapy for the individual patient, but with the increasing problem of acquired antibiotic resistance, they were extended with the intention of minimising this problem. As more and more agents reached the marketplace, an attempt was made to simplify prescribing while in more cost-concious days, economy of use was added. The more sensitive among us saw the need not to hamper clinical freedom unduly. All this was in place by the 1970s, and in terms of concept, little has been added.

Without doubt, the application of the principles of rational antimicrobial chemotherapy has made a major contribution to human health. Unfortunately, this was at a cost, since even the most rational therapy can lead to the emergence of resistance. It seems impossible to dissect out in retrospect, the relative contributions of rational and irrational use, and to determine the overall effect of antibiotic policies. It could be argued that the proper application of policies led to the solving of many particular problems—battles have been won—but it must also be acknowledged that on a universal scale, the problem of resistance intensifies—the war is being lost. Whether this is because the concept of antibiotic policies as a means of avoiding or minimising resistance was wrong, or because the application of a fundamentally sound concept was inadequate, also seems impossible to determine in retrospect. What was once a problem for large hospitals has now spread to the whole community, and shows no sign of abating. We should now ask the question whether what we did—and continue to do—was and is, appropriate. It is to be hoped that the answer to the question will be sought by the rigorous application of sound microbiological and epidemiological science.

REFERENCES

Ayliffe, G. A., Lilly, H. A., and Lowbury, E. J. L., 1979, Decline of the hospital staphylococcus. Incidence of multiresistant *Staphylococcus aureus* in three Birmingham hospitals. *Lancet*, **ii**, 538–544.

Barber, M. and Burston, J., 1955, Antibiotic-resistant staphylococcal infection. *Lancet*, **ii**, 578–583.

Barber, M., Csillag, A., and Medway, A. J., 1958, Staphylococcal infection resistant to chloramphenicol, erythromycin and novobiocin. *Brit. Med. J.*, **2**, 1377–1380.

Barber, M. and Dutton, A. A. C., 1958, Antibiotic-resistant staphylococcal outbreaks in a medical and a surgical ward. *Lancet*, **ii**, 64–68.

Barber, M., Dutton, A. A. C., Beard, M. A., Elmes, P. C., and Williams, R., 1960, Reversal of antibiotic resistance in hospital staphylococcal infections. *Brit. Med. J.*, **1**, 11–17.

Barber, M. and Garrod, L. P., 1963, *Antibiotic and Chemotherapy*. Livingstone, Edinburgh.

Barber, M. and Rozwadowska-Dowzenko, M., 1948, Infection by penicillin-resistant staphylococci. *Lancet*, **ii**, 641–.

Bint, A. J. and Reeves, D. S., 1978, A guide to new antibiotics. *Brit. J. Hosp. Med.*, **4**, 335–342.

Bulger, R. and Sherris, J. C., 1968, Decreased incidence of antibiotic resistance among *Staphylococcus aureus*: A study in a university hospital over a nine-year period. *Ann. Intern. Med.*, **69**, 1099–1108.

Casewell, M. W., Pugh, S., and Dalton, M. T., 1981, Correlation of antibiotic usage with an antibiotic policy in a urological ward. *J. Hosp. Infect.*, **2**, 55–61.

Cooke, D. M., Salter, A. J., and Phillips, I., 1980, Antimicrobial misuse, antibiotic policies and information resources. *J. Antimicrob. Chemother.*, **6**, 435–443.

Cooke, D. M., Salter, A. J., and Phillips, I., 1983, The impact of antibiotic policy on antibiotic prescribing in a London teaching hospital. A one-day prevalence survey as an indicator of antibiotic use. *J. Antimicrob. Chemother.*, **11**, 447–453.

Cooke, M. M., Salter, A. J., and Phillips, I., 1985, Irrational antibiotic prescribing and its prevention. In M. Brittain (ed.). British library research and development report 5842.

Cookson, B. and Phillips, I., 1988, Epidemic methicillin-resistant *Staphylococcus aureus*. *J. Antimicrob. Chemother.*, **21**(Suppl. C), 57–65.

Counts, G. W., 1977, Review and control of antimicrobial usage in hospitalized patients: A recommended collaborative approach. *J. Amer. Med. Ass.*, **238**, 2170.

Eykyn, S., Phillips, I., and Evans, J., 1970, The use of vancomycin for treatment of staphylococcal shunt site infections in patients on regular haemodialysis. *Brit. Med. J.*, **iii**, 80–82.

Finland, M., 1970, Changing ecology of bacterial infections as related to antibacterial chemotherapy. *J. Infect. Dis.*, **122**, 419.

Finland, M. and Jones, W. F., 1956, Staphylococcal infections currently encountered in a large municipal hospital: Some problems in evaluating antimicrobial therapy in such infection. *Ann. New York Acad. Sci.*, **65**, 191–205.

Finland, M., Jones, W. F., and Barnes, M. W., 1959, Occurrence of serious bacterial infections since the introduction of antibacterial agents. *J. Amer. Med. Ass.*, **170**, 2188.

Fleming, A. (ed.) 1946a, Penicillin, its practical application. Butterworth & Co., London.

Fleming, A., 1946b, History and development of penicillin. In A. Fleming (ed.), *Penicillin, its Practical Application.* Butterworth & Co., London, pp. 1–23.

Fleming, A., 1946c, Bacteriological control of penicillin therapy. In A. Fleming (ed.), *Penicillin, its Practical Application.* Butterworth & Co., London, pp. 76–92.

Garrod, L. P., 1946, Pharmacology of penicillin. In A. Fleming (ed.), *Penicillin, Its Practical application.* Butterworth & Co., London, pp. 59–75.

Garrod, L. P., 1955, Present position of the chemotherapy of bacterial infections. *Brit. Med. J.*, **ii**, 756–.

Garrod, L. P. and O'Grady, F., 1971, *Antibiotic and Chemotherapy*, 3rd edn. Livingstone, Edinburgh.

Geddes, A. M., 1977, Good antimicrobial prescribing. *Lancet*, **ii**, 82.

Goodier, T. W. and Parry, W. R., 1959, Sensitivity of clinically important bacteria to six common antibacterial substances. *Lancet*, **i**, 356–357.

Gould, J. C., 1960, The laboratory control of antibiotic therapy. *Br. Med. Bull.*, **16**(1), 29–34.

Gransden, W. R., Atkinson, D., Stead, K. C., and Phillips I., 1983, Antibiotic resistance of *Staphylococcus aureus* 1969–1981, In Proceedings of 13th International Congress of Chemotherapy, Vienna, 1983 Part 74, 36–39.

Gransden, W. R., Eykyn, S. J., and Phillips, I., 1990, The computerised documentation of septicaemia. *J. Antimicrob. Chemother.*, **25**(Suppl. C), 31–39.

Grüneberg, R. N., 1980, Antibiotic prescribing policies: A personal view. In R. N. Grüneberg (ed.), *Antibiotics and Chemotherapy: Current Topics.* MTT Press Ltd, Lancaster, pp. 203–211, ISBN 0-85200-218-1.

Hinton, N. A. and Orr, J. H., 1957, Studies on the incidence and distribution of antibiotic-resistant staphylococci. *J. Lab. Clin. Med.*, **49**, 566–572.

Jepsen, O. B., 1986, The demise of the old methicillin-resistant *Staphylococcus aureus*. *J. Hosp. Infect.*, **7**(Suppl. A), 13–17.

Kayser, F. H., 1978, Use of antimicrobial drugs in general hospitals and in general practice. In current chemotherapy: Proceedings of the 10th Int. Congress of chemotherapy, Zurich, 1977. *Am. Soc. Microbiol.*, **2**, 36.

Kirby, W. M. M. and Ahern, J. A., 1953, Changing patterns of resistance of staphylococci to antibiotics. *Antibiot. Chemother.*, **3**, 831–835.

Kislak, J. W., Eickhoff, T. C., and Finland, M., 1964, Hospital-acquired infections and antibiotic usage in Boston City Hospital, January 1964. *New Engl. J. Med.*, **271**, 834–835.

Kunin, C. M., Tupasi, T., and Craig, W. A., 1973, Use of antibiotics: A brief exposition of the problem and some tentative solutions. *Ann. Intern. Med.*, **79**, 555–560.

Kunin, C. M., 1977a, Audits of antimicrobial usage. *J. Emer. Med. Ass.*, **237**, 1001–1008, 1134–1137, 1241–1245, 1366–1369, 1482–1484, 1605–1608, 1723–1725, 1859–1860, 1967–1970.

Kunin, C. M., 1977b, Guidelines and audits for use of antimicrobial agents in hospitals. *J. Infect. Dis.*, **135**, 335.

Lawson, D. H. and MacDonald, S., 1977, Antibacterial therapy in general medical wards. *Postgrad. Med. J.*, **53**, 306–309.

Lepper, M. H., Moulton, B., Dowling, H. F., Jackson, G. G., and Kofman, S., 1954, Epidemiology of erythromycin-resistant staphylococci in a hospital population. *Antibio. Ann.* 308–319, 1953–1954.

Lowbury, E. J. L., 1955, Cross-infection of wounds with antibiotic-resistant organisms. *Brit. Med. J.*, **i**, 985–990.

Lowbury, E. J. L., 1957, Chemotherapy for *Staphylococcus aureus*. Combined use of novobiocin and erythromycin and other methods in the treatment of burns. *Lancet*, **2**, 305.

Lowbury, E. J. L., 1960, Clinical problems of drug-resistant pathogens. *Brit. Med. Bull.*, **16**(1), 73–78.

McGowan, J. E. and Finland, M., 1974a, Infection and antibiotic usage at Boston City Hospital: Changes in prevalence during the decade 1964–73. *J. Infect. Dis.*, **129**, 421–428.

McGowan, J. E. and Finland, M., 1974b, Usage of antibiotics in a general hospital: Effect of requiring justification. *J. Infect. Dis.*, **130**, 165–168.

McGowan, J. E. and Finland, M., 1976, Effects of monitoring the usage of antibiotics: An interhospital comparison. *South. Med. J.*, **69**, 193.

Modr, Z., 1978, Antibiotic policy in Czechoslovakia. *J. Antimicrob. Chemother.*, **4**, 305.

Moss, F. M., McNichol, M. W., McSwiggan, D. A., and Miller, D. L., 1981, Survey of antibiotic prescribing in a district general hospital. *Lancet*, **ii**, 461–462.

Mouton, R. P., Glerum, J. H., and van Loenen, A. C., 1976, Relationship between antibiotic consumption and frequency of antibiotic resistance of four pathogens—a seven year survey. *J. Antimicrob. Chemother.*, **2**, 9–19.

Neu, H. C., and Howrey, S. P., 1975, Testing the physician's knowledge of antibiotic use. *New Engl. J. Med.*, **293**, 1291–1295.

O'Brien, T. F., Acar, J., Medeiros, A. A., Norton, R. A., Goldstein, F., and Kent, R. L., 1978, International comparison of prevalence of resistance to antibiotics. *J. Amer. Med. Ass.*, **239**, 1515–1523.

Parker, M. T., 1971, Current national patterns: Great Britain. In P. S. Brachman and T. C. Eickhoff (eds.), *Nosocomial Infections*. American Hospital Association, Chicago, IL.

Phillips, I., 1978, The computer in a microbiology department. In F. Siemaszko (ed.), *Computing in Clinical Laboratories*. Pitma Medical, London, pp. 265–268.

Phillips, I., 1979, Antibiotic policies. In D. Reeves and A. Geddes (eds.), *Recent Advances in Infection*, Churchill-Livingstone, Edinburgh, pp. 151–163.

Phillips, I., 1981, The place of cephalosporins in the treatment of infections. In R. P. Mouton (ed.), *Antibiotic Treatment of Infections in the Hospital. Introduction of a New Cephalosporin, Cefotaxime*. Exerpta Medica, pp. 30–42.

Phillips, I., 1998a, The 1997 Garrod Lecture: The subtleties of antibiotic resistance. *J. Antimicrob. Chemother.*, **42**, 5–12.

Phillips, I., 1998b, Lessons from the past: A personal view. *Clin. Infect. Dis.*, **27**(Suppl. 1), S2–S4.

Phillips, I. and Cooke, D., 1982, The control of antibiotic prescribing in a London teaching hospital. In C. H. Stuart-Harris and D. M. Harris (eds.), *The Control of Antibiotic-resistant Bacteria*. Academic Press, London, pp. 201–209.

Phillips, I., King, A., Gransden, W. R., and Eykyn, S. J., 1990, The antibiotic sensitivity of bacteria isolated from the blood of patients in St Thomas' Hospital 1969–1988. *J. Antimicrob. Chemother.*, **25**(Suppl. C), 59–80.

Pollock, M. R., 1960, Drug resistance and mechanisms for its development. *Brit. Med. Bull.*, **16**(1), 16–22.

Porritt, A. E. and Mitchell, G. A. G., 1946, Prophylactic use of penicillin. In A. Fleming (ed.), *Penicillin, its Practical Application*. Butterworth & Co., London, pp. 105–115.

Price, D. J. E. and Sleigh, J. D., 1970, Control of infection due to *Klebsiella aerogenes* in a neurosurgical unit by withdrawal of all antibiotics. *Lancet*, **ii**, 1213–1215.

Ridley, M., 1970, *A Pocket Guide to Antimicrobial Chemotherapy*. Medical Illustrations and Publications, London.

Ridley, M., Barrie D., Lynn, R., and Stead, K. C., 1970, Antibiotic-resistant *Staphylococcus aureus* and hospital antibiotic policies. *Lancet*, **i**, 230–233.

Rosendal, K., Jessen, O., Betzon, M. W., and Bülow, P., 1977, Antibiotic policy and spread of *Staphylococcus aureus* in Danish hospitals 1969–1974. *Acta Path. Microbiol. Scand. Sec. B.* **85**, 143.

Selkon, J. B., 1980, Antibiotic policies, In R. N. Grüneberg (ed.), *Antibiotics and Chemotherapy: Current Topics*. MTT Press, Lancaster, pp. 193–201.

SETRHA, 1979, A guide to the use of antimicrobial drugs. Regional Microbiology Subcommittee.

Sheckler, W. E. and Bennett, J. V., 1970, Antibiotic usage in seven community hospitals. *J. Amer. Med. Ass.*, **213**, 264–267.

Shooter, R. A., 1957, The problem of resistant organisms and chemotherapeutic sensitivity in surgery. *Proc. Roy. Soc. Med.*, **50**, 158–160.

Shooter, R. A., 1981, Evolution of the hospital staphylococcus, In A. MacDonald and G. Smith (eds.), *The Staphylococci: Proceedings of the Alexander Ogston Centennial Conference*. Aberdeen University Press, Aberdeen, pp. 149–155.

Swindell, P. J., Reeves, D. S., Bullock, D. W., Davies, A. J., and Spence, C. E., 1983, Audits of antibiotic prescribing in a Bristol hospital. *Brit. Med. J.*, **286**, 118–122.

Wallmark, G. and Finland, M., 1961, Phage types and antibiotic susceptibility of pathogenic staphylococci. Results at Boston City Hospital 1949–1960 and comparison with previous years. *J. Amer. Med. Ass.*, **175**, 886–897.

Williams, J. D., 1984, Antibiotic guidelines. *Brit. Med. J.*, **288**, 343–344.

Williams, R. E. O., Blowers, R., Garrod, L. P., and Shooter, R. A., 1960, *Hospital Infection, Causes and Prevention*. Lloyd-Luke, London.

Wise, R., 1977, Rational choice of antibiotics. *Prectitioner*, **2**, 449.

Chapter 2

Guideline Implementation: It is Not Impossible

Peter A. Gross
Department of Internal Medicine, Hackensack University Medical Center, 30 Prospect Avenue, Hackensack, New Jersey 07601, USA

In an effort to implement appropriate practices for antibiotic prescribing, changing physician behavior is often cited as one of the key approaches.

I would like to begin this chapter by reviewing the science of changing physician and other provider behavior. Interestingly, much has been written about this over the years. A review of the subject is contained in a summary of a meeting on guideline implementation held at Leeds Castle, England in 1999 by Gross *et al*. The findings are based on a number of Cochrane Collaboration reviews summarized by Grol and Grimshaw as well as other reviews cited in Further Reading at the end of this chapter. We will consider which methods of behavior change are generally ineffective, variably effective and generally effective.

1.   GENERALLY INEFFECTIVE STRATEGIES

The generally ineffective measures include several types of passive educational efforts. Ironically, these are the main methods that are still used to inform providers and change their behavior. Passive educational approaches include *publication of research findings* and *dissemination of guidelines*. While they are important to raise awareness of the diagnosis and management

Antibiotic Policies: Theory and Practice. Edited by Gould and van der Meer
Kluwer Academic / Plenum Publishers, New York, 2005

of diseases, they typically do not change behavior, as would be manifest by using newer approaches to diagnosis and management of diseases. *Lectures* and so-called *Grand Rounds* are also passive educational approaches. In these didactic sessions, usually a single speaker describes an approach and renders an opinion. Because of lack of significant interchange with the audience, these didactic sessions are ineffective in changing provider performance. It is now clear that we should not rely on passive educational approaches to effect change, although they can still be used to inform. We need other methods to effect change.

2.   VARIABLY EFFECTIVE STRATEGIES

There are a number of implementation strategies that have been shown to be variably effective in changing behavior. First, *audit and feedback* may be helpful. In this instance, an individual provider's performance is monitored and that performance is compared confidentially with that of a peer group. This type of provider profiling is most effective when it is used for prescribing and test ordering. The feedback is most likely to be accepted without controversy when it is confidential, though some investigators have compared the individual to other peers by name. The latter approach is more likely to cause political problems and undermine the whole process of improvement.

The use of *local opinion leaders* and *local consensus conferences* often has been shown to be effective. Persons who are considered to be educationally influential by their peers would serve as local opinion leaders and they may encourage others to emulate their behavior. Local consensus conferences for adapting and adopting guidelines will be successful as they most often bring together the local opinion leaders and other major players. The other people to include in such groups are providers who are expected to be early adopters of change. Occasionally, providers who have shown resistance to change may be favorably affected by being included in a consensus conference. If it does not compromise the quality of care, *adapting* a new guideline or policy to the local medical climate will encourage local adoption.

Another variably effective approach is *consumer education*. Providing patients with information about their healthcare needs has been shown to have a positive effect. The effect, however, is small and varies from one clinical condition to another.

Finally, nothing replaces an inquisitive patient to help drive the healthcare provider toward a new, perhaps more appropriate direction. By involving the patient in the clinical decision-making, the care is more *patient centered*. By acknowledging the patient's wishes and needs, the provider is more likely to adapt his approach to management to meet these needs and the patient's satisfaction is likely to be greater.

3.   GENERALLY EFFECTIVE STRATEGIES

There are a number of implementation strategies for behavioral change that are effective most of the time. *Reminders* to healthcare providers are one such strategy. These reminders, however, have to be used sparingly; otherwise, they will be ignored if used too frequently.

Computer information systems can be helpful in a number of ways. *Computerized physician order entry* (CPOE) will avoid many of the problems associated with trying to decipher physicians' illegible handwriting and thereby make what the doctor has ordered clear to the nurse, pharmacist, and other healthcare providers. Errors in medication ordering and ordering of other tests should be significantly reduced.

In addition, *computer checks* can be programmed into a hospital's information system to provide reminders, warnings, and other suggestions to facilitate appropriate ordering of therapeutic and preventive treatments. Importantly, this type of feedback will offer the provider an incentive to use the system if the computer checks are not too numerous and annoying.

Educational outreach is another effective strategy. For example, in *academic detailing*, there is a one-on-one dialog that occurs between the expert detailer and the provider to discuss a new form of therapy or a new procedure. The detailer may be another provider or a pharmaceutical representative. The exchange tends to be *interactive* (i.e., the flow of information is in both directions) rather than didactic. The issue of asking "foolish questions" is usually not present. The provider being detailed can discuss the matter with the academic detailer until the provider feels he understands the issue.

Barrier-oriented interventions are critical. They must be tailored to specific local barriers. Examples of possible local barriers are:

- disagreement among experts,
- availability of alternative practices,
- inapplicability of guidelines to certain patient subgroups,
- patient refusal to comply,
- ceiling and power effects that relate to already high levels of compliance,
- institutional inertia,
- vested interests,
- ineffective continuing medical education, and
- uncertainty about when and how to apply evidence-based medicine measures.

The last generally effective strategy to consider is *multifaceted interventions*. This is probably the most important intervention strategy. Whenever change is being attempted, more than one strategy should be applied. Multiple strategies are likely to be more successful than one.

Inadequate studies have been done to permit us to describe what combination of strategies to use in different clinical situations. The key point is to use a number of the above-described change strategies to assure success.

4.   THEORIES OF FACILITATING CHANGE

Insight into the theories of managing change will help us apply the above strategies more effectively as outlined by Grol and Grimshaw.

1. *Educational theories* point out that change is driven by one's desire to learn and be professionally competent. Learning should be interactive. When a local group meets to reach local consensus, "buy-in" is facilitated.

2. *Epidemiological theories* purports that we are rational human beings who will arrive at a rational decision when the best evidence is presented. Creation of evidence-based guidelines by national professional organizations is an example.

3. *Marketing theories* assume that we will be favorably affected by an attractive marketing package. Which media channel is used will depend on who we want to influence—innovators, early adopters, or late adopters. In a healthcare organization, the media channels may be local opinion leaders, academic detailers, or mass media advertising.

4. *Behaviorist theories* propose that change is influenced by external factors applied before, during, and/or after the targeted objective. The above strategies that apply are audit and feedback as well as reminders before, during, or after ordering a medication. Financial or recognition incentives and sanctions are other behaviorist approaches that could be applied.

5. *Social influence theories* emphasize the importance of social group recognition for implementing change. The group is composed of peers and/or opinion leaders. By feeding back the individual's performance in comparison with that of peers, the effect on changing behavior can be significant because a provider wants to successfully compete with his peers.

6. *Organizational theories* promote improvement by changing the system of care. The emphasis here is on the "bad system," not the "bad apple." These theories view healthcare as a series of interrelated processes and view the participants—the providers—as members of a team that can reach the predetermined goal. This point of view becomes more and more important as efforts to improve care repeatedly come back to the point that it is the system that has

to be changed and the provider must work as part of a larger team of many healthcare professionals.

7. *Coercive theories* promote the idea that exerting pressure and control will accomplish the desired changes. Healthcare regulations from external reviewing agencies or rules imposed internally are the change tools.

Comparisons with the airline industry are relevant to an extent. The airline industry uses checklists to assure that the right thing is done at the right time in the right place. The healthcare industry can certainly learn from this example. Medical care has become complicated enough that one physician cannot remember everything that has to be done for a particular problem. A healthcare checklist, therefore, should be helpful. For example, it can be a set of pre-printed orders for a particular disease such as acute myocardial infarction or community-acquired pneumonia. The checklist can still be individualized to the particular patient because each order on the order sheet has to be checked off before it is executed. This approach is not "cookbook medicine," but "team medicine" using a collaborative approach. The dilemma is getting all providers to accept the fact that they are not all knowing and will not perform perfectly all of the time. The airline pilot's acceptance of imperfection is a lesson for healthcare providers to emulate. The healthcare challenge of this decade and beyond is for the provider to come to the realization that using external resources such as a checklist or a professional team of different healthcare providers will allow them all to perform more perfectly.

An example of using the above information follows:

1. A medical or surgical disorder is selected where improvement in care would have a high impact on disease management, outcome, or cost containment. Community-acquired pneumonia (CAP) is a frequently used example where quality improvement efforts have been applied successfully.

2. A national CAP guideline is selected. In this case, there are at least three or four available from national professional organizations.

3. A local consensus group is gathered together. Champions who are local opinion leaders are selected. Other members of the group may include innovators and early adopters in care changes. Late adopters can be included to get the full spectrum of opinions. One or more of the guidelines is selected or in the case of CAP, performance measures from Medicare (i.e., The Centers for Medicare and Medicaid Services, CMS) and the Joint Commission on Accreditation of Healthcare Organizations (JCAHO) are considered.

4. Performance measures can be viewed as the best part of guidelines where the measures represent the best of evidence-based medicine. Performance measures should leave out those recommendations that are based on expert opinion alone or poorly conducted research studies.

5. The consensus group should then decide which recommendations to implement. The group may have a problem with one or more of the recommendations. In that case, unless the science behind it is incontrovertible, the group should make the change so that the group feels ownership of the outcome by locally adapting the recommendation.

6. The CMS and JCAHO measures used which improve outcome are:

(a) Antibiotics should be administered within 4 hr of admission to the healthcare facility.
(b) The recommended antibiotics should be used and no others.
(c) Blood cultures, if drawn, should be drawn before antibiotics are given.
(d) The patients should be screened for receipt of influenza and pneumococcal vaccine before discharge.

7. Other measures can be added that improve efficiency, patient satisfaction, and reduce length of stay. For example:

(a) Intravenous (IV) antibiotics should be switched to oral antibiotics as the patient is stabilizing according to the Ramirez criteria.
(b) The patient can be discharged to home when the social conditions permit and no other acute conditions need hospital management.

8. An admission order sheet can include the CMS and JCAHO performance measures written in the form of orders that the provider checks off.

9. Nurse practitioners or other personnel as case managers can be made part of the team and help physicians implement the measures. The nurse practitioners or discharge planners can be made aware of the switch therapy recommendations from IV to oral antibiotics and the discharge criteria, thereby helping to facilitate their implementation.

10. Once all these approaches are determined by the consensus group and approved by the medical staff, the medical and nursing staff should be informed of the plans and the reasoning behind them. Their assent to the program is critical. Then the program can be implemented.

11. Additional personnel who have to be hired to implement this program would most likely easily be paid for by the financial gains from a reduction in length of stay and the more efficient utilization of resources. Consequently, awareness of the financial impact of these improvement efforts is important.

The multifaceted intervention described for CAP takes advantage of many of the implementation strategies described and uses most of the theories of facilitating change described. In addition, adding measures that assure the business case for quality improvement will help gain administrative support for the additional resources, such as personnel, required to accomplish the task. In the end, patient care is drastically improved, patient satisfaction is higher and the providers learn that working as part of a healthcare team is significantly easier and more effective.

FURTHER READING

Afridi, S. A., Jafri, S. F., and Marshall, J. B., 1994, Do gastroenterologists themselves follow the American Cancer Society recommendations for colorectal cancer screening? *Am. J. Gastroenterol.*, **89**, 2184–2187.

Bartlett, J. G., Dowell, S. F., Mandell, L. A., File, Jr, T. M., Musher, D. M., and Fine, M., 2000, Practice guideline or the management of community-acquired pneumonia in adults. *Clin. Infect. Dis.*, **31**, 347–382.

Bero, L. A., Grilli, R., Grimshaw, J. M., Harvey, E., Oxman, A. D., and Thomson, M. A., 1998, Closing the gap between research and practice: An overview of systematic reviews of interventions to promote the implementation of research findings. *BMJ*, **317**, 465–468.

Davis, D., O'Brien, M. A., Freemantle, N., Wolf, F. M., Mazmanian, P., and Taylor-Valsey, A., 1999, Impact of formal continuing medical education: Do conferences, workshops, rounds, and other traditional continuing education activities change physician behavior or health care outcomes? *JAMA*, **282**, 867–874.

Davis, D. A., Thomson, M. A., Oxman, A. D., and Haynes, R. B., 1995, Changing physician performance: A systematic review of the effect of continuing medical education strategies. *JAMA*, **274**, 700–705.

Freemantle, N., Harvey, E. L., Wolf, F., Grimshaw, J. M., Grilli, R., and Bero, L. A., 2000, Printed educational materials: Effects on professional practice and health care outcomes. *Cochrane Database Syst. Rev.*, (2), CD000172.

Greco, P. J. and Eisenberg, J. M., 1993, Changing physicians' practice. *N. Engl. J. Med.*, **329**, 1271–1273.

Grimshaw, J. M. and Russell, I. T., 1993, Effect of clinical guidelines on medical practice: A systematic review of rigorous evaluations. *Lancet*, **342**, 1317–1322.

Grimshaw, J. M., Shirran, L., Thomas, R. E. *et al.*, 2001, Changing provider behaviour: An overview of systematic reviews of interventions. *Med. Care*, **39**(8)(Suppl. 2), 2–45.

Grol, R. and Grimshaw, J., 1995, Evidence-based implementation of evidence-based medicine. *Jt. Comm. J. Qual. Improv.*, **25**, 503–521.

Gross, P. A., Greenfield, S., Cretin, S., Ferguson, J., Grimshaw, J., Grol, R. *et al.*, 2001, Optimal methods for guideline implementation: Conclusions from Leeds Castle meeting. *Med. Care*, **39**(8)(Suppl. 2), II-85–92.

Gross, P. A. and Pujat, D., 2001, Implementing practice guidelines for appropriate antimicrobial usage. *Med. Care*, **39**(8)(Suppl. 2), II-55–69.

Hunt, D. L., Haynes, R. B., Hanna, S. E., and Smith, K., 1998, Effects on computer-based clinical decision support systems on physician performance and patient outcomes: A systematic review. *JAMA*, **280**, 1339–1346.

Laffel, G. and Blumenthal, D., 1989, The case for using industrial quality management science in health care organizations. *JAMA*, **262**, 2869–2873.

NHS Centre for Reviews and Dissemination, 1999, Getting evidence into practice. *Effective Health Care*, **5**(1), 1–16 (also available from http://www.york.ac.uk/inst/crd/ehc51.pdf).

Philipchalk, R. P., 1995, *Invitation to Social Psychology*. Harcourt Brace College Publishers, Fort Worth, TX.

Ramirez, J. A., Vargas, S., Ritter, G. W., Brier, M. E., Wright, A., Smith, S. *et al.*, 1999, Early switch from intravenous to oral antibiotics and early hospital discharge: A prospective observation study of 200 consecutive patients with community-acquired pneumonia. *Arch. Intern. Med.*, **159**, 2449–2454.

Redelmeier, D. A. and Tversky, A., 1990, Discrepancy between medical decisions for individual patients and for groups. *N. Engl. J. Med.*, **322**, 1162–1164.

Shortell, S. M., Bennett, C. L., and Byck, G. R., 1998, Assessing the impact of continuous quality improvement on clinical practice: What it will take to accelerate progress. *Milbank Q.*, **76**, 593–624.

Solberg, L., 2001, Guideline implementation: What the literature doesn't tell us. *Jt. Comm. J. Qual. Improv.*, **26**, 525–537.

Thomson, O'Brien, M. A., Oxman, A. D., Haynes, R. B., Davis, D. A., Freemantle, N. *et al.*, 2000, Local opinion leaders: Effects on professional practice and health care outcomes. *Cochrane Database Syst. Rev.*, (2), CD000125.

Wensing, M. and Grol, R., 1994, Single and combined strategies for implementing changes in primary care: A literature review. *Int. J. Qual. Health Care*, **6**, 115–132.

West, S. G. and Wicklund, R. A., 1980, *A Primer of Social Psychological Theories*. Brooks/Cole Publishing Co., Monterery, CA.

Wyatt, J. C., Paterson-Brown, S., Johanson, R., Altman, D. G., Bradburn, M. J., and Fisk, N. M., 1998, Randomised trial of educational visits to enhance use of systematic reviews in 25 obstetrical units. *BMJ*, **317**, 1041–1046.

Chapter 3

UK Guidelines: Methodology and Standards of Care

Dilip Nathwani
Infection Unit (Ward 42), Ninewells Hospital and Medical School, Dundee DD1 9SY, UK

1.   INTRODUCTION

The culture of searching for evidence in medicine has revolved around three areas—the systemic-pathophysiological, the individual-clinical, and statistico-analytic approach (Trohler, 2001). The origin of the first lies with Socrates (*c*. 400 BC) and Galen (*c*. 200 AD), the second with the long evolution of the process we call "medical judgement" and the last with the work of a number of pioneers such as James Lind, a Scottish Naval Surgeon (1716–94) who validated his intuition about the best treatment for scurvy by undertaking a prospective randomised study on board a British Naval Ship (Trohler, 1981). This and other works, were the beginnings of systematic and analytic evaluation of evidence as a basis for clinical practice. The analytic and evidence-based culture has been prevalent and ascendant in UK practice of the last decade and forms the basis of the clinical effectiveness cycle (Figure 1). This process underpins the core components of ensuring good quality clinical practice. This chapter is based on a paper published in the *Journal of Antimicrobial Chemotherapy* (Nathwani, 2003). It aims to synthesise issues related to clinical effectiveness by addressing the development of evidence-based guidelines and the standards used to evaluate the quality of care, with emphasis on antimicrobial prescribing in hospitals. Although some aspects of this chapter may overlap with other chapters, they are in general complementary and provide the readers with a distinctly UK and Scottish perspective on the subject.

Antibiotic Policies: Theory and Practice. Edited by Gould and van der Meer
Kluwer Academic / Plenum Publishers, New York, 2005

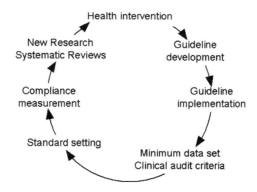

Figure 1. The Clinical effectiveness cycle.

2.   EVIDENCE-BASED GUIDELINES

In 1992, the proponents of modern day evidence-based medicine (EBM) published their recommendations (EBM Working Group, 1992). EBM de-emphasises institutional, unsystematic clinical, and pathophysiological rationale as sufficient grounds for clinical decision making and stresses the examination of evidence from clinical research. Since this one key publication, evidence-based practice has grown exponentially and has stimulated evidence-based clinical practice guidelines. They have been defined as systematically developed statements to assist practitioner and patient decisions about appropriate health-care for specific clinical circumstances (Field and Lohr, 1990). Importantly, they must also reflect the routine working practices of most doctors for them to be accepted as a gold standard that most doctors will accept or admire. In the United Kingdom, clinical guidelines supported by the NHS Management Executive (1993), have been highlighted as a key tool to promote best practice and improve clinical effectiveness, although they cannot be used to mandate, authorise, or outlaw treatment options (Hurwitz, 1999). The criteria for successful development and implementation have subsequently been published by the Nuffield Institute for Health Publication (1994) and the Scottish Intercollegiate Guideline Network (SIGN) in 1993 (http://www.sign.ac.uk; SIGN 50) (SIGN, 2001).

3.   TYPE AND SOURCES OF INFECTION GUIDELINES IN THE UK

In the last decade, the numbers and sources of guidelines have proliferated to an unimaginable level, including those in infection. For example, in the field of infection, UK-based therapeutic guidelines exist for a range of clinical infections, for chemoprophylaxis and immunisation and for infection control. In order for the guideline to be effective, the rigour of development should be transparent and subject to scrutiny. It is important for the reader to appreciate current criteria for the quality of guideline development. This is discussed below. To enhance the dissemination of information related to the effectiveness of treatments or interventions, many sources are available. The following web-sites, amongst many, are particularly useful for guidelines on the web from the UK: National Institute for Clinical Excellence (http://www.nice.org.uk); SIGN (http://www.sign.ac.uk); and others http://www.healthcentre.org.uk/hc/library/guidelines.htm; http://www.sghms.ac.uk/phs/hceu/nhsguide.htm; Eguidelines (http://www.eguidelines.co.uk/link.acgi?index.htm) is a useful source that summarises a range of clinical guidelines, albeit for primary care, but also provides information regarding quality such as the development sources and commendation by various independent and respected authorities. The latter is particularly useful to the busy clinician who does not have the time to appraise the quality of a particular guideline.

4.   RECOMMENDED GUIDELINE DEVELOPMENT METHODOLOGY

One of the key components of a successful or good guideline is the rigour of the development process (Feder, 1998). Unfortunately, most of the guidelines produced by specialist societies do not meet the basic principles of guideline development. One important study reported that 67% did not report any description of the type of stakeholders, 88% gave no information on searches for published studies, and 82% did not give any explicit grading of the strength of the recommendations (Grilli *et al.*, 2000). Of the 431 eligible guidelines considered between January 1988 and July 1998, very few appeared to be for infection although the exact numbers were not explicit in the paper. Another study (Shaneyfely *et al.*, 1999) revealed that the methodological quality of guidelines is generally poor and often exhibits great variation and conflicting recommendations. This study of 279 guidelines published from 1985 to 1997 revealed a mean (standard deviation, SD) adherence to methodological standards on guideline development of 51.1%. There is clearly a need for a common, international, valid and transparent approach to develop good clinical practice guidelines (Cluzeau, 2000; SIGN, 2001). The Appraisal of Guidelines, Research and Evaluation for Europe (AGREE; http://www.agreecollaboration.org) collaboration have developed a European agreed generic methodology to assess the quality of guidelines and the guideline development process. The SIGN (http://www.sign.ac.uk) has long adopted the majority of these criteria and has recently updated many aspects of SIGN methodology which are consistent with the agree criteria. SIGN 50 is an excellent and commendable resource to the guideline process for any group wishing to embark upon this (SIGN, 2001). All information is available on these websites. A more recent comparison of 18 clinical guideline programmes has confirmed a significant improvement in guideline development from around the globe but once again the data emphasises the importance of a continuing international effort to globalise this process (Burgers *et al.*, 2003). Guideline developers in the infection community worldwide need to be encouraged to broadly embrace this process and consider these issues when planning future guidelines. The current guidelines for hospital and ventilator-associated pneumonia being undertaken by a working group of the British Society for Antimicrobial Chemotherapy (BSAC) appear to have taken on board these important considerations.

5.   OTHER APPROACHES TO GUIDELINE DEVELOPMENT

Unfortunately, old habits die hard and many clinicians or organisations persist with supporting nonstandardised *ad hoc* statements or reviews, usually from expert bodies—this undesirable methodology or GOBSAT (Good Old Boys Sat Around a Table) should no longer be valid or encouraged (Miller and Petrie, 2000). These discussions are based on received wisdom, clinical judgement, and experience rather than current scientific evidence; lack an explicit decision-making process; and may be biased by undeclared conflicts of interest (Miller and Petrie, 2000). A good example of this was the guidelines for improving the use of antimicrobial agents in hospitals—a statement by the Infectious Diseases Society of America (IDSA) published in 1988 (Marr *et al.*, 1988). Although commendable in their efforts at aiming to promote good quality prescribing and that it remains a well-respected document, the development process was not evidence based as it was drafted by three authors with subsequent review by 43 multidisciplinary members of the IDSA. Other deficiencies of this document were the rather unhelpful format, lack of summary of the evidence, and no linkage of this to the ultimate recommendations. This chapter highlights a typical example of development methodological flaw before appreciation of EBM methodology became widespread.

"Consensus"-based statements (Murphy *et al.*, 1998) are popular and involve a broad-based panel which listens to the scientific data presented by experts, weighs the information, and then composes a consensus statement that addresses a set of questions previously posed to the panel. Once again within a relatively small group, the interactions are such that some members will have a significant impact on the overall decisions. Other sources of bias include the type of questions set, the composition of the panel, and the selection of the experts and literature. Examples of such a process is the consensus panel recommendations for managing serious candidaemia (Edwards *et al.*, 1997), or synthesising a consensus strategy for combating the prevention and control of antimicrobial resistance microorganisms in hospitals (Goldman *et al.*, 1996). The latter document is worthy of recognition as a good consensus statement—it aims to synthesise a strategy from expert opinion, experience, and key existing evidence. It is explicit in its intention and development process and recognises its inherent deficiencies. This document is widely acknowledged as pivotal in the process of building a subsequent evidence-based approach to preventing antimicrobial resistance in hospitals. This is recognised in the subsequent guidelines on this subject by the Society for Healthcare Epidemiology of America (SHEA) and the IDSA Joint Committee on the Prevention of Antimicrobial Resistance (Shlaes *et al.*, 1997). This approach, although valuable, does not follow the AGREE or SIGN developmental methodology. When the AGREE Instrument for appraising the quality of this guideline is applied to this report (www.agreecollaboration, June 2001), one identifies a number of deficiencies in the areas of stakeholder involvement, rigour of development, and clarity of presentation. This guideline takes much more the form of a specialist but not systematic review, with only one table presented linking four recommendations to evidence. Greater uniformity with AGREE by such North American approach to guidelines would be welcomed. A more structured consensus approach to gathering expert opinion is the classic delphi panel. This technique has merits in that the process is structured, has a number of stages that lead ultimately to a convergence of opinion, and is coordinated by a central person who summarises and feeds the information back from the panel whose responses are anonymous (Evans, 1999).

6. VARIABILITY OF QUALITY OF INFECTION PRACTICE GUIDELINES

In infection practice over many years, we have seen a plethora of documents from speciality societies or other well-respected bodies which have claimed "guideline status" despite a significant absence of methodological rigour. Although one may argue that some of these documents have been well accepted by the broad infection community, for example, the first British Thoracic Society Community Acquired Pneumonia Guidelines (BTS, 1993), and the BSAC Endocarditis Guidelines (Shanson, 1998), both of which have considerably influenced "standard" practice (Woodhead and McFarlane, 2000), one has to question the evidence base of some of these recommendations. This has recently been highlighted for the management of bacterial endocarditis (Graham and Gould, 2002). However, more encouraging signs are appearing. In the last 3 to 4 years, prominent infection and associated groups are showing signs of embracing the EBM approach. These have been highlighted in the IDSA guideline development standards (Kish, 2001) for a plethora of evidence-based guidelines, the updated BTS guidelines for community-acquired pneumonia (CAP) (BTS, 2001) and the SIGN guidelines on acute sore throat and tonsillectomy (SIGN 34), management of genital Chlamydia trachomatis infection (SIGN 42), surgical antibiotic prophylaxis guidelines (SIGN 45), and those for management of lower respiratory tract infection (SIGN 59). Guidelines for the management of adult urinary tract infections are in the developmental phase. The follow up IDSA guideline for managing candidiasis confirmed the adoption of this approach in North America (Rex *et al.*, 2000). This more evidence-based approach has been augmented by greater appreciation amongst clinicians of their value. For example, in an important study by Farquhar *et al.* (2002), clinicians attitudes to clinical practice guidelines were found to be generally favourable as they believed they were educational and would improve quality.[9] A recent survey of Australian Intensivists and ID physicians revealed similar positive attitudes towards a more evidence-based antibiotic use approach in critical care (Sintchenko *et al.*, 2001). However, in the Farquhar systematic review (Farquhar *et al.*, 2002), significant concerns were expressed about guideline practicality, their role in cost-cutting and their potential for increasing litigation. Indeed, Italian physicians (Formoso *et al.*, 2001) in secondary and primary care, particularly share these concerns and appear to suggest poor attitude towards guidelines in general and particularly the proposed methodology for guideline development. One fears that this scepticism may not be confined to Italian physicians only.

7.   GUIDELINE IMPLEMENTATION

How best EBM guidelines are implemented is a crucial question so that a desired change in practice is made and sustained. The determinants of changing physician behaviour are multifactorial and appear to be dependent on an improvement in knowledge, a change in attitude, and a number of organisational, social, and personal factors (Cabana *et al.*, 1999). The implementation strategies which may be most effective at bringing about a positive response to antibiotic guidelines have been the subject of two recent reviews (Brown, 2002; Davey *et al.*, 2003).

Strategies that aim to provide frequent reminders supported by education, provision of interactive and meaningful educational workshops and undertaking a multifaceted approach to guideline implementation have shown to be the most effective (Grol and Grimshaw, 2003; Moulding *et al.*, 1999).

Ultimately, regardless of the optimal implementation strategy, if we believe guidelines improve the clinical and cost effectiveness of healthcare (Davey *et al.*, 2003), then we need to ensure good compliance using internal and external quality assurance systems within any organisation. Therefore, one may conclude that clear implementation strategies with audit criteria should be considered in parallel with the development of the guideline and should be given equal importance.

8. QUALITY ASSURANCE AND DEVELOPMENT OF STANDARDS

Practice guideline programmes, are one of many types of quality programme, which are increasingly being used to improve the quality of care (Ovretveit and Gustafson, 2003). One of the key components of a guideline programme is identification of or setting of key standards and criteria for audit. These clinical standards, based on existing evidence, would be used subsequently as the criterion for evaluating the quality of care provided by an organisation or individual unit or department. Such quality assurance may be undertaken through an internal or external peer review or through an accreditation process (Del Mar, 2001; Steel, 2001). In the United Kingdom, as in Australia, guidelines and clinical standards underpin much of the quality, or more recently, the clinical governance agenda (Scally and Donaldson, 1998). This process aims to make it a statutory responsibility for each organisation to be accountable for ensuring the monitoring and improvement of the quality of healthcare it provides (Scally and Donaldson, 1998).

Clinical standards primarily enable identification of the essentials that need to be right in the treatment of particular conditions if outcomes for the patients are to be optimised. Standards can be set at several levels: minimal, normative, and exemplary or may be deemed essential or desirable (Del Mar, 2001). It is important to recognise which level should be applied to any standard as minimal standards are primarily aimed at promoting basic levels of care by identifying those areas or professionals who perhaps require remedial, or in rare cases even punitive action. Outcome-related standards are deemed as the "gold standard" of performance measurement but in reality they are difficult to capture, particularly in the short term (Davies and Crombie, 1997; Goddard *et al.*, 2002). Increasingly, process or to a lesser extent structure measures are deemed more attractive especially if they are linked through evidence to outcomes (Crombie and Davies, 1998). Indeed, guidelines or care pathways will outline intervention or processes of care that lead to a desired outcome (Nathwani *et al.*, 2001). The timely (within 4–8 hr of admission) administration of appropriate intravenous antibiotics for patients with severe CAP is regarded as a key quality indicator (Nathwani *et al.*, 2001, 2002). This intervention is regarded as an important, validated, credible, consistent, simple, and measurable process standard based on evidence that is valued by clinicians, quality administrators, and patients (Nathwani *et al.*, 2002). A CAP audit in Tayside used this as one of the key performance indicators in prospectively evaluating pneumonia care. This study revealed that a significant 39% of patients admitted with severe CAP did not receive antibiotics within the appropriate timeframe. Indeed, 29% did not receive intravenous antibiotics within 24 hr of hospital admission (Marrie *et al.*, 2000). Poor performance on a process measure gives a clear indication of the remedial action that is required and this can be linked to an incentive to bring about positive change. This audit in Dundee stimulated broad educational feedback and the development and implementation of a care pathway for CAP (Marrie *et al.*, 2000) which will be subject to further evaluation (G. Barlow, personal communication). The pathway implementation is supported by a number of proactive educational and feedback interventions. On the other hand, a commonly used crude outcome marker of death is more difficult to interpret as it is insensitive to the quality of healthcare received and can be influenced by a range of other factors (Mant and Hicks, 1995). In our CAP audit (Nathwani *et al.*, 2002), compliance with the unit protocol did appear to correlate to a reduction in mortality but the association was by no means robust.

9. DEVELOPMENT OF CLINICAL STANDARDS IN SCOTLAND

In Scotland, the development of Clinical Standards has been undertaken by the NHS Quality Improvement Scotland (www.nhshealthquality.org) are subject to audit by means of an internal self-assessment followed by external review. This process aims to increase and promote greater public confidence in the overall standard of care (Steel, 2001).

Standards represent an agreed level of performance and this level should be determined by those who are involved in delivery or receipt of the service. The criteria attached to each standard provide more detailed and practical information on how to achieve the standard and can be described as structure, process, and outcome criteria. These standards focus on the patient journey, the care and treatment the patient receives, and relate very closely to patient outcome; are underpinned by evidence base; may be specific or generic; and are categorised as being desirable or essential (Steel, 2001). An evaluation of national performance against healthcare acquired infection standards has recently been published (www.nhshealthquality.org). The methodology adopted in developing these standards is worthy of greater attention and should form the core principles of any future standards developmental methodology. This methodology (Figure 2) has three distinct phases of setting and describing the standards, measuring the performance followed by recording, analysing, and disseminating the findings. This model may form the "gold standard" methodology but is clearly resource intensive. Therefore, other methods of developing a consensus of recommendations based on evidence and/or good practice are also popular and valued. An example of such work is the standards of care for patients with invasive fungal infections recently published by the British Society for Medical Microbiology (Dennig *et al.*, 2003). The Scottish Infection Standards and Strategy Group (SISS; available on the www.rcpe.ac.uk site), a broad infection specialty subgroup (microbiologists, ID physicians, genito-urinary medicine physicians, public health physicians, epidemiologists, infection control nurses, and pharmacists) of the Bicollegiate Quality of Care Committee, has recognised this and other methodology and has developed good practice recommendations for optimising quality prescribing in hospitals. These recommendations adopt and adapt work based on existing evidence and national strategies aimed at optimising antibiotic prescribing in hospitals (Goldman *et al.*, 1996; Shlaes *et al.*, 1997). They represent a mixture of good practice statements supported by verification criteria aimed at evaluating the process and organisational interventions related to hospital antibiotic prescribing. Further development and validation of "traditional" standards and introducing more "patient based" outcomes are clearly desirable and necessary in the future (Barlow *et al.*, 2003). The SISS good practice guidance statements by no means represent the final product but form the basis of a process that will continue to evolve and adapt in line with validated standard development

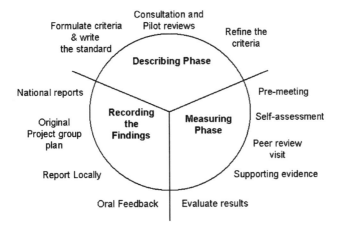

Figure 2. Setting standards: The model.

methodology. We hope that such standards will become an established national tool for benchmarking the quality of antibiotic prescribing in hospitals with the objective of encouraging change where suboptimal practice is identified. This process is a key requirement of the UK Antimicrobial Resistance Strategy (see references).

10.   CONCLUSION

The infection community needs to embrace evidenced-based guideline methodology as a means of ensuring one of the key components of the validity and success of guidelines. However, one must not underestimate the pivotal role of guideline dissemination and implementation in bringing about positive change (Finch and Low, 2002). Indeed, a recent survey (Implementation of SIGN Guidelines in NHS Scotland; Clinical Resources and Audit Group [CRAG], June 2002) of the implementation of a range of noninfection SIGN guidelines in Scotland, revealed significant variability and only 52% implementation at best. This study identified a number of factors and barriers behind successful implementation. These include local/national priority, high level of evidence or recommendations, local champion/lead, resources, intended but not done, and "too complex." The value of such surveys cannot be underestimated as they identify targets for change related to specific disease areas or geographical locations. The discovery of such critical factors for successful change has also been identified as an important research strategy for informing quality programmes (Ovretveit and Gustafson, 2003). Consensus-based approaches may be useful in specific situations, especially where adequate evidence does not exist, but one must avoid the continuing attraction of GOBSAT statements regardless of the stature of the specialist body or its members. The use of consensus as a means of gathering expert opinion and experience has proven to be valuable in informing good practice statements or guidance. Implementation strategies ought to be developed in parallel with this process so that they are effective at the clinical coal face. Standards of care emerging from evidence-based clinical practice guidelines or consensus-based good practice recommendations for antibiotic prescribing in hospitals are a helpful basis for quality assurance nationally and within each organisation. Audit of antibiotic policies in South East England, using such standards, recently revealed considerable variation in content and quality across policies and a clear lack of evidence base (Wiffen and Mayon-White, 2001). The infection community needs to develop existing standards further so they are applicable nationally, if not internationally, but with the important caveat that they ought to be realistic and achievable locally. Measuring compliance with standards ought to be promoted through audit, education, and feedback, but may require to be enforced through internal or external systems of governance. The development of an outcome-based national hospital prescribing database or registry (McKee, 1993) using readily available clinical and microbiological datasets for this purpose should underpin this process.

REFERENCES

Barlow, G. D., Lamping, D., Davey, P. G., and Nathwani, D., 2003, Evaluating outcomes in community-acquired pneumonia: A guide for patients, physicians and policy-makers. *Lancet Infect. Dis.*, submitted for publication.

British Thoracic Society, 1993, British Thoracic Guidelines for the management of community-acquired pneumonia in adults admitted to hospital. *Br. J. Hosp. Med.*, **48**, 346–350.

British Thoracic Society Standards of Care Committee, 2001, British Thoracic Guidelines for the management of community-acquired pneumonia in adults. *Thorax*, **56**, Supplement 4, IV1–64.

Brown, E. M., 2002, Guidelines for antibiotic usage in hospitals. *J. Antimicrob. Chemother.*, **49**, 587–592.

Burgers, J. S., Fervers, B., Haugh, M., Brouwers, M., Browman, G., Philip, T., and Cluzeau, F. A., 2004, International assessment of the quality of clinical practice guidelines in oncology using the appraisal of guidelines and research and evaluation instrument, *J. Clin. Oncol.*, **22**(10), 2000–2007.

Cabana, M. D., Rand, C. S., Powe, N. R., Wu, A. W., Wilson, M. H., Abboud, P. A. C. *et al.*, 1999, Why don't physicians follow clinical practice guidelines? *JAMA*, **282**, 1458–1465.

Cluzeau, F., for the AGREE Collaborative Group, 2000, Guideline development in Europe: An international comparison. *Int. J. Technol. Assessment Health Care*, **16**, 1036–1046.

Evidence Based Medicine Working Group, 1992, Evidence-base medicine. A new approach to teaching the practice of medicine. *JAMA*, **268**, 2420–2425.

Crombie, I. K. and Davies, H. T. O., 1998, Beyond health outcomes: The advantages of measuring process. *J. Eval. Clin. Pract.*, **4**, 31–38.

Davey, P., Nathwani, D., and Rubenstein, E., 2003, Antibiotic policies. In R. G. Finch, D. Greenwood, S. R. Norrby, and R. Whitley, (eds.), *Antibiotic and Chemotherapy*, 8th edn. Churchill Livingstone, London and Edinburgh, pp. 123–138.

Davies, H. T. O. and Crombie, I. K., 1997, Interpreting health outcomes. *J. Eval. Clin. Pract.*, **3**, 187–200.

Del Mar, C., 2001, Guiding guidelines into practice. *Aust. Prescriber*, **24**, 50–51.

Dennig, D. W., Kibbler, C. C., and Barnes, R. A., on behalf of the British Society for Medical Mycology, 2003, British Society for Medical Mycology proposed standards of care for patients with invasive fungal infections. *Lancet Infect. Dis.*, **3**, 230–240.

Edwards, J. E., Bodey, G. P., and Bowden, R. A., 1997, International conference for the development of a consensus on the management and prevention of severe candidal infections. *Clin. Infect. Dis.*, **25**, 43–59.

Evans, C., 1999, The use of consensus methods and expert panels in pharmacoeconomic studies. *Pharmacoeconomics*, **12** (2 Pt 1), 121–129.

Evidence-Based Medicine Working Group, 1992, Evidence-based medicine: A new approach to teaching the practice of medicine. *JAMA*, **268**, 2420–2425.

Farquhar, C. M., Kofa, E. W., and Slutsky, J. R., 2002, Clinicians' attitudes to clinical practice guidelines: A systematic review. *Med. J. Aust.*, **177**, 502–506.

Feder, G., 1998, Guidelines for clinical guidelines. *BMJ*, **317**, 427–428.

Field, M. J. and Lohr, K. N., eds., 1990, Institute of Medicine Committee to Advise the Public Health Service on Clinical Practice Guidelines. Clinical practice guidelines: Directions for a new program. National Academy Press, Washington, DC.

Finch, R. G. and Low, D. E., 2002, A critical assessment of published guidelines and other decision support systems for the antibiotic treatment of community-acquired respiratory tract infections. *Clin. Microbiol. Infect.*, **8**(Suppl. 2), 69–92.

Formoso, G., Liberati, A., and Magrini, N., 2001, Practice guidelines: Useful and "participative" method? Survey of Italian Physicians by Professional Setting. *Arch. Intern. Med.*, **161**, 2037–2042.

Goddard, M., Davies, H. T., Dawson, D., Mannion, R., and McInnes, F., 2002, Clinical performance measurement: Part 1—getting the best out of it. *J. R. Soc. Med.*, **95**, 508–510.

Goldman, D. A., Weinstein, R. A., Wenzel, R. P., Tablan, O. C., Duma, R. J., Gaynes, R. P. *et al.*, 1996, Strategies to prevent and control the emergence and spread of antimicrobial-resistant microorganisms in hospitals. A challenge for hospital leadership. *JAMA*, **275**, 234–240.

Good practice guidance for antibiotic prescribing in hospital. Prepared by the Scottish Infection Standards and Strategy (SISS) Group. *J R Coll Physicians Edinb* 2003; **33**, 281–284.

Graham, J. C. and Gould, F. K., 2002, Role of aminoglycosides in the treatment of bacterial endocarditis. *J. Antimicrob. Chemother.*, **49**, 437–444.

Grilli, R., Magrini, N., Penna, A., Mura, G., and Liberati, A., 2000, Practice guidelines developed by specialty societies: The need for critical appraisal. *Lancet*, **355**, 103–106.

Grol, R. and Grimshaw, J., 2003, From best evidence to best practice: effective implementation of change in patient care. *Lancet*, **362**, 1225–1230.

Hurwitz, B., 1999, Legal and political considerations of clinical practice guidelines. *BMJ*, **318**, 661–664.

Kish, M. A., 2001, Guide to development of practice guidelines. *Clin. Infect. Dis.*, **32**, 851–854.

Mant, J. and Hicks, N., 1995, Detecting differences in quality of care: The sensibility of measures of process and outcomes in treating myocardial infarction. *BMJ*, **311**, 793–797.

Marr, J. J., Moffet, H. L., and Kunin, C. M., 1988, Guidelines for improving the use of antimicrobial agents in hospitals: A statement by the Infectious Diseases Society of America. *J. Infect. Dis.*, **157**, 869–876.

Marrie, T. J., Lau, C. Y., Wheeler, S. L., Wong, C. J., Vanderwoort, M. K., and Feagan, B. G., for the CAPITAL Study Investigators, 2000, A controlled trial of a critical pathway for treatment of community-acquired pneumonia. *JAMA*, **283**, 749–755.

McKee, M., 1993, Routine data: A resource for clinical audit. *Qual. Health Care*, **2**, 104–111.

Miller, J. and Petrie, J., 2000, Development of practice guidelines. *Lancet*, **355**, 82–83.

Moulding, N. T., Silagy, C. A., and Weller, D. P., 1999, A framework for effective management of change in clinical practice: Dissemination and implementation of clinical practice guidelines. *Qual. Health Care*, **8**, 177–183.

Murphy, M. K., Black, N. A., and Lamping, D. L., 1998, Consensus development methods and their use in clinical guideline development. *Health Technol. Assessment*, **2**.

Nathwani, D., 2003, From evidence-based guideline methodology to quality of care standards. *J. Antimicrob. Chemother.*, **51**, 1103–1107.

Nathwani, D., Rubenstein, E., Barlow, G., and Davey, P., 2001, Do guidelines for community acquired pneumonia improve cost-effectiveness of hospital care? *Clin. Infect. Dis.*, **32**, 728–741.

Nathwani, D., Williams, F., Winter, J., Winter, J., Ogston, S., and Davey, P., 2002, Use of indicators to evaluate the quality of community acquired pneumonia management. *Clin. Infect. Dis.*, **34**, 318–329.

NHS Management Executive, 1993, *Improving Clinical Guidelines, EL* (93), 115. Department of Health, Leeds.

Nuffield Institute for Health, 1994, Effective health care. Implementing clinical guidelines, Bulletin No. 8. University of Leeds, Leeds.

Ovretveit, J. and Gustafson, D., 2003, Using research to inform the quality of programmes. *BMJ*, **326**, 759–761.

Rex, J. H., Walsh, T. J., Sobel, J. D., Filler, S. G., Pappas, P. G., Dismukes, W. E. *et al.*, 2000, Practice guidelines for the treatment of candidiasis. *Clin. Infect. Dis.*, **30**, 662–678.

Scally, G. and Donaldson, L. J., 1998, Clinical governance and the drive for quality improvement in the new NHS England. *BMJ*, **317**, 61–65.

Scottish Intercollegiate Guidelines Network (SIGN), 2001, SIGN 50—A guideline developers' handbook. *SIGN*, Edinburgh.

Shaneyfely, T. M., Mayo-Smith, M. F., and Rothwangl, J., 1999, Are guidelines following guidelines? The Methodological Quality of Clinical Practice Guidelines in the Peer Reviewed Medical Literature. *JAMA*, **281**, 1900–1905.

Shanson, D. C., 1998, New guidelines for the antibiotic treatment of streptococcal, enterococcal and staphylococcal endocarditis. *J. Antimicrob. Chemother.*, **42**, 292–296.

Shlaes, D. M., Gerding, D. N., John, J. F., Craig, W., Bornstein, D. L., Duncan, R. A. *et al.*, 1997, Society for Healthcare Epidemiology of America and Infectious Diseases Society of America Joint Committee on the Prevention of Antimicrobial Resistance: Guidelines for the prevention of antimicrobial resistance in hospitals. *Clin. Infect. Dis.*, **25**, 584–599.

Sintchenko, V., Iredell, J. R., Gilbert, G. L., and Coiera, E., 2001, What do physicians think about evidence-based antibiotic use in critical care? A survey of Australian intensivists and infectious disease practitioners. *Intern. Med. J.*, **31**, 462–469.

Steel, D. R., 2001, From evidence-based medicine to clinical standards. *Proc. R. Coll. Physicians Edinb.*, **31**(Suppl. 9), 74–76.

Trohler, U., 1981, Towards clinical investigation on a numerical basis: James Lind at Haslar Hospital 1758–1783. In *Proc XXVII. Int. Congr. Med Barcelona. Barcelona: Academia de Ciences Mediques de Catalunya I Balears*, **1**, 414–419.

Trohler, U., 2001, The history of clinical effectiveness. *Proc. R. Coll. Physicians Edinb.*, **31**, (Suppl. 9), 42–45.

Wiffen, P. J. and Mayon-White, P. T., 2001, Encouraging good antimicrobial prescribing practice: A review of antibiotic prescribing policies used in the South East Region of England. *BMC Public Health*, **1**, 4 (http://www.biomedcentral.com/147-2458/1/4).

Woodhead, M., McFarlane, J., for the BTS CAP Guideline Committee, 2000, Local antibiotic guidelines for adult community acquired pneumonia (CAP): A survey of UK hospital practice in 1999. *J. Antimicrob. Chemother.*, **46**, 141–143.

UK Antimicrobial Resistance Strategy and Action Plan. Department of Health, http://www.doh.gov.uk/arbstrat.htm.

Chapter 4

Pneumonia Guidelines in Practice

Gavin Barlow
*Infection Unit, Tayside University Hospitals NHS Trust,
Dundee DD1 9SY, UK*

1.   INTRODUCTION

Community-acquired pneumonia (CAP) is a common reason for admission to hospitals worldwide and is associated with considerable mortality, morbidity, and use of healthcare resources. For example, it has been estimated that the direct healthcare costs of CAP in the United Kingdom and the United States are in the order of £441 million (Guest and Morris, 1997) and $8.4 billion (Niederman *et al.*, 1998) per year, respectively. Over the last decade there has been a global proliferation in national CAP guideline development and in local and regional quality improvement (QI) initiatives. It is likely that this has been due to a combination of factors, such as the recognition of unexplained variation in clinical practice, the drive towards cost-effective healthcare provision, concern regarding litigation and both the guideline and evidence-based medicine movements.

During this period there has also been increased research activity relating to CAP. This has led to: an improved understanding of aetiology, for example, the emergence of *Chlamydia pneumoniae* as an independent and co-infecting pathogen (Lim *et al.*, 2001); prognostic and clinical-decision support tools (Fine *et al.*, 1997), such as the Pneumonia Severity Index (PSI); understanding of the link between process of care and clinical (Meehan *et al.*, 1997) and economic outcomes (Gleason *et al.*, 1997); evidence regarding the safety of intravenous (IV) to oral switch therapy (Castro-Guardiola *et al.*, 2001); markers of clinical stability to guide hospital discharge (Halm *et al.*, 2002); and new antibiotics, such as fluoroquinolones with enhanced activity against

Antibiotic Policies: Theory and Practice. Edited by Gould and van der Meer
Kluwer Academic / Plenum Publishers, New York, 2005

Streptococcus pneumoniae (File *et al.*, 1997). We have also seen the emergence of the spectre of penicillin and multidrug resistant pneumococcal infection (Pallares *et al.*, 1995). To accommodate this knowledge, changes in the framework of healthcare provision and to counter concerns about the overuse of antibiotics in respiratory tract infections, most of the early CAP guidelines (Bartlett *et al.*, 1998; British Thoracic Society, 1993; Mandell and Niederman, 1993; Niederman *et al.*, 1993), which were largely based on expert opinion, have been updated using evidence-based methodology (American Thoracic Society, 2001; Bartlett *et al.*, 2000; British Thoracic Society, 2001; Mandell *et al.*, 2000). Despite this expanding evidence and guidance base, many important questions remain unanswered. One important question is:

How should healthcare professionals and managers implement national recommendations in order to optimise process of care and outcomes that are important to patients, physicians, and policymakers?

This chapter will review: (1) the QI interventions that have been studied in CAP, (2) the importance of the link between process and outcomes in QI, (3) the evidence that QI interventions can impact on process of care and clinical, patient-centred, and economic outcomes in CAP, and (4) the important steps to consider in the development, implementation, and evaluation of a CAP QI initiative.

2.   WHICH QI INTERVENTIONS HAVE BEEN STUDIED IN CAP?

There is little evidence to suggest that any of the specialist society guidelines developed over the past decade have improved the management of CAP at a national level, although a number of studies have demonstrated that adherence with certain aspects of guidelines is associated with improved outcomes (Battleman *et al.*, 2002; Gordon *et al.*, 1996; Menendez *et al.*, 2002). In the United Kingdom, for example, it is unlikely that the British Thoracic Society (BTS) guidelines will have a nationwide impact on the quality of CAP care until they are widely and appropriately implemented in primary and secondary care. Researchers have used a range of interventions, incorporating various aspects of specialist society guidelines, at a local or regional level, however, to try and improve the quality of CAP healthcare. The majority of these studies are from North America and many have been published in the health management or QI literature (e.g., Florida Medical Quality Assurance, Inc., 1998; Fortune *et al.*, 1996; Gottlieb *et al.*, 1996; Halley, 2000; McGarvey and Harper, 1993; Phillips and Crain, 1998; Reddy *et al.*, 2001; Rollins *et al.*, 1994; Ross *et al.*, 1997; Sperry and Birdsall, 1994), which is generally inaccessible to practising European clinicians. Details of the studies published in the English language medical literature are summarised in Tables 1 and 2.

The objective of most of these studies has been to optimise healthcare process while maintaining or improving clinical or economic outcomes. Only a few studies have measured the impact of QI interventions on patient-centred outcomes and none have assessed cost-effectiveness. All appear to have used a multidisciplinary development process and complex (multifaceted) interventions. While interventions and implementation strategies have varied, common threads are seen in the use of: (1) practice guidelines/protocols, care pathways, or decision support and (2) presentations to healthcare professionals. Research nurses have sometimes been used to support interventions, which is unlikely to be sustainable in the long term. Audit and feedback, reminders and academic detailing (Solomon *et al.*, 2001) have also been used in some studies.

Unfortunately, these studies provide little data on the clinical effectiveness and costs of the single components of multifaceted interventions and there is little information therefore about which single interventions and implementation strategies provide the best value for money. Additionally, the health policy implications of many of these studies are limited because of the common use of uncontrolled observational study designs. For example, the positive results seen in many studies could simply be due to regression to the mean or prevailing temporal trends (American College of Physicians–American Society of Internal Medicine, 1999), such as the known decreasing trend in the length of hospital stay for CAP patients in the United States (Metersky *et al.*, 2000). Because of the predilection of authors and journals to publish positive data, there is also likely to be publication bias.

Particularly in North America, care pathways are commonly used interventions to try and optimise the management of CAP. Care pathways were first developed as an industry tool to identify and manage the rate limiting steps of industrial processes. Over the past two decades they have increasingly been used in healthcare, initially to improve the efficiency of resource use in managed care systems and more recently to improve the quality of healthcare (Pearson *et al.*, 1995). The National Pathway Association (2001) of the United Kingdom defines a care pathway as:

An integrated care pathway determines locally agreed, multidisciplinary practice based on guidelines and evidence where available, for a specific patient/user group. It forms all or part of the clinical record, documents the care given and facilitates the evaluation of outcomes for continuous quality improvement.

The main distinguishing feature from guidelines is that care pathways inform the healthcare professional about an optimal timing and sequence of key healthcare events rather than purely instructing on, for example, the indications for ordering a specific test.

Table 1. A summary of controlled CAP QI studies published in the English language medical literature

Study year (country) Objective	Intervention and implementation	Study design (total sample size)	Results
Marrie *et al.*, 2000 (Canada) To evaluate the safety and effectiveness of a CAP critical pathway	1. Educational plan to reinforce adherence 2. Critical pathway using the PSI, levofloxacin, and IV-oral switch and discharge criteria 3. Study nurse prompts of switch/discharge criteria	Cluster randomised controlled trial of 19 hospitals (1743 patients)	1. Lower admission of low-risk patients in intervention hospitals (31% vs 49% $P = 0.01$) 2. Mean bed-days per patient managed lower in pathway arm (4.4 vs 6.1, $P = 0.04$) 3. No difference in patient quality of life
Dean *et al.*, 2001 (USA) To assess the impact of a pneumonia guideline on mortality	1. ATS-based decision support 2. Presentations 3. Letters and reminders 4. Academic detailing 5. Outpatient and admission order sheets 6. Feedback of outcomes	Controlled before–after (28661 patients)	1. 30-day mortality in hospitalised patients decreased from 13.4% to 11% (pathway) vs 13.2% to 14.2% (non-pathway) (OR = 0.69, 0.49–0.97) 2. Overall use of guideline antibiotics increased from 28% to 56%
Chu *et al.*, 2003 (USA) To demonstrate that an external QI project can lead to an improvement in care not accounted for by temporal trends	1. Feedback of quality indicators 2. The use of either: (a) Clinical pathways *or* (b) Standing orders	Controlled before–after with control hospitals subsequently being exposed to the intervention (2087 patients)	1. Patients receiving antibiotics within 4 hr increased from 57% to 69% ($P < 0.001$) in the intervention group, but decreased in controls 2. Sputum and blood culture procurement within 4 hr also significantly improved 3. Following exposure to the intervention, care at control hospitals also significantly improved

Note: ATS—American Thoracic Society.

Table 2. A summary of uncontrolled CAP QI Studies published in the English language medical literature

Study year (country) Objective	Intervention and implementation	Study design (Total sample size)	Results
Atlas et al., 1998 (USA) To safely increase the % of patients managed as outpatients	1. Identification of low-risk patients (using PSI and nurse support) 2. Enhanced outpatient services 3. Oral clarithromycin 4. Protocol presented to physicians	Uncontrolled before–after (313) with a PORT "control" cohort for patient satisfaction (206)	1. Patients managed as outpatients increased from 42% to 57% (P = 0.01) 2. Eight intervention patients were subsequently hospitalised within 4 weeks compared to no controls 3. Patients in the intervention cohort were less satisfied with care (71% vs 90%, P = 0.04)
Rhew et al., 1998 (USA) To examine the effect of physician decision support for IV-oral switch and discharge	1. Physician and nurse champions 2. Educational sessions 3. Provision of "real-time" decision support about the risk of IV-oral switch and discharge	Uncontrolled before–after (152)	1. No differences in quality of life (SF-36), patient satisfaction or readmission to hospital 2. Underpowered to detect clinically important differences
Benenson et al., 1999 (USA) To determine if a clinical pathway resulted in a shorter antibiotic delivery time and length of hospital stay and improved mortality	1. Clinical pathway (implementation strategy not detailed)	Uncontrolled before–after with two post-intervention cohorts (281)	1. Time to antibiotics decreased from 315 to 175 and 171 minutes in the first and second post-intervention periods, respectively (P < 0.0001) 2. Length of stay decreased from 9.7 to 8.9 (first) and 6.4 (second) days (P < 0.0001)
Al-Eidan et al., 2000 (Northern Ireland) To examine the impact of a community-acquired LRTI treatment protocol	1. Protocol for inpatient LRTI care presented to medical staff 2. Distribution of a written summary to medical staff 3. Posting of the protocol algorithm on wards 4. Pharmacy involvement	Uncontrolled before–after (227)	1. IV antibiotic duration decreased (6–2 mean days (P < 0.001) 2. Treatment failure decreased, 31% to 8% (P < 0.001) 3. Mean length of stay decreased, 9 to 4.5 mean days (P < 0.001) 4. Total costs decreased from £2024 to £1020 per patient (P < 0.001). The only UK-based study of a LRTI protocol

Table 2. *Continued*

Study year (country) Objective	Intervention and implementation	Study design (Total sample size)	Results
Suchyta et al., 2001; Dean et al., 2000 (USA) To study the use of a practice guideline in urban urgent care clinics	1. A practice guideline for CAP in an outpatient setting 2. Use of 1993 ATS guidelines 3. Presented to physicians and nurses 4. Distributions of literature and guideline forms	Uncontrolled before–after (463)	1. Hospital admission decreased from 14% to 6% ($P = 0.02$) 2. Use of recommended antibiotics increased from 45% to 72% ($P < 0.001$) 3. Decreases in mean antibiotic costs ($186 to $141, $P = 0.009$) and direct costs per patient ($678 to $319, $P = 0.008$) 4. Decreased length of stay (154–89 hr, $P = 0.04$)
Meehan et al., 2001 (USA) To study the effects of a statewide quality improvement project	A statewide pathway with locally customised implementation focusing on rapid antibiotic delivery, blood culture collection, and O_2 assessment	Uncontrolled before–after (2388)	1. Antibiotics within 8 hr of admission increased from 83% to 89% ($P = 0.001$) 2. O_2 assessment within 24 hr of admission increased from 94% to 95% ($P < 0.001$) 3. Mean length of stay decreased from 6.6 to 5.5 days ($P < 0.001$)
Dobbin et al., 2001 (Australia) To compare outcomes between hospitals using penicillin-based and cephalosporin-based CAP antibiotic protocols	An antibiotic protocol aimed at limiting use of 3rd generation cephalosporins in CAP (implementation strategy not detailed)	Retrospective cohort study comparing two hospitals post-intervention (158)	1. No difference in length of stay, time to defervescence, and in-hospital mortality 2. IV antibiotic cost per patient lower ($18 vs $95) in penicillin-based protocol using hospital
Lawrence et al., 2002 (USA) To improve time to antibiotics and sputum procurement	1. Protocol presented to physicians and nurses 2. Continuing education, audit and feedback	Uncontrolled before–after (119)	1. Time to antibiotics decreased from 413–291 minutes ($P < 0.01$) 2. Sputum procurement increased from 11% to 25% ($P = 0.06$)

Notes: ATS—American Thoracic Society; LRTI—Lower respiratory tract infection; PORT—Pneumonia Outcomes Research Team; PSI—Pneumonia Severity Index.

Often embedded in care pathways, or provided to healthcare professionals in other formats, is decision support. There are now a number of evidence-based decision support "rules" to guide important clinical management decisions in CAP. These include severity prediction rules, for example, PSI (Fine *et al.*, 1997), to guide the decision about an appropriate site of care (i.e., inpatient vs outpatient), criteria for admission to intensive care (Ewig *et al.*, 1998), and markers of clinical stability for IV to oral switch and discharge from hospital (Halm *et al.*, 2002). As demonstrated by their moderate sensitivity and specificity, however, none of these rules are perfect and they should all be used in combination with clinical judgement (Angus *et al.*, 2002). Further details are provided in Table 3.

3. LINKING PROCESS OF CARE TO OUTCOMES IN QI

"Process of care" is the collective term given to all the activities that contribute to healthcare (Crombie and Davies, 1998). For a QI intervention to improve outcomes, it must first impact on process. But which aspects of process are important in CAP? The process of managing any acute medical condition can be described in the form of a pathway or patient journey. Each journey can be broken down into a series of critical control points or "crossroads," at which important healthcare decisions (i.e., process of care) are made (Figure 1). Critical control points for CAP can be determined using hazard analysis methodology (Bryan, 1981; Macdonald and Engel, 1996). QI programmes should target processes that are either known to or that are likely to impact on outcomes. Identified critical control points should therefore be "tested" using relevant criteria of causality: (1) strength of association, (2) consistency with other studies and evidence, (3) biological or economic credibility, (4) temporal association, and (5) biological or economic gradient (Hill, 1965).

For example, the antibiotic prescription is an important component in the management of CAP. Although there are many potential hazards in antibiotic prescribing (e.g., anaphylaxis), two important hazards that are likely to impact on outcomes for most patients, and that are potentially remediable by an educational intervention, are: (1) delay in administration and (2) inappropriate therapy. A number of studies have shown a statistical association between the timely administration of appropriate antibiotics and lower mortality and length of hospital stay in CAP (Battleman *et al.*, 2002; Meehan *et al.*, 1997). Both of these aspects of process (i.e., delivery and appropriateness) have biological credibility in that they are consistent with the current understanding of the pathogenesis of CAP. There is a temporal association in that the process must occur first to impact on outcomes. Finally, there is a biological gradient in that the longer

Table 3. Outcome and process of care measures for clinical CAP research or QI

Measure	Comments
Patient-based outcome measures	
CAP-Sym (Lamping et al., 2002)	CAP-Sym is the only validated CAP-specific patient-based outcome measure. It is validated for use in outpatients only and could be used in audit and research
Clinically based outcome measures	
Time to clinical stability: One or more of temperature $> 37.8°C$, pulse > 100/min, RR > 24/min, sBP < 90 mmHg, $SaO_2 < 90\%$, lack of availability of the oral route and abnormal mental status = clinical instability (Halm et al., 2002)	Not validated as an outcome measure, but the only evidence-based definition of clinical stability in CAP. Potentially, it could be used as an outcome measure in CAP research
30-day post-admission mortality (Mortenson et al., 2002)	Not validated as an outcome measure, but clearly of importance to all and the evidence-based time point for evaluating mortality in CAP
Processes of care relating to clinical effectiveness	
Door to antibiotic time (Battleman et al., 2002; Meehan et al., 1997)	An evidence-based process of care measure for use in QI programmes or research
% of patients receiving appropriate antibiotic therapy according to a recognised guideline (Battleman et al., 2002; Gordon et al., 1996; Menendez et al., 2002)	An evidence-based process of care measure for use in QI programmes or research
% of hospitalised patients receiving an antibiotic regimen with "atypical" activity (Stahl et al., 1999; Waterer et al., 2001)	An evidence-based process of care measure for use in QI programmes or research

% of patients with ICU admission criteria (and without a DNR order) admitted to ICU: Two of three minor criteria (sBP $\leq$ 90 mmHg, multi-lobar disease, PaO_2/FiO_2 ratio < 250) or one of two major criteria (need for mechanical ventilation or septic shock) predicted the need for ITU admission (Ewig et al., 1998)	An evidence-based process measure for admission to ICU. However, the sensitivity, specificity etc. are suboptimal. On this basis, it would not be adequate for population screening, but it could be used as a physician aid in combination with clinical judgement. It should not be used as an outcome measure
Processes of care relating to economic costs	
% of low-risk patients (PSI 1–3) managed as outpatients (Atlas et al., 1998; Fine et al., 1997; Marrie et al., 2002)	Not all low-risk patients can be managed as outpatients. The proportion is likely to depend on local factors, such as the availability of social support and outpatient services
% of patients switched from IV to oral therapy within 24-hr of clinical stability (Halm et al., 2002; Paladino et al., 2002; Rhew et al., 2001)	An evidence-based process of care measure for use in QI programmes or research
% of patients discharged within 24-hr of reaching clinical stability (Halm et al., 2002; Rhew et al., 2001)	Discharge is also dependent on local factors as outlined above
Economic measures	
Cost-effectiveness (see: www.epoc.uottawa.ca/checklist2002.doc)	The costs of the intervention (including development), direct healthcare costs and indirect costs should be quantified and linked with provider and patient outcomes
Other measures	
Adverse events	Not validated as an outcome measure, but of importance to patients, physicians, policy-makers, and drug regulators

Notes: ICU = intensive care unit; DNR = do not resuscitate; PSI = Pneumonia Severity Index; IV = intravenous.

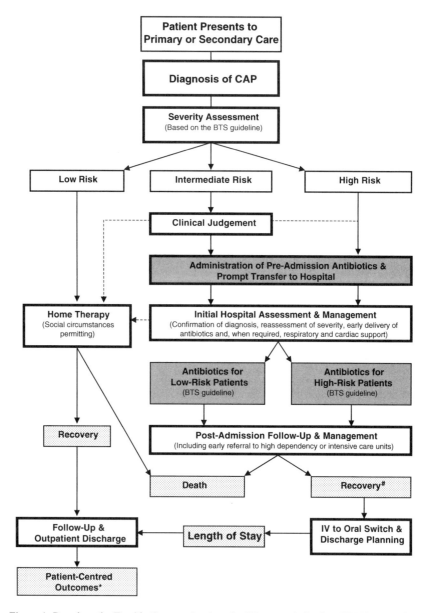

Figure 1. Based on the Tayside Community-Acquired Pneumonia Project (TAYCAPP), the pathway shows important critical control points and outcomes for CAP. Critical control points are shown in bold boxes. Processes of care measures are shown in shaded boxes. Outcomes are shown in patterned boxes. [#]Time to clinical stability. *Patient-centred outcomes should be measured at a fixed-point post-admission (e.g., at 30 days).

the time to administer antibiotics (Battleman *et al.*, 2002; Meehan *et al.*, 1997) and the more inappropriate the therapy (e.g., the extent of under-dosing, lack of "atypical" cover [Stahl *et al.*, 1999] or the use of oral therapy in a severely ill patient), the poorer the outcomes are likely to be.

The failure to draw a blood culture prior to antibiotic therapy is also a potential hazard that is potentially remediable by an educational intervention. While this process is undoubtedly of importance, for example, to detect bacteraemia and rationalise antibiotic therapy, there is little biological credibility to the hypothesis that the act of drawing a blood culture is likely to lead to improved outcomes. Additionally, there is no obvious biological gradient and there is doubt about the impact of microbiological investigations on outcomes (Campbell *et al.*, 2003; Sanyal *et al.*, 1999). The association found in some observational studies (Meehan *et al.*, 1997) is likely to be due to confounding by the overall higher quality of care, including the efficient delivery of appropriate antibiotics and other therapy, provided by physicians who decide to take a blood culture prior to administering antibiotics. For a QI intervention to impact on outcomes therefore, it is important to target processes of care that determine outcome.

3.1.   What are the advantages of process of care measures in QI?

Achievement of important aspects of process (e.g., door to antibiotic time) can be measured as part of QI initiatives instead of or as well as outcome measures, such as 30-day mortality and quality of life. There are a number of advantages in this approach. In contrast to many outcome measures, process of care measures are: easier to measure and interpret; less prone to the effects of case-mix; indicate the aspects of care that need to change; and are more sensitive to the quality of care (Crombie and Davies, 1998; Goddard *et al.*, 2002; Mant and Hicks, 1995). For example, the lack of sensitivity of mortality as an indicator of quality has previously been demonstrated for myocardial infarction (Mant and Hicks, 1995) and is illustrated in CAP by the following example. Waterer *et al.* recently found single effective therapy (SET) to be less effective than dual effective therapy (DET) in the treatment of severe bacteraemic pneumococcal pneumonia (patients receiving SET had a mortality of 18% vs 7% for those receiving DET). If this data is used in a hypothetical model with the assumption that DET is the "standard of care" and that hospital A is 100% adherent with DET (mortality = 7%) and hospital B is 50% adherent with DET (mortality = 13%), and that case-mix and process in hospitals A and B are otherwise identical, to statistically detect ($P \leq 0.05$ with 80% power) the difference in quality of care by using mortality (i.e., 7% vs 13%), both hospitals would need to have data from 338 patients with severe bacteraemic pneumococcal pneumonia. In contrast, to statistically detect the difference in quality of care by measuring the process, that is, adherence with DET (100% vs 50%), both hospitals would require data from only 8 patients (Sample sizes calculated using *CLINSTAT*, Martin Bland, St. George's Hospital Medical School, London, UK, 1996).

4.   WHAT IS THE EVIDENCE THAT QI INITIATIVES IMPROVE PROCESS OF CARE IN CAP?

Process of care measures may be more pragmatic endpoints than "traditional" outcomes (e.g., mortality) in CAP research or QI. Process measures, however, should be linked to clinical, patient-centred, or economic outcomes. This section will review the evidence that QI interventions improve process of care in CAP.

4.1.   Aspects of process that are likely to impact on clinical and patient-centred outcomes

In a cleverly designed controlled before–after study, Chu *et al.* (2003) showed that feedback from an external QI agent resulted in a higher proportion of patients receiving antibiotics within 4 hr of admission at intervention compared to control hospitals (69% vs 57%). This change appears to have been mediated through the use of either clinical pathways or standing orders. When control hospitals were subsequently exposed to the intervention, delivery of antibiotics at these sites also improved (66% vs 53%). This evidence is supported by three uncontrolled studies. Benenson *et al.* (1999) used a before–after study with the use of two post-implementation cohorts to evaluate the impact of a care pathway that emphasised the importance of the early diagnosis and delivery of antibiotic therapy. Door to antibiotic time decreased from 315 to 175 min and 171 min at 1 and 3 years post-implementation, respectively. Although uncontrolled, this is an interesting finding as there has been concern about the sustainability of QI interventions. Meehan *et al.* (2001) found that a locally customised statewide care pathway improved the delivery of antibiotics within 8 hr of admission to hospital from 83% to 89%. Lawrence *et al.* (2002) used an educational intervention with monthly audit and feedback to reduce door to antibiotic time from 413 to 291 min. In some of these studies (Chu *et al.*, 2003; Lawrence *et al.*, 2002), sputum and blood culture procurement have also increased. As discussed above, however, it is debatable if these processes of care have an impact on outcomes.

In another uncontrolled study, Suchyta *et al*. (2001) used a practice guideline incorporating the American Thoracic Society's (ATS's) antibiotic guidance to increase the use of recommended antibiotics from 45% to 72%. Using a wide range of interventions and in the largest published controlled CAP QI study to date; Dean *et al*. showed an improvement in overall use of guideline antibiotics in study hospitals from 28% to 56% (Dean *et al*., 2001). In the only UK study, Al-Eidan *et al*. (2000) described a reduction in the diversity of antibiotics used in CAP (from 12 to 3) following implementation of a practice guideline.

4.2. Aspects of process that are likely to impact on economic outcomes

In an uncontrolled before–after study, Atlas *et al*. (1998) found that the use of the PSI as admission decision support, increased low-risk patients managed as outpatients from 42% to 57%. The intervention included a study nurse during weekday working hours, enhanced outpatient services, and the use of oral clarithromycin. These findings were subsequently confirmed in the only randomised CAP QI trial. A care pathway, which incorporated the PSI with levofloxacin, reduced the proportion of low-risk patients admitted to hospital from 49% to 31% without impacting on patients' quality of life or clinical outcomes (Marrie *et al*., 2000). Using an adapted version of the ATS admission guidance, Suchyta *et al*. (2001) also safely reduced hospital admission from 14% to 6%.

An early switch from IV to oral therapy and subsequent discharge from hospital has been shown to be a safe and cost-effective strategy for CAP in randomised controlled trials (Castro-Guardiola *et al*., 2001; Paladino *et al*., 2002). Only a few studies have evaluated this strategy, however, as part of a QI intervention. Al-Eidan *et al*. (2000) demonstrated a reduction in IV antibiotic administration from 6 to 2 mean days. In contrast, Rhew *et al*. (1998) were unable to demonstrate a significant change in guideline adherence following the implementation of physician decision support for this aspect of care. This study was underpowered, however, to detect clinically important changes. In another small study (not shown in the tables), Weingarten *et al*. (1996) found similar results.

5. WHAT IS THE EVIDENCE THAT QI INITIATIVES IMPROVE OUTCOMES IN CAP?

While it may be necessary to extrapolate improvements in evidence-based process of care measures to clinical, patient-based and economic outcomes, most patients, physicians, and policy-makers would prefer to see "hard evidence."

This section will review the evidence that QI interventions improve outcomes in CAP.

5.1. Clinical outcomes

Relatively few studies have shown any change in clinical outcomes following the implementation of a CAP QI initiative. This is probably due to the large amounts of data required to demonstrate clinically important changes in "traditional" clinical endpoints, for example, mortality. An exception to this was a study that used statewide databases to include 28,661 patients (Dean *et al.*, 2001). Following implementation of a care pathway, 30-day mortality decreased from 13.4% to 11% in hospitalised patients managed by physicians affiliated to pathway using hospitals. During the corresponding period, mortality in patients managed by physicians affiliated to non-pathway hospitals increased from 13.2% to 14.2%. Al-Eidan *et al.* (2000) also demonstrated a significant reduction in "treatment failure" from 31% to 8%. However, this result may have been biased by the use of a non-objective outcome measure. Mortality in the same study also decreased from 8% to 3%, but this was not statistically significant.

5.2. Patient-based outcomes

Clinical outcome measures provide important information to clinicians and policy-makers, but do not necessarily reflect aspects of outcome that are important to patients. Patient-based outcomes refer to measures of subjective well-being, such as quality of life, symptoms and satisfaction with care. The widespread recognition of the need to evaluate the "complete state of physical, mental and social well-being and not merely the absence of disease or infirmity" (World Health Organisation, 1947) means that clinical audit and research is now considered inadequate if patients' experiences of outcome have not been assessed. Few studies have used patient-based outcomes, however, to measure the impact of a CAP QI intervention. Although Atlas *et al.* (1998) showed an increase in low-risk patients managed as outpatients, patients appeared to be less satisfied with care, as measured by a non-validated outcome measure, when compared to a comparable cohort from the Pneumonia Outcomes Research Team (PORT) study. In contrast, the cluster-randomised trial by Marrie *et al.* (2000) did not show any difference in patients' quality of life (as measured by SF-36) between intervention and control hospitals, despite a higher number of low-risk patients being managed as outpatients at intervention-using hospitals. There is currently no evidence therefore, to suggest that CAP QI initiatives improve patient-based outcomes.

The best evidence is that care pathways can be used to optimise resource use, without affecting patients' quality of life.

5.3. Economic outcomes

There have been no cost-effectiveness analyses of CAP QI initiatives. A number of studies have shown decreases in either surrogate markers of resource use or direct healthcare costs. In the cluster-randomised trial by Marrie *et al.* (2000), bed-days per patient managed (BDPM), a surrogate marker of resource use, decreased from 4.4 to 6.1 days. In a subsequent economic paper, Palmer *et al.* (2000) found a cost saving per patient of between $457 and $994. Suchyta *et al.* (2001) demonstrated a reduction in mean antibiotic costs (from $186 to $141) and direct costs per patient (from $678 to $319) as a result of an outpatient QI intervention. Al-Eidan *et al.* (2000) calculated dramatic total cost savings, from £2024 to £1020 per patient, in the United Kingdom. In a small Australian study, which compared two hospitals using penicillin and cephalosporin-based antibiotic regimens, the antibiotic costs per patient were found to be lower in the hospital using penicillin, without an apparent impact on clinical outcomes (Dobbin *et al.*, 2001).

The main problem with all of these studies, however, is that they fail to link the costs of developing the intervention, the direct costs of care (e.g., antibiotics) and the indirect costs of CAP (e.g., time of work) with improvements in either provider (e.g., length of hospital stay) or patient status (e.g., quality of life). If the costs of the intervention and/or the direct costs as a result of it are high therefore, it is possible that the gain in terms of, for example, the cost per additional low-risk patient managed as an outpatient, may not be worthwhile. There is also concern that costs may simply be shifted from secondary to primary care.

6. IMPROVING THE MANAGEMENT OF CAP: EXPERIENCES FROM TAYCAPP

A number of publications provide useful overviews of QI approaches applicable to CAP. In particular, the Medical Research Council (MRC) framework for the development and evaluation of RCTs for complex interventions to improve health (www.mrc.ac.uk), the Cochrane Effective Practice and Organisation of Care checklist (www.epoc.uottawa.ca/checklist2002.doc) and a recent series of articles in the British Medical Journal (Campbell *et al.*, 2003; Øvretveit and Gustafson, 2003; Wensing and Elwyn, 2003) are recommended reading. Although some of these are written from a research perspective, the described principles can be adapted to routine QI. This section will discuss the development, implementation and evaluation of QI initiatives for CAP by reflecting on the Tayside Community-Acquired Pneumonia Project (TAYCAPP). This ongoing initiative is a multifaceted quality improvement programme, which aims to improve the delivery and appropriateness of antibiotic prescribing in CAP patients requiring hospitalisation. The intervention has primary care, accident and emergency (A&E) department, and acute medical admissions unit components.

6.1.   Establishing the problem

There is little point in attempting to improve a process of care that is already highly achieved. For example, is trying to improve oxygenation assessment within 24 hr of admission likely to be cost-effective, when it is already performed in 94% of patients? Before embarking on QI therefore, it is essential to consider the question: "What is currently happening and do we need to improve it?" Surprisingly, this is not always done. In many UK hospitals, the clinical governance agenda is largely being achieved by the local development and implementation of treatment algorithms that have not been informed by pre-implementation audit. Even if a multidisciplinary approach to development is used, which is not always the case, there is a risk that the resulting intervention may focus on areas of ongoing good practice, while neglecting important processes of care that are sub-optimally achieved. The decision about which aspects of process to measure in CAP has been discussed above and is helped by the expanding evidence-base linking process of care to outcomes; evidence-based process of care measures are shown in Table 3. In TAYCAPP, pre-implementation audit of process of care measures showed that half of all patients received appropriate antibiotics within 4 hr (the BTS recommend within 2 hr) of hospital admission. Of patients with severe CAP (as defined by the BTS guidelines), 30% did not receive IV antibiotics within 24 hr. In contrast, 17% of all patients were receiving unnecessarily intensive or expensive antibiotic therapy, an important consideration given current concerns about antibiotic resistance and cost-efficient healthcare.

It is also important to understand how patients pass through the healthcare system (i.e., the patient journey). In TAYCAPP, identification of the pre-implementation pathway of care showed that one-third of CAP patients are admitted via the A&E department. A&E involvement in the development and implementation process was therefore deemed essential. An A&E pathway (Figure 2) and educational session were subsequently developed with the aim of increasing the proportion of CAP patients receiving antibiotics before transfer to the acute medical admissions unit. We were also able to identify that relatively few patients have a delayed discharge because of non-medical reasons (e.g., the need for physiotherapy). We concluded that length of hospital stay was more likely to be reduced for the majority of patients by focusing on the early management of CAP, where deficits clearly existed, rather than addressing issues (e.g., rehabilitation services provision) that were unlikely to be changed by an initiative for a single acute medical condition.

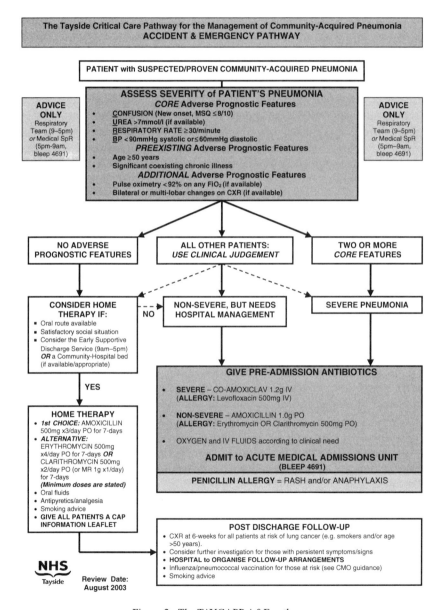

Figure 2. The TAYCAPP A&E pathway.

6.2.   Understanding the problem

Understanding why things go wrong is subtly different to knowing how things go wrong. None of the published QI initiatives for CAP appear to have been informed by studies of why certain aspects of process are not achieved. While multidisciplinary teams may think that they understand a problem and know how to correct it, such teams usually consist of local experts or senior members of staff who may be removed from, or not participating in, the day-to-day delivery of acute medical care. In the United Kingdom, for example, the "gate-keepers" of the initial delivery of appropriate antibiotics in acute medicine are usually junior members of the medical and nursing team, who may not have the necessary knowledge and experience to achieve satisfactory process of care. Involving these healthcare professionals in the development process is therefore essential to understanding why process is not achieved and to designing successful interventions. Failure to explore this question is more likely to lead to an unsuccessful intervention and, in the future, routine or research-based QI programmes are less likely to be funded if this question is not studied or already understood.

A variety of methods are available to explore why things go wrong in healthcare process. Prior to TAYCAPP, we suspected that there was a lack of understanding of the link between severity assessment and appropriate antibiotic therapy amongst physicians, which then resulted in suboptimal process of care. We initially explored this by performing a structured survey of junior and middle grade physicians' attitudes and knowledge. A previously published model of clinician adherence with clinical practice guidelines (Cabana *et al.*, 1999) and other published evidence (Halm *et al.*, 2000; Tunis *et al.*, 1994), informed the design and content of this survey. This showed that suboptimal undergraduate and postgraduate training experiences, poor working environment, lack of familiarity with the BTS guidelines, and lack of knowledge regarding severity assessment and antibiotic prescribing were all potential barriers to the efficient delivery of appropriate antibiotics. In contrast, attitudes towards guidelines in general and CAP appeared to be positive.

We went on to explore the identified barriers in more detail by performing a small number of qualitative in-depth interviews. Qualitative methods are ideally suited to exploring complex phenomena, such as the efficient delivery of antibiotics. In order to ascertain a range of views and experiences, physicians were purposively sampled from the cohort of physicians who had completed the quantitative survey. The aim was to gain a better understanding of phenomena that were related to the efficient delivery of appropriate antibiotics and to identify reasons for non-adherence with local and national guidance and potential interventions. A number of other qualitative methods (e.g., focus groups or observational methods) could have been used. For example, it would have been more appropriate to perform a case-based or critical-incident analysis (i.e., the interviewing of healthcare professionals involved in the care of patients receiving delayed or inappropriate antibiotic therapy). However, probably because of the long and intensive shifts on the acute medical admissions ward, physicians working in this environment were reluctant to be interviewed. In contrast, physicians who had recently finished this part of their post, and who were working in less busy areas of the Medical Unit, were more enthusiastic.

It is also legitimate to use qualitative enquiry to inform the domains and design of questions for a quantitative survey, which is then administered to a larger number of respondents. If qualitative methods are to be used, because of the interviewing and analysis skills required, it is vital to seek expert advice. Good starting points include the Scottish Consensus Statement on Qualitative Research in Primary Health Care (Dowell *et al.*, 1995), Greenhalgh (2001), Pope and Mays (2000), and Pope *et al.* (2000).

The combination of quantitative and qualitative enquiry allowed the development of a local model of the delivery of antibiotic therapy for CAP. One of the interesting and surprising findings was that respondents' confidence and perception in their ability to manage CAP appeared to relate to their previous clinical experience. For example, when starting as a house officer confidence in their ability to manage CAP was low. As their experience of seeing CAP increased so did respondents' confidence and their perception that they were able to manage it well. However, this is a double-edged sword. While respondents' confidence may lead to a timely diagnosis and early delivery of antibiotics, the appropriateness of therapy depends on the knowledge acquired during their clinical experience. For example, some respondents' believed themselves to be managing CAP according or close to recommended practice. When their answers in the quantitative survey, which were performed before the in-depth interviews, were checked, this was not always the case.

This finding suggests that CAP guidance is most likely to be used by junior or inexperienced physicians and highlights the importance of ongoing reflection in clinical practice. These data were subsequently used in interactive educational sessions with physicians and other healthcare professionals to emphasise these issues. Additionally, we have been able to concentrate on other identified correctable barriers (e.g., knowledge about the link between severity assessment and appropriate antibiotic therapy) rather than barriers that we perceived to be important. The model has also identified barriers generic to acute medicine (e.g., ward organisation) that cannot be resolved as part of TAYCAPP. This means that a "ceiling effect" is likely to exist on the extent to which certain process of care measures can be achieved.

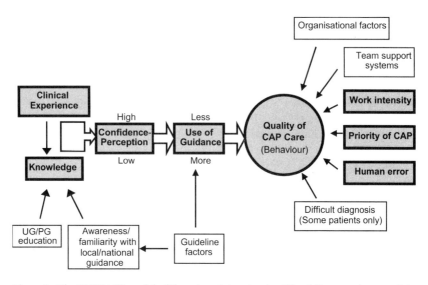

Figure 3. The TAYCAPP model of the micro-determinants of the delivery and appropriateness of antibiotic therapy in the acute medical admissions unit. The model is based on the qualitative and quantitative surveys described above. It should be noted, however, that sample sizes were small and the model was designed to inform TAYCAPP alone. Boxes show barriers to either the delivery or appropriateness of antibiotic therapy. Shaded boxes appear to be the more important barriers in Dundee.

6.3. Designing and implementing a CAP intervention

A multidisciplinary team should oversee the design and implementation of the intervention. The team will need to include representation from local opinion leaders and all the major stakeholders (including patients), and may need to include persons with specific expertise, such as in behavioural psychology and statistics. Prior to deciding on which interventions are to be used and how they are to be implemented, the potential impact and costs of each component of any complex intervention and how these components interrelate should be understood. This can be achieved by review of the literature, modelling (e.g., of the economic costs), and small qualitative and/or quantitative studies, as described above. Ideally, components that appear likely to succeed at reasonable cost should then be tested in small intervention studies to establish, for example, the incremental effectiveness of each component, actual costs, and appropriate control groups (MRC framework for the development and evaluation of RCTs for complex interventions to improve health; www.mrc.ac.uk). The latter is unlikely to be feasible, however, for routine QI or small research projects.

In TAYCAPP, the results of the described quantitative and qualitative studies suggested that multiple and complex barriers to any new intervention were likely to exist and that a multifaceted intervention would be most likely to succeed. For example, we recognised that an educational programme targeting the link between severity assessment and antibiotic therapy may well improve clinicians' knowledge. However, the qualitative interviews identified the presence of a potentially negative attitude towards CAP, in that it is considered less important than some other acute medical problems. For example:

I guess the same for myocardial infarction, we can do a lot for it and minutes mean muscle to coin a phrase, whereas in pneumonia it's just pneumonia ... It's the attitude to it that I see. (Verbatim text from an in-depth interview)

We recognised that potentially this attitudinal barrier could prevent a change in physician behaviour and thereby process of care, even if knowledge did improve. We therefore attempted to upgrade the importance of CAP by emphasising the high mortality and morbidity and marketing a "door to needle time" ethos in all aspects of the programme. The use of a multifaceted approach was supported by a previously published systematic review that found that multifaceted interventions (i.e., two or more of audit and feedback, reminders, a local consensus process, and marketing) and interactive educational meetings (participation of healthcare providers in workshops that include discussion or practice) were more consistently effective in promoting behaviour change in healthcare professionals than, for example, single interventions, such as paper-based educational materials (Bero *et al.*, 1998). In light of the literature review and emerging local evidence, interventions, and implementation strategies subsequently included: (1) strategically sited management pathways based on the BTS recommendations and adapted to local practice, (2) posters marketing the BTS severity assessment criteria and a "door to needle time" ethos, (3) dissemination of implementation packs, which contained an explanation of the programme, a pocket-sized laminated management algorithm and an interactive educational workbook focusing on the link between severity assessment and appropriate therapy, (4) interactive educational meetings emphasising the severity assessment–appropriate antibiotic therapy link, and (5) continuous audit and monthly feedback.

6.4. Evaluation

Cluster-randomisation is the "gold-standard" methodology for research-based evaluations of organisation level interventions. This methodology, however, requires enough hospitals willing to participate and considerable resources. The Cochrane Effective Practice and Organisation of Care (EPOC) Review Group suggest alternative robust methodologies, controlled before–after studies and interrupted time series (ITS) analysis; methodological considerations for these designs are considered at: www.epoc.uottawa.ca/checklist2002.doc. A good example of how a controlled before–after design can be used to evaluate a QI intervention can be seen in the paper by Chu *et al.* (2003). To evaluate an intervention by interrupted time series, process or outcomes are measured longitudinally before and after an intervention (for a short time series at least three data periods are required before and after). The longitudinal nature of this methodology is particularly suited to clinical performance measurement, but can also be used to evaluate research-based projects (Bloor *et al.*, 2003; Wagner *et al.*, 2002). Although not essential, results can be compared with control data from other hospitals and/or national databases to increase external validity. Providing the internal validity (e.g., case selection and recording at the intervention site) is maintained throughout data collection and there are no relevant interventions at the control sites, any statistically significant changes in the level and gradient of the slope (see Figure 4), not mirrored in the control cohorts, are likely to be due to the intervention.

Another important consideration in evaluating a QI initiative is the outcome measures to be used. Most of the outcome measures that have been

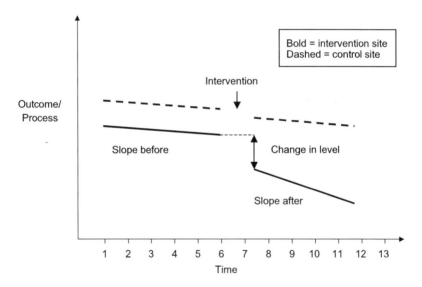

Figure 4. Longitudinal measurement of outcome by ITS analysis. At the intervention site there is a change in both the level and gradient of the slope post-intervention, indicating a change in outcome due to the intervention. In contrast, at the control site the gradient of the slope continues unchanged. The downward gradient at the control site indicates change due to a prevailing temporal trend. The slopes can be statistically evaluated using regression analyses.

used in CAP QI initiatives have not been formally validated for reliability, validity, responsiveness, and acceptability. Additionally, "traditional" clinical endpoints, such as 30-day mortality and length of hospital stay, although important, have fundamental flaws. For example, as discussed above, mortality is an insensitive indicator of healthcare quality and length of hospital stay is subject to factors (e.g., clinician practice style and social services provision) that vary from patient to patient (McCormick *et al.*, 1999) and, if measurable, require risk adjustment during statistical analyses. Alternatives to traditional outcomes include process of care and patient-based measures; a patient-based symptom questionnaire (CAP-Sym) has recently been validated for use in out-patient CAP (Lamping *et al.*, 2002). An important missing link in outcome measurement, however, is the lack of a validated educational outcome measure for CAP. Such a tool would allow the assessment of the impact of an intervention on, for example, clinicians' knowledge and attitudes. A similar educational outcome measure to assess training in evidence-based medicine has recently been validated (Ramos *et al.*, 2003).

The Outcome Measures for Arthritis Clinical Trials (OMERACT) (1998) group have developed a useful paradigm that could be used to assess the appropriateness of non-validated outcome measures during project development:

1. Truth (validity)—Is the measure truthful? Does it measure what is intended? Is it unbiased and relevant?
2. *Discrimination* (reliability and sensitivity to change)—Does the measure discriminate between situations of interest? The situations can be states at one time (for classification and prognosis) or states at different times (to measure change).
3. *Feasibility* (acceptability)—Can the measure be applied easily, given constraints of time, money, and interpretability?

Table 3 shows validated and evidence-based process of care and outcome measures for use in CAP research or clinical performance measurement and that are likely to be of importance to patients, clinicians, policy-makers, and drug regulators. How these measures are used is likely to depend on the objectives of the project and the application of the results (e.g., to assess patients, to evaluate practice at an organisational level, or to determine research or resource priorities), resources, and data availability.

The evaluation of TAYCAPP is currently ongoing. Process of care data will be compared against a local hospital, which has not been exposed to an intervention (i.e., a controlled before–after design). Mortality will be compared to an ICD-10 selected cohort from national databases using ITS analysis.

Cost-effectiveness will be established using recommended methodology. While this design is not "gold-standard," it is pragmatic, relatively inexpensive and we have carefully attempted to minimise the known validity threats to quasi-experimental observational studies, for example, regression to the mean and temporal trends (American College of Physicians–American Society of Internal Medicine, 1999).

Although the statistical evaluations are still awaited, the continuous audit component of the study shows some evidence of success. For example, before implementing TAYCAPP few patients received pre-admission antibiotics, as recommended by the BTS guidelines, either from an admitting general practitioner or in the A&E department. Post-implementation, 20% of patients now receive pre-admission antibiotics from the admitting general practitioner and 80% of patients admitted via A&E are receiving antibiotics prior to transfer to the acute medical admissions unit. Additionally, the appropriateness of antibiotic therapy prescribed on the acute medical admissions ward has increased from 67% in the first month post-implementation to 100% in the fifth month.

6.5.   Long-term surveillance

There have been concerns about the long-term sustainability of QI initiatives in the "real-world" environment. In the United Kingdom, this is due, at least in part, to the lack of QI infrastructure and supporting resources in NHS trusts. For example, many United Kingdom hospitals rely on the "good-will" of healthcare staff to collect and analyse audit data in their own time. It is therefore desirable to embed both the interventions and the evaluation process in "usual practice." For example, the educational sessions for junior house officers in TAYCAPP have been incorporated into an ongoing cycle of protected teaching sessions. Additionally, the developed management algorithms have been included in the undergraduate curriculum and, hopefully, medical students will therefore be aware of this guidance prior to commencing the pre-registration year. The monthly audit and feedback process, which requires considerable effort, will be converted to a simpler postal reminder system. Poster-based educational materials and management algorithms have all been laminated to limit long-term "wear and tear" and are also available in electronic format. These materials will also be distributed in the induction packs given to new members of the Medical Unit staff. Finally, evaluation will continue on a yearly basis when a small set of case-notes (e.g., 40) will be randomly selected for audit. This data will be added to an ongoing ITS analysis to assess the long-term impact.

7. CONCLUSIONS

There have been many studies of complex QI interventions to improve the management of CAP. Many of these have used uncontrolled designs and the health policy implications of these studies are therefore limited. The best evidence comes from three large controlled studies. One study showed a positive impact on the delivery of antibiotics and the procurement of microbiological tests. The only randomised trial demonstrated reduced resource utilisation without compromising clinical or patient-centred outcomes. The largest study showed a positive impact on mortality. There is little evidence, however, about what single interventions work and at what cost. Additionally, can QI initiatives improve patient-centred outcomes and what is the link between educational outcomes, process of care, and patient-based, clinical, and economic outcomes?

In the development of CAP QI initiatives, considerable efforts should be given to first establishing and then understanding the problem. This information is vital in the cost-effective design, implementation and evaluation of targeted interventions. However, in the resource-limited environment of many healthcare systems, the challenge to healthcare professionals and policy-makers is to incorporate and evaluate developed interventions as part of everyday practice and on a long-term basis.

REFERENCES

American College of Physicians-American Society of Internal Medicine, 1999, Compendium of primers. A primer on before-after studies: Evaluating a report of a "successful" intervention. *Effect. Clin. Pract.*, **2**, 5–6.

American Thoracic Society, 2001, Guidelines for the management of adults with community-acquired pneumonia. *Am. J. Respir. Crit. Care Med.*, **163**, 1730–1754.

Al-Eidan, F. A., McElnay, J. C., Scott, M. G., Kearney, M. P., Corrigan, J., and McConnell, J. B., 2000, Use of a treatment protocol in the management of community-acquired lower respiratory tract infection. *J. Antimicrob. Chemother.*, **45**, 387–394.

Angus, D. C., Marrie, T. J., Obrosky, D. S., Clermont, G., Dremsizov, T. T., Coley, C. *et al.*, 2002, Severe community-acquired pneumonia: Use of intensive care services and evaluation of American and British Thoracic Society diagnostic criteria. *Am. J. Resp. Crit. Care Med.*, **166**, 717–723.

Atlas, S. J., Benzer, T. I., Borowsky, L. H., Chang, Y., Burnham, D. C., Metlay, J. P. *et al.*, 1998, Safely increasing the proportion of patients with community-acquired pneumonia treated as outpatients: An interventional trial. *Arch. Intern. Med.*, **158**, 1350–1356.

Bartlett, J. G., Breiman, R. F., Mandell, L. A., and File, T. M. Jr, 1998, Community-acquired pneumonia in adults: guidelines for management. Infectious Diseases Society of America. *Clin. Infect. Dis.*, **26**, 811–838.

Bartlett, J. G., Dowell, S. F., Mandell, L. A., File, T. M. Jr, Musher, D. M., and Fine, M. J., 2000, Practice guidelines for the management of community-acquired pneumonia in adults. *Clin. Infect. Dis.*, **31**, 347–382.

Battleman, D. S., Callaghan, M., and Thaler, H. T., 2002, Rapid antibiotic delivery and appropriate antibiotic selection reduce length of hospital stay of patients with community-acquired pneumonia. *Arch. Intern. Med.*, **162**, 682–688.

Bero, L. A., Grilli, R., Grimshaw, J. M., Harvey, E., Oxman, A. D., and Thompson, M. A., 1998, Closing the gap between research and practice: An overview of systematic reviews of interventions to promote the implementation of research findings. *BMJ*, **317**, 465–468.

Benenson, R., Magalski, A., Cavanaugh, S., and Williams, E., 1999, Effects of a pneumonia clinical pathway on time to antibiotic treatment, length of stay, and mortality. *Acad. Emerg. Med.*, **6**, 1243–1248.

Bloor, K., Freemantle, N., Khadjesari, Z., and Maynard, A., 2003, Impact of NICE guidance on laparoscopic surgery for inguinal hernias: Analysis of interrupted time series. *BMJ*, **326**, 578.

British Thoracic Society, 1993, British Thoracic Society guidelines for the management of community-acquired pneumonia in adults admitted to hospital. *Br. J. Hosp. Med.*, **49**, 346–350.

British Thoracic Society, 2001, BTS guidelines for the management of community-acquired pneumonia in adults. *Thorax*, **56**(Suppl. 4), 1–64.

Bryan, F., 1981, Hazard analysis critical control point approach: epidemiological rationale and application to foodservice operations. *J. Environ. Health*, **44**, 7–14.

Cabana, M. D., Rand, C. S., Powe, N. R., Wu, A. W., Wilson, M. H., and Abboud, P.-A. C., 1999, Why don't physicians follow clinical practice guidelines. A framework for improvement. *JAMA*, **282**, 1458–1465.

Campbell, S. M., Braspenning, J., Hutchinson, A., and Marshall, M. N., 2003, Research methods used in developing and applying quality indicators in primary care. *BMJ*, **326**, 816–819.

Castro-Guardiola, A., Viejo-Rodríguez, A.-L., Soler-Simon, S., Armengou-Arxé, A., Bisbe-Company, V., Peñarroja-Matutano *et al.*, 2001, Efficacy and safety of oral and early-switch therapy for community-acquired pneumonia: A randomised controlled trial. *Am. J. Med.*, **111**, 367–374.

Chu, L. A., Bratzler, D. W., Lewis, R. J., Murray, C., Moore, L., Shook, C. *et al.*, 2003, Improving the quality of care for patients with pneumonia in very small hospitals. *Arch. Intern. Med.*, **163**, 326–332.

Crombie, I. K. and Davies, H. T. O., 1998, Beyond health outcomes: the advantages of measuring process. *J. Eval. Clin. Pract.*, **4**, 31–38.

Dean, N. C., Silver, M. P., Bateman, K. A., Brent, J., Hadlock, C. J., and Hale, D., 2001, Decreased mortality after implementation of a treatment guideline for community-acquired pneumonia. *Am. J. Med.*, **110**, 451–457.

Dean, N. C., Suchyta, M. R., Bateman, K. A., Aronsky, D., and Hadlock, C. J., 2000, Implementation of admission decision support for community-acquired pneumonia. A pilot study. *Chest*, **117**, 1368–1377.

Dobbin, C. J., Duggan, C. J., and Barnes, D. J., 2001, The efficacy of an antibiotic protocol for community-acquired pneumonia. *MJA*, **174**, 333–337.

Dowell, J., Huby, G., and Smith, C., 1995, *Scottish Consensus Statement on Qualitative Research in Primary Health Care*. Tayside Centre for General Practice, University of Dundee, Dundee, UK.

Ewig, S., Ruiz, M., Mensa, J., Marcos, M. A., Martinez, J. A., Arancibia, F. *et al.*, 1998, Severe community-acquired pneumonia: assessment of severity criteria. *Am. J. Respir. Crit. Care Med.*, **158**, 1102–1108.

File, T. M. Jr, Segreti, J., Dunbar, L., Player, R., Kohler, R., Williams, R. R. *et al.*, 1997, A multicenter, randomized study comparing the efficacy and safety of intravenous and/or oral levofloxacin versus ceftriaxone and/or cefuroxime axetil in treatment of adults with community-acquired pneumonia. *Antimicrob. Agents Chemother.*, **41**, 1965–1972.

Fine, M. J., Auble, T. E., Yealy, D. M., Hanusa, B. H., Weissfeld, L. A., Singer, D. E. *et al.*, 1997, A prediction rule to identify low risk patients with community-acquired pneumonia. *N. Engl. J. Med.*, **336**, 243–250.

Florida Medical Quality Assurance, Inc., 1998, Quality of care improvements for patients with pneumonia. *Eval. Health Prof.*, **21**, 514–524.

Fortune, G., Elder, S., Jaco, D., Bentivegna, P., Luebbering, T., and Boechler, M., 1996, Opportunities for improving the care of patients with community-acquired pneumonia. *Clin. Perform. Qual. Health Care*, **4**, 41–43.

Gleason, P. P., Kapoor, W. N., Stone, R. A., Lave, J. R., Obrosky, D. S., Schulz, R. *et al.*, 1997, Medical outcomes and antimicrobial costs with the use of the American Thoracic Society guidelines for outpatients with community-acquired pneumonia. *JAMA*, **278**, 32–39.

Goddard, M., Davies, H. T. O., Dawson, D., Mannion, R., and McInnes, F., 2002, Clinical performance measurement: Part 1—getting the best out of it. *J. R. Soc. Med.*, **95**, 508–510.

Gordon, G. S., Throop, D., Berberian, L., Niederman, M., Bass, J., Ale-Mayehu, D., *et al.*, 1996, Validation of the therapeutic recommendations of the American Thoracic Society (ATS) guidelines for community acquired pneumonia in hospitalized patients. *Chest*, **110**, 55S.

Gottlieb, L. D., Roer, D., Jega, K., D'arc, St. Pierre, J., Dobbins, J., Dwyer, M. *et al.*, 1996, Clinical pathway for pneumonia: Development, implementation, and initial experience. *Best Pract. Benchmarking Healthc.*, **1**, 262–265.

Greenhalgh, T. (ed.), 2001, Papers that go beyond numbers (qualitative research). In *How to Read a Paper*, Chapter 11. BMJ Publishing Group, London, UK.

Guest, J. F. and Morris, A., 1997, Community-acquired pneumonia: The annual cost to the National Health Service in the United Kingdom. *Eur. Respir. J.*, **10**, 1530–1534.

Halley, H. J., 2000, Approaches to drug therapy, formulary, and pathway management in a large community hospital. *Am. J. Health Syst. Pharm.*, **57**(Suppl. 3), S17–S21.

Halm, E. A., Atlas, S. J., Borowsky, L. H., Benzer, T. I., Metlay, J. P., Chang, Y. C. *et al.*, 2000, Understanding physician adherence with a pneumonia practice guideline. Effects of patient, system and physician factors. *Arch. Intern. Med.*, **160**, 98–104.

Halm, E. A., Fine, M. J., Kapoor, W. N., Singer, D. E., Marrie, T. J., and Siu, A. L., 2002, Instability on hospital discharge and the risk of adverse outcomes in patients with pneumonia. *Arch. Intern. Med.*, **162**, 1278–1284.

Hill, A. B., 1965, The environment and disease: Association or causation? *Proc. R. Soc. Med.*, **58**, 295–300.

Lamping, D. L., Schroter, S., Marquis, P., Marrel, A., Duprat-Lomon, I., Sagnier, P.-P., 2002, The community-acquired pneumonia symptom questionnaire: a new, patient-based outcome measure to evaluate symptoms in patients with community-acquired pneumonia. *Chest*, **122**, 920–929.

Lawrence, S. J., Shadel, B. N., Leet, T. L., Hall, J. B., and Mundy, L. M., 2002, An intervention to improve antibiotic delivery and sputum procurement in patients hospitalised with community-acquired pneumonia. *Chest*, **122**, 913–919.

Lim, W. S., Macfarlane, J. T., Boswell, T. C. J., Harrison, T. G., Rose, D., Leinonen, M. *et al.*, 2001, Study of community-acquired pneumonia aetiology (SCAPA) in adults admitted to hospital: implications for management guidelines. *Thorax*, **56**, 296–301.

Macdonald, J. D. and Engel, D., 1996, *A Guide to HACCP Hazard Analysis for Small Businesses*. Highfield Publications, Goncaster.

Mandell, L. A., Marrie, T. J., Grossman, R. F., Chow, A. W., Hyland, R. H., and the Canadian Community-Acquired Pneumonia Working Group, 2000, Canadian guidelines for the initial management of community-acquired pneumonia: An evidence-based update by the Canadian Infectious Diseases Society and the Canadian Thoracic Society. The Canadian Community-Acquired Pneumonia Working Group. *Clin. Infect. Dis.*, **31**, 383–421.

Mandell, L. A. and Niederman, M., 1993, Antimicrobial treatment of community-acquired pneumonia in adults: A conference report. Canadian Community-Acquired Pneumonia Consensus Conference Group. *Can. J. Infect. Dis.*, **4**, 25.

Mant, J. and Hicks, N., 1995, Detecting differences in quality of care: the sensitivity of measures of process and outcome in treating acute myocardial infarction. *BMJ*, **311**, 793–796.

Marrie, T. J., Lau, C. Y., Wheeler, S. L., Wong, C. J., Vandervoort, M. K., and Feagan, B. G., 2000, A controlled trial of a critical pathway for treatment of community-acquired pneumonia. CAPITAL Study Investigators. Community-Acquired Pneumonia Intervention Trial Assessing Levofloxacin. *JAMA*, **283**, 749–755.

McCormick, D., Fine, M. J., Coley, C. M., Marrie, T. J., Lave, J. R., Obrosky, D. S. *et al.*, 1999, Variation in length of hospital stay in patients with community-acquired pneumonia: Are shorter stays associated with worse medical outcomes? *Am. J. Med.*, **107**, 5–12.

McGarvey, R. N. and Harper, J. J., 1993, Pneumonia mortality reduction and quality improvement in a community hospital. *QRB Qual. Rev. Bull.*, **19**, 124–130.

Meehan, T. P., Fine, M. J., Krumholz, H. M., Scinto, J. D., Galush, D. H., Mockalis, J. T. *et al.*, 1997, Quality of care, process, and outcomes in elderly patients with pneumonia. *JAMA*, **278**, 2080–2084.

Meehan, T. P., Weingarten, S. R., Holmboe, E. S., Mathur, D., Wang, Y., Petrillo, M. K. *et al.*, 2001, A statewide initiative to improve the care of hospitalised pneumonia patients: The Connecticut pneumonia pathway project. *Am. J. Med.*, **111**, 203–210.

Menendez, R., Ferrando, D., Valles, J. M., and Vallterra, J., 2002, Influence of deviation from guidelines on the outcome of community-acquired pneumonia. *Chest*, **122**, 612–617.

Metersky, M. L., Tate, J. P., Fine, M. J., Petrillo, M. K., and Meehan, T. P., 2000, Temporal trends in outcomes of older patients with pneumonia. *Arch. Intern. Med.*, **160**, 3385–3391.

Mortenson, E. M., Coley, C. M., Singer, D. E., Marrie, T. M., Obrosky, D. S., Kapoor, W. N. *et al.*, 2002, Causes of death for patients with community-acquired pneumonia. Results from the Pneumonia Patient Outcomes Research Team cohort study. *Arch. Intern. Med.*, **162**, 1059–1064.

National Pathways Association, 2001, Developing care pathways. In K. du Lac (ed.), The Handbook. Radcliffe Medical Press, Abingdon, Oxon, UK.

Niederman, M. S., Bass, J. B. Jr, Campbell, G. D., Fein, A. M., Grossman, R. F., Mandell, L. A. *et al.*, 1993, Guidelines for the initial management of adults with community-acquired pneumonia: Diagnosis, assessment of severity, and initial antimicrobial therapy. American Thoracic Society. *Am. Rev. Respir. Dis.*, **148**, 1418–1426.

Niederman, M. S., McCombs, J. S., Unger, A. N., Kumar, A., and Popovian, R., 1998, The cost of treating community-acquired pneumonia. *Clin. Ther.*, **20**, 820–837.

Øvretveit, J. and Gustafson, D., 2003, Using research to inform quality programmes. *BMJ*, **326**, 759–761.

Paladino, J. A., Gudgel, L. D., Forrest, A., and Niederman, M. S., 2002, Cost-effectiveness of IV-oral switch therapy: Azithromycin versus cefuroxime with or without erythromycin for the treatment of community-acquired pneumonia. *Chest*, **122**, 1271–1279.

Pallares, R., Linares, J., Vadillo, M., Cabellos, C., Manresa, F., Viladrich, P. F. *et al.*, 1995, Resistance to penicillin and cephalosporin and mortality from severe pneumococcal pneumonia in Barcelona, Spain. *N. Engl. J. Med.*, **333**, 474–480.

Palmer, C. S., Zhan, C., Elixhauser, A., Halpern, M. T., Rance, L., Feagan, B. G. *et al.*, 2000, Economic assessment of the community-acquired pneumonia intervention trail employing levofloxacin. *Clin. Ther.*, **22**, 250–264.

Pearson, S. D., Goulart-Fisher, D., and Lee, T. H., 1995, Critical pathways as a strategy for improving care: Problems and potential. *Ann. Intern. Med.*, **123**, 941–948.

Phillips, K. F. and Crain, H. C., 1998, Effectiveness of a pneumonia clinical pathway: Quality and financial outcomes. *Outcomes Manag. Nurs. Pract.*, **2**, 16–22.

Pope, C. and Mays, N., 2000, Qualitative research in health care: Assessing quality in qualitative research. *BMJ*, **320**, 50–52.

Pope, C., Ziebland, S., and Mays, N., 2000, Qualitative research in health care: Analyzing qualitative data. *BMJ*, **320**, 114–116.

Ramos, K. D., Schafer, S., and Tracz, S. M., 2003, Validation of the Fresno test of competence in evidence based medicine. *BMJ*, **326**, 319–321.

Reddy, J. C., Katz, Goldman, L., and Wachter, R. M., 2001, A pneumonia practice guideline and a hospitalist-based reorganization lead to equivalent efficiency gains. *Am. J. Manag. Care*, **7**, 1142–1148.

Rhew, D. C., Riedinger, M. S., Sandhu, M., Bowers, C., Greengold, N., and Weingarten, S. R., 1998, A prospective, multicenter study of a pneumonia practice guideline. *Chest*, **114**, 115–119.

Rhew, D. C., Tu, G. S., Ofman, J., Henning, J. M., Richards, M. S., and Weingarten, S. R., 2001, Early switch and early discharge strategies in patients with community-acquired pneumonia: A meta-analysis. *Arch. Intern. Med.*, **161**, 722–727.

Rollins, D., Thomasson, C., and Sperry, B., 1994, Improving antibiotic delivery to pneumonia patients: Continuous quality improvement in action. *J. Nurs. Care Qual.*, **8**, 22–31.

Ross, G., Johnson, D., Kobernick, M., and Pokriefka, R., 1997, Evaluation of a critical care pathway for pneumonia. *J. Healthc. Qual.*, 22–29,36.

Sanyal, S., Smith, P. R., Saha, A. C., Gupta, S., Berkowitz, L., and Homel, P., 1999, Initial microbiologic studies did not affect outcome in adults hospitalized with community-acquired pneumonia. *Am. J. Respir. Crit. Care Med.*, **160**, 346–348.

Solomon, D. H., Van Houten, L. V., Glynn, R. J., Baden, L., Curtis, K., Schrager, H. *et al.*, 2001, Academic detailing to improve use of broad-spectrum antibiotics at an academic medical center. *Arch. Intern. Med.*, **161**, 1897–1902.

Sperry, S. and Birdsall, C., 1994, Outcomes of a pneumonia critical path. *Nurs. Econ.*, **12**, 332–339,345.

Stahl, J. E., Barza, M., DesJardin, J., Martin, R., and Eckham, M. H., 1999, Effect of macrolides as part of initial empiric therapy on length of stay in patients hospitalized with community-acquired pneumonia. *Arch. Intern. Med.*, **159**, 2511–2512.

Suchyta, M. R., Dean, N. C., Narus, S., and Hadlock, C. J., 2001, Effects of a practice guideline for community-acquired pneumonia in an outpatient setting. *Am. J. Med.*, **110**, 306–309.

Tunis, S. R., Hayward, R. S. A., Wilson, M. C., Rubin, H. R., Bass, E. B., Johnston, M. *et al.*, 1994, Internists' attitudes about clinical practice guidelines. *Ann. Intern. Med.*, **120**, 956–963.

Waterer, G. W., Somes, G. W., and Wunderrink, R. G., 2001, Monotherapy may be sub-optimal for severe bacteraemic pneumococcal pneumonia. *Arch. Intern. Med.*, **161**, 1837–1842.

Wagner, A. K., Soumerai, S. B., Zhang, F., and Ross-Degnan, D., 2002, Segmented regression analysis of interrupted time series studies in medication use research. *J. Clin. Pharm. Ther.*, **27**, 299–309.

Wensing, M. and Elwyn, G., 2003, Methods for incorporating patients' views in health care. *BMJ*, **326**, 877–879.

Weingarten, S. R., Riedinger, M. S., Hobson, P., Noah, M. S., Johnson, B., Giugliano, G. *et al.*, 1996, Evaluation of a pneumonia practice guideline in an interventional trial. *Am. J. Resp. Crit. Care Med.*, **153**, 1110–1115.

World Health Organisation, 1947, The constitution of the World Health Organisation. *WHO Chronicles*, **1**, 29.

Chapter 5

Collecting, Converting, and Making Sense of Hospital Antimicrobial Consumption Data

Stephanie Natsch
Clinical Pharmacologist, Department of Clinical Pharmacy, University Medical Center Nijmegen, The Netherlands

1. INTRODUCTION

Various methods are available to quantify drug use in hospitals. Sound data on the use of antibiotics are crucial for the interpretation of prescribing habits, the evaluation of compliance with clinical guidelines, and the linkage with antimicrobial resistance data. Quantification of drug use does not only comprise collection and registration of data. Before starting this process, it should be considered in which way the data will be described and interpreted. The goal of the use of the data should be established beforehand. Clear decisions have to be taken on the methodology applied to this epidemiological surveillance. The data must provide the ability to measure variation in quality and quantity of use. Care has to be taken on the reliability and completeness of the data as well as the feasibility of the data collection. Methods of assessment need to be simple, rapid, and inexpensive.

Requirements for the establishment of an efficient surveillance programme include authority from the highest level of the hospital management, complete and free access to all relevant information, and clarity as to the ownership of the data generated.

Facilities must tailor surveillance systems to balance the availability of resources with priorities for data collection, population needs, and institutional objectives. Integrating the surveillance system within the framework of the

Antibiotic Policies: Theory and Practice. Edited by Gould and van der Meer
Kluwer Academic / Plenum Publishers, New York, 2005

institution's other quality improvement efforts can facilitate functional collaboration between and among programmes working to improve patient care. Readily available quantitative methods provide significant information on patterns of use and an efficient basis for planning definitive studies. This chapter describes various sources of data, different units of measurement, considerations on the frequency of data collection as well as the level of aggregation of data.

2. SOURCES OF DATA

Several sources of data are available for quantification of hospital drug use, each with its advantages and disadvantages (Chaffee *et al.*, 2000; Eckert *et al.*, 1991).

Pharmacy purchases can be determined from invoices or delivery documents. They allow for an overall assessment of the use in a specific institution. Data can be presented in terms of costs, number of packages purchased, or defined daily doses (DDDs). Care has to be taken as to the accuracy of the data if a pharmacy purchases for more than one institution or if there are other sources of delivery such as trial medication or free samples. Time courses of use are difficult to assess if purchases are done in large quantities at the same time. Data may overestimate use in the case where considerable quantities of drugs are returned to the manufacturer or discarded because the expiry date is reached before use. Nevertheless, purchase data are very easily accessible as they are usually readily available as management data from the finance department. They provide a rough estimate of overall use.

Pharmacy deliveries to wards or units within an institution as a source of data allow for more detailed presentation of the data. If hospital wards reflect particular groups of patients, for example, ICU-unit or oncology, delivery data reflect specific patterns of usage in these patient categories. Data can be presented in terms of costs, number of packages purchased, or DDDs. Time courses of usage are more easily detected as wards usually do not keep large stocks in advance but order drugs when needed. Again, data may not reflect actual use as drugs can be returned to the pharmacy or exchanged with other wards within the institution without administrative correction of the transaction. Also, pharmacies must be able to select specific wards for data presentation, in order to avoid spoiling of the data with deliveries to third parties, nursing homes, psychiatric institutions, or other clients. In most instances, pharmacy delivery data will be the most accessible and most accurate level of data source that is readily available in a timely fashion.

Patient prescription profiles as a source of data give an even more accurate picture of the actual use. They can be performed either as cohort studies or as complete registration of all filled prescriptions. Clearly, data collection is more labour intensive than the above-mentioned methods. However, they provide a lot more detail on patient features and actual patterns of prescribing and use. Linkage with information about indication for use and laboratory values provides even greater insight. Up to date, not many hospitals are ready to provide such detailed data, but rapid advancement of computerisation will allow more easy access to data on prescription levels in due future.

On a regional or national level, other sources of data may be available. Sales data from wholesalers to hospital pharmacies are used, often through commercial databases. Also, in this case, careful evaluation concerning the completeness of the data is necessary. Parallel import of drugs and direct deliveries to hospital pharmacies can lead to underestimation of the total use of drugs. Another useful source of data may be data from reimbursement through insurers. However, the high variability in reimbursement schemes in different countries may make interpretation and comparison of the available data more than cumbersome. Completeness of data will depend on the system and accuracy of declaration of expenses by hospitals.

3. MEASUREMENT UNITS

Different measurement units can be used to evaluate drug use. As numerator, costs, number of packages, volume in grams, number of prescriptions, or DDDs can be used.

Costs allow for an overall analysis of drug expenditures or prescription analysis of one single drug. But there are many disadvantages. Price differences between alternatives confuse the analysis. Comparisons between different institutions or countries are not possible because of different price levels. Each setting may negotiate local prices through direct negotiations. Indexing in the case of long-term studies is necessary.

Number of packages may be independent of sales prices, but may still differ depending on the manufacturer or the country of purchase. This will greatly complicate comparisons.

Assessment of the quantitative volume of use in terms of grams allows only evaluation of the use of one drug at a time. Drugs with a low potency have a larger fraction of the total, not reflecting actual greater use or more frequent prescribing.

The number of prescriptions gives an accurate reflection of the number of patients exposed to a drug, and allows for evaluation of the frequency with which certain drugs are prescribed. It is also a valuable measure when evaluating prophylactic use of antibiotics. A major disadvantage is that, often, the amount of drug used is not known. Also calculations are done more than once for patients receiving more than one prescription or multiple drug regimens.

The use of DDDs as a unit of measurement helps to avoid many of the above-mentioned drawbacks. (Natsch *et al.*, 1998). The system of DDDs has been developed by the World Health Organisation (WHO, 2003). It provides a convenient tool that allows comparisons between different settings, regions, or even countries. A DDD is assigned to every chemical substance, reflecting an international compromise based on the average dosage for the most common indications in adult patients with normal organ function. This is a technical unit of measurement and does not necessarily reflect the recommended or actual dose used. The quality of the results is completely dependent upon strict adherence to the method. Therefore, the system must be used without any adaptations. But the method is independent of sales prices or package sizes and allows for long-term epidemiological studies. Also with DDDs, there are some disadvantages. For some antimicrobials, there are still no DDDs defined; also, for combination preparations, this is a problem. For some antimicrobials, different DDDs are assigned, depending on the route of administration. Conclusions on prescribed dosages and duration of treatment may not be made, particularly not in children or patients with impaired organ function. Also, in the case of prophylactic use of antimicrobials, the use of DDDs has its drawbacks. For amoxicillin and amoxicillin/clavulanic acid, the DDD differs greatly from the actually prescribed doses in inpatients. High consumption of these two antibiotics can, therefore, considerably influence the total use expressed in DDDs. The Collaborating Centre for Drug Statistics of WHO revises the DDDs once in a while. While this allows adjustments in the case of great differences of DDDs from actually prescribed dosages, it makes it difficult to follow data for longer periods of time. Therefore, when using DDDs, it is very important to always state which edition of the DDD system has been applied for calculating the data. The most recent information can be currently found easily on the Internet (http://www.whocc.no/atcddd/).

As denominator, in the community, usually DDDs/1,000 inhabitants/day are calculated. In hospitals, usually DDDs/100 bed-days are used. For better comparison, it is now being discussed that we should calculate in-hospital use per 1,000 bed-days. In either case, the number of bed-days has to be calculated with great care.

There are two ways to do so:

1. Number of beds $\times$ occupancy rate
2. Number of hospitalisations $\times$ length of stay

But also for length of stay, different definitions are used:

1. the day of admission as well as the day of discharge count each for an extra day
2. the day of admission plus the day of discharge count together for one extra day

These calculations give an estimate of the ecological pressure of antimicrobials as they quantify overall use in the population. A simple calculation program named ABC Calc—Antibiotic Consumption Calculator—has been developed and can be downloaded from the Internet free of charge (http://www.escmid.org/sites/index.asp and go to: study groups; ESGAP; News & activities) For measuring patient exposure, the percentage of patients exposed to antimicrobials can be evaluated. Alternatively, the number of prescriptions gives a rough estimate of exposure. Another measure for exposure is the calculation of the number of days that patients are on antimicrobial treatment, calculated per 1,000 patient-days.

Whatever denominator is chosen, the total population has to be defined clearly. For reasons of comparison of data between different settings, it is very important that all variables are defined as carefully as possible.

4. FREQUENCY OF DATA COLLECTION

Whatever source of data is used, a decision has to be taken as to which data are collected as an ongoing process. This is preferably done as part of a wider quality assurance programme. As an alternative, subsets of data could be collected. In this case, it should be carefully evaluated if the dataset is representative of the overall use, allowing for extrapolation of the data to the total population studied. Careful validation of this process is a prerequisite for high quality data.

Another point to consider is the time interval in which data are collected and presented. Most hospitals will only be able to present consumption data on a yearly basis. This is usually the unit, in which management and financial data are presented. Some, however, are able to present data on a more detailed level like quarterly or monthly units. This is more laborious to deal with, but for a close link to resistance data, more detailed data are preferred (Lopez-Lozano *et al.*, 2000; Monnet *et al.*, 2001). A major disadvantage is that sample sizes will become very small and this can affect validity of the data in a negative way. It will allow performance of time-series analysis and modelling and forecasting development of resistance linked to antibiotic consumption. If prescription or consumption data are taken as the basis for data collection, data will primarily be generated on a daily basis. This will need a lot of computer memory to store and handle data. When aggregating the data and discarding the original source, consequences for future analysis have to be carefully analysed well in advance.

5.   LEVEL OF AGGREGATION OF ANTIMICROBIALS

Besides DDDs, the WHO collaborating centre for drug statistics has developed the Anatomical Therapeutical Chemical Classification system (ATC). This is a comprehensive and logical classification system developed to categorise drug substances, which were divided into different groups according to the organ or system on which they act (anatomic), and then according to their therapeutic, pharmacological, and chemical characteristics. It leads to a 5-level hierarchical code assigned to each chemical substance. An example is shown below.

ATC code	ATC level	Description
J	Main anatomic group	General anti-infective agents for systemic use
J01	Therapeutic group	Antibacterial agents for systemic use
J01M	Pharmacological group	Quinolone antibacterial agents
J01MA	Pharmacological subgroup	Fluoroquinolones
J01MA02	Chemical substance	Ciprofloxacin

Computer systems should allow for aggregation of data on different levels according to the research question. If use of a single chemical substance is analysed, analysis on ATC-level 5 should be possible, whereas if a whole pharmacological (sub)group is analysed, aggregation on ATC-level 3 or 4 should be possible.

A major disadvantage of the system is that its first level is based on an anatomical classification. As a consequence, substances used for different disease states are classified in different categories and therefore get more than one ATC code. This is especially relevant in the case of antimicrobials. ATC group J comprises systemic use of antimicrobial agents. Agents specifically used for gastrointestinal or genitourinary diseases are classified in the respective ATC groups. Antimicrobials also used for skin-, eye-, or ear-diseases are also classified in these ATC groups. In general, most interest is focused on systemic use of antimicrobial agents, as this is believed to be most relevant in relation to development of resistance. While performing data collection, it is important to clearly state which categories are excluded from analysis. Despite some disadvantages, and alternatives proposed (Bjornsson, 1996; Pahor *et al.*, 1994), the ATC classification is now widely used and should be chosen as a major classification system.

Besides aggregation based on chemical and anatomical classification, use of antimicrobials can be classified and aggregated according to other criteria. Within a hospital, it is advisable to at least subclassify use per ward or department, preferably linked to special patient categories. Allocation according to the medical specialist gives even more insight into the prescription patterns. Other classifications may be done into prophylactic and therapeutic use, or into parenteral, oral, and local use of antimicrobials. The most sophisticated classification is a registration per indication. But this asks for a linkage with an electronic patient profile, which is not yet widely and easily available in hospitals. In outpatients, these systems become more easily available, often through prescription registration of pharmacies, general practitioners, or insurers. Hospitals should focus on rapid development of these systems in the future.

6. CONCLUSIONS

Data will be used for surveillance of use and feedback to prescribers. They can support development of policies and guidelines. Thereafter, data collections can help monitoring adherence to them and help identify weaknesses in the system or the implementation process.

Increasingly, healthcare institutions are being asked to benchmark or compare their performance data to other similar institutions. This may be a more complex and difficult undertaking than is immediately obvious, because the data may be affected by a variety of factors, some of which, such as the underlying health status of the population served by the institution, are outside the control of the institution. However, ongoing monitoring and benchmarking have been used to implement quality improvement activities. The influence of case-mix and severity-of-illness of the population being studied has to be controlled. Variations in length of stay can also have great influence on interpretation of data, as it can reflect less severely ill-patients but also more intense treatment and earlier discharge in case of shortage of hospital beds. In the latter case, use will increase per bed-day, but not if calculated per patient or admission. Possibly risk stratification can be applied to correct for these influences. The consequences of these effects for the ecological pressure of antimicrobials and selection of resistance are discussed in other chapters.

ACKNOWLEDGEMENTS

I would like to thank all the members of the Working Group on Surveillance of Antimicrobial Use of the Dutch Working Party on Antibiotic Policy (SWAB) for the valuable discussions they have contributed to this chapter and especially their coordinator, P. M. G. Filius, PharmD, for her critical comments on the manuscript.

REFERENCES

Bjornsson, T. D., 1996, A classification of drug action based on therapeutic effects. *J. Clin. Pharmacol.*, **36**, 669–673.

Chaffee, B. W., Townsend, K. A., Benner, T., and DeLeon, R. F., 2000, Pharmacy database for tracking drug costs and utilization. *Amer. J. Health-Syst. Pharm.*, **57**, 669–676.

Eckert, G. M., Ioannides-Demos, L. L., and McLean, A. J., 1991, Measuring and modifying hospital drug use. *Med. J. Aust.*, **154**, 587–592.

Lopez-Lozano, J. M., Monnet, D. L., Yague, A., Burgos, A., Gonzalo, N., Campillos, P. *et al.*, 2000, Modelling and forecasting antimicrobial resistance and its dynamic relationship to antimicrobial use: A time series analysis. *Int. J. Antimicrob. Agents*, **14**, 21–31.

Monnet, D. L., Lopez-Lozano, J. M., Campillos, P., Burgos, A., Yague, A., and Gonzalo, N., 2001, Making sense of antimicrobial use and resistance surveillance data: Application of ARIMA and transfer function models. *Clin. Microbiol. Infect.*, **7**(Suppl. 5), 29–36.

Natsch, S., Hekster, Y. A., de Jong, R., Heerdink, E. R., Herings, R. M. C., and van der Meer, J. W. M., 1998, Application of the ATC/DDD methodology to monitor antibiotic drug use. *Eur. J. Clin. Microbiol. Infect. Dis.*, **17**, 20–24.

Pahor, M., Chrischilles, E. A., Guralnik, J. M., Brown, S. L., Wallace, R. B., and Carbonin, P., 1994, Drug data coding and analysis in epidemiologic studies. *Eur. J. Epidemiol.*, **10**, 405–411.

World Health Organisation, 2003, *Guidelines for ATC Classification and DDD Assignment.* WHO Collaborating Centre for Drug Statistics Methodology, Norwegian Institute of Public Health, Oslo.

Chapter 6

How Do Measurements of Antibiotic Consumption Relate to Antibiotic Resistance?

Roger L. White
Department of Pharmaceutical Sciences, Medical University of South Carolina,
280 Calhoun Street, Charleston, SC, USA

1. INTRODUCTION

Currently, we are confronting an epidemic of resistance to antimicrobials. Although the spread of antimicrobial resistance is often due to lack of adherence to infection control measures, selection of resistance due to inappropriate use of antimicrobials may also play a large role. However, antimicrobial resistance is not a new phenomenon, with *Staphylococcus aureus* demonstrating resistance to penicillin soon after its introduction in the 1940s. Initially, resistance was usually related to only a few antimicrobials; however, multidrug resistance is increasingly common. The societal costs of antimicrobial resistance, in terms of morbidity and mortality, are substantial (Lucas *et al.*, 1998; Meyer *et al.*, 1993). In monetary costs, antibiotics are commonly prescribed drugs and the annual costs to the US healthcare system exceed $7 billion.

Antimicrobial resistance is an incredibly complex problem with no simple solutions. Clonal spread of resistance is facilitated by increased use of day-care centers, international travel, and the transfer of patients to and from hospitals and nursing homes. Antibiotics contribute to selection pressure for resistance due to overuse and misuse of antimicrobials in both inpatient and outpatient settings. In addition, routine use of antimicrobials in the animal husbandry industry is also a factor in resistance. (Smith *et al.*, 2002). Multiple drug resistance and delays in development of resistance add to the complexity

Antibiotic Policies: Theory and Practice. Edited by Gould and van der Meer
Kluwer Academic / Plenum Publishers, New York, 2005

of assessing relationships between antimicrobial use and resistance (Friedrich *et al.*, 1999).

Proving a causal relationship between the use of antimicrobials and development of resistance is very difficult; however they have been linked by a substantial amount of evidence. Unfortunately, there are many confounding factors and most of the studies that have examined relationships between antibiotic use and resistance have been hampered by inability to control confounding variables (Austin *et al.*, 1999a; Levin, 2001; Lipsitch *et al.*, 2000a). Therefore, changes in resistance patterns seen after changes in antimicrobial usage patterns may be due to other factors such as changes in infection control measures that may prevent detection of these relationships. Although direct evidence is lacking, there is compelling evidence that resistance is proportional to antimicrobial usage. McGowan, (1983) and Archibald *et al.* (1997) found resistance to be proportional to antimicrobial usage for Enterococci, Staphylococci and Pseudomonads in studies that compared antibiotic use and resistance in an intensive care unit (ICU) to other patient-care areas of the institution. Moreover, controlling antibiotic use has been shown to reverse trends in resistance (Arason *et al.*, 1996; McNulty *et al.*, 1997; Rahal *et al.*, 1998; Seppala *et al.*, 1997).

Although one might attempt to prevent or reverse antimicrobial resistance by interventions in infection control and antimicrobial use as separate entities, this approach may be too simplistic. In modeling relationships between antimicrobial use and resistance with vancomycin-resistant enterococci (VRE) in an ICU, Austin *et al.* (1999a) found infection control measures such as handwashing and cohorting of nursing staff closely linked with antimicrobial use control. This analysis suggested that the impact of infection control measures might be negated by inappropriate antibiotic use. Furthermore, antibiotic use is related to infection control since antibiotics can affect the transmission of organisms.

Thus, it is not surprising that it has been recommended that more attention be paid to monitoring antibiotic usage. Indeed, the Society for Healthcare Epidemiology of America (SHEA) suggests measuring antibiotic usage to compare usage trends, to provide a benchmark for costs analyses and to facilitate the assessment of relationships with both adverse events and the development of resistance. Guidelines by SHEA and the Infectious Diseases Society of America (IDSA) (Shlaes *et al.*, 1997) suggest monitoring use of antimicrobials by hospital location or prescribing service as well as monitoring the relationship between antibiotic use and resistance.

Historically, as resistance has developed to an antimicrobial, we have been able to depend on the development of new antimicrobials. However, we are now in an era of minimal development of new antimicrobials, especially those directed at Gram-negative aerobic infections. Therefore, we must pay close attention to the proper use of infection control measures and the appropriate use of antimicrobials.

Quantitative relationships between antibiotic use and resistance have not been well studied. However, defining these relationships is important because antibiotic use is one variable upon which we may be able to exert some control. Various measures of antimicrobial use such as the defined daily dose (DDD), grams purchased or administered, days of antimicrobial therapy, the mean daily dose, and the number of doses administered have been utilized in assessing drug use-susceptibility relationships. Differences in these measures and evidence of their relationships with resistance will be discussed.

2. ANTIMICROBIAL USAGE DATA

Most investigations of antimicrobial use have been performed in hospitals, where access to usage data may be readily available. However, some studies have examined antimicrobial usage in the community, and yet others have involved nationwide antibiotic use. In hospitals, antibiotic usage data is usually obtained from either hospital purchase or pharmacy dispensing records, or drug administration records on individual patients. Although dispensing or administration records may record patient-specific information, these data are often aggregated for a specific drug. This is referred to as aggregate or "group level" data. When related to cumulative susceptibility data, relationships between antimicrobial use and resistance are often referred to as ecological studies. Purchase records, which have no patient-specific information, always fall into this category. However, non-aggregated dispensing or preferably, administration records are patient-specific data. This distinction is important since studies have revealed divergent results when these two types of data have been evaluated. Harbarth *et al.* (2001) studied both individual patient data such as days of antibiotic exposure and the average number of doses per day and aggregate data, reported as DDDs over a 5-year period in a single institution. The studies evaluated whether resistance of nosocomial isolates of *Enterobacteriaceae* or *Pseudomonas* species was related to in-hospital exposure to fluoroquinolones, third-generation cephalosporins, ampicillin/ sulbactam, or imipenem. With aggregate level data, increases in DDDs of fluoroquinolones, third generation cephalosporins and ampicillin/sulbactam were noted; however, the proportion of isolates that were susceptible was stable. Relationships between antibiotic use and resistance were weak and only significant for ampicillin/sulbactam at the specific hospital ward level. In contrast, with patient level data, each drug or drug class evaluated was found to be a strong risk factor for resistance to that drug class. In studies involving patient level data, selection of the proper control group may be important (Harris,

2002) and reliance on patients with susceptible isolates as a control group may be insufficient.

With either type of data, one must decide if all use of a specific drug will be included. For example, only antibiotics selected for therapeutic use may be evaluated or those used for prophylaxis may be included. Furthermore, non-systemic uses such as antibiotics used in surgical repair materials or for topical administration may be included. Most evaluations have assumed that all of an administered drug is available systemically; however, exposure to antibiotics with low oral bioavailability may be overestimated unless corrected for bioavailability. Furthermore, those drugs with poor bioavailability may maintain higher concentrations in the gastrointestinal tract and have a significant impact on alterations of flora. Thus, depending on the type and location of organisms, the systemic availability of antimicrobials should be considered. In addition, high protein binding can limit the concentration of the unbound, pharmacologically active concentration of an antimicrobial. Thus, correction for protein binding may be important for highly bound drugs.

2.1.   Patient-specific data

Patient-specific or patient-level antimicrobial use data, used in case control studies, includes collection of the dose, dosing interval and the length of the dosing regimen (Table 1). Depending on the purpose of the study this information may be collected before or after the event of interest. For example, one might study the impact of antecedent antimicrobial use for a fixed period of time prior to detection of a resistant organism. With these data, one could make the distinction between empirical antimicrobial use and that used for directed therapy. Most studies have gathered these data from retrospective chart review, review of pharmacy dispensing records, or medication administration records. With the increasing use of electronic medical records, collection of these data should be less time consuming.

Patient-specific data has several advantages over aggregate data. Most importantly, it allows analysis of the development of resistance that may be due to antecedent and/or concurrent antimicrobial therapy. When other data are collected such as patient demographics, underlying disease states, and the infecting pathogen, one can assess the appropriateness of antimicrobial therapy. Patient-specific data also allows the evaluation of multiple antimicrobial therapy whereas aggregate data usually examines only a single drug. Since a patient's location within a healthcare setting (e.g., an ICU) is known, patient-specific data also facilitates evaluation in specific patient-care areas within an institution. Lastly, since patient-specific data can always be aggregated, comparisons of the relationships with patient-specific and aggregate antimicrobial use data on resistance can be studied. When patient-specific data are analyzed

Table 1. Example of raw antimicrobial use data in specific patients and calculations that may be performed to study aggregate antimicrobial use

Patient	Raw data			Calculations		
	Dose (g)	Dosing interval (hr)	Length of therapy (days)	Doses per day	Grams per day	Grams per course of therapy
1	1	8	10	3	3	30
2	2	8	7	3	6	42
3	1	12	8	2	2	16
Mean	1.3	9.3	8.3	2.7	3.7	29.3
Total			25.0			88.0

without concomitant evaluation of aggregate data, there are some potential disadvantages when compared to aggregate data. In a case control study in which patients are selected based on the occurrence of a specific pathogen, the impact of an antimicrobial on emergence of resistance in other patients, and especially with other organisms, will be missed. In addition, the total amount of selective pressure exerted by an antimicrobial within an institution may be missed if use of that antimicrobial in other patients is not analyzed. Antimicrobial use outside of the patient ward, often referred to as floor stock, such as surgical prophylaxis and dosing in patients undergoing hemodialysis may be excluded if the dispensing records are separate from the primary pharmacy dispensing records. Furthermore, evolving concerns about patient privacy may limit the collection and analysis of patient-specific data (U.S. Office of the Federal Register, 1996).

2.2. Aggregate data

Antimicrobial use data that has been summarized for a group of patients, in which the patient-specific dose, dosing interval, and length of therapy is no longer discernable are referred to as aggregate or group-level data. Aggregate data is most often derived from antimicrobial purchases, but may also be summarized from patient-specific data.

There are several advantages of aggregate antimicrobial use data when compared to patient-specific data. Since aggregate use should capture all of the exposure to an antimicrobial in a healthcare setting, it may better relate to the total microbiological ecology of that setting. Thus, development of resistance to commensal organisms, often called "collateral damage" related to antimicrobial use might be detected with aggregate ecological studies. Purchase data, the most common source of aggregate antimicrobial use data, are usually readily available, and thus, easier to benchmark vs use in other institutions. Furthermore, when simultaneous pharmacoeconomic studies are a consideration, purchase data may be desirable.

A number of disadvantages exist for aggregate antimicrobial use data. A major limitation is that no distinction can be made between antecedent use prior to the development of a resistant isolate and that administered thereafter, thus precluding assessment of resistance due to poor infection control rather than inappropriate antimicrobial use. Aggregate data, if obtained at the nationwide level, may not allow distinction of antimicrobial use that is not used for treatment of infection (e.g., antimicrobials used in the animal husbandry and horticultural industries) from that used in patients. Since all antimicrobials purchased or dispensed are not administered, aggregate data derived from purchases data may overestimate exposure of an antimicrobial in a specific healthcare setting. When aggregate data are derived from patient-specific data, one can evaluate the impact of antimicrobial use within a specific patient-care area such as an ICU; however, purchase data would preclude such assessment. Lastly, aggregate data apportions antimicrobial use over an entire population of patients, potentially masking the true intensity of antimicrobial exposure.

2.3.   Sources of antimicrobial use data

2.3.1.   Antimicrobial purchases

Purchase records to estimate antimicrobial use may be obtained from a specific institution, a wholesale distributor, or from government records in countries with nationalized health plans. Through use of purchase records, antimicrobial use in an institution, a community, or a larger geographic region may be assessed. Purchase data may have an advantage over other types of data collection in that total antimicrobial use within a healthcare setting may be captured. Purchase data, often expressed as grams purchased, can also be linked to acquisition costs of antimicrobials; thus, economic analyses will capture total acquisition costs. The major disadvantage of using purchase data to estimate drug exposure in patients is that the drug may never be administered to the patient. Antimicrobials that are purchased may not be dispensed or administered due to wastage after preparation, changes in a prescriber's orders, or destruction of a drug that has exceeded its expiration date. Therefore, purchase data represent the least sophisticated, but most readily available, method of estimating antimicrobial exposure.

2.3.2.   Antimicrobial dispensing records

Antimicrobial pharmacy dispensing records may represent a more accurate assessment of antimicrobial use than drug purchases since all antimicrobials that are purchased are not dispensed. Dispensing records, in which antimicrobials that are dispensed but not administered to a patient are recorded properly, may accurately assess antimicrobial use in a patient population. However, since some of these "returns" may not be recorded, dispensing records are likely to overestimate actual antimicrobial exposure in patients. In some settings, these data are readily available and are preferred over purchase records as a measure of drug use.

2.3.3. Patient administration records

Within an institution, patient administration records may be readily available as an estimate of antimicrobial use. Theoretically, these records should exactly mirror drug use; however, the accuracy of the records should be verified. For example, a dose of an antimicrobial may be charted, yet the drug is still not administered to the patient due to discovery of a drug allergy or difficulty in establishing venous access. In some cases, a partial dose might be administered before an adverse event (e.g., extravasation) is recognized. Although administration records have been difficult to obtain in many settings, increasing use of electronic medical record data may make this information more readily available.

3. ANTIMICROBIAL USAGE MEASURES

From the records described above, one may use or calculate a variety of measures, or metrics, to quantify antimicrobial use. There may be no single ideal measure of antimicrobial use. Indeed, one may choose a measure over another depending on the purpose of the analysis. There may be no single ideal measure that can be used for each drug, drug class, and relationship with resistance. Measures based on patient-specific data allow more flexibility in the analysis of relationships between antimicrobial use and resistance and can also be used to calculate aggregate data.

3.1. Patient-specific measures

Depending on the type of study relating antimicrobial use and resistance, several patient-specific measures have been utilized. In some case control studies, these measures are non-quantitative. For example, one may assess as a binomial variable whether a patient received a specific antimicrobial during a specified period of time without regard to quantity. However, most patient-specific studies assess quantitative antimicrobial use. These measures include the mean daily dosage, the number of antibiotic orders, doses administered, days of therapy, or the number of grams administered to an individual patient.

Measures that include the intensity of the dose, such as mean daily dose or grams administered may be preferred over measures such as the number of doses or number of prescriptions that fail to account for dose intensity. Thus, the mean daily dose, also referred to as the prescribed daily dose (PDD) may be useful.

Other measures such as the number of days of antimicrobial therapy may be independent of dose; thus, allowing assessment of dosage independent of length of therapy. Integrated, or hybrid, measures such as the total number of grams administered to a patient may account for both of these independent factors, daily dosage and length of therapy (calculation in Table 1). This may be useful since it is plausible that development of resistance with specific organisms may be due primarily to either the length of antimicrobial exposure or the intensity of the daily dose rather than a hybrid measure such as total grams. If one calculates the average length of therapy from days of therapy in individual patients, antibiotic stop order policies may create a problem. For example, if an institution uses a stop order policy that could discontinue a planned 10-day regimen after only 5 days, continuation of the drug regimen may result in a new antibiotic order. In this situation, it would appear that there were two drug orders, each with a 5-day length of therapy. Many hospital information systems would not indicate that the length of therapy in that patient was actually 10 days. Although total grams of an antimicrobial administered to a patient can be derived from patient-specific data, grams of use are usually obtained from purchase records and will be discussed as an aggregate marker of drug usage.

3.2.   Aggregate measures

As mentioned above, one may derive patient-specific data from dispensing or administration records and then use either averages or totals of aggregate data to describe antimicrobial use. Although the average daily dose and average length of therapy have been used as antimicrobial use measures, most aggregate data involve total measures.

3.2.1.   Number of prescriptions

Aggregate measures such as the total number of prescriptions, doses, vials, or packages have been used to estimate antimicrobial use; however, these measures provide no direct information on antimicrobial exposure in patients. In limited situations however, these measures may sufficiently estimate use when all patients are given a fixed dose and/or fixed dosing interval. In an analysis by White (2002) over a 9-year period in a large teaching hospital, the correlation between the number of orders for ceftazidime with grams administered

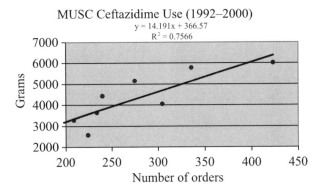

Figure 1. Plot of the number of grams used vs the number of orders for ceftazidime between 1992–2000 at a large teaching institution (White, 2002).

was strong ($R^2 = 0.757$) (see Figure 1). Although correlations of these measures and a more quantitative measure such as grams may be strong, grams as a measure of use provide more quantitative information.

3.2.2.   Expenditures

Historically, antimicrobial expenditures were used to estimate antimicrobial usage. Although this measure may still be a reasonable approximation of use in some circumstances, it may differ significantly from actual usage (Rifenburg *et al.*, 1999). The major limitation with expenditures is that the acquisition cost of an antimicrobial may change over time. In a longitudinal analysis of antimicrobial use, or when use is related to the development of resistance, substantial changes in acquisition costs during the period of analysis significantly limit the usefulness of this measure. However, expenditure may be one of the easiest measures to obtain.

3.2.3.   Grams

Although total grams of an antimicrobial used in a healthcare setting in a fixed period of time can be derived from patient-specific data, grams used are usually obtained from purchase data. As mentioned above, grams can be considered to be a hybrid, rather than independent measure of use since the daily dosage and the length of therapy are either used in the calculation of grams or inherent in the measure if purchased grams are used. If one is analyzing use of only a single drug over time, total grams is a valid measure of use. However,

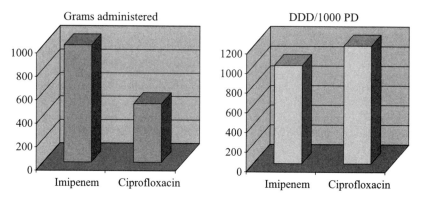

Figure 2. Comparison of grams administered to DDD/1,000 patient-days (PD) of Imipenem and Ciprofloxacin (DDD for Imipenem = 2 g/day, Ciprofloxacin = 0.8 g/day, patient-days = 500) (White, 2002).

when one compares the use of two or more antimicrobials with different daily gram dosages, problems arise. For example, if a drug that is given in a higher gram dosage (e.g., imipenem) is compared to a drug with a lower gram dosage (e.g., ciprofloxacin), differences in use are difficult to discern (Figure 2). Thus, a measure to normalize drugs with different gram daily dosages was needed, especially when evaluating total antimicrobial use for a class of antimicrobials. The DDD was established to alleviate this problem. When one compares grams from one institution to another in an attempt to benchmark antimicrobial use, the source of the data for the grams used may be important. In a comparison of antimicrobial use among 10 hospitals, the source of the grams reported varied from grams dispensed (5 hospitals) to grams purchased (4 hospitals) to grams removed based on storeroom records (Lesch *et al.*, 2001). Obviously collection of data from these varied sources makes meaningful surveillance of drug use among multiple institutions more difficult.

3.2.4.   Defined daily dose (DDD) method

The DDD method is used to measure and compare antimicrobial use within a population of patients. It has primarily been used to assess antibiotic consumption within an institution, but has also been used to estimate non-hospital consumption within a specific geographic region or country (Ruiz-Bremon *et al.*, 2000). In the calculation of DDDs as a measure of antimicrobial use, the total grams used are divided by the DDD, which represents a typical adult daily dosage. This calculation is usually reported as a normalized value of DDD/1,000 patient or inhabitant-days.

Example calculation of the defined daily dose (DDD) assuming the following:

Grams of use	= 600 g
DDD	= 3 g per day
Patient days during that time period	= 2,000 days
Then, DDD/1,000 patient-days	= (600 g/3 g per day)/2,000 days $\times$ 1,000
	= 100

The DDD has been in use since the 1970s, when it was originated by Norwegian researchers. Those researchers, who collaborated with the Norwegian Medical Depot (NMD), developed a system known as the Anatomic Therapeutic Chemical (ATC) classification. They developed a unit of measurement called the DDD that was intended to be used in drug utilization evaluations (World Health Organization, 2002). In 1981, the ATC/DDD system was recommended as a drug use measure by the WHO office in Europe. In 1982, the WHO Collaborating Centre for Drug Statistics Methodology was established to oversee the use of the DDD method. In 1996, this responsibility was transferred to WHO world headquarters in Geneva to promote the DDD as an international standard. Currently, the WHO International Working Group for Drug Statistics advises the WHO Collaborating Centre for Drug Statistics Methodology on use of DDD methodology. In this method, the DDD is the assumed average maintenance dose per day (in grams) for a drug used for its main indication in adults. DDDs are usually based on the monotherapy dosage and on treatment rather than prophylaxis. Furthermore, if a drug is used for more than one indication, different DDDs may be assigned for each indication, thus introducing some potential confusion regarding the appropriate DDD value. DDDs are assigned only for drugs that have been given an ATC code and are reviewed periodically. Antimicrobials are not reviewed and assigned a DDD by the WHO Collaborating Centre until requested from a user of the system (manufacturers, regulatory agencies, and researchers), so DDDs of antimicrobials may not be assigned in a timely manner. Also, since access to DDD values has been expensive, some organizations and authors have utilized non-WHO defined daily doses in DDD calculations (CDC, 2001). With DDD values becoming more accessible, perhaps these problems will subside.

The major advantage of the DDD method is that comparisons of antimicrobial consumption in a population are more meaningful than when simple comparisons of grams are used (Figure 2).

In recent years, the DDD method has become the standard for benchmarking antimicrobial use among institutions or geographic areas. Since DDDs are additive, one may examine total antibiotic exposure within and between drug classes, institutions, regions, and countries.

Although the DDD methodology has much utility in evaluating antimicrobial use, there are several drawbacks that make it less than ideal. A criticism of the method is that the DDD may represent a dose that is seldom used in common clinical practice. Indeed, this can be the case since the DDD may actually represent an average of two or more dosages that are commonly used. Also, since adult doses are the basis for the DDD, these calculations are not as meaningful when used with pediatric antimicrobial use data. Since the DDD is based on the typical adult dose, there is no provision for drug dosages that need to be altered in patients with reduced renal function. When investigators have compared the average daily dosages of antimicrobials (also known as the prescribed daily dose, or PDD) in their institutions, there have been large differences between those dosages and those recommended for calculation of DDDs. Although the DDD is a standard that is useful for benchmarking, the PDD may better represent local usage patterns. Moreover, if the PDD is lower than the DDD, then DDD calculations will underestimate the number of doses or days of therapy (Resi *et al.*, 2001).

In a study to compare DDD values to PDD values, Paterson *et al.* (2002) studied piperacillin/tazobactam, levofloxacin, and cefepime over a 2-month period in a large medical center with over 100 ICU beds. Mean PDD were 60% of the DDD for cefepime (2.4 vs 4 g), 85% for intravenous levofloxacin (0.4 vs 0.5 g), 203% for oral levofloxacin (0.4 vs 0.2 g), and 92% for piperacillin/tazobactam (12.35 vs 13.5 g). PDDs were lower in the ICU than the non-ICU setting for piperacillin/tazobactam. The author suggested that the differences observed were likely due to dosage adjustments for renal dysfunction and the DDD values for oral use. Similar variations between the PDDs and DDDs were reported by White (2002), see Table 2.

Although DDD changes over time are not common, there have been some changes that have occurred with antimicrobials (Table 3). Obviously, this

Table 2. Comparison of DDD and mean daily dose values at a large teaching hospital

Drug	DDD (g/day), NNIS	Mean daily dose (g), MUSC	Mean daily dose/DDD (%)
Ceftazidime	3.0	3.7	123
Cefotaxime	3.0	4.4	146
Imipenem	2.0	1.7	85
Nafcillin	4.0	9.6	240
Piperacillin/tazobactam	13.5	11.9	88
Vancomycin	2.0	1.6	80

Note: NNIS = National Nosocomial Infection Surveillance System, MUSC = Medical University of South Carolina.
Source: CDC (2001); White (2002).

Table 3. Example of some antimicrobials for which the DDD has been changed over time

Drug	WHO DDDs (g/day)		
	Pre-1992	1992–2000	Post-2000
Cefoperazone	2	6	4
Ceftazidime	4	6	4
Cefuroxime IV	2	4	3

Source: WHO (2002).

makes assessment of trends in antimicrobial use more difficult (Ronning *et al.*, 2000). Since DDD calculations do not take repeated courses of treatment into account, Resi *et al.* (2001) have proposed a "therapeutic course" metric as a complement to the DDD system. In this system, the therapeutic course system would account for all prescriptions within a given time frame as the same course of therapy. Although this may be a useful adjunct to DDD calculations, selection of the time frame that would constitute a therapeutic course would likely be a point of much debate.

3.2.5. Pharmacokinetic estimates of antimicrobial exposure

Given that there are known relationships between antimicrobial exposure and both clinical outcome and resistance (Craig and Andes, 1996; Drusano, 2003; Thomas *et al.*, 1998), it would be desirable to measure or estimate antimicrobial exposure in individual patients. The measures of antimicrobial use that have been reviewed fail to account for differences in drug exposure in patients due to intrapatient and interpatient variability. For example, when grams are used (or DDDs as a derivation of grams), the calculations assume that any gram of an antimicrobial administered to a patient will have the same impact on the development of resistance as any gram administered at a different time to that same patient. Furthermore, those calculations assume that the same dose of an antimicrobial will have the same impact on resistance in different patients. These assumptions are invalid due to known interpatient and intrapatient variability in pharmacokinetic profiles. Ultimately, estimations of drug use may be replaced by estimates of drug exposure in patients. To this end, one needs either direct measurements of drug concentrations in individual patients or precise estimations of drug exposure using population pharmacokinetic estimates. These estimates are unlikely to be made in large populations since assays of antimicrobial concentrations sufficient to estimate drug exposure are invasive and costly. Estimates of drug exposure from population pharmacokinetic values is more likely to occur, but may be prone to error if population pharmacokinetic studies are not representative of the underlying populations of interest. A potential disadvantage of this method is that drug concentrations at sites of contact with organisms that may develop resistance may be poorly characterized. For example, if a drug has low systemic concentrations due to poor oral absorption and serum concentrations are measured, drug contact with organisms in the gastrointestinal tract may be underestimated.

4.   NORMALIZATION OF USE DATA

It is common practice to normalize aggregate antimicrobial use data to account for differences in census within an institution or geographic area. This is usually accomplished by correcting the use data so that it reflects a rate of use per unit of time, for example, 1,000 days. This day term is often referred to as patient-days since one calculates this denominator by multiplying the number of patients within the institution in a given period of time by the length of hospitalization (in days) and correcting it to a value such as 1,000. Patient-days are also referred to as occupied bed-days. Correction for changes in the population in this manner allows one to see "true" changes in rates of antimicrobial use rather than fluctuations that may simply reflect changes in the population over time or differences in two or more populations. Although it is always useful to normalize the data to detect the true rate of use, normalization is of little value when small changes in census occur. For example, census changes in smaller patient-care areas may be very important whereas institution-wide fluctuations in census may be minimal; thus, normalization may have only a minimal impact (Figure 3). Obviously, if there are no changes in the number of patient-days over time, non-normalized measures (e.g., grams) and the normalized measures (e.g., grams/patient-day) would perfectly correlate. These calculations mathematically spread the use over the entire population rather than the population that actually received the antimicrobial of interest. Thus normalized values such as grams per patient-day will not reflect average grams per day doses of an antimicrobial.

The most common denominator for normalizing antimicrobial use within an institution in the United States is 1,000 patient-days. In Europe, the European Surveillance of Antimicrobial Consumption (www.ua.ac.be/main.asp?c=*ESAC) recommends use of bed-days, which are calculated by multiplying the number of beds by the occupancy by the length of time of the study. The denominator for use in primary healthcare settings and thus, within a geographic region, is usually inhabitant-days per unit of time, which is calculated by multiplying the number of inhabitants in an area by the number of days studied. For example, 7 DDD/inhabitant/year is equivalent to each inhabitant of an area receiving a 7-day course of that antimicrobial during a 1-year period. If certain

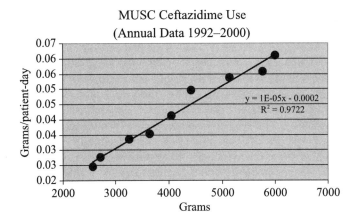

Figure 3. Example of the impact of small changes in patient-days when normalizing grams of use data (White, 2002).

antimicrobials are used only in patients of specific ages, one can utilize the number of inhabitants in that age range when those data are available. The number of admissions or discharges can also be used to estimate the percentage of patients exposed to antimicrobials. Other factors that have been used to calculate rates of use are use per number of beds or occupied beds. However, factors that fail to take length of stay into account are probably less useful; thus patient-days, bed-days, and inhabitant-days are preferred. When patient-specific antimicrobial use data are available, there is no need to normalize the data since the antimicrobial exposure reflects only the patients who received antimicrobials.

5. RELATIONSHIPS BETWEEN MEASURES OF ANTIMICROBIAL USAGE AND RESISTANCE

Establishing a causal relationship between antimicrobial exposure and development of resistance is very difficult since a myriad of factors other than antimicrobial use may contribute to the emergence of resistance (McGowan, 1983). Although the spread of resistant organisms from patient to patient is important, selection pressure from antimicrobials may largely contribute to the emergence of resistance. Levy (1994) suggests that the intensity of antimicrobial use in a population is the most important factor in the selection of resistance and that a threshold may exist that may differ for a specific patient as compared to a population. Furthermore, this threshold may differ among various populations. Since prospective randomized controlled trials of the development of resistance are not performed and would be unethical, most studies are observational and retrospective. Thus, lack of control of confounding variables is a major limitation in these studies and most studies involve a small number of patients and are performed in a single institution. Furthermore, these evaluations have utilized different measures of antibiotic use such as DDDs, grams, days of therapy, and the number of prescriptions over various periods of time. Although antimicrobial use is usually evaluated in discrete time intervals, cumulative use data may be more likely to detect relationships with changes in susceptibility (White and Bosso, 2003).

Statistical evaluation of antimicrobial use-susceptibility relationships has included linear, multiple and logistic regression analysis (Bonapace *et al.*, 2000a, b; Burgess and Jones, 2002; Enzweiler *et al.*, 2000a; Friedrich *et al.*, 1999; Johnson *et al.*, 2002a, b; Polk *et al.*, 2002) as well as time-series analysis (Monnet and Lopez-Lozano, 2002). Mathematical modeling of relationships between antimicrobial use and resistance and the relative contribution of infection control measures has greatly enhanced our understanding of the complexity of the relationships. (Austin and Anderson, 1999; Austin *et al.*, 1999b; Levin *et al.*, 1997; Lipsitch *et al.*, 2000b; Lipsitch and Samore, 2002).

Although we often refer to studies of resistance, many studies evaluate the relationship between antimicrobial use and changes in the percentage of isolates categorized as susceptible rather than resistant. This is not surprising given that antibiograms, the basis for most institutional surveillance studies, report percentage susceptibility rather than percentage resistance. Due to the intermediate category, susceptibility and resistance do not necessarily trend in opposite directions and evaluations with one category may not result in the same conclusions as using another category. Aggregate studies using antibiogram data undoubtedly combine data from pathogens as well as colonizing organisms, which may complicate evaluation of relationships between antimicrobial use and resistance. In an evaluation of 10 years of data in a large teaching hospital, Enzweiler *et al.* (2002c) found that changes in percentage resistance were often more useful than tracking of percentage susceptibility. In that study, clinically relevant changes were detected earlier, in some cases, years earlier, when using percentage resistance rather than percentage susceptibility. Moreover, institution-wide rather than unit-specific data are usually used to assess relationships between antimicrobial use and resistance. Although clinically relevant relationships may be detected, institution-wide data may mask important relationships occurring in specific patient-care areas within the institution (White *et al.*, 2000).

Adding to the complexity of detection of relationships between antimicrobial use and resistance may be the pattern of use in a specific institution. For example, one institution may use an antimicrobial as monotherapy while another consistently uses it as part of combination therapy. This may be important since in mathematical models, combination therapy is better than antimicrobial cycling in prevention of resistance. Typically, the rise in resistance is faster than the decline when selection pressure is removed since the costs of resistance in the absence of antimicrobial pressure is less than the benefit of resistance when pressure is present. Thus, declining susceptibility due to increasing antimicrobial use may be easier to detect than increasing susceptibility when antimicrobial use declines (Bonhoeffer *et al.*, 1997). Further complicating detection of relationships between use and susceptibility is simultaneous resistance to multiple antimicrobials. In a study evaluating the relationship between multiple antimicrobials and resistance in Gram-negative aerobes, Friedrich *et al.* (1999) found that with relationships involving increasing antimicrobial use with declining susceptibility, more than one antimicrobial was statistically associated with the changes in susceptibility (mean 1.7, range 1–14) to any single agent. When combinations of resistance mechanisms are present, associations between antimicrobial use and resistance may be quite complex. Ryan *et al.* (2002) reported on the relationships between fluoroquinolone and carbapenem use with resistance of *P. aeruginosa*. Meropenem and ciprofloxacin, but not imipenem, susceptibility patterns were associated with carbapenem and fluoroquinolone use ($p < 0.0001$). Although not directly evaluated in this study, the authors attributed these effects to a combination of mutations, changes in porins, or drug efflux mechanisms.

5.1. Relationships based on patient-specific data

Patient-specific data are used in case control studies in which antecedent antimicrobial use is associated with the development of resistance. In the case control studies, resistant isolates are identified and risk factors are examined. In other prospective studies, the antimicrobials that may increase the risk of colonization with resistant isolates are examined. In these studies, antimicrobial use may be quantitative (e.g., number of antibiotic courses, grams, number of days of therapy, etc.) or non-quantitative (e.g., binomial data regarding drug exposure of a certain intensity or length of therapy). Since the length of drug exposure prior to development of resistance is not known and whose ascertainment may be the purpose of the study, the length of time for the evaluation of antecedent drug use is not standardized.

Numerous studies have evaluated prior antibiotic use and development of resistance. In investigations of the association of antecedent vancomycin use and development of vancomycin-resistant enterococci (VRE), several studies found relationships between vancomycin and development of VRE colonization or infection (Ena *et al.*, 1993; Frieden *et al.*, 1993; Yates, 1999). In a meta-analysis of 20 studies of vancomycin use and the selection of resistance in enterococci, there was however, no statistically significant relationship. The author suggested that the design of the individual studies and differences in control group selection, length of hospital stay, and publication bias, likely had a great impact on whether such an association was detected (Carmeli *et al.*, 1999). Exposure to broad-spectrum antimicrobials without activity against enterococci was found to be strongly related to colonization with VRE. In this study, the total number of days of antimicrobial exposure was correlated with the prevalence of VRE (Tokars *et al.*, 1999). Yet others have suggested that antimicrobials with significant activity against anaerobes, rather than vancomycin use, may be important in selection of VRE (Donskey *et al.*, 2000; Yates, 1999).

In a prospective, observational study, Chow *et al.* (1991) evaluated the emergence of resistance during antibiotic therapy in 129 patients with Enterobacter bacteremia. Previous administration of a third generation cephalosporin was more likely than other antimicrobials to be associated with multiresistant Enterobacter isolates in an initial blood culture ($p < 0.001$). Emergence of resistance to a third generation cephalosporin was more frequent than to aminoglycosides ($p = 0.001$) or other β-lactam agents ($p = 0.002$). In a 5-year case control study of piperacillin-tazobactam resistant *P. aeruginosa*, Harris *et al.* (2002a) found a number of factors and antimicrobials associated with resistance. The length of time a patient was at risk for the development of resistance, a transfer from one patient-care area to another, ICU stay, and the number of admissions in the previous year were risk factors for the development of *P. aeruginosa* resistant to piperacillin-tazobactam. The odds ratio (OR) for several antimicrobials including piperacillin-tazobactam, OR 6.82, imipenem, OR 2.42, broad-spectrum cephalosporins, OR 2.38, aminoglycosides, OR 2.18, and vancomycin, OR 1.87 indicated an association with resistance. Interestingly, in almost half of the cases of piperacillin-tazobactam resistant *P. aeruginosa*, the patient did not receive piperacillin-tazobactam. In contrast, in an evaluation of the impact of broad-spectrum antibiotics on detection of resistant isolates, Richard *et al.* (2001) found that fluoroquinolones were associated with the development of fluoroquinolone-resistant Gram-negative bacilli in gastrointestinal flora. This study illustrate the complexity of relationships between antimicrobial use and resistance and the value in examining patient-specific antecedent antimicrobial use.

5.2.   Relationships based on aggregate usage

Many studies have examined the relationship between aggregate antimicrobial use and resistance. (Arason *et al.*, 1996; Ballow and Schentag, 1992; Chen *et al.*, 1999; Dahms *et al.*, 1998; Enzweiler *et al.*, 2002a, b; Janoir *et al.*, 1996; Lopez-Lozano *et al.*, 2000; McNulty *et al.*, 1997; Polk *et al.*, 2001;

Rahal *et al.*, 1998; Rice *et al.*, 1996; Seppala *et al.*, 1997; Tokars *et al.*, 1999; Tornieporth *et al.*, 1996). As one might expect, most of these have been conducted in hospital settings; however, several have been analyses of nationwide data. In these studies, various measures of antimicrobial use have been utilized; however, the DDD is the most common metric.

In a nationwide analysis conducted in Finland, resistance of Group A Streptococci was associated with macrolide use expressed in DDD/1,000 patient-days. Erythromycin resistance increased from 5% in 1988 to 19% in 1993 while macrolide use increased 3-fold. Upon reduction of macrolide use by 50%, resistance to Group A Streptococci also declined by approximately 9% (Seppala *et al.*, 1997). In a similar analysis in Canada, Chen *et al.* (1999) evaluated the resistance of *Streptococcus pneumoniae* to fluoroquinolones. Fluoroquinolone use, rose from 0.8 prescriptions/person/year in 1988 to 5.5 prescriptions/person/year in 1997. During that time, *S. pneumoniae* with reduced susceptibility to fluoroquinolones (ciprofloxacin MIC $\geq$ 4 mg/L) increased from 0% in 1988–93 to 1.7% in 1997–8. Janoir *et al.* (1996) also evaluated the relationship between fluoroquinolone use, as DDD/1,000 inhabitants and fluoroquinolone resistance with *S. pneumoniae*. As fluoroquinolone use increased from 0.9 DDD/1,000 inhabitants in 1985 to 2.2 DDD/1,000 inhabitants in 1997, ciprofloxacin-resistant *S. pneumoniae* increased from 0.9% to 3%. In another evaluation of resistance to *S. pneumoniae*, Arason *et al.* (1996) examined the relationships between total antibiotic use, expressed as DDD/inhabitant and drug-resistant *S. pneumoniae* (DRSP) after the percent of DRSP increased to 20% and total antibiotic use increased to 23.2 DDD/inhabitant. From 1992 to 1995, antibiotic use decreased to 20.2 DDD/inhabitant while the percentage DRSP decreased by 5%.

In a study involving 18 hospitals, Ballow and Schentag (1992) studied the relationships between ceftazidime use and susceptibility of *Enterobacter cloacae* to ceftazidime. There was covariance in susceptibility of *E. cloacae* to ceftazidime, piperacillin, mezlocillin, cefotaxime, and ceftriaxone. Although only 10 of the 18 hospitals individually showed a linear relationship between ceftazidime use, expressed as grams/quarter/bed and susceptibility of *E. cloacae* to ceftazidime, overall the relationship was significant ($p < 0.02$). In two hospitals there was a relationship between declining ceftazidime use and increasing susceptibility of *E. cloacae*. Using a multiple hospital database, Polk *et al.* (2001) evaluated total fluoroquinolone use and resistance. Using total fluoroquinolone use as DDD/1,000 patient-days, there was a strong relationship with the prevalence of ciprofloxacin-resistant *P. aeruginosa* ($r = 0.54, p = 0.01$).

In a study of ceftazidime-resistant *Klebsiella pneumoniae*, ceftazidime use was found to be a risk factor for development of resistance. In this study, there was a strong association between ceftazidime use in a specific patient-care area, in grams, and prevalence of ceftazidime-resistant *K. pneumoniae*

(Rice *et al.*, 1990). In response to an outbreak of ceftazidime-resistant extended-spectrum β-lactamase (ESBL) producing *K. pneumoniae*, the use of cephalosporins was restricted. Although the restriction program was successful in reducing cephalosporin use and reducing ceftazidime-resistant ESBL *K. pneumoniae* infections by 44%, imipenem use increased and imipenem-resistant *P. aeruginosa* increased by 57% (Rahal *et al.*, 1998). In an evaluation of antimicrobial use over a 10-year period in a single institution, Enzweiler *et al.* (2002a) found numerous strong linear relationships between antimicrobial use, in DDD/1,000 patient-days, and percentage susceptibility. Of these relationships, most occurred when antimicrobial use was increasing while susceptibility was decreasing; however, relationships were also found where antimicrobial use was declining while susceptibility was increasing.

In the examples above, relationships with resistance were demonstrated with various measures of antimicrobial use. Since many of these measures are highly correlated, it is not surprising that various measures may lead one to the same conclusions. However, these studies did not evaluate which measure may have resulted in the strongest relationship with resistance.

5.2.1. Comparison of various measures of antimicrobial use

Although there have been numerous studies using both patient-specific and aggregate antimicrobial use measures, only a few have compared these measures to each other or with respect to relationships to resistance (Bonapace *et al.*, 2000b; Burgess and Jones, 2002; Enzweiler *et al.*, 2001, 2002b; Johnson *et al.*, 2002b; Polk *et al.*, 2002). Each of these studies is discussed below.

Using 8 years of data from a single institution, Bonapace *et al.* (2000b) evaluated the relationships between four measures of antimicrobial use for 19 antimicrobials (13 β-lactams, 3 aminoglycosides, 2 fluoroquinolones, and trimethoprim/sulfamethoxazole) and changes in susceptibility to eight Gram-negative aerobes (*A. baumannii, E. coli, E. aerogenes, E. cloacae, P. mirabilis, P. aeruginosa, S. marcescens, K. pneumoniae*). Using hospital-wide patient-specific antimicrobial use data, aggregate measures of antimicrobial use were calculated and included total grams, grams/patient-day, days of antimicrobial therapy, and the mean daily dose. Relationships between each of these measures and percentage susceptibility for each organism were assessed by linear regression and only the strongest relationships ($R^2 \geq 0.5$) were further evaluated. Discordance was defined as a regression line slope in the opposite direction from the relationships found with the other measures of drug use (Figure 4). Of the 142 relationships that met the study criteria, in 39% of instances, there was concordance (agreement in the regression line slope) among all four measures of use. When one of the four measures was discordant with the others, it most frequently occurred with the mean daily dose (57% of discordant occurrences). Interestingly, although the mean daily dose was most frequently

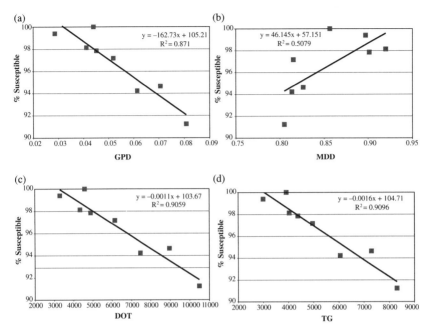

Figure 4. Example of relationships between different aggregate measures of antimicrobial use and percentage susceptibility. (GPD = grams/patient-day, MDD = mean daily dose, TG = total grams, DOT = days of antibiotic therapy) (Bonapace *et al.*, 2000b).

discordant, it was most strongly correlated with changes in susceptibility. There were no apparent trends of discordance among specific antimicrobials or organisms. As expected, some measures of use were highly correlated with each other. Positive correlations between measures of use were strong for total grams vs total grams/patient-day ($r = 0.985$) and days of antibiotic therapy ($r = 0.943$). Days of antibiotic therapy were also highly correlated with total grams/patient-day ($r = 0.928$). However, the mean daily dose was negatively and poorly correlated to each of the other measures of use.

In another study assessing differences in measures of antimicrobial use, Enzweiler *et al.* (2001) evaluated correlations among five measures of use. From patient-specific antimicrobial use data collected from 1992–9 in seven ICUs, the combined ICU, the non-ICU area, and hospital-wide data, the following aggregate measures of use were calculated for 34 antimicrobials: grams, grams/patient-day, DDD/1,000 patient-days, days of antibiotic therapy, and the mean daily dose. Using DDD/1,000 patient-days as a standard, the correlation of the other measures to DDD/1,000 patient-days was assessed. In all instances, grams, grams/patient-day, and days of antibiotic therapy were positively correlated with DDD/1,000 patient-days. Since DDD calculations are derived from grams and then normalized for patient-days, the correlation between DDD/1,000 patient-days and grams/patient-day was 1.00. Similarly, with grams, although not normalized to patient-days, the correlation with DDD/1,000 patient-days was high ($r = 0.8-1.0$) with only 3% of correlations with $r < 0.9$. Days of antibiotic therapy were not as highly correlated to DDD/1,000 patient-days ($r = 0.23-0.99$); however, only 7% of the correlations were lower than 0.7. Correlations between the mean daily dose and DDD/1,000 patient-days were poor and were negatively correlated in 33% of the comparisons.

The relationship between four different measures of antimicrobial use and percentage susceptibility were evaluated using 10 years of data in a large teaching hospital. Enzweiler *et al.* (2002a) evaluated 32 antimicrobials with 7 Gram-positive and 8 Gram-negative organisms. From patient-specific antimicrobial use data, the following aggregate use measures were evaluated: grams/1,000 patient-days, days of antibiotic therapy/1,000 patient-days, DDD/1,000 patient-days, and the mean daily dose. Relationships between each of these measures of use and hospital-wide percentage susceptibility calculated on a quarterly, semi-annual, and an annual basis were assessed by linear regression. Only relationships with $R^2 \geq 0.5$ were further analyzed. Using relationships with DDD/1,000 patient-days as a standard and the slope of the regression line as the basis for agreement, there was agreement of the four measures with DDD/1,000 patient-days in less than 50% of the relationships assessed. With both grams/1,000 patient-days and days of antibiotic therapy/1,000 patient-days, there was good agreement with DDD/1,000 patient-days (100% and 77% of occurrences, respectively); however, the mean daily dose agreed poorly with DDD/1,000 patient-days (44% of occurrences).

Johnson *et al.* (2002b) examined the relationships between aggregate fluoroquinolone use in communities surrounding 35 hospitals in a surveillance network with the prevalence of ciprofloxacin-resistant *P. aeruginosa* in 1999 and 2000. Antimicrobial use measures included prescriptions/1,000 population, total grams/1,000 population, and DDD/1,000 population. Significant linear relationships were found using only DDD/1,000 population for total fluoroquinolone use ($R^2 = 0.25$, $p = 0.02$) and levofloxacin use ($R^2 = 0.33$, $p = 0.01$) in 1999 and levofloxacin use ($R^2 = 0.17$, $p = 0.04$) in 2000. Although this study did not directly compare the measures of use, only DDD/1,000 population was found to correlate with resistance.

In a hospital surveillance network, Polk *et al.* (2002) evaluated the relationship between fluoroquinolone use and the prevalence of ciprofloxacin-resistant *P. aeruginosa*. Linear regression was used to assess the relationship between use and resistance. Antimicrobial use, expressed as DDD/1,000 patient-days, was more strongly associated with resistance ($R^2 = 0.486$, $p < 0.001$) than were grams/1,000 patient-days ($R^2 = 0.237$, $p = 0.017$).

In a large teaching hospital, Burgess and Jones (2002) examined biannual drug usage data from 1998–2001 in 12 hospital units for 18 antibiotics and 6 organisms. Linear regression was used to assess relationships between antimicrobial use and percentage susceptibility. Antimicrobial use measures for each drug included: total milligrams, total milligrams/patient-day, DDD, and DDD/1,000 patient-day. Clinically relevant relationships were defined as having an R value of greater than 0.7 with greater than 70% susceptibility. Using these criteria, there were 143 clinically relevant relationships using total milligrams, 136 using DDD, 138 using total milligram/patient-day, and 141 using DDD/1,000 patient-days. In 47% of occurrences, clinically significant relationships were found by all four measures of use.

6.   SUMMARY AND RECOMMENDATIONS

It is evident from the comparisons of the antimicrobial use measures discussed above, that several measures are highly correlated. Since DDD calculations are derived from grams, it is not surprising that there is a high degree of covariance among these measures when a specific antimicrobial is evaluated. However, the advantage of the DDD calculation over the use of grams is the ability to calculate antimicrobial use within a drug class, a patient-care area, an institution, a region, or a country. With regard to normalization of data use of patient-days or inhabitant days, may help to establish the true rate of antimicrobial use.

In several of the reports described above, the mean daily dose, the number of days of antimicrobial therapy, grams, and DDDs were compared. Grams and thus DDDs, since they are derived from grams, should be considered hybrid measures of antimicrobial use since they may be calculated from the dose, the dosing interval (comprising the daily dose), and the number of days of antimicrobial therapy. Although it seems intuitive that one would want to use a measure that was derived from both drug dose intensity (daily dose) and the length of therapy, it is very plausible that, with some antimicrobial/organism combinations, one of the measures, daily dose, or length of therapy may be more closely associated with the development of resistance than the other. In that scenario, use of a hybrid measure such as grams or DDDs could potentially mask these relationships. In the reports above, the mean daily dose consistently disagreed with other measures. This is likely due to the amount of relative influence that the daily dose and length of therapy (or days of antibiotic therapy) have on the calculation of total grams. Since the number of days that a patient receives an antimicrobial usually numerically exceeds the daily dose, it should be expected that days of therapy would be in closer agreement with grams and DDDs than would the mean daily dose.

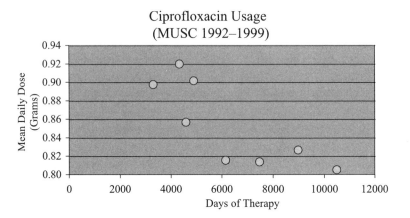

Figure 5. Relationship between total days of therapy with ciprofloxacin and the mean daily dose of ciprofloxacin using 8 years of data at a large teaching hospital (White, 2002).

Interestingly, for some antimicrobials, the correlation between the mean daily dose and days of therapy may be inversely related (White, 2002; Figure 5). This is what one might expect with an antimicrobial in which efficacy is related to dose intensity and thus the area under the serum concentration–time curve (AUC)/MIC relationship (Craig and Andes, 1996). Importantly, it may suggest a relationship that clinicians can manipulate to reduce the length of antimicrobial therapy. Relationships such as this would go undetected if only hybrid measures were analyzed. Therefore, it would be prudent to continue to evaluate these measures as separate entities while also evaluating hybrid measures such as the DDD. Ultimately, the ideal marker may be either direct measurement or an estimate of drug exposure (e.g., AUC).

The ideal marker(s) of antimicrobial use will be selected on the basis of the relative strength of the relationships between various markers and development of resistance. To determine this, improvements in antimicrobial usage data collection and additional research are required. It is likely that no single marker will be optimal for all purposes and may depend on whether one uses it for surveillance of a specific antimicrobial or comparisons with other antimicrobials and whether associations with toxicity or changes in susceptibility are being investigated.

ACKNOWLEDGMENT

The author would like to thank Dr. Deanna S. Jackson for her invaluable assistance on this manuscript.

REFERENCES

Arason, V. A., Kristinsson, K. G., Sigurdsson, J. A., Stefansdottir, G., Molstad, S., and Gudmundsson, S., 1996, Do antimicrobials increase the carriage rate of penicillin resistant pneumococci in children? Cross sectional prevalence study. *BMJ*, **313**, 387–391.

Archibald, L., Phillips, L., Monnet, D., McGowan, Jr., J. E., Tenover, F., and Gaynes, R., 1997, Antimicrobial resistance in isolates from inpatients and outpatients in the United States: Increasing importance of the intensive care unit. *Clin. Infect. Dis.*, **24**, 211–215.

Austin, D. J. and Anderson, R. M., 1999, Studies of antibiotic resistance within the patient, hospitals and the community using simple mathematical models. *Phil. Times R. Soc. Lond. B*, **354**(1384), 721–738.

Austin, D. J., Bonten, M. J. M., Weinstein, R. A., and Slaughter, S., 1999a, Vancomycin-resistant enterococci in intensive-care hospital settings: Transmission dynamics, persistence, and the impact of infection control programs. *Proc. Natl. Acad. Sci., USA*, **96**, 6908–6913.

Austin, D. J., Kristinsson, K. G., and Anderson, R. M., 1999b, The relationship between the volume of antimicrobial consumption in human communities and the frequency of resistance. *Proc. Natl. Acad. Sci., USA*, **96**, 1152–1156.

Ballow, C. H. and Schentag, J. J., 1992, Trends in antibiotic utilization and bacterial resistance: Report of the National Nosocomial Resistance Surveillance Group. *Diagn. Microbiol. Infect. Dis.*, **15**, 37S–42S.

Bonapace, C. R., Lorenz, K. R., Bosso, J. A., and White, R. L., 2000a, Quantitation of increasing resistance to fluoroquinolones in N. America, Europe and Asia, 1982–2000. In, Program and Abstracts of the 40th Interscience Conference on Antimicrobial Agents and Chemotherapy (Toronto, Ontario), Washington, DC, American Society for Microbiology, abstract 99, 403.

Bonapace, C. R., Lorenz, K. R., Bosso, J. A., and White, R. L., 2000b, Use of different measures of drug usage in assessing antibiotic use/susceptibility relationships. In, Program and Abstracts of the 40th Interscience Conference on Antimicrobial Agents and Chemotherapy (Toronto, Ontario), Washington, DC, American Society for Microbiology, abstract 98, 403.

Bonhoeffer, S., Lipsitch, M., and Levin, B. R., 1997, Evaluating treatment protocols to prevent antibiotic resistance. *Proc. Natl. Acad. Sci., USA*, **94**; 12106–12111.

Burgess, D. S. and Jones, B. L., 2002, Effect of different antimicrobial usage markers on clinically significant susceptibility vs drug use relationships. In, 2002 Spring Practice and Research Forum Program and Abstracts, (Savannah, GA), Kansas City, MO, American College of Clinical Pharmacy, abstract 61, 345.

Carmeli, Y., Samore, M. H., and Huskins, C., 1999, The association between antecedent vancomycin treatment and hospital-acquired vancomycin-resistant Enterococci. A meta-analysis. *Arch. Intern. Med.*, **159**, 2461–2468.

CDC (Centers for Disease Control) and Prevention NNIS System, 2001, National Nosocomial Infections Surveillance (NNIS) system report, data summary from January 1992-June 2001, Issued August 2001. *AJIC Am. J. Infect. Control*, **29**, 404–421.

Chen, D. K., McGeer, A., Azavedo, J. C., and Low, D. E., 1999, Decreased susceptibility of *Streptococcus pneumoniae* to fluoroquinolones in Canada. *N. Engl. J. Med.*, **341**, 233–239.

Chow, J. W., Fine, M. J., Shlaes, D. M., Quinn, J. P., Hooper, D. C., Johnson, M. P. *et al.*, 1991, *Enterobacter* bacteremia: Clinical features and emergence of antibiotic resistance during therapy. *Ann. Intern. Med.*, **115**, 585–590.

Craig, W. A. and Andes, D., 1996, Pharmacokinetics and pharmacodynamics of antibiotics in otitis media. *Pediatr. Infect. Dis. J.*, **15**, 255–259.

Dahms, R. A., Johnson, E. M., Statz, C. L., Lee, J. T., Dunn, D. L., and Beilman, G. J., 1998, Third-generation cephalosporins and vancomycin as risk factors for postoperative vancomycin-resistant *Enterococcus* infection. *Arch. Surg.*, **133**, 1343–1346.

Donskey, C. J., Chowdhry, T. K., Hecker, M. T., Hoyen, C. K., Hanrahan, J. A., Hujer, A. M., Hutton-Thomas, R. A., Whalen, C. C., Bonomo, R. A., and Rice, L. B., 2000, Effect of antibiotic therapy on the density of vancomycin-resistant enterococci in the stool of colonized patients. *New Engl. J. Med.*, **343**, 1925–1932.

Drusano, G. L., 2003, Prevention of resistance: A goal for dose selection for antimicrobial agents. *Clin. Infect. Dis.*, **36**(Suppl. 1), S42–S50.

Ena, J., Dick, R. W., Jones, R. N., and Wenzel, R. P., 1993, The epidemiology of intravenous vancomycin usage in a university hospital. A 10-year study. *JAMA*, **269**, 598–602.

Enzweiler, K. A., Lorenz, K., Bosso, J., and White, R., 2001, Agreement among different markers of antibiotic use over an eight-year period. In, Program and Abstracts of the 41st Interscience Conference on Antimicrobial Agents and Chemotherapy (Chicago, IL), Washington, DC, American Society for Microbiology, abstract K-1199, 410.

Enzweiler, K. A., Bosso, J. A., and White, R. L., 2002a, Assessment of relationships between antibiotic use and susceptibility rates over a 10-year period. In, Abstracts of the 40th Annual Meeting of the Infectious Diseases Society of America (Chicago, IL), Alexandria, VA, Infectious Diseases Society of America, abstract 105, 63.

Enzweiler, K. A., White, R. L., and Bosso, J. A., 2002b, Qualitative and quantitative differences among antimicrobial usage markers in assessing antibiotic use/susceptibility relationships. In, Program and Abstracts of the 42nd Interscience Conference on Antimicrobial Agents and Chemotherapy (San Diego, CA), Washington, DC, American Society for Microbiology, abstract O-1001, 418.

Enzweiler, K. A., Bosso, J. A., and White, R. L., 2002c, Percent susceptible versus percent resistant: should both markers be used in surveillance of susceptibility. In, Proceedings of the Annual Meeting of the American College of Clinical Pharmacy, (Albuquerque, NM), Kansas City, MO, American College of Clinical Pharmacy, abstract 166, 63.

European Surveillance of Antimicrobial Consumption (ESAC). Universiteit Antwerpen. http://www.ua.ac.be/main.asp?c=*ESAC

Frieden, T. R., Munsiff, S. S., Williams, G., Faur, Y., Kreiswirth, B., Low, D. E. *et al.*, 1993 Emergence of vancomycin-resistant enterococci in New York City. *Lancet*, **342**; 76–79.

Friedrich, L. V., White, R. L., and Bosso, J. A., 1999, Impact of use of multiple antimicrobials on changes in susceptibility of gram-negative aerobes. *Clin. Infect. Dis.*, **28**, 1017–1024.

Harbarth, S., Harris, A. D., Carmeli, Y., and Samore, M. H., 2001, Parallel analysis of individual and aggregated data on antibiotic exposure and resistance in gram-negative bacilli. *Clin. Infect. Dis.*, **33**, 1462–1468.

Harris, A. D., Samore, M. H., Lipsitch, M., Kaye, K. S., Perencevich, E., and Carmeli, Y., 2002a, Control-group selection importance in studies of antimicrobial resistance: Examples applied to *Pseudomonas aeruginosa*, Enterococci, and *Escherichia coli*. *Clin. Infect. Dis.* **34**, 1558–1563.

Harris, A. D., Perencevich, E., Roghmann, M. C., Morris, G., Kaye, K. S., and Johnson, J. A., 2002b, Risk factors for piperacillin-tazobactam-resistant *Pseudomonas aeruginosa* among hospitalized patients. *Antimicrob. Agents Chemother.*, **46**(3), 854–858.

Janoir, C., Zeller, V., Kitzis, M. D., Moreau, N. J., and Gutmann, L., 1996, High-level fluoroquinolone resistance in *Streptococcus pneumoniae* requires mutations in *parC* and *gyrA*. *Antimicrob. Agents Chemother.*, **40**, 2760–2764.

Johnson, C. K., Polk, R., Edmond, M., and Wenzel, R., 2002a, Association between fluoroquinolone use and prevalence of methicillin-resistant *Staphylococcus aureus* in U.S.

Hospitals: A SCOPE-MMIT report. In, Program and Abstracts of the 42nd Interscience Conference on Antimicrobial Agents and Chemotherapy (San Diego, CA), Washington, DC, American Society for Microbiology, abstract 399.

Johnson, C. K., Tomes, H., and Polk, R. E., 2002b, Community fluoroquinolone use and its relationship to hospital rates of ciprofloxacin resistant *Pseudomonas aeruginosa*: A SCOPE-MMIT report. In: Program and Abstracts of the 42nd Interscience Conference on Antimicrobial Agents and Chemotherapy (San Diego, CA), Washington, DC, American Society for Microbiology, abstract K-1088, 321.

Lesch C. A., Itokazu, G. S., Danziger, L. H., and Weinstein, R. A., 2001, Multi-hospital analysis of antimicrobial usage and resistance trends. *Diagn. Microbiol. Infect. Dis.*, **41**, 149–154.

Levin, B. R., Lipsitch, M., Perrot, V., Schrag, S., Antia, R., Simonsen, L. *et al.*, 1997, The population genetics of antibiotic resistance. *Clin. Infect. Dis.*, **24**(Suppl. 1), S9–16.

Levy, S. B., 1994, Balancing the drug-resistance equation. *Trends Microbiol.*, **2**, 341–342.

Lipsitch, M., Bergstrom, C. T., and Levin, B. R., 2000a, The epidemiology of antibiotic resistance in hospitals: Paradoxes and prescriptions. *Proc. Natl. Acad. Sci., USA*, **97**(4), 1938–1943.

Lipsitch, M., Bacon, T. H., Leary, J. J., Antia, R., and Levin, B. R., 2000b, Effects of antiviral usage on transmission dynamics of *Herpes simplex* virus Type 1 and on antiviral resistance: Predictions of mathematical models. *Antimicrob. Agents Chemother.*, **44**(10), 2824–2835.

Lipsitch, M. and Samore, M. H., 2002, Antimicrobial use and antimicrobial resistance: A population perspective. *Emerg. Infect. Dis.*, **9**(4), 347–354.

Lopez-Lozano, J. M., Monnet, D. L., Yague, A., Burgos, A., Gonzalo, N., Campillos, P. *et al.*, 2000, Modelling and forecasting antimicrobial resistance and its dynamic relationship to antimicrobial use: A time series analysis. *Int. J. Antimicrob. Agents*, **14**, 21–31.

Lucas, G. M., Lechtzin, N., Puryear, D. W., Yau, L. L., Flexner, C. W., and Moore, R. D., 1998, Vancomycin-resistant and vancomycin-susceptible enterococcal bacteremia: Comparison of clinical features and outcomes. *Clin. Infect. Dis.*, **26**, 1127–1133.

McGowan, Jr, J. E., 1983, Antimicrobial resistance in hospital organisms and its relation to antibiotic use. *Rev. Infect. Dis.*, **5**, 1033–1048.

McNulty, C., Logan, M., Donald, I. P., Ennis, D., Taylor, D., Baldwin, R. N. *et al.* 1997, Successful control of *Clostridium difficile* infection in an elderly care unit through use of a restrictive antibiotic policy. *J. Antimicrob. Chemother.*, **40**, 707–711.

Meyer, K. S., Urban, C., Eagan, J. A., Berger, B. J., and Rahal, J. J., 1993, Nosocomial outbreak of *Klebsiella* infection resistant to late-generation cephalosporins. *Ann. Int. Med.*, **19**, 353–358.

Monnet, D. L. and Lopez-Lozano, J. M., 2002, Application of time series analysis to antimicrobial resistance and other surveillance data. In, Program and Abstracts of the 42nd Interscience Conference on Antimicrobial Agents and Chemotherapy (San Diego, CA), Washington, DC, American Society for Microbiology, abstract 395, 447.

Paterson, D. L., Muto, C. A., Gross, P., Ndirangu, M. W., Kuznetsov, D., and Harrison, L. H., 2002, Does the "Defined Daily Dose" adequately measure antibiotic utilization in hospitalized patients in tertiary medical centers? In, Program and Abstracts of the 42nd Interscience Conference on Antimicrobial Agents and Chemotherapy (San Diego, CA), Washington, DC, American Society for Microbiology, abstract O-1002, 418.

Polk, R. E., Johnson, C., Edmond, M., and Wenzel, R., 2001, Trends in fluoroquinolone prescribing in 35 U.S. hospitals and resistance for *P. aeruginosa*: A SCOPE-MMIT report. In, Program and Abstracts of the 41st Interscience Conference on Antimicrobial Agents and Chemotherapy (Chicago, IL), Washington, DC, American Society for Microbiology, abstract UL-1, 1 ("up-to date" late-breaker poster sessions).

Polk, R., Johnson, C., Edmond M, and Wenzel, R., 2002, Relationship of ciprofloxacin-resistant *P. aeruginosa* and quinolone exposure, measured by DDD/1000PD A Grams/1000PD, in U.S. hospitals. In, The 12th Annual Meeting of The Society for Healthcare Epidemiology of America (Salt Lake City, UT), Mt. Royal, NJ.

Rahal, J. J., Urban, C., Horn, D., Freeman, K., Segal-Maurer, S., Maurer, J. *et al.*, 1998, Class restriction of cephalosporin use to control total cephalosporin resistance in nosocomial *Klebsiella*. *JAMA*, **280**, 1233–1237.

Resi, D., Castelvetri, C., Vaccheri, A., and Montanaro N., 2001, The therapeutic course as a measure complementary to defined daily doses when studying exposure to antibacterial agents. *Eur. J. Clin. Pharmacol.*, **57**, 177–180.

Rice, L. B., Eckstein, E. C., DeVente, J, and Shlaes, D.M, 1996, Ceftazidime-resistant *Klebsiella pneumoniae* isolates recovered at the Cleveland Department of Veterans Affairs Medical Center. *Clin. Infect. Dis.*, **23**, 118–124.

Rice, L. B., Willey, S. H., Papanicolaou, G. A., Medeiros, A. A., Eliopoulos, G.M., Moellering, R.C. *et al.* 1990, Outbreak of ceftazidime resistance caused by extended-spectrum β-lactamases at a Massachusetts chronic-care facility. *Antimicrob. Agents Chemother.*, **34**, 2193–2199.

Richard, P., Delangle, M. H., Ralfi, F., Espaze, E., and Richet, H., 2001, Impact of fluoro-quinolone administration on the emergence of fluroquinolone-resistant gram-negative bacilli from gastrointestinal flora. *Clin. Infect. Dis.*, **32**, 162–166.

Rifenburg, R. P., Paladino, J. A., Bhavnani, S. M., Den Haese, D., and Schentag, J. J., 1999, Influence of fluoroquinolone purchasing patterns on antimicrobial expenditures and *Pseudomonas aeruginosa* susceptibility. *Am. J. Health-Syst. Pharm.*, **56**, 2217–2223.

Ronning, M., Blix, H. S., Harbo, B. T., and Strom, H., 2000, Different versions of the anatomical therapeutic chemical classification system and the defined daily dose—are drug utilisation data comparable? *Eur. J. Clin. Pharmacol.*, **56**, 723–727.

Ruiz-Bremon, A., Ruiz-Tovar, M., Gorricho, B. P., Diaz de Torres, P., and Rodriguez, R. L., 2000, Non-hospital consumption of antibiotics in Spain: 1987–1997. *J. Antimicrob. Chemother.*, **45**, 395–400.

Ryan, K. A., Steward, C. D., Tenover, F. C., and McGowan Jr., J. E., 2002, Association between carbapenem and fluoroquinolone resistance among project ICARE (Intensive Care Antimicrobial Resistance Epidemiology) *Pseudomonas aeruginosa* isolates. In, Program and Abstracts of the 42nd Interscience Conference on Antimicrobial Agents and Chemotherapy (San Diego, CA), Washington, DC, American Society for Microbiology, abstract C2-306, 91.

Seppala, H., Klaukka, T., Vuopio-Varkila, J., Muotiala, A., Helenius, H., Lager, K. *et al.*, 1997, The effect of changes in the consumption of macrolide antibiotics on erythromycin resistance in Group A Streptococci in Finland. *N. Engl. J. Med.*, **337**, 441–446.

Shlaes, D. M., Gerding, D. N., John, J. F., Craig, W. A., Bornstein, D. L., Duncan, R. A. *et al.*, 1997, Society for Healthcare Epidemiology of America and Infectious Diseases Society of America joint committee on the prevention of antimicrobial resistance: Guidelines for the prevention of antimicrobial resistance in hospitals. *Clin. Infect. Dis.*, **25**, 584–599.

Smith, D. L., Harris, A. D., Johnson, J. A., Silbergeld, E. K., and Morris Jr., J. G., 2002, Animal antibiotic use has an early but important impact on the emergence of antibiotic resistance in human commensal bacteria. *Proc. Natl. Acad. Sci., USA*, **99**(9), 6434–6439.

Thomas, J. K., Forrest, A., Bhavnani, S. M., Hyatt, J. M., Cheng, A., Ballow, C. H. *et al.*, 1998, Pharmacodynamic evaluation of factors associated with the development of bacterial resistance in acutely ill patients during therapy. *Antimicrob. Agents Chemother.*, **42**, 521–527.

Tokars, J. I., Satake, S., Rimland, D., Carson, L., Miller, E. R., Killum, E. *et al.*, 1999, The prevalence of colonization with vancomycin-resistant *Enterococcus* at a Veterans' Affairs Institution. *Infect. Control Hosp. Epidemiol.*, **20**, 171–175.

Tornieporth, N. G., Roberts, R. B., John, J., and Riley, L. W., 1996, Risk factors associated with vancomycin-resistant *Enterococcus faecium* infection or colonization in 145 matched case patients and control patients. *Clin. Infect. Dis.*, **23**, 767–772.

United States Office of the Federal Register, National Archives and Records Administration, 1996, Health Insurance Portability and Accountability Act of 1996, Public Law 104–191, Washington, D. C.

White, R. L., 2002, What is the best way to express antibiotic use? In, Program and Abstracts of the 42nd Interscience Conference on Antimicrobial Agents and Chemotherapy (San Diego, CA), Washington, DC, American Society for Microbiology, abstract 1038, 460.

White, R. L. and Bosso J. A., 2003, Assessment of antimicrobial usage vs. susceptibility relationships using cumulative vs. non-cumulative antimicrobial data. In, Program and Abstracts of the 43rd Interscience Conference on Antimicrobial Agents and Chemotherapy (Chicago, IL), Washington, DC, American Society for Microbiology, abstract C2-1497, 140.

White, R. L., Friedrich, L. V., Mihm, L. B., and Bosso, J. A., 2000, Assessment of the relationship between antimicrobial usage and susceptibility: differences between the hospital and specific patient-care areas. *Clin. Infect. Dis.*, **31**, 16–23.

World Health Organization, 2002, *Guidelines for ATC Classification and DDD Assignment*, 5th edn, WHO Collaborating Centre for Drug Statistics Methodology. http://www.whocc.no/atcddd, Oslo, Norway, pp. 155–172.

Yates, R.R., 1999, New intervention strategies for reducing antibiotic resistance. *Chest*, **115**, 24S–27S.

Chapter 7

Quantitative Measurement of Antibiotic Use

Fiona M. MacKenzie and Ian M. Gould
Department of Medical Microbiology, Aberdeen Royal Infirmary,
Foresterhill, Aberdeen AB25 2ZN, UK

1. INTRODUCTION

If the assumption that increased antimicrobial use correlates with an increase in antimicrobial resistance, then information on the consumption of these drugs is essential to explore the dynamics of resistance. Any endeavours to contain and reduce the development of resistance must include a reduction in the use of antibiotics and in particular, more appropriate prescribing. Accurate antibiotic consumption data are therefore also essential to evaluate the impact of such intervention studies.

As very few data on antibiotic consumption are published in the literature and these data are reported using various measurement units, it is almost impossible for a hospital or country to benchmark its consumption with other hospitals/countries. Although several studies are currently collating consumption data, very few publications exist placing consumption in an international context. For example, no current publication details UK hospital antibiotic consumption and the most often cited data on UK community consumption dates back to 1997 in the seminal papers of Cars and co-workers (Cars *et al.*, 2001; Mölstad *et al.*, 2002).

Although the greatest use in human medicine is in the community, it is the intensive use of antibiotics in our hospitals that has the greatest impact on resistance. Hospital use will therefore be specifically addressed in this chapter. In order to provide the evidence base to tackle this enormous issue, fundamental

Antibiotic Policies: Theory and Practice. Edited by Gould and van der Meer
Kluwer Academic / Plenum Publishers, New York, 2005

experimental design is essential. So—how do we best measure antibiotic consumption?

2. ANTIBIOTIC CONSUMPTION; ALTERNATIVE UNITS OF MEASUREMENT

2.1. Units used but not recommended

Various units of measurement have been used to express antibiotic consumption, with data coming from numerous sources. Most meaningful units are made up of both a numerator and a denominator, where the numerator measures the amount of antibiotic used and the denominator controls for the size of the population studied. It is the choice of numerator however, which is fundamental to accurately express and compare antibiotic use.

Antibiotic use data are commonly presented in terms of financial expenditure (Silber *et al.*, 1994). Although costs allow for an overall analysis of drug expenditure, there are many disadvantages. Specific hospitals negotiate local prices and different suppliers offer different prices therefore, prices are not comparable between either hospitals or countries. Even at a local level, prices often change periodically and are therefore not appropriate to monitor use over time. Furthermore, data based on costs supplies little, or no, information on indication, route of administration, dose, dosing regimen, and duration of treatment. Even if numbers of packages sold/used may be independent of sale prices they may vary with the manufacturer of the country of purchase.

Number of packages of antibiotic sold, is also often quoted in literature. The reason for this is that the data are relatively easily accessed via organisations such as Intercontinental Marketing Services (IMS). IMS is pharmaceutical industry-based and carries out syndicated market research studies. Their data are obtained from drug manufacturers, wholesalers, retailers, pharmacies, mail order, long-term care facilities, and hospitals. IMS claims to be the world's leading source of information and data analysis for anyone within the pharmaceutical and healthcare industries. The organisation uses the European Pharmaceutical Market Research Association classification of medicinal products, which is not compatible with professional classification schemes used by those who work with antibiotics. The IMS data do not take into account that packages of individual antibiotics may vary with respect to number of unit doses per package, dosages, and route of administration. Again, this means that the data are not appropriate for comparing data between hospitals and countries.

Notably, the oft-quoted data presented by Cars *et al.* (2001) was originally obtained from IMS, although it was converted into a more appropriate unit of measurement. Cars and co-workers subsequently published further data from the same set of countries (Mölstad *et al.*, 2002). Again, the majority of the data were obtained from IMS. They highlighted that the IMS national data were extrapolated from samples of data collected in the individual countries and concluded that it was not possible to validate the data as the IMS data collection method was not fully transparent.

At a local level, it is more often than not the case that data are available at the whole hospital level or, at best, at ward level. Currently, only a minority of hospitals possess a database of patient-level antibiotic prescriptions even though this should be the most accurate data possible. The majority of hospitals can supply antibiotic data expressed as amounts purchased by the pharmacy for the whole hospital or amount of antibiotics distributed to specific wards although this does not take into account the amounts disposed of due to expiry past the use-by date. The number of patients receiving an antibiotic prescription has been quoted in the past. This is most appropriately quoted for patients in the community, as it is relatively safe to assume that the number of people exposed to antibiotics can be calculated from the number of antibiotic prescriptions dispensed. Unlike hospitalised patients, relatively few patients in the community receive more than one prescription and few receive combination therapy. The number of patients receiving an antibiotic is seldom quoted for hospitalised patients. In addition to the reasons stated above, it is not recommended as no quantitative data on dosing and duration of treatment are known and once again, the data from these sources are not comparable. Although number of patients exposed to an antibiotic has been quoted in a few studies, it is really of use only in prevalence studies of infection (Gastmeier *et al.*, 2000). From an ecological standpoint, it is conceivable that it may be an interesting measurement, but it is not recommended for publication of comparative data.

2.2. The recommended unit of antibiotic consumption

A unit of measurement and method of data handling independent of sales prices and package sizes is preferable. Reliable data on antibiotic consumption should ideally be based on individual patient prescriptions—but these data are generally not available. An acceptable compromise is the provision of hospital data which can be broken down by ward/discipline/prescriber. Despite the use of various unsuitable units for measuring antibiotic use in the past, we are moving close to a consensus unit of measurement and data collection. One system has gained a legitimacy and objectiveness—over and above the rest. The consensus numerator for measuring antibiotic use is the "defined daily dose" or "DDD." The consensus denominators are 1,000 inhabitant-days and 100/1,000 bed-days for community and hospitalised patients, respectively. The World Health Organisation (WHO) has met the challenge of standardising the presentation of antibiotic use data in order to monitor and benchmark use of this important class of drugs and to correlate this with the emerging problem of antibiotic resistance.

3.   WHO RECOMMENDATIONS

At a symposium in Oslo in 1969 entitled, The Consumption of Drugs, it was agreed that an internationally accepted classification system for drug consumption studies (not just antibiotics) was needed. At the same symposium the Drug Utilisation Research Group (DURG) was established by the WHO and charged with the task of developing internationally applicable methods for drug utilisation research.

In collaboration with the Norwegian Medicinal Depot (NMD), DURG updated the existing European Pharmaceutical Market Research Association classification system and developed the system now known as the Anatomical Therapeutic Chemical (ATC) classification system. From there, the unit of measurement known as the defined daily dose, or DDD, was developed for use in conjunction with the ATC system. In 1982, the WHO recommended the ATC/DDD system for international drug utilisation studies and introduced the WHO Collaborating Centre for Drug Statistics Methodology, based in Oslo, as a central body responsible for coordinating the use of the methodology. This centre is now responsible for classifying drugs according to the ATC system and setting DDDs.

3.1.   ATC classification

In the ATC system, drugs are divided into 14 main groups according to the organ or system on which they act and their chemical, pharmacological, and therapeutic properties (WHO Collaborating Centre, Oslo, 2002a, b). ATC class "J" includes "general anti-infectives for systemic use" and class "J01" deals specifically with "antibacterials for systemic use." Details of the J01 subclasses can be found in Table 1. Users of the ATC classification system tend only to quote antibiotic consumption for the J01 class. This however, does not necessarily reflect total consumption, as a few antibiotics have been placed in classes other than J01. For example, antibiotics given orally for decontamination of the digestive tract (including vancomycin) or intestinal anti-infectives have been placed in ATC group A07A and nitroimidazole derivatives to treat protozoal diseases are in ATC group P01AB. This is a relatively minor drawback of the system and is currently being addressed by the WHO.

Table 1. Anatomical Therapeutic Chemical (ATC) classification of antibiotics

J ANTIINFECTIVES FOR SYSTEMIC USE
J01 ANTIBACTERIALS FOR SYSTEMIC USE

J01A **TETRACYCLINES**

J01B **AMPHENICOLS**

J01C **BETA-LACTAM ANTIBACTERIALS, PENICILLINS**
  J01CA Penicillins with extended spectrum
  J01CE Beta-lactamase sensitive penicillins
  J01CF Beta-lactamase resistant penicillins
  J01CG Beta-lactamase inhibitors
  J01CR Combinations of penicillins, incl. beta-lactamase inhibitors

J01D **OTHER BETA-LACTAM ANTIBACTERIALS**
  J01DA Cephalosporins and related substances
  J01DF Monobactams
  J01DH Carbapenems

J01E **SULFONAMIDES AND TRIMETHOPRIM**
  J01EA Trimethoprim and derivatives
  J01EB Short-acting sulfonamides
  J01EC Intermediate-acting sulfonamides
  J01ED Long-acting sulfonamides
  J01EE Combinations of sulfonamides and trimethoprim, incl. derivatives

J01F **MACROLIDES, LINCOSAMIDES AND STREPTOGRAMINS**
  J01FA Macrolides
  J01FF Lincosamides
  J01FG Streptogramins

J01G **AMINOGLYCOSIDE ANTIBACTERIALS**
  J01GA Streptomycins
  J01GB Other aminoglycosides

J01M **QUINOLONE ANTIBACTERIALS**
  J01MA Fluoroquinolones
  J01MB Other quinolones

J01R **COMBINATIONS OF ANTIBACTERIALS**

J01X **OTHER ANTIBACTERIALS**
  J01XA Glycopeptide antibacterials
  J01XB Polymyxins
  J01XC Steroid antibacterials
  J01XD Imidazole derivatives
  J01XE Nitrofuran derivatives
  J01XX Other antibacterials

3.2.   Defined daily doses (numerator)

The DDD (expressed in grams) is the assumed average maintenance dose per day for a drug used for its main indication in a 70 kg adult and will only be assigned to drugs with an ATC code. The WHO emphasises that the DDD is a unit of measurement and does not necessarily reflect the recommended or prescribed daily dose—although in many cases it does. The WHO also states that drug consumption data presented in DDDs only gives an estimate of consumption and not an exact picture of actual use. DDDs do however, provide an internationally fixed unit of measurement independent of price and formulation enabling investigators to assess trends in drug consumption and to perform comparisons between groups. An in-depth discussion of the assignment of DDDs can be found on the WHO Collaborating Centre for Drug Statistics Methodology website and in their publications (WHO Collaborating Centre, Oslo, 2002a, b). It should be noted that DDDs are continuously being revised and anybody can request changes to the ATC and DDD systems. Justified requests are fully discussed and changes made if warranted. New and updated ATC/DDD lists are subsequently published at the start of each year. One of the main downsides of the DDD system is that it cannot be applied to children, given that the DDDs relate to an average adult.

3.3.   Denominator data

Antibiotic consumption data expressed as total DDDs used is of little value unless expressed with an appropriate denominator. It is conventional (and useful) to standardise the denominator for measurement of antibiotic use to further facilitate comparisons with other publications. In the community, consumption is commonly expressed as DDDs/1,000 inhabitant-days whereas the WHO recommends DDDs/100 bed-days be used in the hospital. It has however, been suggested that hospital data be expressed as DDDs/1,000 bed-days in order to allow for easier comparison between hospital and community. Inevitably, the bed-days are an approximation as days of admission and discharge are not full days but we believe this has become the international convention. Number of admissions may be a better measurement of hospital activity in view of the shortening length of hospital stay but it is not generally used.

3.4.   Theory into practice

It cannot be overemphasised that for effective data comparison and benchmarking, the ATC/DDD classification system must be used without modification. The process of expressing the consumption of different antibiotics and classes of antibiotics as DDD/100 bed-days could be a complicated one if the investigator were to work up the system of calculation from first principles using the basic information supplied by the WHO Collaborating Centre for Drug Statistics Methodology. Fortunately, all the hard work has already been undertaken and the Antibiotic Consumption Calculator (ABC Calc) simplifies the whole procedure (Monnet, 2003).

4.   ANTIBIOTIC CONSUMPTION CALCULATOR

ABC Calc (Monnet, 2003) is a simple computer tool utilising the ATC/ DDD system to measure antibiotic consumption at both the hospital and ward level as DDDs/100 bed-days (Figure 1). It transforms aggregated data provided by hospital pharmacies (generally as a number of packages or vials) into meaningful antibiotic utilisation rates. It was originally developed in the Department of Antimicrobial Resistance & Hospital Hygiene, Statens Serum Institut (Copenhagen, Denmark), by Dr Dominique Monnet as part of the Danish Integrated Antimicrobial Resistance Monitoring and Research Programme (DANMAP) (Bager, 2000). ABC Calc is freely available as a Microsoft Excel® file and is modified annually to incorporate any changes made to the ATC/DDD system. It can be downloaded from the European Study Group on Antibiotic Policies (ESGAP) page on the European Society of Clinical Microbiology and Infectious Diseases (ESCMID) website (www.escmid.org). To obtain data measured in DDDs/100 bed-days requires the input of data to calculate both the numerator and denominator.

Figure 1. Monnet DL ABC Calc—Antibiotic consumption calculator (Microsoft Excel® application). Version 1.9 Copenhagen (Denmark): Statens Serum Institut; 2003.

4.1.   ABC Calc numerator data

ABC Calc transforms aggregated data provided by hospital pharmacies (generally as a number of packages or vials) into meaningful antibiotic utilisation rates. For each product issued from the pharmacy, ABC Calc prompts the user for information on (1) the name of the product, (2) number of grams per unit dose, (3) the number of unit doses per package, and (4) the number of packages used in a defined time period. The spreadsheet then automatically calculates the total number of grams and DDDs used for each individual antibiotic, each subclass and class as well as total antibiotic consumption. The spreadsheet has the capacity to allow for multiple products of a single antibiotic. Care must be taken to enter each product in the appropriate line when prompted, as for some antibiotics the DDD varies depending on the route of administration.

For a given antibiotic, the number of DDDs are calculated as follows:

$$\frac{(\text{grams per unit dose}) \times (\text{number of unit doses per pack}) \times (\text{number of packs used})}{\text{Defined daily dose (supplied)}}$$

Depending on the product, a unit dose corresponds to one tablet, capsule, vial, etc. Again, depending on the product, a pack corresponds to, for example, a box of 10 tablets, and in some instances the pack may be equal to the unit dose, for example, individually distributed infusion vial.

ABC Calc has gone one step further than the ATC system, in the classification of antibiotics. The ATC system has not subgrouped some classes of antibiotics, whereas ABC Calc has. For example, the ATC system only goes as far as class J01DA "cephalosporins and related substances," whereas ABC Calc has grouped the cephalosporins by generation.

4.2.   ABC Calc denominator data

In order to measure consumption as DDDs/100 bed-days it is crucial to provide accurate data on bed-days and enter it into the ABC Calc spreadsheet.

Bed-days (during a specific time period) are calculated as follows:

$$(\text{Number of beds}) \times (\text{Occupancy index}) \times (\text{Number of days})$$

If the bed occupancy is 85%, then the occupancy index is 0.85. For example if a hospital has 1,200 beds and an occupancy of 85% in 1 year (365 days), then the number of bed-days for that year $= 1,200 \times 0.85 \times 365 = 372,300$.

4.3.   ABC Calc in practice

The authors have had direct experience of using ABC Calc to collate antibiotic consumption data from European hospitals. They coordinate the European Commission Concerted Action project entitled "Antibiotic Resistance, Prevention, and Control" (ARPAC) which runs from January 1, 2002 to December 31, 2004. ARPAC aims to lay the foundations for a better understanding of emergence and epidemiology of antibiotic resistance in human pathogens. It also aims to harmonise strategies for prevention and control of antibiotic resistant pathogens in European hospitals. The project uses ABC Calc to collate and compare antibiotic consumption from European hospitals and to model consumption with antibiotic resistance data as well as infection control and antibiotic policy data.

Although the ABC Calc spreadsheet comes with comprehensive instructions, any problems encountered in the authors' experience have been due to a failure in following the instructions. Most commonly, the bed-days have not been calculated correctly. Table 2 illustrates how such errors translate into errors in total DDDs/100 bed-days presented.

In example 1, the number of bed-days were stated, but no figures were supplied to indicate how they were calculated. Once the raw data were supplied, the bed-days value was modified and the total DDDs/100 bed-days changed from 174 to 37. In example 2, it would appear that the bed occupancy was presented as a percentage rather than an index value and the number of days and the number of bed-days were both entered manually, having been transposed

Table 2. Examples of errors in calculating bed-days and total DDDs/100 bed-days for 1 year (365 days)

	Number of beds	Bed occupancy index	Number of days	Number of bed-days	Total DDD/100 bed-days
Example 1					
Raw data	Not stated	Not stated	Not stated	87,309	174
Corrected data	1,248	0.9	365	410,204	37
Example 2					
Raw data	454	78.26	129,684	355	19,964
Corrected data	454	0.7826	365	129,685	55
Example 3					
Raw data	574	71.7	150,294	6.18×10^9	0
Corrected data	574	0.717	365	150,219	64
Example 4					
Raw data	1,181	84.5	8.7	8,68,212	25
Corrected data	1,181	0.845	365	364,250	60

giving an automatic calculation of a total of 19,964 DDDs/100 bed-days. Once the raw data were corrected, the total consumption became more believable at 55 DDDs/100 bed-days. In example 3, again the bed occupancy was originally presented as a percentage and the number of bed-days calculated manually and entered into the wrong box. The automatic calculation then gave 0 DDDs/100 bed-days instead of 64 DDDs/100 bed-days which was correctly calculated after modifying the entries in the spreadsheet.

Other errors in using ABC Calc have included users disabling the macros used to make the automatic calculations thereby underestimating total consumption, changing the spreadsheet cell formatting, and typographical errors. For example, in one particular spreadsheet the grams per unit dose for the combination of sulfamethoxazole plus trimethoprim were quoted as 480 and 960 g, resulting in total annual consumption of 572 DDDs/100 bed-days. Once the doses were corrected to 0.48 and 0.96 g, the total antibiotic use was reduced to a more realistic 41 DDDs/100 bed-days.

5.   BEWARE OF PSEUDO DDDs!

Although investigators are increasingly quoting antibiotic use expressed in terms of DDDs, not all are using the system appropriately and not all are completely transparent in describing exact methodology used. Ronning and co-workers have looked at the use of the ATC/DDD system in the literature and have also described their own experience of collecting antibiotic use data from European countries using the system (Ronning *et al.*, 2000, 2003).

Ronning *et al.* (2000) carried out a literature search to investigate if drug utilisation studies contain adequate references to the ATC/DDD versions used, thus making them suitable for comparative purposes. They targeted publications in 1996 and 1998 and found 73 papers on drug use. 46% of the articles gave proper references to the DDDs used either by referencing the ATC Index consulted or by listing the DDDs used. 54% of the papers did not give adequate information and it was not possible to identify the specific DDDs used. Thus, comparisons between different datasets could not be made. Ronning *et al.* (2000) concluded that it was not common to supply suitable information on ATC codes and DDDs used in drug utilisation studies, possibly because the authors did not know that the ATC/DDD system is an ever-changing one, which is updated annually.

Ronning *et al.* (2003) then set about requesting retrospective antibiotic use data from European countries, measured in DDDs/1,000 inhabitants per day for the years 1994–9. They distributed spreadsheets for data collection and asked for data presented using the 1999 ATC/DDD version. The authors of this study identified the main methodological problems as use of divergent

ATC/DDD versions, divergent assignment of DDDs for combination products, and the use of unauthorised/national DDDs. Their study also highlighted that most countries could not supply hospital data. Community data were more readily available although most of it was based on total sales from whole-sale data.

Units of antibiotic use, other than those recommended by the WHO have also been used and some published. Because the WHO DDDs do not always correspond to the dose actually used, some investigators have customised DDDs by changing them to the dose prescribed locally. In such cases, it is no longer appropriate to use the term DDDs; the term prescribed daily doses (PDDs) should then be used. Although hospital data using the same PDDs can sometimes be compared (Crowcroft *et al.*, 1999), in general, it is not possible to compare PDDs. An alternative nomenclature for the PDD is the defined daily administration or DDA.

A flurry of additional, unofficial acronyms have recently been added to the arena, most of which serve merely to confuse. For example, DDDs/1,000 inhabitants per day has been abbreviated to DID; DPP is the DDD per package; and DDS stands for DDDs consumed per area/time.

6.   ACTIVITIES COLLATING ANTIBIOTIC USE DATA

As already mentioned, the formation of comparable, local and national databases of antibiotic-use data are essential to inform those exploring the problem of antibiotic resistance. Use of the ATC/DDD system, as the leading consensus method, is essential for benchmarking and evaluating the impact of intervention studies. Numerous projects and surveillance programmes appear to be using the system appropriately, including the following:

1. The ATC/DDD system has been in use for some time by the DANMAP (Monnet *et al.*, 2000).
2. The European Study Group on Antibiotic Policies uses the system via use of ABC Calc, which is freely available via the website of the ESCMID (www.ESCMID.org).
3. The VIRESIST project aims to study relationships between antibiotic consumption and resistance and to forecast future behaviour. VIRESIST is a Spanish acronym, which stands for "Resistance Surveillance using Time Series Analysis Techniques." This is a hospital-level project which models monthly antibiotic consumption data broken down by ward against monthly antibiotic resistance data. Through its participation in this project, Aberdeen Royal Infirmary has shown that its MRSA outbreak is driven by

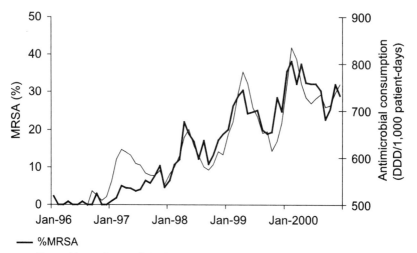

Figure 2. Aberdeen Royal Infirmary; Relationship between %MRSA and sum of lagged consumption of third generation cephalosporins, quinolones, and macrolides (VIRESIST project).

previous use of third generation cephalosporins, quinolones, and macrolides with varying lag times (Figure 2).

4. As previously mentioned, the European Commission DG Research funded project ARPAC—has used the ATC/DDD system and specifically ABC Calc to collate whole hospital and intensive care unit antibiotic-use data from individual European hospitals (www.abdn.ac.uk/arpac).

5. DG Sanco of the European Commission funds the European Surveillance of Antibiotic Consumption (ESAC) project. ESAC is a monitoring programme which aims at collecting standardised, harmonised, and comparable data on antibiotic consumption. Its goal is to document variations in antibiotic consumption and translate them into quality indicators for Public Health monitoring over time and place in order to target interventions and to assess the effectiveness of prevention programmes.

7.   MONITORING ANTIBIOTIC USE

To investigate the epidemiology of antibiotic use and its relationship with antibiotic resistance, antibiotic use and resistance data must be monitored over time to take account of seasonal variations and to be able to accurately monitor the effects of interventions. Data collection should extend to as long a period of time as possible, sampled over short periods of time to take into account small variations. By applying time-series analysis to series of data, it is possible to model the relationships between antibiotic use and resistance, and to demonstrate the temporal relationships between the two (Lopez-Lozano *et al.*, 2000; MacKenzie *et al.*, 2004). It has been suggested by Monnet *et al.* (2001) that time-series analysis to investigate the relationships between use and resistance require a minimum of 60 observations or time intervals. This can be 15 years of trimester-level data, 5 years of monthly data, or about a year of weekly data. Monnet and co-workers suggest that monthly data are preferred as a compromise between getting a maximum number of intervals while collecting a sufficient number of isolates tested in the time interval. The Cochcrane Effective Practice and Organisation of Core Group (EPOC) now considers this one of the best ways to evaluate healthcare interventions (Eccles *et al.*, 1997).

Because patient mix is crucial to antibiotic use, it is best to analyse local antibiotic use by discipline or unit. This is especially true for ICUs (Fridkin *et al.*, 1999; White *et al.*, 2000). The VIRESIST project demonstrates the usefulness of such monitoring schemes (Lopez-Lozano *et al.*, 2000; MacKenzie *et al.*, 2004; Monnet *et al.*, 2001), allowing the statistical analysis of trends in use, the effects of interventions, the study of relationships to antibiotic resistance, and the prediction of antibiotic susceptibilities based on predicted antibiotic use and previous rates of resistance. Comparison of trends within similar units of a hospital and between hospitals is probably useful although there are limited publications on this (see Chapter 8 by H. Westh and Chapter 6 by R. White). The frequency and type of feedback to units and how it can be used to improve prescribing is also unknown. At the moment, as part of the VIRESIST project, we have quarterly updates of data in our institution but only use the data routinely to monitor any major rises in consumption, particularly of limited list drugs and to study associations with resistance. Currently there is no routine feedback to prescribers, although it is our intention to provide this through the ward pharmacists and at annual medical and surgical meetings.

8. CONCLUSIONS

In conclusion, we are only really at the start of learning how best to measure and monitor antibiotic consumption. Computer programmes are increasingly useful but computerised prescribing for the majority, allowing proper audit of quality of use is still on the distant horizon.

REFERENCES

Bager, F., 2000, DANMAP: Monitoring antimicrobial resistance in Denmark. *Int. J. Antimicrob. Agents*, **14**, 271–274.

Cars, O., Mölstad, S., and Melander, A., 2001, Variation in antibiotic use in the European Union. *Lancet*, **357**, 1851–1853.

Crowcroft, N. S., Ronveaux, O., Monnet, D. L., and Mertens, R., 1999, Methicillin-resistant *Staphylococcus aureus* and antimicrobial use in Belgian hospitals. *Infect. Control Hosp. Epidemiol.*, **20**, 31–36.

Eccles, M. and Grimshaw, J., November 1997, Common study design issues. Changing Professional Practice Newsletter. Available from: http://www.dsi.dk/projects/cpp/ Newsletters/News-3.html.

Fridkin, S. K., Steward, C. D., Edwards, J. R., Pryor, E. R., McGowan, J. E. M. Jr., Archibald, L. K. *et al.*, 1999, Surveillance of antimicrobial use and antimicrobial resistance in United States hospitals: Project ICARE phase 2. *Clin. Infect. Dis.*, **29**, 245–252.

Gastmeier, P., Sohr, D., Forster, D., Schulgen, G., Schumacher, M., Daschner, F. *et al.*, 2000, Identifying outliers of antibiotic usage in prevalence studies on nosocomial infections. *Infect. Control Hosp. Epidemiol.*, **21**, 324–328.

Silber, J. L., Paul, S. M., Crane, G., Kupersmit, A., and Spitalny, K., 1994, Influence of hospital antibiotic policy and usage on the incidence of vancomycin-resistant enterococcal (VRE) bacteremia [abstract]. *Infect. Control Hosp. Epidemiol.*, **15**, 32.

Lopez-Lozano, J. M., Monnet, D. L., Yague, A., Burgos, A., Gonzalo, N., Campillos, P. *et al.*, 2000, Modelling and forecasting antimicrobial resistance and its dynamic relationship to antimicrobial use: A time series analysis. *Int. J. Antimicrob. Agents*, **14**, 21–31.

MacKenzie, F. M., Lopez-Lozano, J. M., Beyaert, A., Monnet, D., Camacho, M., Stuart, D. *et al.*, 2004, The role of antimicrobial use in the Aberdeen MRSA outbreak 1996–2000. *Emerg. Infect. Dis.* (in press).

Mölstad, S., Lundborg, C. S., Karlsson, A.-K., and Cars, O., 2002, Antibiotic prescription rates vary markedly between 13 European countries. *Scand. J. Infect. Dis.*, **34**, 366–371.

Monnet, D. L., 2003, *ABC Calc—Antibiotic Consumption Calculator [Microsoft Excel® Application]. Version 1.9.* Statens Serum Institut, Copenhagen (Denmark).

Monnet, D. L., Lopez-Lozano, J. M., Campillos, P., Burgos, A., Yague, A., Gonzalo, N., 2001, Making sense of antimicrobial use and resistance surveillance data: application of ARIMA and transfer function models. *Clin Microbiol Infect.*, **7** (Suppl 5), 29–36. Review.

Monnet, D. L., Soerensen, T. L., and Johansen, H. L., 2000, Comparison of the level of antimicrobial use in hospitals and in primary health care, Denmark, 1997 [abstract]. *Infect. Control Hosp. Epidemiol.*, **21**, 91.

Ronning, M., Salvesen Blix, H., Harbo, B. T., and Strom, H., 2000, Different versions of the anatomical therapeutic chemical classification system and the defined daily dose—are drug utilisation data comparable? *Eur. Clin. Pharmacol.*, **56**, 723–727.

Ronning, M., Salvesen Blix, H., Strom, H., Skovlund, E., Anderson, M., and Vander Stichele, R., 2003, Problems in collecting coparable national drug use data in Europe: The example of antibacterials. *Eur. J Clin. Pharmacol.*, **58**, 843–849.

WHO Collaborating Centre for Drug Statistics Methodology (Norway), 2002a, *Guidelines for ATC Classification and DDD Assignment.* WHO Collaborating Centre, Oslo.

WHO Collaborating Centre for Drug Statistics Methodology (Norway), 2002b, *ATC Index with DDDs.* WHO Collaborating Centre, Oslo.

White, R. L., Friedrich, L. V., Mihm, L. B., and Bosso, J. A., 2000, Assessment of the relationship between antimicrobial usage and susceptibility: Differences between the hospital and specific patient-care areas. *Clin. Infect. Dis.*, **31**, 16–23.

Chapter 8

Benchmarking

Henrik Westh
KMA 445, Hvidovre Hospital, Kettergård Alle 30
DK-2650 Hvidovre, Denmark

Benchmark is a technical term used by surveyors to indicate a point of reference from which measurements may be made. For our purposes, benchmarking is a method for improving operations in our organization typically achieved through systematic comparison with other organization(s) recognized as best in the field. The object of benchmarking is not to compare key figures but to compare how tasks are performed. Learning can only be achieved by looking at those who are better than you. Properly applied benchmarking is an extremely effective way of improving your organization.

Before you try to benchmark your organization, there must be a perceived need for improvement and a willingness to improve. This is very important because otherwise results will only end up as a report. The benchmarking study and the results must be supported by those who later will have to implement changes and improvements.

A number of different definitions of benchmarking are found in Table 1. These definitions have in common keywords such as performance, comparison, measuring, outstanding, best, improvement, process, and practice.

1.   DIFFERENT TYPES OF BENCHMARKING

Before you enter a benchmarking process, you will have to decide what to compare and whom to compare with, bearing in mind that the goal is to

Antibiotic Policies: Theory and Practice. Edited by Gould and van der Meer
Kluwer Academic / Plenum Publishers, New York, 2005

Table 1. Examples of definitions of benchmarking

- A process of finding, adapting, and implementing outstanding practices
- A process of identifying and importing best practice to improve performance
- Comparing the performance of your organization with that of others with outstanding performance to find fresh approaches and new ideas
- The process of comparing the performance of an individual organization against a benchmark, or ideal, level of performance. Benchmarks can be set on the basis of performance over time or across a sample of similar organizations, or against some externally set standard
- A continuous, systematic process for evaluating the products, services, and work processes of organizations that are recognized as representing the best practices for the purpose of organizational improvement

achieve improvements. When you have decided what to compare, three types of benchmarking are typically described.

- Process benchmarking
- Performance benchmarking
- Strategic benchmarking

When you have decided whom to compare with, three types of benchmarking are typically described.

- Internal benchmarking
- External benchmarking
- Generic benchmarking

Process benchmarking is learning from the best to improve one's own processes (comparison of methods and practices). Performance benchmarking is determining how good you are compared to others by comparing performance measures (either financial or operational). Strategic benchmarking is collecting information from other companies to improve one's own strategic planning and positioning.

Internal benchmarking is used to compare different units in the same organization. External benchmarking is the comparison with companies outside your organization that have similar or identical operations and processes. In external benchmarking, you will usually look for noncompeting organizations within your own field. In the private industry, one can run into problems of sensitive or confidential information, but this is rarely a problem in the public sector. Generic benchmarking involves comparison with unrelated industries that are worth learning from.

The benchmarking process is one of many tools for improving your institution or department. But it should be recognized that to be properly performed, it requires many resources. A good benchmarking process includes five major areas of activity.

1. Study and understand one's own process.
2. Find benchmarking partners.
3. Study the partner's process.
4. Analyse the differences between one's own process and that of the partner's.
5. Implement improvements based on what is learned.

It is important to spend enough time on the different elements of benchmarking. Typically 50% should be used in the planning phase, 30% on the study, and 20% in analysing the results. The timeframe for the implementation phase can only be estimated after the results are in.

2. PLANNING FOR BENCHMARKING

- Select a process to benchmark—think company strategy
- Identify needs
- Form a benchmarking team

Benchmarking is a method of improving performance. The goal is to identify and implement improvements by comparing your own operations with those of others who perform better. Almost anything that can be observed or measured can be benchmarked. This is often called gap. The planning and organizing of benchmark activities is extremely important. There must be a clear idea of goals and adequate resources allowing the benchmarking team time to fulfil its assignment. Tools must be found or developed for information and data gathering. Try to keep the goals specific (set limits), as too broad programmes will lead to an enormous amount of information and a lack of specific recommendations. This can often be achieved by identifying critical success factors (CSFs). A CSF could be patient satisfaction or adherence to hospital clinical guidelines or problems that have surfaced in audits. When you have identified CSFs, you have to evaluate your performance and also identify the processes that impact most on your CSFs. Look carefully at your organization and select indicators to benchmark against. Examples of potential indicators are listed in Table 2. It is also important to have a plan for how the results of the benchmarking will be used. The people in the benchmarking team structure must reflect your organization. Typically you will need a process owner, a process worker, a manager, and a user (customer).

3. FIND BENCHMARKING PARTNERS

- Look for long-lasting relationship
- Look for world champions or Best Practice guidelines
- Look out for differences in the scope of operation and in market conditions

Table 2.

Input indicators	Output indicators	Outcome indicators
Hospital type	DDD total antibiotics	Discharges
Antibiotic policies	DDD/1,000 beds	Operations
Drug formulary	Operation types	Surgical site infections, risk-adjusted infection rates
Budgets	Costs	Bed-days
Infection control policies	Surgical prophylaxis	Timing of surgical prophylaxis, length of operations
Staffing	Choice of prophylactic antibiotic regimes for	Survival rates for septicaemia, meningitis, community-
Best practice guidelines	orthopaedic surgery, etc.	acquired or hospital-acquired pneumonia
Microbiology service	Use of key antibiotics	Appropriateness of antibiotics usage
	3. gen cephalosporins	
	Betalactam-combinations, aminoglycosides,	
	Quinolones, monobactams, Vancomycin, others	
	Dosing	
	Time to first dose in meningitis, pneumonia, etc.	
	Administration route iv/oral	
	Resistance levels for key microorganisms	
	S. aureus	
	Pneumococci	

E. coli
Klebsiella
Coagulase negative staphylococci
Acinetobacter
Stenotrophomonas maltophilia
Rate of Candida sepsis
Nosocomial infection rates
Device-associated infection rates   By type of device and type of ICU
Gloves bought
Litres of hand disinfectant bought
Hand disinfection bought   By hospital unit
Microbiology lab in hospital
Infection control audits
Blood cultures per 1,000 bed-days
Positive rate of blood cultures
CNS positive rate in blood cultures   Vancomycin consumption

It is very important to have an open exchange of information so it is important that you find a few partners who are better than you than go for a comparison of a large number of companies leading to superficial conclusions. You should also look outside your network. Partners can be found by searching scientific journals, books, newspapers, the Internet, etc.

4.   STUDY PARTNERS

- Information gathering
- Questionnaires
- Definitions and explanations
- Document
- Get approval from managers

The data gathering process starts with obtaining a deep knowledge of your own organization. This leads to the development of a data gathering model. This model should be validated before using it on your study partner. What you are looking for in your partners is their performance level—how good they are and their practice—how do they do it.

5.   ANALYSE

- Find differences, that is, performance gaps
- Quality control data
- Prepare report

Before you analyse, you must be satisfied that all information is correct. The goal is to find performance gaps that can lead to improvement of CSFs. It is therefore important to check and correct for comparability and quality control your data collection. You will often need to normalize data. For example, in antibiotic consumption, knowing the amount of antibiotics used in Defined Daily Doses (DDDs) is uninteresting if it is not corrected/normalized, typically by per 100 bed-days. The information accumulation leads to knowledge gathering, allowing you to understand why there are performance gaps. This allows you to prepare a report that can be used for the real reason for benchmarking which is to change practices. Several models for identifying causes for the gaps are available, for example, comparison of flow charts, relations diagrams, and root cause analysis (breaking a problem into smaller problems). After identifying gaps, ranking can be relevant, as the closing of gaps will have different costs, improvement potentials, and applicability to your organization. Remember that your conclusions should be adapted to your own organization.

6.   IMPLEMENT IMPROVEMENTS

This is the final great challenge. The momentum of the project must be maintained. In effective implementation, you need to have a clear commitment from those involved and full management support. Factors that could block best practices must be identified (institutional culture, traditional cooperation, technology restraints, user groups, etc.)—all these elements can change best practice back to usual practice. The implementation process needs targets and always needs monitoring. The organization needs continuous orientation on progress and achievements. Remember that benchmarking results cannot be used as a carbon copy manual where best practices are copied directly into your own organization.

Also remember that benchmarking is not a one-time event but a continuous process for improving your own processes. One of the many pitfalls in benchmarking is that it is not successfully integrated into the way the organization solves problems.

7.   ETHICS IN BENCHMARKING

- Request only information that you would give out
- Respect confidentiality
- Full disclosure

As benchmarking is a continuous process, it is very important to have good ethics so you can sustain long-term relations with your benchmarking partners. You must be willing to provide the same information that you seek, to your benchmarking partners. Benchmarking must conform to legislation and moral codes. Otherwise, aspects such as industrial espionage, price fixing, customer allocation schemes, etc. can cloud issues.

8.   BENCHMARKING AND PUBMED

Looking in PubMed on 1 October 2003, the keyword benchmarking gave 3,535 hits. The use of benchmarking and antibiotic as keywords gave 35 hits and benchmarking and microbiology gave 25 hits. Let us briefly look at two of the best of these papers and discuss their study design.

9.   BENCHMARKING FOR REDUCING VANCOMYCIN USE AND VANCOMYCIN-RESISTANT ENTEROCOCCI IN US ICUs

This study was performed to improve compliance with process-of-care guidelines. The goal of the process was to reduce vancomycin consumption

Table 3. Prescribing practice changes implemented in response to benchmark data intervention, and mean rate of vancomycin use[a] before and after intervention, 50 Project ICARE ICUs, January 1996 to July 1999[b]

Vancomycin use prescribing practice change	No. of ICUs (%) ($n=50$)	Change absent		Change present		p value[c]
		Before	After	Before	After	
Hospitalwide[d]	22 (44)					
Drug use evaluation	19 (38)	74.2	80.5	105.3	94.1	0.62
Redistributed HICPAC guidelines on VRE	9 (18)	79.4	84.6	116.0	90.6	0.34
Prior approval of vancomycin required	3 (6)	87.2	84.7	67.2	99.4	0.25
Unit specific[d]	11 (22)					
ICU-specific education on appropriate vancomycin use	9 (18)	75.9	83.3	132.1	96.3	0.01
Removed vancomycin from surgical prophylaxis	3 (6)	82.0	85.9	149.1	82.2	0.01

[a]DDDs per 1,000 patient-days.
[b]ICARE, Intensive Care Antimicrobial Resistance Epidemiology; ICU, intensive care units; HICPAC, Healthcare Infection Control Practices Advisory Committee; VRE, vancomycin-resistant enterococci.
[c]Paired t-test.
[d]Components of each major category are not mutually exclusive, so one ICU may be represented in several components of each category.
Source: Data from Fridkin et al. (2002).

(Fridkin et al., 2002) (Table 3). The study is an example of benchmarking on the basis of performance across a sample of similar organizations. External benchmarking is the comparison with companies outside your organization that have similar or identical operations and processes.

In this Project ICARE study, preintervention data were collected during 1996–7. In part one of the project, the data collected were used to create a national benchmark defined as the aggregate summary data from 113 ICUs. The report included pooled means, medians, and key percentile distributions of the prevalence of VRE and MRSA and vancomycin use as DDDs/1,000 patient-days. This feedback report was presented in October 1997 to the participating hospitals (primarily to the infection-control committees).

In part two of this project, 50 ICUs in 20 hospitals participated in the postintervention period from April 1998 through July 1999 with at least 6 months of data collection. How the 1997 feedback report had been used by the hospitals was surveyed in September 1999. The feedback report survey looked for prescribing practice changes implemented in response to the feedback report. These prescribing practice changes were assessed and compared with vancomycin use before and after intervention.

The ICU-specific use of vancomycin in the 50 ICUs at the 20 study hospitals after the intervention was 89.1 DDD/1,000 patient-days, a 2.8% increase over the preintervention rate of use. This increase in consumption could in part have been caused by an increasing median MRSA prevalence of 33.5% (preintervention) and 39% during the postintervention period.

The only prescribing practice changes that led to a significant reduction in vancomycin use were ICU-specific education on appropriate vancomycin use and not surprisingly the removal of vancomycin from cardiac surgical prophylaxis. A risk-adjusted analysis was performed taking into consideration the ICU type and changes in MRSA prevalence (normalization). ICUs in which unit-specific practices were identified for improvement reported a 35–37% decrease in median vancomycin use (from median 132 to 96 DDD/1,000 patient-days for unit-specific education [9 units] and 149 to 82 DDD/1,000 patient-days for removal of prophylaxis [3 units]).

During the preintervention period, these ICUs reported a median VRE prevalence of 11.7% increasing to 14% during the postintervention period. However, when compared by type of practice change, the difference in VRE prevalence was significantly lower in ICUs in which unit-specific practice changes occurred, compared with other ICUs. Although many of the ICUs with decreases in vancomycin use reported increases in per cent VRE, all the ICUs noting a unit-specific practice change reported decreases in both per cent VRE and vancomycin use.

This study suggests that only focused efforts (i.e., ICU specific) were effective means of reducing excessive vancomycin use. The external benchmarks used were risk adjusted (i.e., stratified by ICU type) to account for the different rates of vancomycin used by different types of ICUs. This made comparison of local data more relevant (and more believable) to the ICU staff responsible for prescribing and other patient-care activities. The ICUs that used unit-specific changes had the highest prestudy rates of vancomycin use, and this excessive use may have made the ICU staff more amenable.

9.1.   Comments

This study analysed performance and measured key figures such as DDDs and resistance levels to methicillin in *Staphylococcus aureus* and to vancomycin in enterococci. This focus on performance gives little information on how to improve or close the gap between the different departments. Most people would accept that there is an over-usage of antibiotics in hospitals. The usage of antibiotics is regulated by clinical guidelines and clinical practices.

The very wide range of vancomycin usage in this study suggests that the true benchmark for vancomycin usage lies somewhere within the range of usages found. If the aim of the study was to reduce vancomycin usage, benchmarking should have been used to analyse the ICUs with low vancomycin consumption and show that their outcome results were as good as best practice. By analysing the methods and practices of these ICUs, one might learn, from the best, to improve one's own processes. This could have lead to a best practice definition of the correct usage of vancomycin in the ICU. This information could be clinical guidelines, training, educational efforts, etc. An analysis of one's own performance would have led to the findings of gaps that then could be corrected. In this study the benchmark was defined as the current median usage of vancomycin in the studied ICUs. With this choice of a benchmark, current practices were accepted as the benchmark, or, in other words, a median/average was defined as best practice. The results of the study were that outliers changed practices and regressed to the mean.

10.   THE HARVARD EMERGENCY DEPARTMENT QUALITY STUDY

This study was performed to improve compliance with process-of-care guidelines and patient-reported measures of quality. The study is an example of benchmarking on the basis of performance across a sample of similar organizations with use of some externally set standards (best practices). External benchmarking is the comparison with companies outside your organization that have similar or identical operations and processes.

Five Harvard teaching hospitals collaborated to improve quality in their emergency departments. The five areas chosen to improve were in patients presenting with abdominal pain, shortness of breath, chest pain, hand laceration, head trauma, or vaginal bleeding. A working group of experts reviewed the medical literature and existing guidelines and developed complaint-specific process-of-care data forms for medical record review. The goal was to improve compliance with process-of-care guidelines, patient satisfaction, and patient-reported problems with care.

In the preintervention phase, 4,876 medical records were evaluated, 2,327 patients completed onsite questionnaires, and 1,386 patients completed a 10-day follow-up questionnaire.

In the postintervention phase, 6,005 medical records were reviewed, 2,899 patients completed onsite questionnaires, and 2,326 patients completed a 10-day follow-up questionnaire.

Physician compliance with the process-of-care guidelines was the medical record based quality measure for the study and was evaluated by physician-reviewers unaware of the purpose of the study. Patients were asked to report problems during their emergency department visit and patient satisfaction was evaluated through the follow-up telephone interview.

One year later, the results of the baseline investigation were provided and preintervention phase process-of-care criteria were distributed to all the emergency departments as clinical guidelines. Based on the preintervention data, the hospitals found 27 different quality improvement interventions. From this list, each hospital chose 8–10 quality improvement efforts for implementation in their hospital.

In multivariate analyses, adjusting for site, age, urgency, and chief complaint, the mean compliance with guidelines for all complaints increased from 55.9% to 60.4% after interventions (see Table 4). For all sites combined, compliance with guidelines was significantly improved for abdominal pain, shortness of breath, and head trauma. There was no significant change in compliance with guidelines for chest pain, hand laceration, or vaginal bleeding. There were significant variations in intersite improvement rates in compliance with guidelines.

Changes in patient-reported problems were investigated by multivariate analyses adjusting for site, age, urgency, and chief complaint. The rate of

Table 4. Hospital specific (hospitals A–E) and total compliance with process-of-care guidelines

Complaint	Mean (95% confidence interval)[a]		p value
	Preintervention	Postintervention	
All complaints			
A ($n = 3,291$)	57.2 (55.2–59.2)	60.3 (58.5–62.1)	0.02
B ($n = 2,903$)	57.4 (55.4–59.4)	60.2 (58.4–61.9)	0.04
C ($n = 1,881$)	54.5 (52.1–56.9)	61.7 (59.5–63.9)	0.0001
D ($n = 2,405$)	52.7 (50.5–54.9)	58.6 (56.6–60.6)	0.0001
E ($n = 456$)	63.4 (58.7–68.1)	62.6 (58.3–66.9)	0.83
Total	55.9 (54.9–56.9)	60.4 (59.4–61.4)	0.0001
Abdominal pain			
A ($n = 1,149$)	57.2 (53.9–60.5)	60.0 (56.9–63.1)	0.23
B ($n = 752$)	58.4 (54.1–62.7)	60.6 (56.5–64.7)	0.45
C ($n = 499$)	53.8 (48.7–58.9)	62.5 (57.8–67.2)	0.02
D ($n = 704$)	55.4 (50.7–60.1)	57.9 (54.2–61.6)	0.42
E ($n = 160$)	71.3 (62.7–79.9)	65.7 (58.3–73.1)	0.34
Total	57.0 (55.0–59.0)	60.5 (58.7–62.3)	0.01
Shortness of breath			
A ($n = 527$)	72.0 (66.5–77.5)	70.0 (64.9–75.1)	0.61
B ($n = 384$)	31.6 (25.1–38.1)	52.1 (45.0–59.2)	0.0001
C ($n = 332$)	58.3 (49.1–67.5)	59.6 (50.4–68.8)	0.96
D ($n = 417$)	37.7 (30.8–44.6)	54.9 (48.6–61.2)	0.005
E ($n = 100$)	56.3 (35.1–77.5)	75.6 (54.4–96.8)	0.29
Total	52.1 (48.8–55.4)	60.9 (57.6–64.2)	0.0002

Table 4. *Continued*

Complaint	Mean (95% confidence interval)[a]		*p* value
	Preintervention	Postintervention	
Chest pain			
A (*n* = 636)	65.5 (62.4–68.6)	61.9 (59.0–64.8)	0.10
B (*n* = 701)	70.7 (67.6–73.9)	69.3 (66.4–72.2)	0.54
C (*n* = 437)	68.0 (64.5–71.5)	64.7 (61.2–68.2)	0.18
D (*n* = 503)	61.3 (57.8–64.8)	63.8 (60.7–66.9)	0.32
E (*n* = 117)	65.9 (59.8–72.0)	62.9 (56.6–69.2)	0.51
Total	66.7 (65.1–68.3)	65.0 (63.4–66.6)	0.13
Hand laceration			
A (*n* = 176)	58.5 (50.7–66.3)	57.6 (50.3–64.9)	0.86
B (*n* = 293)	55.0 (49.9–60.1)	56.9 (50.8–63.0)	0.65
C (*n* = 196)	55.7 (49.0–62.4)	68.1 (62.2–74.0)	0.008
D (*n* = 178)	66.2 (59.5–72.9)	67.8 (60.7–74.9)	0.76
E (*n* = 31)	68.9 (56.4–81.4)	68.2 (58.8–77.6)	0.94
Total	58.7 (55.6–61.8)	62.6 (59.5–65.7)	0.09
Head trauma			
A (*n* = 589)	40.5 (36.6–44.4)	52.7 (49.4–56.0)	0.0001
B (*n* = 728)	48.3 (44.8–51.8)	53.4 (50.0–56.7)	0.04
C (*n* = 348)	31.7 (26.8–36.6)	52.9 (48.6–57.2)	0.0001
D (*n* = 441)	32.0 (28.1–35.9)	46.2 (42.3–50.1)	0.0001
E (*n* = 42)	24.5 (10.6–38.4)	46.0 (29.1–62.9)	0.08
Total	40.0 (38.0–42.0)	51.4 (49.6–53.2)	0.0001
Vaginal bleeding			
A (*n* = 206)	64.0 (57.4–70.7)	70.2 (62.0–78.4)	0.47
B (*n* = 34)	81.4 (63.0–99.8)	72.9 (56.2–89.6)	0.52
C (*n* = 52)	66.6 (52.3 80.9)	70.7 (54.4–87.0)	0.66
D (*n* = 152)	73.3 (65.7–80.9)	73.3 (66.0–80.6)	0.8
E (*n* = 0)[b]			
Total	68.6 (64.1–73.1)	70.2 (65.3–75.1)	0.64

[a]Adjusted for age, urgency, and chief complaint. Total also adjusted for site.
[b]Hospital E had no patients with vaginal bleeding.

problems decreased overall from 24% to 20% and significant improvements were seen in four of the five sites. No improvements were seen in patient satisfaction.

10.1.   Comments

This is a fine study where the use of benchmarking resulted in some improvement in emergency department quality of care. However, the focus on performance gives little information on how to improve or close the gap between the different departments.

This study used best practices as defined by the Harvard Emergency Department Quality Study team and internal benchmarking as the analyses were performed comparing preintervention and postintervention phases for each hospital. The benchmarking partners chosen for the study seem chosen for geographical reasons. A best practice Emergency Department was not part of the analysis; therefore, this is predominantly a study of equals. Looking at the Table 4, it can be seen that total compliance for "shortness of breath" was 72% in hospital A and 31.6% in hospital B. This difference is what benchmarking calls a gap. An analysis of the practices in hospital A that made it possible for them to reach this higher level of performance would have been perfect benchmarking. Although an analysis of differences between one's own and the partner's process was not described, the unblinding of the project results must have allowed for good opportunities to discuss the impacts of the different quality improvement interventions chosen by the different departments. However, the study team felt that the use of multiple interventions did not allow the team to evaluate which initiatives lead to improvement.

Lacking in this chapter is a discussion of the use of CSFs. An example of a CSF was that each unit in the study wanted to fulfil the American College of Emergency Physicians criteria for administration of thrombolytic therapy and achieved 100% compliance with the guideline (up from 65.3%).

11.   CONCLUSION

In conclusion, benchmarking can become a valuable tool for improvements in healthcare. While key performance figures can be used to find gaps in performance, only an analysis of processes will allow one to understand the differences and plan for improvement.

SUGGESTED READING

Andersen, B. and Pettersen, P.-G., 1996, *The Benchmarking Handbook. Step-by Step Instructions.* Chapman & Hall, London, UK. ISBN 0 412 73520 2.

Benchmarking in the Public Sector. Some Methods and Experiences. March 2000. Ministry of Finance. ISBN 87-7856-331-3 www.fm.dk.

Burstin, H. R., Conn, A., Setnik, G., Rucker, D. W., Cleary, P. D., O'Neil, A. C. *et al.*, 1999, Benchmarking and quality improvement: The Harvard Emergency Department Quality Study. *Am. J. Med.*, **107**(5), 437–449.

Fridkin, S. K., Lawton, R., Edwards, J. R., Tenover, F. C, McGowan, J. E., Jr, Gaynes, R. P., 2002, Intensive Care Antimicrobial Resistance Epidemiology Project; National Nosocomial Infections Surveillance Systems Hospitals. Monitoring antimicrobial use and resistance: Comparison with a national benchmark on reducing vancomycin use and vancomycin-resistant enterococci. *Emerg. Infect. Dis.*, **8**(7), 702–707.

Chapter 9

Experiences with Antimicrobial Utilisation Surveillance and Benchmarking

Catherine M. Dollman
Infection Control Service, Communicable Disease Control Branch, Department of Human Services, South Australian Government, 162 Grenfell St, Adelaide SA 5000, Australia

1.   INTRODUCTION

An antimicrobial utilisation surveillance programme was established in Adelaide, South Australia (SA), in November 2001 as an initiative of the Infection Control Service, Communicable Disease Control Branch (CDCB) of the South Australian Department of Human Services. This voluntary surveillance programme incorporates antimicrobial usage data submitted on a monthly basis by major Adelaide metropolitan public and private hospitals. The aim of the programme is to provide SA hospitals with an ongoing overview of their antimicrobial usage rates over time, to enable intervention programmes or policy changes to be planned and implemented where high or increasing rates are identified, and to assess the effectiveness of such programmes or policy changes. Concomitant surveillance of multiresistant organisms within the same institutions is also conducted by the CDCB and when sufficient data are available, links between antimicrobial usage rates and the incidence of particular organisms will be examined. Published data suggests that concomitant surveillance of both antibiotic resistance and antimicrobial use is helpful in interpreting resistance patterns within a particular unit or hospital (Monnet *et al.*, 1998) and may assist in the development of programmes to complement improved infection control procedures in reducing infection rates.

Antibiotic Policies: Theory and Practice. Edited by Gould and van der Meer
Kluwer Academic / Plenum Publishers, New York, 2005   133

This programme was initiated in response to recommendations arising from a report prepared by the Joint Expert Technical Advisory Committee on Antibiotic Resistance (JETACAR), formed by the Australian Government in 1997, in response to the increasing incidence of antimicrobial resistance. The JETACAR made 22 recommendations and, in response, the Commonwealth supported surveillance of both antibiotic resistant organisms and antibiotic utilisation (Commonwealth Department of Health and Aged Care, Commonwealth Department of Agriculture, Fisheries and Forestry—Australia, 1999 and 2000). To date, the SA programme is the only state programme within Australia conducting surveillance of antimicrobial usage. No national programme has yet been established, although preliminary discussions have taken place.

During 2002, 11 SA metropolitan hospitals were involved in the programme. Complete 2002 usage data are available for eight of the eleven hospitals, with three commencing data contribution during the 12-month period. One additional hospital joined the programme in 2003, with at least one more to join in 2004. The hospitals contributing data in 2002 included six public and five private hospitals, ranging in size from approximately 100–650 beds. Stratification by case-mix, size, or other parameter has been avoided due to the limited number of institutions involved; however, submitted data are stratified to provide separate usage rates for intensive care units (ICUs) where applicable, provided these data can be accurately provided by the pharmacy service provider. Intensive care usage data were submitted by six hospitals during 2002, involving four public and two private hospitals.

The pooling of data for hospital areas other than ICUs has some disadvantages, and the reporting of usage rates for particular clinical services or units would assist in identifying units with high antimicrobial usage rates within individual hospitals, and enable more appropriate comparison between similar hospitals. This level of stratification, however, is not currently possible, although antimicrobial consumption by some specialised services or units such as haematology or transplant units, where there is a high usage of antimicrobials, is a future target for collection and analysis where the usage data can be accurately provided. In most Australian hospitals, however, accurate usage data for many wards or units, such as general surgical or medical wards, is difficult to obtain due to patient mix within these areas and the lack of pharmacy resources to enable identification of use by individual patients. No hospital participating in the SA programme is currently able to provide complete, accurate data for antimicrobial consumption at individual patient level.

A survey of contributing hospitals during 2002 found that eight have a hospital formulary which limits to some extent the choice of antimicrobial agents which can be routinely prescribed; however, specific restrictions on antimicrobial use for various indications or patient groups apply in only six of the eleven institutions. These restrictions vary considerably with respect to both the range of agents involved and the limitations imposed. Comprehensive restriction programmes, with the requirement for prior authorisation by infectious diseases staff for some agents, currently operate only in large teaching hospitals. The extent to which these policies are enforced, however, varies significantly between institutions. Restrictions applying to the intensive care setting also vary widely between the contributing hospitals, with these units subject to fewer restrictions or unrestricted use in most cases. Restriction programmes requiring prior authorisation, although labour-intensive, can have major impact on antimicrobial usage patterns and also expenditure. One study has demonstrated no compromise to clinical outcomes from a prior authorisation requirement for a range of antimicrobials which was applied throughout the hospital, including the ICU (White, 1997).

One difficulty encountered during establishment of the SA surveillance programme was the diversity of computer systems used by hospital pharmacy departments or pharmacy service providers for contributing hospitals. Antimicrobial consumption data is submitted to the Infection Control Service in a variety of formats, necessitating the design of a computer program to centrally accept and analyse these data. This programme has the facility to produce automated monthly reports for each hospital, detailing antimicrobial usage density rates within that hospital. Corresponding rates calculated from aggregation of all contributed data are also supplied for comparison, although hospitals are encouraged to use caution when making such comparisons, and should consider inter-institutional differences in multiresistant organism burden and case-mix complexity. All details relating to individual hospital usage rates are kept confidential and only provided to that hospital, unless specific approval is obtained for publication of data. Currently, individual hospital and aggregate usage rates for six antimicrobial classes, and individual agents within those classes, are routinely reported to each contributor. Separate rates for ICU usage are provided where appropriate. Usage rates for other classes or agents can be calculated as required.

Usage data contributed by Adelaide's specialist paediatric hospital are not included in the automated reporting programme, as consumption by a paediatric population cannot be translated into a standard usage density rate. Consumption data is collected, however, and reported separately. The establishment of a national network of paediatric hospitals is planned to allow benchmarking between these institutions, with a particular focus on antimicrobial use in neonatal ICUs, using an agreed unit of measurement.

The establishment of a rural hospital antimicrobial utilisation surveillance network is currently underway, with approximately 30 rural hospitals expected to be involved in this programme. The range of antimicrobials reported routinely in this programme will differ from the current metropolitan network and may provide information on usage patterns in rural areas as a focus for programmes aimed at general practitioner prescribing.

2.   DEFINITIONS USED IN ANTIMICROBIAL
         SURVEILLANCE

Defined daily dose (DDD) has been used as the unit for measurement of antimicrobial consumption in this surveillance programme. The DDD for any drug is defined as the average dose per day to treat an average adult patient. The World Health Organisation (WHO) has determined standard DDDs for most drugs (WHO Collaborating Centre for Drug Statistics Methodology, 2002), and these values have been used in calculating all displayed usage rates. Use of this internationally accepted standard enables the consumption of antimicrobial agents with differing doses to be compared, aggregation of data to assess usage of antimicrobial classes, and comparisons with data from other surveillance programmes or studies. Because DDDs are based on adult dosing, this parameter cannot be used to measure antimicrobial usage in paediatric populations.

The number of DDDs used is calculated as follows:

$$\text{No. of DDDs} = \frac{\text{Total grams used}}{\text{WHO assigned DDD value}}$$

The usage density rate used in the SA surveillance programme is defined as the number of DDDs used per 1,000 occupied bed-days (OBDs). This rate has been widely used as an appropriate measurement of usage in the non-ambulatory setting, and has been adopted by a number of international programmes (DAN-MAP, 2003; Fridkin *et al.*, 1999), although to ensure comparability of data with other centres, the WHO DDD values should always be used. Antimicrobial usage data for outpatient areas, including hospital-in-the-home, day treatment centres, day surgery, and dialysis clinics are excluded from the SA programme to ensure that the denominator corresponds to that used by the concomitant multiresistant organism surveillance programmes conducted by the Infection Control Service.

The rate is calculated as follows:

$$\text{Usage density rate} = \frac{\text{No. of DDDs/time period} \times 1{,}000}{\text{OBD/time period}}$$

3.   DATA COLLECTION AND REPORTING

Numerator data representing antimicrobial consumption, in terms of number of units or packs of individual antimicrobial formulation, are submitted by

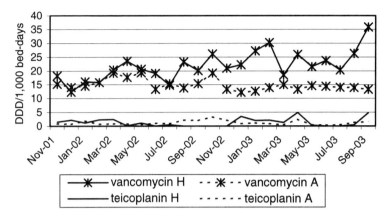

Figure 1. Example of reporting to individual hospitals showing hospital (H) and aggregate (A) rates.

pharmacies on a monthly basis. In the case of public hospitals, this information is supplied by hospital pharmacy departments, while for private hospitals this may be provided by either a contracted individual community pharmacy or a larger hospital pharmacy service provider. Data are stratified into "Intensive Care Unit" (ICU) and pooled usage by other hospital areas (non-ICU), or "total hospital" if there is no ICU, or if ICU data cannot be provided separately. Denominator data representing OBDs are supplied by contributing hospitals.

All contributed datasets are loaded into a custom written database with the facility to calculate usage density rates and produce automated monthly reports for individual hospitals, as well as a report based on aggregate data from all contributors. Rates are routinely calculated for six antimicrobial classes and the individual antimicrobial agents within those classes. Routine reports include charts showing both the usage rate for the institution and the aggregate rate. An example is shown in Figure 1.

4.   OVERALL TRENDS IN ANTIMICROBIAL USAGE RATES

Analysis of aggregate SA data for 2002 suggests a slight overall increasing trend in total antimicrobial usage; however, trends for individual hospitals usage over the year varied considerably. Ongoing monitoring and analysis over a longer time period is required to confirm changes in utilisation rates.

The overall antimicrobial usage rates in contributing SA metropolitan hospitals, for both total hospital use and ICU use, are shown in Figure 2.

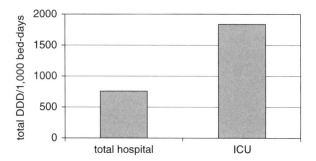

Figure 2. Aggregate antimicrobial usage for all classes and all contributors for 2002 (total hospital and ICU).

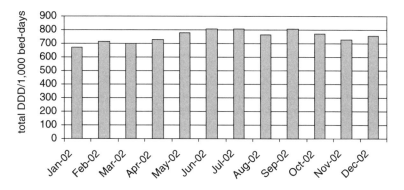

Figure 3. Total monthly aggregate antimicrobial usage for all classes and all contributors.

Aggregate data has been used to calculate these rates. The aggregate rate for total antimicrobial usage in the contributing hospital group for 2002 was 757 DDD/1,000 OBDs. For individual hospitals, total antimicrobial usage ranged from 455 to 966 DDD/1,000 OBDs, with a median rate of 686. For ICU use, the aggregate rate was 1,838 DDD/1,000 OBDs, with a range 1,545–2,252 and a median rate of 1,837.

Monthly rates for aggregate total hospital use are shown in Figure 3. This chart suggests a small upward trend in usage; however, this may be attributable to normal monthly variation, with increased usage of many antimicrobial agents during the winter months. Detailed analysis of usage rates for each contributor and antimicrobial class has demonstrated a seasonal increase in the use of a range of agents, particularly in public hospitals. These include third generation cephalosporins, fluoroquinolones, tetracyclines, amoxicillin and amoxicillin/clavulanate, and particularly macrolides, although significant variation

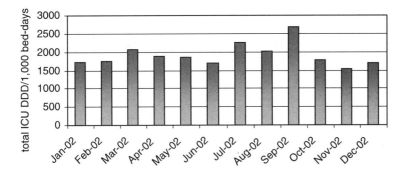

Figure 4. ICU monthly aggregate antimicrobial usage for all classes and all ICU contributors.

was noted between individual hospitals. Only two hospitals demonstrated an increase in benzylpenicillin usage during this time. Analysis of monthly total antimicrobial usage rates for the eight contributors with complete data for the 12-month period demonstrates a seasonal increase in usage for four of these hospitals. Analysis of data collected over a longer time period will be necessary to fully assess both seasonal changes and annual trends in utilisation rates.

Corresponding aggregate monthly usage rates for intensive care units are shown in Figure 4 and do not demonstrate any obvious trend. Analysis of usage for individual units has demonstrated large variations in usage rates, both between and within individual units. Although an increase in usage rates during the winter months was noted for a number of antimicrobial classes, including third generation cephalosporins, macrolides, and fluoroquinolones in some hospitals, this pattern was not consistent for all ICUs. This may reflect the differing proportion of surgical and medical patients within the small number of ICUs submitting data to this surveillance programme.

5.   ANALYSIS OF USAGE BY ANTIMICROBIAL CLASS

Data have been analysed by antimicrobial class to allow assessment of relative use of particular classes, as well as changes occurring over time, to provide aggregate class-specific antimicrobial usage rates as benchmarks for comparison by individual contributing hospitals, and for comparison with other Australian and international data.

Routine monthly reports distributed to contributing hospitals currently include six antimicrobial classes: third/fourth generation cephalosporins, glycopeptides, carbapenems, fluoroquinolones, aminoglycosides, and anti-pseudomonal penicillin/β-lactamase inhibitor combinations. The third and

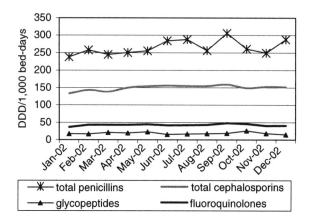

Figure 5. Total hospital usage by antimicrobial class.

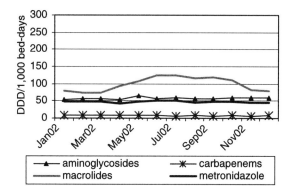

Figure 6. Total hospital usage by antimicrobial class.

fourth generation cephalosporins have been grouped together to simplify reporting; however, usage details for individual agents are also identified. For this
summary, total penicillin, total cephalosporin, macrolide, and metronidazole use
has also been analysed. These rates for monthly total hospital use are displayed
in Figures 5 and 6, using the same scale to allow comparison of relative usage.
Corresponding monthly rates for ICU use are shown in Figures 7 and 8. The
relative frequency of total hospital and ICU usage of the various antimicrobial
classes for 2002 are shown in Figure 9. A breakdown of usage of penicillin and
cephalosporin classes into smaller groups is provided in Figures 10 and 11.

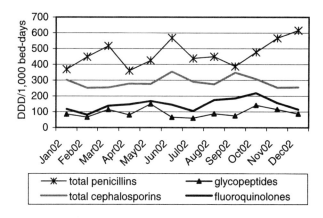

Figure 7. Total ICU use by antimicrobial class.

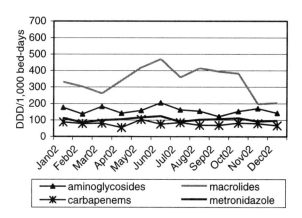

Figure 8. Total ICU use by antimicrobial class.

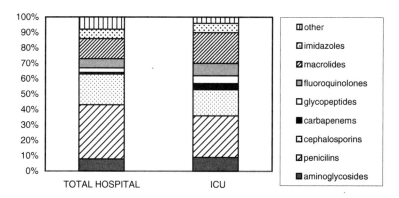

Figure 9. Total hospital and ICU usage by antimicrobial class.

142 Catherine M. Dollman

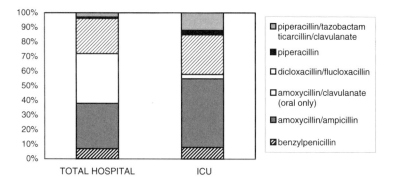

Figure 10. Total hospital and ICU penicillin usage.
Note: The parenteral form of amoxicillin/clavulanate is not available in Australia.

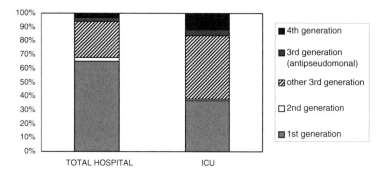

Figure 11. Total hospital and ICU cephalosporin usage.

6.   ANALYSIS BY INDIVIDUAL ANTIMICROBIAL AGENT

The utilisation rates for various individual antimicrobial agents within classes are displayed below in Figures 12–27. Both total hospital and ICU rates are displayed.

6.1.   Cephalosporins

Overall total hospital use of the routinely reported cephalosporins (ceftriaxone/cefotaxime, ceftazidime, cefepime) showed no significant trends during

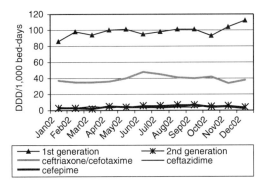

Figure 12. Total hospital cephalosporin use.

Note: First and second generation agents have been grouped to simplify these charts.

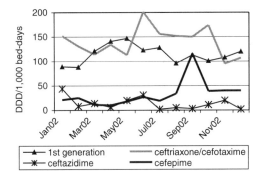

Figure 13. ICU cephalosporin use.

Note: First and second generation agents have been grouped to simplify these charts.

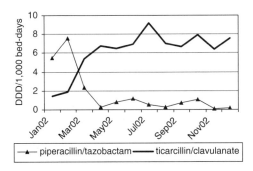

Figure 14. Total hospital use.

Note: These are the only penicillins included in routine monthly reports to contributing hospitals.

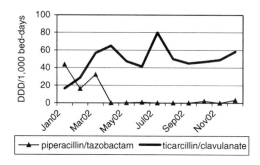

Figure 15. ICU use.

Note: These are the only penicillins included in routine monthly reports to contributing hospitals.

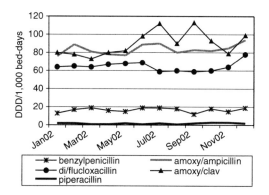

Figure 16. Total hospital use—other penicillins.

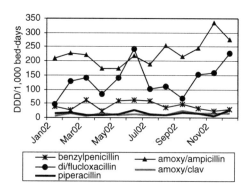

Figure 17. ICU use—other penicillins.

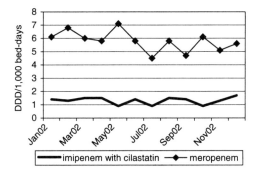

Figure 18. Total hospital carbapenem use.

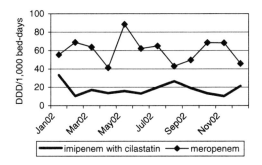

Figure 19. ICU carbapenem use.

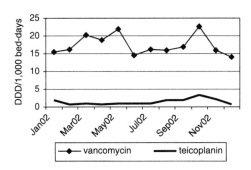

Figure 20. Total hospital glycopeptide use.

2002, although use of ceftriaxone during the latter part of the year may have been influenced by stricter control policies instituted at one teaching hospital during July. Assessment of trends with such limited data is complicated by the normal seasonal increase in usage for the treatment of respiratory infections

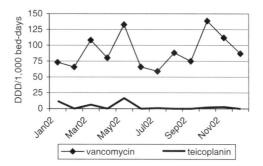

Figure 21. ICU glycopeptide use.

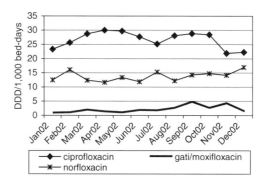

Figure 22. Total hospital fluoroquinolone use.

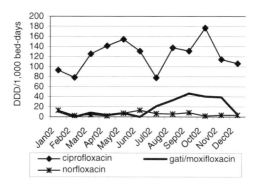

Figure 23. ICU fluoroquinolone use.

during the winter months and ongoing surveillance may show a greater change in use of this agent over time. In the ICU setting, ceftriaxone also displayed the expected seasonal variation. An unexpected peak in cefepime use in three of the six contributing ICUs during September is currently being investigated to

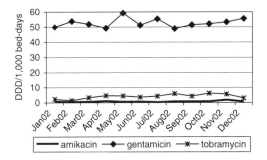

Figure 24. Total aminoglycoside use.

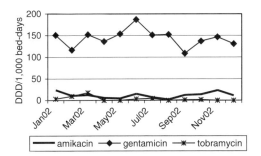

Figure 25. ICU aminoglycoside use.

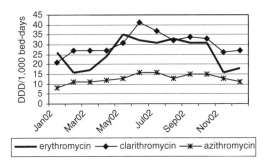

Figure 26. Total hospital macrolide use.

determine whether this was related to an outbreak of a multiresistant organism within these units or a change in prescribing patterns. A slight increasing trend in the use of first generation cephalosporins was noted during the year. Usage of second-generation agents was minimal, with negligible use in the ICU setting.

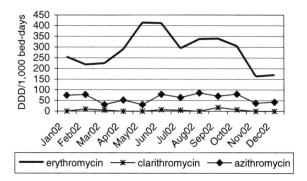

Figure 27. ICU macrolide use.

Use of these agents was predominantly in the private hospital sector, with for-mulary restriction limiting use in the larger public hospitals. Cefoxitin and cefotetan are the only second-generation agents currently available in a par-enteral form in Australia.

6.2.   Penicillins

6.2.1.   Antipseudomonal penicillin/β-lactamase inhibitor combinations

Overall usage of these two agents remained stable during 2002, although there are wide variations in usage between individual institutions contributing to the programme. In most hospitals, ticarcillin/clavulanate has replaced piperacillin/tazobactam due to the variable availability of the latter agent within Australia.

6.2.2.   Other penicillins

Of the other agents in the penicillin class, total hospital amoxicillin/clavulanate usage has increased during 2002; however, ICU use of this combination is low as the IV form is not marketed in Australia. Both amoxicillin/ampicillin and dicloxacillin/flucloxacillin have shown an increasing trend in ICU usage. The use of other agents in this class has remained relatively stable (Figures 16 and 17).

6.3.   Carbapenems

A slight downward trend in total hospital meropenem use was noted during 2002, with no significant change in imipenem use. As expected, ICU use was variable for both of these agents which would normally be limited to use for severe or difficult to treat infections. Most contributing hospitals have made a formulary change to meropenem in recent years. Usage of the newer carbapenem, ertapenem, was negligible overall, with limited use in the private sector.

6.4.   Glycopeptides

Total hospital and ICU use of vancomycin was variable during 2002, with no definite trend in usage shown. A number of hospitals, particularly in the private sector, demonstrated various peaks in teicoplanin usage during the period. Prolonged treatment of a single patient intolerant of vancomycin was involved in most cases, with ensuing discussion with pharmacists to ensure that appropriate vancomycin administration protocols are in place to minimise adverse reactions to this agent. Teicoplanin use in the public hospitals was low.

6.5.   Fluoroquinolones

Use of ciprofloxacin and norfloxacin remained relatively stable during 2002, although high use of norfloxacin in the private setting compared to public hospitals use warrants further investigation and intervention. ICU ciprofloxacin use varied considerably over the period. An increase in use of gatifloxacin and moxifloxacin was noted during the later months, with likely seasonal use for the treatment of community-acquired pneumonia. A significant increase occurred in one public hospital ICU in particular, with a corresponding fall in ceftriaxone use.

6.6.   Aminoglycosides

Total hospital and ICU usage of gentamicin showed no significant trend over 2002. While amikacin use remains very low outside the intensive care setting, ICU use has shown a slight increase, although this is variable between hospitals. This may reflect increasing resistance to gentamicin or altered prescribing patterns within some units, particularly where no antimicrobial restriction policies are in place. Although the aggregate rate is low, tobramycin usage increased over the period, with almost all use contributed by one public hospital with an adult cystic fibrosis unit.

6.7.   Macrolides

Seasonal variation is evident in the monthly usage rates for all macrolides, although no significant overall trend is evident for the period. There is significant variation in usage between hospitals, reflecting individual hospital protocols for the treatment of respiratory infections, in particular, community-acquired pneumonia. As predicted, a fall in macrolide use was noted during 2003. Using of this class fell by 20% overall during this period, with a 33% fall in erythromycin use and a 23% fall in azithromycin use. The other macrolides used, clarithromycin and roxithromycin, also showed a fall in use during 2003. A 73% increase in doxycycline use occurred. Usage of this agent had previously been low. Parenteral azithromycin was introduced into most hospitals during late 2003, and has replaced parenteral erythromycin in some hospital formularies.

7.   BENCHMARKING WITH OTHER ANTIMICROBIAL UTILISATION DATA

While SA is the first Australian state to develop an antimicrobial utilisation surveillance network involving individual public and private hospitals, a number of individual hospitals throughout Australia have been monitoring and analysing their own antimicrobial usage rates for a number of years. Some of these data are available for comparison with SA rates. A number of large programmes conducting surveillance of antimicrobial consumption have been established in Europe and the United States during the last decade, and some of these also provide suitable data for comparison with SA rates. In particular, the DANMAP programme in Denmark has published antimicrobial usage rates for both the primary healthcare sector and hospitals since 1997. A number of charts have been included below to provide an overview of utilisation rates within SA in 2002 compared to rates published in the DANMAP 2002 report.

Figure 28 shows comparative total antimicrobial usage rates for the 11 adult hospitals that contributed data during 2002. Also shown is the total usage rate for the group of contributing SA hospitals calculated from aggregate data (total DDDs and total bed-days) from the 11 hospitals, and the comparative rate for Denmark for 2002. Higher usage is demonstrated for SA overall and for 10 of the 11 hospitals contributing data to the SA surveillance programme.

Comparison with Danish data, and some recently released data from other European countries (European Surveillance of Antimicrobial Consumption, ESAC, 2003), also highlights differences in relative frequency of usage of particular antimicrobial classes. Although limited ESAC data is available relating to hospital use, the frequency of use of various cephalosporin groups has been shown to vary considerably between European countries. For most countries,

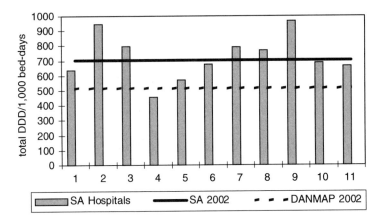

Figure 28. Total antimicrobial usage for 2002 for 11 contributing hospitals.

the significant use of second-generation agents is notable in comparison with SA data. This reflects the difference in availability of these agents, with the parenteral form of cefuroxime not currently marketed in Australia. Danish data (Monnet, July 2003, personal communication. Usage rates for 1st, 2nd, 3rd, and 4th generation cephalosporins) however, suggests that while the low use of third and fourth generation cephalosporins may partly reflect the availability of second generation formulations not available in Australia, use of the cephalosporin class overall is significantly lower than in Australia. Higher total penicillin usage rates are also seen in some European countries, as shown by the Danish data below, with significantly higher use of β-lactamase sensitive penicillins. The Danish usage rate for the extended spectrum penicillin group is slightly higher than that for SA, with a wider range of agents available in Europe. Of this group, only piperacillin, amoxicillin and ampicillin are available in Australia. There is negligible use of β-lactamase inhibitor combinations in Denmark, while these agents are widely used in Australia.

Figures 29 and 30 show comparative usage of the four "generations" of cephalosporins and the different penicillin groups in Denmark and SA.

Figure 31 shows comparative Danish and SA usage rates for glycopeptides, fluoroquinolones, aminoglycosides, macrolides, imidazoles, and carbapenems. For each class, except carbapenems, the SA rate is significantly higher than the corresponding Danish rate.

There are at present limited opportunities for benchmarking between Australian hospitals. Figure 32 shows comparative usage of different antimicrobial classes in the 11 contributing adult SA hospitals and one other teaching hospital located in New South Wales (NSW), another Australian state. Although high use of the penicillin class at the NSW hospital is evident compared to SA

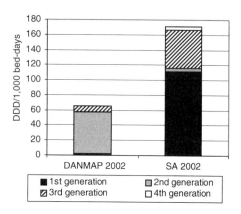

Figure 29. Comparative use of cephalosporins.

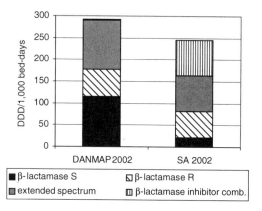

Figure 30. Comparative use of penicillins.

hospitals, usage rates for cephalosporins, particularly for the third generation agents, are lower.

ICU usage rates for parenteral antimicrobial use for the above NSW teaching hospital, one Queensland teaching hospital, and the six SA hospitals with ICUs are displayed in Figure 33. There is a significant variation in the rates for the SA hospitals, partly explained by the diversity of these six units, as previously mentioned. This small number of contributors unfortunately prevent stratification into groups with similar case-mix. Both interstate hospitals show lower usage rates for third generation cephalosporins than most SA hospitals. Fluoroquinolone usage is also significantly lower in the two interstate hospitals. For the penicillin class, usage rates are higher for the NSW hospital than the other Australian centres presented here.

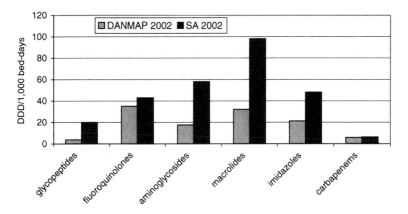

Figure 31. Comparative usage rates for other classes.

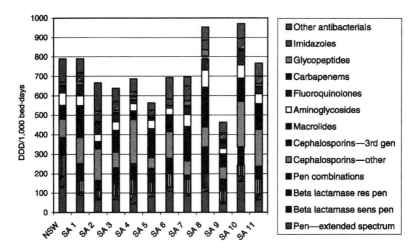

Figure 32. Comparative usage rates for SA hospitals and one NSW hospital.

While there is limited ICU data available for benchmarking locally and internationally using standard WHO DDDs, rates for most of these Australian units appear to be high in comparison with rates published from Scandinavian studies (Petersen *et al.*, 1999; Walther *et al.*, 2002) involving 38 Swedish units and 30 Danish units.

Benchmarking with DANMAP data, as well as that from other Scandinavian countries, clearly demonstrates the high comparative usage rates in SA for many antimicrobial classes and sets a goal for reduction in usage

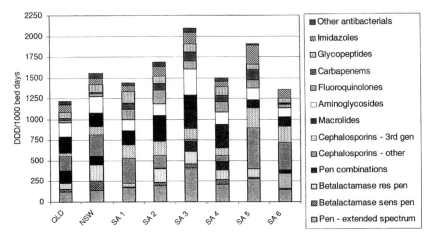

Figure 33. Comparative intensive care usage rates for SA hospitals and 2 other Australian hospitals.

Note: includes parenteral use only.

through improved prescribing and infection control procedures. The wider availability of Australian antimicrobial utilisation surveillance data in future, however, will enable comparison between centres where a similar range of antimicrobial agents is available, and may lead to the sharing of successful intervention models or the development of large-scale intervention programmes. Stratification of a larger pool of contributors by hospital size, case-mix, or other parameter may enable more appropriate benchmarking than currently possible. The availability of comparative data from ICUs and other specialised units of larger hospitals throughout Australia may also provide an opportunity to identify and investigate high antimicrobial use and institute programmes to improve antimicrobial prescribing and infection control within these areas.

8.   ADDENDUM

Subsequent to the surveillance period covered by the preceding report, Flinders Medical Centre, a 430-bed metropolitan teaching hospital contributing to the SA surveillance programme, has implemented a successful programme aimed at promoting a more rational, evidence-based approach to antimicrobial prescribing. This programme has involved the introduction of a reserved antibiotic policy in conjunction with the implementation of revised treatment algorithms based on nationally accepted guidelines (Therapeutic Guidelines Ltd, 2003). Antimicrobial agents are divided into three categories involving either unrestricted use, the requirement for prior Infectious Diseases approval for all use, or use restricted to a range of specified exempt indications. In the last case, use for other indications requires Infectious Diseases approval. Restrictions apply to all hospital areas excluding the ICU. Close cooperation between Infectious Diseases and Pharmacy staff, both before and since the policy implementation, has seen significant changes in antimicrobial usage across a range of agents.

Modification of prescribing patterns for ceftriaxone, the most widely used third generation cephalosporin, was a particular focus of the new policy, with a high rate of inappropriate use demonstrated by a review of ceftriaxone use conducted during late 2002. This agent is widely used for the treatment of respiratory infections in hospitalised patients in many Australian hospitals, and this was shown to be the most common inappropriate indication for ceftriaxone use during the review period. This agent was previously available without Infectious Diseases approval for the treatment of community-acquired pneumonia; however, under the new restriction policy, prior approval is required. Figures 34 and 35 show some changes in ceftriaxone usage rates since the introduction of the restricted antibiotic policy and revised treatment algorithms at this hospital in February 2003.

A significant and sustained fall in ceftriaxone use has been successfully achieved through this policy change. Seasonal variation in ceftriaxone usage rates, suggesting high use for the treatment of respiratory infection during the winter period, has not been evident during 2003. Usage rates for benzylpenicillin

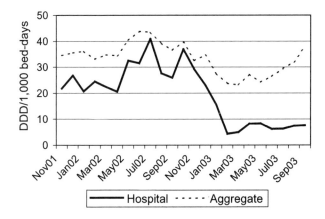

Figure 34. Changes in ceftriaxone use since surveillance commenced in November 2001.

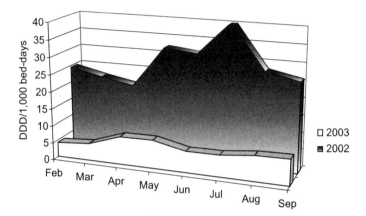

Figure 35. Comparative ceftriaxone usage during the 8-month period February–July 2002 and 2003.

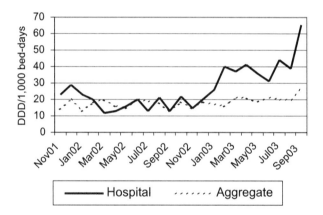

Figure 36. Changes in benzylpenicillin usage since the introduction of antimicrobial restrictions and revised treatment algorithms.

have increased correspondingly in accordance with the hospital algorithm for the treatment of community-acquired pneumonia. A smaller and less sustained increase in gentamicin use has also been noted since the introduction of the policy.

Significant changes in the usage of oral agents have also been noted, in line with the policy changes. Azithromycin usage rates have fallen significantly since this agent was replaced by doxycycline as first line therapy for mild community-acquired pneumonia. All azithromycin use now requires prior Infectious Diseases approval.

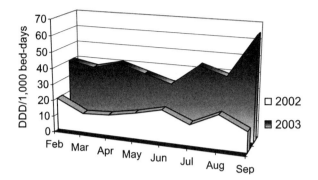

Figure 37. Comparative benzylpenicillin usage during the 8-month period February–July 2002 and 2003.

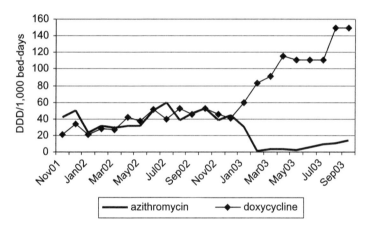

Figure 38. Changes in azithromycin and doxycycline usage since the introduction of antimicrobial restrictions and revised treatment algorithms.

Although limited data are available to date, the concurrent implementation of a reserved antimicrobial policy and revised treatment algorithms has been successful in significantly altering prescribing patterns, particularly for the treatment of community-acquired pneumonia.

REFERENCES

Commonwealth Department of Health and Aged Care, Commonwealth Department of Agriculture, Fisheries and Forestry, Canberra, 1999, The use of antibiotics in food-producing animals: Antibiotic-resistant bacteria in animals and humans.

Commonwealth Department of Health and Aged Care, Commonwealth Department of Agriculture, Fisheries and Forestry, Canberra, August 2000, Commonwealth Government Response to the Report of the JETACAR.

DANMAP, 2003, DANMAP, 2002—Use of antimicrobial agents and occurrence of antimicrobial resistance in bacteria from food animals, foods and humans in Denmark, Copenhagen.

European Surveillance of Antimicrobial Consumption, 2003, Results of the ESAC retrospective data collection (http://www.ESAC.ua.ac.be).

Fridkin, S. K., Steward, C. D., Edwards, J. R. *et al.*, 1999, Surveillance of antibiotic use and antibiotic resistance in United States hospitals: Project ICARE Phase 2. *Clin. Infect. Dis.*, **29**, 245–254.

Monnet, D. L., Archibald, L. K., Phillips, L., Tenovar, F. C., McGowan, J. E., Jr, and Gaynes, R. P., 1998, Antibiotic use and resistance in eight US hospitals: Complexities of analysis and modeling. Intensive care antimicrobial resistance epidemiology project and national nosocomial infections surveillance system hospitals. *Infect. Control Hosp. Epidemiol.*, **19**, 388–394.

Petersen, I. S., Hesselbjerg, L., Jorgensen, L., Renstrup, J., Barnung, S., Schierbeck, J. *et al.*, 1999, High antibiotic consumption in Danish intensive care units? *Acta Path. Microbiol. Immunol. Scand.*, **107**(11), 989–996.

Therapeutic Guidelines Ltd, 2003, *Therapeutic Guidelines: Antibiotic*, 12th ed.

Walther, S. M., Erlandsson, M., Burman, L. G., Cars, O., Gill, H., Hoffman, M. *et al.*, 2002, The ICU-STRAMA Group. Antibiotic prescription practices, consumption and bacterial resistance in a cross section of Swedish intensive care units. *Acta Anaesthesiol. Scand.*, **46**(9), 1075–1081.

White, A. C., Jr., Atmar, R. L., Wilson, J., Cate, T. R., Stager, C. E., and Greenberg, S. B., 1997, Effects of requiring prior authorization for selected antimicrobials: Expenditures, susceptibilities, and clinical outcomes. *Clin. Infect. Dis.*, **25**, 230–239.

WHO Collaborating Centre for Drug Statistics Methodology, 2002, ATC/DDD index 2002 (http://www.whocc.no/atcddd).

Chapter 10

Interventions to Optimise Antibiotic Prescribing in Hospitals: The UK Approach

Erwin M. Brown
Department of Medical Microbiology, Frenchay Hospital, Bristol BS16 1LE, UK

1. INTRODUCTION

A large body of evidence supports a correlation between antibiotic usage and resistance, but confirmation of a causal relationship has proved illusory, the evidence for such a relationship having been exclusively circumstantial. Recently, however, Lopez-Lozano *et al.* (2000), using time-series analysis, a form of mathematical modelling, have demonstrated convincingly that antibiotic prescribing is the driving force behind the emergence of resistance to these drugs. This is an important development because it holds out hope that reducing antibiotic prescribing will lead to a corresponding reduction in the incidences of resistance—a theory that remains unconfirmed as investigators who have evaluated the efficacies of interventions to reduce antibiotic prescribing have, for the most part, used cost, prescribing levels or, less frequently, appropriateness of prescribing, rather than resistance rates, as outcome measures. In the few cases where resistance rates have been employed as an outcome measure, efforts to determine the impact of the interventions have been undermined by the effects of confounding variables, most notably infection control measures. Finally, even if it were possible to totally eliminate inappropriate antibiotic prescribing, resistance rates will continue to be driven upwards by appropriate prescribing.

Notwithstanding uncertainties regarding the effects of interventions to optimise antibiotic prescribing on rates of resistance, several government and

Antibiotic Policies: Theory and Practice. Edited by Gould and van der Meer
Kluwer Academic / Plenum Publishers, New York, 2005

other authoritative bodies (Government Response to the House of Lords Select Committee on Science and Technology Report, 1998; House of Lords Select Committee on Science and Technology, 1998; NHS Executive, 1999; Report from the Invitational EU Conference on the Microbial Threat, 1998; Sub-group on Antimicrobial Resistance of the Standing Medical Advisory Group, 1998) have called for reductions in inappropriate antibiotic usage. This challenge, albeit well motivated, reveals a lack of understanding of the complex relationships between antibiotic usage and antibiotic resistance and the obstacles associated with successfully controlling the prescribing of these drugs. More importantly, while all of these agencies have specified the outcome, none has identified the process by which it is to be achieved. The failure to provide hospitals with guidance on interventions to reduce inappropriate antibiotic prescribing explains, at least in part, why, 4 years after the challenge was issued, many hospitals have not yet implemented a formal antibiotic control programme. One could be forgiven for expecting that the paper produced by the Society for Healthcare Epidemiology of America and Infectious Diseases Society of America Joint Committee on the Prevention of Antimicrobial Resistance, entitled *Guidelines for the Prevention of Antimicrobial Resistance in Hospitals* (Shlaes *et al.*, 1997), might contain clear evidence-based guidelines/ recommendations for optimising antibiotic prescribing, but this is not the case. Indeed, to date, no such guidelines have been published. In 1999, the British Society for Antimicrobial Chemotherapy and the Hospital Infection Society convened a joint working party on optimising antibiotic prescribing in hospitals in order to address this issue.

2.   JOINT BRITISH SOCIETY FOR ANTIMICROBIAL CHEMOTHERAPY/ HOSPITAL INFECTION SOCIETY WORKING PARTY ON OPTIMISING ANTIBIOTIC PRESCRIBING IN HOSPITALS

2.1.   Membership of the Working Party

The membership of the Working Party comprises five medical microbiologists (one of whom is a trainee), three infectious diseases physicians, one surgeon, and one pharmacist; there are also five members from outside the United Kingdom (three from Europe and two from the United States) who are recognised for their expertise in the field of optimising antibiotic usage and who serve as corresponding advisors.

2.2.   Literature search

The Working Party began its deliberations by carrying out a systematic review of the published literature on interventions aimed at optimising antibiotic prescribing in hospitals. Relevant publications were identified by three independent electronic searches. In the first, MEDLINE, EMBASE, and the Cochrane database of clinical trials were searched from 1980 onwards using a broad range of search terms. The second search was conducted in MEDLINE (1966–2000), using PubMed and OVID, and the Cochrane database, and employed terms which differed from those used in the first search. Finally, the third search was of the Cochrane Effective Practice and Organisation of Care (EPOC) specialised register which, itself, was compiled by searching MEDLINE (from 1966), Health STAR (from 1975), and EMBASE (from 1980). There were no language limitations. In addition, the references section of each paper was reviewed and any articles not identified by electronic search were obtained and the process repeated; failing to supplement the electronic searches with a manual search would have resulted in a failure to identify at least one third of the articles.

2.3.   Systematic review

The electronic and manual searches yielded 670 articles, of which 306 published from 1980 onwards contained original data about interventions in hospitals. These 306 studies were then evaluated for eligibility for inclusion in a Cochrane EPOC review. The principal criteria for inclusion related to: study design (only randomised controlled trials, RCTs, controlled clinical trials, CCTs, controlled before and after studies, CBAs, and interrupted time series, ITS, with ≥3 data points before and after the intervention being eligible); minimum methodological criteria (a study must involve objective measurement of provider performance/behaviour or patient outcome(s), and relevant and interpretable data must be presented or obtainable from the investigators); and EPOC scope (a study must involve the evaluation of the effect(s) of behavioural/educational, financial, organisational, or regulatory intervention(s)). Of the 306 studies, 80 (26.1%), comprising 38 ITS, 24 RCTs or CCTs, 11 CBAs, and 7 others, fulfilled the inclusion criteria. Two hundred and twenty-six (74%) studies were excluded for the following reasons: uncontrolled before and after studies (141, 62.4%); inadequate ITS (75, 33.2%); and inadequate CBAs or CTs (10, 5%). There was a significant upward trend with time in terms of the percentage of studies with robust designs, but, even in the final 4-year period (2000–3), only 36.2% of studies were eligible for inclusion in the review.

ITS, which accounted for approximately 48% of the included studies, was the most common type of design. In most cases, the results of these studies were analysed by comparing the means of the pre- and postintervention data points. However, this format can lead to inappropriate, indeed erroneous, conclusions. For example, an upward trend in the data points before the intervention has been implemented may cause the effect of the intervention to be underestimated. Conversely, a downward trend before the intervention has been implemented may lead to the effect of the intervention being overestimated. The EPOC group has recommended that segmented regression analysis be used to estimate the magnitude of the effects of interventions. However, of the 38 ITS studies, the investigators in 23 (60.5%) determined the means of the pre- and postintervention data points, but failed to subject the data to statistical analysis thereby precluding a conclusion regarding whether or not the effect of the intervention was statistically significant, and those in 11 (29%) determined the means of the data points and subjected the results to statistical analysis, but reached the wrong conclusion, leaving only four (10.5%) groups of investigators who employed segmented regression analysis (two of the studies having been carried out by the same group). The advantage of adopting segmented regression analysis to assess the data is that it provides information about the speed of the impact of the intervention and whether or nor the effect is sustained. For example, in a study carried out by Belliveau *et al.* (1996), in which vancomycin prescribing was restricted in an attempt to reduce the incidence of vancomycin-resistant enterococci, the volume of vancomycin dosing was used as the outcome measure. As soon as the intervention was implemented, there was a significant reduction in the number of doses prescribed. However, almost immediately thereafter, dosing levels gradually increased until, by the eleventh month, they had returned to preintervention levels. This trend would not have been detected if the results had been analysed by comparing the means of the pre- and postintervention data points. It is clear, therefore, that the method of analysing the data has a profound influence on the way in which the results are interpreted.

It would be reasonable to expect that the 80 studies which fulfilled the criteria for inclusion in the review were robust in terms of their design and execution. This is, however, not the case: a high proportion of these studies suffering from one or more serious methodological flaws. Although the review has not yet been completed, several conclusions can be drawn from the findings to date.

1. Both the quality and quantity of the evidence in the published literature which supports the efficacies of the interventions are disappointing.
2. The majority of published studies used inadequate control methods, thereby precluding efforts to determine whether any change in practice/outcome was attributable to the intervention.

3. ITS accounted for almost 50% of the studies included in the review; in most of these, inappropriate statistical analysis of the results of the studies overestimated the magnitudes of the effects of the interventions.

4. Many of the studies which fulfilled the criteria for inclusion in the review suffer from methodological flaws.

5. Suboptimal methods of analysing the data generated by some studies led the investigators to reach incorrect conclusions.

6. For some interventions, not even a single study fulfilled the inclusion criteria, there being, therefore, no published evidence to support them.

7. In several studies, up to four interventions were implemented simultaneously, thereby precluding efforts to discern the relative contribution of each measure to the outcome.

8. Only a very small minority of studies employed resistance rates as an outcome measure.

9. Only a very small minority of studies compared the efficacies of interventions.

10. In some cases, assessment of the impact of an intervention was undermined by the effects of confounding variables, most frequently, infection control measures.

11. Owing to the absence of robust evidence from published studies, most, if not all, of the recommendations made by the Working Party will probably represent a consensus of expert opinion.

3.   INTERVENTIONS

As the Working Party had not completed the systematic review of the literature at the time of writing, it is not possible to make evidence-based recommendations regarding interventions to optimise antibiotic prescribing in hospitals. However, it is likely that some or all of the measures described below will be recommended for implementation.

3.1.   Educational/persuasive vs restrictive/ coercive interventions

Interventions fall into two categories, educational or persuasive, and restrictive or coercive. Educational interventions, for example, pharmacy bulletins and newsletters, lectures, conferences, and handbooks are preferable, but it is perceived that, alone, they are of limited value in terms of facilitating judicious antibiotic usage. Moreover, without constant reinforcement to maintain their impact, their effects will be only temporary.

Although there is little evidence to support the efficacies of restrictive interventions specifically to control antibiotic usage, such measures, in general, have consistently been shown to be more effective than educational strategies (not surprisingly as prescribers can only rarely be relied upon to demonstrate goodwill) and their impacts are more enduring. Bamberger and Dahl (1992) compared the impact of voluntary restriction of selected antibiotics (ceftazidime and ceftriaxone) with that of a strict control policy. When restriction was voluntary, only 24.2% of the usage of these drugs was in compliance with local guidelines, compared with 85.4% when restriction was enforced; coincidently, expenditure on these agents was reduced significantly and susceptibility rates among isolates of *Enterobacter cloacae* and *Pseudomonas aeruginosa* increased. In another study (Himmelberg *et al.*, 1991), removal of a restrictive policy led to a 158% increase in usage of previously restricted drugs and a 103% increase in expenditure on these agents. Nonetheless, restrictive interventions are effective only if they are enforced, and enforcement may lead to adversarial relations between prescribers, who have been shown to perceive them as dictatorial and to prefer less coercive measures, and those healthcare workers (usually pharmacists) who have accepted responsibility for enforcing them. Moreover, controlling antibiotic prescribing may be more difficult in smaller hospitals where support mechanisms are not usually available; such hospitals tend to rely less on restrictive interventions than on education to influence prescribing practices.

3.2. Core interventions

The following interventions represent the minimum measures that should be implemented in all hospitals.

3.2.1. Antibiotic Control Plan

The Antibiotic Control Plan (ACP) should be the cornerstone of a hospital's efforts to influence the volume and appropriateness of antibiotic usage. The measures which comprise the plan should be devised, implemented, promoted, enforced, and their efficacies monitored by the Antibiotic Control Committee, a subcommittee of the Drugs and Therapeutics Committee. The Committee should have executive powers and should be chaired by a senior consultant with specialised knowledge of infectious diseases and antibiotics (either a microbiologist or an infectious diseases physician), although this is not essential. The membership should also comprise a microbiologist (if not the chairman), a physician, a surgeon, a trainee doctor, and a pharmacist. The responsibilities of the committee can be summarised as follows:

1. To devise an antibiotic formulary.
2. To produce guidelines for antibiotic prescribing.
3. To develop and implement educational programmes.
4. To develop and implement other interventions for controlling and promoting prudent antibiotic prescribing.
5. To monitor, through the audit process, the efficacies of and compliance with the interventions implemented locally.
6. To undertake surveillance of antibiotic usage within each speciality, providing feedback of prescribers' own antibiotic practices in relation to those of peers or a standard.
7. To undertake regular (2-yearly) reviews of the interventions which have been implemented.

3.2.2.   Antibiotic formulary

An antibiotic formulary is simply a list of drugs available for use within a hospital. Formulary control has been shown to be the most direct and effective means of influencing antibiotic prescribing and reducing antibiotic expenditure and resistance rates, without adversely affecting patient care; it can also have a positive educational impact on prescribers. The following standards apply to formulary development and implementation:

1. The antibiotics included in the formulary should be limited to the minimum necessary to provide effective prophylaxis and therapy, thereby enabling pharmacies to negotiate favourable prices. Ideally, only one antibiotic in each class should be included, thereby eliminating duplicate agents and reducing the number of drugs stocked by the pharmacy. Each drug should be chosen on the basis of efficacy, propensity to promote the development of resistance, pharmacokinetic properties, pharmacodynamic properties, side-effect and safety profiles, tolerability, and cost.
2. The choice of agents should be influenced by local susceptibility patterns.
3. The drugs included in the formulary should be placed into categories, with restrictions on the use of certain agents, based on special indications, breadth of spectrum, toxicity, cost, potential to be misused, and propensity to promote the development of resistance.
4. The formulary should be reviewed periodically, specifically regarding the need to include antibiotics which have recently become available or to delete redundant agents and to determine whether drugs which have been

subject to abuse or to which there have been marked increases in rates of resistance should be reassigned to a category to which restrictions apply.

5. Compliance with the formulary should be audited.

3.2.3.   Enforcing formulary restrictions

The drugs which are included in the formulary can be classified as unrestricted (i.e., can be prescribed by any prescriber without the need for prior approval) or restricted (i.e., available only if usage conforms with guidelines that have been developed by the Antibiotic Control Committee or following discussion with a designated "expert"). The success of a formulary will depend on how rigorously compliance with it is enforced as prescribers frequently fail to adhere to formulary restrictions. There are two methods of enforcing compliance with formulary restrictions.

The first of these is the antibiotic order form which requires written justification to prescribe, or to continue prescribing, drugs included on the restricted list. The information sought on the order form has varied from centre to centre, but has included the following: whether the prescription is for prophylaxis or treatment (empirical or definitive); site of infection; clinical criteria on which the diagnosis of infection is based; the suspected or confirmed cause(s) of the infection; patient-related information, such as age, weight, underlying disease, renal and hepatic function, and known allergies; and drug-related information, such as dosage, frequency, route of administration, and duration. The forms are evaluated by a pharmacist and approval of the use of the antibiotic granted or withheld according to guidelines devised by the Antibiotic Control Committee. This intervention has been shown to be effective in terms of controlling antibiotic usage, limiting the duration of prophylaxis and treatment and facilitating audit. However, it is labour-intensive and sophisticated information technology is required in order for it to be implemented. Moreover, prescribers often fail to complete the forms or the quality of the information is poor or inadequate or both. The lack of sufficient resources in most UK hospitals would make this strategy impracticable.

The alternative method is the requirement to seek approval for the use of antibiotics on the restricted list from an "expert" who is usually either a medical microbiologist or an infectious diseases physician. Normally, approval will be granted if the proposed usage falls within predetermined guidelines. In the United States, approval is sought, during or out of normal working hours, from an infectious diseases physician. In theory, it would be feasible to implement such a strategy in hospitals in the United Kingdom that have large numbers of infectious diseases physicians and/or microbiologists on staff. However, in most hospitals, this will not be the case and a compromise would be necessary. During routine working hours, a pharmacist would refer the prescription to a medical microbiologist or infectious diseases physician who, following discussion with the prescriber, would either approve it or recommend an alternative regimen. If the drug has been prescribed out of hours, it would be issued for a finite period (e.g., 24–48 hr), at the end of which pharmacy staff would notify a local expert who would either approve the prescription or not. This strategy is more suited to use in the United Kingdom, although its implementation would be facilitated by computerised prescribing.

A third, noncoercive method of controlling the use of restricted drugs has been proposed by Williams *et al.* (1985). It involves prospective monitoring of patients prescribed the targeted drugs by an expert. If the prescription is considered appropriate, the prescriber is not contacted, but if the prescription is considered inappropriate, an informal approach is made to the prescriber who is advised either to discontinue the antibiotic or to switch to an alternative, equally effective, less expensive regimen. This strategy was well received by the prescribers in the hospital in which it was evaluated, involved them in negligible additional effort, was educational for those who had prescribed inappropriately, and led to substantial cost savings. It was, however, time- and labour-expensive for the individuals monitoring the prescriptions.

3.2.4. Automatic antibiotic stop-order policy

The aim of an automatic antibiotic stop-order policy is to limit the durations of unnecessarily prolonged prescriptions for therapy and prophylaxis. In the United States, implementation of such a policy is a requirement of the Joint Commission on Accreditation of Hospitals, while in the United Kingdom, a survey carried out by a Working Party of the British Society for Antimicrobial Chemotherapy (1994) revealed that only 26% of the 539 respondents employed such an intervention.

In its simplest form, the policy requires prescribers to specify a duration for each antibiotic prescription, regardless of whether it is for prophylaxis or treatment. In theory, this could be done on a voluntary basis. However, as prescribers can rarely be relied upon to comply voluntarily, it must be enforced in order to be effective. This responsibility is usually devolved to pharmacists who are empowered to discontinue those prescriptions for which durations have not been specified after an agreed period, usually from 48 to 72 hr. Those drugs which have been discontinued must be re-ordered if patients are to continue receiving them, although the policy can be overridden if the prescriber stipulates a duration. An educational programme should always precede the introduction of a stop-order policy and, ideally, prescribers should be notified 24 hr before the stop-order is implemented, thereby preventing lapses in continuous treatment. Stop-order policies have been shown to be effective means of ensuring that antibiotics are not inadvertently administered for excessive periods and have enabled the durations of treatment of patients with some types of infections to be standardised. Problems include the potential risks to patients associated with premature discontinuation of therapy (although this may be more theoretical than real), the need for human resources (pharmacists) to enforce the policy, reluctance by pharmacists to act as 'policemen' and antagonism from prescribers whose prescriptions have been discontinued. In addition, 48–72 hr is widely perceived to be an excessive duration for surgical prophylaxis. Many of these difficulties can be overcome by computerised prescribing which obviates the need for pharmacists to enforce the policy.

A variation of the stop-order policy is the pre-printed antibiotic prescription/order form on which prescribers must specify the indication for the antibiotic(s) they want a patient to receive, that is, as prophylaxis or empirical or definitive therapy. If an antibiotic has been designated as prophylaxis, administration is discontinued after 24 hr; alternatively, the pharmacy dispenses only enough doses to cover a 24-hr period. If the prescriber indicates that the antibiotic is to be given as empirical therapy, the prescription is discontinued after 48–72 hr. This stimulates the prescriber to reassess the need for antibiotic therapy and the appropriateness of the antibiotic regimen in the light of additional clinical information, as well as the results of laboratory and radiological investigations which should be available by that time, rather than simply continuing with a regimen that is ineffective or inappropriate. Finally, if a prescription is designated as definitive therapy, it will be administered, commonly for between 5 and 7 days. The policy is enforced by pharmacy staff or implemented through a computer-assisted antibiotic order entry programme. All of the above can be overridden if the prescriber specifies a duration. Again, the introduction of an antibiotic order form should be preceded by an educational programme and prescribers should be given 24 hr notice if a prescription is to be discontinued. The disadvantages of this type of intervention include the need for adequate human resources and the potential for antagonism between prescribers and pharmacists.

3.2.5.   Guidelines for antibiotic prescribing

In general, the implementation of clinical guidelines can lead to improvements in clinical practice by reducing variations in the methods and standards of care, improving the appropriateness and quality of care, reducing the cost of care, improving the cost-effectiveness of care, serving as educational tools, and promoting evidence-based decision making. Specifically in relation to antibiotic usage, clinical guidelines have been shown to promote more prudent use of these drugs and to reduce expenditure on them; they may also lead to reductions in the incidences of antibiotic-resistant pathogens.

Guidelines for antibiotic prescribing are becoming increasingly popular as a means of influencing clinicians' practice. In a survey of consultant microbiologists and hospital pharmacists in the United Kingdom (Working Party of the British Society for Antimicrobial Chemotherapy, 1994), 62% of respondents indicated that antibiotic guidelines were available in their hospitals. More recently, a 1998 survey of hospitals in the United States participating in Project ICARE (Intensive Care Antimicrobial Resistance Epidemiology) revealed that 70% of these institutions had introduced clinical guidelines for antibiotic usage (Lawton *et al.*, 2000). The success of the guidelines will depend on many factors but, most importantly, the rigour and commitment used in developing, disseminating, implementing, and evaluating them.

3.2.5.1.   Guideline development
There have been concerns about the quality of many of the clinical guidelines produced by specialty societies on the grounds that they do not conform to the basic principles of guideline development (Grilli *et al.*, 2000); the implementation of inappropriate recommendations may compromise patient care. A detailed description of the methodology by which guidelines are developed is outside the remit of this chapter. Readers, and particularly those contemplating the development of guidelines, should refer to the websites of guideline development groups, such as the Appraisal of Guidelines Research and Evaluation in Europe (AGREE) collaboration (http://www.agreecollaboration.org) or the Scottish Intercollegiate Guideline Network (SIGN) (http:// www.sign.ac.uk), as well as reviews by Brown (2002), Finch and Low (2002), Kish (2001), Natsch and van der Meer (2003), Peetermans and Ramaekers (2002), and Thomson *et al.* (1995). However, the essential features of guideline development can be summarised as follows:

1. There should be guidelines for prophylaxis and both empirical and definitive therapy.
2. The group developing the guidelines should be multidisciplinary and there should be a sufficient number of members (6–10) with expertise and experience in the subject of the guidelines in order to allow it to be adequately explored and to ensure that the guidelines are credible. The group should comprise at least one individual with the skills necessary to conduct literature and systematic reviews. There should be input from all stakeholders, including trainees, who are the prescribers who are most likely to use them, in order to ensure that there is 'ownership' of the guidelines, thereby increasing the likelihood that they will be implemented.
3. The development group should determine whether or not evidence-based guidelines on the same topic already exist. If so, they can be adopted as they are or adapted to suit local circumstances.

4. The guidelines must be based on a systematic review of the scientific evidence. In order to minimise the risk of bias, the literature should be identified according to an explicit search strategy, selected according to defined inclusion criteria, and assessed against consistent methodological standards. The method by which the literature is obtained, along with the search terms and the period of the search, should be specified.

5. As scientifically robust evidence is not always available it is likely that many guidelines will be hybrids of varying degrees of evidence and expert opinion. To ensure transparency of the recommendations that comprise the guidelines, the recommendations should be graded according to the strength of the evidence supporting them. The grading system should be validated, with the grading based on an objective measurement of the study design and quality and of the consistency, clinical relevance, and external validity of the evidence.

6. The guidelines should not be excessively long, that is, no more than 20–25 pages.

7. They should be simple, clear, non-controversial, clinically relevant, flexible, applicable to day-to-day practice, and available in a user-friendly format.

8. The antibiotics recommended in the guidelines should take account of the pathogens encountered locally and their susceptibility patterns.

9. As well as providing recommendations for optimal selection, the guidelines should include information regarding dosage, route of administration, duration, alternatives for patients who are allergic to first-line agents, and adjustments of dosages for patients with impaired renal function.

10. For prophylactic use, the guidelines should specify the procedures for which antibiotics are needed (or not needed) and the optimal agents, their dosages, and the timing, route, and duration of administration.

11. The guideline development group should identify evidence that is lacking and areas for further research.

12. The development group should identify sample outcome measures that would form the basis for auditing both the process and outcome of the guidelines.

13. The guidelines should be reviewed by respected peers who are not members of the guideline panel, but who are experts in the relevant field.

14. Guidelines are not static. They should be reviewed at periodic intervals that should be specified (e.g., 2-yearly) and updated to take account of advances in medical knowledge, changes in clinical practice and local circumstances, and the outcome of guideline evaluations. Any modifications of the guidelines must be the result of the same rigour and commitment as the original recommendations.

Guidelines can be developed nationally or locally. Those developed locally by the clinicians who will use them are less likely to be scientifically valid than those developed nationally by Royal Colleges and working parties of specialty societies because local groups lack the clinical, managerial, and technical skills, as well as the time and financial resources, needed for the task. Moreover, expertise at the local level is unlikely to be sufficiently broad, and personal opinions may introduce bias into the decision-making process. Locally produced guidelines must be no less robust than those produced nationally if patients are to receive optimal care. On the other hand, prescribers may disagree with or distrust guidelines written by remote national 'experts.' Guidelines are more likely to be implemented if users have participated in their development. Consequently, fewer resources are needed for effective dissemination and to promote implementation, compared with national guidelines for which greater emphasis must be placed on these phases of the process. A reasonable compromise would be to adapt national evidence-based guidelines (where such guidelines exist) for local use, a strategy that may be adequate to ensure prescribers' compliance.

3.2.5.2.   Guideline dissemination

One reason why guidelines are ineffective is that the target prescribers are often unaware of their existence. Dissemination then is the process of bringing guidelines to the attention of their intended users with the aim of increasing awareness and influencing knowledge, attitudes, and behaviour.

Dissemination can be achieved in a variety of ways: publication in journals, newsletters, local reports or documents, junior doctors' handbooks, configuration into a brief and portable format that is readily accessible to clinicians, posters on wards and in relevant departments, patient literature, group educational programmes, and personal visits. The optimal strategy has not been determined. Publication in medical journals, especially general medical journals, has, to date, been the most commonly used method, but is regarded as a poor means of disseminating guidelines and has a low likelihood of implementation. Direct mailing to relevant practitioners is seen as a more effective measure, but it is still of limited efficacy, although the impact of this intervention can be enhanced by making the guidelines visually attractive and/or by staging their delivery in manageable 'chunks' of information. In general, however, passively delivered interventions, such as written communications, have minimal abilities to achieve even temporary changes in behaviour. Grimshaw and Russell (1994) have claimed that the more overtly educational the dissemination strategy, the greater the likelihood that the guidelines will be adopted and the more lasting their impact, provided that dissemination is linked to an effective implementation strategy.

3.2.5.3.   Guideline implementation

Simply developing and disseminating guidelines, irrespective of how well they are done, is of limited value in terms of affecting improvements in healthcare unless the guidelines are implemented. Implementation is the process of ensuring that guidelines are introduced into clinical practice. Regrettably, the resources dedicated to developing guidelines have not been matched by those to promote compliance with them and, consequently, there is strong evidence that guidelines are often not adopted. Surveys have shown that compliance can vary from 20% to >90%, depending on the nature of the guideline, the specific clinical problem it is designed to address, the patient group being targeted, the mode of implementation and the definition of adherence. The most experienced practitioners may be the least likely to comply with guidelines. Cabana *et al.* (1999) identified three domains of barriers to implementation which related to: knowledge (lack of awareness or familiarity with the guidelines); attitudes (lack of agreement with the guidelines, lack of trust in the guidelines, i.e., low outcome expectancy, lack of self-confidence, i.e., self-efficacy, or the inertia of previous practice); and behaviour (external barriers which may be guideline-, patient- or environment-related). Others have suggested the following additional explanations for practitioners' failure to adhere to guidelines:

1. Guidelines may not be written for practising clinicians, but merely represent a summary of the current state of knowledge, that is, they lack scientific validity.
2. Important stakeholders may not have been represented on the group that developed the guidelines.
3. Clinicians may choose to ignore guidelines for nonclinical reasons, such as financial incentives or fear of litigation.
4. Guidelines may lack applicability to individual patients.
5. Local opinion leaders may not have endorsed the guidelines.
6. There may be inefficiencies of the healthcare system.

Guidelines should facilitate changes in practice, but if the changes are to be sustained, measures designed to promote implementation of guidelines must also change clinicians' knowledge, attitudes, and beliefs. Active educational interventions, such as seminars that are devoted exclusively to the guidelines and where potential users are given the opportunity to discuss them, are more likely to be effective than didactic lectures or simply including the guidelines as part of an educational programme. However, education alone is insufficient to ensure compliance. Other interventions that have been shown in at least some studies to promote adoption of guidelines and to lead to improvements in practice behaviour and clinical outcome include the following:

1. Endorsement by local and national professional organisations.
2. Incorporation into routine practice by local opinion leaders.

3. Dissemination of guidelines by department heads.
4. Audit of compliance with guidelines, with feedback of results to clinicians.
5. Peer review.
6. Printed patient-specific reminders at the time of consultations to prompt clinicians to use guidelines, for example, by attaching the guidelines to clinical notes or by including them on desktop computers.
7. General reminders of guidelines.
8. Making guidelines available to prescribers when they are making clinical decisions. This process has been facilitated by computer-assisted decision support programmes such as that described by Pestotnik *et al.* (1996), although the efficacy of this intervention has not been validated independently and the effects on patient outcomes have not been adequately assessed.
9. Promoting 'ownership' of guidelines by involving potential users in their development; alternatively, local adaptation of national guidelines may be sufficient to convey a sense of ownership.
10. Incorporation of guidelines into service contracts between purchasers and providers.
11. Educational outreach visits ('academic detailing'), that is, pre-arranged face-to-face discussions between a detailer (a trained educator such as a pharmacist) and a practitioner at the latter's place of work with the aim of persuading the practitioner to change behaviour through information and evidence (Soumerai and Avorn, 1990). To date, this has been the most effective and most lasting method of promoting compliance and has the advantage of allowing those clinicians who most need to change their practices to be targeted. On the other hand, it is expensive and labour-intensive and concerns have been raised about whether or not it is effective outside the research setting.

Any one or a combination of interventions improves compliance with guidelines to varying degrees. However, because most studies of the efficacies of these interventions have involved multiple strategies, it has not been possible to discern the relative contribution of each one. For this reason, and because many of the studies suffered from methodological flaws and because there have been very few comparative studies, efforts to identify the most effective intervention(s) have been frustrated. In general, multiple measures have proved more effective than single interventions and a combination of strategies is, therefore, most likely to have the maximum impact on guideline implementation.

3.2.5.4.   Evaluation

Evaluation is the assessment of the efficacy of the guidelines, with the aim of ensuring that they have produced the intended changes in both practice and outcome. Audit is the most effective means of achieving this objective, but it is essential to evaluate all of the components of the guideline process, not simply outcome, as they are inextricably linked. In other words, improvements in clinical outcome will not be realised unless guidelines are received, read, and adopted.

3.2.6. Laboratory control and the role of the medical microbiologist/infectious diseases physician

The clinical microbiology laboratory and specialists in infectious diseases (clinical microbiologists and infectious diseases physicians) can make important contributions to a hospital's programme to optimise antibiotic prescribing. Laboratory control can be achieved in a variety of ways:

1. By promoting optimal usage of diagnostic services, ensuring that specimens are appropriate, clinically relevant, and timely. The submission of inappropriate specimens, in particular, those taken from sites that are not clinically infected and those obtained after antibiotic therapy has been initiated, should be discouraged as they may lead to inappropriate treatment.
2. By undertaking selective susceptibility testing, that is, including only those antibiotics which are listed in the hospital formulary.
3. By appending clinical interpretations to laboratory reports (e.g., casting doubts on the significance of laboratory isolates) when such comments are appropriate.
4. By not determining, or by withholding, the susceptibilities of clinical isolates when there is inadequate clinical information to enable an informed opinion about significance or when there are doubts about the significance of these isolates. Failure to do so will, in at least some cases, cause inexperienced prescribers to assume that the results have been interpreted by the laboratory as being clinically significant and to initiate antibiotic therapy inappropriately.
5. By selective reporting of antibiotic susceptibility test results, that is, reporting the susceptibility patterns of only a limited number of agents which are appropriate treatment of the patient from whom the specimen has been obtained; ideally, these drugs should be the least expensive and most narrow-spectrum available.
6. By undertaking rapid identification and susceptibility testing of clinical isolates. It has been demonstrated that rapid provision of the results of susceptibility testing is more likely than conventional testing to lead to timely changes to appropriate treatment and to have a demonstrable impact on the care and outcome of hospitalised patients with infections. However, this will mean that laboratories will be required to adopt aggressive reporting

strategies in order to bring the results to the attention of prescribers for appropriate action. Furthermore, rapid methods have reduced abilities to detect some types of inducible resistance, thereby leading to false reports of susceptibility.

7. By collecting local surveillance data and reporting trends and susceptibility patterns in order to guide optimal empirical therapy.

As well as playing a pivotal role in the development and implementation of the various interventions which comprise a hospital's ACP, clinical microbiologists and infectious diseases physicians provide timely advice to colleagues regarding diagnosis, the most appropriate specimens which should be submitted for microbiological investigations and optimal empirical and definitive therapy. Compared with nonspecialists in the management of patients with infectious diseases, they have been shown to distinguish more accurately between infected and noninfected patients, to prescribe appropriate empirical and definitive therapy more often and at an earlier stage and to be associated with higher survival and cure rates. They also prescribe fewer antibiotics overall and fewer broad-spectrum antibiotics specifically and are more likely to convert from intravenous (iv) to oral treatment and from broad- to narrow-spectrum agents when culture and susceptibility test results are available. Patients treated by such specialists experience shorter mean lengths of hospital stay, fewer relapses and readmissions, higher satisfaction scores, and shorter times to return to regular activities compared with patients under the care of nonspecialists. Yet, in a survey conducted by a Working Party of the British Society for Antimicrobial Chemotherapy (1994), only 75% of respondents, who were consultant medical microbiologists, indicated that they provided a clinical consultative service. Not every patient about whom an opinion is sought needs to be seen directly and much useful advice can be given over the telephone. However, it is only at the bedside that a patient's clinical status can be accurately assessed and it is at the bedside where the greatest influence over antibiotic prescribing can be exerted. Face-to-face contact between microbiologists/infectious diseases physicians and prescribers promotes confidence in the former, increases the likelihood of future consultations, and is educational.

3.2.7. Educational interventions

As stated previously, educational interventions have had only minimal effects on antibiotic prescribing and the impacts of those which have been shown to be effective were short-lived unless they were constantly reinforced. On the other hand, education complements the effects of other interventions, including those which are more restrictive or coercive, and must be regarded as the foundation of a hospital's efforts to optimise antibiotic prescribing; it is potentially the only means by which prescribers can be persuaded to accept ownership of the problem of antibiotic resistance. In order to have a sustained effect on prescribing behaviour it is necessary to change prescribers' underlying attitudes and beliefs. The interventions which have been shown in several studies to be the most effective in terms of changing practice are: audit and feedback, computer-assisted decision support, educational outreach visits, local opinion leaders, mass media interventions, and printed educational material. However, prescribers tend to revert to preintervention practices once the study has been terminated. Those who have had experience of trying to change prescribers' practices will empathise with the views of Sbarbaro (2001): "Changing physician behaviour is considered by many to be an exercise in futility—an unobtainable goal intended only to produce premature ageing in those seeking the change. The more optimistic might describe the process as uniquely challenging."

A long-term strategy which might be more effective than changing the behaviour of existing prescribers is to 'mould' the behaviour of future prescribers, that is, medical students. As well as understanding the need for prudent prescribing, medical students must be taught how to use the services of the diagnostic laboratory effectively, inappropriate investigations leading to inappropriate prescribing.

3.2.8.   The role of the hospital pharmacist

Although clinical pharmacists in the United States have for many years occupied high-profile roles and have been extremely effective in terms of controlling anti-biotic usage, this resource has not, to date, been adequately utilised in the United Kingdom, despite the obvious benefits of doing so. Indeed, the costs of employing one or more pharmacists to fulfill the role of antibiotic utilisation coordinator/infectious diseases pharmacist can be offset by savings on antibiotic expenditure. As well as being a member of the Antibiotic Control Committee and enforcing the interventions implemented by the Committee (such as formulary restrictions and automatic antibiotic stop-order policies), the pharmacist has an educational role (promoting good and cost-effective prescribing practices), monitors compliance with clinical guidelines and other interventions, monitors antibiotic consumption (to highlight inappropriate antibiotic usage), and undertakes audit initiatives (including evaluating the effects of clinical guidelines on outcome and antibiotic resistance patterns). Pharmacists should be provided with modern computer facilities in order to enable them to expedite these functions and should promote the introduction of electronic prescribing. For a more detailed discussion of the role of the pharmacist in antimicrobial management, the reader is referred to Chapter 13, this volume, by Knox *et al.*

3.3.   Other interventions

Several other strategies for optimising antibiotic prescribing in hospitals have been proposed and/or evaluated. Although there is a paucity of robust evidence in the literature to support the efficacies of most of these interventions, and some are controversial, at least a few may have benefits and may eventually find places in the antibiotic control programmes of some hospitals.

3.3.1.   'Streamlining'

'Streamlining' is the conversion of initial therapy, based on the results of culture and susceptibility testing and clinical response, from a broad- to a narrow-spectrum regimen, from combination therapy to monotherapy or from newer, expensive drugs to older, less-expensive drugs with equivalent efficacies. Too often, patients are left to complete initial courses of therapy because they are responding to them and because prescribers are reluctant to change the regimens. Although there is little evidence of the efficacy of streamlining, the collective experience in many centres suggests that it is feasible, effective, and safe. Its implementation has been associated with substantial cost savings, lower incidences of toxicity, and reduced selective pressures for resistance and it has been shown to have a marked educational impact on prescribers.

3.3.2.   Intravenous (iv)-oral switch therapy

The conversion from a parenteral to an oral antibiotic regimen, also known as sequential antibiotic therapy, is a form of streamlining. The oral alternative may simply be a different formulation of the same drug, a drug belonging to the same class of antibiotics or a drug belonging to a different class of antibiotics. Regardless, the most important criteria are that the oral agent has therapeutic efficacy that is comparable to that of the iv drug, that it is active against the cause of the infection and that it has good oral bioavailability. The conversion should be implemented in accordance with recognised criteria.

Treatment by the oral route has several advantages. Oral formulations are easier to administer and less expensive (in terms of both acquisition and administration costs) and are associated with lower incidences of complications (phlebitis and catheter-related bloodstream infections). Perhaps most importantly, they facilitate early discharge from hospital, thereby reducing the cost of care and the likelihood of patients being exposed to or transmitting antibiotic-resistant potential pathogens. The implementation of such a programme is not without its difficulties, its success depending upon the collaborative efforts of the parental clinical team, the hospital pharmacy, members of the microbiology department, and nursing staff.

3.3.3.   Combination therapy

The practice of using combination therapy is an extension of effective antituberculous and antihuman immunodeficiency virus therapy, that is, the administration of two or more antibiotics reduces the likelihood of the emergence of resistant strains. However, while some investigators outside of the setting of tuberculosis have demonstrated a trend towards less frequent emergence of resistance in patients given combinations of drugs, and others have reported higher clinical and bacteriological cure rates (the latter, in principle, helping to reduce transmission of antibiotic-resistant strains), most of those who have compared the efficacy of combination therapy with that of monotherapy (usually a β-lactam/aminoglycoside combination and a β-lactam alone respectively) have failed to show that the former is superior to the latter in terms of preventing the emergence of resistant strains. The combination approach also leads to considerable hidden costs and may be associated with drug interactions (antagonism) at the receptor sites, an increased frequency of superinfection (secondary to greater disruption of the normal flora) and a greater likelihood of adverse drug reactions. With the exception of antituberculous therapy, there is currently insufficient evidence to justify the routine use of combination treatment as a general means of minimising the emergence of antimicrobial resistance.

3.3.4.   Therapeutic substitution

Therapeutic substitution involves replacing a prescribed antibiotic with one having a different chemical structure, but belonging to the same therapeutic class and having comparable pharmacokinetic and pharmacodynamic properties and clinical efficacy. This intervention has been applied broadly to those therapeutic classes having little diversities among constituent drugs or large disparities in drug prices. Examples of antibiotics to which the strategy might apply are cephalosporins, aminoglycosides, and quinolones. A therapeutic substitution is initiated by a hospital's drug and therapeutic committee and is implemented within the context of the formulary system. The principal motivation is cost savings. The practice has been widely adopted throughout both the United States and the United Kingdom.

The challenges of therapeutic substitution include identifying appropriate therapeutic alternatives, obtaining prescriber approval before making a therapeutic substitution, adequately monitoring the effects of therapeutic substitution on patient outcome, dealing with toxic reactions and drug interactions, and identifying true savings after taking account of the costs of implementing and administering the intervention, adverse events, and drug administration.

3.3.5. 'Cycling' (rotation)

'Cycling' is the scheduled withdrawal of a class of antibiotics (or a specific member of a class) and substitution with a different class (or a specific member of that class). This may be followed after a specified interval by a third or a fourth substitution, but, in order to fulfil the definition, the initial regimen must be re-introduced at a later stage and the cycle repeated. The duration of each cycle is based on either local susceptibility patterns or a predetermined time period. Cycling has normally involved substitution of one class of antibiotics with another, as opposed to substitution with a member of the same class (which shares resistance mechanisms), although, in some studies, one aminoglycoside was replaced with another. Cycling is not the same as simply withdrawing one drug and replacing it with another. The rationale behind the intervention is that the more frequently an antibiotic is prescribed, the more likely resistance to it will develop. Withdrawal of an antibiotic for a proscribed period of time will limit the selective pressures exerted by that agent, thereby allowing rates of resistance to it to stabilise or decrease during the period of restriction and ensuring that its efficacy is intact when it is re-introduced at a later date in place of a substitute. Each cycle is timed to occur before the emergence of significant levels of resistance to the substitute drug. The objective, therefore, is to maintain the total mass of any drug below the critical level that leads to the emergence of resistance to it.

Notwithstanding the current popularity of cycling, data supporting its efficacy are limited. Most of the investigators who claimed to have evaluated cycling assessed withdrawal/substitution; the initial regimen was not re-introduced. Of the studies that actually investigated cycling, most did not fulfil the criteria for inclusion in a systematic review, the majority being uncontrolled before-and-after studies. Of the three which fulfilled these criteria, interpretation of the effects of cycling on resistance rates was undermined, owing to a lack of standardisation, the impact of confounding variables, in particular, infection control interventions, the failure to differentiate clinical isolates which were simply colonising patients from those causing infection and the administration of "off-cycle" drugs to as many as 50% of patients. Furthermore, each of the studies published to date involved only a single intensive care unit, thereby precluding efforts to make generalisations. Finally, the results of a study which used mathematical models suggest that cycling will always be inferior to "mixed" antibiotic use (the simultaneous prescribing of alternative drugs belonging to different classes) at the population level (Bonhoeffer *et al.*, 1997).

In conclusion, the efficacy of cycling, in terms of preventing or reversing the trend towards increasing antibiotic resistance, has not been demonstrated. Indeed, the investigators in four studies described the rapid re-emergence of strains resistant to the initial antibiotic when it was re-instated. There remains a need for large, well-designed, CCTs employing high-quality epidemiological tools, sophisticated resistance mechanism and molecular typing analyses, and effective and consistent infection control interventions. A great many issues relating to cycling need to be resolved before undertaking such trials, let alone implementing this intervention on a routine basis.

3.3.6. Computer-assisted decision support

Computer-assisted decision support provides prescribers with information relevant to individual patients at the bedside when the decision to administer antibiotics is made, this being the most critical period in terms of influencing the choice of treatment. Recommendations on prophylaxis and empirical and definitive therapy are based on patient data, local susceptibility patterns, local practice guidelines and costs of formulary drugs, all of which must be programmed into the hospital information system. As well as advising on the choice of antibiotics, the system recommends dosages and durations and alerts prescribers to incorrect dosages, routes of administration and intervals between doses, resistant pathogens, cost-effective alternatives, drug incompatibilities, the need to monitor serum drug concentrations, etc. In its most highly developed form, this computer-driven aid has been shown to lead to reduced antibiotic usage and expenditure, increased appropriateness of antibiotic prescribing, improved clinical outcome, and reduced incidences of adverse drug reactions, without leading to increased incidences of resistance (Evans et al., 1998; Pestotnik et al., 1996). However, its efficacy has not been confirmed by well-designed clinical trials and the benefits in terms of patient outcome have not yet been adequately assessed. Moreover, there is a requirement for highly sophisticated information technology systems which are not widely available in hospitals in the United Kingdom, although they are currently under development in some centres.

3.4.   Outcome measures

Measuring the impact of the interventions implemented in a hospital in order to optimise antibiotic prescribing is fraught with problems. For example, simply determining the incidences of antibiotic-resistant organisms before and after the introduction of an intervention does not allow the relative contribution of the intervention to be distinguished from that of infection control measures. However, the following parameters might be used as a basis for assessing the efficacies of the strategies which have been introduced.

1. Auditing compliance with the intervention.
2. Monitoring changes in total drug usage, expressed in terms of Defined Daily Doses (DDDs), before and after implementation and annually.

3. Monitoring changes in the usage of targeted drugs (in DDDs) before and after implementation and annually.
4. Monitoring changes in the mean durations of antibiotic prescriptions.
5. Monitoring changes in the mean durations of hospital stay.
6. Monitoring changes in the appropriateness of prescriptions.
7. Monitoring changes in the antibiotic susceptibilities of target organisms before and after implementation.

4. CONCLUSIONS

1. Most, if not all, of the recommendations which will be made by the Working Party, once the systematic review of the literature has been completed, are likely to be based on a consensus of expert opinion, rather than robust published evidence.
2. A multifaceted approach, that is, one involving a combination of interventions, will be needed to achieve a maximum impact on prescribing behaviour.
3. It may not be feasible or practicable to implement all of the interventions described above (indeed, it may not even be possible to implement all of the core interventions) in each hospital and it will, therefore, be necessary to develop a programme of interventions that suits local circumstances and needs.
4. The interventions that are implemented will need to be enforced and their efficacies monitored through the audit process.
5. Hospital management will need to demonstrate support for the programme developed by the Antibiotic Control Committee by making available to it adequate resources to enable the interventions introduced to control antibiotic usage to be implemented and enforced.

REFERENCES

Bamberger, D. M. and Dahl, S. L., 1992, Impact of voluntary vs enforced compliance of third-generation cephalosporin use in a teaching hospital. *Arch. Intern. Med.*, **152**, 554–557.

Belliveau, P. P., Rothman, A. L., and Maday, C. E., 1996, Limiting vancomycin use to combat vancomycin-resistant *Enterococcus faecium*. *Am. J. Health Syst. Pharm.*, **53**, 1570–1575.

Bonhoeffer, S., Lipsitch, M., and Levin, B. R., 1997, Evaluating treatment protocols to prevent antibiotic resistance. *Proc. Natl. Acad. Sci. USA*, **94**, 12106–12111.

Brown, E.M., 2002, Guidelines for antibiotic usage in hospitals. *J. Antimicrob. Chemother.*, **49**, 587–592.

Cabana, M. D., Rand, C. S., Powe, N. R., Wu, A. W., Wilson, M. H., Abboud, P.-A. C. *et al.*, 1999, Why don't physicians follow clinical practice guidelines? A framework for improvement. *JAMA*, **282**, 1458–1465.

Evans, R. S., Pestotnik, S., Classen, D. C., Clemmer, T. P., Weaver, L. K., and Orme, J. F., Jr., 1998, A computer-assisted management program for antibiotics and other antiinfective agents. *N. Engl. J. Med.*, **338**, 232–238.

Finch, R. G. and Low, D. E., 2002, A critical assessment of published guidelines and other decision-support systems for the antibiotic treatment of community-acquired respiratory tract infections. *Clin. Microbiol. Infect.*, **8**(Suppl. 2), 69–92.

Government Response to the House of Lords Select Committee on Science and Technology Report, 1988, *Resistance to Antibiotics and Other Antimicrobial Agents*, CM 4172. The Stationery Office, London.

Grilli, R., Magrini, N., Penna, A., Mura, G., and Liberati, A., 2000, Practice guidelines developed by specialty societies: The need for a critical appraisal. *Lancet*, **355**, 103–106.

Grimshaw, J. M. and Russell, I. T., 1994, Achieving health gain through clinical guidelines. II: Ensuring guidelines change medical practice. *Qual. Health Care*, **3**, 45–52.

Himmelberg, C. J., Pleasants, R. A., Weber, D. J., Kessler, J. M., Samsa, G. P., Spivey, J. M. *et al.*, 1991, Use of antimicrobial drugs in adults before and after removal of a restriction policy. *Am. J. Hosp. Pharm.*, **48**, 1220–1227.

House of Lords Select Committee on Science and Technology, 1998, *Resistance to Antibiotics and Other Antimicrobial Agents*. The Stationery Office, London.

Kish, M. A., 2001, Guide to development of practice guidelines. *Clin. Infect. Dis.*, **32**, 851–854.

Lawton, R. M., Fridkin, S. K., Gaynes, R. P., McGowan, J. E., and the Intensive Care Antimicrobial Resistance Epidemiology (ICARE) Hospitals, 2000, Practices to improve antimicrobial use at 47 US hospitals: The status of the 1997 SHEA/IDSA position paper recommendations. *Infect. Cont. Hosp. Epidemiol.*, **21**, 256–259.

Lopez-Lozano, J.-M., Monnet, D. L., Yague, A., Burgos, A., Gonzalo, N., Campillos, P. *et al.*, 2000, Modelling and forecasting antimicrobial resistance and its dynamic relationship to antimicrobial use: A time series analysis. *Int. J. Antimicrob. Agents*, **14**, 21–31.

Natsch, S. and van der Meer, J. W. M., 2003, The role of clinical guidelines, policies and stewardship. *J. Hosp. Infect.*, **53**, 172–176.

NHS Executive, 1999, *Resistance to Antibiotics and Other Antimicrobial Agents*. HSC 1999/049, NHS Executive, Leeds, UK.

Peetermans, W. E. and Ramaekers, D., 2002, Clinical practice guidelines in infectious diseases. *Neth. J. Med.*, **60**, 343–348.

Pestotnik, S. L., Classen, D. C., Evans, R. S., and Burke, J. P., 1996, Implementing antibiotic practice guidelines through computer-assisted decision support: Clinical and financial outcomes. *Ann. Intern. Med.*, **124**, 884–890.

Report from the Invitational EU Conference on The Microbial Threat, 1998, *The Copenhagen Recommendations*. Ministry of Health, Ministry of Food, Agriculture and Fisheries, Copenhagen, Denmark.

Sbarbaro, J. A., 2001, Can we influence prescribing patterns? *Clin. Infect. Dis.*, **33**(Suppl. 3), S240–S244.

Shlaes, D. M., Gerding, D. N., John, J. F., Jr., Craig, W. A., Bornstein, D. L., Duncan, R. A. *et al.*, 1997, Society for Healthcare Epidemiology of America and Infectious Diseases Society of America Joint Committee on the Prevention of Antimicrobial Resistance: Guidelines for the prevention of antimicrobial resistance in hospitals. *Clin. Infect. Dis.*, **25**, 584–599.

Soumerai, S. B. and Avorn, J., 1990, Principles of educational outreach ('academic detailing') to improve clinical decision making. *JAMA*, **263**, 549–556.

Sub-Group on Antimicrobial Resistance of the Standing Medical Advisory Group, 1998, *The Path of Least Resistance*. Department of Health, London.

Thomson, R., Lavender, M., and Madhok, R., 1995, How to ensure that guidelines are effective. *BMJ*, **311**, 237–242.

Williams, R. R., Gross, P. A., and Levine, J. F., 1985, Cost containment of second-generation cephalosporins by prospective monitoring at a community teaching hospital. *Arch. Intern. Med.*, **145**, 1978–1981.

Working Party of the British Society for Antimicrobial Chemotherapy, 1994, Hospital antibiotic control measures in the UK. *J. Antimicrob. Chemother.*, **34**, 21–42.

Chapter 11

Improving Prescribing in Surgical Prophylaxis

Jos W. M. van der Meer and Marjo van Kasteren
Department General Internal Medicine, UMC St Radboud, PO Box 9101, 6500HB Nijmegen, The Netherlands

1. INTRODUCTION

Surgical wound infections have occurred ever since the early days of surgery. Although the concept of asepsis and antisepsis, as proposed by Lister was a major step forward, surgical wound infections have remained very common. With the advent of antibiotics it was felt that these infections would be preventable. However, it took decades until optimal usage of antibiotics for surgical prophylaxis became a more or less settled problem.

In fact, one could state that the use of antibiotics for the prevention of surgical wound infection has become the area in antibiotic use nowadays that is least controversial. This chapter on antimicrobial prophylaxis in surgery is divided into two parts. The first part reviews the state of the art of antibiotic prophylaxis in surgery, the second deals with auditing and improving antibiotic prophylaxis in practice.

2. STATE OF THE ART OF ANTIBIOTIC PROPHYLAXIS IN SURGERY

2.1. Historical background

Based on a set of seminal studies, the background of the current policies of antimicrobial prophylaxis in surgery has been settled. The first studies to be

Antibiotic Policies: Theory and Practice. Edited by Gould and van der Meer
Kluwer Academic / Plenum Publishers, New York, 2005

mentioned within this context are—without doubt—the animal experiments by Burke. In a series of elegant investigations, he investigated the timing of antibiotic prophylaxis (Burke, 1961). He produced surgical wounds in rats which he contaminated with *Staphylococcus aureus*. At various points in time he administered antibiotics and discovered that wound infection could only be prevented if the antibiotics were given close to the time of inoculation.

A second landmark study was that of Weinstein and collaborators (Weinstein *et al.*, 1975). In rats, these investigators showed that prophylaxis for abdominal surgery should be directed to both anaerobic bacteria and aerobic bacteria. When rats were intra-abdominally inoculated with intestinal flora, it was found that animals treated with antibiotics that were only effective against anaerobic bacteria (i.e., clindamycin), developed potentially lethal sepsis caused by aerobes; those animals treated with antibiotics that were selectively effective against aerobes, in contrast, would not develop sepsis but suffer from intra-abdominal abscesses (apparently mainly caused by anaerobes). Treatment with combined antibiotics prevented both types of problems. The importance of coverage of anaerobes and aerobes in clinical surgery was proposed by Nichols, who investigated regimens of oral antibiotic for prevention of wound infection after colonic surgery (Nichols *et al.*, 1972). Thus, from these key studies the concept emerged that prophylaxis after abdominal surgery has to cover both anaerobes and aerobes.

The third group of landmark studies was that of Cruse and Foord. These investigators scrutinised the occurrence of surgical wound infections and determined the standards for the "acceptable" frequencies of wound infections after dirty surgery, contaminated surgery, so called clean-contaminated surgery, and clean surgery (Cruse and Foord, 1980). These investigations have set the stage for investigations looking into the effect of antibiotic prophylaxis in contaminated and clean-contaminated surgical procedures. The surgical classification, given by Mayhall (1993) is presented here (Table 1). The percentages of post-operative wound infection without antibiotics range between 2% and 5% for clean wounds, around 10% for clean-contaminated wounds, and >20% for contaminated wounds.

2.2.   Principles of antibiotic prophylaxis in surgery

Taken together the studies discussed in the previous paragraphs have laid the foundations for the following principles in surgical prophylaxis:

- Antibiotic prophylaxis in surgery is mainly indicated for (clean-)contaminated procedures
- The antibiotics to be selected for surgical prophylaxis should cover the microorganisms that predictably cause wound infection

Table 1. Surgical wound classification

Class	Description of the wound
Clean	Elective, primary closure, without drains Non-traumatic, not infected, not inflamed Adequate asepsis Airways, GI tract, or genitourinary system not opened
Clean-contaminated	Airways, GI tract, or genitourinary system opened under controlled conditions or without extraordinary contamination Oropharynx opened Vagina opened Urinary tract opened without positive urine culture Biliary tract opened without infected bile
Contaminated	Open, traumatic wound (not older than 6 hr) Visible leakage from the GI tract GI tract or genitourinary system opened with infected bile or urine Break in aseptic instrumentation Incision in area of acute, non-purulent inflammation
Dirty	Surgery through traumatic wounds with necrotic tissue, foreign material or (faecal) contamination Traumatic wound with delayed treatment Perforated viscus detected Acute bacterial infection with pus detected at operation

Source: Adapted from Mayhall (1993).

- The antibiotics selected for prophylaxis should be present in adequate concentrations in the wound at the time that bacterial contamination occurs.

  Commonly five more principles are added:

- The prophylactic antibiotics should be given for a short duration
- The antibiotics selected for this indication should not be used therapeutically
- The antibiotics selected should not readily lead to emergence of microbial resistance
- The antibiotics used for surgical prophylaxis should be free of side effects
- The antibiotics used for surgical prophylaxis should be relatively cheap.

Each of these principles will be discussed in a little more detail.

Antibiotic prophylaxis mainly in (clean-)contaminated procedures. It is generally accepted that the main classes within the wound classification (Table 1) for which antibiotic prophylaxis is indicated are clean-contaminated and contaminated wounds. However, there are clean surgical procedures in which a wound infection is such a disaster, that antibiotic prophylaxis is considered to be indicated. These are especially those procedures in which prosthetic material is implanted. Some surgical procedures with clean wounds meet with frequencies of wound infection in the range of 8–20% (e.g., craniotomy and coronary bypass surgery). For these procedures antibiotic prophylaxis is considered acceptable.

Areas of uncertainty are abdominal hysterectomy and pulmonary surgery. For both procedures the clinical trials are equivocal (Boldt *et al.*, 1999; Hemsell, 1991). Guidelines for prevention of infections in surgery have been published by several organsations, for example, the Centers for Disease Control in the United States (Mangram *et al.*, 1999). In Table 2, the Dutch

Table 2. Surgical procedures for which perioperative antibiotic prophylaxis is indicated

Clean wounds	
ENT surgery	Stapedectomy
	Cochlear implant
	Implant/bone transplant nose
Neurosurgery	Craniotomy
Vascular surgery	Application of prosthetic material
	Aorta reconstruction
	Vascular surgery with inguinal incision
Cardiac surgery	Open-heart surgery, coronary bypass surgery, implantation of prosthetic valve
Bone and joint surgery	Implantation of prosthetic joints
	Osteosynthesis
	Amputation in ischaemic area
Clean-contaminated/contaminated	
Head and neck surgery	Incision of pharynx/oesophagus
Neurosurgery	Surgery through naso-/oropharynx
Thoracic surgery	Lobectomy en pulmonary resection
Abdominal surgery	Stomach and duodenum surgery with hpochorhydria, disturbed motility or extreme overweight
	Biliary surgery for acute cholecystitis, stone in the common duct, age >70 years, or obstructive jaundice
	Colonic or rectal surgery
	Appendectomy for appendicitis
Urogenital surgery	Surgery of the urinary tract with non-sterile urine
	Vaginal/abdominal hysterectomy
	Secondary Caesarean section
	Manual removal of the placenta
	Abortion in second trimester or in first trimester after PID
	Vulvectomy
Trauma	Open fracture
	Penetrating abdominal or thoracic trauma

Source: Adapted from van Kasteren *et al.* (2000).

consensus (issued by the Working Party on Antibiotic Policy, SWAB) regarding antibiotic prophylaxis in surgery is reproduced (van Kasteren *et al.*, 2000).

Prophylaxis to cover microorganisms that predictably cause wound infection. Surgical wound infections are caused by endogenous and exogenous microflora. The endogenous microorganisms, the patient's own flora, are dependent on the body site where the surgical procedure will take place. The antimicrobial susceptibility of the endogenous flora is dependent on the patient's history (has the patient been exposed to resistant microorganisms, for example, methicillin-resistant *S. aureus* [MRSA], has the patient been using antibiotics recently?). This implies that antibiotic prophylaxis should be individualised, especially when there is a risk of resistant endogenous flora.

Exogenous microflora (transferred by medical personnel or by inanimate material) is generally not taken into account for the selection of the antimicrobial drugs for prophylaxis: infections by these microorganisms should be prevented by hygienic measures. The choice of antibiotics should be aimed at optimal prophylaxis, not maximal.

Adequate antibiotic concentrations at the time of bacterial contamination. Since bacterial contamination of a surgical wound almost exclusively occurs between incision and closure, effective antibiotic concentrations should be present in the wound during that period. The literature on surgical prophylaxis has concentrated more on the presence of effective concentrations at the beginning of surgery than at time of closure (see next paragraph).

There has been quite some debate regarding the optimal time of giving the first antibiotic dose. Studies have provided evidence that administration within 2 hr before incision is effective (Classen *et al.*, 1992). Given the rapidity of diffusion of antibiotics into tissues with adequate blood supply, it is probably better to shorten this time period to approximately 30 min.

The prophylactic antibiotics should be given for a short duration. There is general agreement that antibiotic prophylaxis in surgery should not be given for longer than 24 hr. A number of studies have compared administration of a single dose versus 24 hr of administration and met with equal efficacy (Rowe-Jones *et al.*, 1990; Wymenga *et al.*, 1992). Despite the still insufficient power of these (large) studies, most authorities feel that a single dose of an antibiotic will suffice for surgical procedures that do not exceed 3–4 half-lives of the drug, provided there is no substantial blood loss and no use of extra-corporal circulation. Under the latter circumstances an extra dose of antibiotic shortly before the end of the operation is indicated.

Antibiotics selected for this indication should not be used therapeutically. This principle is of a somewhat lesser order than the others. It is mainly put forward for practical reasons. If an antibiotic is only used for prophylaxis, its use and compliance with guidelines is relatively easily monitored. In addition, since many antibiotics are effective in prophylaxis—provided they cover the microorganisms that are likely to cause the infection—it is unnecessary to use new antibiotics. A drug like cefazolin is very suitable for the coverage of aerobic pathogens: its spectrum is sufficient, it has a relatively long half-life (90 min) and it is safe. It should be noted that this principle does not hold up as easily for anaerobic coverage. Here, metronidazole is often used and this drug cannot be avoided for therapy of anaerobic infections.

Antibiotics selected should not readily lead to emergence of microbial resistance. This general principle not only pertains to the effect of the antibiotic on the microflora of the patient, but also on that of the hospital environment (i.e., the operating theatre). Antibiotics that may readily induce resistance, for instance, through one-step mutation (macrolides, fucidic acid) are better avoided. Also, antibiotics with a very broad spectrum (such as carbapenems) should not be used in this setting: by destroying most of the endogenous flora, they pave the way for colonisation and subsequent infection by resistant hospital bacteria and also by fungi.

Antibiotics used for surgical prophylaxis should be free of side effects. One would preferably select antibiotics with which allergic reactions are rare. For this reason, penicillins are less suitable. Cephalosporins that are clearly less allergenic in general, can be used safely (Kelkar and Li, 2001). Although toxicity for most antibiotics is not a great problem if only a single dose is given, it is still wise to avoid aminoglycosides here. Wrong estimations of renal clearance and lean body mass are easily made and volume depletion and deterioration of renal function can occur during operation, all of which may enhance toxicity when more than one dose is given.

Antibiotics used for surgical prophylaxis should be relatively cheap. Cost containment is another reason to give preference to old ("off-patent") drugs. Indirect costs also have to be taken into account and for this reason oral bowel preparations (containing for instance neomycin and erythromycin) have become less used.

3. AUDITING AND IMPROVING THE QUALITY OF ANTIBIOTIC PROPHYLAXIS IN SURGERY

3.1. Surveying the quality of antibiotic prophylaxis in practice

Assessing the quality of the actual practice of antibiotic prophylaxis in surgery is relatively easy. In fact, if one considers performing studies on quality of antimicrobial prescribing in hospital, it is a sensible decision to start with this particular area. Before embarking on an extensive investigation of the whole chain of events preceeding the actual administration of the antibiotics, it is wise to perform a pilot investigation. Such a pilot should start with collecting the relevant data of a small series of postoperative patients (e.g., ten patients that are currently admitted) on a surgical ward. The data needed are:

- Age, gender, bodyweight, and length
- Diagnosis/indication for surgery
- Kind of surgery
- Comorbidity (including recent use of antibiotics; known drug allergy)
- Liver tests and renal function (preoperative)
- Antibiotic(s) given for prophylaxis
- Dosis
- Time of administration
- Administered by whom
- Starting time of operation
- Duration of operation
- Blood loss
- Additional dosages of antibiotic.

Collection of these data should be performed in a casual way, so as not to induce a change in prescribing behaviour. When these data have been collected, the actual quality review can start. To this end, use of the algorithm of Gyssens *et al.* (1992) is recommended (Figure 1). In addition, the prevailing local hospital guideline on antibiotic prophylaxis should be available.

The following patient vignette may serve as an example on how to perform the review.

A 72 year old female, weight 83 kg, length 163 cm
Diagnosis: stenosis of sigmoid colon due to recurrent diverticulitis
Surgical procedure: resection of sigmoid colon
Comorbidity: non-insulin dependent type 2 diabetes; no antibiotic usage; no drug
allergy; normal liver tests, estimated creatinine clearance 70 ml/min
Antibiotic prophylaxis: ceftazidime; Dose: 1 g
Time of administration: 11:10 am
Administered iv by a nurse on the ward
Start of operation: 1:20 pm
Duration of operation: 2 hours and 45 min
Blood loss 400 ml.

3.2. No additional antibiotics

Review of the case history described above, with the help of the algorithm will yield the following results.

- The data are adequate. It concerns a procedure with a contaminated wound, so antibiotics are indicated.

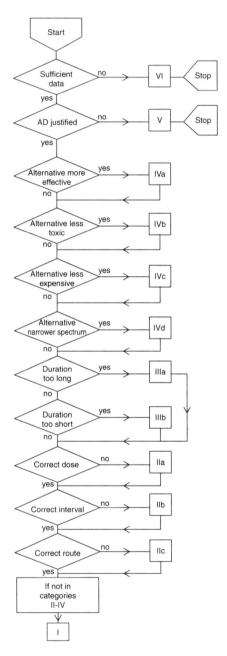

Figure 1. Flow chart for quality-of-use assessment of antimicrobial drug prescriptions according to Gyssens *et al.* (1992). Reprinted with permission.

- The patient only received ceftazidime and no anti-anaerobic coverage; thus there is an alternative which is more effective; we score **IVa**.
- Let us assume that the hospital guideline recommends cefazolin with metronidazole for this type of surgical procedure.
- Although metronidazole adds a little toxicity to cefazolin, toxicity is not an issue here.
- Ceftazidime is considerably more expensive than cefazolin: we score **IVc**.
- Also, the spectrum of ceftazidime is too wide: we score **IVd**.
- Since only one dose was given, the duration is correct.
- The dose is correct. Since there was only one dose given, the dosis interval is not applicable. The administration of the drug is correct.
- The patient received the prophylaxis on the ward. This is of course not ideal, since the surgical procedure may be delayed, as in this case. The time elapsed between the injection and the operation is more than 2 hr. We score **I**.
- This case already points to serious problems that need urgent attention.

First of all, the administration of a third generation cephalosporin does raise the question, "Is this structural or incidental? Is the mistake with the person who has been giving the order or with the nurse administering the drug (e.g., insufficient knowledge of different cephalosporins)?" More cases from the same ward are needed to see what is wrong.

3.3.   Further quality assessment

If the pilot survey of antibiotic prophylaxis in surgery appears to meet with poor quality, it is necessary to describe the sequence of events that has preceded the actual prescribing practice. Within that context, barriers that interfere with this chain of events and with the actual implementation of the antibiotic policy should be looked for.

The sequence starts with the formulation of the guideline (often by a multidisciplinary hospital committee that makes use of already existing national or international guidelines to formulate local guidelines). In respect of the acceptance of the guideline, it is important to check whether key players of the surgical disciplines were involved in the decision-making at this stage. An important next step is the issuing of the guideline, its acceptance by the professionals and its implementation in daily practice.

It is an important question whether barriers (defined as factors that limit or restrict complete physician adherence to a guideline [Cabana *et al.*, 1999]) are present. Such barriers may be physician related, patient related, or system related. Physician-related barriers have to do with knowledge, attitude, or behaviour. According to Cabana *et al.* (1999) lack of knowledge comes down to lack of awareness and lack of familiarity with the guideline. The amount of information, the time needed to become and stay informed, and the accessibility to the guidelines are critical issues here. Attitude has to do with lack of agreement with the guideline, lack of outcome expectancy, and perhaps worst of all, inertia. With regard to attitude there is the crucial influence of supervisors and local opinion leaders: their involvement and endorsement is essential. Behaviour, which is a consequence of knowledge and attitude is strongly influenced by external factors, for example, patient demands and factors in the system such as financial constraints and bureaucracy.

For the assessment of quality, each of these factors that may pose a barrier should be given critical attention.

3.4.   How to intervene?

Once the quality is assessed and the barriers that interfere with it have been identified, a plan can be made for intervention. As an example, the intervention studies as published by Gyssens *et al.* (1996, 1997) may serve. In that study, rather poor timing of the preoperative antibiotic dose was found. In the analysis a critical factor was that nobody really felt responsible for the administration of the antibiotic. This was due to lack of knowledge of the surgeons (not sufficiently aware of the importance of antibiotic timing [Classen *et al.*, 1992]) and a lack of involvement on the side of the anaesthesiologists and a lack of communication between surgeons and anaesthesiologists. The successful intervention consisted of the education of surgeons, anaesthesiologists, and nurses.

In a recent multicentre study of surgical antimicrobial prophylaxis in the Netherlands (van Kasteren *et al.*, 2003), it was found that the most important barriers to local guideline adherence were lack of awareness due to ineffective distribution of the most recent version of the guidelines, lack of agreement by surgeons with the local hospital guidelines, and environmental factors, such as organizational constraints in the surgical suite and in the ward. Especially, adherence to guidelines on dosing interval and timing needs improvement.

In conclusion, it is clear that meticulous analysis of all steps is necessary to design and install effective intervention. Most often a set of interventional measures is necessary. The effect of the intervention should be assessed. Such assessment should be performed relatively shortly after the intervention, but also later, to see whether the effect of the intervention lasts. After all, it is human to fall back to one's old routines and mistakes.

REFERENCES

Boldt, J., Piper, S., Uphus, D., Fussle, R., and Hempelmann, G., 1999, Preoperative microbiologic screening and antibiotic prophylaxis in pulmonary resection operations. *Ann. Thorac. Surg.*, **68**, 208–211.

Burke, J. F., 1961, The effective period of preventive antibiotic action in experimental incision and dermal lesions. *Surgery*, **50**, 161–168.

Cabana, M. D., Rand, C. S., Powe, N. R., Wu, A. W., Wilson, M. H., Abboud, P. A., Rubin, H. R. *et al.*, 1999, Why don't physicians follow clinical practice guidelines? A framework for improvement. *JAMA*, **282**, 1458–1465.

Classen, D. C., Evans, R. S., Pestotnik, S. L., Horn, S. D., Menlove, R. L., and Burke, J. P., 1992, The timing of prophylactic administration of antibiotics and the risk of surgical-wound infection. *N. Engl. J. Med.*, **326**, 281–286.

Cruse, P. J. and Foord, R., 1980, The epidemiology of wound infection. A 10-year prospective study of 62,939 wounds. *Surg. Clin. North Am.*, **60**, 27–40.

Gyssens, I. C., Geerligs, I. E., Nannini-Bergman, M. G., Knape, J. T., Hekster, Y. A., and van der Meer, J. W. M., 1996, Optimizing the timing of antimicrobial prophylaxis in surgery: An intervention study. *J. Antimicrob. Chemother.*, **38**, 301–308.

Gyssens, I. C., Knape, J. T., Van Hal, G., and van der Meer, J. W. M., 1997, The anesthetist as determinant factor of quality of surgical antimicrobial prophylaxis. A survey in a university hospital. *Pharm. World Sci.*, **19**, 89–92.

Gyssens, I. C., van den Broek, P. J., Kullberg, B. J., Hekster, Y., and van der Meer, J. W. M., 1992, Optimizing antimicrobial therapy. A method for antimicrobial drug use evaluation. *J. Antimicrob. Chemother.*, **30**, 724–727.

Hemsell, D. L., 1991, Prophylactic antibiotics in gynecologic and obstetric surgery. *Rev. Infect. Dis.*, **13**(Suppl. 10), S821–S841.

Kelkar, P. S. and Li, J. T. C., 2001, Cephalosporin Allergy. *N. Engl. J. Med.*, **345**, 804–809.

Mangram, A. J., Horan, T. C., Pearson, M. L., Silver, L. C., and Jarvis, W. R., 1999, The Hospital Infection Control Practices Advisory Committee. Guideline for prevention of surgical site infection 1999. *Infect. Control Hosp. Epidemiol.*, **20**, 247–280.

Mayhall, C. G., 1993, Surgical infections including burns. In R.P. Wenzel (ed.), *Prevention and Control of Nosocomial Infections.* Williams and Perkins, Baltimore, MD.

Nichols, R. L., Condon, R. E., Gorbach, S. T., and Nyhus, L. M., 1972, Efficacy of preoperative antimicrobial preparation of the bowel. *Ann. Surg.*, **176**, 227–232.

Rowe-Jones, D. C., Peel, A. L., Kingston, R. D., Shaw, J. F., Teasdale, C., and Cole, D. S., 1990, Single dose cefotaxime plus metronidazole versus three dose cefuroxime plus metronidazole as prophylaxis against wound infection in colorectal surgery: Multicentre prospective randomised study. *BMJ*, **300**, 18–22.

van Kasteren, M. E. E., Kullberg, B. J., de Boer, A. S., Mintjes-de Groot, J., and Gyssens, I. C., 2003, Adherence to local hospital guidelines for surgical antimicrobial prophylaxis: A multicentre audit in Dutch hospitals. *J. Antimicrob. Chemother.*, **51**, 1389–1396.

van Kasteren, M. E. E., Gyssens, I. C., Kullberg, B. J., Bruining, H. A., Stobberingh, E. E., and Goris, R. J. A., 2000, Optimalisering van het antibioticabeleid in Nederland. V. SWAB-richtlijnen voor preoperatieve antibiotische profylaxe. *Ned. Tijdschr. Geneesk.*, **144**, 2049–2055.

Weinstein, W. M., Onderdonk, A. B., Bartlett, J. G., Louie, T. J., and Gorbach, S. L., 1975, Antimicrobial therapy of experimental intraabdominal sepsis. *J. Infect. Dis.*, **132**, 282–286.

Wymenga, A., van Horn, J., Theeuwes, A., Muytjens, H., and Slooff, T., 1992, Cefuroxime for prevention of postoperative coxitis. One versus three doses tested in a randomized multicenter study of 2,651 arthroplasties. *Acta Orthop. Scand.*, **63**, 19–24.

Chapter 12

Audits for Monitoring the Quality of Antimicrobial Prescriptions

Inge C. Gyssens
Department of Internal Medicine, and Department of Medical Microbiology and Infectious Diseases, Erasmus University Medical Center, Dr. Molewaterplein 40, 3000 CA Rotterdam, The Netherlands

1.   INTRODUCTION

Antibiotic therapy differs from all other types of pharmacotherapy. It is based on the characteristics of not only the patient and the drug but also on the nature of the infection and the microorganism causing the infection. There is a complex relationship between the host, the pathogens, and the anti-infective agents. The rational use of antimicrobial drugs is based on an understanding of the many aspects of infectious diseases. Factors relating to host defence, the identity, virulence, and susceptibility of the microorganism and the pharmacokinetics and pharmacodynamics of antimicrobial drugs have to be considered. Antimicrobial use is the major determinant of microbial resistance. To guarantee the long-term efficacy of antimicrobial drugs, the quality-of-use should be maximised and overconsumption (inappropriate use) eliminated. There are major differences in antimicrobial consumption in different parts of the world (Cars *et al.*, 2001). However, much less is known about the quality of antimicrobial use. An optimal treatment for an infection is obtained when a maximum efficacy is combined with a minimal toxicity for the host, at a reasonable cost and with a minimal development of microbial resistance. In healthcare facilities, antimicrobial drugs are used in three types of situations (Table 1).

The quality of empiric therapy and antimicrobial prophylaxis is largely determined by the availability of local surveillance data on microbial resistance

Antibiotic Policies: Theory and Practice. Edited by Gould and van der Meer
Kluwer Academic / Plenum Publishers, New York, 2005

Table 1. Classification of the different types of antimicrobial therapy and definitions

Empiric therapy: Administration of antibiotics to treat an active infection in a blind approach before the causative microorganism has been identified and its antibiotic susceptibility determined
Definitive therapy: Administration of antibiotics targeted at a specific microorganism causing an active or latent infection
Prophylaxis: Administration of antibiotics to prevent a possible infection (which is not yet present or incubating)

and by the information that prescribers have on the local epidemiology of infections and the causative organisms. The microbiology laboratory plays a major role in the aggregation, analysis, and reporting of surveillance data and provides a major contribution towards the choice of empiric therapy ("well-educated guess") or prophylaxis. Guidelines for empiric therapy and prophylaxis that are based on this surveillance should be available in every healthcare facility. The accessibility of microbiology laboratory facilities is crucial for the identification of a pathogen and determination of its susceptibility to facilitate and streamline a definitive therapy with a spectrum of action that is less broad than the blindly chosen empiric therapy. When the patient is in a stable condition, sequential therapy or step down therapy from parenteral to oral administration is preferable and allows for outpatient therapy (Eron and Passos, 2001). Antibiotic therapy should be streamlined at the earliest opportunity. Recent studies have shown that the duration of antimicrobial therapy of some infections can be shortened. This chapter reviews the different methods of evaluation of quality of use at the patient level. It cites the evidence supporting the principles of prudent prescribing of antimicrobial (antibacterial and antifungal) drugs.

2.   OUTCOME PARAMETERS

Outcome measures of audits can be categorised in process outcome, patient outcome, and microbiological outcome parameters.

2.1.   Process outcome: prescribing behaviour

2.1.1.   Definition

Traditionally, quality is measured by an in-depth analysis of medical records, also called audits of practice. An audit of antimicrobial drug use is defined as the analysis of appropriateness of individual prescriptions (Gould *et al.*, 1994). Although this approach is costly in manpower, an audit is certainly the most complete method to judge all aspects of therapy. Moreover, the evaluation process (see below) can be used as an educational tool (Gyssens *et al.*, 1992). The feedback of the results of an audit can be part of an intervention to improve prescribing (Gyssens *et al.*, 1996a, b, 1997a, b).

2.1.2. Criteria

To evaluate the quality of prescribing of antimicrobial drugs by audits, the criteria developed by Kunin *et al.* (Kunin, 1973) have traditionally been used. In the past, the classification was mainly based on the authority of infectious diseases specialists who performed the evaluation. Usage was categorised as appropriate, probably appropriate, inappropriate due to less costly alternatives, dose adaptation needed, or totally inappropriate. Because the formulation of the criteria was rather aspecific, these original criteria have been modified by several authors. They have adapted or extended these criteria in order to be able to judge specific parameters, for example, dose (Byl *et al.*, 1999; Dunagan *et al.*, 1989; Evans *et al.*, 1998), dose interval (Volger *et al.*, 1988), way of administration (Byl *et al.*, 1999; Maki and Schuna, 1978), obtaining the necessary serum concentrations for monitoring (Dunagan *et al.*, 1989; Maki and Schuna, 1978), monitoring allergic reactions (Dunagan *et al.*, 1989; Maki and Schuna, 1978), cost (Dunagan *et al.*, 1991), broadness of spectrum (Byl *et al.*, 1999; Maki and Schuna, 1978), difference between empiric or definitive treatment (failing to adapt after the laboratory results are known) (Maki and Schuna, 1978; Parret *et al.*, 1993), records insufficient for categorisation (Volger *et al.*, 1988), timing of administration of surgical prophylaxis, postoperative administration of prophylaxis (Gyssens *et al.*, 1996a, b).

2.1.3. Algorithm

Based on the original criteria of Kunin, we developed in 1992, an algorithm to facilitate the classification of prescriptions in different categories of inappropriate use (Gyssens *et al.*, 1992). This algorithm allows an evaluation of each parameter of importance associated with prescribing antimicrobial drugs. Since 1996, the algorithm has been modified to include most criteria (Figure 1).

In order to obtain a complete evaluation, the series of questions in Table 2 have to be asked in a fixed order so that no parameter of importance is omitted. The questions in the algorithm are classified in categories of good use in order to structure and accelerate the process of evaluation. By utilising the algorithm, experts can categorise individual prescriptions. Prescriptions can be inappropriate for different reasons at the same time and can be placed in more than one category. During the evaluation procedure, the algorithm is read from top to bottom in order to evaluate each parameter associated with process

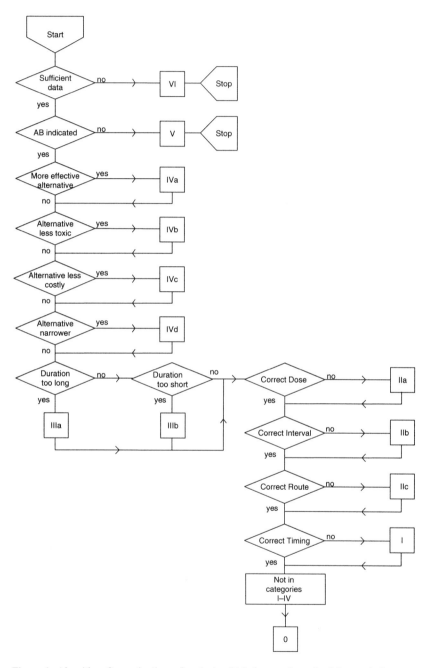

Figure 1. Algorithm for evaluation of antimicrobial therapy (van der Meer and Gyssens, 2001).

Table 2. Quality evaluation criteria of antimicrobial therapy

---

- Is there enough information to allow for categorisation?
- Is the clinical picture compatible with an infection? Is there an INDICATION for treatment with antibiotics?
- Is the CHOICE of the antimicrobial drug adequate?
  (a) EFFICACY: Is the (suspected) agent active?
  (b) TOXICITY/ALLERGY: Is there a less toxic alternative?
  (c) COST: Is there a less costly alternative at equal efficacy and toxicity?
  (d) BROADNESS OF SPECTRUM: is the spectrum unnecessarily broad?
- Is the DURATION of treatment appropriate?
  TOO LONG
  TOO SHORT
- Is the DOSAGE correct?
  DOSE
  INTERVAL
  MODE of ADMINISTRATION
- Is the TIMING appropriate?
  TOO EARLY
  TOO LATE

---

outcome. These parameters are explained below and are illustrated with a few examples.

Category VI: Are the data sufficient for categorization?

If the information concerning the treatment is insufficient, the evaluation can of course not be performed. In our own audits, 4% of prophylaxis prescriptions and 10% therapeutic prescriptions were not evaluable due to insufficient data in the medical records (Gyssens *et al.*, 1996a, 1997; Van Kasteren *et al.*, 2003). This corresponds with 5% in a Swiss study (Parret *et al.*, 1993). The presence or absence of sufficient data to prescribe antimicrobial drugs is associated with quality (Gyssens *et al.*, 1997; Maki and Schuna, 1978; Nathwani *et al.*, 1996). Maki has correlated appropriateness of therapy with the comprehensiveness of the notes of the prescriber in the medical record (Maki and Schuna, 1978).

Category V: Is the clinical picture compatible with an infection? Is there an indication for antimicrobial treatment?

A severe infection often starts with fever. On the other hand, fever is not always caused by an infection. A knowledge of infectious diseases and good microbiology facilities enable prescribers to discriminate between patients who need and those who do not need antibiotics. Distinguishing an infection from inflammation and the difference between bacterial sepsis and SIRS remains a field of continuous research. Apart from the clinical signs, laboratory parameters can guide the clinician, for example, C reactive protein (CRP). Overconsumption of antimicrobial drugs (prophylaxis not indicated) is traditionally a problem in surgical prophylaxis. Inappropriate prescribing of 40–75% has been reported in the United States for more than 15 years (Everitt *et al.*, 1990; Gorecki *et al.*, 1999; Silver *et al.*, 1996). Audits in Canada (Girotti *et al.*, 1990); the United Kingdom (Griffiths *et al.*, 1986), Italy (Motola *et al.*, 1998; Mozillo *et al.*, 1988), Belgium (Sasse *et al.*, 1998), The Netherlands (Gyssens *et al.*, 1996a), Switzerland (Parret *et al.*, 1993), Israel (Finkelstein *et al.*, 1996), and Australia (Johnston *et al.*, 1992) show similar problems. Empirical treatment for (suspected) nosocomial pneumonia often leads to overconsumption because diagnostic criteria are lacking (Singh *et al.*, 2000). Therapeutic prescriptions have been judged unnecessary (no signs of infection) in 9% in the United States (Maki and Schuna, 1978), 4% in France (Thuong *et al.*, 2000), and more than 35% in the United Kingdom (Swindell *et al.*, 1983), and in 4% of patients with bacteraemia (Dunagan *et al.*, 1991). In the Netherlands, therapeutic prescriptions were judged unnecessary in 16% of surgical cases and 5% in patients in internal medicine wards. After an intervention, this inappropriate use was reduced to 8% and 3%, respectively (Gyssens *et al.*, 1996a, 1997).

Category IV: Is the choice of the antimicrobial drug appropriate?

a. *Efficacy*: Is the causative microorganism susceptible?

For severely ill patients, treatment is inevitably started in a situation of uncertainty about the identity of the causative microorganism and its susceptibility. Blind empiric therapy often consists of large doses of a broad-spectrum drug or with a combination of drugs. A rational choice of an antimicrobial agent can only be expected from a prescriber if he is aware of the most likely causative organism and the prevailing susceptibility patterns. Since bacteria become more and more resistant, an empiric therapy with an even broader spectrum will be necessary. The distribution of resistant strains varies between countries, between hospitals, and even between services within one hospital. Local surveillance data should be available. Empiric therapy with vancomycin for a staphylococcal infection can be appropriate in one hospital or in a particular ward, but not in others where the prevalence of MRSA is virtually zero (Struelens, 1998). A standard of treatment for bacteraemias has been developed by the Society of Infectious Diseases of America (IDSA) to ensure that the administration of an antimicrobial drug is appropriate for the susceptibility of the isolated microorganism. The implementation of this standard of therapy of bacteraemias has been the subject of audits in different countries (Byl *et al.*, 1999; Fowler *et al.*, 1998; Nathwani *et al.*, 1996). The susceptibility has been studied for cases of bacteraemia (Dunagan *et al.*, 1989) and cases of bacter-aemic pneumococcal pneumonia (Meehan *et al.*, 1997). Many audits have reported on susceptibility mismatches. Maki found 9% of inappropriate use due to susceptibility mismatches (Maki and Schuna, 1978) and Wilkins 25% (Wilkins *et al.*, 1991). In Israel, 7.5% of empiric treatments of bacteraemia were inappropriate (Elhanan *et al.*, 1997); in Ireland, this represented 44% (Cunney *et al.*, 1997). The susceptibility for empiric therapy of 69% increased to 90% in internal medicine in the Netherlands (Gyssens *et al.*, 1997). In Germany, antibiotics empirically chosen with the help of a computer program were appropriate in 74% of cases (Heininger *et al.*, 1999). The management computer program of Evans *et al.* has significantly reduced susceptibility mismatches (Evans *et al.*, 1998). The clinical impact of rapid identification and *in vitro* susceptibility techniques on the care and outcome parameters of hospitalised patients has been reported by Doern *et al.* (1994).

b. *Toxicity/Allergy*: Is there a less toxic alternative?

The majority of antimicrobial drugs are eliminated by the kidneys. For drugs with a narrow therapeutic index, for example, aminoglycosides, adapta-tion of the dose is necessary in case of renal failure. Recently, a randomised double blind controlled trial demonstrated that once daily dosing had a lower probability of nephrotoxicity in patients with a normal renal function at base-line (Rybak *et al.*, 1999).

Several authors have analysed the potential toxicity of aminoglycoside use. Failure to monitor serum concentrations or to adapt regimens has been reported (Dunagan *et al.*, 1989; Gyssens *et al.*, 1997; Li *et al.*, 1989). In a British teaching hospital, control of serum concentrations were omitted in 14%; incorrect requests (21%) and incorrect sampling were frequent (Shrimpton *et al.*, 1993). Fear of toxicity can lead to underdosing of aminogly-cosides (Gyssens *et al.*, 1997). A monitoring programme of aminoglycosides has been evaluated by a randomised controlled study: improved response rate (60–48%) and a reduced length of stay but no difference in toxicity were observed (Burton *et al.*, 1991).

c. *Cost*: Can costs be cut without impairing quality?

The cost of antimicrobial therapy is considered as a quality indicator. Oral drugs are much less costly than their parenteral equivalents. The use of older drugs, reduction in the frequency of administration of parenteral drugs, bolus injections instead of infusions, and avoidance of drugs that require monitoring of serum concentrations result in cost containment (Gyssens *et al.*, 1991). Several audits have analysed cost aspects and many interventions were directed at cost savings (Briceland *et al.*, 1988; Evans *et al.*, 1990; Raz *et al.*, 1989). Some authors analysed costs as sole process outcome (Destache *et al.*, 1990). An early switch (after 72 hr) from parenteral to oral can cut costs

(Byl *et al.*, 1999; Ehrenkranz *et al.*, 1992; Evans *et al.*, 1990; Grasela *et al.*, 1991; Nathwani *et al.*, 1996). Shortening the duration of prophylaxis through interventions or the use of older drugs has been shown cost-effective (Evans *et al.*, 1990; Everitt *et al.*, 1990; Gyssens *et al.*, 1996a, 1997).

d. *Broadness of spectrum*: Is the spectrum unnecessarily broad?

Prolonged administration of a broad-spectrum antibiotic has important ecologic consequences. A comparison of two empiric antibiotic policies with a different spectrum in a neonatal ward has demonstrated that the combination of cefotaxime with amoxicillin favoured the selection of resistant Enterobacter strains compared to a regimen with penicillin and tobramycin (De Man *et al.*, 2000). Replacing an antibiotic by another antibiotic with a narrower spectrum, but also active against the isolated microorganism is a classic strategy used by infectious disease consultants. The rationale behind this strategy is to avoid selective pressure by blindly using broad-spectrum antimicrobial drugs. This strategy has not been well documented by prospective randomised studies. Some pathogens are still susceptible to narrow spectrum antibiotics, for example, group A streptococci are still ubiquitously susceptible to penicillin (Macris *et al.*, 1998).

In the Netherlands and Scandinavia, the tailored narrow spectrum definitive therapy with older drugs has been taught for years and "prudent prescribing" has resulted from this strategy (van der Meer and Gyssens, 2001). In Denmark, where there is the lowest incidences in Europe of resistance against antibiotics (DANMAP, 2000), older antibiotics with narrow spectrum are most often prescribed (Cars *et al.*, 2001; Røder *et al.*, 1993). In contrast to Spain and France, British practitioners also prescribe older antibiotics with narrow spectrum (Halls, 1993). In Maki and Schuna's study, continuation of definitive therapy with unnecessarily broad spectrum drugs was considered inappropriate in less than 10%, by Wilkins *et al.* in less than 16% (Wilkins *et al.*, 1991), by Parret in 14% (Parret *et al.*, 1993), and by Gyssens *et al.* in 4–7% (Gyssens *et al.*, 1997). Overconsumption of broad spectrum drugs was frequent in patients with bacteraemia in Israel (Elhanan *et al.*, 1997). In Belgium, this amounted to 29% of prescriptions for definitive therapy prescribed by clinicians lacking additional training in infectious diseases; for patients prescribed antibiotics by infectious diseases specialists, this was still relatively high, that is, 19% (Byl *et al.*, 1999). The strategy of streamlining was applied more frequently for patients with ID consultation than for controls (Fluckiger *et al.*, 2000).

Category III: Is the duration of treatment appropriate?

a. *Too long*

Studies on surgical prophylaxis have demonstrated that a single dose is sufficient for the majority of procedures. Inappropriate use of prophylaxis is often due to prolonged administration (Gould and Jappy, 1996; Gyssens *et al.*, 1996a; Moss *et al.*, 1981; Parret *et al.*, 1993; Van Kasteren *et al.*, 2003), and many intervention studies have been successful in reducing this practice (Evans *et al.*, 1990; Gyssens *et al.*, 1996a). There is a lack of evidence-based information on the required duration of treatment of the majority of infectious diseases. Even the duration of treatment of common infectious diseases is often based on tradition. There are also cultural differences. An European inquiry reported that the shortest mean duration of treatment with antimicrobial drugs was in the United Kingdom (8 days) and the longest in France (12 days) (Halls, 1993). The duration of treatment is often defined by the absence of relapse after an arbitrarily chosen number of treatment days, for example, 7 or 10 days. A recent randomised study in children comparing a duration of treatment of 7 days for acute bacterial infections with a treatment of 4 days found no difference in outcome (Peltola *et al.*, 2001). For many indications, the minimal duration of treatment is not known. The influence of prolonged administration on colonisation with resistant pneumococci in patients in the community has been documented by an observational study (Guillemot *et al.*, 1996) and a randomised controlled study (Schrag *et al.*, 2001). In the hospital setting, Harbarth *et al.* have shown that prolonged prophylaxis is correlated with a risk of acquired resistance (Harbarth *et al.*, 2000). Prolonged duration of treatment has been evaluated in patients with staphylococcal bacteraemia (Fowler *et al.*, 1998). In the Netherlands, an excessive duration was not frequently encountered (Gyssens *et al.*, 1997).

b. *Too short*

This parameter deserves more attention. An insufficient duration of treatment of oesophageal candidiasis or disseminated candidiasis has been reported in 40–60% by Natsch *et al.* (2001).

Category II: Is the dosage correct?

a. *The dose*

The dose of an antimicrobial drug has to be calculated in order to attain optimal serum concentrations in relation to the Minimum Inhibitory Concentration (MIC) of the drug against the (presumed) pathogens. Optimal therapy requires concentrations well above the MIC. In immunocompromised patients and for infections in body sites that are difficult to reach (meningitis, abscesses), it is necessary that concentrations at multiples of MICs are reached. For concentration dependent drugs, for example, aminoglycosides, the most efficient strategy is to administer a large dose (6 or 7 mg/kg) to all patients and adjust doses (or intervals) with the help of individual pharmacokinetic monitoring as soon as possible (Kashuba *et al.*, 1999). Furthermore, data from *in vitro* and *in vivo* studies in animal models suggest that the risk of development of resistance is reduced when the maximum concentration of a quinolone exceeds the MIC (Peak/MIC ratio) for the targeted microorganism at least 8- or 10-fold and the 24-hr AUC/MIC ratio is at least 100 for Gram-negative bacillae (Thomas *et al.*, 1998). Lower doses of betalactams were found to be a risk for colonisation with penicillin-resistant pneumococci in French children (Guillemot *et al.*, 1996). A large variation has been found in the dosage and duration of treatment of similar clinical syndromes in children presenting with an infectious disease (Van Houten *et al.*, 1998). Underdosing has been found in audits in the United Kingdom and in the Netherlands (Gyssens *et al.*, 1997; Natsch *et al.*, 2001). The height of the serum concentration peak (Cmax) has been described as a quality indicator for prescribing aminoglycosides (Burton *et al.*, 1991; Destache *et al.*, 1990).

c. *Dosing interval*

Optimal dosing frequency depends on the half-life and mechanism of action of the drug. By using aminoglycosides in a once daily regimen, the optimal pharmacodynamic conditions are combined with minimal toxicity (Rybak *et al.*, 1999). On the other hand, continuous infusions have been used based on mechanisms of time-dependent action of betalactams (Visser *et al.*, 1993). A reduced frequency of parenteral administration results in cost containment (Tanner, 1984). Moreover, parenteral drugs that can be administered once daily allow for an outpatient treatment of serious infections. For this reason, ceftriaxone has been preferred to penicillin, for example, for the treatment of streptococcal endocarditis, even in case of susceptibility of the microorganism to penicillin (Sexton *et al.*, 1998). Using unnecessarily broad spectrum drugs with long half-lifes, for reasons of convenience, resulted in an increase of resistance in healthcare facilities (Conus and Francioli, 1992).

d. *Oral or parenteral administration*

Parenteral administration should be used for empiric therapy in serious infections, for patients with gastrointestinal disturbances, and for drugs with reduced bioavailability. In practice, cultural factors seem to play an important role in the choice of administration route. Although the sites and severity of the infections were probably comparable in several European hospitals, in the United Kingdom, 60% of hospitalised patients were treated with oral antibiotics, while in Italy more than 80% of patients were treated with intramuscular injections (Halls, 1993). In the United States, iv administration has been considered as a standard of care for a long time. Sequential therapy is now more frequently used for patients in a stable clinical condition, mostly for economic reasons (Ehrenkranz *et al.*, 1992; Paladino *et al.*, 1991; Schentag, 1993). To reach sufficient serum concentration is the primary requirement for oral therapy. The switch from parenteral to oral is only optimal when oral therapy is limited to drugs with an excellent bioavailability. Infrequent dosing increases patient compliance. To allow for twice daily dosing, oral antibiotics should have a half-life of at least 1 hr. Unfortunately, some of these potent antibiotics which also have a low toxicity and low cost, for example, first generation cephalosporins (Gyssens *et al.*, 1996a; Kunin, 1973; Raz *et al.*, 1989; Seligman, 1981), ciprofloxacin (Frieden and Mangi, 1990; Seligman, 1981), and flucona-zole (Natsch *et al.*, 2001) have been overconsumed for years. Finally, a critical review on quality aspects of parenteral to oral switches has warned that a switch to oral antibiotics should not be postponed until the treatment could be stopped altogether (Davey and Nathwani, 1998). An intervention to promote parenteral to oral switches shortened hospital stay for patients with pneumonia (Ehrenkranz *et al.*, 1992), or patients with acute infections (Eron and Passos, 2001). An intervention consisted of protocols to reduce parenteral treatment for community-acquired pneumonia (Al-Eidan *et al.*, 2000).

Category I: Is the timing appropriate?

a. *Too late*
    The timing of surgical prophylaxis has been considered as optimal within 30 min before the incision, that is, at induction of anaesthesia. Administration within 2 hr before incision is considered correct. Failing to comply with this schedule seems to be a ubiquitous problem. Incorrect timing has been reported in 54% of US hospitals (Silver *et al.*, 1996) and in 46% of cases in Israel (Finkelstein *et al.*, 1996). The timing of prophylaxis—depending mainly on logistics—is also relatively easy to correct. The timing improved from 40% in 1985 to 99.1% in 1994 with the help of a computer-assisted prescription pro-gram in Salt Lake City (Evans *et al.*, 1998); in the Netherlands, interventions have succeeded in improving the timing into the optimal range (within 30 min before incision) from 39% to 64% and from 70% to 80%, respectively (Gyssens *et al.*, 1996b). The timing of therapy in the emergency room for patients admitted with a serious infection has been audited in a Dutch inter-vention study (Natsch *et al.*, 2000). Other temporal relationships have been identified as quality indicators; in a study of elderly patients with pneumonia, a lower 30-day mortality was associated with administration of antibiotics within 8 hr after hospitalisation (Meehan *et al.*, 1997).

b. *Too early*
    Antimicrobial therapy can be administered too early, for example, before blood and/or another sample has been drawn for culture (Gyssens *et al.*, 1997).
    In conclusion, by using the algorithm, the whole process of prescribing can be analysed systematically. Evaluation has to be performed by two or more independent experts in infectious diseases. Agreement can be calculated by kappa test. Ratings of 0.8 have been described for prophylaxis, for which much evidence is available (Gyssens *et al.*, 1996a). Partial agreement was found for evaluation of therapy in internal medicine wards (Gyssens *et al.*, 1997).

2.2.   Patient outcome

Recently there has been a trend in studying the quality of antimicrobial therapy by applying an intervention, and measuring the effect of an intervention on the patients and causative microorganisms. In patients suffering from bacteraemia, mortality related to septicaemia was lower in patients to whom appropriately chosen antibiotics had been administered (Byl *et al.*, 1999; Weinstein *et al.*, 1997), and the complications less frequent in cases of pneumonia (Metlay *et al.*, 2000). Often the intervention consists of the implementation of a clinical practice guideline or a clinical pathway partially based on evidence or on expert opinion. Several authors have shown potential cost savings of parenteral to oral switches (Paladino *et al.*, 1991; Quintiliani *et al.*, 1987). Ehrenkranz *et al.* and Weingarten *et al.* have described the advantages of a switch for patients with pneumonia (Ehrenkranz *et al.*, 1992; Weingarten *et al.*, 1996). Eron *et al.* have shown in a matched case control study that a strategy of early discharge and outpatient therapy conducted by an infectious diseases physician had a favourable outcome, although in this particular study, the cost was the only process outcome (Eron *et al.*, 2001). Singh *et al.* have studied in a randomised and controlled way that a 3-day course of ciprofloxacin was sufficient in cases in which a score indicated a low risk of pneumonia (Singh *et al.*, 2000). Another randomised controlled study by Marrie *et al.* has shown the value of a critical pathway containing levofloxacin to treat pneumonia (Marrie *et al.*, 2000). It has to be mentioned that in these two studies, the quality of antibiotic treatment is difficult to evaluate, since the choice was fixed, often independent of the identity of the causative microorganism and was probably determined by the financing of pharmaceutical companies. An uncontrolled study has demonstrated an impact of a rotating empiric antibiotic schedule on mortality in intensive care (Raymond *et al.*, 2001). In Table 3 are listed selected patient outcome variables from recently published intervention studies.

2.3.   Microbiological outcome, emergence of resistance, spread of resistance

Few interventions to evaluate and improve the quality of prescription have analysed microbiological outcomes. An intervention which consisted of an authorisation for drugs on a restricted list was followed by an increase in the susceptibility of causative microorganisms while mortality rates remained unchanged (White *et al.*, 1997). An awareness of the selection potential of resistance of different groups of antibiotics or of different individual antibiotics is still minimal in the clinical setting. The ecological effects of different prescribing policies have been documented in neonates (De Champs *et al.*, 1994; De Man *et al.*, 2000), in a geriatric ward (Bendall *et al.*, 1986), in a haematology

Table 3. Patient outcome variables as indicators of quality-of-use in intervention studies

Indicator	References
Mortality related to infection (30 days)	Fraser *et al.* (1997), Meehan *et al.* (1997), Fowler *et al.* (1998), Byl *et al.* (1999), Singh *et al.* (2000), Lemmen *et al.* (2001), Raymond *et al.* (2001)
Length of stay	Destache *et al.* (1990), Al-Eidan *et al.* (2000), Eron *et al.* (2001), Lemmen *et al.* (2001)
Prolongation of stay related to infection	Burton *et al.* (1991), Ehrenkranz *et al.* (1992), Fraser *et al.* (1997), Raymond *et al.* (2001)
Length of stay in intensive care	Singh *et al.* (2000)
Readmission or admission of non-outpatients (12 weeks)	Ehrenkranz *et al.* (1992), Fraser *et al.* (1997), Fowler *et al.* (1998), Eron *et al.* (2001)
Nosocomial infection rate	De Champs *et al.* (1994), Frank *et al.* (1997), Taylor (2000)
"Cure" (or lack of relapse)	Burton *et al.* (1991), Fraser *et al.* (1997), Fowler *et al.* (1998), Al-Eidan *et al.* (2000), Singh *et al.* (2000)
Side effects (allergies, toxicity)	Frequency of allergic reactions (Dunagan *et al.*, 1991; Evans *et al.*, 1998; Maki *et al.*, 1978). Aminoglycosides, markers of renal toxicity (Burton *et al.*, 1991; Li *et al.*, 1989), or general toxicity (Fraser *et al.*, 1997)

ward (Bradley *et al.*, 1999), or even an entire hospital (Landman *et al.*, 1999). It should be noted that most published studies have been conducted in wards or hospitals with a high antibiotic consumption and that have problems with selected resistant strains. The antibiotic policy intervention is intended to eliminate these strains (Bendall *et al.*, 1986; Bradley *et al.*, 1999; De Man *et al.*, 2000; Landman *et al.*, 1999; Meyer *et al.*, 1993). Risk factors for the development of resistance are not yet well known or have not been properly studied and the impact of infection control measures is often not well documented.

3.   TYPES OF STUDIES TO OBTAIN QUALITY DATA ON A PATIENT LEVEL

3.1.   Prevalence or incidence studies?

3.1.1.   Prevalence studies

The first evaluation studies were conducted at Harvard Medical School in the United States (Adler *et al.*, 1971; Barrett *et al.*, 1968) and in the United Kingdom (Cooke *et al.*, 1983). The data collection took place during one day per ward. Only a limited number of parameters on use can be collected by this method. This type of study is generally part of a larger study on the prevalence of nosocomial infections. A large European epidemiologic study on nosocomial infections and antibiotic use in intensive care units (EPIIC) was conducted according to this method in 1993 (Vincent *et al.*, 1995). Recently, an intervention programme on the quality of use of vancomycin used repeated point prevalence studies before and after an intervention (Hamilton *et al.*, 2000).

3.1.2.   Incidence studies

The majority of audits have measured the incidence of antibiotic treatment courses. The data have been collected during longer periods: one month (Gyssens *et al.*, 1996a; Moss *et al.*, 1981; Swindell *et al.*, 1983), two months (Durbin *et al.*, 1981; Van Houten *et al.*, 1998), three months (Parret *et al.*, 1993), four months (Dunagan *et al.*, 1991), five months (Fluckiger *et al.*, 2000), and 2 years (Fowler *et al.*, 1998; Quintiliani *et al.*, 1987). These studies gave more precise data for determining the relationship between use and quality, for example, excessive consumption in grams per 100 patient-days.

3.2.   Simple audits or intervention audits?

The aim of an audit is to ultimately improve practices. In the spirit of a strategy of continuous improvement, studies that combine audits with interventions are preferred above simple audits.

3.2.1.   Simple audits "Case Review" or "Recommendations audit"

Simple audits are evaluations of quality of prescription without intervention. Often these studies are based on a selection of cases in a nonquantitative approach. Early reports were retrospective, for example, in the Netherlands (Sturm, 1988), in Switzerland (Parret *et al.*, 1993), in Denmark (Røder *et al.*, 1993), in France (Roger *et al.*, 2000). This retrospective approach of clinical cases has also been utilised recently for a multicentre study of cohorts of patients with pneumonia (Meehan *et al.*, 1997).

An evaluation of an infectious diseases service with a control group of patients for whom no consultation was requested has been conducted in Belgium (Byl *et al.*, 1999) and in Switzerland (Fluckiger *et al.*, 2000).

3.2.2.   Audits with intervention (before and after) without a control group

A popular method since 1990 is the before and after intervention audit. These are mainly incidence studies which measure the effect of an intervention aimed at improving prescribing practices. The first studies in the United States were aimed at controlling costs and numerous studies have been conducted by pharmacists for this purpose (Quintiliani *et al.*, 1987; Seligman, 1981). In general, there is a global "before and after" analysis, the data of the initial audit are compared to the data of an identical study during (Al-Eidan *et al.*, 2000; Durbin *et al.*, 1981; Lemmen *et al.*, 2001; Raymond *et al.*, 2001) or after intervention (Bamberger and Dahl, 1992; Bradley *et al.*, 1999; Drori-Zeides *et al.*, 2000; Frank *et al.*, 1997; Gyssens *et al.*, 1996a, b; White *et al.*, 1997). The analysis can also be conducted over several well-determined periods (Everitt *et al.*, 1990). In this type of analysis, the data of a number of periods of several weeks are gathered and checked. Everitt *et al.* have compared the choice of surgical prophylaxis during 34 periods of time before intervention and 20 periods after intervention. An intervention study by introducing a protocol for the administration of antibiotics at all caesarean sections found a compliance rate of over 97% with a reduction of the infection rate to around 3% (Taylor, 2000).

3.2.3.   Audits with intervention (before and after or simultaneous) with a nonrandomised control group

According to the Cochrane Effective Practice and Organisation of Care (EPOC) group, www.epoc.uottawa.ca, invention studies are considered to have a rigorous design only if they have a parallel control group (simultaneous). This study design has to correct for the error introduced by the choice of a historical control, that is, for the possibility that a spontaneous trend, but not the intervention, was responsible for the behavioural change of the prescriber. The control group could be a ward where the intervention has not been implemented (Bartlett *et al.*, 1991; Gyssens *et al.*, 1997). A series of controlled studies has been conducted to demonstrate the effect of consultations by infectious diseases physicians on the quality of treatment. The comparison has been made with a group of patients without consultation (sequential analysis) (Gómez *et al.*, 1996) or with a patient group for whom the recommendations by infectious diseases physicians were not followed (Fowler *et al.*, 1998). Some of the controlled studies are retrospective and have a case control design (Classen *et al.*, 1992; Eron *et al.*, 2001).

3.2.4.   Randomised controlled clinical trials

Certain authors have conducted randomised controlled clinical trials in order to compare the effects of their interventions on the quality of prescribing (Ehrenkranz *et al.*, 1992; Fraser *et al.*, 1997). In certain cases, the evaluation of the evaluation process itself was sometimes completely replaced by the evaluation of patient outcome (Burton *et al.*, 1991; Destache *et al.*, 1990; Fraser *et al.*, 1997; Singh *et al.*, 2000).

4.   STRATIFICATION LEVEL

4.1.   Healthcare facility

The first studies on the quality of antibiotic use evaluated entire hospitals (Moss *et al.*, 1981). When the study target is a single antibiotic or a single indication, the evaluation can be performed relatively easily on the hospital as a whole (Drori-Zeides *et al.*, 2000; Dunagan *et al.*, 1991; Fluckiger *et al.*, 2000; Fowler *et al.*, 1998; Hamilton *et al.*, 2000; White *et al.*, 1997). Computerised methods also allow for the evaluation of an entire hospital (Echols *et al.*, 1984).

4.2.   Hospital units, medical disciplines

The main disciplines (internal medicine, general surgery) (Achong *et al.*, 1977; Bartlett *et al.*, 1991; Durbin *et al.*, 1981; Fraser *et al.*, 1997; Gyssens *et al.*, 1996a; Kunin, 1973; Lemmen *et al.*, 2001) are generally sufficient to provide a global picture of the quality of consumption of a hospital. Studies on paediatric use are scarce (Peltola *et al.*, 2001; Van Houten *et al.*, 1998). Certain disciplines can be selected on the basis of high consumption (Parret *et al.*, 1993). In Germany (Heininger *et al.*, 1999), Denmark, and France (Roger *et al.*, 2000), antibiotic use has been evaluated in intensive care units and in the United States, intervention studies have been conducted in intensive care units (Evans *et al.*, 1998; Raymond *et al.*, 2001; Singh *et al.*, 2000). Haematology departments are also great consumers of antimicrobial drugs (Bradley *et al.*, 1999).

4.3.   Patient level

Empiric therapy is often broad spectrum to avoid complications and mortality. Patients with severe, well-defined diseases have been studied: evaluation of these populations allows the detection of differences in patient outcome. Mostly patients with bacteraemia have been studied (Byl *et al.*, 1999; Dunagan *et al.*, 1991; Fowler *et al.*, 1998; Fluckiger *et al.*, 2000; White *et al.*, 1997), or patients with bacterial pneumonia. Pneumonia has been studied most often (Grasela *et al.*, 1990), including pneumonias in older patients (Meehan *et al.*, 1997) or patients suspected of pneumonia (Singh *et al.*, 2000). Another category of patients that has been studied is immunocompromised patients: patients with neutropenia caused by chemotherapy (Bradley *et al.*, 1999), premature infants with fever without apparent cause (De Champs *et al.*, 1994; De Man *et al.*, 2000). Some specific aspects of surgical prophylaxis have been studied in patients undergoing specific operations, for example, caesarean section (Everitt *et al.*, 1990).

4.4. Drug level

Some authors have chosen to analyse consumption of some antimicrobial drugs for reasons of cost, overconsumption, or because of high toxicity.

5. ANTIMICROBIAL DRUG CLASSES

Traditionally, antibacterial drugs were audited for economic reasons. They have been considered to contribute 75% of the total drug cost of an institution's drug bill (Achong *et al.*, 1977). Often these drugs have a broad spectrum and are on a restricted drug list (Fraser *et al.*, 1997) including vancomycin (Drori-Zeides *et al.*, 2000; Evans *et al.*, 1999; Fowler *et al.*, 1998) ciprofloxacin, imipenem, piperacillin-tazobactam (Thuong *et al.*, 2000). Certain drugs of different classes are considered to have a high resistance potential, such as ceftazidime, ciprofloxacin, imipenem, and gentamicin (Cunha, 2000). There is still little information available on the basic mechanisms that account for the differences in resistance selection potential (see Section 3.2 on Microbiological outcome). A particular problem encountered with some drugs is the combination of excellent activity, broad spectrum of action and good bioavailability. Combined with excellent tolerability, these characteristics paint a portrait of an agent that is identified by Kunin as "a drug of fear" (Kunin, 1973; Seligman, 1981). Prescribed for fever without diagnosis (cefazolin [Ma *et al.*, 1979] and cephalexin [Seligman, 1981]) and for all indications (ciprofloxacin) (Frieden *et al.*, 1990), these drugs have been overconsumed for years in the United States. Overconsumption of ciprofloxacin was also frequent in France: up to 60% of inappropriate prescriptions (Thuong *et al.*, 2000). Aminoglycosides have been studied for reasons of toxicity by many authors (Dunagan *et al.*, 1991; Li *et al.*, 1989). The value of dosage programmes and monitoring has been demonstrated by randomised controlled trials (Burton *et al.*, 1991; Destache *et al.*, 1990). Apart from antibiotics, antifungal therapy is now complicated by development of resistance by yeasts. Antifungal and antiviral drugs have been audited, mainly in referral hospitals. Well before the appearance of liposomal amphotericin B, antiviral and antifungal drugs contributed 42% of total antimicrobial drug costs (Gyssens *et al.*, 1997). The problems caused by resistant *Candida albicans* strains and the selection of resistant yeasts such as *Candida krusei* have been reported and are associated with the high utilisation of azoles. Considering the exorbitant high cost of liposomal amphotericin B compared to regular amphotericin B, there is room for development of an antifungal antibiotic policy. Two audits have been conducted on the utilisation of antifungal drugs (Gutierrez *et al.*, 1996; Natsch *et al.*, 2001). Fluconazole has all the potential to become another overconsumed "drug of fear" (Natsch *et al.*, 2001).

6.   DATA COLLECTION TECHNIQUES

6.1.   Abstracts of the medical record

Manual collection. Medical and nursing records are analysed in the wards. The following demographic data should be collected: identification of the patient (code number), age, sex, weight, identification of the prescriber (code number). The (presumed) diagnosis of the infection. The generic and brand name of antibiotics prescribed, the route of administration, the unit dose, the dates (and hours) of the start and stop of treatment. If the exact time of administration is not known, the hours of the nursing rounds can be used as an approximation. For the audit of surgical prophylaxis, the diagnosis and type of surgical procedure can be retrieved from medical records or from surgical intervention reports. Details supporting the suspected or confirmed diagnosis of infection (history, clinical findings) and laboratory data are needed in order to study empiric or definitive treatments. Depending on the degree of computerisation of the hospital, much of this information can be retrieved from the central hospital computer.

6.2.   Interviews

Generally there is no contact between the prescribers and the researchers. Few authors have used the interview technique (Moss *et al.*, 1981). Interviews with prescribers can in itself act as an intervention.

6.3.   Antibiotic Order Forms

This prescription sheet, filled in by the prescriber, forces the prescriber to review clinical information and laboratory results in order to specify whether therapy is started on an empiric, a definitive, or prophylactic basis; to name a (suspected) causative microorganism, the spectrum needed, the dose, the frequency, and the duration (Durbin *et al.*, 1981). This approach has mainly been used for some specific restricted drugs (Thuong *et al.*, 2000). By filling in the Antibiotic Order Form, the prescriber provides the data for the audit. In return, the pre-printed drug information on the form facilitates prescribing by providing information on formulary antibiotics, standardised dosing regimens at the time the prescription is written. In this way, an antibiotic form acts in itself as an intervention. The antibiotic order form may be compulsory (Durbin *et al.*, 1981; Echols *et al.*, 1984) or voluntary (Gyssens *et al.*, 1997). Some centres have reported successes with such a form (Echols *et al.*, 1984; Gyssens *et al.*, 1997; Lipsy *et al.*, 1993; Soumerai *et al.*, 1993; Thuong *et al.*, 2000) while others have not. Recently, authors have developed a "Vancomycin Continuation Form" in order to monitor the use of vancomycin (Evans *et al.*, 1999). Another advantage of the form is automatic stop order, for example after 24–48 hr in case of prophylaxis (Echols *et al.*, 1984; Lipsy *et al.*, 1993) or after 72 hr of empiric therapy (Lipsy *et al.*, 1993).

6.4. Automated methods—computer-assisted prescribing

For 15 years, in the university hospital LDS of Salt Lake City, a computer program "Clinical-decision-support" has been in use for assisting clinicians with antimicrobial drug prescribing (Evans *et al.*, 1986). Clinicians prescribe drugs with the help of a computer (Evans *et al.*, 1998). The "errors" of prescribing are recorded and they allow a continuous evaluation of process parameters (allergies, dosages, costs . . .) and patient outcome parameters such as side effects, length of stay. The program is linked to microbiology laboratory results and provides alerts to identify patients with inappropriate therapy due to susceptibility mismatches (Pestotnik *et al.*, 1996). The LDS pharmacy database has provided several evaluations of quality interventions in intensive care (Evans *et al.*, 1998) and on the timing of surgical prophylaxis (Classen *et al.*, 1992). A similar program, "Computer-based expert system for quality assurance of antimicrobial therapy," which produces reports of the discrepancies between prescribed therapy and microbiology results was described in 1993 by hospital pharmacists of a US reference hospital (Morrell *et al.*, 1993). Other reports on automated prescribing programs are scarce. German authors have published their experience in intensive care with a program "Computer-assisted infection monitoring program" developed with the support of Bayer to improve the choice of empiric therapy (Heininger *et al.*, 1999). Another group has reported an "Antimicrobial prescribing program" with a historic control (Frank *et al.*, 1997).

6.5. Use of surveillance data as a quality measure

Data collected at individual patient level provide the most reliable information on antibiotic use regarding the population exposed. However, since data at the individual level are not often available, the majority of studies on antibiotic consumption present data that have been collected at the collective level. These data are now preferentially expressed as defined daily dose (DDD) (World Health Organization, 2003) per 100 or 1,000 patient-days. On the other hand, in quality audits, when individual patient data are retrieved from medication charts at the wards, consumption is traditionally expressed in grams or exactly prescribed daily doses (PDD). To allow comparison of quantitative audit data with surveillance reports, consumption measured in audits should also be converted in DDD. Combining quantitative and qualitative methods of evaluation allows one to calculate inappropriate or excessive consumption either in grams (Parret *et al.*, 1993), in DDD/100 patient-days (Gyssens *et al.*, 1996a, 1997; Røder *et al.*, 1993) or both (Hamilton *et al.*, 2000). Quantitative surveillance with interrupted time-series analysis can be used as a method for identifying wards with changes in use and for targeting more detailed audits (Ansari *et al.*, 2003). Finally, by calculating the PDD/DDD ratio, audits can reveal to what extent the local dosage in a particular ward differs from the DDD.

Example: The DDD of cefazolin is 3 g. In a surgical department, an audit reveals that cefazolin is uniquely used for prophylaxis in a single preoperative dose of 1 g. The PDD/DDD ratio in this ward is 0.33. The PDD/DDD ratio shows that three times more patients are exposed to cefazolin than would be expected from the number of DDD/100 patient-days.

7.   MANPOWER

7.1.   Personnel for data collection

Demographic and quantitative data from the medical and nursing record can be retrieved by data collectors. Data collectors can be trained for that purpose. Pharmacy technicians (Gyssens *et al.*, 1996a), medical assistants, infection control practitioners (Van Kasteren *et al.*, 2003) have performed this task successfully. Qualitative data from the medical record should be recorded by medical doctors in order to provide clinical abstracts for an evaluation (Gyssens *et al.*, 1992; Parret *et al.*, 1993). Clinical pharmacists have also been active in this field in the United States (Achong *et al.*, 1977) and in the Netherlands (Natsch *et al.*, 2001).

7.2.   Experts for quality evaluation

Because of the need for an understanding of the available evidence in clinical infectious diseases, infectious diseases physicians are considered the experts to perform quality evaluations. Traditionally they have performed the first audits

Table 4. Who are the experts that evaluate quality of antimicrobial therapy?

One expert infectious diseases physician + one pharmacist, ID prevails	Natsch *et al.* (2001)
Two infectious diseases physicians + one surgeon	Parret *et al.* (1993)
Two independent infectious diseases physicians, comparison of agreement with kappa test	Volger *et al.* (1988) Gyssens *et al.* (1996a), Gyssens *et al.* (1997)
Two infectious diseases physicians + one pharmacist, discussion till agreement	Dunagan *et al.* (1991)
Clinical pharmacologist	Achong *et al.* (1977)
Senior physicians with expertise in infectious diseases	Thuong *et al.* (2000)
Senior infectious diseases physician, panel of local infectious diseases physicians, or other centres	Maki *et al.* (1978), Fluckiger *et al.* (2000)
Infectious diseases physician, not specified	Kunin (1973)
Infectious diseases physicians, microbiologists, not specified	Byl *et al.* (1999)
Two independent microbiologists, till agreement	Swindell *et al.* (1983)

and publications (Kunin, 1973; Marr *et al.*, 1988). Hospital pharmacists, specialised in infectious diseases, have performed many analyses and interventions in the United States. The evaluations have often targeted pharmacokinetic and economic aspects and they have evaluated compliance with recommendations (Bamberger *et al.*, 1992; Burton *et al.*, 1991; Lipsy *et al.*, 1993; Schentag *et al.*, 1993). The profiles of the different experts are shown in Table 4.

Quality evaluation can be performed by (1) experts handling authoritative criteria or (2) comparison of agreement with local, national, or international guidelines or standards. When the quality of prescription is evaluated only by local experts, the evaluation is in fact a local audit. The disadvantage of this system is the involvement of the expert(s) limiting objectivity of judgement. The evaluation can be influenced by local habits instead of based on evidence. To obtain an independent evaluation, experts of other centres can be consulted (Gyssens *et al.*, 1997).

8.   PROCEDURES OF AUDIT

8.1.   Evaluation by experts

The quality is evaluated for process outcome. Criteria are listed in Table 3. The experts listed in Table 4 perform an evaluation according to criteria based on evidence from the literature and their own experience. The detailed procedure is described in Section 2. In the United States, Medicare has selected a committee of experts to develop a series of quality indicators in order to evaluate the process care of patients hospitalised with pneumonia (Meehan *et al.*, 1997). In this retrospective study on 14,069 patients of more than 65 years with pneumonia, the only indicator related to antibiotic therapy was the timing of the administration of antibiotics (see timing).

8.2.   Audits for adherence to guidelines

Antibiotic prescribing can also be evaluated by studying the degree of concordance with (inter) national guidelines. Concordance with guidelines have been studied, for example, "Sanford Guide to Antimicrobial Therapy" (Fluckiger *et al.*, 2000), the standard of surgical prophylaxis by the CDC (Dellinger *et al.*, 1994) or "The Medical Letter" (Anonymous, 1992), a standard of treatment for bacteraemias by the CDC (Gross *et al.*, 1994), and the CDC recommendations on the use of vancomycin (HICPAC, 1995). Some centres have studied antimicrobial use by comparison with recommendations developed locally: treatment of *Staphylococcus aureus* bacteraemias (Fowler *et al.*, 1998), recommendations to use a restricted list of drugs (Thuong *et al.*, 2000), recommendations on the use of vancomycin (Drori-Zeides *et al.*, 2000), regulation of prescription of some third generation cephalosporins (Bamberger *et al.*, 1992), recommendation of standardized treatments (Bartlett *et al.*, 1991).

9.   FREQUENCY OF AUDITS

9.1.   Quantitative audit

With a surveillance system in place, it is probably sufficient to repeat the in-depth audit from time to time on a sample of each antibiotic in the formulary. The actual daily dose is interesting if one wants to evaluate the dose in a qualitative analysis. Computerised systems allow for continuous surveillance of these differences between DDD and PDD.

9.2.   Qualitative audit

The majority of authors have published a single audit of their hospital or a before-and-after intervention study. Follow-up reports are rare (Adler *et al.*, 1971; Everitt *et al.*, 1990). Publication bias probably plays an important role. Success stories are preferentially reported. The frequency of evaluations at collective level should be dictated by the problems that have been detected by screening surveillance data of antibiotic use or resistance.

10.   CONCLUSIONS

The most informative studies are those which provide incidence data, by a prospective or concurrent approach. The point prevalence approach gives an underestimation compared to longitudinal studies of several weeks (Cooke *et al.*, 1983). The prevalence approach is less costly in manpower and can serve as an indicator for future in-depth studies. Point prevalence studies can be repeated at regular intervals and become more interesting. This approach is probably more easily applicable in the long run. Studies considered rigorous by the Cochrane collaboration have a controlled design, preferentially controlled randomised trials, or at least a controlled before-and-after study. Without a control group, studies with an interrupted time-series analysis are necessary to determine trends. Moreover, outcomes should include patient outcome parameters and microbiological outcome parameters besides process outcomes. Feedback of audits can be used as an intervention. The purpose of the evaluation of use is change of prescribing behaviour. Peer review audits are more accepted by clinicians (Cooke *et al.*, 1983).

Many approaches are described in the international literature to evaluate the quality of use at the individual patient level in healthcare facilities. The technique of audit chosen and the frequency with which it has to be repeated is dictated by the available budget and manpower. The structure and organisation differs between countries and hospitals.

REFERENCES

Achong, M. R., Wood, J., Theal, H. K., Goldberg, R., and Thompson, D. A., 1977, Changes in hospital antibiotic therapy after a quality-of-use study. *Lancet*, **ii**, 1118–1122.

Adler, J. L., Burke, J. P., and Finland, M., 1971, Infection and antibiotic usage at Boston City Hospital, January 1970. *Arch. Intern. Med.*, **127**, 460–465.

Al-Eidan, F. A., McElnay, J. C., Scott, M. G., Kearney, M. P., Corrigan, J., and McConnell, J. B., 2000, Use of a treatment protocol in the management of community acquired lower respiratory tract infection. *J. Antimicrob. Chemother.*, **45**, 387–394.

Anonymous, 1992, Antimicrobial prophylaxis in surgery. *Med. Lett. Drug Ther.*, **34**, 5–8.

Ansari, F., Gray, K., Nathwani, D., Phillips, G., Ogston, S., Ramsay, C. *et al.*, 2003, Outcomes of an intervention to improve hospital antibiotic prescribing: Interrupted time series with segmented regression analysis. *J. Antimicrob. Chemother.*, **52**, 842–848.

Bamberger, D. M. and Dahl, S. L., 1992, Impact of voluntary vs enforced compliance of third-generation cephalosporin use in a teaching hospital. *Arch. Intern. Med.*, **152**, 554–557.

Barrett, F. F., Casey, J. I., and Finland, M., 1968, Infections and antibiotic use among patients at Boston City Hospital, February, 1967. *N. Engl. J. Med.*, **278**, 5–9.

Bartlett, R. C., Quintiliani, R. D., Nightingale, C. H., Platt, D., Crowe, H., Grotz, R. *et al.*, 1991, Effect of including recommendations for antimicrobial therapy in microbiology reports. *Diagn. Microbiol. Infect. Dis.*, **14**, 157–166.

Bendall, M. J., Ebrahim, S., Finch, R. G., Slack, R. C. B., and Towner, K. J., 1986, The effect of an antibiotic policy on bacterial resistance in patients in geriatric medical wards. *Q. J. Med.*, New series **60**(233), 849–854.

Bradley, S. J., Wilson, A. L. T., Allen, M. C., Sher, H. A., Goldstone, A. H., and Scott, G. M., 1999, The control of hyperendemic glycopeptide-resistant *Enterococcus* spp. on a haematology unit by changing antibiotic usage. *J. Antimicrob. Chemother.*, **43**, 261–266.

Briceland, L. L., Nightingale, C. H., Quintiliani, R., Cooper, B. W., and Smith, K. S., 1988, Antibiotic streamlining from combination therapy to monotherapy utilizing an interdisciplinary approach. *Arch. Intern. Med.*, **148**, 2019–2022.

Burton, M. E., Ash, C. L., Hill, D. P. J., Handy, T., Shepherd, M. D., and Vasko, M. R., 1991, Controlled trial of bayesian aminoglycoside dosing. *Clin. Pharmacol. Ther.*, **49**, 685–694.

Byl, B., Clevenberg, P., Jacobs, F., Struelens, M., Zech, F., Kentos, A. *et al.*, 1999, Impact of infectious diseases specialists and microbiological data on the appropriateness of antimicrobial therapy for bacteremia. *Clin. Infect. Dis.*, **29**, 60–66.

Cars, O., Mölstad, S., and Melander, A., 2001, Variation in antibiotic use in the European Union. *Lancet*, **357**, 1851–1853.

Classen, D. C., Evans, R. S., Pestotnik, S. L., Horn, S. D., Menlove, R. L., and Burke, J. P., 1992, The timing of prophylactic administration of antibiotics and the risk of surgical-wound infection. *N. Engl. J. Med.*, **326**, 281–286.

Conus, P. and Francioli, P., 1992, Relationship between ceftriaxone use and resistance of Enterobacter species. *J. Clin. Pharm. Ther.*, **17**, 303–305.

Cooke, D. M., Salter, A. J. and Philips, I., 1983, The impact of antibiotic policy on prescribing in a London teaching hospital. *JAC*, **11**, 447–453.

Cunha, B. A., 2000, Antibiotic resistance. *Med. Clin. N. Amer.*, **84**, 1407–1429.

Cunney, R. J., McNamara, E. B., Alansari, N., Loo, B., and Smyth, E. G., 1997, The impact of blood culture reporting and clinical liaison on the empiric treatment of bacteraemia. *J. Clin. Pathol.*, **50**, 1010–1012.

DANMAP (DANMAP 2000). Consumption of antimicrobial agents and ocurrence of antimicrobial resistance in bacteria from food animals, foods and humans in Denmark.

Davey, P. and Nathwani, D., 1998, Sequential antibiotic therapy: The right patient, the right time and the right outcome. *J. Infect.*, **37**(Suppl 1), 37–44.

De Champs, C., Franchineau, P., Gourgand, J.-M., Loriette, Y., Gaulme, J., and Sirot, J., 1994, Clinical and bacteriological survey after change in aminoglycoside treatment to control an epidemic of Enterobacter cloacae. *J. Hosp. Infect.*, **28**, 219–229.

De Man, P., Verhoeven, B. A. N., Verbrugh, H. A., Vos, M. C., and van den Anker, J. N., 2000, An antibiotic policy to prevent emergence of resistant bacilli. *Lancet*, **355**, 973–978.

Dellinger, E. P., Gross, P. A., Barrett, T. A., Krause, P. J., Martone, W. J., McGowan, J. E., Sweet, R. L., and Wenzel, R. P., 1994, Quality standard for antimicrobial prophylaxis in surgical procedures. *Infect. Control Hosp. Epidemiol.*, **15**, 182–188.

Destache, C. J., Meyer, S. K., Bittner, M., and Hermann, K. G., 1990, Impact of a clinical pharmacokinetic service on patients treated with aminoglycosides: A cost-benefit analysis. *Ther. Drug. Monit.*, **12**, 419–426.

Doern, G. Y., Vautour, R., Gaudet, M., and Levy, B., 1994, Clinical impact of rapid in vitro susceptibility testing and bacterial identification. *J. Clin. Microbiol.*, **32**, 1757–1762.

Drori-Zeides, T., Raveh, D., Schlesinger, Y., and Yinnon, A. M., 2000, Practical guidelines for vancomycin usage, with prospective drug-utilization evaluation. *Infect. Control Hosp. Epidemiol.*, **21**, 45–47.

Dunagan, W. C., Woodward, R. S., Medoff, G., Gray, J. L., Casabar, E., Lawrenz, C. *et al.*, 1991, Antibiotic misuse in two clinical situations: Positive blood culture and administration of aminoglycosides. *Rev. Infect. Dis.*, **13**, 405–412.

Dunagan, W. C., Woodward, R. S., Medoff, G., Gray, J. L., III, Casabar, E., Smith, M. D. *et al.*, 1989, Antimicrobial misuse in patients with positive blood cultures. *Am. J. Med.*, **87**, 253–259.

Durbin, W. A. J., Lapidas, B., and Goldmann, D. A., 1981, Improved antibiotic usage following introduction of a novel prescription system. *JAMA*, **246**, 1796–1800.

Echols, R. M., and Kowalsky, S. F., 1984, The use of an antibiotic order form for antibiotic utilization review: Influence on physicians' prescribing patterns. *J. Infect. Dis.*, **150**, 803–807.

Ehrenkranz, N. J., Nerenberg, D. E., Shultz, J. M., and Slater, K. C., 1992, Intervention to discontinue parenteral antimicrobial therapy in patients hospitalized with pulmonary infections: Effect on shortening hospital stay. *Infect. Control. Hosp. Epidemiol*, **13**, 21–32.

Elhanan, G., Sarhat, H., and Raz, R., 1997, Empiric antibiotic treatment and the misuse of culture results and antibiotic sensitivities in patients with community-acquired bacteraemia due to urinary tract infection. *J. Infect.*, **35**, 283–288.

Eron, L. J., and Passos, S., 2001, Early discharge of infected patients through appropriate antibiotic use. *Arch. Intern. Med.*, **161**, 61–65.

Evans, M. E., Millheim, E. T., and Rapp, R. P., 1999, Vancomycin use in a university medical center: Effect of a vancomycin continuation form. *Infect. Control Hosp. Epidemiol.*, **20**, 417–420.

Evans, R. S., Larsen, R. A., Burke, J. P., Gardner, R. M., Meier, F. A., Jacobson, J. A. *et al.*, 1986, Computer surveillance of hospital-acquired infections and antibiotic use. *JAMA*, **256**, 1007–1011.

Evans, R. S., Pestotnik, S. L., Burke, J. P., Gardner, R. M., Larsen, R. A., and Classen, D. C., 1990, Reducing the duration of prophylactic antibiotic use through computer monitoring of surgical patients. *DICP Ann. Pharmacother.*, **24**, 351–354.

Evans, R. S., Pestotnik, S. L., Classen, D. C., Clemmer, T. P., Weaver, L. K., Orme, J. F. *et al.*, 1998, A computer-assisted management program for antibiotics and other anti-infective agents. *N. Engl. J. Med.*, **338**, 232–238.

Everitt, D. E., Soumerai, S. B., Avorn, J., Klapholz, H., and Wessels, M., 1990, Changing surgical antimicrobial prophylaxis practices through education targeted at senior department leaders. *Infect. Control. Hosp. Epid.*, **11**, 578–583.

Finkelstein, R., Reinhertz, G., and Embom, A., 1996, Surveillance of the use of antibiotic prophylaxis in surgery. *Isr. J. Med. Sci.*, **32**, 1093–7.

Fluckiger, U., Zimmerli, W., Sax, H., Frei, R., and Widmer, A. F., 2000, Clinical impact of an infectious diseases service on the management of bloodstream infection. *Eur. J. Clin. Microbiol. Infect. Dis.*, **19**, 493–500.

Fowler, V. G., Jr., Sanders, L. L., Sexton, D. J., Kong, L., Marr, K. A., Gopal, A. K. *et al.*, 1998, Outcome of Staphylococcus aureus bacteremia according to compliance with recommendations of infectious diseases specialists: Experience with 244 patients. *Clin. Infect. Dis.*, **27**, 478–486.

Frank, M. O., Batteiger, B. E., Sorensen, S. J., Hartstein, A. I., Carr, J. A., McComb, J. S. *et al.*, 1997, Decrease in expenditures and selected nosocomial infections following implementation of an antimcrobial-prescribing improvement program. *Clin. Perform. Qual. Health Care*, **5**, 180–188.

Fraser, G. L., Stogsdill, P., Dickens Jr, J. D., Wennberg, D. E., Smith, R. P., and Prato, B. S., 1997, Antibiotic optimization. *Arch. Intern. Med.*, **157**, 1689–1694.

Frieden, T. R., and Mangi, R. J., 1990, Inappropriate use of oral ciprofloxacin. *JAMA*, **264**, 1438–1440.

Girotti, M. J., Fodoruk, S., Irvine-Meek, J., and Rotstein, O. D., 1990, Antibiotic handbook and pre-printed perioperative order forms for surgical antibiotic prophylaxis: Do they work? *CJS*, **33**, 385–388.

222   *Inge C. Gyssens*

Gómez, J., Conde Cavero, S. J., Hernández Cardona, J. L., Núnez, M. L., Ruiz Gómez, J., Canteras, M. *et al.*, 1996, The influence of the opinion of an infectious diseases consultant on the appropriateness of antibiotic treatment in a general hospital. *J. Antimicrob. Chemother.*, **38**, 309–314.

Gorecki, P., Schein, M., Rucinski, J. C., and Wise, L., 1999, Antibiotic administration in patients undergoing common surgical procedures in a community teaching hospital: The chaos continues. *World J. Surg.*, **23**, 429–432.

Gould, I. M., Hampson, J., Taylor, E. W., Wood, M. J., for the Working Party of the British Society for Antimicrobial Chemotherapy, 1994, Hospital antibiotic control measures in the UK. *J. Antimicrob. Chemother.*, **34**, 21–42.

Gould, I. M., and Jappy, B., 1996, Trends in hospital antibiotic prescribing after introduction of an antibiotic policy. *J. Antimicrob. Chemother.*, **38**, 896–904.

Grasela, T. H., Jr., Paladino, J. A., Schentag, J. J., Huepenbecker, D., Rybacki, J., Purcell, J. B. *et al.*, 1991, Clinical and economic impact of oral ciprofloxacin as follow-up to parenteral antibiotics. *DICP Ann. Pharmacother.*, **25**, 857–862.

Grasela, T. H., Welage, L. S., Walawander, C. A., Timm, E. G., Pelter, M. A., Poirier, T. I. *et al.*, 1990, A nationwide survey of antibiotic prescribing patterns and clinical outcomes in patients with bacterial pneumonia. DICP *Ann. Pharmacother.*, **24**, 1220–1225.

Griffiths, L. R., Bartzokas, C. A., Hampson, J. P., and Ghose, A. R., 1986, Antibiotic cost and prescribing patterns in a recently commissioned Liverpool teaching hospital. Part I: Antimicrobial therapy. *J. Hosp. Infect.*, **8**, 159–167.

Gross, P. A., Barrett, T. L., Patchen Dellinger, E., Krause, P. J., Martone, W. J., McGowan, J. E. *et al.*, 1994, Quality standard for the treatment of bacteremia. *Clin. Infect. Dis.*, **18**, 428–430.

Guillemot, D., Carbon, C., Balkau, B., Geslin, P., Lecoeur, H., Vauzelle-Kervroedan, F. *et al.*, 1996, Low dosage and long treatment duration of beta-lactam. Risk factors for carriage of penicillin-resistant *Streptococcus pneumoniae*. *JAMA*, **279**, 365–370.

Gutierrez, F., Wall, P. G., and Cohen, J., 1996, An audit of the use of antifungal agents. *J. Antimicrob. Chemother.*, **37**, 175–185.

Gyssens, I. C., Blok, W. L., Van den Broek, P. J., Hekster, Y. A., and Van der Meer, J. W. M., 1997, Implementation of an educational program and an antibiotic order form to optimize quality of antimicrobial drug use in a department of internal medicine. *Eur. J. Clin. Microbiol. Infect. Dis.*, **16**, 904–912.

Gyssens, I. C., Geerligs, I. E. J., Dony, J. M. J., Van der Vliet, J. A., Van Kampen, A., Van den Broek, P. J. *et al.*, 1996a, Optimising antimicrobial drug use in surgery: An intervention study in a Dutch university hospital. *J. Antimicrob. Chemother.*, **38**, 1001–1012.

Gyssens, I. C., Geerligs, I. E. J., Nannini-Bergman, M. G., Knape, J. T. A., Hekster, Y. A., and van der Meer, J. W. M., 1996b, Optimizing the timing of antimicrobial drug prophylaxis in surgery: An intervention study. *J. Antimicrob. Chemother.*, **38**, 301–308.

Gyssens, I. C., Lennards, C. A., Hekster, Y. A., and Van der Meer, J. W. M., 1991, The cost of antimicrobial chemotherapy. A method for cost evaluation. *Pharm. Weekbl. Sci.*, **13**(6), 248–253.

Gyssens, I. C., Smits-Caris, C., Stolk-Engelaar, M. V., Slooff, T. J. J. H., and Hoogkamp-Korstanje, J. A. A., 1997, An audit of microbiology laboratory utilization. The diagnosis of infection in orthopedic surgery. *Clin. Microb. Infect.*, **3**, 518–522.

Gyssens, I. C., Van den Broek, P. J., Kullberg, B. J., Hekster, Y. A., and Van der Meer, J. W. M., 1992, Optimizing antimicrobial therapy. A method for antimicrobial drug evaluation. *J. Antimicrob. Chemother.*, **30**, 724–727.

Halls, G. A., 1993, The management of infections and antibiotic therapy: A European survey. *J. Antimicrob. Chemother.*, **31**, 985–1000.

Hamilton, D. C., Drew, R., Janning, S. W., Kure Latour, J., and Hayward, S., 2000, Excessive use of vancomycin: A successful intervention strategy at an academic medical center. *Infect. Control Hosp. Epidemiol.*, **21**, 42–45.

Harbarth, S., Samore, M. H., Lichtenberg, D., and Carmeli, Y., 2000, Prolonged antibiotic prophylaxis after cardiovascular surgery and its effect on surgical site infections and antimicrobial resistance. *Circulation*, **101**, 2916–2921.

Heininger, A., Niemetz, A. H., Keim, M., Fretschner, R., Doering, G., and Unertl, K., 1999, Implementation of an interactive computer-assisted infection monitoring program at the bedside. *Infect. Control Hosp. Epidemiol.*, **20**, 440–447.

Hospital Infection Control Practices Advisory Committee, (HICPAC), 1995, Recommendations for preventing the spread of vancomycin resistance. *Infect. Control Hosp. Epidemiol.*, **16**, 105–113.

Johnston, J., Harris, J., and Hall, J. C., 1992, The effect of an educational intervention on the use of peri-operative antimicrobial agents. *Austr. Clin. Rev.*, **12**, 53–56.

Kashuba, A. D., Nafziger, A. N., Drusano, G. L., and Bertino, J. S. J., 1999, Optimizing aminoglycoside therapy for nosocomial pneumonia caused by gram-negative bacteria. *Antimicrob. Agents Chemother.*, **43**, 623–629.

Kunin, C. M., 1973, Use of antibiotics: A brief exposition of the problem and some tentative solutions. *Ann. Intern. Med.*, **79**, 555–560.

Landman, D., Chockalingam, M., and Quale, J. M., 1999, Reduction in the incidence of methicillin-resistant S. aureus and Ceftazidime-resistant Klebsiella pneumoniae following changes in a hospital antibiotic formulary. *Clin. Infect. Dis.*, **28**, 1062–6.

Lemmen, S. W., Becker, G., Frank, U., and Daschner, F. D., 2001, Influence of an infectious disease consulting service on quality and costs of antibiotic prescriptions in a university hospital. *Scand. J. Infect. Dis.*, **33**, 219–221.

Li, S. C., Ioannides-Demos, L. L., Spicer, W. J., Berbatis, C., Spelman, D. W., Tong, N. *et al.*, 1989, Prospective audit of aminoglycoside usage in a general hospital with assessments of clinical processes and adverse clinical outcomes. *Med. J. Aust.*, **151**, 224–232.

Lipsy, R. J., Smith, G. H., and Maloney, M. E., 1993, Design, implementation, and use of a new antimicrobial order form: A descriptive report. *Ann. Pharmacother.*, **27**, 856–861.

Ma, M. Y., Goldstein, J. C., and Meyer, R. D., 1979, Effect of control programs on cefazolin prescribing in a teaching hospital. *Am. J. Hosp. Pharm.*, **36**, 1055–1058.

Macris, M. H., Hartman, N., Murray, B., Klein, R. F., Roberts, R. B., Kaplan, E. L. *et al.*, 1998, Studies of the continuing susceptibility of group A streptococcal strains to penicillin during eight decades. *Ped. Infect. Dis. J.*, **17**, 377–381.

Maki, D. G. and Schuna, A. A., 1978, A study of antimicrobial misuse in a university hospital. *Am. J. Med. Sci.*, **275**, 271–282.

Marr, J. J., Moffet, H. L., and Kunin, C. M., 1988, Guidelines for improving the use of antimicrobial agents in hospitals: A statement by the Infectious Disease Society of America. *J. Infect. Dis.*, **157**, 869–876.

Marrie, T. J., Lau, C. Y., Wheeler, S. L., Wong, C. J., Vandervoort, M. K., and Feagan, B. G., 2000, A controlled trial of a critical pathway for treatment of community-acquired pneumonia. *JAMA*, **283**, 749–755.

Meehan, T. P., Fine, M. J., Krumholz, H. M., Scinto, J. D., Galusha, D. H., Mockalis, J. T. *et al.*, 1997, Quality of care, process, and outcomes in elderly patients with pneumonia. *JAMA*, **278**, 2080–2084.

Metlay, J. P., Hoffmann, J., Cetron, M. S., Fine, M. J., Farley, M. M., Whitney, C. *et al.*, 2000, Impact of penicillin susceptibility on medical outcomes for adult patients with bacteremic pneumococcal pneumonia. *Clin. Infect. Dis.*, **30**, 520–528.

Meyer, S. K., Urban, C., Eagan, J. A., Berger, B. J., and Rahal, J. J., 1993, Nosocomial outbreak of *Klebsiella* infection resistant to late-generation cephalosporins. *Ann. Intern. Med.*, **119**, 353–358.

Morrell, R., Wasilauskas, B., and Winslow, R., 1993, Personal computer-based expert system for quality assurance of antimicrobial therapy. *Am. J. Hosp. Pharm.*, **50**, 2067–2073.

Moss, F., McNicol, M. W., McSwiggan, D. A., and Miller, D. L., 1981, Survey of antibiotic prescribing in a district general hospital I. Pattern of use. *Lancet*, **ii**, 349–352.

Motola, G., Russo, F., Mangrella, M., Vacca, C., Mazzeo, F., and Rossi, F., 1998, Antibiotic prophylaxis for surgical procedures: A survey from an Italian university hospital. *J. Chemother.*, **10**, 375–380.

Mozillo, N., Greco, D., Pescini, A., and Formato, A., 1988, Chemoprophylaxis in the surgical ward: Results of a national survey in Italy. *Eur. J. Epidemiol.*, **4**, 357–359.

Nathwani, D., Davey, P., France, A. J., Phillips, G., Orange, G., and Parratt, D., 1996, Impact of an infection consultation service for bacteraemia on clinical management and use of resources. *Q. J. Med.*, **89**, 789–797.

Natsch, S., Kullberg, B. J., Meis, J. F. G. M., and Van der Meer, J. W. M., 2000, Earlier initiation of antibiotic treatment for severe infections after intervention to improve the organization and specific guidelines in the emergency room. *Arch. Intern. Med.*, **160**, 1317–1320.

Natsch, S., Steeghs, M. H. M., Hekster, Y. A., Meis, J. F. G. M., van der Meer, J. W. M., and Kullberg, B. J., 2001, Use of fluconazole in daily practice: Still room for improvement. *J. Antimicrob. Chemother.*, **48**, 303–310.

Paladino, J. A., Sperry, H. E., Backes, J. M., Gelber, J. A., Serriann, D. J., Cumbo, T. J. *et al.*, 1991, Clinical and economic evaluation of oral ciprofloxacin after an abbreviated course of intravenous antibiotics. *Am. J. Med.*, **91**, 462–470.

Parret, T., Schneider, R., Rime, B., Saghafi, L., Pannatier, A., and Francioli, P., 1993, Evaluation de l'utilisation des antibiotiques: une étude prospective dans un centre hospitalier universitaire. *Schweiz med Wschr*, **123**, 403–413.

Peltola, H., Vuori-Holopainen, E., Kallio, M. J. T., and Group, S.-T. S., 2001, Successful shortening from seven to four days of parenteral beta-lactam treatment for common childhood infections infections: A prospective and randomized study. *Int. J. Infect. Dis.*, **5**, 3–8.

Pestotnik, S. L., Classen, D. C., Evans, R. S., and Burke, J. P., 1996, Implementing antibiotic practice guidelines through computer-assisted decision support: Clinical and financial outcomes. *Ann. Intern. Med.*, **124**, 884–90.

Quintiliani, R., Cooper, B. W., Briceland, L. L. *et al.*, 1987, Economic impact of streamlining antibiotic administration. *Am. J. Med.*, **82**(suppl. 4A), 391–395.

Raymond, D. P., Pelletier, S. J., Crabtree, T. D., Gleason, T. G., Hamm, L. L., Pruett, T. L. *et al.*, 2001, Impact of rotating empiric antibiotic schedule on infectious mortality in an intensive care unit. *Crit. Care Med.*, **29**, 1101–1108.

Raz, R., Sharir, R., Ron, A., and Laks, N., 1989, The influence of an infectious disease specialist on the antimicrobial budget of a community teaching hospital. *J. Infect.*, **18**, 213–219.

Røder, B. L., Nielsen, S. L., Magnussen, P., Engquist, A., and Frimodt-Møller, N., 1993, Antibiotic usage in an intensive care unit in a Danish university hospital. *J. Antimicrob. Chemother.*, **32**, 633–642.

Roger, P. M., Hyvernat, H., Verleine-Pugliese, S., Bourroul, C., Giordano, J., Fosse, T. *et al.*, 2000, Consultation systématique d'infectiologie en réanimation médicale. *Presse Méd,* 29.

Rybak, M. J., Abate, B. J., Kang, S. L., Ruffing, M. J., Lerner, S. A., and Drusano, G. L., 1999, Prospective evaluation of the effect of an aminoglycoside dosing regimen on rates of observed nephrotoxicity and ototoxicity. *Antimicrob. Agents Chemother.*, **43**, 1549–1555.

Sasse, A., Mertens, R., Sion, J. P., Bossens, M., De Mol, P., Goossens, H. *et al.* 1998, Surgical prophylaxis in Belgian hospitals: Estimate of costs and potential savings. *J. Antimicrob. Chemother.*, **41**, 267–272.

Schentag, J. J., 1993, The results of a targeted pharmacy program. *Clin. Ther.*, **15**(Suppl. A), 29–36.

Schentag, J. J., Ballow, C. H., Fritz, A. L., Paladino, J. A., Williams, J. D., Cumbo, T. J. *et al.*, 1993, Changes in antimicrobial agent usage resulting from interactions among clinical pharmacy, the infectious diseases division, and the microbiology laboratory. *Diagn. Microbiol. Infect. Dis.*, **16**, 255–264.

Schrag, S. J., Pena, C., Fernandez, J., Sanchez, J., Gomez, V., Perez, E. *et al.*, 2001, Effect of short-course, high-dose amoxillin therapy on resistant pneumococcal carriage: A randomized trial. *JAMA*, **286**, 49–56.

Seligman, S. J., 1981, Reduction in antibiotic costs by restricting use of an oral cephalosporin. *Am. J. Med.*, **71**, 941–944.

Sexton, D. J., Tenenbaum, M. J., Wilson, W. R., Steckelberg, J. M., Tice, A. D., Gilbert, D. *et al.*, 1998, Ceftriaxone once daily for four weeks compared with ceftriaxone plus gentamicin once daily for two weeks for treatment of endocarditis due to penicillin-susceptible streptococci. *Clin. Infect. Dis.*, **27**, 1470–1474.

Shrimpton, S. B., Milmoe, M., Wilson, A. P. R., Felmingham, D., Drayan, S., Barrass, C. *et al.*, 1993, Audit of prescription and assay of aminoglycosides in a U.K. teaching hospital. *J. Antimicrob. Chemother.*, **31**, 599–606.

Silver, A., Eichorn, A., Kral, J., Pickett, G., Barie, P., Pryor, V. *et al.*, 1996, Timeliness and use of antibiotic prophylaxis in selected inpatient surgical procedures. *Am. J. Surg.*, **171**, 548–552.

Singh, N., Rogers, P., Atwood, W., Wagener, M. W., and Yu, V. L., 2000, Short-course empiric antibiotic therapy for patients with pulmonary infiltrates in the intensive care unit. *Am. J. Respir. Crit. Care Med.*, **162**, 505–511.

Soumerai, S. B., Avorn, J., Taylor, W. C., Wessels, M., Maher, D., and Hawley, S. L., 1993, Improving choice of prescribed antibiotics through concurrent reminders in an educational order form. *Med. Care*, **31**, 552–558.

Struelens, M. J., 1998, The epidemiology of antimicrobial resistance in hospital acquired infections: Problems and possible solutions. *BMJ*, **317**, 652–654.

Sturm, A. W., 1988, Rational use of antimicrobial agents and diagnostic microbiology facilities. *J. Antimicrob. Chemother.*, **22**, 257–260.

Swindell, P. J., Reeves, D. S., Bullock, D. W., Davies, A. J., and Spence, C. E., 1983, Audits of antibiotic prescribing in a Bristol hospital. *BMJ*, **286**, 118–122.

Tanner, D. J., 1984, Cost containment of reconstituted parenteral antibiotics: Personnel and supply costs associated with preparation, dispensing, and administration. *Rev. Infect. Dis.*, **6**(Suppl. 4): S924–S937.

Taylor, G. M., 2000, An audit of the implementation of guidelines to reduce wound infection following caesarean section. *Health Bull. (Edinb)*, **58**, 38–44.

Thomas, J. K., Forrest, A., Bhavnani, S. M., Hyatt, J. M., Cheng, A., Ballow, C. H. *et al.*, 1998, Pharmacodynamic evaluation of factors associated with the development of bacterial resistance in acutely ill patients during therapy. *Antimicrob. Agents Chemother.*, **42**, 521–527.

Thuong, M., Shortgen, F., Zazempa, V., Girou, E., Soussy, C. J., and Brun-Buisson, C., 2000, Appropriate use of restricted antimicrobial agents in hospitals: The importance of empirical therapy and assisted re-evaluation. *J. Antimicrob. Chemother.*, **46**, 501–508.

Van der Meer, J. W. M. and Gyssens, I. C., 2001, Quality of antimicrobial drug prescription in hospital. *Clin. Microbiol. Infect.*, **7**(Suppl. 6), 12–15.

Van Houten, M. A., Luinge, K., Laseur, M., and Kimpen, J. L., 1998, Antibiotic utilisation for hospitalised paediatric patients. *Int. J. Antimicrob. Agents*, **10**, 161–164.

Van Kasteren, M. E. E., Kullberg, B. J., De Boer, A. S., Mintjes-de Groot, J., and Gyssens, I. C., 2003, Adherence to local hospital guidelines for surgical antimicrobial prophylaxis: A multicentre audit in Dutch hospitals. *J. Antimicrob. Chemother.*, **51**, 1389–1396.

Vincent, J.-L., Bihari, D. J., Suter, P. M., Bruining, H. A., White, J., Nicolas-Chanoin, M.-H. *et al.*, 1995, The prevalence of nosocomial infection in Intensive Care Units in Europe. *JAMA*, **274**, 639–644.

Visser, L. G., Arnouts, P., van Furth, R., Mattie, H., and van den Broek, P. J., 1993, Clinical pharmacokinetics of continuous intravenous administration of penicillins. *Clin. Infect. Dis.*, **17**, 491–495.

Volger, B. W., Ross, M. B., Brunetti, H. R., Baumgartner, D. D., and Therasse, D. G., 1988, Compliance with a restricted antimicrobial agent policy in a university hospital. *Am. J. Hosp. Pharm.*, **45**, 1540–1544.

Weingarten, S. R., Riedinger, M. S., Hobson, P., Noah, M. S., Johnson, B., Giugliano, G. *et al.*, 1996, Evaluation of a pneumonia practice guideline in an interventional trial. *Am. J. Respir. Crit. Care Med.*, **153**, 1110–1115.

Weinstein, M. P., Towns, M. L., Quartey, S. M., Mirrett, S., Reimer, L. G., Parmigiani, G. *et al.*, 1997, The clinical significance of positive blood cultures in the 1990s: A prospective comprehensive evaluation of the microbiology, epidemiology, and outcome of bacteraemia and fungemia in adults. *Clin. Infect. Dis.*, **24**, 584–602.

White, A. C., Atmar, R. L., Wilson, J., Cate, T. R., Stager, C. E., and Greenberg, S. B., 1997, Effects of requiring prior authorization for selected antimicrobials: Expenditures, susceptibilities, and clinical outcomes. *Clin. Infect. Dis.*, **25**, 230–239.

Wilkins, E. G. L., Hickey, M. M., Khoo, S., Hale, A. D., Umasankar, S., Thomas, P. *et al.*, 1991, Northwick Park Infection Consultation Service. Part II. Contribution of the service to patient management: An analysis of results between September 1987 and July 1990. *J. Infect.*, **23**, 57–63.

World Health Organization (2003). ATC index with DDDs and Guidelines for ATC classification and DDD assignment. WHO Collaborating centre for drugs statistics methodology, Oslo. www.whocc.no.

Chapter 13

Multidisciplinary Antimicrobial Management Teams and the Role of the Pharmacist in Management of Infection

Karen Knox[1], W. Lawson[2], and A. Holmes[3]

[1]*Department of Microbiology, Surrey and Sussex Healthcare NHS Trust, Crawley Hospital, West Sussex, RH11 7DH, UK*
[2]*Department of Pharmacy, Hammersmith Hospital NHS Trust, Hammersmith Hospital, Du Cane Road, London W12 ONN, UK;* [3]*Department of infectious Diseases and Microbiology, Imperial College, Hammersmith Hospital NHS Trust, Du cane Road, London W12 ONN, UK*

1. INTRODUCTION

1.1. Background

Multidisciplinary or integrated care networks are used widely and successfully within healthcare systems for the delivery of patient care. Examples of such networks include, amongst others, pain control, diabetes, and cancer services. In the United Kingdom (UK), the model for healthcare reform proposes the expansion of multidisciplinary team working (Department of Health, 2000). The application of multidisciplinary team working to improve antimicrobial prescribing has been shown to be successful and is advocated by a number of bodies representing medical and allied professions. The inappropriate or suboptimal use of antimicrobials both in hospital and community settings remains a huge problem, despite the potential benefits of prudent, targeted prescribing practices. Antimicrobial costs account for at least 30% of the drug expenditure of most hospitals, with 30–50% of patients receiving antibiotics at any one time (Berman *et al.*, 1992). Studies from the United States of America (USA) estimate that in excess of 50% of all antimicrobial

Antibiotic Policies: Theory and Practice. Edited by Gould and van der Meer
Kluwer Academic / Plenum Publishers, New York, 2005

prescriptions may be inappropriate, either in terms of drug choice, route of administration, dose, or length of treatment (Jarvis, 1996; Marr *et al.*, 1988). In the Netherlands, surveys on antimicrobial use in the hospital setting specifically, suggest that this figure is at least 15% (Van der Meer and Gyssens, 2001). Prescribing practices should be targeted for improvement via a multidisciplinary team approach.

The reason for controlling antimicrobial use is to encourage responsible prescribing in order to (1) increase the quality of patient care, (2) contain costs, and (3) attempt to minimise the emergence of microbial resistance. There is currently a paucity of robust published data reporting the effects of antimicrobial prescribing control on such outcome measures and quality research using sound methodology is encouraged. Although cost containment has often been considered to be the overriding priority when instituting antimicrobial control measures, more emphasis must be placed upon measuring effects on aspects of patient outcome and upon the epidemiology of microbial pathogens (BSAC Working Party Report, 1994; Department of Health NHS Executive, 2000; Goldmann *et al.*, 1996; House of Lords Select Committee Report, 1997; Shlaes *et al.*, 1997).

Programmes to manage or control antimicrobial use are built upon antibiotic policies. These policies are based upon local epidemiology of prescribing practices and antimicrobial resistance patterns (BSAC Working Party Report, 1994; Shlaes *et al.*, 1997) and include protocols and guidelines for the treatment and prevention of infection. Where available and appropriate, national guidelines and data should be taken into consideration. The preparation and implementation of such policies must involve prescribers who have ultimate responsibility for individual patient care (Knox and Holmes, 2002).

1.2.   Understanding prescribing practices

The epidemiology of prescribing practices at a local level needs to be assessed before attempting to alter it. This will identify areas of inappropriate antimicrobial use where interventions could be targeted and have most impact. In addition, it is important to define key process or outcome indicators by which the impact of any intervention may be assessed (Nathwani *et al.*, 2002). The factors that determine antimicrobial prescribing are multiple and complex and need to be fully realised and addressed (Avorn and Solomon, 2000). Changing prescribing practice equates to changing human behaviour, which can be extremely difficult to achieve. Studies addressing this are lacking. Factors such as clinician education, lack of local expertise, bed and staffing shortages, and consultation time constraints contribute to inappropriate antimicrobial prescribing. Furthermore, the rapid turnover of medical and nonmedical staff, particularly at junior level, necessitates a sustained interventional effort to maintain improved prescribing habits. These issues can only be addressed if there is strong support from hospital management (Avorn and Solomon, 2000; Goldmann *et al.*, 1996; Swindell *et al.*, 1983). Other factors, such as changes in patient epidemiology over time are important to recognise as these may appreciably affect both antimicrobial prescribing and the epidemiology of pathogens within individual heathcare settings (Gould and Jappy, 2000).

1.3.   Antimicrobial management programmes

Effective implementation of antibiotic management programmes is a complex task and as such, cannot be performed by an individual or individual discipline. Instead, a multidisciplinary team approach is required. Any control strategy that targets clinical practice requires the cooperation of senior members of clinical staff from the start if it is to be successfully implemented. Interventions aimed at improving antimicrobial prescribing are intended to contribute actively and positively to patient care. This needs to be emphasised in order to prevent interpretation of the exercise as being largely corrective or a curb on clinical freedom. To this end, the appointment of a key healthcare practitioner to lead or oversee such a programme will lend credibility to it and increase its chances of success (Ibrahim *et al.*, 2001; Marr *et al.*, 1988; Schentag *et al.*, 1993). The support of hospital administration is also essential from the outset, to ensure that the appropriate administrative infrastructure, financial backing, and information technology (IT) is made available. Strategies to implement antimicrobial policies include passive and interactive prescriber education, standardised antimicrobial order forms, formulary restrictions, prior approval to start or continue antimicrobials, protocolised antimicrobial streamlining, prescribing feedback, computerised decision support, and online ordering. The use of passive education alone is not effective in altering prescribing habits. It is only when some form of antimicrobial restriction, combined with educational efforts is used, that success in altering prescribing practices is shown (Evans *et al.*, 1998; Lipsky *et al.*, 1999; Schiff and Rucker, 1998). The exception to this is one-on-one educational outreach (also called "academic detailing") which has been shown to be effective (Avorn and Soumerai, 1983). This strategy is used by the pharmaceutical industry, but is unlikely to be cost-effective or achievable in the majority of healthcare settings.

In the UK, under the supervision of the Effective Practice and Organisation of Care group of the Cochrane Collaboration, a review of international published studies of antimicrobial control programmes in the hospital setting has been conducted by a working party of the British Society of Antimicrobial Chemotherapy (BSAC) and the Hospital Infection Society (Davey *et al.*, 2002). Preliminary results to date show that published studies concentrate primarily on reporting economic outcomes, with the effect on patient outcomes generally being measured by length of inpatient stay and mortality. Studies assessing effects on antimicrobial resistance patterns are few. The major inadequacy of many published studies is the use of flawed methodology such as the use of uncontrolled before and after intervention data. In these instances, confounding factors such as patient case-mix, elements of healthcare provision, and purchasing agreements are not easy to control for, thus making interpretation of conclusions difficult. This review confirms that there is a paucity of and need for further well-designed and conducted studies to assess the impact of different models of antimicrobial control programmes.

2.   MULTIDISCIPLINARY ANTIMICROBIAL MANAGEMENT TEAMS

The idea of a multidisciplinary approach for the improvement of antimicrobial prescribing practice is not new. In 1988, the Infectious Diseases Society of America (IDSA) published guidelines that advocated such an approach for the improvement of use of antimicrobials in hospitals (Marr *et al.*, 1988). Likewise in Europe, as an outcome of the European Union Conference on "the Microbial Threat," it was recommended that every hospital introduce a multidisciplinary Antimicrobial Management (or Review) Team (AMT), and that this team be given both the authority to modify antimicrobial prescribing practices as well as the responsibility for ensuring compliance with guidelines (The Copenhagen Recommendations, 1998). Authority implies the support of hospital administration, which is essential for the implementation of control programmes. This is succinctly put by Goldmann *et al.* (1996) who advocate a "multidisciplinary, systems-oriented approach, catalysed by hospital leadership." In addition, fundamental to the design of antimicrobial management programmes and the function of a multidisciplinary AMT is the careful consideration and design of an IT strategy to facilitate their implementation. Inadequate resources and a lack of manpower are limiting factors in many healthcare settings, and well-conducted studies are needed to confirm the most cost-effective methods of implementing a team approach.

2.1.   Structure of Antimicrobial Management Teams

Members of an AMT should include professionals with expertise and interest in dealing with the management and prevention of infection. As a minimum this will include an Infectious Diseases (ID) Physician or Clinical Microbiologist and a clinical pharmacist with experience in infection management (Infectious

Diseases Pharmacist) (Fraser *et al.*, 1997; Gross *et al.*, 2001; Hirschman *et al.*, 1988; Jenney *et al.*, 1999; Lee *et al.*, 1995; Schentag *et al.*, 1993), but may also include any or all of an Infection Control professional or hospital epidemiologist, a drug-utilisation review pharmacist, members from the microbiology laboratory staff, and colleagues representing medical and surgical specialties, and an IT expert (Berman *et al.*, 1992; Cook and Sanchez, 1992; Gentry *et al.*, 2000; Gums *et al.*, 1999; Hayman and Crane, 1993; Minooee and Rickman, 2000; Prado *et al.*, 2002).

A team approach requires strong leadership, good management and organisation, and strict role definitions for each member. This is to avoid potential role conflict between members of the team, which may be encountered when professionals from multiple disciplines work together (Barriere *et al.*, 1989; Burke *et al.*, 1996; Marr *et al.*, 1988). Crucially, as with any multidisciplinary team working approach, clear lines of accountability for each member of the team need to be defined from the start. This is not always emphasised in examples in the published literature. It is recommended that there should be a nominated individual to take the lead in overseeing antimicrobial management programmes (Department of Health, 1999). Although results of a recent postal survey of North American ID pharmacists' perspectives on antibiotic control programmes highlighted ID physician leadership as one of the most important factors in predicting the success of a programme (Garey *et al.*, 2000), the authors believe that a successful lead may be either clinical (ID physician or Clinical Microbiologist) or non-clinical (Pharmacist). However, recognised support from a key clinician for the team leader will serve to enhance the profile and is necessary to increase the chance of success of any scheme aimed at influencing clinical practice (Ibrahim *et al.*, 2001; Marr *et al.*, 1988; Schentag *et al.*, 1993). Other factors identified in the survey by Garey *et al.* (2000) which increased intervention success included a multidisciplinary approach and adequate allotment of time and resources.

2.2. Functions and responsibilities of AMTs

The AMT can be viewed as the driving force behind the formulation and implementation of antimicrobial policies with the aim to ensure prudent, appropriate antimicrobial prescribing within healthcare settings. IT systems must be specifically adapted to the needs of the AMT in order to ensure efficient functioning, and members of the team should be involved with development of these where possible. Responsibilities of the team include selection of antimicrobial agents for empirical and individual uses, development of protocols and guidelines for antimicrobial use, and establishment and regular review of an antimicrobial formulary (including consideration of new drugs to be included). This is done in collaboration with clinical colleagues who are encouraged to take ownership of such policies relevant to their speciality. In addition, the team has educational responsibilities and is active in developing and updating continuing educational programmes for other professional staff. The maintenance of self-education is of paramount importance, and research must be high on the agenda. Surveillance activities as well as audit and feedback functions are equally important. Examples of these functions include the monitoring and reporting of compliance with published protocols and guidelines for antimicrobial use. Reasons for breaches in protocols should be identified and addressed, and protocols may need to be adapted accordingly.

2.3.   AMTs—models of delivery

As noted above, there is a paucity of adequate published literature on the effects of AMTs on antimicrobial prescribing. The majority of published studies primarily report on economic outcomes (Hayman and Crane, 1993; Hirschman *et al.*, 1988; Lee *et al.*, 1995; Schentag *et al.*, 1993) with some in addition revealing modest positive patient outcome, measured as trends towards decreasing inpatient stay and mortality (Fraser *et al.*, 1997; Gentry *et al.*, 2000; Gross *et al.*, 2001; Gums *et al.*, 1999). Occasional studies report solely on effects on prescriber compliance with local recommendations for antimicrobial use (Berman *et al.*, 1992; Cook and Sanchez, 1992; Feucht and Rice, 2003; Jenney *et al.*, 1999; Prado *et al.*, 2002). None of the above-mentioned studies assess the effect on the epidemiology of antimicrobial resistance. Examples that illustrate how individual institutions implement AMTs may not be directly applicable universally as they have been developed with local frameworks of staffing organisation and clinical approach. For example, historically, hospital pharmacists in North America, UK, and Australasia have had primarily a clinical, ward-based role, which is in contrast with many other countries in Europe, where hospital pharmacists have less of a ward-based, clinical presence. This will obviously influence how AMTs and the role of the pharmacist can be developed in these different countries. However, published illustrations are useful in providing a framework to guide the development and application of AMTs worldwide.

2.4.   Operational aspects of AMTs

The specific way in which the AMT operates from a day-to-day basis can be tailored to individual circumstances. The team as a whole is responsible for identifying key areas for intervention through surveillance activities, as well as overseeing the formulation of appropriate protocols and guidelines for treatment and prophylaxis. Ways in which the team communicates appropriate interventions to antimicrobial prescribing will differ depending upon the available workforce, expertise, and IT. Consults may be conveyed telephonically, in written format, by direct bedside consultation on individual patients or by formal directorate-based consultations (e.g., specialist unit ward rounds). Where suitable IT is in place, advice can be given at the point of prescribing (e.g., electronic prescribing) (Table 1).

The following are examples of primarily telephonic and/or ward-based consults, given as a team approach either by a pharmacist with variable degrees of expertise in infection management and/or members of an ID team or a microbiologist.

In one of the first published randomised trials to evaluate whether antibiotic choices could be influenced favourably by a multidisciplinary AMT, patients were reviewed by both an ID fellow and a clinical pharmacist and suggested changes to therapy were placed in the medical progress note section of the medical records (Fraser *et al.*, 1997). In a study by Gums *et al.* (1999), antibiotic consults were undertaken by an ID physician and/or clinical pharmacy fellow. The consult was in the form of a simple one-page format and was either left on the patient's chart or communicated directly to the attending physician depending upon urgency. In both models, pertinent information regarding rationale for changing antimicrobial therapy was included. The multidisciplinary nature of the interventions was considered to be important in encouraging physician acceptance, which occurred in over 85% of cases. Gross *et al.* (2001) reported successful interventions by an AMT comprising an ID physician and a clinical pharmacist with post-graduate training in anti-infective therapy. Consults in this study, which largely comprised approval of restricted antimicrobials, were conveyed telephonically. Similarly, in a report by Jenney *et al.* (1999), a combination of telephonic and ward-based consultation by ID registrars and/or pharmacy staff was successfully used to implement an antibiotic control programme.

Further examples highlight the role of the clinical pharmacist in operational aspects of the AMT. Lee *et al.* (1995) describe a system where patients

Table 1. AMT activities—effecting interventions

- Telephoned consults
- Written consults
  - Clinical notes
  - Attached to prescription chart
  - Attachment of stickers/notes to prescription chart
- Automated computer-assisted decision support at the point of prescribing
- Attendance at clinical unit- or specialty-based ward rounds
- Formal ward-based review of individual patients
- Participation in educational events

---

suitable for intervention were identified by ward-based pharmacists, according to protocol. These patients were presented to the AMT for a decision upon changes to therapy. Subsequent evaluation of patient's clinical progress was made by the pharmacists and reported to the ID physician only when a new development or complication arose. In a prospective evaluation by Barenfanger *et al.* (2001), interventions involving antimicrobial agents by pharmacists were either written or telephonic. In this instance, microbiology expertise was available to the pharmacists as necessary. Uniquely, pharmacists intervening in patient care for the study group in this report were given prior specific in-service training sessions on microbiological topics to enable more informed interventions to be made. Such topics included guidelines for determination of contamination vs colonisation; interpretations of Gram-staining; and guidelines for interpretation of results from sterile and non-sterile sites. In addition, pharmacists in the intervention group had the use of a computer software program (TheraTrac 2) to allow for more timely access to patient data. This software program served as an electronic link between data generated in the microbiology laboratory and data available in the pharmacy department, such as current antimicrobial therapy and patient allergies.

A similar computer software program was used in an example by Schentag *et al.* (1993). In this illustration, optimisation of antimicrobial therapy was undertaken by clinical pharmacy antimicrobial specialists. These pharmacists worked in conjunction with members of the Clinical Infectious Diseases Division. The latter advised primarily on empirical antimicrobial therapy and consulted on complex cases. This was done either telephonically or directly by ward-based review. The roles of the two specialist teams were seen to be complementary rather than conflicting, which was vital in ensuring success in implementing the programme.

Feucht and Rice (2003) describe the use of monthly educational conferences, directed at medical residents, to reinforce previously disseminated local hospital prescribing guidelines. Information on aspects of antimicrobial resistance was also highlighted as part of this interventional programme.

There are many further examples of specialist pharmacist-led antimicrobial control programmes (Cradle *et al.*, 1995; McMullin *et al.*, 1999). Gentry *et al.* (2000) describe the role of a clinical pharmacist specialist in infectious diseases who was appointed to lead an antimicrobial control programme. In this example, consultations for change were conveyed directly at ward level by both the pharmacist and an ID physician depending upon complexity of the case.

2.5.   Methods of identifying targets for intervention

A variety of methods can be used to identify patients who are suitable for intervention by an AMT (Table 2). A commonly used point of contact to identify

Table 2. AMT activities—methods of identifying targets for intervention

---

- Attendance of clinical unit- or speciality-based ward rounds
- Response to formal request for review
- Review of antimicrobial requests
  - Ward-based chart review
  - Computer-generated order review
  - Antibiotic order forms
  - Analysis of consumption/expenditure data
  - Use of restricted antimicrobials
- Review of significant microbiological data
- Therapeutic drug monitoring
- Renal function
- Reports of adverse drug reactions
- Review of compliance to standard protocols

---

patients who may benefit from tailoring of antimicrobial therapy is that of formal ward rounds either of specialist units or on review of newly admitted patients. Attendance on these rounds by a member of an AMT is invaluable. Patients are also regularly identified in most healthcare settings by formal requests for review from the clinician ultimately responsible for individual patient care. In published examples by Cradle *et al.* (1995), Fraser *et al.* (1997), Gentry *et al.* (2000), Gross *et al.* (2001), Feucht and Rice (2003), and Wyllie *et al.* (2003), use of prescription chart review, pharmacy records, and/or computer-generated antimicrobial orders identify specific antimicrobial use likely to benefit from consult by an AMT. Other reports describe a variety of integrated methods to identify patients. These include pertinent microbiological data (culture and sensitivity results), chart reviews, antibiotic levels, renal dysfunction, use of restricted antimicrobials, and reviews of compliance to standard protocols (Gums *et al.*, 1999; Lee *et al.*, 1995).

Suitable computerised systems as described by Lee *et al.* (1995) and Barenfanger *et al.* (2001) can facilitate identification processes by linking up microbiological and pharmacy data.

2.6.   Intervention activities

One of the key factors that allow for successful intervention is that certain alterations in antimicrobial use can be protocolised. This is important, as it will enable each member of the AMT to give consistent advice in those specific situations. Examples include dose alterations, intravenous to oral switching of antibiotics (sequential therapy), streamlining or narrowing of empirical therapy based on culture results and clinical diagnosis, advice on therapeutic drug monitoring and interpretation of levels, limiting antimicrobial prophylaxis,

Table 3. AMT activities—interventions

- Antimicrobial dose or regimen alteration
- Streamlining and sequential therapy
- Discontinuation of antimicrobials
- Advice on and as a result of therapeutic drug monitoring
- Automatic stop orders for:
  - ○ antimicrobial prophylaxis
  - ○ restricted antimicrobials
  - ○ empirical antimicrobials
- Approval of restricted antimicrobials
- Assistance in interpretation of laboratory results
- Indications for specific antimicrobial use
- Suggestions for additional laboratory test ordering
- Formal educational events

indications for use of specific antimicrobials, approval of restricted antimicrobials, and additional laboratory test ordering (Table 3). In situations where protocols cannot be applied or where more complex clinical advice is required, appropriate expertise in the form of an ID physician or clinical microbiologist should be sought.

In the example by Lee *et al.* (1995) and Gentry *et al.* (2000), treatment and surgical prophylaxis guidelines were developed in order to maintain consistency in the team's recommendations. In the former, specific interventional activities included rationalising of intravenous cephalosporin use, rapid identification of patients eligible for home therapy, switch from intravenous to oral agents, and discontinuing extended use of intravenous antibiotics. Appropriate patients were identified at a ward level via these protocols, and decision for changing therapy was approved at team level. The focus of the model described by Gentry *et al.* (2000) was upon modifying the approval process of non-formulary and restricted antibiotics. The clinical pharmacy specialist was given authority and primary responsibility to approve restricted and non-formulary drugs within these guidelines. In addition, the pharmacist assisted the clinical team with clinical follow-up. In both examples, an ID physician consult was advised for complex cases or those falling outside of approved protocols.

The focus of the education-based example given by Feucht and Rice (2003) was to improve the use of intravenous vancomycin and fluoroquinolone prescribing practices in line with locally produced guidelines for appropriate use of these agents. A clinical pharmacist prospectively reviewed new orders for these drugs and intervened where appropriate, with the aim of reducing unnecessary duplication of anti-Gram-negative agent use and reducing the duration of inappropriate empirical antibiotic cover.

A team approach to improve antibiotic turnaround times has been used successfully on a medical intensive care unit (Watling *et al.*, 1996), although effect on patient outcomes was not measured. In this example, the use of a pre-printed antibiotic order form was instrumental in eliminating errors in prescription interpretation, and improved communication between all members of the healthcare staff facilitated improvement in timeliness of therapy.

Extended intervention activities were employed in the studies by Schentag *et al.* (1993) and Barenfanger *et al.* (2001). Defined interventions included antimicrobial dose adjustments, early discontinuation of intravenous antibiotics, sequential therapy, and protocol-driven early conversion from empiric to targeted therapy. In both examples, the clinically specialised pharmacists were given primary responsibility for implementing the programme but worked closely with ID physicians and microbiologists in a complementary fashion.

A number of common conclusions can be drawn from the published literature on AMTs. First, where an approach to addressing the issue of antimicrobial prescribing is multidisciplinary, it is more likely to gain acceptance from clinical colleagues. Second, leadership of such a team (or recognised support of a non-clinical lead) by a respected clinician lends credibility to it. Third, roles within the team should be well defined in order to avoid potential conflict. Fourth, a sustained effort is required to improve antimicrobial prescribing. Fifth, educational benefits are seen where clinical AMT intervention is employed, which in turn serve to sustain improved prescribing practices.

3. THE ROLE OF THE PHARMACIST IN INFECTION MANAGEMENT

There is a global precedence to promote the role of pharmacists in the prevention and treatment of infection in order to enhance prudent antimicrobial prescribing (ASHP, 1998; Audit Commissions Report NHS England and Wales, 2001; Bosch, 2000; BSAC Working Party Report, 1994; IDSA, 1997; Shlaes *et al.*, 1997). It is important to establish a role for pharmacists complementary to those of other specialists in infection management. An excellent review published by Dickerson *et al.* (2000) discusses the active contribution made by pharmacists towards promoting optimal antimicrobial use in both the hospital and community settings by providing education; developing and implementing clinical practice guidelines; and audit and feedback activities (see also Table 4). Pharmacists with specific training in infection management may be referred to as Infectious Diseases (ID), Microbiology or Antibiotic Use Review (AUR) Pharmacists. It is notable, however, that pharmacists working in other specialised clinical areas, for example, Critical Care Units, Oncology,

Table 4. The role of the pharmacist in infection management

- Clinical role in conjunction with colleagues on AMT
  - ○ Member of antimicrobial review committee—policy making, clinical practice guideline development, new drugs review
  - ○ Identification of patients for intervention activities
  - ○ Initiation of streamlining or sequential therapy
  - ○ Dose adjustments
  - ○ Therapeutic drug monitoring
  - ○ Approval of restricted antimicrobials
- Provision of expert advice on antimicrobial use
- Surveillance of antimicrobial use
  - ○ Collection and analysis of local consumption and expenditure
  - ○ Compliance with policies
  - ○ Prescribing errors
- Audit and feedback
  - ○ Includes evaluation of impact of clinical guidelines on process of care, patient outcomes, financial outcomes, and antimicrobial resistance patterns
- Educational role directed at
  - ○ Clinicians at the point of prescribing and generally
  - ○ Nursing and technical staff
  - ○ Patients—UK NHS medicines information patients' helpline
  - ○ Pharmacists
- Infection Control activities—integrating antibiotic control with infection control
  - ○ Member of hospital/community infection control committees
- Provision of outpatient or community parenteral therapy programmes
- Coordination and implementation of immunisation programmes
  - ○ Community and hospital

Haematology and Transplant Units, Renal Units, and Human Immunodeficiency Virus (HIV) medicine will have a good working knowledge of likely infections and antimicrobial use in these settings.

3.1.   Lead role in AMTs

Where available, a dedicated ID pharmacist should play a lead role in and be viewed as co-therapist with other members of the AMT (Barriere *et al.*, 1989; Lee *et al.*, 1995). The above examples by Schentag *et al.* (1993), Cradle *et al.* (1995), Lee *et al.* (1995), McMullin *et al.* (1999), Gentry *et al.* (2000), and Barenfanger *et al.* (2001) illustrate the possible role expansion of an appropriately trained pharmacist working within a multidisciplinary AMT in a variety of hospital settings. These illustrations and the other examples given above, demonstrate how, as members of a team dedicated to improving antimicrobial use in the hospital setting or with access to such a resource, clinical pharmacists with differing levels of expertise in infection management can make specific contributions towards optimising antimicrobial prescribing.

3.2.   Specialist advice and education

The educational role of the pharmacist is extremely valuable, with opportunities to inform prescribing clinicians at the point of care and in general about prudent antimicrobial prescribing (Fraser *et al.*, 1997; Gentry *et al.*, 2000). Pharmacists involved in education use a variety of methods to improve prescribing knowledge. In the examples of antimicrobial control programmes described above, communicating the rationale for changing antimicrobials to clinicians at the point of prescribing increased the likelihood that the suggested changes were made. This also contributed to the maintenance of positive alterations in prescribing practice.

A major specific role of clinical pharmacists in infection management is to provide specialist advice on aspects of antimicrobial use such as appropriate initial dose of antimicrobial and dose alterations according to renal or hepatic function, therapeutic drug monitoring, and information on drug interactions and side effects. This information can be supplied directly at the point of prescribing, on review of drug charts or antimicrobial order forms, within a formulary, via computer-assisted support programmes, or as a telephonic service. Formal pharmacokinetic consultation services are commonly established in teaching hospitals and tertiary referral facilities and could be further expanded in smaller community hospitals (Bedard and McLean, 1994).

Therapeutic drug monitoring programmes are successfully led by appropriately trained pharmacists and result in improved, consistent prescribing of antimicrobials such as aminoglycosides and glycopeptides, with fewer associated adverse drug reactions (Lynch *et al.*, 1992) as well as having financial benefits (Ariano *et al.*, 1995).

Other opportunities for disseminating information to other medical and allied professionals include formal lectures and teaching sessions within undergraduate and postgraduate training schemes, interactive educational meetings, participation in Grand Rounds, clinical ward rounds, interactive computerised educational activities (e.g., available on hospital intranet), and where resources are available, academic detailing. The reasons for the need for responsible antimicrobial use should be emphasised at clinical staff induction programmes.

In addition to providing education to medical and nursing professionals, pharmacists also provide a valuable resource for patients. A variety of pharmacist-run, telephonic-based medicines information services to provide advice to patients on aspects of their medication are available worldwide (Raynor *et al.*, 2000). In addition, appropriate patient counselling given at the point of discharge or at outpatient dispensing is designed to empower patients and is likely to improve compliance.

Pharmacists are involved in teaching other members of their professional group and, in addition, as part of continuing professional development, must maintain their knowledge- and skill-base. Adequate resources and time should be allocated for this within healthcare settings.

3.3.   Surveillance and audit activities

Surveillance activities include monitoring of antimicrobial use, generating meaningful expenditure and consumption data, and monitoring the occurrence of adverse drug reactions. Pharmacy-based monitoring services are a valuable tool for reviewing hospital prescribing and have been shown to have a positive impact (Berman *et al.*, 1992; Burke, 2001; Dean *et al.*, 2002a; Fletcher *et al.*, 1990). Pharmacists should play an increasing role in monitoring compliance to clinical guidelines as well as evaluating their impact on the process of care and patient outcomes (Dickerson *et al.*, 2000). Regular application of relatively simple collection methods such as point prevalence studies can provide a wealth of information on local prescribing practices as well as providing a means by which to monitor and feedback the effects of interventional activities (Dean *et al.*, 2002a). Formal pharmacy-oriented drug surveillance networks have been shown to be successful in collecting drug experience data generated during the routine clinical care of patients (Grasela *et al.*, 1987). As an integral part of these processes, suitable information technology and substantial effort are needed so that data may be standardised and pooled across healthcare institutions to aid in addressing important public health issues (Grasela *et al.*, 1993).

Prescribing errors are an important target for improvement, and initiatives to reduce such errors have been proposed and implemented both in the UK and in the USA (Dean *et al.*, 2002b). Errors include drug overdosing or underdosing, inappropriate dosing interval, incorrect route of delivery, prescription of agents to patients known to be allergic, and delay or omission to give a prescribed drug. Prescribing errors may result in serious adverse patient outcomes. In a review by Lesar *et al.* (1997), antimicrobials were associated with almost 40% of all medication-prescribing errors. Ward-based pharmacists who routinely examine drug charts on a daily basis are ideally situated to identify errors, as well as to gather information on possible reasons for them. Antimicrobial review systems have the potential to reduce prescribing errors and hence their associated adverse events in hospitalised patients (Guglielmo *et al.*, 1999).

Audit functions are an integral part of any process of care. Results of surveillance activities should be actively reported back to relevant parties. Examples include evaluation of compliance to clinical guidelines as well as of their impact on patient outcomes, impact of antimicrobial management and educational programmes on defined key indicators. As seen in the examples cited above, feedback to clinicians on prescribing practices serves to effect positive prescribing changes.

3.4.   Implementation of streamlining and sequential therapy

Antimicrobial streamlining is the conversion of broad-spectrum empirical therapy to a narrower spectrum agent, intravenous to oral switch (sequential therapy), as well as controlling the use of "redundant" combinations of antimicrobials. Such interventions have been shown to be safe and efficacious, and contribute to substantial cost savings (Ramirez, 1996). Specifically, the use of sequential therapy as a single measure of modifying antimicrobial use is an accepted method by which to improve the quality of patient care, achieve cost savings, and reduce drug administration time (Hamilton-Miller, 1996; Lelekis and Gould, 2001). Pharmacists, in many instances, may take the lead in initiating streamlining and sequential therapy (Allen *et al.*, 1992; Cairns, 1998; Chawla and Slayter, 1996; Frighetto *et al.*, 1992; Kuti *et al.*, 2002; Pastel *et al.*, 1992).

3.5.   Provision of outpatient or home parenteral therapy services

Outpatient or home parenteral anti-infective therapy (OHPAT) for a variety of specific infectious conditions can lead to improved patient care and has become an accepted and growing practice worldwide (Nathwani and Zambrowski, 2000; Williams *et al.*, 1997). The key element for successful delivery of such a service is a team approach (the patient, nurse, pharmacist, and clinician). In the USA and Canada (where this is usually referred to as community-based parenteral anti-infective therapy or CoPAT), a well-developed infrastructure for delivery of this service exists and practice guidelines have been published. Outside of these counties, the service is generally less well developed, although in the UK, national guidelines are also available (Nathwani and Conlan, 1998) and several centres have successful working programmes. Aside from the responsibility of preparing and supplying the anti-infective agent, specific roles for the specialist pharmacist in ensuring the successful running of such programmes include the evaluation of the patient for suitability for OHPAT, development of a treatment plan, provision of education (to patients and healthcare workers), and collection of outcome data.

3.6.   Infection control activities

Infection control and antibiotic control should be more formally integrated in the hospital setting. An ID pharmacist should be encouraged to be a member of the Infection Control Team and attend infection control committee meetings on a regular basis. In the USA, both the American Society of Consultant Pharmacists (ASCP) and the American Society of Health-System Pharmacists (ASHP) have published statements regarding the contribution by pharmacists towards promoting infection control activities (ASCP, 1997; ASHP, 1998). Participation in policy-making, education, surveillance, and quality assurance activities are some of the areas where a pharmacist with a background in infection management can be a valuable resource.

3.7. Adult immunisation programmes

Vaccination campaigns against infectious diseases are a key part of community health initiatives. In the community setting, pharmacist-led adult immunisation programmes for pneumococcal and influenza vaccination have long been advocated by ASHP (1993). Within both community and hospital settings, pharmacists can facilitate identification of patients and staff to be targeted for vaccination, provide relevant education, and supply the vaccine. In addition, in select instances, pharmacists may be in the best position to administer the vaccine (Grabenstein and Bonasso, 1999; Sanchez *et al.*, 2003).

3.8. Formulary development

Antibiotic formulary development, collation, and distribution, as well as regular review and update, require a collaborative effort in which the pharmacist can take the lead. The duties of an infectious diseases pharmacist in this regard, are guided by an antimicrobial review committee. Cook and Sanchez (1992) describe a model of multidisciplinary approach to development and implementation of an effective antibiotic formulary. During this process, pharmacist/physician working relationships were strengthened and the result was unanimous formulary acceptance.

4. TRAINING AND SUPPORT IN INFECTION MANAGEMENT FOR PHARMACISTS

4.1. Promoting the role of the pharmacist in infection management

In Europe, North America, and Australia, aside from government initiatives, the case for promoting the role of pharmacists in infection management is advocated by such authorities as the United Kingdom Clinical Pharmacy Association (UKCPA), the European Society of Clinical Pharmacy (ESCP), the ASHP, the

Society of Infectious Diseases Pharmacists (SIDP), the American College of Clinical Pharmacy (ACCP), and the Canadian Society of Hospital Pharmacists (CSHP). In addition, other professional bodies worldwide support an extended role for pharmacists (e.g., BSAC, IDSA, the Society for Healthcare Epidemiology of America, the Canadian Infectious Disease Society, Australian Society for Antimicrobials [ASA]).

4.2.   Information networking and specialised practice interest groups

Networking groups for pharmacists with the aim of disseminating information and expertise in management of infections are facilitated and supported by some of the above-mentioned organisations (Table 5). As an example, as a result of a survey of NHS Trusts, the UK (Lawson *et al.*, 2000), a national network for pharmacists involved in antimicrobial prescribing has been established. The Infection Management Practice Interest Group (now called the Infection Management Group) under the auspices of the UKCPA, aims to provide and exchange information regarding clinical experience, evidence, and best practice. Its remit also includes encouraging and supporting practice research, in addition to providing education events. This information can be further networked amongst other existing UKCPA practice interest groups for pharmacists representing other hospital specialties who have a role in antimicrobial prescribing. Other organisations that have special interest groups in infectious diseases specifically for pharmacists, with active e-mail discussion

Table 5. Resources and support networks for pharmacists in infectious disease and antimicrobial management

- Europe
  - United Kingdom Clinical Pharmacy Association (UKCPA)—Practice Interest Group in Infection Management: www.ukcpa.org
  - European Society of Clinical Pharmacy (ESCP)—Special Interest Group in Infectious Diseases: www.escpweb.org
- USA
  - Society of Infectious Diseases Pharmacists (SIDP): www.sidp.org
  - American College of Clinical Pharmacists (ACCP)—Practice and Research Network in Infectious Diseases: www.accp.com
  - American Society of Health-System Pharmacists (ASHP): www.ashp.org
  - American Society of Consultant Pharmacists (ASCP): www.ascp.org
- Canada
  - Canadian Society of Hospital Pharmacists (CSHP)—Pharmacy Specialty Network in Infectious Diseases: www.cshp.ca
- Australia
  - Australian Society for Antimicrobials (ASA): www.asainc.net.au

facilities include the ESCP, SIDP, ACCP, and CSHP. Active discussion and sharing of expertise by all professionals with an interest in infectious disease management is encouraged and facilitated by other organisations such as the IDSA, Hospital Infection Society, and the ASA.

4.3.   Post-graduate training opportunities in infection management

Although many post-graduate courses for pharmacists contain modules relating to infectious diseases and antimicrobial management, current opportunities for dedicated training in these specific areas are limited. In the USA, there are structured full-time residency programmes in Infectious Diseases pharmacy practice which are defined as organised, directed, post-graduate programmes that centre on developing the competencies necessary to provide pharmaceutical care to patients with infectious diseases (minimum 12 months). Baseline standards and learning objectives to be met by these programmes have been prepared jointly by the ASHP and the SIDP. These can be found on the ASHP website (www.ashp.org). Fellowships in research or practice are also available. More recently, in the UK, a part-time MSc in Infection Management for pharmacists is now available. This is a collaborative venture between the Health Protection Agency (HPA), Imperial College Faculty of Medicine (London), and the Academic Pharmacy Unit (Hammersmith Hospitals NHS Trust, London). In Belgium, there is a recently introduced training course in Hospital Management of Anti-Infectives open to healthcare professionals including pharmacists (Professor M. Struelens, personal communication), however, to the authors' best knowledge, there are no other post-graduate training programmes in infection management designed specifically for pharmacists in any other countries.

5.   THE FUTURE

A multidisciplinary effort is required to ensure prudent, responsible, antimicrobial prescribing practice in healthcare settings. The way forward is via the formation of antimicrobial management teams which facilitate the formulation, implementation, and auditing of antimicrobial management programmes. More research is urgently needed to produce a solid evidence-base to direct the way in which antimicrobial use can be positively influenced. This includes monitoring financial, clinical, and antimicrobial resistance outcomes of such interventions. Substantial ongoing administrative, financial, and IT support is crucial to pave the way and to ensure the success of these initiatives.

In the UK and elsewhere, it is apparent that the contribution of the pharmacist in promoting prudent antimicrobial prescribing and infection management is under-recognised and under-utilised. The misguided view of the pharmacist as "prescribing policemen" in implementing prescribing restrictions and enforcing drug approval needs to be actively discouraged. Instead, their profile and status as professionals who can contribute actively and effectively to the prevention and treatment of infectious disease, within a multidisciplinary team framework, should be promoted. In countries where there has been under-investment in this role, improved opportunities for training for pharmacists, and the creation of new posts must be high on the agenda. Over the past 3 years, the UK has seen the creation of at least 14 new hospital posts for infectious diseases/antimicrobial pharmacists, as well as the development of a post-graduate training course in infection management specifically for pharmacists who wish to practice in this area. In addition, specific funding from the Department of Health has been provided to support initiatives led by hospital-based pharmacists to promote prudent antimicrobial prescribing.

Inappropriate antibiotic prescribing must remain prominent on the research agenda. In this era of accountability and antibiotic resistant "superbugs," as members of the healthcare profession, we all have a duty of care to ensure responsible antimicrobial prescribing. Success of initiatives to achieve this remains a strategic priority, which is dependent upon hospital leadership and administrative support. The formation and deployment of multidisciplinary AMTs can successfully bring together the necessary expertise to effect relevant antimicrobial control programmes to positively influence use of antimicrobial agents.

REFERENCES

Allen, B., Naismith, N. W., Manser, A. J., and Moulds, R. F. W., 1992, A campaign to improve timing of conversion from intravenous to oral administration of antibiotics. *Aust. J. Hosp. Pharm.*, **12**, 343–439.

American Society of Consultant Pharmacists, 1997, *Guidelines on the Role of the Consultant Pharmacist in Infection Control in Long-term Care Facilities*. www.ascp.com/public/pr/ guidelines/infection.shtml

American Society of Hospital Pharmacists, 1993, ASHP technical assistance bulletin on the pharmacists' role in immunization. *Am. J. Hosp. Pharm.*, **50**, 501–505.

American Society of Health-System Pharmacists, 1998, ASHP statement on the pharmacist's role in infection control. *Am. J. Health-Syst. Pharm.*, **55**, 1724–1726.

Ariano, R. E., Demianczuk, R. H., Danziger, R. G., Richard, A., Milan, H., and Jamieson, B., 1995, Economic impact and clinical benefits of pharmacist involvement on surgical wards. *Can. J. Hosp. Pharm.*, **48**, 284–289.

Audit Commission for Local Authorities and the National Health Service in England and Wales. *A Spoonful of Sugar*. Medicines management in NHS Hospitals. Audit Commissions Publications, December 2001. www.audit-commission.gov.uk/publications/ spoonfulsugar.shtml.

Avorn, J. and Solomon, D. H., 2000, Cultural and economic factors that (mis)shape antibiotic use: The nonpharmacologic basis of therapeutics. *Ann. Intern. Med.*, **133**, 128–135.

Avorn, J. and Soumerai, J. A., 1983, Improving drug therapy decisions through educational outreach. A randomised trial of academically based "detailing." *N. Engl. J. Med.*, **308**, 1457–1463.

Barenfanger, J., Short, M. A., and Groesch, A. A., 2001, Improved antimicrobial interventions have benefits. *J. Clin. Microbiol.*, **39**, 2823–2828.

Barriere, S. L., Dudley, M. N., Kowalsky, S. F., Kreter, B., Polk, R. E., Prince, R. A. *et al.*, 1989, The role of the pharmacist in antimicrobial agent therapy. *J. Infect. Dis.*, **159**, 593–594.

Bedard, M. and McLean, W., 1994, A regional pharmacokinetic consultation service. *Can. J. Hosp. Pharm.*, **47**, 268–276.

Berman, J. R., Zaran, F. K., and Rybak, M. J., 1992, Pharmacy-based antimicrobial-monitoring service. *Am. J. Hosp. Pharm.*, **49**, 1701–1706.

Bosch, X., 2000, New role proposed for spanish pharmacists (Letter). *Lancet*, **356**, 52.

Burke, J. P., 2001, Maximizing appropriate antibiotic prophylaxis for surgical patients: An update from LDS Hospital, Salt Lake City. *Clin. Infect. Dis.*, **33**(Suppl 2), S78–S83.

Burke, J. D., Ahkee, S., Ritter, G. W., and Ramirez, J. A., 1996, Development of an interdisciplinary antimicrobial team: Elements for success. *Hosp. Pharm.*, **31**, 361–366.

Cairns, C., 1998, Implementation of sequential therapy programmes—a pharmacist's view. *J. Infect.*, **37**(Suppl 1), 55–59.

Chawla, P. and Slayter, K., 1996, Pharmacy initiated antibiotic streamlining. *Can. J. Hosp. Pharm.*, **49**, 128–129.

Cook, A. A. and Sanchez, M. L., 1992, A multidisciplinary process to determine, communicate, and manage an antibiotic formulary. *Hosp. Pharm.*, **27**, 867–869, 872–874, 882.

The Copenhagen Recommendations. Report from the Invitational EU Conference on the Microbial Threat. September 9–10, 1998, Copenhagen, Denmark.

Cradle, R. M., Darouiche, R. O., Tibbetts, C. S., and Graviss, E., 1995, Pharmacist's impact on antimicrobial drug therapy. *Am. J. Health-Syst. Pharm.*, **52**, 1544–1546.

Davey, P., Brown, E., Hartman, G., and Ramsay, C., 2002, Room for improvement: Poor quality of evaluation of interventions for improving antibiotic prescribing to hospital inpatients. Poster presentation at 42nd Interscience Conference on Antimicrobial Agents and Chemotherapy; September 26–30, 2002, San Diego, USA.

Dean, B., Lawson, W., Jacklin, A., Rogers, T., Azadian, B., and Holmes, A., 2002a, The use of serial-point prevalence studies to investigate hospital anti-infective prescribing. *Int. J. Pharm. Prac.*, **10**, 121–125.

Dean, B., Schachter, M., Vincent, C., and Barber, N., 2002b, Causes of prescribing errors in hospital inpatients: A prospective study. *Lancet.*, **359**, 1373–1378.

Department of Health, March 1999, Health Services Circular 1999/049. Resistance to antibiotics and other antimicrobial agents. Crown Publishing.

Department of Health, July 2000, The NHS Plan; A plan for investment—A plan for reform. The Stationary Office. www.nhs.uk/nhsplan

Department of Health NHS Executive, June 2000, UK Antimicrobial Resistance Strategy and Action Plan. Crown Publishing. www.doh.gov.uk/arbstrat.htm

Dickerson, L. M., Mainous, A. G., and Carek, P. J., 2000, The pharmacist's role in promoting optimal antimicrobial use. *Pharmacotherapy*, **20**, 711–723.

Evans, R. S., Pestonik, S. L., Classen, D. C., Clemmer, T. P, Weaver, L. K., Orme, J. F. *et al.*, 1998, A computer-assisted management program for antibiotics and other antiinfective agents. *N. Engl. J. Med.*, **338**, 232–238.

Feucht, C. L. and Rice, L. B., 2003, An interventional program to improve antibiotic use. *Ann. Pharmacother.*, **37**, 646–651.

Fletcher, C. V., Metzler, D., Borchardt-Phelps, P., and Rodman, J. H., 1990, Patterns of antibiotic use and expenditure during 7 years at a university hospital. *Pharmacotherapy*, **10**, 199–204.

Fraser, G. L., Stogsdill, P., Dickens, J. D., Wennberg, D. E., Smith, R. P., and Prato, B. S., 1997, Antibiotic optimisation: An evaluation of patient safety and economic outcomes. *Arch. Intern. Med.*, **157**, 1689–1694.

Frighetto, L., Nickoloff, D., Martinusen, S. M., Mamdani, G. S., and Jewesson, P. J., 1992, Intravenous- to oral stepdown programme: Four years of experience in a large teaching hospital. *Ann. Pharmacother.*, **26**, 1447–1451.

Garey, K. W., Liang, P., Itokazu, G., Schwartz, D., Rydman, R., Danziger, L. H. *et al.*, 2000, A North American (NA) survey of Infectious Diseases Pharmacists (IDP) perspectives on Antibiotic Control Programs (ACP). Abstracts of the 40th Interscience Conference on Antimicrobial Agents and Chemotherapy, September 17–20, 2000, Toronto, Canada.

Gentry, C. A., Greenfield, R. A., Slater, L. N., Wack, M., and Huycke, M. M., 2000, Outcomes of an antimicrobial control program in a teaching hospital. *Am. J. Health-Syst. Pharm.*, **57**, 268–274.

Goldmann, D. A., Weinstein, R. A., Wenzel, R. P., Tablan, O. C., Duma, R. J., Gaynes, R. P. *et al.*, 1996, Strategies to prevent and control the emergence and spread of antimicrobial-resistant microorganisms in hospitals: A challenge to hospital leadership. *J. A. M. A.*, **275**, 234–240.

Gould, I. M. and Jappy, B., 2000, Antimicrobial practice. Trends in antimicrobial prescribing after 9 years of stewardship. *J. Antimicrob. Chemother.*, **45**, 913–917.

Grabenstein, J. D. and Bonasso, J., 1999, Health-system pharmacists' role in immunizing adults against pneumococcal disease and influenza. *Am. J. Health-Syst. Pharm.*, **56**(17 Suppl 2), S3–S22.

Grasela, T. H., Edwards, B. A., Raebel, M. A., Sisca, T. S., Zarowitz, B. J., and Schentag, J. J., 1987, A clinical pharmacy-oriented drug surveillance network: II. Results of a pilot project. *Drug Intell. Clin. Pharm.*, **21**, 909–914.

Grasela, T. H., Walawander, C. A., Kennedy, D. L., and Jolson, H. M., 1993, Capability of hospital computer systems in performing drug-use evaluations and adverse drug event monitoring. *Am. J. Hosp. Pharm.*, **50**, 1889–1895.

Gross, R., Morgan, A. S., Kinky, D. E., Weiner, M., Gibson, G. A., and Fishman, N. O., 2001, Impact of a hospital-based antimicrobial management program on clinical and economic outcomes. *Clin. Infect. Dis.*, **33**, 289–295.

Guglielmo, B. J., Lubner, A. D., Corelli, R. L., Flaherty, J. F., and Jacobs, R. A., 1999, Prevention of adverse events in hospitalised patients using an antimicrobial review program. *West. J. Med.*, **171**, 159–162.

Gums, J. G., Yancey, R. W., Hamilton, C. A., and Kubilis, P. S., 1999, A randomised, prospective study measuring outcomes after antibiotic therapy intervention by a multidisciplinary consult team. *Pharmacotherapy*, **19**(12), 1369–1377.

Hamilton-Miller, J. M., 1996, Switch therapy: The theory and practice of early change from parenteral to non-parenteral antibiotic administration. *Clin. Microbiol. Infect.*, **2**(1), 12–19.

Hayman, J. N. and Crane, V. S., 1993, Multidisciplinary task force for controlling drug expenses. *Am. J. Hosp. Pharm.*, **50**, 2343–2347.

Hirschman, S. Z., Meyers, B. R., Bradbury, K., Mehl, B., Gendelman, S., and Kimelblatt, B., 1988, Use of antimicrobial agents in a university teaching hospital. Evolution of a comprehensive control program. *Arch. Int. Med.*, **148**, 2001–2007.

House of Lords Select Committee on Science and Technology, 1997, *Resistance to Antibiotics and Other Antimicrobial Agents*. 7th Report. Session 1997–8. HM Stationary Office, London.

Ibrahim, K. H., Gunderson, B., and Rotschafer, J. C., 2001, Intensive care unit antimicrobial resistance and the role of the pharmacist. *Crit. Care Med.*, **29**(Suppl), N108–N133.

Infectious Diseases Society of America Position Statement, 1997, Hospital pharmacists and infectious diseases specialists. *Clin. Infect. Dis.*, **25**, 802.

Jarvis, W. R., 1996, Preventing the emergence of multidrug-resistant microorganisms through antimicrobial use controls: The complexity of the problem. *Infect. Cont. Hosp. Epidemiol.*, **17**, 490–495.

Jenney, A., O'Reilly, M., Meagher, D., and Corallo, C. E., 1999, Interventions of an antibiotic management team. *Aust. J. Hosp. Pharm.*, **29**, 36–39.

Knox, K. L. and Holmes, A. H., 2002, Regulation of antimicrobial prescribing practices—a strategy for controlling nosocomial antimicrobial resistance. *Int. J. Infect. Dis.*, **6**(Suppl 1), S8–S13.

Kuti, J. L., Le, T. N., Nightingale, C. H., Nicolau, D. P., and Quintilliani, R., 2002, Pharmacoeconomics of a pharmacist managed program for automatically converting levofloxacin route from i.v. to oral. *Am. J. Health-Syst. Pharm.*, **59**, 2209–2215.

Lawson, W., Ridge, K., Jacklin, A., and Holmes, A., 2000, Infectious diseases pharmacists in the UK: Promoting their role and establishing a national network. *J. Infect.*, **40**, A31.

Lee, J., Carlson, J. A., and Chamberlain, M. A., 1995, A team approach to hospital formulary replacement. *Diagn. Microbiol. Infect. Dis.*, **22**, 239–242.

Lelekis, M. and Gould, I. M., 2001, Sequential therapy for cost containment in the hospital setting: Why not? *J. Hosp. Infect.*, **48**, 249–257.

Lesar, T. S., Briceland, L., and Stein, D. S., 1997, Factors related to errors in medication prescribing. *J. A. M. A.*, **277**, 312–317.

Lipsky, B. A., Baker, C. A., McDonald, L. L., and Suzuki, N. T., 1999, Improving the appropriateness of vancomycin use by sequential interventions. *Am. J. Infect. Control*, **27**, 84–90.

Lynch, T. J., Possidente, C. J., Cioffi, W. G., and Hebert, J. C., 1992, Multidisciplinary protocol for determining aminoglycoside dosage. *Am. J. Hosp. Pharm.*, **49**, 109–115.

Marr, J. J., Moffet, H. L., and Kunin, C. M., 1988, Guidelines for improving the use of antimicrobial agents in hospitals: A statement by the Infectious Diseases Society of America. *J. Infect. Dis.*, **157**, 869–876.

McMullin, S. T., Hennenfent, J. A., Ritchie, D. J., Huey, W. Y., Lonergan, T. P., Schaiff, R. A. *et al.*, 1999, A prospective, randomised trial to assess the cost impact of pharmacist-initiated interventions. *Arch. Intern. Med.*, **159**, 2306–2309.

Minooee, A. and Rickman, L. S., 2000, Expanding the role of the infection control professional in the cost-effective use of antibiotics. *Am. J. Infect. Control*, **28**, 57–65.

Nathwani, D. and Conlon, C., on behalf of the OHPAT Workshop, 1998. Outpatient and home parenteral antibiotic therapy (OHPAT) in the UK: A consensus statement by a working party. *Clin. Microbiol. Infect.*, **4**, 537–551.

Nathwani, D., Gray, K., and Borland, H., 2002, Quality indicators for antibiotic control programmes. *J. Hosp. Infect.*, **50**, 165–169. doi:10.1053/jhin.2001.1171.

Nathwani, D. and Zambrowski, J. J., on behalf of the AdHOC Workshop, 2000, Advisory group on home-based and outpatient care: An international consensus statement on non-inpatient parenteral therapy. *Clin. Microbiol. Infect.*, **6**(9), 464–476.

Pastel, D. A., Chang, S., Nessim, S., Shane, R., and Morgan, M. A., 1992, Department of pharmacy-initiated program for streamlining empirical antibiotic therapy. *Hosp. Pharm.*, **27**, 596–603, 614.

Prado, M. A., Lima, M. P., Gomes, I., and Bergsten-Mendes, G., 2002, The implementation of a surgical antibiotic prophylaxis program: The pivotal contribution of the hospital pharmacy. *Am. J. Infect. Control*, **30**, 49–56.

Ramirez, J. A., 1996, Antibiotic streamlining: Development and justification of an antibiotic streamlining program. *Pharm. Pract. Manag. Q.*, **16**, 19–34.

Raynor, D. K., Sharp, J. A., Ratenbury, H., and Towler, R. J., 2000, Medicine Information Help Lines: A survey of hospital pharmacy-based services in the UK and their conformity with guidelines. *Ann. Pharmcother.*, **34**, 106–110.

Sanchez, D., Breland, B. D., Pinkos, L., Eagle, A., Nowlin, D., and Duty, L., 2003, Pharmacist-run influenza immunization clinic for health workers. *Am. J. Health-Syst. Pharm.*, **60**, 241–243.

Schentag, J. J., Ballow, C. H., Fritz, A. L., Paladino, J. A., Williams, J. D., Cumbo, T. J. *et al.*, 1993, Changes in antimicrobial agent usage resulting from interactions among clinical pharmacy, the infectious disease division and the microbiology laboratory. *Diagn. Microbiol. Infect. Dis.*, **16**, 255–264.

Schiff, G. D. and Rucker, T. D., 1998, Computerized prescribing. Building up electronic infrastructure for better medication usage. *J. Am. Med. Assoc.*, **279**, 1024–1029.

Shlaes, D. M., Gerding, D. N., John, J. F., Craig, W. A., Bornstein, D. L., Duncan, R. A. *et al.*, 1997, Society for Healthcare Epidemiology of America and Infectious Diseases Society of America Joint Committee on the Prevention of Antimicrobial Resistance: Guidelines for the prevention of antimicrobial resistance in hospitals. *Infect. Control Hosp. Epidemiol.*, **18**, 275–291.

Swindell, P. J., Reeves, D. S., Bullock, D. W., Davies, A. J., and Spence, C. E., 1983, Audits of antibiotic prescribing in a Bristol hospital. *Br. J. Med.*, **286**, 118–122.

Van der Meer, J. W. and Gyssens, I. C., 2001, Quality of antimicrobial drug prescription in hospital. *Clin. Microbiol. Infect.*, 7(Suppl 6), 12–15.

Watling, S. M., Harter, P. J., Lee, S. M., and Yanos, J., 1996, Multidisciplinary approach to improving antibiotic turnaround time in a medical intensive care unit. *J. Pharm. Technol.*, **12**, 280–283.

Williams, D. N., Rehm, S. J., Tice, A. D., Bradley, J. S., Kind, A. C., and Craig, W., 1997, Practice guidelines for community-based parenteral anti-infective therapy. *Clin. Infect. Dis.*, **25**, 787–801.

Working Party of the British Society for Antimicrobial Chemotherapy, Working Party Report, 1994, Hospital antibiotic control measures in the UK. *J. Antimicrob. Chemother.*, **34**, 21–42.

Wyllie, S., Weeks, C., Khachi, H., Vickers, M., and Jones, G., 2003, Letter to the editor. *J. Hosp. Infect.*, **53**, 85–90.

Chapter 14

Antibiotic Policy—Slovenian Experiences

Milan Čižman and Bojana Beović
Department of Infectious Diseases, University Medical Centre Ljubljana, Japljeva 2, 1525 Ljubljana, Slovenia

1.   INTRODUCTION

Antibiotic resistance is increasing worldwide. Infections with resistant organisms have been associated with treatment failures, higher morbidity and mortality, and increased cost. In Slovenia, antibiotic resistance has become a serious healthcare issue both in hospitals and in the community. *Streptococcus pneumoniae* is becoming increasingly resistant to penicillin (20% in 2001) and erythromycin (13% in 2001). The resistance of *Streptococcus pyogenes* to erythromycin (7% in 2001) is increasing year by year. High resistance rates to fluoroquinolones have been observed recently in *Escherichia coli* and *Campylobacter* sp. isolates (10–15% and 45%, respectively in 2001). The problem of methicillin resistant *Staphylococcus aureus* in hospitals appears to be moderate and is stable or decreasing (20% in 2001). In 2001, the resistance of *Pseudomonas aeruginosa* to fluoroquinolones and to carbapenems was 30% and 10%, respectively (Mueller-Premru *et al.*, 2002). Antibiotic utilization patterns and the impact of antibiotic policy measures in the community and in hospitals in Slovenia are presented and discussed.

2.   COUNTRY AND REIMBURSEMENT

Slovenia is a small central European country with 1,992,035 million inhabitants according to the census in 30 June 2001. Almost all inhabitants (>99%)

Antibiotic Policies: Theory and Practice. Edited by Gould and van der Meer
Kluwer Academic / Plenum Publishers, New York, 2005

have compulsory insurance. Over 1 million people have additional insurance. Medicinal products are grouped in three categories: the so-called positive list, the intermediary (semi-reimbursed) list, and the negative list. Generally, compulsory insurance covers 75% of the price of a medicinal product from the positive list and 25% of the price of a product from the intermediary list. Compulsory insurance also covers the full cost of medicinal products from the positive list for children, young people up to 18 years of age, students, handicapped individuals, and persons suffering from contagious diseases, as well as the full cost of drugs and other medicinal products from the intermediary list for children, young people, students, and handicapped individuals. Additional insurance covers the difference to 100% of the price for medicinal products from the positive and intermediary lists for other citizens. Most antibiotics are on the positive list.

3.   NATIONAL ANTIBIOTIC POLICY

Slovenia does not have a national expert committee on antibiotic policies, in contrast with some other Eastern European countries (Krcmery *et al.*, 2000). A prescription is needed for every antibiotic purchase, and in human medicine, antibiotics may only be prescribed by physicians. Since there is no national expert committee, the existing guidelines, both for hospitals and primary care, have been developed by hospital committees and/or individuals (Čižman and Beović, 2002; Čižman and Marolt, 1998).

3.1.   Ambulatory care

The consumption of antibiotics in ambulatory care has been monitored in Slovenia since 1974. All pharmacies in the country are involved in the monitoring process. Data on the number of packages and the cost of antibiotics purchased, the age and gender of the patients, and the identity numbers of the physicians and healthcare institutions prescribing antibiotics are collected and analysed (Oražem and Milovanovič, 1996).

In the 1980s, the number of prescriptions for all antimicrobial agents for systemic use, including antibacterials, antifungals, antivirals, and antiparasitic drugs, was between 770 and 860 per 1,000 inhabitants per year. Antibacterial drugs represented 95% of all antimicrobial products. Antimicrobial agents in turn accounted for 12.2–14.5% of all prescriptions and were the second most commonly prescribed group of medicinal products. Extended-spectrum penicillins were the most commonly prescribed class of antibiotics (36–39%), followed by narrow-spectrum penicillins (18–24%), trimethoprim/sulfamethoxasol

(TMP/SMX) and sulfonamides (13–19%), and tetracyclines (9–12%). Over the decade, a decline in the use of narrow-spectrum penicillins, tetracyclines, and TMP/SMX was observed, accompanied by an increase in the use of extended-spectrum penicillins, cephalosporins, and at the end of the decade, macrolides (roxithromycin, azithromycin) (Marolt-Gomišček and Čižman, 1992).

In the first half of the 1990s, the proportion of antimicrobial prescriptions increased to 16% (Oražem and Milovanovič, 1996). The consumption of macrolides, quinolones, cephalosporins, and extended-spectrum penicillins increased steadily, while the number of prescriptions for sulfonamides (including combinations), narrow-spectrum penicillins and tetracyclines declined (Oražem and Milovanovič, 1996). Amoxicillin (20%) was the most commonly prescribed antibacterial agent, followed by amoxicillin/clavulanic acid (16.5%), narrow-spectrum penicillins (16.2%), TMP/SMX (11.7%), and azithromycin (7.2%). Cefaclor (2.7% of all antibacterials) was the most commonly prescribed cephalosporin and ciprofloxacin the most common fluoroquinolone (1.0%).

From 1996 to 1999, the total consumption of antibacterials in Slovenia increased by 39%, attaining 19.8 DDD/1,000 inhabitants/day in 1999 (Čižman *et al.*, 2003). The consumption of tetracyclines, narrow-spectrum penicillins, TMP/SMX, cephalosporins, lincosamides, and extended-spectrum penicillins decreased, but the consumption of combinations of penicillins with β-lactamase inhibitors (amoxicillin/clavulanic acid), macrolides, and fluoroquinolones increased significantly. The increased use of macrolides and fluoroquinolones was associated with the emergence of resistance in *S. pyogenes, S. pneumoniae*, and *E. coli* (Čižman *et al.*, 1999; Čižman *et al.*, 2000; Čižman *et al.*, 2001a, b; Čižman *et al.*, 2002).

In 1990, the Health Insurance Company in collaboration with infectious diseases (ID) specialists decided to restrict the use of amoxicillin/clavulanic acid and cephalosporins to cases where penicillin or TMP/SMX had proved ineffective or the prescription was based on susceptibility testing (Fürst, 2001). Fluoroquinolones were to be used only as sequential therapy in patients after discharge from hospital. Because of a steady increase in the consumption of amoxicillin/clavulanic acid and fluoroquinolones, the health insurance company imposed further restrictions on the use of these two antibacterial classes in 1999. Amoxicillin/clavulanate could no longer be prescribed for patients with *S. pyogenes* infections, and fluoroquinolones could only be given as an alternative treatment for acute respiratory and urinary tract infections (after clinical failure of other antibiotics), or on the basis of susceptibility tests showing sensitivity to quinolones and resistance to other antibiotics.

In 2001, the total outpatient consumption of antibiotics declined to 17.4 DDD/1,000 inhabitants/day, but still remained 36% higher than in Denmark and 21% higher than in Sweden over the same period (ESAC, 2003). The pattern of use of antibiotics in Slovenia in 2001 is shown in Table 1. The decline in

Table 1. The pattern of antimicrobial use in Slovenia in 2001

Antibiotic group	Ambulatory care %	Hospital care %
Tetracyclines	4.42	1.13
Chloramphenicols	0.00	0.01
Penicillins	58.79	43.24
Cephalosporins	2.98	18.55
Macrolides	18.16	6.57
Lincosamides	0.63	2.90
TMP/SMX	6.89	3.42
Aminoglycosides	0.00	4.68
Quinolones	7.58	13.65
Others	0.00	5.18
Total use in DDD/1,000 inhabitants/day	17.4	1.80

consumption may be attributed to the measures imposed by the health insurance company as well as to the educational efforts of ID physicians aiming to raise the awareness of general practitioners about appropriate prescribing of antibiotic drugs.

3.2. Hospital antibotic policy

In 2001, a total of 335,557 patients were admitted to 27 hospitals (1 tertiary care centre with 4 hospitals, 11 general hospitals and 12 special hospitals including psychiatric hospitals and rehabilitation centres). The mean length of hospital stay was 8.3 days and the total number of bed-days was 2,773,164 (Institute of Public Health of the Republic of Slovenia, 2002).

Almost every hospital in Slovenia has a therapeutic committee (TC). Antibiotic committees, composed of ID physicians and other specialists, have been founded in all larger hospitals over the past 20 years. The multidisciplinary composition of antibiotic committees ensures that the antibiotic policy is influenced and accepted by the different specialities in the institution. In hospitals, all antibiotics registered in the country are usually available. Unregistered antibiotics can only be purchased by request to the TC of the University Medical Centre Ljubljana (UMC), which is the only tertiary care centre in the country. In 1998, the antibiotic committee of the UMC published guidelines, which were later adopted by many other hospitals in Slovenia (Čižman and Marolt, 1998). They include recommendations on empiric therapy, documented therapy, and prophylaxis of common community- and hospital-acquired infections. The dosage, length of treatment, and cost are included as well.

Since 1998, lists of so-called restricted antibiotics have been drawn up in several hospitals. In General Hospital Celje, antimicrobial agents are divided into four groups with different levels of restriction (Šibanc *et al.*, 2002).

At the University Medical Centre Ljubljana, a list of restricted antimicrobials has been maintained since 1999. At the beginning it included 12 antimicrobial agents: ceftazidime, cefoperazone, cefpirome, imipenem, meropenem, aztreonam, vancomycin, teicoplanin, amikacin, tobramycin, chloramphenicol, and lipid forms of amphotericin B. In all hospital departments except intensive care, hemato-oncology, and surgical infection units, the use of any drug from the list is subject to authorization by an ID specialist. Piperacillin-tazobactam and cefepime were added to the list after being registered in 2001 and 2002. Several other hospitals follow the pattern of restriction applied in Ljubljana.

Slovenia participates in the ESAC (European Surveillance of Antimicrobial Consumption) project. Table 2 shows the consumption of antibacterials in Slovenian hospitals from 1985 to 2001, including the data collected for the ESAC project (Čižman *et al.*, 2003).

The data presented in Table 2 show an increase in the total use of antibacterials and specifically an increase in the use of combinations of penicillins and β-lactamase inhibitors (especially amoxicillin/clavulanic acid), cephalosporins, carbapenems, macrolides, lincosamides, fluoroquinolones, and some other antibacterials including glycopeptides. On the other hand, the consumption of narrow-spectrum penicillins, tetracyclines, amphenicols, and TMP/SMX decreased as in ambulatory care. Excluding all psychiatric hospitals and rehabilitation centres, the total consumption of antibacterials in Slovenia in the years 1998–2001 was between 48.28 and 50.60 DDD/100 bed-days (data covering annually 89–100% of bed-days).

A trend towards an increased use of systemic antibacterials has been observed in many countries. The consumption per 100 bed-days in the Netherlands increased from 37.2 DDD in 1991 to 42.5 DDD in 1996 (Janknegt, *et al.*, 2000), and in Denmark from 39.24 in 1997 to 42.8 DDD in 1999 (Danmap, 2000). In the Netherlands, the data covered over 70% of all hospital bed-days and in Denmark 95% of all bed-days (excluding psychiatric hospitals, private hospitals, and one rehabilitation centre).

In Slovenian hospitals, the consumption of antibacterials for systemic use in 2001 was 20% higher than in Denmark and Sweden (Sørensen and Monnet, 2002; Strama, 2001). The total consumption per 100 bed-days varied considerably between hospitals: in maternity hospitals it was between 23 and 29 DDD, in general hospitals between 38 and 69 DDD, in the tertiary care centre around 65 DDD, in special hospitals (orthopaedic surgery, oncology, pulmonology) it ranged from 20 to 100 DDD, in psychiatric hospitals from 3 to 13 DDD, and in rehabilitation centres from 5 to 13 DDD. In general hospitals, the utilization of

Table 2. Use of antibacterials in Slovenian hospitals (in DDD/100[a] bed-days)

ATC group	Therapeutic group	1985	1990	1995	2000	2001
J01AA	Tetracyclines	3.972	2.139	1.073	0.5853	0.5384
J01B	Amphenicols				0.0075	0.0057
J01C	Penicillins	25.138	21.801	22.844	20.9451	20.5057
J01CA	Penicillins with extended spectrum	13.374	0.076	3.386	1.6420	1.5414
J01CE	β-lactamase sensitive penicillins	10.500	8.529	6.443	2.4295	2.6498
J01CF	β-lactamase resistant penicillins	1.264	0.916	0.512	1.3941	1.5416
J01CR	Combinations of penicillins, including β-lactamase inhibitors	0.000	3.280	12.500	15.4795	14.7729
J01DA	Cephalosporins	4.672	6.546	8.576	9.5900	8.7981
J01DF	Monobactams	0.000	0.000	0.002	0.0003	0.0004
J01DH	Carbapenems	0.000	0.071	0.189	0.2495	0.2931
J01EE	Combinations of sulfonamides and trimethoprim, including derivatives	3.572	2.738	2.051	2.0204	1.6251
J01FA	Macrolides	1.195	1.959	2.981	3.5181	3.1178
J01FF	Lincosamides	0.355	0.440	0.913	1.2902	1.3753
J01GB	Aminoglycosides	2.841	2.318	2.718	2.2910	2.2192
J01MA	Fluoroquinolones	0.000	2.597	3.537	5.8517	6.4751
	Others	0.128	0.0509	1.246	2.3520	2.4579
J01	Total antibacterials for systemic use	42.148	41.326	46.345	48.4123	47.4119

[a]ATC/DDD classification, WHO version, 2001.

individual antibiotic classes varied significantly from one hospital to another. As it seems unlikely that these big differences between hospitals only reflect differences in morbidity from bacterial infections, other explanations must be sought.

In Slovenia, penicillins are the most commonly prescribed class of antibiotics (43%), followed by cephalosporins (18.5%), fluoroquinolones (13.6%), macrolides (6.5%), and other drugs, mainly metronidazole (1.8 DDD/100 bed-days), and glycopeptides (0.6 DDD/100 bed-days) (Table 2). In the Netherlands and Denmark, penicillins are prescribed more widely than in Slovenia (57%), followed by cephalosporins (8–12%), fluoroquinolones (5–8%), and aminoglycosides (4–5%). The situation in the United Kingdom is similar to that in the Netherlands and Denmark (Standing Medical Advisory Committee, 1998). In Sweden, after penicillins, the most widely used group of antibiotics in 2001 were cephalosporins with 0.24 DDD/1,000 inhabitants per day (17%), followed by fluoroquinolones with 0.19 DDD/1,000 inhabitants per day (14.6%) (Strama, 2002).

4.   ANTIBIOTIC POLICY IN THE TERTIARY CARE CENTRE

The University Medical Centre Ljubljana (UMC) is the only tertiary care centre in Slovenia. In 2001 the UMC had 2,455 beds and admitted 82,594 patients, who stayed in hospital for a total of 582,745 bed-days, the average length of stay being 7.0 days. An open drug formulary is used for all drugs in the centre.

In the period from 1995 to 1997, the consumption of antibacterial drugs in the UMC (including psychiatric units which accounted for approximately 25% of bed-days) increased by 5.9%, attaining 43.68 DDD/100 bed-days in 1997 (Vižintin and Čižman, 1998). β-lactam agents were the most commonly used antimicrobials (53%), followed by macrolide and lincosamide antibiotics (13%), and quinolones (12%) (Vižintin and Čižman, 1998). In most units, a trend towards increasing use of macrolides, lincosamides, and quinolones was associated with a marked decline in the use of tetracyclines and amphenicols.

The utilization of some problem antibiotics in the UMC has been regulated since 1998. The original list of so-called restricted agents, drawn up by the antibiotic committee, included 11 antibacterials and 1 antifungal drug (lipid associated forms of amphotericin B). Any drug from the list may be prescribed only on approval of ID or a few members of the antibiotic committee. A special order form, including data on the patient, type of infection or prophylaxis, and dosage is used for these drugs. Exceptionally, the use of an antibiotic from the list may be approved by telephone on the basis of previous consultation. The utilization of these antibiotics in individual hospital departments is monitored by a team of two ID doctors. Several departments may be covered by one team. Other common measures such as stop orders, systematic education of physicians, auditing, computer guided prescription, or rotation of antibiotics (Gould, 1999, 2002; Keuleyan and Gould, 2001; Struelens *et al.*, 1999; van der Meer and Gyssens, 2001; Wilton *et al.*, 2002) have not been used in the UMC. Unfortunately, the hospital management shows inadequate understanding of problems of antibiotic consumption and bacterial resistance and so most of the work in this field is done on a voluntary basis by a handful of enthusiasts. The consumption of antibacterials in the UMC from 1998 to 2002 is presented in Figure 1.

The data in Figure 1 show a 9% increase in the total consumption of antibacterials (from 58.91 to 64.31 DDD/100 bed-days) in the period from 1998 to 2002. The highest increase was observed in fluoroquinolones (64%),

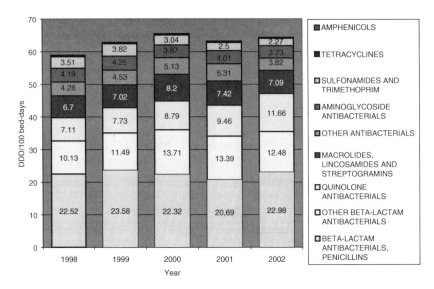

Figure 1. Consumption of antibiotics in the UMC Ljubljana (DDD/100 bed-days)

followed by macrolides and lincosamides (9%), and other β-lactam antibacterials (33%). The consumption of penicillins was stable, whereas the consumption of other antibacterials, aminoglycosides, TMP/SMX, tetracyclines, and amphenicols decreased. The consumption of restricted antibiotics varied from 2.93 DDD/100 bed-days in 1998 to 2.77 DDD/100 bed-days in 1999, 3.76 DDD/100 bed-days in 2001, and 3.39 DDD/100 bed-days in 2002. In the same period, the average length of hospital stay declined from 7.9 days in 1998 to 7.3 days in 2002. Since restricted antibacterials accounted for only 5% of all antibiotic consumption in the centre, the institution of restrictive measures for this group of antibiotics could not influence significantly the total consumption of antibitiotics in the hospital. Consequently, a decrease in the total use of antibiotics was observed only in a few departments where ID specialists were responsible for the treatment of all bacterial infections (Beović *et al.*, 2003). On the other hand, restricted antimicrobials represent approximately a third of the total cost of antimicrobial agents in the centre. Therefore the restrictions may be expected to have a significant financial impact, besides helping to reduce the development of resistant organisms (White *et al.*, 1997).

5.   CONCLUSION

The consumption of antibiotics in Slovenia is moderate. At the national level, antibiotic consumption has decreased over the past 2 years both in outpatient and inpatient settings. The resistance rates in some common community pathogens are moderate but show a tendency to increase. The problem of methicillin resistant *S. aureus* is stable and decreasing. The restrictive measures imposed by the health insurance company, along with the educational efforts of ID specialists have had a major role in reducing the consumption in outpatient settings. In hospitals, effective collaboration of ID specialists with microbiologists, hospital pharmacists, and departmental staff is often impossible due to inadequate awareness of the problem on the part of the management team.

REFERENCES

Beović, B., Matos, B., Bošnjak, R., Seme, K., Mueller-Premru, M., Hergouth-Križan, V. *et al.*, 2003, Prevention of nosocomial lower respiratory tract infections in patients after intracranial artery aneurysm surgery with a short course of antimicrobials. *Int. J. Antimicrob. Agents*, **21**.

Černelč, V., 2002, Utilization review of systemic antimicrobials in Slovene hospitals. *Med. Razgl.*, **41**(Suppl.), 43–42.

Čižman, M. and Beovič, B., 2002, Priročnik za ambulantno predpisovanje protimikrobnih zdravil. Arkadia.

Čižman, M. and Marolt, M., 1998, Priporočila za uporabo protimikrobnih zdravil KC 1998. Komisija za antibiotike v KC, Klinični center, Sekcija za kemoterapijo SZD.

Čižman, M., Oražem, A., Križan-Hergouth, V., and Kolman, J., 2001a, Correlation between increased consumption of fluoroquinolones in outpatients and resistance of *Escherichia coli* from urinary tract infections. *J. Antimicrob. Chemother.*, **47**, 502.

Čižman, M., Pečar-Čad, S., Stefancic, M., Elseviers, M., Ferech, M., Van der Stichele, R. *et al.* 2003, Consumption of antibiotics in Slovenia, first results of the ESAC retrospective data collection. ESCMID Meeting, Glasgow.

Čižman, M., Pokorn, M., Paragi, M., and the Slovenian Meningitis Study Group, 2002, Antimicrobial resistance of *invasive Streptococcus pneumoniae* in Slovenia from 1997 to 2000. *J. Antimicrob. Chemother.*, **49**, 582–584.

Čižman, M., Pokorn, M., Seme, K., and Oražem, A., 2001b, The relationship between trends in macrolide use and resistance to macrolides of common respiratory pathogens. *J. Antimicrob. Chemother.*, **47**, 475–477.

Čižman, M., Pokorn, M., Seme, K., Paragi, M., and Oražem, A., 1999, Influence of increased macrolide consumption on macrolide resistance of common respiratory pathogens. *Eur. J. Clin. Microbiol. Infect. Dis.*, **18**, 522–524.

Danmap, Danmap 2000-Consumption of antimicrobial agents and occurrence of antimicrobial resistance in bacteria from food animals, food and humans in Denmark. ISSN 1600-2032.

ESAC—European Surveillance of Antimicrobial Consumption, 2002, Interim report, November 2001-2, 51–76.

Fürst, J., 2001, Primerjava uporabe zdravil v otroški dobi v Sloveniji in Skandinaviji. In C. Kržišnik and T. Battelino (eds.), *Izbrana poglavja iz pediatrije*. Medicinska fakulteta Univerze v Ljubljani, Katedra za pediatrijo, pp. 71–76.

Gould, I. M., 1999, Stewardship of antibiotic use and resistance surveillance: the international scene. *J. Hosp. Infect.*, **43**(Suppl.), 253–260.

Gould, I. M., 2002, Antibiotic policies and control of resistance. *Curr. Opin. Infect. Dis.*, **15**, 395–400.

Institute of Public Health of the Republic of Slovenia. Health Statistical Year Book, 2002, Hospital treatment of diseases.

Janknegt, R., Lashof, A. O., Gould, I. M., and van der Meer, J. W. M., 2000, Antibiotic use in Dutch hospitals 1991–1996. *J. Antimicrob. Chemother.*, **45**, 251–256.

Keuleyan, E. and Gould, I. M., 2001, Key issues in developing antibiotic policies: From an institutional level to Europe-wide, European Study Group on Antibiotic Policy (ESGAP), Subgroup III. *Clin. Microbiol. Infect.*, **7**, 16–21.

Krcmery, V., Jeljaszewicz, J., Grzesiowski, P., Hryniewicz, W, Metodiev, K., Stratchounski, L. *et al.*, 2000, National and local antibiotic policies in central and eastern Europe. *J. Chemother.*, **12**, 471–474.

Marolt-Gomišček, M. and Čižman, M., 1992, The usage of antibiotics for outpatients and recommendations for the treatment of bacterial infections. *Zdrav. Vestn.*, **61**, 403–406.

Mueller-Premru, M., Seme, K., Križan-Hergouth, V., Andlovic, A., Kolman, J., and Gubina, M., 2002, Trends of bacterial resistance at the University Medical Center in Ljubljana. *Med. Razgl.*, **41**(Suppl. 2), 25–34.

Oražem, A. and Milovanovič, M., 1996, Out-patient prescribing of antibacterials for systemic use in Slovenia. *Med. Razgl.*, **35**(Suppl. 1), 1–11.

Sørenson, L. T., and Monnet, D., 2002, Control of antibiotic use in the community: The Danish experience. *Infect. Control Hosp. Epidemiol.*, **21**, 387–389.

Standing Medical Advisory Committee, 1998, Subgroup on Antimicrobial Resistance, the path of least resistance. Department of Health http://www.doh.gov.uk/smac.thm

Struelens, M. J., Byl, B., and Vincent, J. L., 1999, Antibiotic policy: A tool for controlling resistance of hospital pathogens. *Clin. Microbiol. Infect.*, **5**(Suppl.), 19–24.

Šibanc, B., Lešničar, G., and Tratar, F., 2002, Results of controlled antimicrobial treatment at the Celje general hospital between 1995–2000. *Med. Razgl.*, **41**(Suppl.), 53–60.

The Swedish Strategic Programme for the rational use of antimicrobial agents (Strama), 2001, O. Cars K. Ekdahl (eds.) Swedres 2001, A report on Swedish antibiotic utilisation and resistance in human medicine. The Swedish Strategic Programme for the Rational Use of Antimicrobial Agents (STRAMA) and the Swedish Institute for Infectious Disease Control, 23–26.

Van der Meer, J. W. M. and Gyssens, I. C., 2001, Quality of antimicrobial drug prescription in hospital. *Clin. Microbiol. Infect.*, **7**(Suppl. 6), 12–15.

Vižintin, T. and Čižman, M., 1998, Antimicrobial usage and trends in the University Medical Centre in Ljubljana from 1995 to 1997. *Zdrav. Vestn.*, **67**, 721–725.

White, A. C. Jr., Atmar, R. L., Wilson, J., Cate, T. R., Stager, C. E., and Greenberg, S. B., 1997, Effects of requiring prior authorization for selected antimicrobials: Expenditures, susceptibilities, and clinical outcomes. *Clin. Infect. Dis.*, **25**, 230–239.

WHO Collaborating Centre for Drug Statistics Methodology, 2001, Guidelines for ATC classification and DDD assignment. Oslo.

Wilton, P., Smith, R., Coast, J., and Millar, M., 2002, Strategies to contain the emergence of antimicrobial resistance: A systematic review of effectiveness and cost-effectiveness. *J. Health Serv. Res. Policy*, **7**, 111–117.

Chapter 15

Intensive Care Unit

Hakan Hanberger[1], Dominique L. Monnet[2], and Lennart E. Nilsson[3]
[1]Division of Infectious Diseases, Department of Molecular and Clinical Medicine, University Hospital, S-581 85 Linköping, Sweden; [2]Antimicrobial Resistance Surveillance Unit, National Center for Antimicrobials and Infection Control, Statens Serum Institut Denmark; [3]Division of Clinical Microbiology, Department of Molecular and Clinical Medicine, University Hospital, S-581 85 Linköping, Sweden

1. INTRODUCTION

Critically ill-patients admitted to intensive care units (ICUs) are highly susceptible to infections because of predisposing illnesses and the use of invasive procedures, and are therefore exposed to high antibiotic pressure. Use of antibiotics in the ICU must follow best clinical practice if the emergence of resistance to antibiotics is to be minimised. Antibiotic resistance is an important factor governing treatment success and mortality (Carmeli *et al.*, 2002; Kollef and Ward, 1998; Kollef *et al.*, 1999). The problem of resistance is greater in ICUs than in other hospital wards or primary care centres (Archibald *et al.*, 1997; Hanberger and Nilsson, 2000; Hanberger *et al.*, 2001a). Control of antibiotic resistance, that is, detecting, monitoring, and fighting the emergence of resistant bacteria is, therefore, especially important in the intensive care environment. According to a recent review of European ICUs, the prevalence of antibiotic resistance in bacteria was, with some exceptions, highest in ICUs in Southern European countries and in Russia, and lowest in Scandinavia (Hanberger *et al.*, 2001a). This was also true for the key organism methicillin-resistant *Staphylococcus aureus* (MRSA) (Regnier, 1996; Vincent *et al.*, 1995). Antimicrobial resistance also varies markedly by region and ward level in the United States, Canada, and Latin America with the highest resistance rates

Antibiotic Policies: Theory and Practice. Edited by Gould and van der Meer
Kluwer Academic / Plenum Publishers, New York, 2005

being found in Latin America (Burwen *et al.*, 1994; Diekema *et al.*, 1997; Edmond *et al.*, 1999). As patterns of resistance change, physicians need to reassess standard therapies to ensure appropriate antibiotic coverage.

2.   ANTIBIOTIC CONSUMPTION IN ICUs

Because data on antimicrobial use are reported using various measurement units, comparisons are only possible among ICUs using the same measurement unit. Several studies have reported antibiotic use expressed as a number of WHO Defined Daily Doses (DDD) per 1,000 patient-days in individual or groups of European ICUs. Depending on the ICU, antibiotic use ranged from 490 to 3,456 DDD per 1,000 patient-days (Erlandsson *et al.*, 1999; Gruson *et al.*, 2000; Hanberger *et al.*, 2004; Kiivet *et al.*, 1998; Lemmen *et al.*, 2000; Naaber *et al.*, 2000; Petersen *et al.*, 1999; Vlahovic-Palcevski *et al.*, 2000; Walther *et al.*, 2002).

In one study, Bergmans *et al.* (1997) used the prescribed daily dose (PDD) as the measurement unit and reported 921 PDD per 1,000 patient-days in two Dutch general ICUs in 1994. The ICARE DDDs developed to report antibiotic use in Centers for Disease Control and Prevention (CDC) Project ICARE represent a form of PDDs (Capellà, 1993). In 40 US hospitals, which participated in Project ICARE during the period 1996–7, antibiotic use ranged from 413 to 927 ICARE DDD per 1,000 patient-days depending on the type of ICU (ICARE Surveillance Report, 1999). It is important to note that this does not correspond to total antibiotic use since Project ICARE did not collect data on all antibiotic classes used in these ICUs.

Other studies have collected data at patient level and expressed antibiotic use as the number of daily antibiotic treatments (all individual antibiotics received on a single day are taken into account) per 1,000 patient-days. In a group of four Danish ICUs, Petersen *et al.* (1999) reported that antimicrobial use ranged from 1,390 to 2,510 daily antimicrobial treatments per 1,000 patient-days. Although the highest use was reported from one ICU that routinely used selective decontamination of the digestive tract (SDD), antimicrobial use in this ICU is likely to have been underestimated because multiple agents for the SDD protocol were recorded as one single antimicrobial.

In the European Strategy for Antibiotic Prophylaxis (ESAP) study, the median antimicrobial use (including antifungals) was 928 daily treatments per 1,000 patient-days (range: 355–1,686) in 21 ICUs that did not use SDD (Monnet *et al.*, 2000). In comparison, two ICUs that routinely used SDD reported 3,753 and 4,794 daily antimicrobial treatments per 1,000 patient-days and two other ICUs that used SDD for very selected indications only reported 997 and 1,085 daily antimicrobial treatments per 1,000 patient-days (Monnet *et al.*, 2000). It seems that, when routinely used, SDD may represent the largest part of overall antimicrobial use in an ICU.

Data collected at patient level also allow the expression of antibiotic use in terms of exposure, either as a number of antibiotic exposure-days (several antibiotics received by a single patient on a single day count for one day of exposure) per 1,000 patient-days or as a percentage of ICU patients who received at least one antibiotic. Fischer *et al.* (2000) reported 546 antibiotic exposure-days per 1,000 patient-days in a Swiss paediatric ICU in 1998–9. The European Prevalence of Infection in Intensive Care (EPIC) one-day multicentre prevalence study performed in 1992 found that 62% of patients in 1,047 ICUs from 17 European countries received antibiotics (Vincent *et al.*, 1995). A German point prevalence survey in 1994 found that 53% of ICU patients in 72 hospitals received antibiotics (Gastmeier *et al.*, 2000). In a 3-month incidence study performed in 49 Spanish ICUs in 1996, 53% of patients received antibiotics (GTEI-SEMIUC, 1996). In the 21 ESAP ICUs that did not use SDD, a median 75% of patients received antimicrobials (range: 23–100%) (ESAP, unpublished data). Similar figures were found in a 2-week prevalence study carried out in 2000 in 23 Swedish ICUs (Hanberger *et al.*, 2001b). Studies performed in individual adult ICUs reported that 68–80% of patients received antibiotics (Bourdain *et al.*, 1999; Kollef *et al.*, 2000; Røder *et al.*, 1993; Tarp and Møller, 1997). In neonatal ICUs, studies performed in individual units showed that 24–46% of patients received antibiotics (Andersen and Meberg, 1999; Borderon *et al.*, 1992; Fonseca *et al.*, 1994; Tullus *et al.*, 1988). However, much higher percentages were reported in selected groups of neonates. For example, the percentage of neonates receiving antibiotics was 92% in premature infants weighing less than 1,500 g at birth (Fonseca *et al.*, 1994) and virtually 100% in preterm neonates (<30 weeks) requiring mechanical ventilation (Gortner, 1993).

Some studies have attempted to compare antibiotic use in ICUs and other hospital wards. In Project ICARE, the median rate of antibiotic use was higher in adult ICUs than in non-ICU areas combined (Fridkin *et al.*, 1999). This was especially true for third-generation cephalosporins, intravenous vancomycin, penicillins with anti-pseudomonal activity, and intravenous fluoroquinolones. In three European hospitals, Kiivet *et al.* (1998) reported that antibiotic use expressed as a number of DDD per 1,000 patient-days was 2–6 times higher in ICUs than in surgical and medical units. In one US hospital, the total number of days of antibiotic therapy and total number of grams of antibiotic per patient-day were 1.5 times greater in the ICU than in non-ICU areas (White *et al.*, 2000). In one Danish hospital, Tarp and Møller (1997), reported that 69% of patients in ICU received antibiotics as compared to only 24% and 17% in surgical and medical wards, respectively. Finally, antibiotic pressure in ICUs is much higher than in primary care. In European Member States, antibiotic use in primary care in 1997 ranged from 8.9 DDD per 1,000 inhabitant-days in the Netherlands to 36.5 DDD per 1,000 inhabitant-days in France (Cars *et al.*, 2001). Similar data from outpatients can be compared to antibiotic use in hospitals (including ICUs), for example, 392 DDD per 1,000 patient-days in Danish hospitals in 1999 (DANMAP, 2001) or in specific hospital areas such as ICUs (see above, data in WHO DDD per 1,000 patient-days) since both one inhabitant-day and one patient-day represent one person on a defined day.

3. ANTIBIOTIC RESISTANCE IN ICU

3.1. *Enterobacteriaceae*

The Gram-negative pathogens most frequently isolated from ICU infections are *Escherichia coli, Klebsiella pneumoniae, Enterobacter cloacae,* and *Pseudomonas aeruginosa.* The number of nosocomial infections caused by *Acinetobacter* spp has increased in recent years because they are intrinsically resistant to many of the commonly used antimicrobial agents. The carbapenems are more active against *E. coli* and *K. pneumoniae* than the "third-generation" cephalosporins, ceftazidime, and cefotaxime/ceftriaxone (Table 1). The prevalence of ESBL-producing strains amongst *E. coli* (0–23%) and *K. pneumoniae* (5–64%) explains the difference in activities observed between these two antimicrobial classes. ESBL production in Gram-negative bacteria may confer resistance to virtually all commonly prescribed β-lactam antimicrobials, with the exception of the carbapenems (Table 1).

A substantial increase in the levels of ciprofloxacin resistance in *E. coli* and *K. pneumoniae* can be seen (Table 1). This is a cause for concern, especially as these species are the most frequently isolated *Enterobacteriaceae* in the ICU and can harbour ESBLs.

The resistance of *E. cloacae* to ceftazidime (19–68%—Table 1) is probably due to either the selection of Enterobacter strains producing stable derepressed constitutive chromosomal class I lactamases which hydrolyse most β-lactam antibiotics (except carbapenems which show 0–6% resistance), or the spread of Enterobacter strains, producing ESBL. The high level of use of β-lactam antibiotics such as amoxicillin, and second- and third-generation cephalosporins probably explains the increased endemic prevalence of Enterobacter producing class I β-lactamases. This endemic situation is also seen in Northern Europe (Table 1).

An alarmingly high resistance to ciprofloxacin (31%) in *Enterobacter* spp was seen in ICUs in Belgium in 1994–5 and in a European study in 1999 (20%)

(Table 1). Quinolone resistance was markedly lower in a European study in 2001 (9%) and in Belgium in 2002 (9%) (Table 1). Ciprofloxacin resistance was lower (0–9%) in Germany, Spain, Sweden, and Turkey (Table 1).

The explanation for this may differ between these countries and may depend on low total quinolone consumption, more appropriate quinolone use, the regional emergence and spread of epidemic multiresistant strains of *Enterobacter aerogenes*, or better hygiene routines in hospitals, thereby preventing outbreaks of quinolone-resistant *Enterobacter* spp. Carbapenems are the most active agents against *Enterobacter* spp (Table 1).

3.2. Non-fermentative Gram-negative bacilli

The non-fermentative Gram-negative bacilli, *Acinetobacter* spp and *P. aeruginosa*, generally show lower levels of susceptibility than the *Enterobacteriaceae* to all antimicrobials. Imipenem and meropenem have a markedly wider spectrum than other antibiotics when tested against *Acinetobacter* spp. Against *P. aeruginosa*, imipenem and meropenem also exhibit relatively high activity, with meropenem having higher activity than imipenem (Table 1). Piperacillin/tazobactam and ceftazidime were also active against many *P. aeruginosa* (Table 1). An increase in quinolone and carbapenem resistance among *P. aeruginosa* was seen in some studies (Table 1).

3.3. Coagulase-negative staphylococci (CoNS)

CoNS are low virulence pathogens. However, over the past two decades, CoNS have been increasingly recognised as a prevalent cause of nosocomial infection. For example, the NNIS and SCOPE studies rank CoNS as the most common cause of nosocomial bloodstream infection in US ICUs (Edmond *et al.*, 1999; Fridkin *et al.*, 1999) and the EPIC study found CoNS to be the fourth most common cause of nosocomial infection when all sites of infection were considered (Vincent *et al.*, 1995). Unfortunately, antimicrobial treatment of CoNS is complicated by very high rates of oxacillin resistance worldwide. The EPIC study, carried out in 1992 in 17 Western European countries, demonstrated a 70% rate of oxacillin resistance in CoNS (Vincent *et al.*, 1995). More recent data from a European study revealed higher rates of oxacillin resistance in CoNS from ICUs (88%) than non-ICUs (74%) (Rodriguez-Villabos *et al.*, 2000). North American data from 2001 revealed 84% of CoNS to be oxacillin resistant (Stephen *et al.*, 2002). In a study performed in 1999–2000 at 16 Nordic centres, 68% of CoNS from ICU patients were oxacillin resistant, but that was the case for only 33% of CoNS collected from patients at primary care centres (Hanberger and Nilsson, 2000). Most oxacillin-resistant CoNS are

Table 1. Surveillance of antibiotic resistance in European ICUs

Escherichia coli

Reference	Hanberger et al. (1999a)	MYSTIC (2003)	Ruckdeschel et al. (1998)	MYSTIC (2003)	Hanberger et al. (1999a)	MYSTIC (2003)
Country	Belgium	Belgium	Germany	Germany	Spain	Spain
Year	1994–5	1997–2002	1996–7	1997–2002	1994–5	1997–2002
NCCLS breakpoints	I + R	I + R	R	I + R	I + R	I + R
Ceftazidime	4	6*	0	8*	1	5*
Ceftriaxone/Cefotaxime	2	5*	0	2*	2	2*
Ciprofloxacin	6	10	6	15	14	20
Gentamicin	4	5	4	7	7	4
Imipenem	1	0	0	0	0	0
Meropenem	—	0	—	0	—	0
Piperacillin–tazobactam	15	2	3	3	4	2

Enterobacter spp

Species	Enterobacter spp	E. cloacae	E. cloacae	E. cloacae	Enterobacter spp	E. cloacae
Year	1994–5	1997–2002	1996–7	1997–2002	1994–5	1997–2002
Ceftazidime	43	20	27	26	31	19
Ceftriaxone/Cefotaxime	37	20	—	26	30	24
Ciprofloxacin	31	3	0	3	4	1
Gentamicin	3	4	—	3	4	0
Imipenem	3	0	—	0	2	0
Meropenem	—	0	—	1	—	0
Piperacillin–tazobactam	51	13	—	14	23	16

Klebsiella spp

Species	K. pneumoniae	K. pneumoniae	Klebsiella spp	K. pneumoniae	K. pneumoniae	K. pneumoniae
Year	1994–5	1997–2002	1996–7	1997–2002	1994–5	1997–2002
Ceftazidime	3	17*	1	19*	4	6*
Ceftriaxone/Cefotaxime	6	15*	1	9*	4	10*
Ciprofloxacin	1	3	4	14	2	3
Gentamicin	2	9	2	10	5	7
Imipenem	0	0	0	0	0	0
Meropenem	—	0	—	0	—	0
Piperacillin–tazobactam	14	7	10	13	3	1

Pseudomonas aeruginosa

Year	1994–5	1997–2002	1996–1997	1997–2002	1994–1995	1997–2002
Ceftazidime	11	14	2	5	16	9
Ciprofloxacin	16	59	13	16	14	10
Gentamicin	23	12	—	23	18	11
Imipenem	16	17	7	5	22	20
Meropenem	—	10	—	3	—	3
Piperacillin–tazobactam	13	14	4	4	8	10

Acinetobacter spp

Species	Acinetobacter spp	A. baumannii	A. baumannii	A. baumannii	Acinetobacter spp	A. baumannii
Year	1994–5	1997–2002	1996–1997	1997–2002	1994–5	1997–2002
Ceftazidime	18	14	3	3	76	80
Ciprofloxacin	18	19	15	5	81	90
Gentamicin	18	14	—	4	81	86
Meropenem	—	2	—	0	—	11
Imipenem	12	2	—	0	16	18
Piperacillin–tazobactam	36	12	0	2	58	79

*ESBL phenotype by NCCLS criteria.

Hanberger et al. (1999a) Sweden 1997 I + R	MYSTIC (2003) Sweden 1997–2002 I + R	Aksaray et al. (2000) Turkey 1997 I + R	MYSTIC (2003) Turkey 1997–2002 I + R	Mathai et al. (2000) Europe 1999 I + R	Garcia-Rodriguez and Jones (2002) Europe 2000 I + R	Mathai et al. (2000) USA 1997–9 R	Stephen et al. (2002) USA 2001 I + R
1	2*	26	23*	9	13	3*	8*
1	0*	25	27*	4	—	2*	4*
2	5	19	38	11	16	3	13
0	4	22	23	6	8	3	8
0	0	1	1	0	<1	0	0
—	0	—	2	—	<1	—	0
4	4	35	31	11	15	5	5

Enterobacter spp 1997	*E. cloacae* 1997–2002	*Enterobacter* spp 1997	*E. cloacae* 1997–2002	*Enterobacter* spp 1999	*E. cloacae* 2000	*Enterobacter* spp 1997–9	*Enterobacter* spp 2001
19	22	68	42	40	53	33	28
30	—	70	39	37	—	30	23
3	1	8	9	20	9	6	11
0	0	44	—	9	14	9	4
0	0	4	0	6	1	< 1	2
—	0	—	0	—	1	—	2
17	18	72	35	37	49	27	26

Klebsiella spp 1997	*K. pneumoniae* 1997–2002	*Klebsiella* spp 1997	*K. pneumoniae* 1997–2002	*Klebsiella* spp 1999	*K. pneumoniae* 2000	*Klebsiella* spp 1997–9	*Klebsiella* spp 2001
1	5*	73	64*	39*	41	10*	16*
1	—	69	57*	37*	—	10*	15*
4	4	30	34	24	22	7	15
2	1	66	55	36	27	6	10
0	0	3	<1	0	1	0	0
—	0	—	4	—	1	—	0
10	3	76	50	32	32	9	10

1997	1997–2002	1997	1997–2002	1999	2000	1997–9	2001
2	11	57	56	29	29	24	23
8	26	56	61	40	45	20	30
14	6	70	81	44	46	15	19
16	18	52	57	38	36	15	22
—	11	—	50	34	31	—	20
0	10	53	44	36	19	12	13

Acinetobacter spp 1994–5	*A. baumannii* 1997–2002	*Acinetobacter* spp 1997	*A. baumannii* 1997–2002	*Acinetobacter* spp 1999	*A. baumannii* 1997–2000	*Acinetobacter* spp 1997–9	*Acin baumannii* 2001
0	5	88	80	79	65	43	43
19	10	67	78	78	66	41	47
0	5	83	94	81	65	41	47
—	0	—	42	—	20	—	21
19	5	44	47	58	18	7	19
56	29	89	85	78	72	36	41

resistant to many other antimicrobial classes (Diekema *et al.*, 2001), which no doubt has contributed to the widespread use of glycopeptides in the hospital setting. Glycopeptide resistance has been well described in CoNS (Schwalbe *et al.*, 1987), but appears to be relatively uncommon in contemporary surveillance studies. Of over 6,000 strains of CoNS collected between 1997 and 1999 from centres worldwide, none were resistant to vancomycin, 1.9% had a vancomycin MIC of 4 μg/ml, and 1.9% were resistant to teicoplanin (MIC 16 μg/ml) (Diekema *et al.*, 2001). Of the isolates collected from European centres, only 9 of 2,068 CoNS strains (0.4%) were resistant to teicoplanin and none were resistant to vancomycin (Diekema *et al.*, 2001). As rates of glycopeptide resistance rise, the use of newer agents for treatment of resistant Gram-positive pathogens will increase. Three years of SENTRY surveillance (1997–9) revealed 99.9% of CoNS to be inhibited by 4 μg/ml of linezolid, while 99% were inhibited by 1 μg/ml of quinupristin/dalfopristin (Diekema *et al.*, 2001).

3.4.   *Staphylococcus aureus*

S. aureus is one of the most virulent of human bacterial pathogens. Since the emergence of the first oxacillin-resistant *S. aureus* (ORSA) strains in the early 1960s, the spread of ORSA has been reported in Europe and throughout the world. In the EPIC study, 60% of *S. aureus* isolates causing ICU infections were oxacillin resistant (Vincent *et al.*, 1995). However, the prevalence of ORSA varied widely from country to country, with national oxacillin resistance rates of approximately 80% in Italy and France, 77% in Greece, 67% in Portugal and Belgium, 54% in Spain, 53% in Austria, 37% in Germany, 14% in Switzerland, 13% in Great Britain, and no oxacillin resistance detected in Norway, Holland, Sweden, or Denmark (Regnier, 1996; Vincent *et al.*, 1995). Voss *et al.* (1994) confirmed that ORSA prevalence in many European ICUs exceeds 50% with the highest resistance rates seen in the countries of Southern Europe. In 25 European centres, mean ORSA prevalence during 1997 in all blood isolates collected from ICU patients was 39% and levels also varied widely by country (Fluit *et al.*, 2001). In the 25 European centres, the mean ORSA prevalence during 1997 in all blood isolates collected from ICU patients was 39% and levels also varied widely by country. Overall, national ORSA rates ranged from <5% in Germany, Switzerland, and the Netherlands to >50% in Italy and Portugal (Fluit *et al.*, 2001). These data are consistent with other data from the Netherlands and Germany that reveal low rates of oxacillin resistance in *S. aureus* (Ruckdeschel *et al.*, 1998).

In other areas of Northern Europe, a recent study performed in 16 Nordic centres in 1999–2000 showed an oxacillin resistance rate in ICUs of only 3% (Hanberger and Nilsson, 2000).

In the United States, ORSA rates in *S. aureus* are high in ICUs—approximately 40% in a study performed in 1994–5 in eight geographically separate hospitals (Archibald *et al.*, 1997). Data collected in 2001 reported a 51% rate of oxacillin resistance in isolates of *S. aureus* from ICU patients (Stephen *et al.*, 2002). Finally, in a European study performed in 2000, Rodriguez-Villalobos *et al.* (2002) showed that 47% of *S. aureus* isolates collected from ICU patients were oxacillin resistant compared to 25% of isolates from non-ICU patients.

Most strains of ORSA are resistant to multiple drugs, but co-resistance patterns vary from region to region. Using representatives of eight different classes of antimicrobial agents (gentamicin, rifampicin, chloramphenicol, ciprofloxacin, tetracycline, clindamycin, erythromycin, and trimethoprim/sulfamethoxazole), all ORSA strains from 1997 to 1999 were examined with respect to the mean number of co-resistances by country and region (Diekema *et al.*, 2001). The highest co-resistance rates were found in Latin America (mean = 4.7) and the Western Pacific (mean = 5.7). ORSA from European centres had a mean of 4.5 co-resistances: over 89% were resistant to ciprofloxacin, 83% to erythromycin, 74% to clindamycin, and 72% to gentamicin. Such high rates of co-resistance in ORSA underscore the importance of developing newer agents to treat serious infections caused by ORSA. Scattered reports of serious infections caused by ORSA with decreased susceptibility to the glycopeptides have been appearing in the literature since 1997 (CDC, 1997).

If glycopeptide resistance becomes widespread in ORSA, additional therapeutic options will be required. Fortunately, glycopeptide resistance in ORSA appears to be neither common nor widespread. Of over 15,000 clinical strains of *S. aureus* collected between 1997 and 1999 from SENTRY centres worldwide (including 3,477 strains from European centres), none were resistant to vancomycin. However, nine strains (or 0.3%) from European centres had an MIC to vancomycin of 4 μg/ml and one strain (0.03%) was resistant to teicoplanin (MIC > 16 μg/ml) (Diekema *et al.*, 2001). In addition, all ORSA in this study were inhibited by 4 μg/ml of linezolid, and 98% were inhibited by 1 μg/ml of quinupristin/dalfopristin. These newer agents appear to be promising alternatives to the glycopeptides for the treatment of strains of ORSA that are resistant to multiple drugs.

The first documented case of infection caused by vancomycin-resistant *S. aureus* (VRSA) was reported in the United States in 2002 (CDC, 2002). The MIC results for vancomycin, teicoplanin, and oxacillin were >128, 32, and >16 μg/ml, respectively. The isolate contained the *vanA* vancomycin resistance gene from enterococci, which is consistent with the glycopeptide MIC profiles. It also contained the oxacillin-resistance gene *mecA*. The isolate was susceptible to chloramphenicol, linezolid, minocycline, quinupristin/dalfopristin, tetracycline, and trimethoprim/sulfamethoxazole.

3.5.   Enterococci

The incidence of resistance among *Enterococcus faecium* was 10% for vancomycin, 7% for teicoplanin, 53% for ampicillin, and 30% for aminoglycoside (high-level resistance) according to a study performed in 1997–8 in 25 European hospitals (Fluit *et al.*, 2001). No glycopeptide resistance, less than 1% ampicillin resistance and 32% high-level aminoglycoside resistance was seen among *Enterococcus faecalis*, which were isolated five times more frequently than *E. faecium* (Fluit *et al.*, 2001). No vancomycin-resistant enterococci (VRE) were seen in *Enterococcus* spp collected in a European study in 1999 (Mathai *et al.*, 2000). Similarly, a study performed in ICUs in 16 Nordic centres in 1999–2000 showed a VRE prevalence below 1% (Hanberger and Nilsson, 2000). Data collected in 2001 from European ICUs showed a 3% rate of VRE in isolates of *E. faecalis* from ICU patients compared to 1% VRE in non-ICUs (Rodriguez-Villabos *et al.*, 2002).

The VRE problem is more widespread in North American ICUs according to the study carried out by Fridkin *et al.* (1999) in 1996–7, showing an overall VRE prevalence in ICUs in the United States of 13% which is higher than that reported in the study carried out in 1994–5 by Archibald *et al.* (1997). In a more recent study performed during 2001 in North American ICUs, Stephen *et al.* (2002) found 28% VRE among *Enterococcus* spp.

4.   IMPROVING ANTIBIOTIC PRESCRIBING

4.1.   The impact of antibiotic policies and antibiotic consumption on antibiotic resistance

Controlling antibiotic resistance requires not only improved antibiotic usage but also better compliance with infection control practices—in particular, hand disinfection. Emergence of antibiotic resistance in the ICU setting may be due to the development of resistance during therapy, or to the selection and overgrowth of preexisting resistant flora. These processes can be prevented by reducing the use of antibiotics, selecting narrow-spectrum drugs with low ecological impact, or by using bactericidal drugs that discourage mutations. However, the spread of resistant clones of, for example, MRSA, ESBL, or VRE from patients already colonised or infected with these resistant bacteria on admission or acquired within the ICU (Bonten and Mascini, 2003) has to be controlled by hygiene measures such as isolation and improved hand hygiene. Various strategies have been tried to limit antibiotic resistance (Diaz and Rello, 2002). However, some basic requirements must first be met and these are: reducing unnecessary use of antibiotics, selecting the proper dose, frequency, route of administration and duration of treatment, and monitoring drug levels, when appropriate.

Adverse outcomes resulting from inadequate antimicrobial treatment of infections caused by antibiotic-resistant bacteria have been shown in studies by Kollef *et al.* (1998, 1999) and Zaidi *et al.* (2002). As resistance patterns change, physicians need to re-evaluate standard therapies to ensure appropriate antibiotic coverage. However, it is important to have quality control of anti-biotic therapy and all ICUs need locally adapted guidelines for the prudent use of antibiotics, including restricted use of prophylactic and therapeutic antibiotics which affect local resistance patterns (Albrich *et al.*, 1999). The use of SDD has been associated with the emergence of antibiotic-resistant bacterial strains, limiting its usefulness. The routine use of SDD has not been advocated because individual trials have failed to demonstrate any reduction in mortality (Bonten *et al.*, 2003; Kollef, 2003). However, a recent Dutch study has shown improved patient survival and lower prevalence of antibiotic resistance in ICU-patients receiving SDD (de Jonge *et al.*, 2002), but the findings are under debate and need to be confirmed. Moreover, as the prevalence of antibiotic resistance is very low in the Netherlands compared to Southern Europe and the Americas, the extrapolation of the resistance findings to ICUs in other countries may not be valid (Bonten *et al.*, 2003).

Several studies have shown that antimicrobial control has a beneficial effect on resistance patterns. Indeed, a recent study has reported the results of a new programme of antibiotic strategy control (Gruson *et al.*, 2000). In that study, rotation and restricted use of antibiotics in a medical ICU reduced the incidence of ventilator-associated pneumonia caused by antibiotic-resistant Gram-negative bacteria (Gruson *et al.*, 2000). In addition, Burke and Pestotnik (1999) showed that a computer-assisted decision support programme for antibiotic prescribing had the potential to stabilise bacterial resistance in the ICU. It is difficult to design a study that can prove that any reduction in colonisation or infections caused by antibiotic-resistant pathogens is due to a change in antibiotic policy, as it would be difficult to allow for improved compliance with hygiene instructions that could also lead to reduced cross-transmission of antibiotic resistant clones (Struelens *et al.*, 1999). In a recent study, Allaouchiche *et al.* (2002) showed concomitant variations of antimicrobial use and the incidence of ICU-acquired infections due to third-generation cephalosporin-resistant Gram-negative bacilli, carbapenem-resistant Gram-negative bacilli, or MRSA over a 5-year period in a French ICU. Interestingly, the same study showed a protective effect of an increase in the use of medicated soaps plus alcoholic hand rubs on the incidence of ICU-acquired infections due to these resistant bacteria (Allaouchiche *et al.*, 2002).

In the ESAP study, having a list of antibiotics subject to restricted use and reporting excellent communication between senior and junior doctors were independent factors associated with low total antimicrobial use (Monnet *et al.*, 2000). Reduction of the duration of therapy is another method of reducing antibiotic resistance (Baughman, 2002; Ibrahim *et al.*, 2001; Singh *et al.*, 2000).

4.2.   Antibiotic cycling and their role in reducing resistance

Antibiotic cycling has been suggested as a strategy for discouraging the emergence of antimicrobial resistance. The concept is to withdraw an antibiotic or class of antibiotic from use in order to allow resistance rates to decrease or stabilise (Bonten, 2002). However, conflicting results have been reported in studies of antibiotic cycling and the results are inconclusive. In an early study of antibiotic cycling, Gerding *et al.* (1991) evaluated cycling of aminoglycosides and could demonstrate that a change to amikacin caused a 50% reduction in gentamicin resistance in Gram-negative bacteria but gentamicin resistance increased when it was reintroduced. The use of aminoglycosides also decreased during the study period. In another more recent antibiotic cycling study, Raymond *et al.* (2001) demonstrated a reduced incidence of antibiotic-resistant bacteria, but the study was not controlled for the relative contribution of decreased emergence of resistance vs control of cross transmission. Mathematical modelling may be used to design prospective cycling studies (Bonten *et al.*, 2001).

4.3.   IT and benchmarking to improve antibiotic prescribing

As most antibiotic use in the ICU is empirical, it is important to know the most prevalent pathogens and their local resistance patterns. These data can be easily provided via the Intranet or Internet if the clinical microbiology laboratory is computerised. Providing physicians with pathogen frequency, susceptibility data by ward level and site of infection, and patient-specific clinical information has been shown to improve antibiotic selection, control antibiotic costs, and slow the emergence of resistance (Evans *et al.*, 1998; Pestotnik *et al.*, 1996). The selection of antibiotics in the hospital setting is still a largely manual task and therefore fraught with potential errors (Bailey and McMullin, 2001). These include overuse of antibiotics, choice of inadequate agents, and dosage regimens. The decision process for antibiotic prescriptions in the ICU setting was investigated in a Swedish study carried out in 2000 (Hanberger *et al.*, 2001b). Three of four ICU patients were treated with antibiotics (see above). Most prescriptions were strictly empirical and only 27% were based on a positive culture with or without an antibiogram, and only 8% of the prescriptions were accompanied by a preliminary discontinuation date. In order to improve antibiotic use in the ICUs, more microbiological information as well as patient-specific clinical information must be made available to the prescriber. Improving antibiotic prescribing by using information systems is technically feasible, but commercial solutions are still suboptimal (Bailey and McMullin, 2001). Another option is to use an antibiotic stewardship team working in concert with critical care specialists in choosing optimal empirical regimens and in streamlining therapy once culture results are available (Paterson, 2003).

Interventions aimed at controlling the use of antibiotics require education and access to local data on antibiotic resistance and consumption. Therefore, a national ICU-surveillance programme, ICU-STRAMA was developed in Sweden in 1999, with the aim of aiding clinicians by providing feedback on local antibiotic consumption data and bacterial resistance patterns (ICU-STRAMA, 1999–2000). Local multidisciplinary ICU groups consisting of specialists in intensive care, infectious diseases, and infection control, as well as pharmacists, microbiologists, and others have formulated local policies using the information in the database which is easily accessible through a website. Person-to-person interactions are likely to be too time-consuming and unsustainable in the long term. By using the Internet, it will be possible to create a sustainable programme for the coordinated collection of information about antibiotic policy, antibiotic use, antibiotic resistance, infection control, and intensive care demography. The susceptibility of clinical isolates to important drugs has been high in Swedish ICUs, despite comparatively high consumption of antibiotics which may be due to the moderate ecological impact of the drugs chosen and the positive impact of hospital hygiene on the resistance rates. It is difficult to measure the effect of a bench programme such as ICU-STRAMA on antibiotic resistance in a low-level resistance ICU setting, but Fridkin *et al.* (2002) showed that monitoring antimicrobial use and resistance and promoting changes of practice in specific ICUs were associated with decreases in ICU vancomycin use and VRE prevalence.

5. INFECTION CONTROL

The effectiveness of infection control measures in the prevention and control of the spread of resistant bacteria has been convincingly demonstrated (Bergogne-Berezin, 1999; Eggimann *et al.*, 2000; Lingnau and Allerberger, 1994; Souweine *et al.*, 2000). Since bacteria can be transmitted on the hands of healthcare workers, the most effective way to prevent patient-to-patient spread of resistant pathogens is by maintaining good hand hygiene (Scott, 2000). Both hand washing and the use of alcohol-based hand disinfectants are effective ways of reducing bacterial carriage on the hands of healthcare workers. However, alcohol-based hand disinfectants may provide superior efficacy and fewer barriers to healthcare worker compliance (Voss and Widmer, 1997). Pittet *et al.* (2000) recently published data suggesting a decline in nosocomial infection rates and ORSA prevalence after the introduction of these products into routine use in a large university hospital. Additional infection control measures (e.g., use of gloves and gowns) are necessary to prevent spread of pathogens like ORSA and VRE, which are known to contaminate the environment around infected or colonised patients (Boyce *et al.*, 1997; Rhinehart *et al.*, 1990; Srinivasan *et al.*, 2002). The Center for Disease Control and Prevention publishes literature-based recommendations for the prevention and control of selected resistant bacterial pathogens (www.cdc.gov).

REFERENCES

Allaouchiche, B., Monnet, D., López-Lozano, J., Müller-Pebody, B., Ayzac, L., Tigaud, S. *et al.*, 2002, Incidence of ICU-acquired infections due to antimicrobial-resistant microorganisms and relationships with antimicrobial use, infection control and workload in a French ICU. *42nd Interscience Conference on Antimicrobial Agents and Chemotherapy.* San Diego, CA, abstr. K-1346.

Aksaray, S., Dokuzoguz, Guvener, E. *et al.*, 2000, Surveillance of antimicrobial resistance among Gram-negative isolates from intensive care units in eight hospitals in Turkey. *J. Antimicrob. Chemother.*, **45**, 695–699.

Albrich, W. C., Angstwurm, M., Bader, L., and Gartner, R., 1999, Drug resistance in intensive care units. *Infection*, **27**(Suppl. 2), S19–S23.

Andersen, C. T., and Meberg, A., 1999, [Drug use in a neonatal unit]. *Tidsskr Nor Lægeforen*, **119**, 197–200.

Archibald, L., Phillips, L., Monnet, D., McGowan, J. E., Tenover, F., and Gaynes, R., 1997, Antimicrobial resistance in isolates from inpatients and outpatients in the United States: Increasing importance of the intensive care unit. *Clin. Infect. Dis.*, **24**, 211–215.

Bailey, T. C. and McMullin, S. T., 2001, Using information systems technology to improve antibiotic prescribing. *Crit. Care Med.*, **29**(Suppl.), 87–91.

Baughman, R. P., 2002, Antibiotic resistance in the intensive care unit. *Curr. Opin. Crit. Care*, **8**, 430–434.

Bergmans, D. C., Bonten, M. J., Gaillard, C. A., van Tiel, F. H., van der Geest, S., de Leeuw, P. W. *et al.*, 1997, Indications for antibiotic use in ICU patients: A one-year prospective surveillance. *J. Antimicrob. Chemother.*, **39**, 527–535.

Bergogne-Berezin, E., 1999, Current guidelines for the treatment and prevention of nosocomial infections. *Drugs*, **58**, 51–67.

Bonten, M. J., Austin, D. J., and Lipsitch, M., 2001, Understanding the spread of antibiotic resistant pathogens in hospitals: Mathematical models as tools for control. *Clin. Infect. Dis.*, **33**, 1739–1746.

Bonten, M. J., 2002, Infection in the intensive care unit: Prevention strategies. *Curr. Opin. Infect. Dis.*, **15**, 401–405.

Bonten, M. J., and Mascini, E. M., 2003, The hidden faces of the epidemiology of antibiotic resistance. *Intensive Care Med.*, **29**, 1–2.

Bonten, M. J., Joore, H. C., de Jongh, B. M. *et al.*, 2003, Selective decontamination of the digestive tract: All questions answered? *Crit. Care*, **7**, 203–205.

Borderon, J. C., Laugier, J., Ramponi, N., Saliba, E., Gold, F., and Blond, M. H., 1992, Surveillance of antibiotic therapy in a pediatric intensive care unit. *Ann. Pediatr.* (Paris), **39**, 27–36.

Bourdain, N., Lepape, A., Galzot, F., Lamy, B., Haessler, D., Chomarat, M. *et al.*, 1999, Peut-on utiliser des protocoles d'antibiothérapie probabiliste en réanimation? *Pathol. Biol.*, **47**, 584–588.

Boyce, J. M., Potter-Bynoe, G., Chenevert, C., and King, T., 1997, Environmental contamination due to methicillinresistant Staphylococcus aureus: Possible infection control implications. *Infect. Control Hosp. Epidemiol.*, **18**, 622–627.

Burke, J. P. and Pestotnik, S. L., 1999, Antibiotic use and microbial resistance in intensive care units: Impact of computer-assisted decision support. *J. Chemother.*, **11**, 530–535.

Burwen, D. R., Banerjee, S. N., Gaynes, R. P., and the National Nosocomial Infections Surveillance System, 1994, Ceftazidime resistance among selected nosocomial Gram-negative bacilli in the United States. *J. Infect. Dis.*, **170**, 1622–1625.

Capellà, D., 1993, Descriptive tools and analysis. *WHO Reg. Publ. Eur. Ser.*, **45**, 55–78.

Carmeli, Y., Eliopoulos, G., Mozaffari, E., and Samore, M., 2002, Health and economic outcomes of vancomycin-resistant enterocci. *Arch. Intern. Med.*, **162**, 2223–2228.

Cars, O., Mölstad, S., and Melander, A., 2001, Variation in antibiotic use in the European Union. *Lancet*, **357**, 1851–1853.

Centers for Disease Control and Prevention, 1997, Update: Staphylococcus aureus with reduced susceptibility to vancomycin—United States, 1997. *MMWR*, **46**, 813–815.

Centers for Disease Control and Prevention, 2002, Staphylococcus areus resistant to vancomycin—United States, 2002. *MMWR*, **51**, 565–567.

DANMAP 2000, Consumption of antimicrobial agents and occurrence of antimicrobial resistance in bacteria from food animals and humans in Denmark. Danish Veterinary Laboratory. Copenhagen, Denmark. 2001. Available from http://www.dfvf.dk/Files/Filer/Zoonosecentret/Publikationer/Danmap/Danmap_2000.pdf

De Jonge, E., Schultz, M. J., Spanjaard, L. *et al.*, 2002, European Society of Intensive Care Medicine, 15th Annual Congress (Barcelona): Effects of selective decontamination of the digestive tract on mortality and antibiotic resistance (abstract 30). *Intensive Care Med.*, **28**(Suppl. 1), S12.

Diaz, E. and Rello, J., 2002, Top ten list in antibiotic policy in the ICU. *Chest*, **122**, 712–714.

Diekema, D. J., Pfaller, M. A., Jones, R. N., Doern, G. V., Winokur, P. L., Gales, A. C. *et al.*, and the SENTRY participants group (Americas), 1997, Survey of bloodstream infections due to gram-negative bacilli: Frequency of occurrence and antimicrobial susceptibility of isolates collected in the United States, Canada and Latin America for the Sentry antimicrobial surveillance program. *Clin. Infect. Dis.*, **29**, 595–607.

Diekema, D. J., Pfaller, M. A., Schmitz, F. J. *et al.*, and the SENTRY Participants Group, 2001, Survey of infections due to Staphylococcus species: Frequency of occurrence and antimicrobial susceptibility of isolates collected in the United States, Canada, Latin America, Europe and the Western Pacific for the SENTRY Antimicrobial Surveillance Program, 1997–1999. *Clin. Infect. Dis.*, **32**(Suppl 2), S114–S132.

Eggimann, P., Harbarth, S., Constantin, M. N., Touveneau, S., Chevrolet, J. C., and Pittet, D., 2000, Impact of a prevention strategy targeted at vascular-access care on incidence of infections acquired in intensive care. *Lancet*, **355**, 1864–1868.

Edmond, M. B., Wallace, S. E., McClish, D. K., Pfaller, M. A., Jones, R. N., and Wenzel, R. P., 1999, Nosocomial bloodstream infections in United States hospitals: A three-year analysis. *Clin. Infect. Dis.*, **29**, 239–244.

Erlandsson, C.-M., Hanberger, H., Eliasson, I., Hoffmann, M., Isaksson, B., Lindgren, S. *et al.*, 1999, Surveillance of antibiotic resistance in ICUs in south-eastern Sweden. *Acta Anaesthesiol. Scand.*, **43**, 815–820.

Evans, R. S., Pestotnik, S. L., Classen, D. C., *et al.*, 1998, A computer-assisted management program for antibiotics and other anti-infective agents. *N. Engl. J. Med.* **338**, 232–238.

Fischer, J. E., Ramser, M., and Fanconi, S., 2000, Use of antibiotics in pediatric intensive care and potential savings. *Intensive Care Med.*, **26**, 959–966.

Fonseca, S. N., Ehrenkranz, R. A., and Baltimore, R. S., 1994, Epidemiology of antibiotic use in a neonatal intensive care unit. *Infect. Control Hosp. Epidemiol.*, **15**, 156–162.

Fluit, A. C., Verhoef, J., Schmitz, F. J., European SENTRY Participants, 2001, Frequency of isolation and antimicrobial resistance of gram-negative and gram-positive bacteria from patients in intensive care units of 25 European university hospitals participating in the European arm of the SENTRY antimicrobial surveillance program 1997–1998. *Eur. J. Clin. Microbiol. Infect. Dis.*, **20**, 617–625.

Fridkin, S. K., Steward, C. D., Edwards, J. R., Pryor, E. R., McGowan, J. E., Jr., Archibald, L. K. *et al.*, 1999, Surveillance of antimicrobial use and antimicrobial resistance in United States hospitals: Project ICARE phase 2. *Clin. Infect. Dis.*, **29**, 245–252.

Fridkin, S. K., Lawton, R., Edwards, J. R., Tenover, F. C., McGowan, J. E., Gaynes, R. P., the Intensive Care Antimicrobial Resistance Epidemiology (ICARE) Project, and the National Nosocomial Infections Surveillance (NNIS) system hospitals, 2002, Monitoring antimicrobial use and resistance: Comparison with a national benchmark on reducing vancomycin use and vancomycin-resistant enterococci. *Emerg. Infect. Dis.*, **8**, 702–707.

Garcia-Rodriguez, J. A. and Jones, R. N., 2002, Antimicrobial resistance in gram-negative isolates from European intensive care units: Data from the Meropenem Yearly Susceptibility Test Information Collection (MYSTIC) programme. *J. Chemother.*, **14**, 25–32.

Gastmeier, P., Sohr, D., Forster, D., Schulgen, G., Schumacher, M., Daschner, F. *et al.*, 2000, Identifying outliers of antibiotic usage in prevalence studies on nosocomial infections. *Infect. Control Hosp. Epidemiol.*, **21**, 324–328.

Gerding, D. N., Larson, T. A., Hughes, R. A. *et al.*, 1991, Aminoglycoside resistance and aminoglycoside usage: Ten years of experience in one hospital. *Antimicrob. Agents. Chemother.*, **35**, 1284–1290.

Gortner, L., 1993, Drug utilisation in preterm and term neonates. *Pharmacoeconomics*, **4**, 437–445.

Grupo de Trabajo de Enfermedades Infecciosas de la Sociedad Española de Medicina Intensiva y Unidades Coronarias (GTEI-SEMIUC). ENVIN-UCI, Estudio Nacional de Vigilancia de Infeccion Nosocomial en Servicios de Medicina Intensiva. Informe 1996.

Gruson, D., Hilbert, G., Vargas, F., Valentino, R., Bebear, C., Allery, A. *et al.*, 2000, Rotation and restricted use of antibiotics in a medical intensive care unit. Impact on the incidence of ventilator-associated pneumonia caused by antibiotic-resistant gram-negative bacteria. *Am. J. Respir. Crit. Care Med.*, **162**, 837–843.

Hanberger, H., Garcia-Rodriguez, J. A., Gobernado, M., Goossens, H., Nilsson, L. E., and Struelens, M. J., 1999a, Antibiotic susceptibility among aerobic Gram-negative bacilli in intensive care units. *JAMA*, **281**, 67–71.

Hanberger, H., Nilsson, L. E., Claesson, B. *et al.*, 1999b, New species-related MIC breakpoints for early detection of development of resistance among Gram-negative bacteria in Swedish intensive care units. *J. Antimicrob. Chemother.*, **44**, 611–619; **35**, 152–160.

Hanberger, H., Nilsson, L. E., and the SCOPE study Group, 2000, May, Higher incidence of antibiotic resistance among nosocomial bacteria in Nordic intensive care units (ICUs) compared to primary care centres. Abstracts of the 3rd European Congress of Chemotherapy, Madrid, Abstr. T 136.

Hanberger, H., Diekema, D., Fluit, A., Jones, R., Struelens, M., Spencer, R. *et al.*, 2001a, Surveillance of antibiotic resistance in European ICUs 2001a, *J. Hosp. Infect.*, **48**, 161–176.

Hanberger, H., Erlandsson, M., Burman, L. G., Cars, O., Gill, H., Lindgren, S., Nilsson, L. E., Olsson-Liljequist, B., Walther, S., and ICU-Strama Study Group, 2004, High Antibiotic Susceptibility among Bacterial Athogens in Swedish ICUs, *Scand. J. Infect. Dis.*, **36**, 24–30.

Ibrahim, E. H., Ward, S., Sherman, G. *et al.*, 2001, Experience with a clinical guideline for the treatment of ventilator-associated pneumonia. *Crit. Care Med.*, **29**, 1109–1115.

ICU-STRAMA 1999–2000, Surveillance of antibiotic consumption and antibiotic resistance and its relationship to severity of illness and infection control in Swedish intensive care units. Available from http://dior.imt.liu.se/icustrama/

Intensive Care Antimicrobial Resistance Epidemiology (ICARE) Surveillance Report, 1999, June, Data Summary from January 1996 through December 1997, A report from the National Nosocomial Infections Surveillance (NNIS) System. *Am. J. Infect. Control*, **27**, 279–284.

Kiivet, R. A., Dahl, M. L., Llerena, A., Maimets, M., Wettermark, B., and Berecz, R., 1998, Antibiotic use in 3 European university hospitals. *Scand. J. Infect. Dis.*, **30**, 277–280.

Kollef, M. H., and Ward, S., 1998, The influences of mini-BAL cultures on patient outcomes: Implications for antibiotic management of ventilator-associated pneumonia. *Chest*, **113**, 412–420.

Kollef, M. H., Sherman, G., Ward, S., and Fraser, V. J., 1999, Inadequate antimicrobial treatment of infections: A risk factor for hospital mortality among critically ill patients. *Chest*, **115**, 462–474.

Kollef, M. H., Ward, S., Sherman, G., Prentice, D., Schaiff, R., Huey, W. *et al.*, 2000, Inadequate treatment of nosocomial infections is associated with certain empiric antibiotic choices. *Crit. Care Med.*, **28**, 3456–3464.

Kollef, M. H., 2003, Selective digestive decontamination should not be routinely employed. *Chest*, **123**(Suppl.), 464S–468S.

Lemmen, S. W., Häfner, H., Kotterik, S., Lütticken, R., and Töpper, R., 2000, Influence of an infectious disease service on antibiotic prescription behavior and selection of multiresistant pathogens. *Infection*, **28**, 384–387.

Lingnau, W. and Allerberger, F., 1994, Control of an outbreak of methicillin-resistant Staphylococcus aureus (MRSA) by hygienic measure in a general intensive care unit. *Infection*, **22**(Suppl. 2), S135–S139.

Mathai, D., Jones, J. N., Stilwell, M., and Pfaller, M. A., 2000, Three year analysis of pathogen occurrence and antimicrobial resistance in 315 intensive care units within 71 participating medical centres (32 nations). Report from the SENTRY antimicrobial surveillance program (1997–99). Presented in part at the 40th Interscience Conference on Antimicrobial Agents and Chemotherapy, Toronto, Session 108, Poster 1027.

Monnet, D. L., Suetens, C., Jepsen, O. B., Burman, L. G., Carsauw, H., Gastmeier, P. *et al.*, and the European Strategy for Antibiotic Prophylaxis (ESAP) Project Team, 2000, Overall antimicrobial use and control strategies in intensive care units from 6 European countries. 4th Decennial International Conference on Nosocomial and Healthcare-Associated Infections, Atlanta, GA, 2000, abstr. P-S2-03. *Infect. Control Hosp. Epidemiol.*, **21**, 88.

MYSTIC (AstraZeneca, data on file), 2003.

Naaber, P., Kõljalg, S., and Maimets, M., 2000, Antibiotic usage and resistance—trends in Estonian University Hospitals. *Int. J. Antimicrob. Agents*, **16**, 309–315.

Paterson, D. L., 2003, Restrictive antibiotic policies are appropriate in intensive care units. *Crit. Care Med.*, **31**(Suppl.), S25–S28.

Pestotnik, S. L., Classen, D. C., Evans, R. S. *et al.*, 1996, Implementing antibiotic practice guidelines through computer-assisted decision support: Clinical and financial outcomes. *Ann. Intern. Med.*, **124**, 884–890.

Petersen, I. S., Hesselbjerg, L., Jørgensen, L., Renstrup, J., Barnung, S., Schierbeck, J. *et al.*, 1999, High antibiotic consumption in Danish intensive care units? *APMIS*, **107**, 989–996.

Pittet, D., Hugonnet, S., Harbarth, S. *et al.*, 2000, Effectiveness of a hospital-wide programme to improve compliance with hand hygiene. *Lancet*, **356**, 1307–1312.

Raymond, D. P., Pelletier, S. J., Crabtree, T. D. *et al.*, 2001, Impact of rotating empiric antibiotic schedule on infectious mortality in an intensive care unit. *Crit. Care Med.*, **29**, 1101–1108.

Regnier, B., 1996, Epidemiology on control of antibiotic multiresistant bacteria in hospitals (in French). *Path. Biol.*, **44**, 113–123.

Rhinehart, E., Smith, N. E., Wennersten, C. *et al.*, 1990, Rapid dissemination of beta-lactamase producing aminoglycoside resistant Enterococcus faecalis among patients and staff on an infant-toddler surgical ward. *N. Engl. J. Med.*, **323**, 1814–1818.

Røder, B. L., Nielsen, S. L., Magnussen, P., Engquist, A., and Frimodt-Møller, N., 1993, Antibiotic usage in an intensive care unit in a Danish university hospital. *J. Antimicrob. Chemother.*, **32**, 633–642.

Rodriguez-Villalobos, H., Jones, R., and Struelens, M. J., 2002, Epidemiology of antibiotic resistance of bacterial pathogens from intensive care units: The SENTRY surveillance program in Europe 2000. *Clin. Microbiol. Infect.*, **8**(Suppl. 1), Abstract: P908.

Ruckdeschel, G., Grimm, H., Machka, K., and Wiedemann, B., 1998, Comparison of antibiotic resistance in clinical bacterial strains from general wards vs. intensive care unit patients. In *Abstracts of the 2nd European Congress of Chemotherapy*. Hamburg, Abstract T253.

Scott, G., 2000, Prevention and control of infections in intensive care. *Intensive Care Med.*, **26**(Suppl. 1), S22–S25.

Schwalbe, R. S., Stapleton, J. T., and Gilligan, P. H., 1987, Emergence of vancomycin resistance in coagulase-negative staphylococci. *N. Engl. J. Med.*, **316**, 927–931.

Singh, N., Rogers, P., Atwood, C. W. *et al.*, 2000, Short-course empiric antibiotic therapy for patients with pulmonary infiltrates in the intensive care unit. A proposed solution for indiscriminate antibiotic prescription. *Am. J. Respir. Crit. Care Med.*, **162**, 505–511.

Souweine, B., Traore, O., Aublet-Cuvelier, B. *et al.*, 2000, Role of infection control measures in limiting morbidity associated with multi-resistant organisms in critically ill patients. *J. Hosp. Infect.*, **45**, 107–116.

Srinivasan, A., Song, X., Ross, T., Merz, W., Brower, R., and Perl, T. M., 2002, A prospective study to determine whether cover gowns in addition to gloves decrease nosocomial transmission of vancomycin-resistant enterococci in an intensive care unit. *Infect. Control Hosp. Epidemiol.*, **23**, 424–428.

Stephen, J., Mutnick, A., and Jones, R. N., 2002, Assessment of pathogens and resistance (R) patterns among intensive care unit (ICU) patients in North America (NA): Initial Report from the SENTRY Antimicrobial Surveillance Program (2001). *42nd Interscience Conference on Antimicrobial Agents and Chemotherapy*. San Diego, Abstract: C2-297.

Struelens, M. J., Byl, B., and Vincent, J. L., 1999, Antibiotic policy: A tool for controlling resistance of hospital pathogens. *Clin. Microbiol. Infect.*, **5**, S19–S24.

Tarp, B. D., and Møller, J. K., 1997 [Utilization of antibiotics at the Arhus Municipal Hospital. A prevalence study]. *Ugeskr Laeger*, **159**, 936–939.

Tullus, K., Berglund, B., Fryklund, B., Kühn, I., and Burman, L. G., 1988, Epidemiology of fecal strains of the family Enterobacteriaceae in 22 neonatal wards and influence of antibiotic policy. *J. Clin. Microbiol.*, **26**, 1166–1170.

Vlahovic-Palcevski, V., Morovic, M., and Palcevski, G., 2000, Antibiotic utilization at the university hospital after introducing an antibiotic policy. *Eur. J. Clin. Pharmacol.*, **56**, 97–101.

Vincent, J. L., Bihari, D. J., Suter, P. M., Bruining, H. A., White, J., Nicolas-Chanoin, M.-H. *et al.*, 1995, The prevalence of nosocomial infection in intensive care units in Europe. *JAMA*, **274**, 639–644.

Voss, A., Milatovic, D., Wallrauch-Schwarz, C. *et al.*, 1994, Methicillin-resistant Staphylococcus aureus in Europe. *Eur. J. Clin. Microbiol. Infect. Dis.*, **13**, 50–55.

Voss, A. and Widmer, A. F., 1997, No time for handwashing!? Handwashing versus alcoholic rub: Can we afford 100% compliance? *Infect. Control Hosp. Epidemiol.*, **18**, 205–208.

Walther, S. M., Erlandsson, M., Burman, L. G., Cars, O., Gill, H., Hoffman, M. *et al.*, and the Icustrama Study Group, 2002, Antibiotic prescription practices, consumption and bacterial resistance in a cross section of Swedish intensive care units. *Acta Anaesthesiol. Scand.*, **46**, 1075–1081.

White, R. L., Friedrich, L. V., Mihm, L. B., and Bosso, J. A., 2000, Assessment of the relationship between antimicrobial usage and susceptibility: Differences between the hospital and specific patient-care areas. *Clin. Infect. Dis.*, **31**, 16–23.

Zaidi, M., Sifuentes-Osornio, J., Rolon, A. L. *et al.*, 2002, Inadequate therapy and antibiotic resistance. Risk factors for mortality in the intensive care unit. *Arch. Med. Res.*, **33**, 290–294.

Chapter 16

The Real Cost of MRSA

Stephanie J. Dancer
*Scottish Centre for Infection and Environmental Health, Clifton House,
Clifton Place, Glasgow G37LN, UK*

1.   INTRODUCTION

Methicillin-resistant *Staphylococcus aureus* (MRSA) has become endemic
in UK hospitals over the last decade. It is primarily responsible for infections
of wounds, chest, and urinary tract, but also causes more serious conditions
such as osteomyelitis, endocarditis, and septicaemia. Patients usually acquire
the organism in hospitals and then return home with it, thus contributing
towards the increasing reservoir in the community. The number of serious
MRSA infections has increased over the last few years so that over 40% of all
staphylococcal bacteraemias in the United Kingdom are now due to MRSA
(EARSS Annual Report, 2002). This constitutes one of the highest rates of
MRSA infection in Europe.

Eradication of the carrier state is difficult due to the unusually high epidemicity of circulating strains. The prevalent strains in the United Kingdom at
present are EMRSA-15 and EMRSA-16, with EMRSA-17 recently described
(Aucken *et al.*, 2002). These strains are able to adhere to skin, wounds, and
mucous membranes, spread between persons, and displace previously resident
strains. The complex epidemiology of staphylococci makes any clearance
strategy unattractive, especially when the organism has spread widely through
an institution to achieve endemic status. Equally difficult is the treatment of
MRSA infections, because there are so few effective antibiotics remaining.
Those agents, which are clinically effective, tend to be toxic and/or expensive.
The overall management of patients with MRSA is therefore perceived as

Antibiotic Policies: Theory and Practice. Edited by Gould and van der Meer
Kluwer Academic / Plenum Publishers, New York, 2005

time-consuming, costly, and liable to fail. These difficulties have prompted comments in the literature such as, "Trying to control MRSA causes more problems than it solves," "MRSA is a suitable case for inactivity," and even that, "Is it time to stop searching for MRSA? Stop the ritual of tracing colonised people" (Barrett *et al.*, 1998; Lacey, 1987; Teare and Barrett, 1997).

Tempting though it is to abandon search and destroy policies for MRSA, there are reasons why continued activity towards control is still strongly recommended. MRSA carriage is more likely than methicillin-susceptible *S. aureus* (MSSA) to lead to infection and MRSA bacteraemia has a worse outcome than MSSA bacteraemia (Muder *et al.*, 1991; Romero-Vivas *et al.*, 1995). The mortality rate attributed to MSSA bacteraemia in critically ill patients was 1.3% in one study, compared to 23.4% for MRSA bacteraemia (Blot *et al.*, 2002). Clinicians may be unconcerned about the possibility that methicillin resistance will become the norm for *S. aureus*, but the very nature of evolution decrees further resistance— notably to glycopeptides (Linares, 2001). It has already been established that patients with glycopeptide-intermediate *S. aureus* (GISA) tend to have an even less favourable outcome than patients with MRSA (Walsh and Howe, 2002).

Such therapeutic difficulties have major implications for the cost of healthcare. A patient found to have infection due to MSSA is usually quickly and easily treated with isoxazolyl penicillins or macrolides. These drugs are relatively nontoxic and cheap and patients can be discharged on oral therapy. Contracting MRSA, however, leads to lengthened hospital stay, prolonged treatment, use of expensive drugs, laboratory costs, and occasionally surgical intervention. Furthermore, successful eradication is not guaranteed, as MRSA can persist for years (Sanford *et al.*, 1994). Hospitals with rising numbers of MRSA patients are experiencing an increasing financial burden, which is now impacting upon health policies and overall societal costs. There are also indefinable costs to human health and well being. This chapter will review the evidence for the costs of MRSA within different healthcare systems, the breakdown of these costs and the potential impact of various control strategies on health service resources.

2.   COST OF HOSPITAL-ACQUIRED INFECTION

There has already been much interest in the excess costs generated by hospital-acquired infection (HAI). The last few years have witnessed an explosion of reports examining the economic impact of these infections and the most significant contributory factors (Emmerson *et al.*, 1996; Haley *et al.*, 1981; Jarvis, 1996; Plowman *et al.*, 2001). Infected patients, on average, incurred hospital costs that were almost three times higher than those of uninfected patients and they remained in hospital 2.5 times longer (Plowman *et al.*, 2001).

It is generally agreed that HAI costs money, which could have been saved if preventative measures had been in place, but there is uncertainty surrounding the most effective control strategies and the best methods by which to measure them (Haley, 1991). Most calculations are based upon the increase in bed-days attributed to HAI (Coello *et al.*, 1993; Pena *et al.*, 1996) but this methodology has to make assumptions and extrapolations from data not always generated for the purposes of measuring HAI (Walker, 2002). More recent studies provide information on the distribution of additional costs incurred between different hospital sectors or dependent upon patient diagnoses (Zoutman *et al.*, 1998). The former includes specific wards such as Intensive Care and Special Care Baby Units, and the latter, conditions such as hospital-acquired bacteraemia, surgical site infection, urinary tract infection, and pneumonia (Hollenbeak *et al.*, 2000; Jarvis, 1996; Khan and Celik, 2001; Mahieu *et al.*, 2001; Pittet *et al.*, 1994; Plowman *et al.*, 2001; Rose *et al.*, 1977; Spengler and Greenough, 1978).

Some HAI's cost more than others—combined analysis reveals two groups that appear to generate higher costs than for other HAI types; in particular, surgical patients who acquire wound infections and medical patients with lower respiratory tract infections (Lynch *et al.*, 1992; Pinner *et al.*, 1982). There have been attempts at economic analyses of surgical wound infections (Fry, 2002; Poulsen *et al.*, 1994; Reilly *et al.*, 2001a). Many studies measure only hospital costs, to varying degrees, and fail to include the costs for community services, follow-up, and social security benefits (Noel *et al.*, 1997; Poulsen *et al.*, 1994; Reilly *et al.*, 2001a). There is little on the direct cost to the patient, in terms of pain, disability, reduction in quality of life, and lost work (Davey and Nathwani, 1998; Poulsen and Gottschau, 1997; Reilly *et al.*, 2001a; Whitehouse *et al.*, 2002). It is known that HAI may only present when the patient has gone home (Reid *et al.*, 2002; Reimer *et al.*, 1987; Stockley *et al.*, 2001) but the potential savings generated from early discharge may well be negated by the subsequent impact of HAI on patients in the community (Jönsson and Lindgren, 1980; Whitehouse *et al.*, 2002).

In the wake of such a drain on healthcare resources, results of various infection control programmes demonstrating cost benefits for both patients and hospital budgets have been presented (Miller *et al.*, 1989). There is evidence that surveillance alone will reduce HAI and associated costs (Haley *et al.*, 1985; Olson *et al.*, 1984), particularly if the results are then fed back to the clinicians involved (Smyth and Emmerson, 2000). Surveillance initiatives should not ignore infected patients presenting post-discharge; for this, the employment of a full-time audit nurse is more than justified in terms of cost-effectiveness (Reilly *et al.*, 2001b). There are also significant cost benefits from treating infected patients in their own homes, using outpatient parenteral antibiotic therapy and support staff (Nathwani, 2001).

Comprehensive infection control programmes pay for themselves and more (Haley et al., 1985; Miller et al., 1989; Wenzel, 1995). There will, however, continue to be uncertainties over the best methods of audit, surveillance, and feedback, which types of HAI require specific attention and which specialities and/or clinical units are most likely to generate the greatest cost benefits for both patients and budgets (Fry, 2002; Haley, 1991). Decisions on where to place control precautions should rest with Infection Control committees in hospital and community, but all require full managerial support (Brachman and Haley, 1981). As with the introduction of any clinical improvement programme, resources must first be released before implementation (Jarvis, 1996; Miller et al., 1989).

3.   THE COST OF ANTIMICROBIAL RESISTANCE

While the cost of HAI can be aptly demonstrated in the literature, evidence for the cost of antimicrobial resistance is not quite so robust (Coast et al., 1996). This is despite the fact that an increasing proportion of HAI is due to resistant organisms (Edmond et al., 1999; Schaberg et al., 1991). Hospitals seek to provide guidance for the use of antibiotics, but an empirical regimen without targeted cover, or dosed too low to provide optimal therapy, will not eradicate a pathogen. It will also encourage the development of resistance, which then increases patient morbidity and mortality (Paladino et al., 2002). Concomitant higher rates of treatment failure and the need for an increase in either the number or duration of hospital admissions will almost certainly be associated with a huge economic burden (Paladino et al., 2002; Singer et al., 2003).

Infections with resistant pathogens lengthen the stay in hospital, likely to be one of the most significant contributors towards the economic consequences of resistant bacteria (Brooklyn Antibiotic Resistance Task Force, 2002; Kollef and Fraser, 2001; Nathwani, 2003). Managerial costs are increased when infection involves these organisms, as well as unnecessary and prolonged therapy (Niederman, 2001a). Treatment of resistant bacteria is associated with increased drug costs, as the newer, broad spectrum and often far more potent drugs are usually far more expensive than the narrower spectrum agents employed for less-resistant pathogens (Janknegt, 1997). These powerful drugs bring their own particular adverse effects, including the problems generated by overgrowth or superinfection of naturally resistant organisms (Drew et al., 2000; Khare and Keady, 2003; Sanyal and Mokaddas, 1999). Other factors to consider are the prohibitive costs required to develop new antimicrobial agents, and the implementation of broader infection control and public health interventions aimed at curbing the spread of resistant pathogens (Kollef and Fraser, 2001).

The economic evaluation of HAI due to resistant organisms is often confounded by the same risk factors that are associated with poor outcomes (Holmberg *et al.*, 1987). Resistance is an important risk factor for inadequate empirical therapy. Such therapy is itself a potent determinant of a number of adverse outcomes, including mortality (Niederman, 2001b). Another practical difficulty in actually measuring the economic impact of resistance is the identification of a suitable control population (Coast *et al.*, 2002; Niederman, 2001b). Nevertheless, for both HAI and community-acquired infections, the mortality, the likelihood of hospitalisation, and the length of hospital stay is usually at least twice as great for patients infected with drug-resistant strains as for those infected with drug-susceptible strains of the same bacteria (Holmberg *et al.*, 1987). The likely cost per capita of healthcare-associated *S. aureus* infection in Denmark and the Netherlands is lower than that in the United States, because MRSA infections, which are rare in both of these countries, cost significantly more than do MSSA infections (Farr and Jarvis, 2002; Farr *et al.*, 2001; Janknegt, 1997). A Danish patient with healthcare-associated *S. aureus* would be treated with an old-fashioned β-lactam antibiotic with faster response, higher cure rate, and quicker hospital discharge at lower overall cost to society (Farr and Jarvis, 2002).

Although the adverse, economic, and health effects of drug-resistant bacterial infections can only be roughly quantified, it is concluded that antimicrobial resistance is an important health problem and an economic burden to society (Holmberg *et al.*, 1987). Future evaluation of interventions aiming to contain resistance might benefit from the use of modelling as a means of measuring optimal policy response, as well as trying to resolve some of the difficulties associated with such interventions (Coast *et al.*, 1996, 2002).

4. GENERAL COSTS OF MRSA

Most authorities would agree that MRSA is probably the most important resistant bacterium associated with HAI. Staphylococci themselves are the most common pathogens causing bacteraemia according to national surveillance of bloodstream infections (Pfaller *et al.*, 1998). Along with *Escherichia coli*, they account for over 55% of all bacteraemias from one recent study (Diekema *et al.*, 2000). The only major change from similar previous studies was the increase in methicillin (oxacillin) resistance in both coagulase negative staphylococci and *S. aureus* (Diekema *et al.*, 2000; Pfaller *et al.*, 1998; Schaberg *et al.*, 1991). *S. aureus* was the most common pathogen referred to the SENTRY Antimicrobial Surveillance Program from 1997 to 1999, where it was found to be the most prevalent bloodstream infection, skin and soft tissue infection, and cause of pneumonia in all geographic areas studied (Diekema *et al.*,

2001). Similar increases in methicillin resistance were also observed in these isolates, originating from both hospital and community (Diekema *et al.*, 2001).

MRSA does not appear to replace MSSA in the overall burden of infection, as the attack rate of methicillin-susceptible strains has not decreased (Stamm *et al.*, 1993). MRSA has simply added to the infection rate, particularly the HAI rate (Herwaldt, 1999; Law and Gill, 1988; Stamm *et al.*, 1993). At some point in the future, when the prevalence of MRSA is such that *S. aureus* can be redefined as "naturally" methicillin resistant, MRSA will replace MSSA in all situations—just as penicillin-resistant *S. aureus* replaced penicillin-susceptible strains in the 1950s and 1960s.

Before pursuing the specific costs of MRSA and individual control strategies, it might be helpful to examine the general costs of MRSA overall, particularly when compared to MSSA. One study modelled estimates of the incidence, deaths, and direct medical costs of *S. aureus* infections in the New York metropolitan area in 1995 (Rubin *et al.*, 1999). The study examined the relative impact of methicillin-resistant vs susceptible strains of *S. aureus* and of community vs hospital-acquired staphylococcal infections. The attributable cost of a patient with MRSA was approximately $2,500 higher than the attributable cost of a patient with MSSA ($34,000 vs $31,500). The higher cost of MRSA infections was due to the cost of vancomycin, longer hospital stay, and the cost of patient isolation procedures. For HAIs alone, the cost attributable to MRSA alone was approximately $3,700 higher on a per patient basis than the cost attributable to MSSA infections ($31,400 vs $27,700). MRSA infections also caused more deaths than MSSA infections (21% vs 8%) (Rubin *et al.*, 1999). The study concluded that reducing the incidence of hospital-acquired MRSA and MSSA would reduce the overall societal costs from *S. aureus* infections.

Nearly 500 patients were included in a study assessing the impact of methicillin resistance on patients with *S. aureus* surgical site infection (Engemann *et al.*, 2003). Patients infected with MRSA had a greater 90-day mortality rate and a longer stay in hospital than patients infected with MSSA. Median hospital charges were nearly $53,000 for MSSA infections and approximately $92,000 for MRSA infections. This resulted in a 1.19-fold increase in hospital charges overall and a mean attributable excess charge of nearly $14,000 for patients with MRSA infections, compared with MSSA patients ($p < 0.001$) (Engemann *et al.*, 2003). Another study on patients with primary bacteraemias due to MRSA and MSSA similarly compared attributable hospital stay and total and variable direct costs of hospitalisation (Abramson and Sexton, 1999). This study showed that the median hospital stay due to MSSA bacteraemia was 4 days, compared with 12 days for MRSA. The median total cost for MSSA bacteraemia was $9,661 vs $27,083 for MRSA bacteraemia, that is, an approximate 3-fold increase in direct cost (Abramson and Sexton, 1999).

Prolongation of hospital stay is often used as a measurable index for the economic impact of acquiring MRSA in hospital (Abramson and Sexton, 1999; Engemann *et al.*, 2003; Kim *et al.*, 2001; Nathwani, 2003; Niederman, 2001b, Walker, 2002). Investigators highlight the importance of increased hospitalisation as a key determinant of the total cost of an episode of infection but it does not encompass all the components of the final bill (Nathwani, 2003). Wakefield *et al.* analysed the relative importance of laboratory, antibiotic, and per diem costs of caring for 58 patients with serious *S. aureus* infections (Wakefield *et al.*, 1988). Laboratory costs accounted for 2%, antibiotic for 21%, and per diem costs for 77% of total infection related costs, with a greater proportion attributable to MRSA infections. Results of another study by Kim *et al.* confirmed the relatively marginal (4%) contribution of antibiotic costs to the overall cost of care in Canadian hospitals, but the final estimate for the cost of MRSA in Canadian hospitals approached $60 million each year (Kim *et al.*, 2001).

The cost of eradicating MRSA with isolation, screening of patients, staff and the environment, topical clearance, and education was found to be approximately half that of treating a single MRSA bacteraemia (Rao *et al.*, 1988). The incidence of hospital-acquired MRSA fell to zero in the 5 months following the implementation of these control strategies (Rao *et al.*, 1988). Others have demonstrated that a strict MRSA policy is financially worthwhile (Hornberg *et al.*, 2003; Pan *et al.*, 2001; Vriens *et al.*, 2002).

5. SPECIFIC COSTS OF MRSA

5.1. Costs of screening/surveillance cultures

Most hospitals have an MRSA screening policy of some description, usually for patients received from other hospitals, or for patients due to be admitted to a high-risk ward or for a high-risk procedure (Boyce, 1991). Some will also routinely screen patients from residential or nursing homes in the community, since these facilities frequently act as a reservoir for colonised patients (Fraise *et al.*, 1997; Jernigan *et al.*, 1995; Muder *et al.*, 1991). Within the hospital, there are usually policies for screening patients involved in a cross-infection incident or outbreak, when new or unexpected cases are identified following routine laboratory investigation (Ayliffe *et al.*, 1998). Such surveillance is deemed to be a useful control measure, since establishing carriage offers more rapid management of patients and their clinical area, including neighbouring patients and healthcare staff (Coello *et al.*, 1994). Furthermore, inappropriate antibiotic therapy can be circumvented and future episodes of cross-infection averted. Regarding the former, prescribing a patient an antibiotic to which MRSA is resistant, encourages proliferation of the organism and increased risk of invasive sepsis (Dancer, 2001). These patients are also more likely to shed the organism into the environment and onto others (Schentag *et al.*, 1998).

Comprehensive screening policies for all admissions, however, are rare due to the increased time, effort, and resources required. Since most microbiology laboratories cannot confirm MRSA in less than 48 hr, clinicians often have to make their own decisions regarding the potential MRSA status of a patient and act accordingly. Screening samples would not generally be regarded as urgent and it is unlikely that they would be processed out-of-hours or over a week-end—thus adding further delay to the time of final report. Despite these difficulties, current opinion on the role of active surveillance cultures has been gaining momentum recently (Arnold *et al.*, 2002; Farr and Jarvis, 2002). Infection control professionals are in agreement that screening patients for MRSA carriage is a useful procedure, because it is not possible to control what you don't know about (Farr and Jarvis, 2002).

While there are obvious benefits for hospitals and the individual from active surveillance, there should be an evaluation regarding cost benefits. One study found that the increase in hospital-acquired MRSA was associated with increased transfer of colonised patients from nursing homes and other hospitals (Jernigan *et al.*, 1995). A subsequent cost–benefit analysis suggested that surveillance cultures of transferred patients would save between $20,000 and $460,000 and prevent from 8 to 41 HAIs due to MRSA (Jernigan *et al.*, 1995). Such a screening programme would also reduce the number of patient-days spent in isolation. A more recent study showed that if early identification of colonised patients prevented transmission of MRSA to as few as six patients, then a screening programme would save money (Papia *et al.*, 1999). The average cost of implementing recommended infection control measures for patients colonised with MRSA was approximately $5,235 per patient (laboratory and nursing costs) (Papia *et al.*, 1999).

When MRSA eradication fails, or is never attempted, endemicity ensues. It is argued that in this situation, control efforts are no longer justified (Barrett *et al.*, 1998). High rates of MRSA in an acute hospital, however, are associated with increased HAI rates, increased use of glycopeptides, higher risk of generating antibiotic-resistant Gram-positive bacteria, and additional healthcare costs (Herwaldt, 1999; Rubinovitch and Pittet, 2001). When the evidence in endemic hospitals is reviewed, containment efforts appear to decrease the incidence of MRSA HAIs (Rubinovitch and Pittet, 2001). Successful programmes are based upon early identification of the MRSA reservoir and prompt implementation of control precautions, particularly via the screening of high-risk patients on admission (Girou *et al.*, 1998; Papia *et al.*, 1999). This strategy has been shown to be cost-effective in a number of different acute-care endemic settings using varied cost indicators (Chaix *et al.*, 1999; Jernigan *et al.*, 1995; Papia *et al.*, 1999; Rubinovitch and Pittet, 2001).

Targeted screening is similarly cost-effective (Girou *et al.*, 2000). In one study of dermatology patients, it was shown that a selective strategy is cheaper because there are fewer samples submitted to the laboratory and less work imposed upon healthcare personnel (Girou *et al.*, 2000). Another study involving neonates in an intensive care unit examined the cost benefits from active surveillance cultures and contact/droplet precautions for control of MRSA in the unit (Karchmer *et al.*, 2002). Estimated costs of controlling a 10-month outbreak that resulted in 18 colonized and 4 infected infants in one unit ranged from over $48,000 to $68,000. The estimated excess cost of 75 MRSA bacteraemias in a second neonatal unit was $1,306,000; this outbreak included 14 deaths and lasted for over 50 months. The study concluded that weekly screening and isolation policies halted the outbreak in the first unit and cost 19–27-fold less than the attributable costs of MRSA bacteraemias in the second (Karchmer *et al.*, 2002).

Recently, a performance improvement task force was set up to report best hospital practices for controlling MRSA (Arnold *et al.*, 2002). The group looked at screening, isolation, cohorting, decolonisation, post-exposure follow-up, microbiology procedures, and surveillance methods, and categorised the recommendations into priority levels according to published data. Evidence for screening patients at risk of having MRSA was particularly strong, with most protocols receiving the top-level priority (strongly recommended and supported by well-designed epidemiological studies and experience) (Arnold *et al.*, 2002). The group emphasised the need to increase patient screening. This was the first time that a public health department in the United States had stated that identification of the MRSA reservoir is necessary for effective control (Farr and Jarvis, 2002).

5.2. Handwashing

There is a huge amount of literature extolling the benefits of handwashing in the control of HAI (Larson, 1995). It is generally agreed that hand hygiene is the single most important activity in an infection control programme, but it is very difficult to get everyone to wash his or her hands routinely, particularly when the ward is busy or short staffed (Afif *et al.*, 2002; Gopal Rao *et al.*, 2002; Grundman *et al.*, 2002). It is also difficult to cost the effectiveness of hand hygiene because it is not usually practised in isolation, but as part of an infection-control package (Harbarth *et al.*, 2000). One recent study examined the effectiveness of a hospital-wide programme to improve compliance with hand hygiene (Pittet *et al.*, 2000). The programme introduced bedside alcohol-based hand disinfection and monitored overall compliance with hand hygiene during routine patient care before, and during, the handwashing campaign. Surveys were performed over a 3-year period and HAIs were measured in parallel. The study demonstrated a significant increase in compliance with hand hygiene (48–66%) over the study period, with concomitant decrease in the overall numbers of HAI and MRSA transmission rate. In particular, MRSA acquisition decreased from 2.16 to 0.93 episodes per 10,000 patient-days ($p < 0.001$). The consumption of alcohol-based hand rub solution represented extra costs of SFr 110,833, an average of SFr 101.15 per 1,000 patient-days. Total crude direct and indirect costs associated with the intervention were estimated as less than SFr 380,000. Given a conservative estimate of SFr 3,500 saved per HAI averted, prevention of 108 infections during the study period would have offset programme costs. In actual fact, it is possible that over 900 infections were prevented by the intervention (Pittet *et al.*, 2000).

Others have noted a reduction in the proportion of hospital-acquired MRSA following the introduction of alcohol-based gel placed at the bedside (Gopal Rao *et al.*, 2002). Different handwash products have similarly been associated with decreased numbers of patients acquiring MRSA (Reboli *et al.*, 1989; Webster *et al.*, 1994). One of these studies introduced the use of triclosan 1%w/v into a neonatal intensive care unit and reported elimination of MRSA from the unit (Webster *et al.*, 1994). In addition, a reduction in the use of vancomycin resulted in a cost saving of approximately $A17,000 (Webster *et al.*, 1994).

A sustained programme of educating clinical staff about MRSA and the benefits of hand hygiene was introduced at another hospital in order to cut costs (Nettleman *et al.*, 1991). The transmission rate of MRSA was halved after assigning responsibility to medical and surgical residents on whose wards an MRSA isolate was identified. The programme was supplemented with handouts, bacteriological screening of staff hands, and feedback. Residents were encouraged to serve as role models for appropriate hand hygiene and other infection control precautions. These inexpensive interventions saved the hospital $42,000, or 115 hospital days, per year (Nettleman *et al.*, 1991).

Despite the obvious benefits to patients, hospital staff do not wash their hands. Published standards for the control of HAI tend to focus upon the managerial and structural aspects of infection control but perhaps it is time for an explicit standard to be set on hand decontamination for all healthcare staff (Teare, 2000). Infection control teams can audit this (Pittet *et al.*, 2000). Senior medical staff are role models and play a significant role in influencing hand hygiene compliance (Lankford *et al.*, 2003; Teare, 2000).

5.3.   Isolation, cohorting, and contact isolation

Patients may need source isolation because they are infected and hazardous to others, or they may need protective isolation because their susceptibility to infection is increased. Such precautions against infection are often costly in money and the time of skilled staff and may be exasperating impediments to routine clinical work (Bagshawe *et al.*, 1978; Barrett *et al.*, 1998). Single room isolation may also precipitate psychological distress, particularly in the elderly (Tarzi *et al.*, 2001). Cross-infection on wards, however, is difficult to prevent by routine measures, particularly if there is a staphylococcal disperser present. Both carriers and infected patients may heavily contaminate their immediate environment and release airborne particles carrying staphylococci (Boyce *et al.*, 1997). Such patients should be isolated in a single room, if possible (Ayliffe *et al.*, 1998).

Isolation is not generally practised without other basic hygienic procedures, so its impact on reducing staphylococcal spread in a hospital has to be assessed as part of an infection control package (Ayliffe *et al.*, 1998). Furthermore, there are few, if any, studies examining the cost benefits of isolation as the single, or predominant, activity in control of MRSA. There is plenty of evidence, however, that isolation can facilitate control of endemic MRSA, including outbreaks and this in itself suggests that significant savings can be made from physically separating MRSA patients from others (Murray-Leisure *et al.*, 1990; Shanson *et al.*, 1985). In contrast, increasing the proximity of patients by adding a fifth bed to four-bedded bays significantly heightens the risk of cross-infection with MRSA (Kibbler *et al.*, 1998). This supports the finding that the relative risk of MRSA acquisition increases with the colonization pressure from imported cases (Talon *et al.*, 2003).

Hospitals can establish the use of specialised units with designated staff for control purposes (Fitzpatrick *et al.*, 2000). An eight-bedded isolation unit in one large hospital was built specifically to control MRSA. The number of infections was more than halved in 2 years and further reductions occurred during the following 4 years (Selkon, 1980). A decrease in the spread of MRSA by use of isolation units was also observed in several London hospitals in the 1980s (Bradley *et al.*, 1985; Dacre *et al.*, 1986; Duckworth *et al.*, 1988; Shanson *et al.*, 1985). In contrast, a statistical model predicts that the risk of MRSA acquisition would increase by 160% per year in the absence of a dedicated cohort facility (Talon *et al.*, 2003).

Hospitals without plentiful isolation facilities, or when overwhelmed with MRSA patients, have to be inventive when faced with control issues. Cohorting known positive patients together in ward bays is an option; if basic infection control practices are reinforced regularly, and there is access to flexible domestic support, spread of MRSA can be curtailed (Duckworth *et al.*, 1988). Risk assessment regarding both patient and ward environment are helpful when considering control scenarios (Ayliffe *et al.*, 1998; Wilson and Dunn, 1996). The establishment of an isolation facility on a temporary basis, however, may be required if the number of cases suddenly increases (Bradley *et al.*, 1985; Cox *et al.*, 1995).

Contact isolation is another control option for MRSA patients (Cohen *et al.*, 1991; Jernigan *et al.*, 1995), with or without cohorting (Arnold *et al.*, 2002; Murray-Leisure *et al.*, 1990). A theoretical cost–benefit analysis demonstrated that contact precautions could decrease the total days of MRSA isolation by 42%, prevent 8–41 infections each year, and decrease hospital costs by over $20,000 (Jernigan *et al.*, 1995). Even the cost of gowns used for contact precautions can be reduced following the introduction of a programme of active surveillance cultures and immediate contact isolation for new ICU admissions (Muto *et al.*, 2003). In the first 3 months of this particular programme, the incidence rate of MRSA decreased from 5.4 per 1,000 patient-days to 1.6 per 1,000 patient-days, while overall gown use decreased by 40% (Muto *et al.*, 2003).

Contact isolation does not always control hospital-acquired MRSA, however, particularly if there is understaffing, unidentified carriers, unusual modes of transmission, or high carriage rates among patients on admission (Herwaldt, 1999). Infection control staff need to identify the reasons for persistent transmission and target their interventions specific to local circumstances (Herwaldt, 1999). Control can be achieved through a multifaceted approach, even in endemic situations. Patient cohorting, respiratory or contact isolation precautions, use of an isolation ward, and a hand hygiene programme significantly reduced the transmission of MRSA in one acute hospital (Herwaldt, 1999). The proportion of MRSA isolates decreased from 32% to 22%, the number of MRSA carriers was halved in 1 year and the rate of MRSA infections decreased from 2% to 0.2%. Furthermore, vancomycin expenditure decreased from $32,000 to $12,500 a year, saving the hospital $19,500 (Herwaldt, 1999).

5.4.   Laboratory costs

Laboratory costs from HAI are often ignored because attention is more likely to be focused on the direct costs relating to patients, that is, treatment costs and extra days spent in hospital (Wakefield *et al.*, 1988). Extra specimens from HAI patients, however, provide an increasing drain on laboratory resources. Both clinical and environmental samples require uplift to the laboratory, which may not be in site, followed by registration, processing, and reporting. Often the finding from one specimen generates a request for more, either from the same patient or from others. HAI are likely to be associated with multiple resistant organisms, which themselves cause extra work due to extended susceptibility testing. They may also require additional tests, including molecular typing at reference laboratories. The cost of such specialist attention is most unlikely to be included in the final total.

Plowman *et al.* examined the cost ratios of microbiology tests from hospital patients with and without HAI (Plowman *et al.*, 2001). The mean cost incurred for non-HAI patients was £6.97, compared with £33.13 for patients with HAI. Thus, microbiology tests from HAI patients were nearly five times more expensive than those from non-HAI patients. Other pathology tests gave a cost ratio of 2.3 between non-HAI and HAI patients, that is, more than twice as expensive for HAI patients. The overall percentage contribution from microbiology tests towards the total cost of HAI was 0.83%; all laboratory tests accounted for nearly 3% (Plowman *et al.*, 2001).

The latter study investigated the rate and cost of all HAI occurring in an English district general hospital, so the proportion of patients with MRSA was unknown. A study specifically examining the cost of serious *S. aureus* infections, including MRSA, analysed the relative importance of laboratory and antibiotic costs and extra days spent in hospital (Wakefield *et al.*, 1988). Per diem costs were 77% of total infection-related costs, antibiotics contributed a further 21%, and laboratory costs accounted for just 2% of the final total. Not unexpectedly, extra days in hospital due to HAI contributed most towards the total.

Laboratory costs included bacterial cultures, antibiotic susceptibility testing, and antibiotic serum levels, with labour and consumables already incorporated (Wakefield *et al.*, 1988). They were calculated by multiplying the number of tests by the direct cost per test. Indirect costs, such as overheads and equipment, were not included. Also ignored were non-microbiological tests, such as white blood cell counts and renal function tests. The final total associated with treating 58 serious *S. aureus* infection gave an overall mean of $66.72. The mean laboratory cost for 48 patients with MSSA infections was $61.74, compared with $92.73 for 10 patients with MRSA infections ($p > 0.05$) (Wakefield *et al.*, 1988).

Other previously mentioned studies have considered laboratory costs as part of an MRSA screening programme (Jernigan *et al.*, 1995; Karchmer *et al.*, 2002; Papia *et al.*, 1999; Rubinovitch and Pittet, 2001), although it is not always clear exactly which components of laboratory processing were used in the final estimate. There are several different MRSA screening methods, with varying degrees of cost-effectiveness (Kunori *et al.*, 2002).

Some laboratories have been driven to incorporate more expensive molecular tests, such as the polymerase chain reaction (PCR), because these tests can identify selected organisms significantly faster that conventional culture (Martineau *et al.*, 1998). They can also detect smaller numbers of these organisms with reliable specificity (Tokue *et al.*, 1992). Such techniques have the potential to markedly improve patient management, while also reducing the risk of cross-infection and even outbreaks. Unfortunately, the cost of introducing and maintaining this technology is prohibitive for most laboratory budgets. PCR testing, however, has been shown to be an accurate and cost-effective method for identifying patients with *S. aureus* (Jayaratne and Rutherford, 1999; Shrestha *et al.*, 2003). The total cost for PCR per test has been quoted as $3.62, compared with $4.77 for conventional culture, with an average turnaround time of 48 hr compared with 82 hr for culture (Jayaratne and Rutherford, 1999).

5.5.   Antibiotic costs

It is generally agreed that the use of antibiotics has selected for resistant organisms from initially susceptible populations and further use has encouraged the subsequent proliferation witnessed today (Schentag *et al.*, 1998). While any antibiotic has the propensity to select for a resistant strain, there are some classes that are more likely to be associated with MRSA (Dancer, 2001; Hill *et al.*, 1998; Hori *et al.*, 2002; Monnet, 1998; Monnet and Frimodt-Moller, 2001; Venezia *et al.*, 2001). The consequence is clinically significant infection with an organism for which there are few treatment options (Dancer, 2003). Those antibiotics, which might be expected to be effective, tend to be expensive and toxic (Dancer, 2003; Janknegt, 1997). It is likely that the cost of using and monitoring such drugs for MRSA provide a significant contribution towards overall HAI costs (Casewell, 1995; Rao *et al.*, 1988; Vriens *et al.*, 2002). Conversely, decreasing overall antimicrobial use and/or improving the quality of antimicrobial prescribing might be expected to lower the cost of pharmaceuticals for HAI (Geissler *et al.*, 2003; Landman *et al.*, 1999).

The study by Plowman *et al.* also examined the antimicrobial costs for HAI (Plowman *et al.*, 2001). Antimicrobials accounted for a mean cost of £13.40 for non-HAI patients, compared to £71.07 for HAI patients. The percentage contribution towards the overall costs of HAI was 1.83%. The contribution from antibiotics towards costs of serious *S. aureus* infection, however, was shown to be 21% in a separate study (Wakefield *et al.*, 1988). The same study examined antibiotic costs for both MSSA and MRSA, and found that mean costs for MSSA accounted for $612.53 compared with $1067.52 for MRSA. Adverse complications of antibiotic therapy were not investigated, but the authors postulated that the requirement for parenteral therapy for MRSA contributed towards the number of additional drugs in hospital for MRSA patients (mean 19.1 days) compared to patients with MSSA (mean 5.9 days) ($p < 0.004$) (Wakefield *et al.*, 1988). MRSA outbreaks also have the potential for driving up drug costs. A 5-week outbreak involving five wards consumed £6,440 worth of teicoplanin, almost half the total cost attributed to the outbreak excluding labour costs and extra days in hospital (Mehtar *et al.*, 1989).

It seems reasonable to examine the potential for savings from infection control policies and programmes regarding the use of antibiotics (Mehtar, 1993, 1995). The role of the medical microbiologist includes giving advice on antimicrobial therapy, encompassing choice, dose, length of course, route, toxicity, combinations, and monitoring. Timely and appropriate advice on the management of infected patients can contribute significantly towards cost savings. Without microbiological guidance, it was shown that antibiotic usage increased by £2,000 per month, compared with a similar period in the previous year (Mehtar, 1995).

The implementation of antibiotic policies has been shown to be cost-effective (Geissler *et al.*, 2003; Mehtar, 1993). Such policies can be extended to formulary restrictions and physician monitoring (Landman *et al.*, 1999; Woodward *et al.*, 1987). Strictly enforced restrictions for aminoglycosides, cephalosporins, and vancomycin generated combined savings of $2.61 per antibiotic day ($p < 0.0046$) and $34,597 per month ($p < 0.0003$) (Woodward *et al.*, 1987). A retrospective analysis of 322 patients with bacteraemia, treated before and after the onset of controls, revealed that antibiotics were more appropriately used afterwards (Woodward *et al.*, 1987). Another study examined both cost savings and the effect on HAI organisms after the introduction of an antimicrobial-prescribing improvement programme (Frank *et al.*, 1997). Over a two-year period, antibiotic prescribing decreased from nearly 160,000 to 140,000 defined daily doses, with concomitant savings of $280,000 in the first year and $390,000 in the second (Frank *et al.*, 1997). These accompanied significant decreases in the rates of enterococcal and selected Gram-negative bacteraemias and in the rates of MRSA and Stenotrophomonas colonisation and infection (Frank *et al.*, 1997).

Antibiotic consumption is particularly high in the Intensive Care Unit. This offers an opportunity for implementing and evaluating policies designed to emphasize more rational use of antibiotics (Gruson *et al.*, 2000). Implementing such a policy in an 11-bedded ICU over a 5-year period resulted in a significant reduction in MRSA and ceftriaxone-resistant Enterobacteriaceae, while costs showed a progressive decrease from €64,500 to €42,000 in the final year (Geissler *et al.*, 2003).

The requirement for parenteral administration of glycopeptides precipitates the extended hospital stay and associated costs for MRSA patients (Janknegt, 1997). Strategies to circumvent this have been introduced, namely outpatient programmes utilising once-daily teicoplanin administration and earlier switching from iv to oral routes if possible (Janknegt, 1997; Nathwani, 2001, 2003; Neiderman, 2001b). An economic evaluation of linezolid, flucloxacillin and vancomycin in the empirical treatment of cellulitis suggested that use of linezolid alone would result in a higher overall success rate and would be less costly than vancomycin across the entire spectrum of the patients' risk of being infected by a resistant pathogen (Vinken *et al.*, 2001).

Other relatively simple policies benefit both patients and budgets (Jewell, 1994; Rubinovitch and Pittet, 2001). A decrease in vancomycin expenditure from $32,000 to $12,500 per year occurred following the introduction of contact precautions, isolation, and a hand hygiene programme (Herwaldt, 1999). Even the targeted application of nasal mupirocin can demonstrate cost savings through the lesser numbers of patients who go on to develop MRSA sepsis (Bloom *et al.*, 1996; van den Burgh *et al.*, 1996). The huge difference in the cost of newer antibiotics for Gram-positive infections will provide a significant impact in prescribing budgets, should glycopeptide-resistant *S. aureus* flourish (Muto *et al.*, 2003). The parenteral streptogramin, Synercid®, and the oxazolidanone, linezolid, cost at least five to seventeen times more than vancomycin (Muto *et al.*, 2003).

5.6. Cleaning

The importance of hospital cleaning in the control of MRSA, as with so many other control components, is still unclear (Dancer, 1999a). Much is known about the epidemiology of *S. aureus* and its potential to contaminate the environment (Boyce *et al.*, 1997; Layton *et al.*, 1993), but it is not known what proportion of patients acquire their MRSA directly, or indirectly, from a contaminated environment. It seems likely that falling standards of hygiene in hospitals have contributed towards increasing rates of MRSA; lack of evidence, however, means that its role in reducing infection remains contentious, let alone evaluated for potential savings within an HAI control programme. One recent study, however, concluded that vigorous environmental cleaning helped to control an MRSA outbreak, along with other control measures (Rampling *et al.*, 2001). Preintervention, nearly 70 patients acquired E-MRSA 16 on a male surgical ward over a period of about 20 months. Environmental cultures provided indistinguishable strains. The domestic cleaning time was then doubled, with emphasis on removal of dirt by vacuum cleaning and allocation of responsibility for the routine cleaning of shared medical equipment. In the 6 months that followed, only three patients acquired the epidemic strain and monthly surveillance cultures failed to detect it in the environment. A cost analysis showed that the cleaning initiative saved nearly £28,000 (Rampling *et al.*, 2001).

It is likely that cleaning is a cost-effective method for MRSA control, but lack of evidence should not be used as an excuse for the continued erosion of domestic services (Dancer, 2002). Neither should the requisite evidence be mandatory before improving cleaning standards in healthcare premises (Talon, 1999). There is little point in educating healthcare workers to wash their hands before and after patient contact, when the first item they touch after washing recontaminates their hands once again (Dancer, 2002).

5.7. Outbreaks

Strains of MRSA may be epidemic in character, affecting two or more patients to cause episodes of cross-infection or even outbreaks. Molecular typing techniques demonstrate transmission within and between healthcare facilities in cities, countries, and across continents (Ayliffe, 1997; Witte *et al.*, 1997). When a clearly defined outbreak occurs in hospital, there is an opportunity for costing the incident; it is much more difficult to estimate the costs of endemicity. Such analyses offer strong support for prompt control measures, since they invariably show significantly increased costs (Mehtar, 1993). Even a limited cross-infection episode can be expensive. One particular incident on an orthopaedic ward was estimated to cost an extra £7,321, excluding physiotherapy and X-ray, following MRSA transmission between a long-term patient with a chronic ulcer and a patient who had just received a prosthetic knee implant (Mehtar, 1993).

Larger outbreaks generate even greater costs. A 5-week outbreak in 1986 involving five wards cost nearly £13,000 but this did not include labour or extra days in hospital (Mehtar *et al.*, 1989). More recently, an outbreak involving more than 400 patients in England was estimated to be in excess of £400,000 of which the provision of isolation wards accounted for a large proportion (Cox *et al.*, 1995).

The costs of not controlling MRSA are much greater than the costs of control (Casewell, 1995; Karchmer *et al.*, 2002; Rao *et al.*, 1988). The analysis of a large outbreak in Madrid suggested that the extra costs incurred exceeded £700,000 (Ayliffe *et al.*, 1998; Casewell, 1995; Coello *et al.*, 1994). It was thought that the extra length of stay and escalating use of vancomycin required for routine prophylaxis were the main contributors towards excess costs.

Even simple procedures such as screening, isolation, and cleaning could potentially save thousands of pounds if an outbreak is averted. A 10-month outbreak in a neonatal SCBU was estimated to cost approximately $49,000–69,000 compared with a similar SCBU, which did not receive any of the basic control interventions implemented in the first SCBU. The second unit witnessed 14 deaths in the outbreak, which lasted more than 50 months and cost $1,307,000 (Karchmer *et al.*, 2002). Doubling the cleaning on a male surgical ward, as already described, eradicated an epidemic strain of MRSA and saved over £28,000 (Rampling *et al.*, 2002).

5.8.   Intensive care unit

ICUs are unique because they house seriously ill patients who require constant hands-on care in a confined environment. These patients are commonly exposed to high concentrations of antibiotics (Kollef and Fraser, 2001). This results in the emergence and spread of antibiotic-resistant bacteria, which create additional costs to the overall total generated by intensive care (Geldner *et al.*, 1999; Niederman, 2001a, b; Pittet *et al.*, 1994). Acquisition of MRSA in this environment is strongly and independently influenced by colonisation pressure (Merrer *et al.*, 2000). Anything that might reverse this pressure, therefore, should have a significant impact on the cost of managing MRSA in the ICU.

In one 26-bedded medical ICU, an infection control programme with selective screening significantly reduced the incidence of ICU-acquired infection or colonisation (from 5.6 to 1.4 per 100 admissions) over a 4-year period despite a persistently high (4%) prevalence of MRSA carriage among newly admitted patients (Girou *et al.*, 1998). A cost–benefit analysis showed that the mean cost attributable to MRSA infection was $9,275, compared to the total costs of a control programme ranging from $340 to $1,480 per patient. The study demonstrated that a 14% reduction in MRSA infection rate resulted in the control programme being beneficial (Chaix *et al.*, 1999). A similar study utilising universal screening was also shown to be cost-effective (Lucet *et al.*, 2003).

Other cost-effective strategies include the rational use of antibiotics in the ICU. In a study already mentioned, an antibiotic-use policy demonstrated a reduction in antibiotic-selective pressure, MRSA and other resistant organisms and a progressive reduction in costs (100% for 1994, 81% for 1995, 65% for 1998) (Geissler *et al.*, 2003). It stands to reason that combining as many of these strategies as possible is likely to have a significant impact on the rates of MRSA colonisation and infection, as well as reducing the high costs required for management. Once again, the costs of control are offset by savings from lesser numbers of infected patients (Chaix *et al.*, 1999; Khan and Celik, 2001; Lucet *et al.*, 2003; Rubinovitch and Pittet, 2001).

5.9.   Community/long-term care

Early discharge of infected or colonised patients to convalescent homes, or to homes for the elderly, has created an expanding reservoir of MRSA in the community (Boyce, 1998; Fraise *et al.*, 1997; Jewell, 1994; Muder *et al.*, 1991). Patients with MRSA in the community may not receive so much attention as those in hospital because their environment is regarded as relatively low risk (Dancer, 1997). Colonised patients in low dependency units, however, have four times the clinical infection rate of uncolonised patients (Muder *et al.*, 1991). The cost of managing long-term care patients with MRSA is almost twice as expensive as managing patients with MSSA (Capitano *et al.*, 2003). There are also treatment difficulties, since active infection with MRSA may require parenteral therapy and there are few oral options (Dancer, 2003; Nathwani, 2001).

Aggressive containment strategies can reduce the MRSA infection rate in nursing homes. In one study, the initial colonisation rate in residents in a 42-bedded extended care unit/nursing home was 52%, but dropped to 2%, with an infection rate of 1.4%, following a programme of screening, contact isolation, topical clearance, and treatment of infected patients

(Jacqua-Stewart *et al.*, 1999). The process was shown to be cost-effective (Jacqua-Stewart *et al.*, 1999).

MRSA is not solely hospital-acquired (Boyce, 1998; Herold *et al.*, 1998; McLaws *et al.*, 1988); new strains are appearing, some characterised by glycopeptide resistance and others by specific virulence determinants such as the PVL gene (Dufour *et al.*, 2002; Hiramatsu, 2001). Such evolutionary changes do not bode well for the future, whereby the costs from untreatable infection could be incalculable.

6.   CONCLUSION

Rising healthcare costs have become an increasing concern to everyone involved in delivering healthcare services (Robinson, 1993). It is no longer appropriate for infection control to be regarded merely as a programme for self-improvement, or as a commendable marker of quality, but as a necessary, critical and cost-effective activity (Duffy, 1985). It should receive strong support from seniors, managers, and any others who might influence healthcare budgets (Brachman and Haley, 1981; Casewell, 1995; Jarvis, 1996). Infection control programmes are, without doubt, cost-effective, but if preventive care is to be encouraged, financial incentives for the value of benefits received require continued emphasis and evaluation (Miller *et al.*, 1989). It is almost as difficult to cost an event that does not happen (Duffy, 1985), as it is to motivate managers faced with bed pressures and waiting lists, especially when they do not understand the epidemiology of infectious organisms or the potential benefits of infection control (Anon, 1985).

There has been some debate on the type and extent of MRSA control measures, particularly in endemic hospitals (Barrett *et al.*, 1998; Boyce, 1991). Accordingly, widely divergent strategies have been employed, from "search and destroy" to almost complete complacency (Barrett *et al.*, 1998; Spicer, 1984). There is, however, mounting evidence that even simple control procedures impact upon the rate of MRSA acquisition and save considerable sums of money (Arnold *et al.*, 2002; Chaix *et al.*, 1999; Pittet *et al.*, 2000; Rampling *et al.*, 2001). Complacency in the face of endemic MRSA is thus unwarranted and, frankly, irresponsible (Dancer, 1999b; Farr and Jarvis, 2002). In addition, there are the societal and human costs of MRSA infection, almost impossible to evaluate and, for the most part, completely ignored (Drummond *et al.*, 1989; Engemann *et al.*, 2003; Jönsson and Lindgren, 1980; Karchmer *et al.*, 2002; Romero-Vivas *et al.*, 1995). Complacency towards MRSA from the human perspective should be regarded as unethical, particularly if the cost of controlling MRSA is balanced against the "cost" of an MRSA death (Dancer, 1999b).

Physicians have an obligation towards their patients' safety (Scolan *et al.*, 2000). Hospital admission, whether for routine or emergency treatment, should not routinely include the acquisition of MRSA. Such an outcome has the potential to elicit legal interest, since there are burgeoning cost implications from malpractice and adverse events (Korin, 1993; Olson, 1981; Scolan *et al.*, 2000). Successful litigation is yet to occur in the United Kingdom, but increasing patient and media interest in hospital hygiene and the "superbug" has already initiated legal activity (BBC Scotland, 2003).

From the evidence presented in this chapter, it is clear that a programme for preventing HAI will not only pay for itself but will also generate other direct and indirect benefits to patients and society as a whole (Khan and Celik, 2001; Rubin *et al.*, 1999). In view of the global threat from multiple resistant organisms, an effective infection control programme could be one of the most cost-beneficial medical interventions available in modern public health (Wenzel, 1995).

ACKNOWLEDGEMENT

Thanks are due to Ms Barbara Nolan for her expert secretarial services.

REFERENCES

Abramson, M. A. and Sexton, D. J., 1999, Nosocomial methicillin-resistant and methicillin-susceptible *Staphylococcus aureus* primary bacteraemia: At what costs? *Infect. Control Hosp. Epidemiol.*, **20**, 408–411.

Afif, W., Huor, P., Brassard, P., and Loo, V. G., 2002, Compliance with methicillin-resistant *Staphylococcus aureus* precautions in a teaching hospital. *Am. J. Infect. Control*, **30**, 430–433.

Anonymous, 1985, In search of efficiency. *Health Soc. Serv. J.*, **95**, 1–8.

Arnold, M. S., Dempsey, J. M., Fishman, M., McAuley, P. J., Tibert, C., and Vallande, N. C., 2002, The best hospital practices form controlling methicillin-resistant *Staphylococcus aureus*: On the cutting edge. *Infect. Control Hosp. Epidemiol.*, **23**, 69–76.

Aucken, H. M., Ganner, M., Murchan, S., Cookson, B. D., and Johnson, A. P., 2002, A new UK strain of epidemic methicillin-resistant *Staphyloccocus aureus* (EMRSA-17) resistant to multiple antibiotics. *J. Antimicrob. Chemother*, **50**, 171–175.

Ayliffe, G. A. J., 1997, The progressive intercontinental spread of methicillin-resistant *Staphylococcus aureus*. *Clin. Infect. Dis.*, **24**(Suppl.), 74–79.

Ayliffe, G. A. J. *et al.*, 1998, Report of combined Working Party of the British Society for Antimicrobial Chemotherapy, The Hospital Infection Society and the Infection Control Nurses Association. Revised guidelines for the control of methicillin-resistant *Staphylococcus aureus* infection in hospitals. *J. Hosp. Infect.*, **39**, 253–290.

Bagshawe, K. D., Blowers, R., and Lidwell, O. M., 1978, Isolating patients in hospital to control infection. Part 1—Sources and routes of infection. *BMJ*, **2**, 609–613.

Barrett, S. P., Mummery, R. V., and Chattopadhyay, B., 1998, Trying to control MRSA causes more problems than it solves. *J. Hosp. Infect.*, **39**, 85–93.

BBC Scotland, 2003, Call to close city hospital wards. Accessed at http://news.bbc.co.uk/l/hi/scotland/2937797.stm

Bloom, B. S., Fendrick, A. M., Chernew, M. E., and Patel, P., 1996, Clinical and economic effects of mupirocin calcium on prevention of *Staphylococcus aureus* infection in haemodialysis patients: A decision analysis. *Am. J. Kidney Dis.*, **27**, 687–694.

Blot, S. I., Vandewoude, K. H., Hoste, E. A., and Colardyn, F. A., 2002, Outcome and attributable mortality in critically ill patients with bacteraemia involving methicillin-susceptible and methicillin-resistant *Staphylococcus aureus*. *Arch. Intern. Med.*, **162**, 2229–2235.

Boyce, J. M., 1991, Should we vigorously try to contain and control methicillin-resistant *Staphylococcus aureus*? *Infect. Control Hosp. Epidemiol.*, **12**, 46–54.

Boyce, J. M., Potter-Bynoe, G., Chenevert, C., and King, T., 1997, Environmental contamination due to methicillin-resistant *Staphylococcus aureus*: Possible infection control implications. *Infect. Control Hosp. Epidemiol.*, **18**, 622–627.

Boyce, J. M., 1998, Are the epidemiology and microbiology of methicillin-resistant *Staphylococcus aureus* changing? *JAMA*, **279**, 623–624.

Brachman, P. S. and Haley, R. W., 1981, Nosocomial infection control: Role of the hospital administrator. *Rev. Infect. Dis.*, **3**, 783–784.

Bradley, J. M., Noone, P., Townsend, D. E., and Grubb, W. B., 1985, Methicillin-resistant *Staphylococcus aureus* in a London hospital, *Lancet*, **1**, 1493–1495.

Brooklyn Antibiotic Resistance Task Force, 2002, The cost of antibiotic resistance: Effect of resistance among *Staphylococcus aureus, Klebsiella pneumoniae, Acinetobacter baumannii,* and *Pseudomonas aeruginosa* on length of hospital stay. *Infect. Control Hosp. Epidemiol.*, **23**, 106–108.

Capitano, B., Leshem, O. A., Nightingale, C. H., and Nicolau, D. P., 2003, Cost effect of managing methicillin-resistant *Staphylococcus aureus* in a long-term care facility. *J. Am. Ger. Soc.*, **51**, 10–16.

Casewell, M. W., 1995, New threats to the control of methicillin-resistant *Staphylococcus aureus. J. Hosp. Infect.*, **30**(suppl), 465–471.

Chaix, C., Durand-Zaleski, I., Alberti, C., and Brun-Buisson, C., 1999, Control of endemic methicillin-resistant *Staphylococcus aureus.* A cost-benefit analysis in an intensive care unit. *JAMA*, **282**, 1745–1751.

Coast, J., Smith, R. D., and Millar, M. R., 1996, Superbugs: Should antimicrobial resistance be included as a cost in economic evaluation? *Health Econ.*, **5**, 217–226.

Coast, J., Smith, R. D., Karcher, A. M., Wilton, P., and Millar, M., 2002, Superbugs II: How should economic evaluation be conducted for interventions, which aim to contain antimicrobial resistance? *Health Econ.*, **11**, 637–647.

Coello, R., Glenister, H., Fereres, J., Bartlett, C., Leigh, D., Sedgwick, J. *et al.*, 1993, The cost of infection in surgical patients: A case-control study. *J. Hosp. Infect.*, **25**, 239–250.

Coello, R., Jimenez, J., Garcia, M., Arroyo, P., Minguez, D., Fernandez, C. *et al.*, 1994, Prospective study of infection, colonization and carriage of methicillin-resistant *Staphylococcus aureus* in an outbreak affecting 990 patients. *Eur. J. Clin. Microbiol. Infect. Dis.*, **13**, 74–81.

Cohen, S. H., Morita, M. M., and Bradford, M., 1991, A seven-year experience with methicillin-resistant *Staphylococcus aureus. Am. J. Med.*, **91**, 233S–237S.

Cox, R. A., Conquest, C., Mallaghan, C., and Marples, R. R., 1995, A major outbreak of methicillin-resistant *Staphylococcus aureus* caused by a new phage-type (EMRSA-16). *J. Hosp. Infect.*, **32**, 73–78.

Dacre, J., Emmerson, A. M., and Jenner, E. A., 1986, Gentamicin-methicillin-resistant *Staphylococcus aureus*: Epidemiology and containment of an outbreak. *J. Hosp. Infect.*, 7, 130–136.

Dancer, S. J., 1997, MRSA in the community: Who is responsible? *SCIEH Weekly Report*, 31(97/01), 2–4.

Dancer, S. J., 1999a, Mopping up hospital infection. *J. Hosp. Infect.*, 43, 85–100.

Dancer, S. J., 1999b, Swinging back the MRSA pendulum? *J. Hosp. Infect.*, 42, 69–71.

Dancer, S. J., 2001, The problem with cephalosporins. *J. Antimicrob. Chemother.*, 48, 463–478.

Dancer, S. J., 2002, Hospital-acquired infection: Is cleaning the answer? *CPD Infect.*, 3, 40–46.

Dancer, S. J., 2003, Glycopeptide resistance in *Staphylococcus aureus*. *J. Antimicrob. Chemother.*, 51, 1309–1311.

Davey, P. G. and Nathwani, D., 1998, What is the value of preventing post-operative infections? *New Horizon*, 6(suppl. 2), S64–S71.

Diekema, D. J., Pfaller, M. A., Jones, R. N., Doern, G. V., Kugler, K. C., Beach, M. L. *et al.*, 2000, Trends in antimicrobial susceptibility of bacterial pathogens isolated from patients with bloodstream infections in the USA, Canada and Latin America. SENTRY Participants Group. *Int. J. Antimicrob. Agents*, 13, 257–271.

Diekema, D. J., Pfaller, M. A., Schmitz, F. J., Smayevsky, J., Bell, J., Jones, R. N. *et al.*, 2001, Survey of infections due to staphylococcus species: Frequency of occurrence and antimicrobial susceptibility of isolates collected in the United States, Canada, Latin America, Europe, and the Western Pacific region for the SENTRY Antimicrobial Surveillance Programme, 1997–1999. SENTRY Participants Group. *Clin. Infect. Dis.*, 32(suppl. 2), S114–S32.

Drew, R. H., Perfect, J. R., Srinath, L., Kurkimilis, E., Dowzicky, M., and Talbot, G. H., 2000, Treatment of methicillin-resistant *Staphyloccocus aureus* infections with quinupristin-dalfopristin in patients intolerant of or failing prior therapy. *J. Antimicrob. Chemother.*, 46, 775–784.

Drummond, M., Stoddart, G., and Torrance, G., 1989, *Methods for Economic Evaluation of Health Care Programmes*. Oxford University Press, Oxford, England, pp. 18–83.

Duckworth, G. J., Lothian, J. L., and Williams, J. D., 1988, Methicillin-resistant *Staphylococcus aureus*: Report of an outbreak in a London teaching hospital. *J. Hosp. Infect.*, 11, 1–15.

Duffy, K. R., 1985, Cost-effective applications of the Centres for Disease Control guidelines for prevention of nosocomial infections. *Am. J. Infect. Control*, 13, 216–217.

Dufour, P., Gillet, Y., Bes, M., Lina, G., Vandenesch, F., Floret, D., Etienne, J., and Richet, H., 2002, Community-acquired methicillin-resistant *Staphylococcus aureus* infections in France: Emergence of a single clone that produces Panton-Valentine leukocidin. *Clin. Infect. Dis.*, 35, 819–824.

EARSS Annual Report, 2002, European Antimicrobial Resistance Surveillance System. Accessed at http://www.earss.rivm.nl.

Edmond, M. B., Wallace, S. E., McClish, D. K., Pfaller, M. A., Jones, R. N., and Wenzel, R. P., 1999, Nosocomial bloodstream infections in United States hospitals—a three-year analysis. *Clin. Infect. Dis.*, 29, 239–244.

Emmerson, A. M., Enstone, J. E., Griffin, M., Kelsey, M. C., and Smyth, E. T. M., 1996, Second national prevalence survey of infections in hospitals—overview of the results. *J. Hosp. Infect.*, 32, 175–190.

Engemann, J. J., Carmeli, Y., Cosgrove, S. E., Fowler, V. G., Bronstein, M. Z., Trivette, S. L. *et al.*, 2003, Adverse clinical and economic outcomes attributable to methicillin resistance among patients with *Staphylococcus aureus* surgical site infection. *Clin. Infect. Dis.*, 36, 592–598.

Farr, B. M. and Jarvis, W. R., 2002, Would active surveillance cultures help control health-care-related methicillin-resistant *Staphylococcus aureus* infections? *Infect. Control Hosp. Epidemiol.*, **23**, 65–68.

Farr, B. M., Salgado, C. D., Karchmer, T. B., and Sherertz, R. J., 2001, Can antibiotic-resistant nosocomial infections be controlled? *Lancet Infect. Dis.*, **1**, 38–45.

Fitzpatrick, F., Murphy, O. M., Brady, A., Prout, S., and Fenelon, L. E., 2000, A purpose built MRSA cohort unit. *J. Hosp. Infect.*, **46**, 271–279.

Fraise, A. P., Mitchell, K., O'Brien, S. J., Oldfield, K., and Wise, R., 1997, Methicillin-resistant Staphylococcus aureus (MRSA) in nursing homes in a major UK city: An anonymised point prevalence survey. *Epidemiol. Infect.*, **118**, 1–5.

Frank, M. O., Batteiger, B. E., Sorensen, S. J., Hartstein, A. I., Carr, J. A., McComb, J. S. *et al.*, 1997, Decrease in expenditures and selected nosocomial infections following implementation of an antimicrobial-prescribing improvement program. *Clin. Perform. Qual. Health Care*, **5**, 180–188.

Fry, D. E., 2002, The economic costs of surgical site infection. *Surg. Infect. (Larchmt)*, **1**, S37–S43.

Geissler, A., Gerbeaux, P., Granier, I., Blanc, P., Facon, K., and Durand-Gassell, 2003, Rational use of antibiotics in the intensive care unit: Impact of microbial resistance and costs. *Intensive Care Med.*, **29**, 49–54.

Geldner, G., Ruoff, M., Hoggmann, H. J., Kiefer, P., Georgieff, M., and Wiedeck, H., 1999, Cost analysis concerning MRSA infection in ICU. *Anaesthesiol. Intensivmed. Notfallmed. Schmerzther.*, **34**, 409–413.

Girou, E., Azar, J., Wolkenstein, P., Cizeau, F., Brun-Buisson, C., and Roujeau, J-C., 2000, Comparison of systematic versus selective screening for methicillin-resistant *Staphylococcus aureus* carriage in a high-risk dermatology ward. *Infect. Control Hosp. Epidemiol.*, **21**, 583–587.

Girou, E., Pujade, G., Legrand, P, Cizeau, F., and Brun-Buisson, C., 1998, Selective screening of carriers for control of methicillin-resistant *Staphylococcus aureus* (MRSA) in high-risk hospital areas with a high level of endemic MRSA. *Clin. Infect. Dis.*, **27**, 543–550.

Gopal Rao, G., Jeanes, A., Osman, M., Aylott, C., and Green, J., 2002, Marketing hand hygiene in hospitals—a case study. *J. Hosp. Infect.*, **50**, 42–47.

Grundmann, H., Hori, S., Winter, B., Tami, A., and Austin, D. J., 2002, Risk factors for the transmission of methicillin-resistant *Staphylococcus aureus* in an adult intensive care unit: Fitting to the data. *J. Infect. Dis.*, **185**, 481–488.

Gruson, D., Hilbert, G., Vargas, F., Valentino, R., Bebear, C., Allery, A. *et al.*, 2000, Rotation and restricted use of antibiotics in a medical intensive care unit. Impact on the incidence of ventilator-associated pneumonia caused by antibiotic-resistant gram-negative bacteria. *Am. J. Respir. Crit. Care Med.*, **162**, 837–843.

Haley, R. W., 1991, Measuring the costs of nosocomial infections: Methods for estimating economic burden on the hospital. *Am. J. Med.*, **91**, 32S–38S.

Haley, R. W., Culver, D. H., White, J. W., Morgan, W. M., Emori, T. G., Munn, V. P. *et al.*, 1985, The efficacy of infection surveillance and control programme in preventing nosocomial infections in US hospitals. *Am. J. Epidemiol.*, **121**, 182–205.

Haley, R. W., Schaberg, D. R., Crossley, K. B., Von Allmen, S. D., and McGowan, J., Jr., 1981, Extra charges and prolongation of stay attributable to nosocomial infections: A prospective inter-hospital comparison. *Am. J. Med.*, **70**, 51–58.

Harbarth, S., Martin, Y., Rohner, P., Henry, N., Auckenthaler, R., and Pittet, D., 2000, Effect of delayed infection control measures on a hospital outbreak of methicillin-resistant *Staphylococcus aureus*. *J. Hosp. Infect.*, **46**, 43–49.

Herold, B. C., Immergluck, L. C., Maranan, M. C., Lauderdale, D. S., Gaskin, R. E., Boyle-Vavra, S. _et al._ 1998, Community-acquired methicillin-resistant _Staphylococcus aureus_ in children with no identified predisposing risk. _JAMA_, **279**, 593–598.

Herwaldt, L., 1999, Control of methicillin-resistant _Staphylococcus aureus_ in the hospital setting. _Am. J. Med._, **106**, 11S–18S.

Hill, D. A., Herford, T., and Parratt, D., 1998, Antibiotic usage and methicillin-resistant _Staphylococcus aureus_: An analysis of causality. _J. Antimicrob. Chemother._, **42**, 676–677.

Hiramatsu, K., 2001, Vancomycin-resistant _Staphylococcus aureus_: A new model of antibiotic resistance. _Lancet Infect. Dis._, **1**, 147–155.

Hollenbeak, C. S., Murphy, D. M., Koenig, S., Woodward, R. S., Dunagan, W., and Fraser, V. J., 2000, The clinical and economic impact of deep chest surgical site infections following coronary artery bypass graft surgery. _Chest_, **118**, 397–402.

Holmberg, S. D., Solomon, S. L., and Blake, P. A., 1987, Health and economic impacts of antimicrobial resistance. _Rev. Infect. Dis._, **9**, 1065–1078.

Hori, S., Sunley, R., Tami, A., and Grundmann, H., 2002, The Nottingham _Staphylococcus aureus_ population study: Prevalence of MRSA among the elderly in a university hospital. _J. Hosp. Infect._, **50**, 25–29.

Hornberg, C., Schafer, T. R., Koller, A., and Wetz, H. H., 2003, The MRSA patient in technical orthopaedics and rehabilitation, Part 2: Hygiene Management. _Orthopade_, **32**, 218–224.

Janknegt, R., 1997, The treatment of staphylococcal infections with special reference to pharmacokinetic, pharmacodynamic and pharmacoeconomic considerations. _Pharm. World Sci._, **19**, 133–141.

Jaqua-Stewart, M. J., Tjaden, J., Humphreys, D. W., Bade, P., Tille, P. M., Peterson, K. G. _et al._, 1999, Reduction in methicillin-resistant _Staphylococcus aureus_ infection rate in a nursing home by aggressive containment strategies. _S. D. J. Med._, **52**, 241–247.

Jarvis, W. R., 1996, Selected aspects of the socio-economic impact of nosocomial infections: Morbidity, mortality, cost, and prevention. _Infect. Control Hosp. Epidemiol._, **17**, 552–557.

Jayaratne, P. and Rutherford, C., 1999, Detection of methicillin-resistant _Staphylococcus aureus_ (MRSA) from growth on mannitol salt oxacillin agar using PCR for nosocomial surveillance. _Diagn. Microbiol. Infect. Dis._, **35**, 13–18.

Jernigan, J. A., Clemence, M. A., Stott, G. A., Titus, M. G., Alexander, C. H., Palumbo, C. M. _et al._, 1995, Control of methicillin-resistant _Staphylococcus aureus_ at a university hospital: One decade later. _Infect. Control Hosp. Epidemiol._, **16**, 686–696.

Jewell, M., 1994, Cost-containment using an outcome-based best practice model for the management of MRSA. _J. Chemother._, **6**(suppl. 2), 35–39.

Jönsson, B. and Lindgren, B., 1980, Five common fallacies in estimating the economic gains of early discharge. _Social Sci. Med._, **14**, 27–33.

Karchmer, T. B., Durbin, L. J., Simonto, B. M., and Farr, B. M., 2002, Cost-effectiveness of active surveillance cultures and contact/droplet precautions for control of methicillin-resistant _Staphylococcus aureus_. _J. Hosp. Infect._, **51**, 126–132.

Khan, M. M. and Celik, Y., 2001, Cost of nosocomial infection in Turkey: An estimate based on the university hospital data. _Health Serv. Man. Res._, **14**, 49–54.

Khare, M. and Keady, D., 2003, Antimicrobial therapy of methicillin-resistant _Staphylococcus aureus_ infection. _Exp. Opin. Pharmacother._, **4**, 165–177.

Kibbler, C. C., Quick, A., and O'Neill, A. M., 1998, The effect of increased bed numbers on MRSA transmission in acute medical wards. _J. Hosp. Infect._, **39**, 213–219.

Kim, T., Oh, P. I., and Simor, A. E., 2001, The economic impact of methicillin-resistant _Staphylococcus aureus_ in Canadian hospitals. _Infect. Control Hosp. Epidemiol._, **22**, 99–104.

Kollef, M. H. and Fraser, V. J., 2001, Antibiotic resistance in the intensive care unit. *Ann. Intern. Med.*, **134**, 298–314.

Korin, J., 1993, Cost implications of malpractice and adverse events. *Hosp. Formul.*, **28**(suppl. 1), 59–61.

Kunori, T., Cookson, B., Roberts, J. A., Stone, S., and Kibbler, C., 2002, Cost-effectiveness of different MRSA screening methods. *J. Hosp. Infect.*, **51**, 189–200.

Lacey, R. W., 1987, Multi-resistant *Staphylococcus aureus*—a suitable case for inactivity? *J. Hosp. Infect*, **9**, 103–105.

Landman, D., Chockalingam, M., and Quale, J. M., 1999, Reduction in the incidence of methicillin-resistant *Staphylococcus aureus* and ceftazidime-resistant *Klebsiella pneumoniae* following changes in a hospital antibiotic formulary. *Clin. Infect. Dis.*, **28**, 1067–1070.

Lankford, M. G., Zembower, T. R., Tricks, W. E., Hacek, D. M., Noskin, G. A., and Peterson, L. R., 2003, Influence of role models and hospital design on hand hygiene of healthcare workers. *Emerg. Infect. Dis.*, **9**, 217–223.

Larson, E. L., 1995, APIC guideline for handwashing and hand antisepsis in health care settings. *Am. J. Infect. Control*, **23**, 251–269.

Law, M. R. and Gill, O. N., 1988, Hospital-acquired infection with methicillin-resistant and methicillin-sensitive staphylococci. *Epidemiol. Infect.*, **101**, 623–629.

Layton, M. C., Perez, M., Heald, P., and Patterson, J. E., 1993, An outbreak of mupirocinresistant *Staphylococcus aureus* on a dermatology ward associated with an environmental reservoir. *Infect. Control Hosp. Epidemiol.*, **14**, 369–375.

Linares, L., 2001, The VISA/GISA problem: Therapeutic implications. *Clin. Microbiol. Infect.*, **7**(suppl. 4), 8–15.

Lucet, J. C., Chevret, S., Durand-Zaleski, I., Chastang, C., and Regnier, B., 2003, Prevalence and risk factors for carriage of methicillin-resistant *Staphylococcus aureus* at admission to the intensive care unit results of a multi-centre study. *Arch. Intern. Med.*, **163**, 181–188.

Lynch, W., Malek, M., Davey, P. G., Byrne, D. J., and Napier, A., 1992, Costing wound infection in a Scottish hospital. *Pharmacoeconomics*, **2**, 163–170.

Mahieu, L. M., Buitenweg, N., Bentels, P., and De Dooy, J. J., 2001, Additional hospital stay and charges due to hospital-acquired infections in a neonatal intensive care unit. *J. Hosp. Infect.*, **47**, 223–229.

Martineau, F., Picard, F. J., Roy, P. H., Ouellette, M., and Bergeron, M. G., 1998, Species-specific and ubiquitous-DNA-based assays for rapid identification of *Staphylococcus aureus*. *J. Clin. Microbiol.*, **36**, 618–623.

Mehtar, S., Drabu, Y. J., and Mayet, F., 1989, Expenses incurred during a 5-week epidemic methicillin-resistant *Staphylococcus aureus* outbreak. *J. Hosp. Infect.*, **13**, 199–200.

Mehtar, S., 1993, How to cost and fund an infection control programme. *J. Hosp. Infect.*, **25**, 57–69.

Mehtar, S., 1995, Infection control programmes—are they cost-effective? *J. Hosp. Infect.*, **30**(suppl), 26–34.

Merrer, J., Santoli, F., Appere de Vecchi, C., Tran, B., De Jonghe, B., and Outin, H., 2000, "Colonisation pressure" and risk of acquisition of methicillin-resistant *Staphylococcus aureus* in a medical intensive care unit. *Infect. Control Hosp. Epidemiol.*, **21**, 718–723.

Miller, P. J., Farr, B. M., and Gwaltney, J. M., Jr., 1989, Economic benefits of an effective infection control program: Case study and proposal. *Rev. Infect. Dis.*, **11**, 284–288.

Monnet, D. L., 1998, Methicillin-resistant *Staphylococcus aureus* and its relationship to antimicrobial use: Possible implications for control. *Infect. Control Hosp. Epidemiol.*, **19**, 552–559.

Monnet, D. L. and Frimodt-Moller, N., 2001, Antimicrobial-drug use and methicillin-resistant *Staphylococcus aureus. Emerg. Infect. Dis.*, 7, 161–163.

Muder, R. R., Brennen, C., Wagener, M. M., Vickers, R. M., Rihs, J. D., Hancock, G. A. *et al.*, 1991, Methicillin-resistant staphylococcal colonisation and infection in a long-term care facility. *Ann. Intern. Med.*, 114, 107–112.

Murray-Leisure, K. A., Geib, S., Graceley, D., Rubin-Slutsky, A. B., Saxena, N., Muller, H. A. *et al.*, 1990, Control of epidemic methicillin-resistant *Staphylococcus aureus. Infect. Control Hosp. Epidemiol.*, 11, 343–350.

Muto, C. A., Jernigan, J. A., Ostrowsky, B. E., Richet, H. M., Jarvis, W. R., Boyce, J. M. *et al.*, 2003, SHEA guideline for preventing nosocomial transmission of multidrug-resistant strains of *Staphylococcus aureus* and enterococcus. *Infect. Control Hosp. Epidemiol.*, 24, 362–386.

McLaws, M. L., Gold, J., King, K., Irwig, L. W., and Berry, G., 1988, The prevalence of nosocomial and community-acquired infections in Australian hospitals. *Med. J. Aust.*, 149, 582–590.

Nathwani, D., 2001, The management of skin and soft tissues infections: Outpatient parenteral antibiotic therapy in the United Kingdom. *Chemotherapy*, 47(suppl. 1), 17–23.

Nathwani, D., 2003, Impact of methicillin-resistant *Staphylococcus aureus* infections on key health economic outcomes: Does reducing the length of hospital stay matter? *J. Antimicrob. Chemother.*, 51(suppl.S2), ii37–ii44.

Nettleman, M. D., Trilla, A., Fredrickson, M., and Pfaller, M., 1991, Assigning responsibility: Using feedback to achieve sustained control of methicillin-resistant *Staphylococcus aureus. Am. J. Med.*, 91, 228S–232S.

Niederman, M. S., 2001a, Cost effectiveness in treating ventilator-associated pneumonia. *Crit. Care*, 5, 243–244.

Niederman, M. S., 2001b, Impact of antibiotic resistance on clinical outcomes and the cost of care. *Crit. Care Med.*, 29(suppl. 4), N114–N120.

Noel, I., Hollyoak, V., and Galloway, A., 1997, A survey of the incidence and care of post-operative wound infections in the community. *J. Hosp. Infect.*, 36, 267–273.

Olson, M. D., 1981, Medicolegal review: Nosocomial infections next target for malpractice suits. *Hosp. Med. Staff.*, 10, 19–24.

Olson, M., O'Connor, M., and Schwartz, M. L., 1984, Surgical wound infections. A five-year prospective study of 20,193 wounds at the Minneapolis VA Medical Centre. *Ann. Surg.*, 199, 253–259.

Paladino, J. A., Sunderlin, J. L., Price, C. S., and Schentag, J. J., 2002, Economic consequences of antimicrobial resistance. *Surg. Infect. (Larchmt)*, 3, 259–267.

Pan, A., Catenazzi, P., Ferrari, L., Tinelli, C., Seminari, E., Ratti, A. *et al.*, 2001, Evaluation of the efficacy of a program to control nosocomial spread of methicillin-resistant *Staphylococcus aureus. Infez. Med.*, 9, 163–169.

Papia, G., Louie, M., Tralla, A., Johnson, C., Collins, V., and Simor, A. E., 1999, Screening high-risk patients for methicillin-resistant *Staphylococcus aureus* on admission to the hospital: Is it cost effective? *Infect. Control Hosp. Epidemiol.*, 20 473–477.

Pena, C., Pujol, M., Pallares, R., Corbella, X., Vidal, T., Tortras, N. *et al.*, 1996, Estimation of costs attributable to nosocomial infection: Prolongation of hospitalisation and calculation of alternative costs. *Med. Clinica*, 106, 441–444.

Pfaller, M. A., Jones, R. N., Doern, G. V., and Kugler, K., 1998, Bacterial pathogens isolated from patients with bloodstream infection: frequencies of occurrence and antimicrobial susceptibility patterns from the SENTRY antimicrobial surveillance program (United States and Canada, 1997). *Antimicrob. Agents Chemother.*, 42, 1762–1770.

Pinner, R. W., Haley, R. W., Blumenstein, B. A., Schaberg, D. R., Von Allmen, S. D., and McGowan, J. E., Jr., 1982, High cost nosocomial infections. *Infect. Control*, **3**, 143–149.

Pittet, D., Hugonnet, S., Harbarth, S., Mourouga, P., Sauvan, V., Touveneau, S. *et al.*, and members of the infection control programme, 2000, Effectiveness of a hospital-wide programme to improve compliance with hand hygiene. *Lancet*, **356**, 1307–1312.

Pittet, D., Tarara, D., and Wenzel, R. P., 1994, Nosocomial blood-stream infection in critically ill patients. Excess length of stay, extra costs, and attributable mortality. *JAMA*, **271**, 1598–1601.

Plowman, R., Graves, N., Griffin, M. A. S., Roberts, J. A., Swan, A. V., Cookson, B. *et al.*, 2001, The rate and cost of hospital-acquired infections occurring in patients admitted to selected specialties of a district general hospital in England and the national burden imposed. *J. Hosp. Infect.*, **47**, 198–209.

Poulsen, K. B., Bremmelgaard, A., Sorensen, A. I., Raahave, D., and Petersen, J. W., 1994, Estimated costs of post-operative wound infections. A case-control study of marginal hospital and social security costs. *Epidemiol. Infect.*, **113**, 283–295.

Poulsen, K. B. and Gottschau, A., 1997, Long-term prognosis of patients with surgical wound infections. *World J. Surg.*, **21**, 799–804.

Rampling, A., Wiseman, S., and Davis, L., 2001, Evidence that hospital hygiene is important in the control of methicillin-resistant *Staphylococcus aureus*. *J. Hosp. Infect.*, **49**, 109–116.

Rao, N., Jacobs, S., and Joyce, L., 1988, Cost-effective eradication of an outbreak of methicillin-resistant *Staphylococcus aureus* in a community teaching hospital. *Infect. Control Hosp. Epidemiol.*, **9**, 255–260.

Reboli, A. C., John, J. F., Jr., and Levkoff, A. H., 1989, Epidemic methicillin-gentamicinresistant *Staphylococcus aureus* in a neonatal intensive care unit. *Am. J. Dis. Child.*, **143**, 34–39.

Reid, R., Simcock, J. W., Chisholm, L., Dobbs, B., and Frizelle, F. A., 2002, Post-discharge clean wound infections: Incidence underestimated and risk factors overemphasised. *ANZ J. Surg.*, **72**, 339–343.

Reilly, J., Twaddle, S., McIntosh, J., and Kean, L., 2001a, An economic analysis of surgical wound infection. *J. Hosp. Infect.*, **49**, 245–249.

Reilly, J. S., Baird, D., and Hill, R., 2001b, The importance of definitions and methods in surgical wound infection audit. *J. Hosp. Infect.*, **47**, 64–66.

Reimer, K., Gleed, C., and Nicolle, L. E., 1987, The impact of post-discharge infection on surgical wound infection rates. *Infect. Control*, **8**, 237–240.

Robinson, R., 1993, Economic evaluation and health care: What does it mean? *BMJ*, **307**, 670–673.

Romero-Vivas, J., Rubio, M., Fernandez, C., and Picazo, J. J., 1995, Mortality associated with nosocomial bacteraemia due to methicillin-resistant *Staphylococcus aureus*. *Clin. Infect. Dis.*, **21**, 1417–1423.

Rose, R., Hunting, K. J., Townsend, T. R., and Wenzel, R. P., 1977, Morbidity/mortality and economics of hospital-acquired bloodstream infections: A controlled study. *S. Med. J.*, **70**, 1267–1269.

Rubin, R. J., Harrington, C. A., Poon, A., Dietrich, K., Greene, J. A., and Moiduddin, A., 1999, The economic impact of *Staphylococcus aureus* infection in New York City hospitals. *Emerg. Infect. Dis.*, **5**, 9–17.

Rubinovitch, B. and Pittet, D., 2001, Screening for methicillin-resistant *Staphylococcus aureus* in the endemic hospital: What have we learned? *J. Hosp. Infect.*, **47**, 9–18.

Sanford, M. D., Widmer, A. F., Bale, M. J., Jones, R. N., and Wenzel, R. P., 1994, Efficient detection and long-term persistence of the carriage of methicillin-resistant *Staphylococcus aureus*. *Clin. Infect. Dis.*, **19**, 1123–1128.

Sanyal, S. C. and Mokaddas, E. M., 1999, The increase in carbapenem use and emergence of *Stenotrophomonas maltophilia* as an important nosocomial pathogen. *J. Chemother.*, **11**, 28–33.

Schaberg, D. R., Culver, D. H., and Gaynes, R. P., 1991, Major trends in the microbial aetiology of nosocomial infection. *Am. J. Med.*, **91**, 72S–75S.

Schentag, J. J., Hyatt, J. M., Carr, J. R., Paladino, J. A., Birmingham, M. C., Zimmer, G. S. *et al.*, 1998, Genesis of methicillin-resistant *Staphylococcus aureus* (MRSA), how treatment of MRSA infections has selected for vancomycin-resistant *Enterococcus faecium*, and the importance of antibiotic management and infection control. *Clin. Infect. Dis.*, **26**, 1204–1214.

Scolan, V., Telmon, N., Rouge, J. C., and Rouge, D., 2000, Medical responsibility and nosocomial infections. *Rev. Med. Intern.*, **21**, 361–367.

Selkon, J. B., 1980, The role of an isolation unit in the control of hospital infection with methicillin-resistant staphylococci. *J. Hosp. Infect.*, **1**, 41–46.

Shanson, D., Johnstone, D., and Midgley, J., 1985, Control of a hospital outbreak of methicillin resistant *Staphylococcus aureus* infections: Value of an isolation unit. *J. Hosp. Infect.*, **6**, 285–292.

Shrestha, N. K., Shermock, K. M., Gordon, S. M., Tuohy, M. J., Wilson, D. A., Cwynar, R. E. *et al.*, 2003, Predicitive value and cost-effectiveness analysis of a rapid polymerase chain reaction for preoperative detection of nasal carriage of *Staphylococcus aureus*. *Infect. Control Hosp. Epidemiol.*, **24**, 327–333.

Singer, M. E., Harding, I., Jacobs, M. R., and Jaffe, D. H., 2003, Impact of antimicrobial resistance on health outcomes in the outpatient treatment of adult community-acquired pneumonia: A probability model. *J. Antimicrob. Chemother.*, **51**, 1269–1282.

Smyth, E. T. and Emmerson, A. M., 2000, Surgical site infection surveillance. *J. Hosp. Infect.*, **45**, 173–184.

Spengler, R. F. and Greenough, W. B., 3rd, 1978, Hospital costs and mortality attributed to nosocomial bacteraemias. *JAMA*, **240**, 2455–2458.

Spicer, W. J., 1984, Three strategies in the control of staphylococci including methicillin-resistant *Staphylococcus aureus*. *J. Hosp. Infect.*, **5**(suppl. A), 45–49.

Stamm, A. M., Long, M. N., and Belcher, B., 1993, Higher overall nosocomial infection rate because of increased attack rate of methicillin-resistant *Staphylococcus aureus*. *Am. J. Infect. Control*, **21**, 70–74.

Stockley, J. M., Allen, R. M., Thomlinson, D. F., and Constantine, C. E., 2001, A district general hospital's method of post-operative infection surveillance including post-discharge follow-up, developed over a five-year period. *J. Hosp. Infect.*, **49**, 48–54.

Talon, D., 1999, The role of the hospital environment in the epidemiology of multi-resistant bacteria. *J. Hosp. Infect.*, **43**, 13–17.

Talon, D., Vichard, P., Muller, A., Bertin, M., Jeunet, L., and Bertrand, X., 2003, Modelling the usefulness of a dedicated cohort facility to prevent the dissemination of MRSA. *J. Hosp. Infect.*, **54**, 57–62.

Tarzi, S., Kennedy, P., Stone, S., and Evans, M., 2001, Methicillin-resistant *Staphylococcus aureus*: Psychological impact of hospitalisation and isolation in an older adult population. *J. Hosp. Infect.*, **49**, 250–254.

Teare, E. L. and Barrett, S. P., 1997, Is it time to stop searching for MRSA? Stop the ritual of tracing colonised people. *BMJ*, **314**, 666–667.

Teare, L., 2000, Lavate vestras manus. *CPD Infect.*, **2**, 51–53.

Tokue, Y., Shoji, S., Satoh, K., Watanabe, A., and Motomiya, M., 1992, Comparison of a polymerase chain reaction assay and a conventional microbiologic method for detection of methicillin-resistant *Staphylococcus aureus*. *Antimicrob. Agents Chemother.*, **36**, 6–9.

Van den Bergh, M. F. Q., Kluytmans, J. A. J. W., van Hout, B. A., Maat, A. P., Seerden, R. J., McDonnel, J. *et al.*, 1996, Cost-effectiveness of perioperative mupirocin nasal ointment in cardiothoracic surgery. *Infect. Control Hosp. Epidemiol.*, **17**, 786–792.

Venezia, R. A., Domaracki, B. E., Evans, A. M., Preston, K. E., and Graffunder, E. M., 2001, Selection of high-level oxacillin resistance in heteroresistant *Staphylococcus aureus* by fluoroquinolone exposure. *J. Antimicrob. Chemother.*, **48**, 375–381.

Vinken, A., Li, Z., Balan, D., Rittenhouse, B., Welike, R., and Nathwani, D., 2001, Economic evaluation of linezolid, flucloxacillin and vancomycin in the empirical treatment of cellulitis in UK hospitals: A decision analytical model. *J. Hosp. Infect.*, **49**(suppl. A), S13–S24.

Vriens, M., Blok, H., Fluit, A., Troelstra, A., Van Der Werken, C., and Verhoef, 2002, Costs associated with a strict policy to eradicate methicillin-resistant *Staphylococcus aureus* in a Dutch University Medic Centre: A ten-year study. *E. J. Clin. Microbiol. Infect. Dis.*, **21**, 782–786.

Wakefield, D. S., Helms, C. M., Massanari, R. M., Mori, M., and Pfaller, M., 1988, Cost of nosocomial infection: Relative contributions of laboratory, antibiotic and per diem costs in serious *Staphylococcus aureus* infections. *Am. J. Infect. Control*, **16**, 185–192.

Walker, A., 2002, Hospital-acquired infection and bed use in NHS Scotland. Accessed at http://www.ageconcernscotland.org.uk/pdf.pl?file=HAI_Rept.pdf

Walsh, T. R. and Howe, R. A., 2002, The prevalence and mechanisms of vancomycin resistance in *Staphylococcus aureus. Ann. Rev. Microbiol.*, **56**, 657–675.

Webster, J., Faoagali, J. L., and Cartwright, D., 1994, Elimination of methicillin-resistant *Staphylococcus aureus* from a neonatal intensive care unit after hand washing with triclosan. *J. Paediatr. Child Health*, **30**, 59–64.

Wenzel, R. P., 1995, The Lowbury Lecture. The economics of nosocomial infections. *J. Hosp. Infect.*, **31**, 79–87.

Whitehouse, J. D., Friedman, N. D., Kirkland, K. B., Richardson, W. J., and Sexton, D. J., 2002, The impact of surgical-site infections following orthopaedic surgery at a community hospital and a university hospital: Adverse quality of life, excess length of stay, and extra costs. *Infect. Control Hosp. Epidemiol.*, **23**, 183–189.

Wilson, P. and Dunn, L. J., 1996, Using an MRSA scoring system to decide whether patients should be nursed in isolation. *Hyg. Med.*, **21**, 465–477.

Witte, W., Kresken, M., Braulke, C., and Cuny, C., 1997, Increasing incidence and widespread dissemination of methicillin-resistant *Staphylococcus aureus* (MRSA) in hospitals in central Europe, with special reference to German hospitals. *Clin. Microbiol. Infect.*, **3**, 414–422.

Woodward, R. S., Medoff, G., Smith, M. D., and Gray, J. L., 3rd, 1987, Antibiotic cost savings from formulary restrictions and physicians monitoring in a medical-school-affiliated hospital. *Am. J. Med.*, **83**, 817–823.

Zoutman, D., McDonald, S., and Vethanayagan, D., 1998, Total and attributable costs of surgical-wound infections at a Canadian tertiary-care centre. *Infect. Control Hosp. Epidemiol.*, **19**, 254–259.

Chapter 17

Antifungal Agents: Resistance and Rational Use

Frank C. Odds
Aberdeen Fungal Group, Institute of Medical Sciences, Foresterhill,
Aberdeen AB25 2ZD, UK

1.   INTRODUCTION

For several years, almost every publication in the field of clinical mycology has begun by stating one or more of the following points. The incidence of invasive fungal disease continues to rise despite judicious antifungal prophylaxis and heightened clinical awareness of the risk of such disease in particular types of patient. There has been a shift among species causing invasive disease away from *Candida albicans* towards other *Candida* species with resistance to triazole antifungal agents such as fluconazole. There is an urgent need for new antifungal agents active against new molecular targets to combat the rising tide of infection and antifungal resistance. While such claims inevitably generate a climate of apprehension about mycoses, they tend to simplify and overstate the reality. This chapter will attempt to evaluate the true extent of antifungal resistance problems and suggest approaches to rational prophylactic and therapeutic use of antifungal agents.

2.   EPIDEMIOLOGY OF INVASIVE
FUNGAL INFECTIONS

Expressed concerns about a rising incidence of invasive *Candida* infection can be traced back at least to the 1950s (Keye and Magee, 1956). There is little doubt, however, that the greatest rise in invasive infections caused by

Antibiotic Policies: Theory and Practice. Edited by Gould and van der Meer
Kluwer Academic / Plenum Publishers, New York, 2005

Candida spp. and many other types of fungi began in the early 1980s, in parallel with rapidly increasing medical and surgical use of immunosuppressive procedures. The AIDS epidemic also began at this same time and AIDS became recognized as a factor predisposing not only to superficial fungal infections, but also commonly to potentially fatal deep-tissue mycoses such as cryptococcal meningitis and *Pneumocystis jiroveci* pneumonia worldwide, and disseminated histoplasmosis and *Penicillium marneffei* infection in geographical areas where these mycoses are endemic. By 1990, the major emphasis of clinical mycology had switched from infections of the skin and mucous membranes to the study of disseminated, invasive, and all too commonly lethal fungal diseases. While seriously immunosuppressed patients remain those most at risk of invasive mycoses, fungal infections have grown as a cause for concern in intensive care and after major surgical procedures.

Since 1990, the epidemiology of mycoses has been far from static, and a number of surveys have illustrated the major trends. These can be characterized as follows. There has been a decreasing incidence of invasive *Candida* infection and a steadily rising incidence of invasive infections caused by *Aspergillus* and other mould species. The species causing *Candida* infection differ between countries, regions, and even individual institutions, as well as between patients with different underlying diseases, making it difficult to generalize about temporal changes in the incidence of *Candida* species. The incidence of all types of mycoses associated with HIV infection has declined dramatically in countries where highly active anti-retroviral therapy (HAART) is used widely. The status of certain fungi with very low susceptibility to existing antifungal agents (*Fusarium* spp., *Scedosporium* spp., *Zygomycota*) has emerged from that of obscure case reports to routine mention in lists of opportunistic fungal risks in haematological malignancy.

2.1.   *Candida* infections

The decrease in invasive *Candida* infections began during the 1990s and has been evidenced from US mortality records (McNeil *et al.*, 2001), incidence data from US intensive care units (Trick *et al.*, 2002) and neutropenic patients (Wisplinghoff *et al.*, 2003), and from Japanese autopsy data (Yamazaki *et al.*, 1999). Most of the decrease is attributable to a marked decline in infections caused by *C. albicans*, so it is not surprising that the proportion of other *Candida* species incriminated in disseminated disease has risen. However, evidence for a rising *incidence* of *Candida* infections due to species other than *C. albicans* is less impressive than data illustrating their increased *prevalence*. Only *Candida glabrata* infections may have increased in incidence in some areas in a manner and extent consistent with a general trend: infections caused by *Candida krusei, Candida parapsilosis,* and *Candida tropicalis* have occurred at fairly consistent rates through the 1990s (Trick *et al.*, 2002). In a particularly thorough analysis of publications detailing the epidemiology of candidaemia, Sandven (2000) showed how, in the United States, the prevalence of *C. glabrata* among *Candida* species isolated from blood cultures has risen from around 10% up to 1990 to around 20% in surveys done since that date; the change is at the expense of *C. tropicalis*, which has had a lower prevalence since 1990. The large survey by Pfaller *et al.* also shows *C. glabrata* representing 18% of *Candida* spp. blood isolations in the United States from 1992 to 1998 (Pfaller *et al.*, 1999b). In European surveys, a similar overall doubling in the prevalence of *C. glabrata* (at the expense of *C. albicans*) is apparent between the 1980s and the 1990s, although the average current prevalence of *C. glabrata* in European surveys is lower (~15%) than in the United States (Sandven, 2000). By contrast, most data from Latin American countries, Japan, and elsewhere in Asia all show *C. glabrata* to be a relatively rare species, with *C. parapsilosis* highly prevalent and second to *C. albicans* as a cause of bloodstream infections (Pfaller *et al.*, 2000; Sandven, 2000). These observations are slightly confused by results from the SENTRY prospective surveillance scheme, which covers the United States, Europe, and Latin America, and which puts *C. parapsilosis* as the second most common species in Europe (Pfaller *et al.*, 1999a). With such mixed messages emerging from large surveys of bloodstream isolates, it is impossible to make confident statements about any particular trends or their causes.

2.2.   *Aspergillus* infections

The almost continually rising incidence of aspergillosis worldwide is far more easy to discern. The same surveys that show a fall in *C. albicans* infections document a steady rise in aspergillosis, mainly due to *Aspergillus fumigatus*, but sometimes caused by other species such as *Aspergillus flavus* (McNeil *et al.*, 2001; Yamazaki *et al.*, 1999). The source of the increased incidence is easy to define: the number of patients undergoing procedures that predispose to invasive aspergillosis (primarily allogeneic haematopoietic stem cell transplantation and major solid organ transplantation) has grown steadily through the 1980s and 1990s (Denning, 1998; K. A. Marr *et al.*, 2002a, b). Mortality in invasive aspergillosis is often very high, exceeding 80% in stem cell transplant recipients and patients with aspergillosis disseminated from a primary pulmonary site (Lin *et al.*, 2001).

2.3.   Other fungal diseases

The same clinical settings that predispose patients to aspergillosis also increase the risk of nosocomial infections by other filamentous fungi.

Groll and Walsh have extensively reviewed the threat posed by uncommon fungal diseases (Groll and Walsh, 2001). Among the fungi they discuss, *Fusarium* spp., *Scedosporium* spp., and members of the *Zygomycota* pose the greatest threat to life since they are commonly refractory to systemic antifungal agents.

The introduction of HAART has reduced HIV burdens so effectively that the incidence of most life-threatening AIDS-related mycoses has declined, sometimes dramatically (Ives *et al.*, 2001; Raffaele *et al.*, 2003). This change particularly affects *Pneumocystis* and *Cryptococcus* infections, where the high incidence and clinical consequences stimulated intensive research into both diseases through the 1980s and 1990s. Both are now encountered only occasionally in countries where HAART is readily available and affordable.

3.   ANTIFUNGAL RESISTANCE: IS IT A GROWING PROBLEM?

The still-rising overall incidence of invasive fungal disease creates a particular concern that is easily expressed. Since the obvious and necessary action to combat a growing fungal infection problem is to increase prophylactic and therapeutic usage of antifungal agents, will this not inevitably lead to a rise in incidence of infections caused by antifungal-resistant strains and species? Should we not take precautionary steps to ensure that resistant fungi do not become a clinical problem comparable with multiresistant bacteria?

3.1.   Antifungal resistance in *Candida* species

These are very reasonable questions, and some authors have already published alarming accounts that have engendered concerns without necessarily delivering accurate detail and evidence to support their claims. For example, the now almost universal use of the ugly term "non-*albicans Candida* species" in publications has created a widespread illusion that only *C. albicans* is susceptible to fluconazole and other azoles. The detailed reality is quite different. The only clinically important *Candida* species regarded as resistant to fluconazole *per se* is *C. krusei*, and this species remains susceptible to most other triazole antifungals, albeit with lower susceptibility than *C. albicans*. *C. glabrata* is less susceptible to triazoles than *C. albicans*, but to characterize this species as azole-resistant is a gross oversimplification of the data; *in vitro*, most isolates of *C. glabrata* fall within the "susceptible" range of triazole minimal inhibitory concentrations (MIC) (Pfaller *et al.*, 2000) and resistance prevalence varies between age groups and geographical locations (Pfaller *et al.*, 2003b). *Candida dubliniensis* can be readily induced to develop resistance to fluconazole *in vitro*, although most fresh isolates of this species are susceptible in the absence of exposure to the drug (Moran *et al.*, 1997; Quindos *et al.*, 2000). Almost all other *Candida* species remain equally or more susceptible to triazoles than *C. albicans*, and occasional reports of widespread azole resistance among isolates of, for example, *C. tropicalis* (St Germain *et al.*, 2001) may represent problems of interpretation of trailing end-points in azole susceptibility tests (Rex *et al.*, 1998) since the majority of surveys have failed to show reduced azole susceptibility in species other than *C. krusei* and *C. glabrata* (Sanglard and Odds, 2002).

Resistance to triazole antifungals can result from alterations in the structure of the protein target for these agents, Erg11p, from upregulation of expression of this protein and from upregulation of multidrug efflux transporters in fungi (Sanglard and Odds, 2002). These mechanisms may be expressed in combination in some isolates (White, 1997). Fluconazole is unique among the triazole antifungal agents because it is a substrate for the major facilitator family of efflux transporters; all triazoles are exported by ABC-family transporters (Sanglard and Odds, 2002). This difference suggests that resistance to fluconazole may arise slightly more commonly than to other triazoles (at least in isolates of *C. albicans* where the mechanisms have been most thoroughly studied).

Regardless of details of resistance mechanisms, it is unquestionable that antifungal resistance can develop in normally susceptible fungal species and that resistance can lead to treatment failure. The high prevalence of resistance development among oral *C. albicans* isolates during the peak of the AIDS epidemic has been well documented (Canuto *et al.*, 2000; Chryssanthou *et al.*, 1995; Milan *et al.*, 1998) and is clearly associated with treatment failures (Rex *et al.*, 1995). Resistance to itraconazole and to voriconazole and concomitant treatment failure has been reported in clinical *A. fumigatus* isolates (Denning *et al.*, 1997; Manavathu *et al.*, 2000), and the inherently low susceptibility to amphotericin B and the older established triazoles of fungi such as *Fusarium* spp., *Scedosporium* spp., and the *Zygomycota* is considered to be the principal reason for high mortality rates when these moulds cause disseminated disease (Groll and Walsh, 2001).

3.2. Antifungal resistance cannot be transmitted by extrachromosomal DNA

However, there is a most important difference between resistance development and transmission among fungi as compared with bacteria; fungi do not, to our knowledge, have any mechanism comparable to bacteria for the transfer of genes encoding resistance from one isolate to another. Antifungal resistance is not encoded in extrachromosomal DNA, and transformation of fungi with DNA is far less easy than with bacteria, even under optimized laboratory conditions.

Current experimental studies with *C. albicans* show that, contrary to long-held opinions, the fungus probably can undergo mating, but does so naturally at a remarkably low frequency (Soll *et al.*, 2003). The same seems likely to apply to *A. fumigatus* (Poggeler, 2002). This inability to transmit antifungal resistance implies that development of clinically relevant resistance, at least to amphotericin B and triazoles where experience with their usage now extends to 20–30 years, is likely to be encountered almost exclusively among patients undergoing active treatment with these agents. How else can it be explained that the fluconazole resistance that developed so readily with oropharyngeal *Candida* infections in HIV-positive patients during the 1990s is now so much reduced in the most recent surveys (Barchiesi *et al.*, 2002; Martins *et al.*, 1998; Tacconelli *et al.*, 2002)? The low prevalence or absence of resistance among patients who have received no prior azole treatment is demonstrated clearly in a large survey of South African patients infected with HIV (Blignaut *et al.*, 2002). The lesson of the pre-HAART HIV era is that, among HIV-positive patients under the pressure of azole therapy, resistance to the agents can develop rapidly among many isolates of *C. albicans* (21% is the highest recorded point prevalence; Martins *et al.*, 1997) and prevalences of *C. dubliniensis* and *C. glabrata* rise unequivocally (Dupont *et al.*, 2000). However, transmission of resistant strains to untreated individuals seems not to occur on any significant scale.

The rapid azole resistance development seen among *Candida* isolates in the HIV-positive patient cohort has not been observed consistently in any other clinical setting, although reports from some institutions attest to obvious increases in prevalence of *C. glabrata* concurrent with the introduction of routine fluconazole prophylaxis (Abi-Said *et al.*, 1997; Price *et al.*, 1994). These reports are balanced by publications showing the *opposite* change in other institutions (Baran *et al.*, 2001; Kunova *et al.*, 1997) and by emergence of *C. glabrata* temporally associated with amphotericin B, not azole prophylaxis (Michel-Nguyen *et al.*, 2000). Warnings of an epidemic of azole resistance among *Candida* isolates are not supported by large surveys showing very low rates of such resistance among recent isolates from blood (Pfaller *et al.*, 1999b, 2003a) nor by reports indicating no change in levels of resistant isolates in a number of settings, including community-acquired mycoses such as vaginal *Candida* infections (Asmundsdottir *et al.*, 2002; Chen *et al.*, 2003; Marrazzo, 2003; Walker *et al.*, 2000).

Among fungi other than *Candida* spp., no publications so far suggest the emergence of antifungal resistant isolates on a large scale, although resistance to agents such as itraconazole undoubtedly *can* develop during treatment, as already mentioned. A survey of 170 isolates of *A. fumigatus* found only three resistant to itraconazole (Verweij *et al.*, 2002).

3.3. Antifungal resistance: conclusion

The most considered response that can be given to questions about the danger of emergence of antifungal-resistant fungi is that the phenomenon definitely occurs and that it has been seen to occur rapidly in oral *Candida* isolates in HIV-infected patients. However, in other clinical settings, the emergence of resistant fungi seems not to be an inevitable corollary of antifungal usage, and recorded changes in incidence and/or prevalence of causative fungal species have been associated with alterations in the type of patient at risk of mycosis and the methods of their management (Husain *et al.*, 2003; Kovacicova *et al.*, 2001; Krcmery and Barnes, 2002; Nucci and Colombo, 2002; Singh *et al.*, 2002; Torres *et al.*, 2003), and by no means exclusively with alterations in antifungal treatments. Prudence to avoid unnecessary use of antifungal agents is reasonable; anxiety about the large-scale emergence of resistant strains is not.

4. RATIONAL USE OF ANTIFUNGAL AGENTS

4.1. Antifungal agents available for prophylaxis and treatment of invasive mycoses

A further factor diminishing concerns about resistance developing when antifungal agents are used is the increased antifungal coverage offered by the current armoury of antifungal drugs. The number of antifungal agents and of antifungal classes have grown remarkably in recent years (Odds *et al.*, 2003). This section provides a very brief review of the agents available. Table 1 lists the main properties of systemic antifungal agents approved for clinical use or soon likely to be approved.

Amphotericin B, a polyene, kills susceptible fungal species by directly damaging their cell membranes. The selective toxicity of amphotericin B for fungal, as opposed to mammalian membranes, is low, and nephrotoxicity is the major hazard associated with use of the drug. The risk of nephrotoxicity has been considerably reduced by the availability of lipid-associated amphotericin B formulations. The cost of the lipid complex and colloidal suspension formulations is higher than that of conventional (deoxycholate-complexed) amphotericin B, and that of liposomal amphotericin B is considerably higher, but

Table 1. Systemic antifungal agents

Antifungal class	Agent	Mode of action	Antifungal spectrum	Comments
Polyenes	Amphotericin B	Damages fungal membranes by binding to ergosterol	Most fungal types but weak vs *Aspergillus flavus*, *Fusarium* spp., *Scedosporium* spp., *Zygomycota*	IV administration only; nephrotoxicity greatly reduced in lipid-associated formulations
Pyrimidine analogue	Flucytosine	Interferes with DNA synthesis after intracellular conversion to 5-fluorouracil	*Candida* spp., *Cryptococcus neoformans*	Toxic to bone marrow when blood levels high; resistant strains fairly common among susceptible species
Triazoles	Fluconazole	Inhibits 14α-sterol demethylase and alters sterol-dependent membrane fluidity	*Candida* spp., *Cryptococcus neoformans*, some filamentous fungi but not *Aspergillus* or *Fusarium* spp.	A safe and effective agent, limited by gaps in its spectrum and by poor activity vs *C. krusei* and some isolates of *C. glabrata*
	Itraconazole	See fluconazole	Many fungal types but not *Fusarium* spp., *Scedosporium* spp., or *Zygomycota*	Poor and variable oral bioavailability from capsules; oral solution has good bioavailability but poor palatability; IV formulation contains high cyclodextrin concentration; many interactions with drugs metabolised by hepatic P450 enzymes

Voriconazole	See fluconazole	Many fungal types including some isolates of *Fusarium* spp., *Scedosporium* spp., and *Zygomycota*	Side effects include high incidence of visual disturbances; many interactions with drugs metabolised by hepatic P450 enzymes
Ravuconazole	See fluconazole	Many fungal types including some isolates of *Fusarium* spp., *Scedosporium* spp., and *Zygomycota*	Agent in clinical development at the time of writing; clinical details not yet known
Posaconazole	See fluconazole	Many fungal types including some isolates of *Fusarium* spp., *Scedosporium* spp., and *Zygomycota*	Agent in clinical development at the time of writing; clinical details not yet known
Echinocandins Caspofungin	Inhibits synthesis of β-1:3-D-glucan in fungal cell walls	Most *Candida* and *Aspergillus* spp., some other filamentous fungi; not *Cryptococcus neoformans*	IV only. Very low toxicity and drug–drug interaction potential
Anidulafungin	Inhibits synthesis of β-1:3-D-glucan in fungal cell walls	Most *Candida* and *Aspergillus* spp., some other filamentous fungi; not *Cryptococcus neoformans*	Agent in clinical development at the time of writing; clinical details not yet known
Micafungin	Inhibits synthesis of β-1:3-D-glucan in fungal cell walls	Most *Candida* and *Aspergillus* spp., some other filamentous fungi; not *Cryptococcus neoformans*	Agent in clinical development at the time of writing; clinical details not yet known

many practitioners regard the higher cost of these formulations as justifiable in view of their considerably improved safety profiles.

Flucytosine inhibits growth of fungi that can actively import the compound and convert it to 5-fluorouracil, which restricts its use principally to *Candida* and *Cryptococcus* infections. The prevalence of yeast isolates resistant to flucytosine was probably overstated in the past, when susceptibility testing was not standardized (Sanglard and Odds, 2002). The current use of flucytosine is mainly as adjunct therapy in combination with other antifungal agents.

Fluconazole is well established as a safe and effective drug that has now been used for many years for the prophylaxis and treatment of fungal infections, particularly *Candida* infections. The agent is essentially inactive against *Aspergillus* spp. and has become regarded increasingly as a drug principally of use for treating yeast (*Candida* and *Cryptococcus*) infections. It has the shortest list of the drug–drug interactions that typify the triazole antifungal family (the fungal cytochrome P450 target is structurally similar to mammalian P450 enzymes) and is the least likely of the class to generate transient changes in serum levels of hepatic enzymes.

Itraconazole has a broad spectrum of antifungal activity that should make it a useful agent for treating many types of invasive mycosis. However, in its capsule formulation, its bioavailability is poor in some patients. Its formulation as an oral solution offers reliable bioavailability but its poor palatability often leads to patient noncompliance and both the oral solution and intravenous solution depend on high hydroxylpropyl-β-cyclodextrin concentrations to dissolve the itraconazole and this substance is a cause of diarrhoea and occasional renal effects. Itraconazole has a long list of drug–drug interactions associated with its use.

Voriconazole has an antifungal activity spectrum and potency similar to, but even better than that of itraconazole. It has proved itself to be first-line therapy for invasive aspergillosis (Herbrecht *et al.*, 2002). Its associated drug–drug interactions are similar to those of itraconazole and approximately 30% of patients given voriconazole orally or IV experience visual disturbances of short duration.

Caspofungin, the first of the echinocandin antifungal family to be registered for clinical use, is available only for intravenous administration. Its antifungal spectrum excludes *Cryptococcus neoformans* but otherwise covers the main pathogenic *Candida* and *Aspergillus* species. The agent has an excellent safety profile and few to no drug–drug interactions. Micafungin has been licensed in Japan for treatment of several types of *Aspergillus* and *Candida* infection. Its antifungal spectrum and IV-only formulation are very similar to those of caspofungin. Anidulafungin, the third echinocandin likely to be close to regulatory approval, so far seems also to have a similar profile to the other agents in the class.

4.2.   Recommending uses of antifungal agents: the limitations

The existence of a diverse range of classes and formulations of antifungal drugs (Table 1) should offer many possibilities for their use in the prevention and treatment of invasive mycoses. In practice, any recommendation is limited by the officially licensed indications for each individual agent (which often vary from country to country) and by the extent to which recommendations can be supported by evidence from well-designed prospective, randomized clinical trials. What follows will include suggestions for antifungal usage that are offered as future possibilities and are not (yet) supported either by licensed drug indications or by evidence-based medicine. Such suggestions are offered in good faith and arise from the consideration that licensing and evidence-based medicine commonly lag substantially behind the available opportunities for therapeutic management.

Agents undergoing major clinical trials but not yet licensed in the United States or Europe have not been included among the recommendations that follow. It is likely highly that posaconazole and ravuconazole will have many properties in common with the licensed triazoles, itraconazole and voriconazole, and that anidulafungin and micafungin will closely match caspofungin. However, their place in clinical practice will depend on the details of their formulations and their adverse event and drug interaction profiles, so it is too early to guess their final place in the antifungal armoury.

Since 2000, several publications have provided guidelines from working parties and other consensus groups for antifungal prophylaxis and treatment in several categories of patients (Bohme *et al.*, 2001; Denning *et al.*, 2003; Dykewicz, 2001; Hughes *et al.*, 2002; K. Marr and Boeckh, 2001; Quilitz *et al.*, 2001; Rex *et al.*, 2000; Saag *et al.*, 2000; Stevens *et al.*, 2000). These recommendations show impressive similarities in their choices of agent and other suggestions for management and should be consulted for the excellence of the detail they provide. The discussion that follows takes account of these publications, but ventures further, as already stated, by suggesting some new approaches for management that are not yet supported by data from appropriate clinical trials.

Two principles are common to most of the published guidelines, as follows:

1. Fluconazole represents a reasonable antifungal choice where the infection under treatment is known or likely to be caused by *C. albicans* or other fluconazole-susceptible *Candida* sp. For infections caused by *C. krusei* or fluconazole-resistant strains of a *Candida* species, a systemic antifungal agent with activity against the infecting yeast is the preferred choice.

2. Amphotericin B is a broad-spectrum, systemically active antifungal agent for use against infections by most fungal types, but a lipid-associated formulation should be used in patients who have impaired renal function, or develop signs of nephrotoxicity under treatment.

This advice, though entirely reasonable, was published before any data were published for the novel systemically active agents. It should, therefore, be supplemented by the general suggestion:

3. New triazoles (itraconazole and voriconazole) and the echinocandin caspofungin all have defined and licensed places in the treatment of aspergillosis and other invasive fungal infections; in some patients they may offer demonstrable benefits over the better-known fluconazole and amphotericin B, and the possibility that they are more appropriate choices needs to be considered in every case where diagnostic evidence suggests a strong possibility of serious fungal disease.

4.3.   Rational prophylaxis against fungal disease

Prevention of fungal disease is an obviously desirable goal for patients at high risk of invasive infection. Each individual patient has to be assessed for the appropriateness of antifungal prophylaxis: the level of risk of mycosis, the extent of immunocompromising factors such as neutrophil count, the ability of the patient to take oral medication, and the other drugs being used to manage the patient are just a few of the detailed factors that need to be considered in deciding whether to use antifungal prophylaxis and, if so, which drug to choose.

Patients at risk of invasive fungal disease fall into one of two very broad categories: (1) patients with neutropenia and (2) patients with normal leukocyte counts but who are at risk because of other forms of debilitation (e.g., low birth weight, abdominal or transplant surgery, chronic granulomatous disease, burns, etc.). Strategies for preventing fungal infections in neutropenia have been worked out over very many years. Many of the clinical trials were subjected to a meta-analysis by Bow and colleagues (Bow *et al.*, 2002), which concluded that prophylaxis of neutropenic patients with azoles or intravenous amphotericin B formulations reduced morbidity and mortality due to fungal infection, but it had no effect on the incidence of aspergillosis and was of much greater benefit in patients with prolonged neutropenia or undergoing stem-cell transplantation than in other neutropenic patients undergoing chemotherapy.

This meta-analysis sets the stage for a rational approach to antifungal prophylaxis in neutropenia. It is logical that the need for preventive anti-infective therapy increases with the level of risk of the infection. Giving prophylaxis to a patient who has previously suffered an invasive mycosis and who is again made neutropenic by subsequent chemotherapy carries the special name "preemptive" therapy. So the decision about *whether* to attempt prophylaxis in a neutropenic patient should be taken on the basis of level of risk. Criteria for assessment of risk of invasive fungal disease in neutropenia have been detailed by others and the recommendation is, therefore, relatively simple: Patients at high risk definitely require prophylactic antifungals, patients at intermediate risk may benefit from antifungal prophylaxis, and patients at low risk do not require prophylaxis.

The choice of agent for prophylaxis in neutropenia is controversial, since many potentially suitable antifungal drugs do not have prophylaxis as a licensed indication. Fluconazole *is* licensed and has been used prophylactically for many years. However, there are trial data to prove that a triazole such as itraconazole, which includes *Aspergillus* spp. in its spectrum, is superior to fluconazole, which does not, when given as prophylaxis to very high-risk patient groups such as allogeneic stem-cell transplant recipients (Boogaerts *et al.*, 2001). The eventual rational choice for prophylaxis in such special subsets of patient is therefore likely to be an agent with proven activity against both *Candida* and *Aspergillus* spp., which will include itraconazole and voriconazole (both can be given orally and IV), caspofungin, and amphotericin B (both IV only).

For patients without neutropenia, the decision to undertake prophylactic antifungal therapy and the choice of agent are more controversial than with neutropenic patients. Few intensive therapy units (ITUs) and even fewer surgical wards would ever consider routine antifungal prophylaxis for all their patients. However, there is respectable evidence to show that—as with neutropenic patients—the subsets of patients at highest risk of invasive mycosis *do* benefit from a prophylactic approach. Fluconazole was shown to prevent *Candida* peritonitis in patients who had undergone invasive intra-abdominal surgery (Eggimann *et al.*, 1999) and to reduce the incidence of invasive mycoses in critically ill post-surgical patients of all types (Pelz *et al.*, 2001). Itraconazole and fluconazole gave results judged as equivalent in preventing invasive mycoses in liver transplant recipients (Winston and Busuttil, 2002). Both nystatin and fluconazole were shown to be effective antifungal prophylaxis when given to very low birth weight neonates receiving intensive care (Kaufman *et al.*, 2001; Kicklighter *et al.*, 2001; Sims *et al.*, 1988).

On the basis of these studies, a conclusion can be drawn that, given adequate and carefully drawn up guidelines to define the sets of non-neutropenic patients at highest risk of invasive fungal disease in any given clinical setting, a prophylactic antifungal regime may be instituted with benefit. The choice of agent will depend on whether a *Candida* or a mould infection is more likely in a high-risk patient.

4.4.   Empirical antifungal therapy in neutropenic patients

By definition, empirical therapy does not fit the description of "rational" therapy (although it would be entirely irrational to *exclude* a persistently febrile neutropenic patient from antifungal treatment!). In clinical trials with antifungal agents, the usual criterion for treatment is fever in a neutropenic patient that persists for 5 days or more despite antibacterial chemotherapy. The clinical trials lean solely on fever as a primary criterion both for admission to the study and for efficacy. Though rigorously scientific, this approach fails to resemble the most common situation in "febrile neutropenia" where, by 5 days after onset of fever, the attending physicians usually have clues as to the nature of any possible fungal infection. This means that agents such as fluconazole can be avoided when the diagnostic evidence, albeit feeble, points to a possible invasive aspergillosis.

Trial data in "empiric" antifungal therapy can be criticized on many fronts and recently have been in a forceful manner (Bennett *et al.*, 2003). The limitations of study designs may have *under*estimated, rather than optimized the performance of the many agents that have been tested clinically. At present, amphotericin B (in various formulations), fluconazole, itraconazole, and voriconazole have all shown efficacy in published trials (though all are not licensed for empirical therapy in, e.g., the United States) and—to judge from meeting abstracts—caspofungin will also demonstrate efficacy as empirical therapy.

In the present author's opinion, the problems with empiric antifungal therapy can be overstated; they result from the well-known difficulties of establishing sound diagnoses of invasive infections due to *Candida* and *Aspergillus* species. It seems unthinkable that agents with efficacy proven against mycoses with a well-established diagnosis should suddenly become impotent against the same mycosis in the absence of diagnostic information. The problem, then, lies with the expectation that any prolonged fever that does not respond to antibacterial therapy in a neutropenic patient must be the result of a fungal infection. In the everyday clinical arena, a best judgement can be and has to be made as to whether a patient's fever might be attributable to a mycosis and, if the possibility is high, the best course is to institute antifungal therapy as rapidly as possible, not to wait for an academically defined period of non-responsiveness to antibacterial agents. Perhaps, in time, people with the appropriate clinical expertise and experience will be able to draw up an algorithm for more rational management of "fever in neutropenia" that will facilitate decisions whether or not to institute antifungal therapy.

4.5. Rational therapy of diagnosed invasive mycosis

Published guidelines for the management of proven candidaemia and other forms of *Candida* infection (Rex *et al.*, 2000) and of proven invasive aspergillosis of all types (Stevens *et al.*, 2000) indicate that the well-established systemically active antifungal agents all have a role to play in appropriate circumstances.

For candidaemia, amphotericin B, liposomal amphotericin B, fluconazole, and caspofungin (but *not* yet voriconazole or itraconazole) are all licensed therapies in the United States. Choice of agent should be determined according to the circumstances of the patient. For invasive aspergillosis, voriconazole and amphotericin B are now regarded as the agents of first choice, with caspofungin and itraconazole available should alternative therapies be required.

From published case reports and small series, voriconazole is developing a reputation as a useful agent for treatment of *Scedosporium* infections (Girmenia *et al.*, 1998; Munoz *et al.*, 2000; Walsh *et al.*, 2002) and the drug may prove to be useful in other infections caused by uncommon mould species. The newest triazoles, posaconazole and ravuconazole, may also prove ultimately to have a role in such infections, and the potential value of the echinocandin class for uncommon fungal diseases remains to be evaluated. Susceptibility testing *in vitro* is likely to be of more value in determining the choice of agents for unusual mycoses than for the more common *Candida* and *Aspergillus* infections.

4.6. Antifungal combinations and therapy changes

The combination of high-dose oral fluconazole with IV amphotericin B is no less effective than amphotericin B alone for treatment of candidaemia in non-neutropenic patients and may be slightly more efficacious (Rex *et al.*, 2003). However, the main stimulus for the use of combinations of antifungals is therapeutic failure of single agents in life-threatening situations such as invasive aspergillosis and diseases caused by unusual fungi. Prospective clinical trials of antifungal combinations in unusual mycoses will be extremely hard to design and implement; all the evidence so far available comes from anecdotal and open studies, much of it so far presented only in meetings abstracts and lectures. Clear evidence that combinations reduce mortality rates when a mycosis has been well diagnosed is hard to find. It is too early to pronounce on the potential future value for antifungal combinations.

When treatment of an invasive mycosis appears to be failing, a commonly raised question is whether to change antifungal treatment and, if so, to what. A patient treated with fluconazole can be usefully switched to one of the broader spectrum triazoles (voriconazole, itraconazole, etc.), but to switch from one broad-spectrum triazole to another would require strong evidence *in vitro* of superior antifungal potency at achievable blood levels of the second azole. Outwith straightforward pharmacological considerations (formulation, route of administration, potential for toxicity, or drug interactions in a given patient), there is no particular reason not to switch from one appropriate antifungal class to another. The advent of the echinocandins into clinical use expands the possibilities for class switching in treatment failure, and may in time generate a database that can offer predictive clues to optimize treatment switches.

REFERENCES

Abi-Said, D., Anaissie, E., Uzun, O., Raad, I., Pinzcowski, H., and Vartivarian, S., 1997, The epidemiology of hematogenous candidiasis caused by different *Candida* species. *Clin. Infect. Dis.*, **24**, 1122–1128.

Asmundsdottir, L. R., Erlendsdottir, H., and Gottfredsson, M., 2002, Increasing incidence of candidemia: Results from a 20-year nationwide study in Iceland. *J. Clin. Microbiol.*, **40**, 3489–3492.

Baran, J., Muckatira, B., and Khatib, R., 2001, Candidemia before and during the fluconazole era: Prevalence, type of species and approach to treatment in a tertiary care community hospital. *Scand. J. Infect. Dis.*, **33**, 137–139.

Barchiesi, F., Maracci, M., Radi, B., Arzeni, D., Baldassarri, I., Giacometti, A. *et al.*, 2002, Point prevalence, microbiology and fluconazole susceptibility patterns of yeast isolates colonizing the oral cavities of HIV-infected patients in the era of highly active antiretroviral therapy. *J. Antimicrob. Chemother.*, **50**, 999–1002.

Bennett, J. E., Powers, J., Walsh, T., Viscoli, C., de Pauw, B., Dismukes, W. *et al.*, 2003, Forum report: Issues in clinical trials of empirical antifungal therapy in treating febrile neutropenic patients. *Clin. Infect. Dis.*, **36**, S117–S122.

Blignaut, E., Messer, S., Hollis, R. J., and Pfaller, M. A., 2002, Antifungal susceptibility of South African oral yeast isolates from HIV/AIDS patients and healthy individuals. *Diagn. Microbiol. Infect. Dis.*, **44**, 169–174.

Bohme, A., Ruhnke, M., Karthaus, M., Einsele, H., Guth, S., Heussel, G. *et al.*, 2001, Treatment of fungal infections in haematology and oncology. Guidelines of the Working Party on Infections in Haematology and Oncology (AGIHO) of the German Society for Haematology and Oncology (DGHO). *Dtsch. Med. Wochenschr.*, **126**, 1440–1447.

Boogaerts, M., Winston, D. J., Bow, E. J., Garber, G., Reboli, A. C., Schwarer, A. P. *et al.*, 2001, Intravenous and oral itraconazole versus intravenous amphotericin B deoxycholate as empirical antifungal therapy for persistent fever in neutropenic patients with cancer who are receiving broad-spectrum antibacterial therapy—a randomized, controlled trial. *Ann. Intern. Med.*, **135**, 412–422.

Bow, E. J., Laverdiere, M., Lussier, N., Rotstein, C., Cheang, M. S., and Ioannou, S., 2002, Antifungal prophylaxis for severely neutropenic chemotherapy recipients—a meta-analysis of randomized-controlled clinical trials. *Cancer*, **94**, 3230–3246.

Canuto, M. M., Rodero, F. G., Ducasse, V. O. D., Aguado, I. H., Gonzalez, C. M., Sevillano, A. S. *et al.*, 2000, Determinants for the development of oropharyngeal colonization or infection by fluconazole-resistant *Candida* strains in HIV-infected patients. *Eur. J. Clin. Microbiol. Infect. Dis.*, **19**, 593–601.

Chen, Y. C., Chang, S. C., Luh, K. T., and Hsieh, W. C., 2003, Stable susceptibility of *Candida* blood isolates to fluconazole despite increasing use during the past 10 years. *J. Antimicrob. Chemother.*, **52**, 71–77.

Chryssanthou, E., Torssander, J., and Petrini, B., 1995, Oral *Candida albicans* isolates with reduced susceptibility to fluconazole in Swedish HIV-infected patients. *Scand. J. Infect. Dis.*, **27**, 391–395.

Denning, D. W., 1998, Invasive aspergillosis. *Clin. Infect. Dis.*, **26**, 781–803.

Denning, D. W., Kibbler, C. C., and Barnes, R. A., 2003, British Society for Medical Mycology proposed standards of care for patients with invasive fungal infections. *Lancet Infect. Dis.*, **3**, 230–240.

Denning, D. W., Venkateswarlu, K., Oakley, K. L., Anderson, M. J., Manning, N. J., Stevens, D. A. *et al.*, 1997, Itraconazole resistance in *Aspergillus fumigatus. Antimicrob. Agents Chemother.*, **41**, 1364–1368.

Dupont, B., Brown, H. H. C., Westermann, K., Martins, M. D., Rex, J. H., Lortholary, O. *et al.*, 2000, Mycoses in AIDS. *Med. Mycol.*, **38**, 259–267.

Dykewicz, C. A., 2001, Summary of the guidelines for preventing opportunistic infections among hematopoietic stem cell transplant recipients. *Clin. Infect. Dis.*, **33**, 139–144.

Eggimann, P., Francioli, P., Bille, J., Schneider, R., Wu, M. M., Chapuis, G. *et al.*, 1999, Fluconazole prophylaxis prevents intra-abdominal candidiasis in high-risk surgical patients. *Crit. Care Med.*, **27**, 1066–1072.

Girmenia, C., Luzi, G., Monaco, M., and Martino, P., 1998, Use of voriconazole in treatment of *Scedosporium apiospermum* infection—case report. *J. Clin. Microbiol.*, **36**, 1436–1438.

Groll, A. H. and Walsh, T. J., 2001, Uncommon opportunistic fungi: New nosocomial threats. *Clin. Microbiol. Infect.*, **7**, 8–24.

Herbrecht, R., Denning, D. W., Patterson, T. F., Bennett, J. E., Greene, R. E., Oestmann, J. W. *et al.*, 2002, Voriconazole versus amphotericin B for primary therapy of invasive aspergillosis. *N. Engl. J. Med.*, **347**, 408–415.

Hughes, W. T., Armstrong, D., Bodey, G. P., Bow, E. J., Brown, A. E., Calandra, T. *et al.*, 2002, 2002 guidelines for the use of antimicrobial agents in neutropenic patients with cancer. *Clin. Infect. Dis.*, **34**, 730–751.

Husain, S., Tollemar, J., Dominguez, E. A., Baumgarten, K., Humar, A., Paterson, D. L. *et al.*, 2003, Changes in the spectrum and risk factors for invasive candidiasis in liver transplant recipients: Prospective, multicenter, case-controlled study. *Transplantation*, **75**, 2023–2029.

Ives, N. J., Gazzard, B. G., and Easterbrook, P. J., 2001, The changing pattern of AIDS-defining illnesses with the introduction of highly active antiretroviral therapy (HAART) in a London clinic. *J. Infect.*, **42**, 134–139.

Kaufman, D., Boyle, R., Hazen, K. C., Patrie, J. T., Robinson, M., and Donowitz, L. G., 2001, Fluconazole prophylaxis against fungal colonization and infection in preterm infants. *N. Engl. J. Med.*, **345**, 1660–1666.

Keye, J. D. and Magee, W. E., 1956, Fungal infections in a general hospital. *Am. J. Clin. Pathol.*, **26**, 1235–1253.

Kicklighter, S. D., Springer, S. C., Cox, T., Hulsey, T. C., and Turner, R. B., 2001, Fluconazole for prophylaxis against candidal rectal colonization in the very low birth weight infant. *Pediatrics*, **107**, 293–298.

Kovacicova, G., Spanik, S., Kunova, A., Trupl, J., Sabo, A., Koren, P. *et al.*, 2001, Prospective study of fungaemia in a single cancer institution over a 10-y period: Aetiology, risk factors, consumption of antifungals and outcome in 140 patients. *Scand. J. Infect. Dis.*, **33**, 367–374.

Krcmery, V. and Barnes, A. J., 2002, Non-*albicans Candida* spp. causing fungaemia: Pathogenicity and antifungal resistance. *J. Hosp. Infect.*, **50**, 243–260.

Kunova, A., Trupl, J., Demitrovicova, A., Jesenska, Z., Grausova, S., Grey, E. *et al.*, 1997, Eight-year surveillance of non-*albicans Candida* spp in an oncology department prior to and after fluconazole had been introduced into antifungal prophylaxis. *Microb. Drug Res. Mech. Epidemiol. Dis.*, **3**, 283–287.

Lin, S. J., Schranz, J., and Teutsch, S. M., 2001, Aspergillosis case-fatality rate: Systematic review of the literature. *Clin. Infect. Dis.*, **32**, 358–366.

Manavathu, E. K., Cutright, J. L., Loebenberg, D., and Chandrasekar, P. H., 2000, A comparative study of the *in vitro* susceptibilities of clinical and laboratory-selected resistant isolates of *Aspergillus* spp. to amphotericin B, itraconazole, voriconazole and posaconazole (SCH 56592). *J. Antimicrob. Chemother.*, **46**, 229–234.

Marr, K. and Boeckh, M., 2001, Practice guidelines for fungal infections: A risk-guided approach. *Clin. Infect. Dis.*, **32**, 321.

Marr, K. A., Carter, R. A., Boeckh, M., Martin, P., and Corey, L., 2002a, Invasive aspergillosis in allogeneic stem cell transplant recipients: Changes in epidemiology and risk factors. *Blood*, **100**, 4358–4366.

Marr, K. A., Carter, R. A., Crippa, F., Wald, A., and Corey, L., 2002b, Epidemiology and outcome of mould infections in hematopoietic stem cell transplant recipients. *Clin. Infect. Dis.*, **34**, 909–917.

Marrazzo, J., 2003, Vulvovaginal candidiasis—over the counter treatment doesn't seem to lead to resistance. *BMJ*, **326**, 993–994.

Martins, M. D., Lozanochiu, M., and Rex, J. H., 1997, Point prevalence of oropharyngeal carriage of fluconazole-resistant *Candida* in human immunodeficiency virus-infected patients. *Clin. Infect. Dis.*, **25**, 843–846.

Martins, M. D., Lozano-Chiu, M., and Rex, J. H., 1998, Declining rates of oropharyngeal candidiasis and carriage of *Candida albicans* associated with trends toward reduced rates of carriage of fluconazole-resistant *C. albicans* in human immunodeficiency virus-infected patients. *Clin. Infect. Dis.*, **27**, 1291–1294.

McNeil, M. M., Nash, S. L., Hajjeh, R. A., Phelan, M. A., Conn, L. A., Plikaytis, B. D. *et al.*, 2001, Trends in mortality due to invasive mycotic diseases in the United States, 1980–1997. *Am. J. Hum. Genet.*, **33**, 641–647.

Michel-Nguyen, A., Favel, A., Azan, P., Regli, P., and Penaud, A., 2000, Dix-neuf années de données épidémiologiques en centre hospitalier universitaire: place de *Candida (Torulopsis) glabrata*; Sensibilité. *J. Mycol. Med.*, **10**, 78–86.

Milan, E. P., Burattini, M. N., Kallas, E. G., Fischmann, O., Costa, P. R. D., and Colombo, A. L., 1998, Azole resistance among oral *Candida* species isolates from AIDS patients under ketoconazole exposure. *Diagn. Microbiol. Infect. Dis.*, **32**, 211–216.

Moran, G. P., Sullivan, D. J., Henman, M. C., McCreary, C. E., Harrington, B. J., Shanley, D. B. *et al.*, 1997, Antifungal drug susceptibilities of oral *Candida dubliniensis* isolates from human immunodeficiency virus (hiv)-infected and non-hiv-infected subjects and generation of stable fluconazole-resistant derivatives in vitro. *Antimicrob. Agents Chemother.*, **41**, 617–623.

Munoz, P., Marin, M., Tornero, P., Rabadan, P. M., Rodriguez-Creixems, M., and Bouza, E., 2000, Successful outcome of *Scedosporium apiospermum* disseminated infection treated with voriconazole in a patient receiving corticosteroid therapy. *Clin. Infect. Dis.*, **31**, 1499–1501.

Nucci, M. and Colombo, A. L., 2002, Risk factors for breakthrough candidemia. *Eur. J. Clin. Microbiol. Infect. Dis.*, **21**, 209–211.

Odds, F. C., Brown, A. J. P., and Gow, N. A. R., 2003, Antifungal agents: Mechanisms of action. *Trends Microbiol.*, **11**, 272–279.

Pelz, R. K., Hendrix, C. W., Swoboda, S. M., Diener-West, M., Merz, W. G., Hammond, J. *et al.*, 2001, Double-blind placebo-controlled trial of fluconazole to prevent candidal infections in critically ill surgical patients, *Ann. Surg.*, **233**, 542–548.

Pfaller, M. A., Diekema, D. J., Messer, S. A., Boyken, L., and Hollis, R. J., 2003a, Activities of fluconazole and voriconazole against 1,586 recent clinical isolates of Candida species determined by broth microdilution, disk diffusion, and Etest methods: Report from the ARTEMIS global antifungal susceptibility program, 2001. *J. Clin. Microbiol.*, **41**, 1440–1446.

Pfaller, M. A., Jones, R. N., Doern, G. V., Fluit, A. C., Verhoef, J., Sader, H. S. *et al.*, 1999a, International surveillance of blood stream infections due to *Candida* species in the European SENTRY program: Species distribution and antifungal susceptibility including the investigational triazole and echinocandin agents. *Diagn. Microbiol. Infect. Dis.*, **35**, 19–25.

Pfaller, M. A., Jones, R. N., Doern, G. V., Sader, H. S., Messer, S. A., Houston, A. *et al.*, 2000, Bloodstream infections due to *Candida* species: SENTRY Antimicrobial Surveillance Program in North America and Latin America, 1997–1998. *Antimicrob. Agents Chemother.*, **44**, 747–751.

Pfaller, M. A., Messer, S. A., Boyken, L., Tendolkar, S., Hollis, R. J., and Diekema, D. J., 2003b, Variation in susceptibility of bloodstream isolates of *Candida glabrata* to fluconazole according to patient age and geographic location. J. *Clin. Microbiol.*, **41**, 2176–2179.

Pfaller, M. A., Messer, S. A., Hollis, R. J., Jones, R. N., Doern, G. V., Brandt, M. E. *et al.*, 1999b, Trends in species distribution and susceptibility to fluconazole among blood stream isolates of *Candida* species in the United States. *Diagn. Microbiol. Infect. Dis.*, **33**, 217–222.

Poggeler, S., 2002, Genomic evidence for mating abilities in the asexual pathogen *Aspergillus fumigatus. Curr. Genet.*, **42**, 153–160.

Price, M. F., Larocco, M. T., and Gentry, L. O., 1994, Fluconazole susceptibilities of *Candida* species and distribution of species recovered from blood cultures over a 5-year period. *Antimicrob. Agents Chemother.*, **38**, 1422–1424.

Quilitz, R. E., Arnold, A. D., Briones, G. R., Dix, S. P., Ippoliti, C., Kennedy, L. D. *et al.*, 2001, Practice guidelines for lipid-based amphotericin B in stem cell transplant recipients. *Ann. Pharmacother.*, **35**, 206–216.

Quindos, G., Carrillo-Munoz, A. J., Arevalo, M. P., Salgado, J., Alonso-Vargas, R., Rodrigo, J. M. *et al.*, 2000, In vitro susceptibility of *Candida dubliniensis* to current and new antifungal agents. *Chemotherapy*, **46**, 395–401.

Raffaele, B., Sacchi, P., and Filice, G., 2003, Overview on the incidence and the characteristics of HIV-related opportunistic infections and neoplasms of the heart: Impact of highly active antiretroviral therapy. *AIDS*, **17**, S83–S87.

Rex, J. H., Nelson, P. W., Paetznick, V. L., Lozanochiu, M., Espinelingroff, A., and Anaissie, E. J., 1998, Optimizing the correlation between results of testing in vitro and therapeutic outcome in vivo for fluconazole by testing critical isolates in a murine model of invasive candidiasis. *Antimicrob. Agents Chemother.*, **42**, 129–134.

Rex, J. H., Pappas, P. G., Karchmer, A. W., Sobel, J., Edwards, J. E., Hadley, S. *et al.*, 2003, A randomized and blinded multicenter trial of high-dose fluconazole plus placebo versus fluconazole plus amphotericin B as therapy for candidemia and its consequences in non-neutropenic subjects. *Clin. Infect. Dis.*, **36**, 1221–1228.

Rex, J. H., Rinaldi, M. G., and Pfaller, M. A., 1995, Resistance of *Candida* species to fluconazole. *Antimicrob. Agents Chemother.*, **39**, 1–8.

Rex, J. H., Walsh, T. J., Sobel, J. D., Filler, S. G., Pappas, P. G., Dismukes, W. E. *et al.*, 2000, Practice guidelines for the treatment of candidiasis. *Clin. Infect. Dis.*, **30**, 662–678.

Saag, M. S., Graybill, R. J., Larsen, R. A., Pappas, P. G., Perfect, J. R., Powderly, W. G. *et al.*, 2000, Practice guidelines for the management of cryptococcal disease. *Clin. Infect. Dis.*, **30**, 710–718.

Sandven, P., 2000, Epidemiology of candidemia. *Rev. Iberoamer. Micol.*, **17**, 73–81.

Sanglard, D. and Odds, F. C., 2002, Resistance of *Candida* species to antifungal agents: Molecular mechanisms and clinical consequences. *Lancet Infect. Dis.*, **2**, 73–85.

Sims, M. E., Yoo, Y., You, H., Salminen, C., and Walther, F. J., 1988, Prophylactic oral nystatin and fungal infections in very-low-birthweight infants. *Am. J. Perinatol.*, **5**, 33–36.

Singh, N., Wagener, M. M., Marino, I. R., and Gayowski, T., 2002, Trends in invasive fungal infections in liver transplant recipients: Correlation with evolution in transplantation practices. *Transplantation*, **73**, 63–67.

Soll, D. R., Lockhart, S. R., and Zhao, R., 2003, Relationship between switching and mating in *Candida albicans*. *Eukaryot. Cell*, **2**, 390–397.

St Germain, G., Laverdiere, M., Pelletier, R., Bourgault, A. M., Libman, M., Lemieux, C. *et al.*, 2001, Prevalence and antifungal susceptibility of 442 *Candida* isolates from blood and other normally sterile sites: Results of a 2-year (1996 to 1998) multicenter surveillance study in Quebec, Canada. *J. Clin. Microbiol.*, **39**, 949–953.

Stevens, D. A., Kan, V. L., Judson, M. A., Morrison, V. A., Dummer, S., Denning, D. W. *et al.*, 2000, Practice guidelines for diseases caused by *Aspergillus*. *Clin. Infect. Dis.*, **30**, 696–709.

Tacconelli, E., Bertagnolio, S., Posteraro, B., Tumbarello, M., Boccia, S., Fadda, G. *et al.*, 2002, Azole susceptibility patterns and genetic relationship among oral *Candida* strains isolated in the era of highly active antiretroviral therapy. *Jaids*, **31**, 38–44.

Torres, H. A., Rivero, G. A., Lewis, R. E., Hachem, R., Raad, II, and Kontoyiannis, D. P., 2003, Aspergillosis caused by non-fumigatus Aspergillus species: Risk factors and in vitro susceptibility compared with *Aspergillus fumigatus*. *Diagn. Microbiol. Infect. Dis.*, **46**, 25–28.

Trick, W. E., Fridkin, S. K., Edwards, J. R., Hajjeh, R. A., and Gaynes, R. P., 2002, Secular trend of hospital-acquired candidemia among intensive care unit patients in the United States during 1989–1999. *Clin. Infect. Dis.*, **35**, 627–630.

Verweij, P. E., Dorsthorst, D. T. A. T., Rijs, A. J. M. M., De Vries-Hospers, H. G., and Meis, J. F. G. M., 2002, Nationwide survey of in vitro activities of itraconazole and voriconazole against clinical *Aspergillus fumigatus* isolates cultured between 1945 and 1998. *J. Clin. Microbiol.*, **40**, 2648–2650.

Walker, P. P., Reynolds, M. T., Ashbee, H. R., Brown, C., and Evans, E. G. V., 2000, Vaginal yeasts in the era of "over the counter" antifungals. *Sex Trans. Infect.*, **76**, 437–438.

Walsh, T. J., Lutsar, I., Driscoll, T., Dupont, B., Roden, M., Ghahramani, P. *et al.*, 2002, Voriconazole in the treatment of aspergillosis, scedosporiosis and other invasive fungal infections in children. *Pediatr. Infect. Dis. J.*, **21**, 240–248.

White, T. C., 1997, Increased mrna levels of erg16, cdr, and mdr1 correlate with increases in azole resistance in *Candida albicans* isolates from a patient infected with human immunodeficiency virus. *Antimicrob. Agents Chemother.*, **41**, 1482–1487.

Winston, D. J. and Busuttil, R. W., 2002, Randomized controlled trial of oral itraconazole solution versus intravenous/oral fluconazole for prevention of fungal infections in liver transplant recipients. *Transplantation*, **74**, 688–695.

Wisplinghoff, H., Seifert, H., Wenzel, R. P., and Edmond, M. B., 2003, Current trends in the epidemiology of nosocomial bloodstream infections in patients with hematological malignancies and solid neoplasms in hospitals in the United States. *Clin. Infect. Dis.*, **36**, 1103–1110.

Yamazaki, T., Kume, H., Murase, S., Yamashita, E., and Arisawa, M., 1999, Epidemiology of visceral mycoses: Analysis of data in Annual of the Pathological Autopsy Cases in Japan. *J. Clin. Microbiol.*, **37**, 1732–1738.

Chapter 18

Strategies for the Rational Use of Antivirals

Sheila M. L. Waugh and William F. Carman
West of Scotland Specialist Virology Centre, Gartnavel General Hospital,
Great Western Road, Glasgow G12 OYN, UK

The treatment and prevention of viral disease is a rapidly evolving field. In only a few years there has been an exponential increase in the availability of effective antiviral compounds, together with major improvements in the diagnosis and monitoring of viral infections. Predictably, development has not kept pace with the ability of these pathogens to adapt, viral resistance has been documented to almost all antivirals in current use. To minimise the emergence of resistant virus, and thus optimise patient care, it is important that antivirals are only used within an evidence-based framework.

Here, a number of issues will be discussed. First the changing profile of viral infections encountered in hospital and community settings is outlined, together with a brief overview of the antivirals currently in common use. In the next section the major problems associated with antiviral treatment are discussed, with particular emphasis on the emergence of viral resistance. The final two sections detail principles that can help to minimise these problems and aid in the rational use of antivirals in both treatment and preventative regimes.

1. THE CHANGING FACE OF VIRAL INFECTIONS AND THEIR MANAGEMENT

Viral infections have often been considered as either minor illnesses not requiring intervention, or serious conditions for which there is no effective treatment. This perspective is changing, with the importance of seemingly

Antibiotic Policies: Theory and Practice. Edited by Gould and van der Meer
Kluwer Academic / Plenum Publishers, New York, 2005

innocuous viruses increasingly recognised. For example, rhinoviruses, a cause of the common cold, are now known to be associated with severe lower respiratory disease in the immunocompromised and exacerbations of asthma (Gern and Busse, 1999; Greenberg, 2003). Similarly, though it is true that many viral infections remain untreatable, effective treatment is now available for several serious conditions, such as aciclovir in herpes simplex virus (HSV) encephalitis, ribavirin for Lassa fever, and combination antiretroviral therapy for human immunodeficiency virus (HIV).

The increase in interest in the treatment of viral disease has been perpetuated both by the emergence of new viral pathogens and by the increasing prevalence of impaired immunity, either as a result of immunosuppressive treatment regimes or AIDS. Changing behaviour patterns have also led to the opportunity for increased spread of pathogens, such as hepatitis C virus (HCV) in intravenous drug users (Mathei *et al.*, 2002).

It is important to remember that the major drive against viral infection remains defensive, based on the use of sound infection control principles and vaccination. Rigorous infection control policies have had significant impact in many situations, such as the transmission of hepatitis B virus (HBV) in renal dialysis units (UK Department of Health, 2002) and the spread of norovirus, the cause of winter vomiting disease, during outbreaks on hospital wards (Chadwick *et al.*, 2000; McCall and Smithson, 2002). The eradication of smallpox and the elimination of poliovirus from large parts of the globe are two of the most striking examples of vaccine preventable disease, but there are many more, including the prevention of influenza virus infection (Nichol, 2003) and vaccination against HBV (Bonanni and Bonaccorsi, 2001).

There remain a limited number of antivirals, though in recent years there has been an explosive increase in licensed drugs. Nowhere is this more obvious than in the development of antiretroviral therapies for the treatment of HIV, with new drugs being licensed every year and many more entering clinical trials (Gulick, 2003). Not only are more antiviral compounds being developed and licensed, but there are also an ever-increasing number of situations where their use is being considered.

Antivirals are generally targeted at a single virus or closely related viruses, rather than a large group of viruses. Amantadine acts well against influenza A virus but has no activity against influenza B virus, and aciclovir is useful against HSV and varicella zoster virus (VZV), but is not effective as treatment for cytomegalovirus (CMV) or Epstein–Barr virus infections, despite these being members of the herpes virus family. There are two available antivirals that could reasonably be described as broad-spectrum: ribavirin (Snell, 2001) and cidofovir (Safrin *et al.*, 1997). However, as their use is limited in many situations by uncertain *in vivo* efficacy and, for the latter especially, a poor safety profile, it is not possible to treat a presumed viral infection empirically.

A specific virus must either be suspected clinically or found to be present on diagnostic testing before antiviral treatment can be considered.

Examples of the currently available antivirals are detailed in Table 1. The majority of antiviral compounds in common current use are against viruses of

Table 1. Examples of currently available antivirals (British National Formulary, 2003; USA Food and Drug Administration, 2003)

Virus	Available antivirals	Main target (Molecular)	Resistance documented
HSV	Aciclovir, penciclovir, valaciclovir[a], famciclovir[a], foscarnet, cidofovir	Viral polymerase	Yes[b]
VZV	Aciclovir, valaciclovir[a], famciclovir[a], foscarnet	Viral polymerase	Yes[c]
CMV	Ganciclovir, valganciclovir[a], foscarnet	Viral polymerase	Yes[d]
Influenza A	Amantadine	Viral fusion protein	Yes[e]
	Zanamivir, oseltamivir	Viral neuraminidase	Yes[f]
Influenza B	Zanamivir, oseltamivir	Viral neuraminidase	Yes[f]
RSV	Ribavirin	Various modes of action	Not yet[g]
HBV	Lamivudine	Viral polymerase	Yes[h]
	Adefovir	Viral polymerase	Not yet[i]
	Interferon	Immune system and direct antiviral effects	
HCV	Ribavirin and interferon/pegylated interferon	Immune system and direct antiviral effects	Yes[j]
HIV 1	Abacavir, didanosine, lamivudine, stavudine, tenofovir disoproxil zalcitabine, zidovudine	Viral reverse transcriptase	Yes[k]
	Efavirenz, nevirapine	Viral reverse transcriptase	Yes[k]
	Amprenivir, indinavir, lopinavir, nelfinavir, ritonavir, saquinavir	Viral protease	Yes[k]

[a]Famciclovir, valaciclovir, and valganciclovir are prodrugs of penciclovir, aciclovir, and ganciclovir, respectively, with higher oral bioavailability.
[b]Morfin and Thouvenot (2003).
[c]Boivin *et al.* (1994).
[d]Limaye *et al.* (2000).
[e]Hayden and Hay (1992).
[f]Gubareva *et al.* (1998, 2001).
[g]Snell (2001).
[h]Dienstag *et al.* (1999).
[i]Marcellin *et al.* (2003).
[j]Pawlotsky (2000).
[k]Pillay *et al.* (2000).

the herpes family (HSV, VZV, CMV), influenza, and HIV 1. The majority are nucleoside analogues which inhibit viral polymerases including reverse transcriptase. Other targets include the influenza neuraminidase and HIV protease. There is currently a great deal of interest in the development of drugs with novel targets, such as the cellular virus receptor; such an approach may help to overcome the problem of cross-resistance between drugs with a similar mode of action.

As the number of antiviral drugs grows, and in particular their increased long-term use in chronic infections such as HBV and HIV, so the problems of resistance and toxicity become more marked and more challenging.

2.   PROBLEMS ASSOCIATED WITH ANTIVIRAL THERAPY

2.1.   Limited experience in antiviral use generally and locally

The majority of antivirals have been available for a relatively short time, such that there is little experience of their use in a range of conditions. Controlled clinical trials have generally been conducted only for one or two major indications. For example, valganciclovir was originally licensed only for CMV retinitis in AIDS patients, but as an oral preparation with equivalent activity to intravenous ganciclovir (Pescovitz *et al.*, 2000), it had obvious advantages in the treatment and prophylaxis of CMV in transplant patients. In addition, trials tend to be carried out in selected populations of patients, which do not necessarily reflect everyday practise, where patients are more heterogeneous, and may differ in terms of coexisting pathologies and degree of adherence to treatment regimes. Experience with the drug therefore still requires to be built up both formally, with further published studies, and at a local level. This is a continuing process as use is expanded to more clinical situations and particular patient groups.

Limited information on a drug often means that it is considered as treatment in some situations where there is no evidence or consensus regarding efficacy. This is an issue particularly in severe life-threatening infections in the immunocompromised, where antivirals are used despite unproven clinical benefit. The use of drugs in these situations should be carefully monitored, and where possible included as part of larger studies.

As data on the use of a particular antiviral becomes available, it is important that information is disseminated to ward level and actual patient treatment.

2.2.   Adverse effects

As viruses use host cell machinery for replication, enzymes such as the viral polymerase have significant homology to the cellular enzyme, thus resulting in an increased potential for inhibition of host cell processes. Many antivirals have a significant side-effect profile, for example, discontinuation of cidofovir is required in 25–30% of patients due to nephrotoxicity (Safrin *et al.*, 1997). Other drugs, such as aciclovir and its derivatives, are relatively free of side effects, in part because they have a requirement for the viral thymidine kinase, and thus will only be metabolised to their active form in an infected cell. The likelihood and range of unwanted effects becomes much greater when treatment is required long term, and where combinations of drugs are used, such as in HIV. In these cases there is also a particular concern that the experience of side effects may effect adherence and thus have further adverse impact on the success of treatment.

2.3.   Resistance

Antiviral drug resistance, due to mutation, has been described for almost all antiviral compounds (Table 1). Mutation is a common event for viruses, due to rapid replication rates. RNA-dependant RNA polymerases and reverse transcriptases are particularly error prone and lack the proofreading ability of DNA polymerases, thus mutation is particularly common in RNA viruses, such as HIV and influenza. Indeed RNA viruses are often described as existing as a quasispecies due to the vast variation which can exist in viral sequence, even within a single infected individual (Holland *et al.*, 1992). Mutations resulting in antiviral drug resistance may arise during treatment with a particular drug, or may pre-exist within the viral quasispecies. In either case drug treatment selects out resistant virus and allows it to dominate. A pre-requisite therefore, for the emergence of resistant virus, is replication of the virus in the presence of the drug. This can be prevented by fully suppressing viral replication, though this is seldom achieved. The avoidance of subtherapeutic drug treatment is vital, as this can significantly increase the risk of resistance (Pillay and Zambon, 1998).

Resistance may take the form of single or multiple point mutations or deletions. These usually occur in sequences either coding for the target enzyme, such as the HIV reverse transcriptase, or for proteins necessary for drug activation, such as the HSV thymidine kinase. There are often a variety of mutations which can confer resistance to a particular drug—for example, both viral polymerase and thymidine kinase mutations can confer aciclovir resistance in HSV infection (Morfin and Thouvenot, 2003). The effect of a given mutation is not always straightforward and there may be variation in the degree of resistance conferred. Unsurprisingly, where many drugs have the same viral target, cross-resistance to related antivirals is a common phenomenon. Some mutations confer resistance to one drug while increasing viral susceptibility to another, as with the M184V lamivudine resistance mutation in HIV reverse transcriptase, which increases viral susceptibility to zidovudine (Pillay *et al.*, 2000).

There is great variation in the frequency with which resistance to a particular antiviral is observed and in the speed with which it emerges in a given treated patient. Amantadine treatment results in the emergence of resistant influenza in 30% of immunocompetent individuals within 5 days (Hayden and Hay, 1992), while to date virus resistant to the influenza neuraminidase inhibitor zanamivir has only been encountered rarely in immune compromised patients after prolonged treatment (Gubareva *et al.*, 1998). In the treatment of HIV infection, resistance can arise rapidly to the non-nucleoside reverse transcriptase inhibitors such as nevirapine; in contrast, the development of resistance to the protease inhibitors is a slower process, requiring the accumulation of a number of different mutations (Pillay *et al.*, 2000).

Antiviral drug resistance is a substantial problem in the immunocompromised and in long-term treatment regimes; these are the two areas where effective treatment is arguably most important. Two commonly encountered viral pathogens in transplant patients are HSV and CMV. Aciclovir resistance in HSV is common in these patients with up to 5% of viral isolates carrying resistance mutations (Morfin and Thouvenot, 2003). Although such viruses are attenuated in animal models, they can cause severe clinical disease in the immunosuppressed. In opportunistic CMV disease, resistance to the front-line antiviral ganciclovir can be observed in association with prolonged drug exposure, such as during prophylactic regimes (Limaye *et al.*, 2000). In chronic HBV, up to a third of patients show resistance after 1 year's treatment with the nucleoside analogue lamivudine (Dienstag *et al.*, 1999). Antiretroviral regimes, which are potentially life-long, with three or more drugs give ample time for resistance mutations to emerge, often to more than one drug (Pillay *et al.*, 2000). Given that a particular viral mutation can confer resistance to several related antiretrovirals, the available effective antivirals can rapidly become limited, particularly after two or three changes in therapy. Resistance is one of the main driving forces in the continuing search for new drugs against HIV.

Resistance is not a common problem in the treatment of most acute infections such as genital herpes simplex, or varicella zoster in immune-competent individuals. This may be explained by the fact that mutant viruses are often attenuated with respect to the wild-type virus and less fit with regard to their replication competency (Nijhuis *et al.*, 2001). Unfit, resistant virus may therefore be more easily dealt with by an intact immune system. In addition, mutations can take some time to emerge, often longer than the time required to complete a course of treatment for an acute infection.

From the public health perspective, a concern is the risk of spread of resistant virus. Resistant viruses are not yet commonly found circulating in the community, perhaps, as lacking selective pressure from an antiviral, the fitter wild-type forms predominate. Whether this situation will change as antiviral use increases is currently unclear. There are, however, many examples of the transmission of resistant viruses between individuals. In particular there is mounting concern regarding the increase in transmission of resistant HIV in western countries since antiretroviral therapy became widely available, including the transmission of multiply drug resistant virus (UK collaborative group on monitoring the transmission of HIV drug resistance, 2001). Another example is the anti-influenza drug amantadine, which is one case where resistant virus commonly emerges in immune competent individuals, the resistant influenza is readily transmitted and spread of resistant virus has been observed during outbreaks (Hayden and Hay, 1992).

As antiviral use increases, it is inevitable that resistance will be encountered more commonly, both in the individual patient undergoing treatment and in the wider community. This again emphasises the need for rational drug use, particularly with a view to preventing the emergence of resistant virus.

3. ANTIVIRAL TREATMENT STRATEGIES

Antiviral policies must ensure that any treatment is indicated, and appropriate with regard to three major goals: successful treatment of the patient's illness, avoidance of treatment-related adverse effects, and prevention of the emergence of resistant virus.

3.1. Sources of information and advice

In a rapidly evolving field such as the treatment of viral disease, the evidence base changes as new antivirals are developed and more studies are completed on existing drugs. National guidelines, produced by professional bodies, are evidence-based recommendations, which are supplemented by expert opinion where necessary. Examples in the United Kingdom include: The British HIV Association (BHIVA) guidelines giving the most recent recommendations for HIV treatment (BHIVA, 2003); The Royal College of Obstetrics and Gynaecology guidelines, which are regularly reviewed and cover the treatment of genital HSV and VZV in pregnancy (RCOG, 2001, 2002); and government associated agency guidance, such as the UK National Institute of Clinical Excellence recommendations covering Influenza (NICE, 2003). In the United States guidelines on a number of virology issues are published by the Centers for Disease Control and Prevention (e.g., Bridges *et al.*, 2003; Dybul *et al.*, 2002). Similar guidelines are available from national bodies in other countries. National guidelines can be adapted to local protocols taking into account local factors such as referral routes, prescribing policies, and the available viral diagnostic and monitoring services. Guidance on the use and monitoring of antiviral treatment is also available from a number of recent reviews (Pillay *et al.*, 2002; Waugh *et al.*, 2002). Guidelines may not exist for situations where there is little information or no clear evidence of efficacy, or where there are conflicting studies, such as ribavirin for parainfluenza virus infections in the immunocompromised (Elizaga *et al.*, 2001) or cidofovir in progressive multifocal leukoencephalopathy (Segarra-Newnham and Vodolo, 2001).

3.2.   Confirm the diagnosis

Good laboratory virology provision is essential to the use of antiviral agents. As antivirals tend to have a limited spectrum of activity and are often associated with significant toxicity and cost, it is important to confirm a suspected diagnosis. Ideally this would be done prior to commencing treatment, although sometimes this may not be possible. Examples of this are where delay might have serious consequences, as in suspected HSV encephalitis, or where early treatment (within 36 hr of onset of disease) is required for therapeutic benefit, as in influenza virus infection. In such cases it is still important that laboratory confirmation is obtained as soon as possible, to ensure the correct treatment or allow cessation of unnecessary treatment. In the past diagnosis was often retrospective, requiring several days or even weeks for virus growth in tissue culture or entailing the testing of convalescent blood samples. Some viruses cannot be grown in standard cultures and sensitivity is often too low to reliably detect low amounts of virus (as with cerebrospinal fluid (CSF) in HSV encephalitis). The arrival of the polymerase chain reaction (PCR) in laboratories has revolutionised the impact of diagnostic virology in clinical practice (Carman, 2001). This technique is a highly sensitive and specific method for the detection of viral nucleic acid, and can be adapted to detect practically any virus in any body fluid or tissue. With classical PCR a result can generally be available within 1–2 days, however the recent advance to real time PCR allows a result to be available within a few hours (Niesters, 2002). This means that in many cases it is possible to await a confirmed diagnosis before starting treatment. Such rapid techniques are particularly useful in serious illnesses where the cause of symptoms cannot be satisfactorily determined clinically, for example, systemic illness in a transplant recipient which could be due to a number of factors including potentially treatable viral infections such as CMV or adenovirus, or where the side effects of unnecessary treatment could compromise other aspects of patient management as with increased neutropoenia with ganciclovir in bone-marrow transplant patients. In all cases adequate samples, usually from the site of infection, should be taken as soon as possible. Where there is any doubt as to the correct sample, sample collection buffer, or method of sample transport, the laboratory should be consulted.

3.3. Limit antiviral use where possible

In the majority of patients most viral infections do not require treatment. This will limit adverse effects and selection of resistant virus. For example, although aciclovir is an effective treatment for primary varicella zoster infection (chickenpox), there is little benefit in treating children with uncomplicated infection, as the illness is generally self-limiting with serious complications being rare (Tarlow and Walters, 1998). Similarly, although the drug pleconaril has recently been shown to reduce the severity of rhinovirus infections (Hayden *et al.*, 2003), its use is not currently indicated in upper respiratory tract infections in otherwise healthy patients. It is also important not to use antivirals in situations where they have been shown not to be of benefit. Thus, although aciclovir can inhibit Epstein–Barr virus replication in vitro, it has not been shown to be effective in cases of glandular fever, probably due to the immune-mediated nature of the illness (van der Horst *et al.*, 1991).

3.4. Correct timing of treatment

It is important to start treatment at the optimum time. For acute infections, most antivirals are only effective if commenced rapidly. In adults with primary varicella zoster, treatment is normally only recommended if it can commence within 24 hr of rash onset, or within 36–48 hr of symptom onset in influenza virus infection, as there is little evidence of efficacy beyond this point (Couch, 2000; Wilkins *et al.*, 1998) and viral replication has usually peaked. This requires systems to be in place to see, diagnose, and treat such patients quickly. In many situations treatment may need to be started on a clinical basis unless very rapid PCR (or direct immunofluorescence) results can be obtained. Unnecessary treatment can be discontinued if the original diagnosis is not confirmed. In the case of recurrences of genital herpes, patients are often provided with antivirals in advance so that they can commence treatment at the first sign of symptoms (Drake *et al.*, 2000).

In chronic illness, factors indicating a need for antiviral treatment are not always present at diagnosis and very often treatment may be not be required for several years, if at all. Delaying treatment prolongs the development of resistance. Thus, in chronic HBV or HCV infection treatment is not normally indicated until liver biopsy, among other factors, shows a certain degree of pathology (Booth *et al.*, 2001; Wai and Lok, 2002). The high incidence of

HBV resistance to lamivudine means that earlier treatment may reduce the value of the drug when liver pathology reaches the stage where treatment really is indicated. In the case of HIV, once treatment is started it requires to be carried on life-long; this has major implications in terms of chronic debilitating side effects such as lipodystrophy (Bodasing and Fox, 2003) and the emergence of viral resistance. The latter necessitates changes in treatment regime that can rapidly reduce the antiviral options available to the patient in the future.

3.5.   Correct drug choice

Having decided that antiviral treatment should be commenced, the next step is to decide on the correct drug, or drug combination. Despite the relatively limited number of antivirals, there are a number of options for most treatable viral infections (see examples in Table 1, or for more detailed information see Waugh *et al.*, 2002). The decision will be based on a number of factors: efficacy, side-effect profile, route of administration, dosage schedule, resistance profile of the drug and the virus, underlying pathology, and cost.

Proven efficacy. Where possible a drug should be chosen in line with evidence-based guidelines or controlled trial data for that particular situation.

Side effect profile. Where alternatives exist an antiviral should be chosen which has the fewest side effects. Thus in the treatment of HSV, foscarnet, which is an effective inhibitor of viral replication, is not the first-line treatment due to a poor safety profile with nephrotoxicity (Balfour *et al.*, 1994).

Route of administration. Many antiviral compounds that are currently available are formulated for only one route of administration, it is therefore important to attempt to use one which can be delivered in a way which ensures adequate drug levels and maximises the likelihood of adherence. Inhalers are often difficult for elderly patients to administer correctly, thus for an elderly patient with influenza, the oral neuraminidase inhibitor oseltamivir may be preferred to inhaled zanamivir (Diggory *et al.*, 2001).

Dosage schedule. If adherence is likely to be a problem, or treatment will be long term or frequent, then an antiviral requiring fewer daily doses may be preferred. In the treatment of recurrent genital herpes simplex infection the oral aciclovir dose is five times a day while valaciclovir requires administration only twice daily (British National Formulary, 2003). In HIV treatment regimes multiple factors need to be considered, such as the dosage interval of the different drugs in the regime, whether or not food should be taken at the same time, and in the patient's own daily routine. Fitting the regime as much as possible to the patient's lifestyle helps improve tolerability and thus adherence (Trotta *et al.*, 2002).

Resistance profile of the drug. In some situations it is particularly desirable to use an antiviral that is less prone to selecting out resistant virus. In nursing homes, for example, a neuraminidase inhibitor may be preferred to amantadine for the treatment of influenza, as the rapid emergence of resistance to the later would result in the spread of resistant virus (Hayden and Hay, 1992). Where the patient is known to have a virus resistant to a particular antiviral, a drug should be used where cross-resistance is unlikely to be a problem.

Underlying patient pathology. The relative importance of a particular side effect will often vary when the patient has other coexisting pathology. Hence, although ganciclovir is normally the first-line choice in CMV infection in the immunocompromised, in patients where myelosuppression is a particular concern, foscarnet may be preferred, despite its nephrotoxicity, as it causes less neutropoenia.

Cost. Antiviral cost is an issue that can never be overlooked. The newer drugs tend to be more expensive than those that are more established. Where two equally effective alternatives exist the cheaper should always be considered. However, a number of factors need to be considered, such as the likelihood of adherence and possible side effects, and the cheapest drug will not necessarily be the most cost-effective in overall terms.

3.6. Correct dose and duration

Achieving an antiviral dose sufficient to maximally suppress viral replication is critical not only in terms of successful treatment of the infection, but also to prevent the development of resistant virus. Subtherapeutic antiviral doses result in viral replication in the presence of a drug, thus allowing the emergence of resistant virus. General recommendations on antiviral doses are available from national and local formularies, but a number of factors need to be considered before deciding on a dose for a patient in a particular situation. One factor is the susceptibility of the virus to the antiviral. The IC_{50} for aciclovir (the concentration required to inhibit viral replication by 50%) is 10-fold greater for VZV than for HSV. The latter therefore requires higher treatment doses. A further factor is the site of the infection and the ease with which the drug can penetrate that compartment of the body. Only 50% of intravenous aciclovir serum concentrations are achieved in the CSF, so again higher doses are recommended for HSV encephalitis.

In acute illness the aim is normally to continue treatment until the viral infection is resolved by a combination of the antiviral and the patient's own immune system. Where the immune system is not acting optimally this often results in a requirement for longer duration of treatment. In chronic infection, where the aim is to control rather than cure the disease, treatment may be continued indefinitely if benefit is still being achieved. In HBV infection lamivudine is often continued for several years, and in HIV treatment it is usually expected to continue life-long once started. This can lead to resistance if treatment is stopped, particularly for drugs with a long half-life, as there is an effective period of under-treatment as antiviral levels drop, so replication continues in the presence of suboptimal drug concentrations. Nevertheless, stopping treatment may be necessary due to side effects or if patient wishes. If possible a replacement regime should be insti-tuted immediately to suppress replication. A controversial therapeutic strategy for HIV, known as a structured drug interruption (Gulick, 2002), is aimed at boosting the patient's own immune response to the virus and allowing them a break from arduous treatment regimes. Although there are potential benefits, one of the major concerns is the risk of promoting the emergence of resistant virus.

3.7.   Adherence

As with all medical treatment the adherence of the patient to the regime is vital to its success. Erratic drug intake results in subtherapeutic drug levels with treatment failure and resistance. In patients on short-term courses of antivirals for acute infections, such as primary herpes simplex, this is less of a concern. In HCV the patients require to undergo treatment with a combination of interferon and ribavirin for 6 months or more. The interferon is given by subcutaneous injection and most patients experience flu-like illness and lethargy, with a sig-nificant percentage suffering depressive symptoms. Especially as the patient may not have felt particularly unwell prior to treatment, the future benefits may be less obvious to them than the acute side effects. In HIV, regimes are complex with multiple drugs and significant side effects, particularly given the long-term nature of treatment. Adherence may be compromised by all these factors. It is important that regimes are kept as simple as possible and that patients are fully informed regarding their treatment and the need for good adherence.

3.8.   Combination antiviral therapy

Combination treatments, which can prevent or delay the development of resistance, are most commonly used for chronic conditions. In HCV, the com-bination of interferon and ribavirin helps achieve viral suppression by a com-bination of boosting host immune responses and direct viral effects. In HIV combination therapy with three or more antiretrovirals from two or three classes of drug (Highly Active Antiretroviral Therapy—HAART) has resulted in regimes of sufficient potency to fully suppress viral replication, which should delay the emergence of resistant virus. There are, however, also obvious disadvantages, with cumulative side effects and drug interactions. Adherence to combination regimes has been greatly helped by the availability of antivirals as compound preparations containing two or three drugs, which reduces pill burden and improves compliance. As with all compound drugs, this does limit the flexibility to change individual drugs and their doses.

3.9. Monitoring treatment

Often, simple clinical improvement is all that is required to monitor antiviral treatment, for example, in acute infection with influenza or herpes simplex in the immunocompetent. For chronic infections, or infections in the immunosuppressed, regular testing is required. This may be qualitative, for example, weekly testing after infection or reactivation of CMV, or quantitative, such as in HBV and HIV infections (Pillay *et al.*, 2002).

Where symptoms fail to resolve or worsen on treatment, or where laboratory testing shows a failure to suppress viral levels or viral rebound, the antiviral regime should be reviewed in terms of dose, duration, and adherence, and other possible confounding patient factors considered. If, despite optimised treatment, symptoms still fail to resolve, resistance should be considered (Pillay *et al.*, 2002). Viral resistance testing is now usually based on sequencing of the virus and identification of known resistance mutations, rather than on viral susceptibility in culture. This is generic technology and is becoming increasingly available in regional virus laboratories. The detection of resistant virus normally indicates the need to change therapy.

Monitoring is particularly important where an antiviral is being used in a situation where experience and evidence for benefit is limited. In such cases both regular clinical and virological review are essential.

4. ANTIVIRAL PROPHYLAXIS STRATEGIES

Many of the considerations above for the use of antivirals in the treatment of viral infections are also relevant when considering the use of antivirals for prevention. One should consider in which situations antiviral prophylaxis is warranted, and whether there are alternatives, such as vaccination or pre-emptive treatment strategies for CMV.

Prophylaxis should only be considered where there is a significant risk of a specific viral infection or of reactivation, where such infection would cause a significant risk to the patient and where effective vaccination is not possible. The need for prophylaxis must also be balanced against the possible disadvantages, such as unwanted side effects and possible selection for resistance.

4.1.   Restrict to necessary situations

Most routine antiviral prophylaxis is used in patients on immunosuppressive regimes, usually post-transplant, and patients with low CD4 counts in HIV infection. The aim is to prevent primary infection or reactivation of latent infection. The main viral targets are CMV and HSV. Not all immunocompromised patients require prophylaxis, and protocols should define those patients for whom it is necessary. The decision should be based on the likelihood of infection or reactivation and the severity of the immunosuppressive regime or level of the CD4 count. In solid-organ transplants, CMV prophylaxis is often only considered where there is a miss-match in serostatus, such that a primary CMV infection will result from transplant of an organ from a seropositive donor. In HIV, CMV reactivation and disease become a problem requiring prophylaxis only when CD4 counts fall below 50 (Kaplan *et al.*, 2002).

Antiviral prophylaxis can also be used to prevent infection post-exposure, such as after significant contact with HIV (Department of Health, 2000). It is important to reserve treatment to situations where there is a real risk of infection, especially for HIV where treatment requires multiple drugs and side effects are common. Thus, a needlestick requires a thorough risk assessment to establish whether significant blood contact occurred and the likely HIV status of the source.

Prophylaxis may also be indicated to prevent spread of infection in outbreaks. Again it should only be considered in high-risk situations, such as an influenza outbreak in a nursing home with elderly patients, or in a hospital ward with immunocompromised patients. As described below vaccination is normally the preferred option in these situations.

4.2.   Alternatives to antiviral prophylaxis

It is always important to consider alternatives to drug therapy. Two options are vaccination and pre-emptive treatment strategies.

Vaccines, where available, are the preferred method for viral prophylaxis pre-exposure and often also post-exposure, for example, the hepatitis B vaccine and influenza vaccines (Department of Health, 1996). Vaccine, or a course of vaccine, need usually only be given once in a lifetime or, in the case of influenza, yearly. They generally have minimal side effects and no further action is required if contact does occur. Antiviral prophylaxis may still be required in some circumstances where a vaccine is available, such as where vaccine is contraindicated or response to vaccination is poor, as in the immunocompromised. Passive vaccination with immunoglobulin may be another option after contact. For example, varicella zoster immunoglobulin is available after contact with VZV and is usually preferred to aciclovir as there is more information on its use in this situation (Department of Health, 1996).

In some transplant units a CMV pre-emptive strategy is used in place of antiviral prophylaxis. In this situation patients at risk of CMV infection or reactivation are monitored by PCR at regular intervals (usually once or twice a week). When the virus is detected in the blood, or levels are seen to be rising in quantitative assays, antiviral drug is started at treatment, rather than prophylaxis, doses. This has the advantage of limiting drug treatment, thus minimising side effects and reducing the emergence of resistant virus. There is still debate on which strategy is best (Emery, 2001; Hart and Paya, 2001). It should be noted that for pre-emptive strategies to succeed there needs to be adequate laboratory funding and expertise and a high degree of organisation regarding the collection, transport, and testing of samples and communication of results.

4.3. Timing and duration of treatment

Ideally a patient should be on prophylaxis before there is significant risk of infection or reactivation. Thus treatment should start before or soon after transplant, or when the CD4 count is seen to be dropping towards the threshold for treatment in HIV. Treatment should continue until immunity is reconstituted to a significant level, often taken as 100 days post-transplant. Various studies have shown that although immune reconstitution on HAART in HIV is not necessarily complete it appears to be safe to discontinue prophylactic treatment when the CD4 count has risen (Macdonald *et al.*, 1998).

Timing of treatment is vital for post-exposure prophylaxis. For example, after a significant needlestick from an HIV positive source, antiretroviral post-exposure prophylaxis should ideally start within 1 hr of contact (Department of Health, 2000). In contrast, following contact with VZV in a high-risk individual, such as a pregnant woman, aciclovir treatment is given from 6 to 10 days post-contact in keeping with the natural history of the infection and timing of viraemia (Asano *et al.*, 1993).

4.4. Drug choice and dose

Drug choice should be guided by the factors mentioned in the discussion on treatment strategies. Avoidance of side effects is particularly important given the fact that the patients are not being treated for an active infection. Drug doses for prophylaxis are often lower than those recommended for treatment. Drug choice is a particular issue for post-exposure prophylaxis in HIV, where knowledge of the source's treatment history, and possible viral resistance profile, may influence the treatment combination chosen.

4.5.   Adherence

Adherence is important for successful antiviral prophylaxis. It is important that the patient is aware of the reasons why the treatment is necessary.

4.6.   Monitoring for infection

Prophylactic regimes do fail. Prophylaxis doses are generally lower than those required for effective treatment, thus it is necessary to react quickly to the development of infection, not only to ensure correct treatment, but also to prevent the development of resistance on regimes that are not fully suppressive. Clinical and virological monitoring is important. Where monitoring detects active viral infection or a patient becomes unwell on prophylaxis, samples should be taken to confirm the diagnosis as quickly as possible, although it may be necessary to start treatment before results are available.

5.   CONCLUSION

Antiviral therapy continues to be a rapidly expanding field. Complex issues need to be addressed such as the effects of long-term use and the emergence of resistance. If these agents are to realise their full potential both now and in the future it is essential for their use to be the subject of rational and well monitored prescribing practises.

REFERENCES

Asano, Y., Yoshikawa, T., Suga, S., Kobayashi, I., Nakashima, T., Yazaki, T. *et al.*, 1993, Postexposure prophylaxis of varicella in family contact by oral acyclovir. *Pediatrics*, **92**, 219–222.

Balfour, H. H., Jr, Benson, C., Braun, J., Cassens, B., Erice, A., Friedman-Kien, A. *et al.*, 1994, Management of acyclovir-resistant herpes simplex and varicella-zoster virus infections. *J. Acquir. Immune Defic. Syndr.*, **7**, 254–260.

BHIVA, 2003, British HIV Association (BHIVA) guidelines for the treatment of HIV-infected adults with antiretroviral therapy. *HIV Med.*, **2**, 276–313.

Bodasing, N. and Fox, R., (2003): HIV-associated lipodystrophy syndrome: Description and pathogenesis. *J. Infect.*, **46**, 149–154.

Boivin, G., Edelman, C. K., Pedneault, L., Talarico, C. L., Biron, K. K., and Balfour, H. H., Jr, 1994, Phenotypic and genotypic characterization of acyclovir-resistant varicella-zoster viruses isolated from persons with AIDS. *J. Infect. Dis.*, **170**, 68–75.

Bonanni, P. and Bonaccorsi, G., 2001, Vaccination against hepatitis B in health care workers. *Vaccine*, **19**, 2389–2394.

Booth, J. C., O'Grady, J., and Neuberger, J., 2001, Clinical guidelines on the management of hepatitis C. *Gut*, **49**(Suppl. 1), I1–I21.

Bridges, C. B., Harper, S. A., Fukuda, K., Uyeki, T. M., Cox, N. J., and Singleton, J. A., 2003, Prevention and control of influenza. Recommendations of the Advisory Committee on Immunization Practices (ACIP). *MMWR Recomm. Rep.*, **52**, 1–34; quiz CE1–4.

British Medical Association and Royal Pharmaceutical Society of Great Britain, 2003, British National Formulary, 45.

Carman, B., 2001: Molecular techniques should now replace cell culture in diagnostic virology laboratories. *Rev. Med. Virol.*, **11**, 347–349.

Chadwick, P. R., Beards, G., Brown, D., Caul, E. O., Cheesbrough, J., Clarke, I. *et al.*, 2000, Management of hospital outbreaks of gastro-enteritis due to small roundstructured viruses. *J. Hosp. Infect.*, **45**, 1–10.

Couch, R. B., 2000, Prevention and treatment of influenza. *N. Engl. J. Med.*, **343**, 1778–1787.

Department of Health, 1996, Immunisation against infectious disease. 'The Green Book'.

Department of Health, 2000, HIV post-exposure prophylaxis: guidance from the UK Chief Medical Officers' Expert Advisory Group on AIDS. PL CO (2000) 4.

Dienstag, J. L., Schiff, E. R., Wright, T. L., Perrillo, R. P., Hann, H. W., Goodman, Z. *et al.*, 1999, Lamivudine as initial treatment for chronic hepatitis B in the United States. *N. Engl. J. Med.*, **341**, 1256–1263.

Diggory, P., Fernandez, C., Humphrey, A., Jones, V., and Murphy, M., 2001, Comparison of elderly people's technique in using two dry powder inhalers to deliver zanamivir: Randomised controlled trial. BMJ, **322**, 577–579.

Drake, S., Taylor, S., Brown, D., and Pillay, D., 2000, Improving the care of patients with genital herpes. BMJ, **321**, 619–623.

Dybul, M., Fauci, A. S., Bartlett, J. G., Kaplan, J. E., and Pau, A. K., 2002, Guidelines for using antiretroviral agents among HIV-infected adults and adolescents. Recommendations of the Panel on Clinical Practices for Treatment of HIV. *MMWR Recomm. Rep.*, **51**, 1–55.

Elizaga, J., Olavarria, E., Apperley, J., Goldman, and J., and Ward, K., 2001, Parainfluenza virus 3 infection after stem cell transplant: Relevance to outcome of rapid diagnosis and ribavirin treatment. *Clin. Infect. Dis.*, **32**, 413–418.

Emery, V. C., 2001, Prophylaxis for CMV should not now replace pre-emptive therapy in solid organ transplantation. *Rev. Med. Virol.*, **11**, 83–86.

Gern, J. E., and Busse, W. W., 1999, Association of rhinovirus infections with asthma. *Clin. Microbiol. Rev.*, **12**, 9–18.

Greenberg, S. B., 2003, Respiratory consequences of rhinovirus infection. *Arch. Intern. Med.* **163**, 278–284.

Gubareva, L. V., Matrosovich, M. N., Brenner, M. K., Bethell, R. C., and Webster, R. G., 1998, Evidence for zanamivir resistance in an immunocompromised child infected with influenza B virus. *J. Infect. Dis.*, **178**, 1257–1262.

Gulick, R. M., 2002, Structured treatment interruption in patients infected with HIV: A new approach to therapy. *Drugs*, **62**, 245–253.

Gulick, R. M., 2003, New antiretroviral drugs. *Clin. Microbiol. Infect.*, **9**, 186–193.

Hart, G. D, and Paya, C. V., 2001, Prophylaxis for CMV should now replace pre-emptive therapy in solid organ transplantation. *Rev. Med. Virol.*, **11**, 73–81.

Hayden F. G and Hay, A. J., 1992, Emergence and transmission of influenza A viruses resistant to amantadine and rimantadine. *Curr. Top Microbiol. Immunol.*, **176**, 119–130.

Hayden, F. G., Herrington, D. T., Coats, T. L., Kim, K., Cooper, E. C., Villano, S. A. *et al.*, 2003, Efficacy and safety of oral pleconaril for treatment of colds due to picornaviruses in adults: Results of 2 double-blind, randomized, placebo-controlled trials. *Clin. Infect. Dis.*, **36**, 1523–1532.

Holland, J. J., De La Torre, J. C., and Steinhauer, D. A., 1992, RNA virus populations as quasispecies. *Curr. Top Microbiol. Immunol.*, **176**, 1–20.

Kaplan, J. E., Masur, H., and Holmes, K. K, 2002, Guidelines for preventing opportunistic infections among HIV-infected persons—2002. Recommendations of the U.S. Public Health Service and the Infectious Diseases Society of America. *MMWR Recomm. Rep.*, **51**, 1–52.

Limaye, A. P., Corey, L., Koelle D. M., Davis, C. L., and Boeckh, M., 2000, Emergence of ganciclovir-resistant cytomegalovirus disease among recipients of solid-organ transplants. *Lancet*, **356**, 645–649.

Macdonald, J. C., Torriani, F. J., Morse, L. S., Karavellas, M. P., Reed, J. B., and Freeman, W. R., 1998, Lack of reactivation of cytomegalovirus (CMV) retinitis after stopping CMV maintenance therapy in AIDS patients with sustained elevations in CD4 T cells in response to highly active antiretroviral therapy. *J. Infect. Dis.*, **177**, 1182–1187.

Marcellin, P., Chang, T. T., Lim, S. G., Tong, M. J., Sievert, W., Shiffman M. L. *et al.*, 2003, Adefovir dipivoxil for the treatment of hepatitis B e antigen-positive chronic hepatitis B. *N. Engl. J Med.*, **348**, 808–816.

Mathei, C., Buntinx, F., and van Damme, P., 2002, Seroprevalence of hepatitis C markers among intravenous drug users in western European countries: A systematic review. *J. Viral Hepat.*, **9**, 157–173.

McCall, J., and Smithson, R., 2002, Rapid response and strict control measures can contain a hospital outbreak of Norwalk-like virus. *Commun. Dis. Public Health*, **5**, 243–246.

Morfin, F., and Thouvenot, D., 2003, Herpes simplex virus resistance to antiviral drugs. *J. Clin. Virol.*, **26**, 29–37.

NICE, 2003, National Institute for Clinical Excellence. Guidance on the use of zanamivir, oseltamivir and amantidine for the treatment of influenza.

Nichol, K. L., 2003, The efficacy, effectiveness and cost-effectiveness of inactivated influenza virus vaccines. *Vaccine*, **21**, 1769–1775.

Niesters, H. G., 2002, Clinical virology in real time. *J. Clin. Virol.* **25**(Suppl. 3), S3–S12.

Nijhuis, M., Deeks, S., and Boucher, C., 2001, Implications of antiretroviral resistance on viral fitness. *Curr. Opin. Infect. Dis.*, **14**, 23–28.

Pawlotsky, J. M., 2000, Hepatitis C virus resistance to antiviral therapy. *Hepatology*, **32**, 889–896.

Pescovitz, M. D., Rabkin, J., Merion, R. M., Paya, C. V., Pirsch, J., Freeman, R. B. *et al.*, 2000, Valganciclovir results in improved oral absorption of ganciclovir in liver transplant recipients. *Antimicrob. Agents Chemother.* **44**, 2811–2815.

Pillay, D., Emery, V. C., Mutimer, D., Ogilvie, M. M., Carman, W., Mutton, K. *et al.*, 2002, Guidelines for laboratory monitoring of treatment of persistent virus infections. *J. Clin. Virol.*, **25**, 73–92.

Pillay, D., Taylor, S., and Richman, D. D., 2000, Incidence and impact of resistance against approved antiretroviral drugs. *Rev. Med. Virol.*, **10**, 231–253.

Pillay, D. and Zambon, M., 1998, Antiviral drug resistance. BMJ, **317**, 660–662.

RCOG, 2001, Royal college of obstetricians and gynaecologists clinical green top guidelines. Chickenpox in pregnancy. 30.

RCOG, 2002: Royal college of obstetricians and gynaecologists clinical green top guidelines. Management of genital herpes in pregnancy. 13.

Safrin, S., Cherrington, J., and Jaffe, H. S., 1997, Clinical uses of cidofovir. *Rev. Med. Virol.*, **7**, 145–156.

Segarra-Newnham, M. and Vodolo, K. M., 2001, Use of cidofovir in progressive multifocal leukoencephalopathy. *Ann. Pharmacother.*, **35**, 741–744.

Snell, N. J., 2001, Ribavirin—current status of a broad spectrum antiviral agent. *Expert Opin Pharmacother.*, **2**, 1317–1324.

Tarlow, M. J., and Walters, S., 1998, Chickenpox in childhood. A review prepared for the UK Advisory Group on Chickenpox on behalf of the British Society for the Study of Infection. *J. Infect.*, **36**(Suppl. 1), 39–47.

Trotta, M. P., Ammassari, A., Melzi, S., Zaccarelli, M., Ladisa, N., Sighinolfi, L. *et al.*, 2002, Treatment-related factors and highly active antiretroviral therapy adherence. *J. Acquir. Immune Defic. Syndr.* **31**(Suppl. 3), S128–S131.

UK collaborative group on monitoring the transmission of HIV drug resistance 2001: Analysis of prevalence of HIV-1 drug resistance in primary infections in the United Kingdom. BMJ *322*, 1087–1088.

UK Department of Health, 2002, Good practise guidelines for renal dialysis/transplantation units.

USA Food and Drug Administration, 2003, Centre for drug evaluation and research. Approved drug products. *Orange Book*, 23rd edn.

van der Horst, C., Joncas, J., Ahronheim, G., Gustafson, N., Stein, G., Gurwith, M. *et al.*, 1991, Lack of effect of peroral acyclovir for the treatment of acute infectious mononucleosis. *J. Infect. Dis.*, **164**, 788–792.

Wai, C. T., and Lok, A. S., 2002, Treatment of hepatitis B. *J. Gastroenterol.*, **37**, 771–778.

Waugh, S. M., Pillay, D., Carrington, D., and Carman, W. F., 2002, Antiviral prophylaxis and treatment (excluding HIV therapy). *J. Clin. Virol.*, **25**, 241–266.

Wilkins, E. G., Leen, C. L., McKendrick, M. W., and Carrington, D., 1998, Management of chickenpox in the adult. A review prepared for the UK Advisory Group on Chickenpox on behalf of the British Society for the Study of Infection. *J. Infect.*, **36**(Suppl. 1), 49–58.

Chapter 19

Disinfection Policies in Hospitals and the Community

Emine Alp and Andreas Voss
University Medical Centre St Radboud, Department of Medical Microbiology
(440 MMB) and Nijmegen University Centre of Infectious Diseases
P.O. Box 9101, 76500 HB Nijmegen, The Netherlands

The prevalance of hospital-acquired infections with multiresistant bacteria has increased in many countries around the world (Boyce, 1990). Nosocomial pathogens originating from colonized or infected patients can contaminate the environment and survive for extended periods. As a result, the hospital environment has become an important source (and or reservoir) of multiresistant bacteria capable of colonizing or infecting patients. Environmental surfaces have been associated with nosocomial outbreaks of multiresistant bacteria (Hayden, 2000; Talon, 1999) and community outbreaks of hepatitits A and acute gastroenteritis due to other viruses (Evans *et al.*, 2002; Leoni *et al.*, 1998; Love *et al.*, 2002). Still these problems do not justify routine disinfection of surfaces and fomites, but targeted use of disinfectants is an important factor in preventing infections in the hospitals (and possibly in the community).

1.   DEFINITIONS

Cleaning is the removal of all visible and invisible organic material (e.g., soil) from objects to prevent microorganisms from thriving, multiplying, and spreading. It is accomplished using water with detergents or enzymatic products. In sanitary facilities (e.g., washbasins, toilets), separate buckets and cloths have to be used and alkaline detergent is recommended for cleaning.

Antibiotic Policies: Theory and Practice. Edited by Gould and van der Meer
Kluwer Academic / Plenum Publishers, New York, 2005

Cleaning must precede disinfection and sterilization, since it reduces the number of microorganisms on contaminated equipment (Rutala, 1996; WIP, 2002 Module 6.1).

Disinfection describes the inactivation of pathogenic microorganisms (vegetative bacteria and/or fungi and/or viruses) on inanimate surfaces as well as intact skin and mucous membranes. It can be accomplished by the use of liquid chemicals or wet pasteurization in healthcare settings. Disinfection is aimed at minimizing the risk of transfer of microorganisms, but this process does not inactivate all microorganisms; bacterial endospores usually survive. Disinfection differs from sterilization by its lack of sporocidal activity, but a few disinfectants (frequently referred to as "chemical sterilants") will kill spores after prolonged exposure times (6–10 hr). At similar concentrations but with shorter exposure periods (<30 min), these disinfectants may kill all microorganisms with the exception of high numbers of bacterial spores and are called "high level disinfectants". Disinfectants that kill only most vegetative bacteria, some fungi, and some viruses (≤10 minutes) are called "low level disinfectants" (Rutala, 1996).

Disinfection (as well as sterilization) can be effected by the prior cleaning of the object, organic load, the type and the level of microbial contamination, the concentration of and exposure time to the germicide, the nature of the object, and the temperature and the pH of the disinfection process.

Quaternary ammonium, iodine, alcohol, aldehyde, organic acid, peroxide, and halogenated compounds are the chemical disinfectants and have proven effective against a broad spectrum of microorganisms (Rutala, 1996; WIP, 2002 Module 6.1).

Based on the risk of infection, items are classified in three risk categories:

1. Critical items enter sterile tissue or the vascular system and if such an item is contaminated there is high risk of infection. Therefore these items must be sterile.
2. Semicritical items come in contact with mucous membranes or non-intact skin and these objects must be correctly cleaned and should undergo a disinfection process that eradicates all microorganisms and most bacterial spores.
3. Noncritical items come in contact with intact skin but not mucous membranes and these items need not be sterile. Environmental surfaces and fomites (e.g., bed rails, linens, bedside tables) in hospital are considered noncritical items (Rutala, 1996).

2.   ENVIRONMENTAL CONTAMINATION

Outbreaks of hepatitis A or acute gastroenteritis can occur in hospitals, but are furthermore major public health problems, especially in schools, military quarters, and nurseries. The aetiological agents of these diseases are excreted in high numbers in the faeces of infected individuals and are able to persist for extended periods of time in the environment (Evans *et al.*, 2002; Kawai and Feinstone, 2000). In many outbreaks, surfaces may act as vehicles for the spread of the infection (Leoni *et al.*, 1998; Lloyd-Evans *et al.*, 1986; Weniger *et al.*, 1983). During viral infections of the respiratory tract, patients shed large amounts of virus into their naso-tracheal secretions and these can contaminate the environment. Respiratory viruses, such as respiratory synscytial virus (RSV), rhinovirus, and parainfluenza virus have been shown to survive for extended periods in suspensions and on surfaces (Brady *et al.*, 1990; Hall *et al.*, 1980; Hendley *et al.*, 1973; Sizun *et al.*, 2000). Transmission of rhinovirus infection by contaminated surfaces was also shown in an experimental study (Gwaltney and Hendley, 1982). Contaminated environmental surfaces are considered to represent a significant vector for viral infections in the community and, also in paediatric units in the hospital. Next to direct droplet transmission, indirect transmission (environment → hands → self-inoculation of mucous membranes) is probably even more important in spreading viral respiratory diseases.

Nosocomial infections result from a patient's endogenous flora, person-to-person transmission, or are linked to contaminated surfaces (Shaikh *et al.*, 2002; Weber and Rutala, 1993). Extensive environmental contamination has been demonstrated in rooms housing patients with multiresistant bacteria (Byers *et al.*, 1995; Dembry *et al.*, 1995; Hargreaves *et al.*, 2001; Rutala *et al.*, 1983). Several investigators have demonstrated that the inanimate environment near an infected patient commonly becomes contaminated with pathogenic microorganisms (Bonten *et al.*, 1996; Boyce *et al.*, 1997; Karanfil *et al.*, 1992; Weber and Rutala, 1997). Furthermore, these microorganisms can survive in the environment—including on working surfaces—for a long time (Ansari *et al.*, 1988; Byers *et al.*, 1998; Getchell-White *et al.*, 1989; Mbithi *et al.*, 1992; Neely and Maley, 2000; Weber and Rutala, 2001). Consequently, CDC guidelines include measures to prevent infection originating from environmental contamination (CDC, 2001, 2003c).

Despite the fact that noncritical items or contact with noncritical surfaces carries little risk of transmitting infectious agents to patients, these items may contribute to secondary transmission by contaminating hands of healthcare workers or medical equipment (Weber and Rutala, 1993). In a survey study, 63% of 369 infection control professionals strongly or somewhat agreed that the inanimate environment plays a critical role in transmission of organisms (Manangan *et al.*, 2001).

Boyce *et al.* (1997) found environmental contamination of rooms in 73% of the rooms harboring patients with methicillin-resistant *Staphylococcus aureus* (MRSA) infections and 69% when patients were colonized with MRSA.

Objects that frequently were contaminated included the floor, bed linens, the patient's gown, tables, and blood pressure cuffs. Even in the absence of direct patient contact 42% of the healthcare workers (HCW) contaminated their gloves by touching contaminated surfaces, thereby proving that contaminated environmental surfaces may serve as a source for MRSA spread. Rightfully, infection control measurements for MRSA outbreaks include decontamination of the environment (Burd *et al.*, 2003; Hails *et al.*, 2003; O'Connell and Humphreys, 2000). While antibiotic-resistant pathogens, including MRSA, so far were a strictly nosocomial problem, it recently became an important and growing threat to the public health. Around the world, cases of serious infections due to community acquired MRSA have been described (Mongkolrattanothai *et al.*, 2003).

Another bacterial genus that has emerged as important nosocomial pathogen with increasing resistance to antibiotics are enterococci. The National Nosocomial Infections Surveillance (NNIS) system has identified enterococci as the second most common nosocomial pathogen. The report also demonstrated an overall increase in the incidence of vancomycin-resistant enterococci (VRE) from 0.3% to 7.9% between 1989 and 1993 (CDC, 1993). In many reports, hospital outbreaks of VRE have been related to environmental contamination and most outbreaks have been controlled with appropriate infection control measures (Armstrong-Evans *et al.*, 1999; Boyce, 1995; Boyce *et al.*, 1995; Calfee *et al.*, 2003; Karanfil *et al.*, 1992; Mayer *et al.*, 2003; Montelcalvo *et al.*, 1999; Noskin *et al.*, 1995; Porwancher *et al.*, 1997; Sample *et al.*, 2002; Smith *et al.*, 1998; Timmers *et al.*, 2002). On the other hand standard disinfection methods, as those used in the United States (using sprays to apply the disinfectant to the surface) may not be sufficient to properly free the environment and surfaces from VRE. In a study by Byers *et al.* (1998) only the use of the "bucket method" successfully achieved room decontamination.

Multi-resistant Gram-negative bacilli have emerged as nosocomial pathogens, especially in intensive care units and became endemic in many hospitals, causing local outbreaks. Since environmental contamination plays an important role in these outbreaks, they frequently have been controlled with simple infection control measures including environmental disinfection (Alfieri *et al.*, 1999; Aygun *et al.*, 2002; Berg *et al.*, 2000; Dijk *et al.*, 2002; Engelhart *et al.*, 2002; Talon, 1999).

The risks of transmission of blood-borne viruses, like human immunodeficiency virus (HIV), hepatitis B virus (HBV), and hepatitis C virus (HCV) have been well documented (Bolyard *et al.*, 1998; CDC, 2003a). Although percutaneous injuries are among the most efficient modes of HBV transmission, these exposures probably account for only a part of these infections among HCWs. In several investigations of nosocomial hepatitis B outbreaks, most infected HCWs could not recall an overt percutaneous injury, although in some studies, up to one third of infected HCWs recalled caring for a patient who was

HBsAg-positive. In addition, HBV demonstrated to survive in dried blood at room temperature on environmental surfaces for at least 1 week. The potential for HBV transmission through contact with environmental surfaces has furthermore been demonstrated in investigations of HBV outbreaks among patients and staff of haemodialysis units. Data on survival of HCV and HIV in the environment are limited (Bolyard *et al.*, 1998; CDC, 2003b; Sattar *et al.*, 2001). Transmission of these viruses from patient to staff or from patient to patient could theoretically be mediated by contaminated surfaces and instruments and avoiding contact with contaminated materials are valuable means of protection. The indirect spread of these viruses, although much less common, can occur when objects that are freshly contaminated with tainted blood enter the body or contact damaged skin (CDC, 1977; Lewis *et al.*, 1992).

2.1.   Cleaning or disinfection of environment

Cleaning is necessary to control environmental contamination and for minimizing the spread of microorganisms. Cleaning also serves aesthetic aspects and a clean environment promotes further hygienic action.

There are two cleaning methods: dry cleaning and wet cleaning. The choice between wet and dry cleaning depends on the nature of the dirt and the room. A dry system is preferred for the cleaning of floors and particular materials. The dry system uses little or no liquid. Floors remain dry during cleaning and can be used by HCWs and patients immediately after cleaning without the danger of slipping. While recent wet cleaning or fluid spillage promotes the growth of Gram-negative organisms, these pathogens are rare in dry cleaned environments (Ayliffe *et al.*, 1990; Dharan *et al.*, 1999). Ballemans *et al.* (2003) conducted an experimental prospective study over a 10-week period and compared a new dry cleaning method, using humidified high-performance cloths with the wet routine cleaning practice. They showed a statistically significant reduction of the total viable counts, for the new dry cleaning method compared with the wet mopping. Unfortunately the study did not look into the impact on infection rates and/or reduced cross-contamination. Despite these results, dry cleaning is not sufficient to remove "stuck-on dirt"; a wet system must be used. Wet cleaning has to be the choice in departments where frequent spilling occurs (e.g., intensive care units). In general, wet cleaning of large surface is always preceded by dry mopping (WIP, 2002 Module 6.4).

Patient areas should be cleaned periodically and after contamination. Tables 1–7 give the advised frequency of routine cleaning according to the national Dutch infection control guidelines (WIP) in the nursing department (Table 1), isolation rooms (Tables 2–4), outpatients' clinic (Table 5), operating department (Table 6), and in "other" rooms within healthcare institutions (Table 7) (WIP, 2002 Module 6.4).

Table 1. Frequency of routine cleaning in the nursing department (including contact isolation and droplet isolation)

	Floor	Furniture/objects
Patient room	Clean daily	Clean daily
Treatment room	Clean daily	Clean daily
Sanitary facilities	Sanitary clean twice a day, every day of the week	Sanitary clean twice a day, every day of the week
Utility room	Clean daily	Clean daily
Kitchen	Clean daily	Clean daily
Administrative area	Clean 5 days in a week	Dusting 3 days in a week
Storage room	Clean 2 days in a week	Dusting 1 day in a week
Cloakroom	Clean 2 days in a week	Dusting 1 day in a week
Hallways and stairs	Clean daily	

Table 2. Frequency of routine cleaning in isolation room: airborne isolation

	Floor	Furniture/objects
Room	Clean daily	Clean daily
Sanitary facilities	Sanitary clean twice a day, every day of the week	Clean daily
After end of isolation; room, sanitary facilities, and sluice	Clean	Clean

Table 3. Frequency of routine cleaning in isolation room: strict isolation

	Floor	Furniture/objects
Room	Clean daily	Clean daily
Sanitary facilities	Sanitary clean twice a day, every day of the week	Clean daily
Sluice	Clean daily	Clean daily
After end of isolation; room, sanitary facilities, and sluice	Disinfection	Disinfection

Table 4. Frequency of routine cleaning in isolation room: protective isolation

	Floor	Furniture/objects
Room	Clean daily	Clean daily
Sanitary facilities	Sanitary clean twice a day, every day of the week	Sanitary clean twice a day, every day of the week

Table 5. Frequency of routine cleaning in the outpatients' clinic

	Floor	Furniture/objects
Treatment room[a]	Clean 5 days in a week	Clean 5 days in a week
Consulting room, hard floor covering[a]	Clean 5 days in a week	Clean (dusting) 5 days in a week
Consulting room, soft floor covering	Vacuum cleaning 2 days in a week	Clean (dusting) 2 days in a week
Examination room[a]	Clean 5 days in a week	Clean 5 days in a week

[a]If in use at the weekend, clean daily.

Table 6. Frequency of routine cleaning in the operating department

	Floor	Furniture/objects
Operating room	Clean daily	Clean daily
Scrub area	Clean daily	Clean daily
Room for immediate pre-operative care	Clean daily	Clean daily
Storage room	Clean daily	Clean (dusting) 1 day in a week
Waste storage room	Clean daily	Clean daily
Dirty linen storage room	Clean daily	Clean daily
Instrument washing room	Clean daily	Clean daily
Office area	Clean daily	Clean (dusting) 1 day in a week
Hallway	Clean daily	Clean (dusting) daily
Sluice	Clean daily	Clean (dusting) daily
Changing room	Clean daily	Clean (dusting) daily
Recovery room	Clean daily	Clean daily

[a]If not in use at the weekend, then clean 5 days in a week.

Table 7. Frequency of routine cleaning in various rooms

	Floor	Furniture/objects
Baby and children's room	Clean 5 days in a week	Clean 5 days in a week
Endoscopy room	Clean 5 days in a week	Clean 5 days in a week
Physiotherapy exercise room	Clean 5 days in a week	Clean 5 days in a week
Radiology, room for invasive examination	Clean 5 days in a week	Clean 5 days in a week
Radiology, room for other examination	Clean 5 days in a week	Clean (dusting) 2 days in a week
Rooms other than operating room in which invasive procedures are performed	Clean 5 days in a week	Clean 5 days in a week
Laboratory	Clean 5 days in a week	Clean 5 days in a week (only workbenches)
Central kitchen	Clean daily	Clean daily

[a]If in use at the weekend clean daily.

Routine cleaning of environmental surfaces with detergents and elimination of heavy dust is sufficient in most circumstances (WIP, 2002 Module 6.4). Detergents, disinfectants, and cleaning equipment itself may become a source of contamination. Werry *et al.* (1988) reported contamination of detergent solutions used for cleaning of surfaces. The contaminants, mainly Gram-negative non-fermentative bacilli, including *Acinetobacter* spp. and *Pseudomonas aeruginosa*. Since contamination may occur during the preparation of fresh solutions, cleaning solutions must be prepared daily or as needed, and frequently be replaced with fresh solution according to facility policies. The mop head should be changed at the beginning of each day and/or after cleaning up large spills of blood or other body substances. The use of disposable materials is preferred for all methods of cleaning. When using non-disposable materials, they have to be sent to the laundry service immediately after completing a cleaning job (WIP, 2002 Module 6.4).

Rutala *et al.* (2000) reviewed the epidemiological and microbiological data regarding the use of disinfectants on noncritical surfaces. They concluded to disinfect housekeeping and noncritical patient care equipment–surfaces given the minimal extra cost and added antimicrobial activity. Still, the routine use of disinfectants to clean hospital floors and other surfaces is controversial and the influence on nosocomial infections unclear. Danforth *et al.* (1987) compared the influence of disinfection vs cleaning using plain soap on nosocomial infection rates during a 6-month period. The combined nosocomial infection rate for the eight acute-care nursing wards did not differ between the disinfectant and detergent groups. No differences in floor contamination were observed. Comparing detergent- and disinfectant-use, Dharan *et al.* (1999) observed no change in the incidence of hospital-acquired infections during a 4-month trial period compared to the preceding 12 months.

When surfaces, furniture, or objects are found to be contaminated with blood or other body fluids, disinfectants must be used. Further indications for disinfection are: rooms of patients infected or colonized with multiresistant microorganisms, for example, MRSA, VRE, multiresistant Gram-negative bacteria (Hayden, 2000; Muto *et al.*, 2003) and during viral epidemics, for example, HAV, coronavirus, rotavirus, etc. (Griffith *et al.*, 2000; Leoni *et al.*, 1998; Lloyd-Evans *et al.*, 1986; Rutala and Weber, 1997). In general, when controlling epidemics, disinfection should be part of the solution to control the spread.

For noncritical instruments and devices as well as for general environmental surfaces high level disinfectants/liquid chemical sterilants should not be used (Rutala, 1996). As mentioned above, pre-disinfection cleaning is necessary, to reduce the biological burden and to remove organic matter (e.g., blood) that can partly inactivate disinfectants. Only a few industrial products offer cleaning and disinfection in one. Furthermore, disinfectants must be applied in the right concentration and the prescribed contact time must be used (WIP, 2002 Module 6.4).

A chlorine-based disinfectant is typically used for disinfection of floors and other large surfaces. It is bactericidal, virucidal, tuberculocidal, and fungicidal. When chlorine reacts with proteinaceous material, such as blood, some of the chlorine combines with proteins and forms N-chloro compounds. The surface should be cleaned before the disinfectant is applied. Otherwise, a high concentration of available chlorine is required to inactivate virus in the presence of undiluted blood. Surfaces contaminated by blood or other body fluids which cannot be physically cleaned before disinfection, should be disinfected with 0.5% (5,000 ppm, i.e., 1:10 dilution) available chlorine or iodine. On the other hand, if the surface is hard and smooth and has been cleaned appropriately, 0.05% (500 ppm, i.e., 1:100) solutions are sufficient. Higher concentrations (1,000 ppm) are required to kill *Mycobacterium tuberculosis*. Once surfaces and objects have been cleaned, they must be in contact with the chlorine ("wet") for at least 5 min. This is the minimum time required to let the disinfectant take effect. After disinfection surfaces should be allowed to air dry (Rutala, 1996). *Clostridium difficile* has been associated with outbreaks of diarrhea and colitis in hospitalized adults, especially those receiving antimicrobial therapy. Although there is evidence of person-to-person transmission in the hospital as well as transmission via contaminated environmental surfaces and transiently colonized hands, control of *C. difficile* is usually achieved by proficient cleaning of environmental surfaces (McFarland *et al.*, 1989). In order to reduce surface contamination and to control outbreaks, the use of sodium hypochlorite solutions (500 and 1,600 ppm) were shown to be more effective than chlorine. Thus, in outbreak situations, sodium hypochlorite should be the disinfectant of choice in reducing the levels of environmental contamination with *C. difficile* (Rutala and Weber, 1997).

Viruses can be transmitted from environmental surfaces either directly to mucous membranes or from surface-to-finger-to-mucous membranes. There may be discontinuous phases of infections between hospitals and the community involving environmental surfaces (Griffith *et al.*, 2000; Rheinbaben *et al.*, 2000; Rutala and Weber, 2000). Apart from good hand hygiene, Ward *et al.* (1991) showed, in an experimental study, that the use of disinfectants is an efficient method of inhibiting the transmission of rotavirus to human subjects. Sattar *et al.* (1993) determined that chlorine, phenolic, and phenol/ethanol products prevented rotavirus transmission from stainless steel disks to fingerpads. Infection occurred in 63% to 100% of volunteers who licked rotavirus-contaminated fingers/surfaces, but no volunteers became infected after licking contaminated surfaces that had been disinfected with the phenolic/ethanol spray. Hypochlorite was shown to be effective in controling outbreaks with coronavirus (1,000 ppm, 1 min), human parainfluenza virus (1,000 ppm, 1 min), coxsackie B virus and adenovirus type 5 (5,000 ppm, 1 min), rotavirus (800 ppm, 10 min), hepatitis A virus (5,000 ppm, 1 min), and rhinovirus type 14 (800 ppm, 10 min) (Abad *et al.*, 1997; Muto *et al.*, 2003).

Also phenolics and quaternary ammonium can be used to disinfect environmental surfaces. Sodium hypochlorite has the additional advantage of being considerably less expensive than commercially available phenolic and quaternary disinfectants. Also, cleaning with sodium hypochlorite solutions may exacerbate respiratory disorders in some patients. When sodium hypochlorite is inappropriate, a phenolic or quaternary ammonium compound may be the preferred alternative. The phenolic detergents are tuberculocidal, fungicidal, virucidal, and bactericidal at their recommended use-dilution. The quaternary ammonium compounds sold as hospital disinfectants are generally fungicidal, bactericidal, and virucidal against lipophilic viruses (HBV, HCV, HIV, HSV-1); they are not sporocidal and generally not tuberculocidal or virucidal against hydrophilic viruses (Rutala, 1996).

For disinfection of smaller surfaces and materials, 70% alcohol can be used. Engelenburg *et al.* (2002) showed that a high concentration alcohol mixture (80% ethanol and 5% isopropanol) has a high virucidal potential in particular for the blood-borne lipid-enveloped viruses HIV, HBV, and HCV. But they are flammable, evaporate quickly, and are not appropriate for large surfaces cleaning (Rutala, 1996).

Also, for fomites (beds, bed linens, tray tables, etc.) in the patient room, cleaning and, if necessary (i.e., blood and/or other body fluids are detected or contamination with multiresistant bacteria), disinfection should be done regularly. The fomites are cleaned after the patient is discharged or in the event of visible soiling. If the fomites are used by the same patient for a long period of time, they must be cleaned at least once every 4 weeks. Shortstay beds and the other fomites for outpatients' treatment are cleaned after each use (WIP, 2002 Module 33.1).

2.2. Susceptibility to disinfectants

The multiple antibiotic-resistant Gram-negative bacilli, MRSA, and VRE have become established as a major problem in many hospitals around the world (Alfieri *et al.*, 1999; Aygun *et al.*, 2002; Berg *et al.*, 2000; Boyce, 1990; Burd *et al.*, 2003; Dijk *et al.*, 2002; Engelhart *et al.*, 2002; Getchell-White *et al.*, 1989; Mongkolrattanothai *et al.*, 2003; Timmers *et al.*, 2002). Repeated exposure of hospital pathogens to antibiotics can lead to resistance, and also a similarly intensive exposure to antiseptics and disinfectants might result in a possible resistance to these agents.

Russell *et al.* (1998) have developed stable chlorhexidine resistance in some strains of *Pseudomonas stutzeri* by exposure to increasing concentrations of the bisbiguanide. The chlorhexidine-resistant strains showed a variable increase in resistance to quaternary ammonium compounds and to trichlosan. Additionally, these chlorhexidine-resistant strains demonstrated a variable increase in resistance to polymyxin B, gentamicin, nalidixic acid, erythromycin, and ampicillin. They concluded that concomitant antibiotic and antiseptic resistance in Gramnegative bacteria may occur. Thomas *et al.* (2000) reported a stable increase in chlorhexidine MICs for *P. aeruginosa* after exposure to subinhibitory concentrations simulating residual levels of this antiseptic in the environment. Increases in resistance may result from exposure of microorganisms to sublethal doses of disinfectants. The high organic load or bioburden protects the bacteria and requires higher concentrations of the disinfectant to reach the disinfecting efficacy on the predominant microflora (Gebel *et al.*, 2002).

However, Martro *et al.* (in press) assessed the bactericidal activity of several antiseptics and disinfectants on *Acinetobacter baumannii* strains obtained from clinical and environmental specimens in an intensive care unit during an outbreak. And observed neither evidence of development of resistance to biocides over time, nor a correlation between resistance to antibiotics and a decreased susceptibility to antiseptics or disinfectants.

The studies about activity of disinfectants against VRE, found no difference in the *in vitro* susceptibility of VRE and vancomycin-susceptible enterococci to standard hospital disinfectants (Block *et al.*, 2000; Saurina *et al.*, 1997). Also, Rutala *et al.* (1997) conducted a study to evaluate the susceptibility of antibiotic-susceptible and antibiotic-resistant hospital bacteria and did not find a correlation between antibiotic resistance and resistance to disinfectants. So, routine disinfection methods do not need to be altered for resistant bacteria.

REFERENCES

Abad, X. F., Pinto, R. M., and Bosch, A., 1997, Disinfection of human enteric viruses on fomites. *FEMS Microbiol. Lett.*, **156**, 107–111.

Alfieri, N., Ramotar, K., Armstrong, P. *et al.*, 1999, Two consecutive outbreaks of *Stenotrophomonas maltophilia* (*Xanthomonas maltophilia*) in an intensive-care unit defined by restriction fragment-length polymorphism typing. *Infect. Control Hosp. Epidemiol.*, **20**, 553–556.

Ansari, S. A., Sattar, S. A., Springthorpe, V. S., Wells, G. A., and Tostowaryk, W., 1988, Rotavirus survival on human hands and transfer of infectious virus to animate and non-porous inanimate surfaces. *J. Clin. Microbiol.*, **26**, 1513–1518.

Armstrong-Evans, M., Litt, M., McArthur, M. A. *et al.*, 1999, Control of transmission of vancomycin-resistant *Enterococcus faecium* in a long-term-care facility. *Infect. Control Hosp. Epidemiol.*, **20**, 312–317.

Aygun, G., Demirkiran, O., Utku, T. *et al.*, 2002, Environmental contamination during a carbapenem-resistant *Acinetobacter baumannii* outbreak in an intensive care unit. *J. Hosp. Infect.*, **52**, 259–262.

Ayliffe, G. A. J., Babb, J. R., Taylor, L. J., 1990, The hospital environment. In *Hospital Acquired Infection*. Bodmin, Cornwall, UK: MPG Books Ltd, pp. 109–120.

Ballemans, C. A. J. M., Blok, H. E. M., Swennenhuis, J., Troelstra, A., and Mascini, E. M., 2003, Dry cleaning or wet mopping: Comparison of bacterial colony counts in the hospital environment. *J. Hosp. Infect.*, **53**, 150–152.

Berg, R. W. A., Claahsen, H. L., Niessen, M., Muytjens, H. L., Liem, K., and Voss, A., 2000, *Enterobacter cloacae* outbreak in the NICU related to disinfected thermometers. *J. Hosp. Infect.*, **45**, 29–34.

Bonten, M. J. M., Hayden, M. K., Nathan, C. *et al.*, 1996, Epidemiology of colonisation of patients and environment with vancomycin-resistant enterococci. *Lancet*, **348**, 1615–1619.

Block, C., Robenshtok, E., Simhon, A., and Shapiro, M., 2000, Evaluation of chlorhexidine and povidone iodine activity against methicillin-resistant *Staphylococcus aureus* and vancomycin-resistant *Enterococcus faecalis* using a surface test. *J. Hosp. Infect.*, **46**, 147–152.

Bolyard, E. A., Tablan, O. C., Williams, W. W., Pearson, M. L., Shapiro, C. N., Deltchman, S. D., and The Hospital Infection Control Practices Advisory Committee, 1998, Guideline for infection control in health care personnel, 1998. http://www.cdc.gov/ncidod/hip/guide/InfectControl98.pdf.

Boyce, J. M., 1990, Increasing prevalence of methicillin-resistant *Staphylococcus aureus* in the United States. *Infect. Control Hosp. Epidemiol.*, **11**, 639–642.

Boyce, J. M., 1995, Vancomycin-resistant enterococci: Pervasive and persistent pathogens. *Infect. Control Hosp. Epidemiol.*, **16**, 676–679.

Boyce, J. M., Mermel, L. A., Zervos, M. J. *et al.*, 1995, Controlling vancomycin-resistant enterococci. *Infect. Control Hosp. Epidemiol.*, **16**, 634–637.

Boyce, J. M., Potter-Bynoe, G., Chenevert, C., and King, T., 1997, Environmental contamination due to methicillin-resistant *Staphylococcus aureus*: Possible infection control implications. *Infect. Control Hosp. Epidemiol.*, **18**, 622–627.

Brady, M. T., Evans, J., and Cuartas, J., 1990, Survival and disinfection of parainfluenza viruses on environmental surfaces. *Am. J. Infect. Control*, **18**, 18–23.

Burd, M., Humphreys, H., Glynn, G. *et al.*, 2003, Control and the prevention of methicillin-resistant *Staphylococcus aureus* in hospitals in Ireland: North/south study of MRSA in Ireland 1999. *J. Hosp. Infect.*, **53**, 297–303.

Byers, K. E., Durbin, L. J., Simonton, B. M., Anglim, A. M., Adal, K. A., Farr, B. M., 1998, Disinfection of hospital rooms contaminated with vancomycin-resistant *Enterococcus faecium*. *Infect. Control Hosp. Epidemiol.*, **19**, 261–264.

Byers, K. E., Simonton, R. M., Anglim, A. M., Adal, K. A., and Farr, B. M., 1995, Environmental contamination with vancomycin-resistant *Enterococcus faecium* (VRE). In The Fifth Annual Meeting of the Society for Healthcare Epidemiology of America; April 2–4, 1995; San Diego, CA. *Infect. Control Hosp. Epidemiol.*, **16**(Suppl.), 18, abstract 17.

Calfee, D. P., Giannetta, E. T., Durbin, L. J., Germanson, T. P., Farr, B. M., 2003, Control of endemic vancomycin-resistant *Enterococcus* among inpatients at a university hospital. *Clin. Infect. Dis.*, **37**, 326–332.

CDC, 1977, Centers for Disease Control and Prevention. Control measures for hepatitis B in dialysis centers. Available at: www.cdc.gov/ncidod/hip/control.htm.

CDC, 1993, Centers for Disease Control. Nosocomial enterococci resistant to vancomycin-United States, 1989–1993. *MMWR Morb. Mortal Wkly. Rep.*, **42**, 597–599.

CDC, 2001, Centers for Disease Control and Prevention. Recommendations for preventing transmission of infections among chronic hemodialysis patients. *MMWR Morb. Mortal Wkly. Rep.*, **50**, 1–43.

CDC, 2003a, Centers for Disease Control and Prevention. Transmission of hepatitis B and C viruses in outpatient settings-New York, Oklahoma, and Nebraska, 2000–2002. *MMWR Morb. Mortal Wkly. Rep.*, **52**(38), 901–906.

CDC, 2003b, Centers for Disease Control and Prevention. Prevention and control of infections with hepatitis viruses in correctional settings. *MMWR Morb. Mortal Wkly. Rep.*, **52**, RR-1.

CDC, 2003c, Centers for Disease Control and Prevention. Guidelines for environmental infection control in health-care facilities, 2003. Available at: http://www.cdc.gov/ncidod/hip/enviro/guide.htm.

Danforth, D., Nicolle, L. E., Hume, K., Alfieri, N., and Sims, H., 1987, Nosocomial infections on nursing units with floors cleaned with a disinfectant compared with detergent. *J. Hosp. Infect.*, **10**, 229–235.

Dembry, L. M., Farrel, P. A., Barrett, C., Bell, A., and Hierholzer, W. J., 1995, Vancomycin-resistant enterococci and contamination of the environment. In The Fifth Annual Meeting of the Society for Healthcare Epidemiology of America; April 2–4, 1995; San Diego, CA. *Infect. Control Hosp. Epidemiol.*, **17**(suppl.), 18, abstract 16.

Dharan, S., Mourouga, P., Copin, P., Bessmer, G., Tschanz, B., and Pittet, D., 1999, Routine disinfection of patients' environmental surfaces. Myth or reality? *J. Hosp. Infect.*, **42**, 113–117.

Dijk, Y. V., Bik, E. M., Hochstenbach-Vernooij, S. *et al.*, 2002, Management of an outbreak of *Enterobacter cloacae* in a neonatal unit using simple preventive measures. *J. Hosp. Infect.*, **51**, 21–26.

Engelenburg, F. A. C., Terpstra, F. G., Schuitemaker, H., and Moorer, W. R., 2002, The virucidal spectrum of a high concentration alcohol mixture. *J. Hosp. Infect.*, **51**, 121–125.

Engelhart, S., Krizek, L., Glasmacher, A., Fischnaller, E., Marklein, G., and Exner, M., 2002, *Pseudomonas aeruginosa* outbreak in a haemotology-oncology unit associated with contaminated surface cleaning equipment. *J. Hosp. Infect.*, **52**, 93–98.

Evans, M. R., Meldrum, R., Lane, W. *et al.*, 2002, An outbreak of viral gastroenteritis following environmental contamination at a concert hall. *Epidemiol. Infect.*, **129**, 355–360.

Gebel, J., Sonntag, H. G., Werner, H. P., Vacata, V., Exner, M., and Kistemann, T., 2002, The higher disinfectant resistance of nosocomial isolates of *Klebsiella oxytoca*: How reliable are indicator organisms in disinfectant testing?. *J. Hosp. Infect.*, **50**, 309–311.

Getchell-White, S. I., Donowitz, L. G., and Groschel, D. H., 1989, The inanimate environment of an intensive care unit as a potential source of nosocomial bacteria: Evidence for long survival of *Acinetobacter calcoaceticus*. *Infect. Control Hosp. Epidemiol.*, **10**, 402–407.

Griffith, C. J., Cooper, R. A., Gilmore, J., Davies, C., and Lewis, M., 2000, An evaluation of hospital cleaning regimens and standards. *J. Hosp. Infect.*, **45**, 19–28.

Gwaltney, J. M., Jr and Hendley, J. O., 1982, Transmission of experimental rhinovirus infection by contaminated surfaces. *Am. J. Epidemiol.*, **116**, 828–833.

Hails, J., Kwaku, F., Wilson, A. P., Bellingan, G., and Singer, M., 2003, Large variation in MRSA policies, procedures and prevalence in English intensive care units: A questionnaire analysis. *Intensive Care Med.*, **29**, 481–483.

Hall, C. B., Douglas, G., and Geiman, J. M., 1980, Possible transmission by fomites of respiratory syncytial virus. *J. Infect. Dis.*, **141**, 98–102.

Hargreaves, J., Shireley, L., Hansen, S. *et al.*, 2001, Bacterial contamination associated with electronic faucets: A new risk for healthcare facilities. *Infect. Control Hosp. Epidemiol.*, **22**, 202–205.

Hayden, M. K., 2000, Insights into the epidemiology and control of infection with vancomycin-resistant enterococci. *Clin. Infect. Dis.*, **31**, 1058–1065.

Hendley, J. O., Wenzel, R. P., and Gwaltney, J. M., 1973, Transmission of rhinovirus colds by self-inoculation. *N. Engl. J. Med.*, **288**, 1361–1364.

Karanfil, L. V., Murphy, M., Josephson, A. *et al.*, 1992, A cluster of vancomycin-resistant *Enterococcus faecium* in an intensive care unit. *Infect. Control Hosp. Epidemiol.*, **13**, 195–200.

Kawai, H. and Feinstone, S. M., 2000, Acute viral hepatitis. In G. L. Mandell, J. E. Bennett, and R. Dolin (eds.), *Principles and Practice of Infectious Diseases*. Philadelphia: Churchill Livingstone; pp. 1279–1297.

Leoni, E., Bevini, C., Degli Esposti, S., and Graziano, A., 1998, An outbreak of intrafamiliar hepatitis A associated with clams epidemic transmission to a school community. *Eur. J. Epidemiol.*, **14**, 187–192.

Lewis, D. L., Arens, M., Appleton, S. S. *et al.*, 1992, Cross-contamination potential with dental equipment. *Lancet*, **340**, 1252–1254.

Lloyd-Evans, N., Springthorpe, V. S., and Sattar, S. A., 1986, Chemical disinfection of human rotavirus-contaminated inanimate. *J. Hyg.*, **97**, 163–173.

Love, S. S., Jiang, X., Barrett, E., Farkas, T., and Kelly, S., 2002, A large hotel outbreak of Norwalk-like virus gastroenteritis among three groups of guests and hotel employees in Virginia. *Epidemiol.Infect.*, **129**, 127–132.

Manangan, L. P., Pugliese, G., Jackson, M., Lynch, P., Sohn, A. H., Sinkowitz-Cochran, R. L., *et al.*, 2001, Infection control dogma: 10 suspects. *Infect. Control Hosp. Epidemiol.*, **22**, 243–247.

Martro, E., Hernandez, A., Ariza, J. *et al.*, (In press). Assessment of *Acinetobacter baumannii* susceptibility to antiseptics and disinfectants. *J. Hosp. Infect.*

Mayer, R. A., Geha, R. C., Helfand, M. S., Hoyen, C. K., Salata, R. A., and Donskey, C. J., 2003, Role of fecal incontinence in contamination of the environment vancomycin-resistant enterococci. *Am. J. Infect. Control*, **31**, 221–225.

Mbithi, J. N., Springthorpe, V. S., Boulet, J. R., and Sattar, S. A., 1992, Survival of hepatitis A virus on human hands and its transfer on contact with animate and inanimate surfaces. *J. Clin. Microbiol.*, **30**, 757–763.

McFarland, L. V., Mulligan, M. E., Kwok, R. Y. Y., and Stamm, W. E., 1989, Nosocomial acquisition of *Clostridium difficile* infection. *N. Engl. J. Med.*, **320**, 204–210.

Mongkolrattanothai, K., Boyle, S., Kahana, M. D., and Daum, R. S., 2003, Severe *Staphylococcus aureus* infections caused by clonally related community-acquired methicillin-susceptible and methicillin-resistant isolates. *Clin. Infect. Dis.*, **37**, 1050–1058.

Montelcalvo, M. A., Jarvis, W. R., Uman, J. *et al.*, 1999, Infection-control measures reduce transmission of vancomycin-resistant enterococci in an endemic setting. *Ann. Intern. Med.*, **131**, 269–272.

Muto, C. A., Jernigan, J. A., Ostrowsky, B. E. *et al.*, 2003, SHEA guideline for preventing nosocomial transmission of multidrug-resistant strains of *Staphylococcus aureus* and *Enterococcus*. *Infect. Control Hosp. Epidemiol.*, **24**, 362–386.

Neely, A. N. and Maley, M. P., 2000, Survival of enterococci and staphylococci on hospital fabrics and plastic. *J. Clin. Microbiol.*, **38**, 724–726.

Noskin, G. A., Stosor, V., Cooper, I., and Peterson, L. R., 1995, Recovery of vancomycin-resistant enterococci on fingertrips and environmental surfaces. *Infect. Control Hosp. Epidemiol.*, **16**, 577–581.

O'Connell, N. H. and Humphreys, H., 2000, Intensive care unit design and environmental factors in the acquisition of infection. *J. Hosp. Infect.*, **45**, 255–262.

Porwancher, R., Sheth, A., Remphrey, S., Taylor, E., Hinkle, C., and Zervos, M., 1997, Epidemiological study of hospital-acquired infection with vancomycin-resistant

Enterococcus faecium: possible transmission by an electronic ear-probe thermometer. *Infect. Control Hosp. Epidemiol.*, **18**, 771–774.

Rheinbaben, F. v., Schunemann, S., GroB, T., and Wolff, M. H., 2000, Transmission of viruses via contact in a hosehold setting: Experiments using bacteriophage φX174 as a model virus. *J. Hosp. Infect.*, **46**, 61–66.

Russell, A. D., Tattawasart, U., Maillard, J. Y., and Furr, J. R., 1998, Possible link between bacterial resistance and use of antibiotics and biocides. *Antimicrob. Agents Chemother.*, **42**, 2151.

Rutala, W. A., 1996, Disinfection and sterilization. In C.G. Mayhall (ed.), *Hospital epidemiology and infection control*. Baltimore: Williams & Wilkins, pp. 913–936.

Rutala, W. A., Katz, E. B. S., Sherertz, R. J., and Sarubbi, F. A., 1983, Environmental study of a methicillin-resistant *Staphylococcus aureus* epidemic in a burn unit. *J. Clin. Microbiol.*, **18**, 683–688.

Rutala, W. A., Stiegel, M. M., Sarubbi, F. A., and David, J. W., 1997, Susceptibility of antibiotic-susceptible and antibiotic-resistant hospital bacteria to disinfectants. *Infect. Control Hosp. Epidemiol.*, **18**, 417–421.

Rutala, W. A. and Weber, D. J., 1997, Uses of inorganic hypochlorite (bleach) in health-care facilities. *Clin. Microbiol. Rev.*, **10**, 597–610.

Rutala, W. A. and Weber, D. J., 2000, Surface disinfection: Should we do it? *J. Hosp. Infect.*, **48**(Suppl. A), S64–S68.

Sample, M. L., Gravel, D., Oxley, C., Toye, B., Garber, G., and Ramotar, K., 2002, An outbreak of vancomycin-resistant enterococci in a hematology-oncology unit: Control by patient cohorting and terminal cleaning of the environment. *Infect. Control Hosp. Epidemiol.*, **23**, 468–470.

Sattar, S. A., Jacopsen, H., Springthorpe, S., Cusack, T. M., and Rubino, J. R., 1993, Chemical disinfection to interrupt transfer of rhinovirus type 14 from environmental surfaces to hands. *Appl. Environ. Microbiol.*, **59**, 1579–1585.

Sattar, S. A., Tetro, J., Springthorpe, V. S., and Giulivi, A., 2001, Preventing the spread of hepatitis B and C viruses: Where are germicides relevant?. *Am. J. Infect. Control*, **29**, 187–197.

Saurina, G., Landman, D., and Quale, J. M., 1997, Activity of disinfectants against vancomycin-resistant *Enterococcus faecium*. *Infect. Control Hosp. Epidemiol.*, **18**, 345–347.

Shaikh, Z. H. A., Osting, C. A., Hanna, H. A., Arbuckle, R. B., Tarrand, J. J., and Raad, I. I., 2002, Effectiveness of a multifaceted infection control policy in reducing vancomycin usage and vancomycin-resistant enterococci at a tertiary care cancer centre. *J. Hosp. Infect.*, **51**, 52–58.

Sizun, J., Yu, M. W. N., and Talbot, P. J., 2000, Survival of homan coronaviruses 229E and OC43 in suspension and after drying on surfaces: A possible source of hospital acquired infections. *J. Hosp. Infect.*, **46**, 55–60.

Smith, T. L., Iwen, P. C., Olson, S. B., and Rupp, M. E., 1998, Environmental contamination with vancomycin-resistant enterococci in an outpatient setting. *Infect. Control Hosp. Epidemiol.*, **19**, 515–518.

Talon, D., 1999, The role of the hospital environment in the epidemiology of multi-resistant bacteria. *J. Hosp. Infect.*, **43**, 13–17.

Thomas, L., Maillard, J. Y., Lambert, R. J. *et al.*, 2000, Development of resistance to chlorhexidine diacetate in *Pseudomonas aeruginosa* and the effect of a residual concentration. *J. Hosp. Infect.*, **46**, 297–303.

Timmers, G. J., Zwet, W. C., Simoons-Smit, I. M. *et al.*, 2002, Outbreak of vancomycin-resistant *Enterococcus faecium* in a haemotology unit: Risk factor assessment and successful control of the epidemic. *Br. J. Haematol.*, **116**, 826–833.

Ward, R. L., Bernstein, D. I., Knowlton, D. R. *et al.*, 1991, Prevention of surface-to-human transmission of rotaviruses by treatment with disinfectant spray. *J. Clin. Microbiol.*, **29**, 1991–1996.

Weber, D. J. and Rutala, W. A., 1993, Environmental issues and nosocomial infections. In R. P. Wenzel (ed.), *Prevention and Control of Nosocomial infections.* Williams & Wilkins, Baltimore, MD: pp. 420–449.

Weber, D. J. and Rutala, W. A., 1997, Role of environmental contamination in the transmission of vancomycin-resistant enterococci. *Infect. Control Hosp. Epidemiol.*, **18**, 306–309.

Weber, D. J. and Rutala, W. A., 2001, The emerging nosocomial pathogens *Cryptosporidium*, *Esherichia coli 0157:H7*, *Helicobacter pylori*, and hepatitis C: Epidemiology, environmental survival, efficacy of disinfection, and control measures. *Infect. Control Hosp. Epidemiol.*, **22**, 306–315.

Weniger, B. G., Ruttenber, A. J., Goodman, R. A., Juranek, D. D., and Wahlquist, S., 1983, Fecal coliforms on environmental surfaces in two day care center. *Appl. Environ. Microbiol.*, **45**, 733–735.

Werry, C., Lawrence, J. M., and Sanderson, P., 1988, Contamination of detergent cleaning solutions during hospital cleaning. *J. Hosp. Infect.*, **11**, 44–49.

WIP, 2002, Module 6.1. Decontamination policy. Dutch Workingparty Infection Prevention (WIP). November 2002. www.wip.nl.

WIP, 2002, Module 6.4. Decontamination. Cleaning and disinfection of rooms, furniture and objects. Dutch Workingparty Infection Prevention (WIP). December 2002. www.wip.nl/UK.

WIP, 2002, Module 33.1. Beds and accessories. Dutch Workingparty Infection Prevention (WIP). November 2002. www.wip.nl/UK.

Chapter 20

The Evolution of Antibiotic Resistance within Patients

Ian R. Booth
Department of Molecular and Cell Biology, Institute of Medical Sciences, University of Aberdeen, Foresterhill, Aberdeen AB25 2ZD, UK

1.   INTRODUCTION

Antibiotic resistance is a major clinical problem that affects both hospital and community populations. The community problems arise mainly from the carriage of resistant organisms that while basically quiescent can alter and initiate disease. In hospitals, the major challenge is the development of multiply resistant organisms in long-stay and severely immunocompromised patients. It is well established that many infections, in otherwise healthy patients, involve penetration of the mucosae by a small group of related organisms that then proliferate (Moxon *et al.*, 1994) (Figure 1). Most individuals cope with this continuous challenge via the immune system, but problems arise either from very vigorous infections or when the immune system is partially disabled (Levin and Antia, 2001). For the majority of infections, antibiotic treatment is a sufficient aid to the immune system to accelerate bacterial clearance. However, when bacteria either occupy or create niches that are difficult to reach with either antibiotics or the immune system, then the potential for the explosive development of high levels of resistance is considerable. This chapter supplies a personal perspective on the current understanding of the issues associated with the development of antibiotic resistance. It is not comprehensive but there are now many excellent reviews in this area that provide coverage of this important topic (Canton *et al.*, 2003; Chopra *et al.*, 2003; Drlica, 2003; Livermore, 2003; Martinez and Baquero,

Antibiotic Policies: Theory and Practice. Edited by Gould and van der Meer
Kluwer Academic / Plenum Publishers, New York, 2005

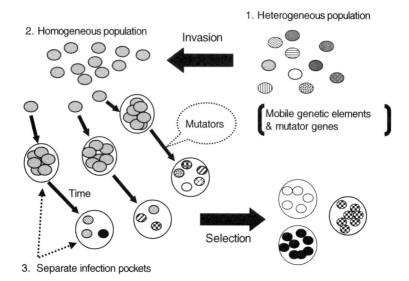

Figure 1. Generation of diversity among microorganisms. A few organisms from a hetero-geneous group infect the patient and are either cleared from the system or find protected niches where they can evolve. The treatment with antibiotics (or interaction with the immune system) establishes the potential for selection of variants. The outcome of each mutagenic/selection regime may differ depending upon the properties of that niche with regard to accessibility to antibiotics, the immune system, and other organisms. The presence of muta-tor organisms and of *strangers* carrying drug resistance elements can accelerate and/or change the direction of evolution.

2002; Ruiz, 2003; Whittle *et al.*, 2002). Rather, I have incorporated papers that report experiments or philosophies that I consider to be illuminating.

1.1.   Acquisition of resistance

Antibiotic resistance can arise by the acquisition of resistance determinants from other organisms by horizontal gene transfer or by the occurrence of muta-tions that either affect the structural genes for target proteins in the cell or alter the rates of penetration or expulsion of the antibiotic. Some mechanisms might be considered hybrids, such as the evolution of β-lactamases that has occurred to create the TEM, SHV, and ESBL categories of enzymes—the initial acquisition of the gene encoding the enzyme is followed by the accumulation of mutations that confer new properties. Changes in the fidelity of DNA replication and repair can potentiate the acquisition of new drug resistance (see Section 4, Mutators), both by enhancing the rate of mutation and by allowing more frequent gene transfer events between species that are normally maintained separate

(Chopra *et al.*, 2003). The degree to which acquired resistance is a significant event in patients depends largely upon the degree of isolation of the evolving organism from other microorganisms. In isolated niches, such as deep seated cysts in the liver (Low *et al.*, 2001), the opportunities for gene exchange with "*strangers*" would be rare, whereas for a uropathogenic organism, the opportunities for illicit gene exchange might be more frequent. It is clear, therefore, that the study of the development of antibiotic resistance in patients is as interesting as a purely scientific problem as it is important as a clinical one.

2. ANTIBIOTIC CONCENTRATIONS AT TARGET SITES

It is well recognised that the efficiency of antibiotics *in vivo* is determined by their pharmacokinetics and pharmacodynamics (Wise, 2003). The former describes relatively simple processes of absorption, distribution, metabolism, and excretion of the antibiotic, which are properties of the molecules themselves. Resistance selection is favoured among the microbial community if the drug does not reach the site of infection at sufficiently high concentrations to kill all of the microorganisms. Variable penetration of tissues will lead to heterogeneity in the antibiotic concentration in different body compartments leading to eradication of organisms at one site but promoting resistance development at another. One of the significant attributes of fluoroquinolones is considered to be the rapidity with which they are absorbed and distributed around the body—but even these antibiotics show low uptake into cerebrospinal fluid and fatty tissues (Wise, 2003). For β-lactams, a more restricted distribution may negatively impact their efficiency. Purulent fluid sampled from infected liver cysts in a patient on intravenous meropenem therapy contained only 1.2–5 mg/L compared with serum levels of 91.7 mg/L 30 min prior to surgery (Low *et al.*, 2001). This may represent an extreme example, but similar ratios have been recorded for salivary concentrations of most β-lactams and sulfamethoxazole (Soriano and Rodriguez-Cerrato, 2002). Most antibiotics shared these characteristics with only the weakly basic antibiotics being concentrated to high levels (>40% of serum concentration), for example, ciprofloxacin and trimethoprim. This study (Soriano and Rodriguez-Cerrato, 2002) sought to correlate the pharmacokinetic properties of different antibiotics with their impact on the nasopharyngeal flora since the mucosa is coated with saliva. With the exception of antibiotics that are known to have low *in vitro* activity on the flora (e.g., ciprofloxacin), there was a reasonable correlation between salivary concentration and impact on the flora (Soriano and Rodriguez-Cerrato, 2002). However, in terms of either reducing the carriage of resistant streptococci or the development of resistance, there was a tendency for β-lactams to eliminate only the sensitive strains creating the potential for the expansion of the resistant and intermediate flora. It was considered that the low salivary concentration of some of these antibiotics, coupled with their low to moderate effectiveness *in vitro*, was a significant factor in determining their effectiveness. Antibiotics with a high *in vitro* activity proved more successful (Soriano and Rodriguez-Cerrato, 2002). The compounding problem is that selection for non-susceptible strains can lead to cross-resistance. In the ARISE project (Soriano and Rodriguez-Cerrato, 2002), it was observed that elimination of 67% of the *Streptococcus pneumoniae* strains by treatment with penicillin (0.06 mg/L) led to an increase in non-susceptible strains but also doubled the incidence of clarithromycin-non-susceptible organisms. A similar finding was made for treatment with clarithromycin. These observations point to the importance of the antibiotic reaching the effective concentration at the target tissue in order to prevent the development, or enrichment, of resistant strains.

2.1. Mutant prevention concentration and mutant selection window

Pharmacodynamics deals with the relationship between antibiotic concentration and bactericidal effects and post-antibiotic effects. Such relationships can only be examined using either animal or *in vitro* models or by collation of data obtained with human subjects. Wise (2003) has pointed out that the MIC is the major *in vitro* parameter studied with clinical isolates, but because this analyses bacterial growth at a fixed concentration, it cannot reflect *in vivo* activity very closely because here the concentration of the antibiotic varies with time. A number of different pharmacodynamic parameters have been defined that relate the persistence of the antibiotic in the serum at concentrations above the MIC to the efficacy of the drug (Hyatt *et al.*, 1995; Schentag *et al.*, 2001). Which of these parameters is the most important depends on the antibiotic class being studied (Hyatt *et al.*, 1995). For β-lactams, oxazolidinones, and macrolides, it has been suggested that the serum concentration needs to be maintained above the MIC for an effective treatment (Wise, 2003). In contrast to the aminoglycosides and fluoroquinolones, the ratio of maximum serum concentration to MIC is considered to more important as a predictor of efficacy of treatment.

It is generally held that if the concentration of the antibiotic at the site of infection exceeds the MIC for the microorganism, then resistance can be avoided. This led Drlica and colleagues to propose the *mutant prevention concentration* (MPC) for an antibiotic (Dong *et al.*, 2000; Drlica, 2003; Sindelar *et al.*, 2000). Based on their analysis of fluoroquinolone potency against *Mycobacterium*, they argued that the selection of mutations could be avoided if the antibiotic concentration could be maintained above one that prevented the growth of cells that had acquired the first step mutation required for increasing the MIC (Dong *et al.*, 2000). The MPC definition was arrived at after *in vitro* studies of resistance development and considerations of extant clinical data and defines the MPC as *the minimum concentration that allows no mutant recovery when more than 10^{10} cells are applied to drug-containing agar* (Dong *et al.*, 2000; Drlica, 2003). Many current antibiotics cannot achieve serum concentrations that are likely to exceed, or even approach, the MPC (Dong *et al.*, 2000), suggesting that the selection of resistant clones will remain a significant issue for the immediate future. However, the definition of MPC does give a basis for the design of antibiotics to reduce the frequency with which mutations to resistance will arise.

The proposal of the MPC was developed from an analysis of the rise in resistance to fluoroquinolones in mycobacteria. These observations fit with other data on the pharmacodynamics of antibiotics, but the concept has been superceded by the *mutant selection window*, which is the idea that there is a range of concentrations of the antibiotic over which more resistant mutants are most frequently selected (Drlica, 2003). As the concentration of antibiotic increased, the sensitive flora died, but then a plateau in the kill was observed due to the presence within the population of resistant mutants. Progressively greater concentrations lead to a further decline in the numbers of survivors as the resistance is overcome. The selection window is that range of concentrations over which survival is essentially constant and represents the interval between the lower boundary, at which concentration-sensitive organisms are killed, and the upper boundary, which represents the concentrations that kill cells that possess only a single mutation conferring resistance (Drlica, 2003). Wise (2003) documents a number of cases where high doses of antibiotics given for short periods resulted in eradication of the pathogen in a high percentage of patients and prevented the appearance of resistance. Thus, the pharmacokinetics and pharmacodynamics of antibiotics offer opportunities for limiting the development of resistance in patients.

3. SELECTIVE PRESSURE

The ideal environment for the selection of antibiotic-resistant bacteria is one in which the concentration of the antibiotic is too low to kill all the organisms but high enough to present an opportunity for a slightly more resistant organism to outgrow the rest of the population. Over the last 40 years, since the introduction of ampicillin, more than 60 variants of the TEM-1 β-lactamase have been characterised. Genetic experiments have shown that many more variants can be isolated in the laboratory, but that selection in the patient involves more complex parameters than simply the maximum activity against a particular β-lactam (Blazquez *et al.*, 2000). Similar observations have been made for ciprofloxacin resistance determinants (Ruiz, 2003). In particular the "patient-environment" represents a spatially- and temporally-variable context within which organisms must evolve to survive. The reason for the observation of a narrower spectrum of mutations in the patient isolates than in the laboratory represents the more varied challenges that are posed in the former, particularly with respect to variations in antibiotic exposure. A patient who is failing treatment with one β-lactam will be rapidly switched to another—the microorganism that specialises in survival of the first antimicrobial at the expense of dealing with a related molecule will be short-lived!

3.1.   The principle that not all change is good

A demonstration of this principle was achieved through the analysis of the evolution of TEM-1 β-lactamase *in vitro* (Figure 2). *Escherichia coli* cells

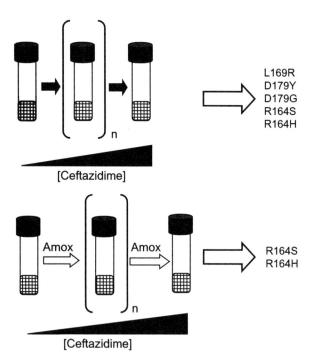

Figure 2. Fluctuating β-lactam pressure alters the pattern of evolution. Blazquez and colleagues (2000) devised an experiment in which cells were serially sub-cultured at increasing concentrations of ceftazidime and in a parallel experiment the cells were also subjected to an overnight incubation with a fixed concentration of amoxicillin. The outcome of the evolution of a TEM-1 β-lactamase was investigated by sequencing the resulting plasmids when the cultures had achieved growth in the presence of 32 mg/L ceftazidime.

expressing TEM-1 were subjected to serial passage through medium containing increasing, doubling concentrations of ceftazidime until growth occurred at 32 mg/L. In parallel, a second series of similar serial passages on ceftazidime was conducted but an overnight challenge with amoxicillin was performed between each growth cycle. The protein sequences of the resulting TEM enzymes were deduced by sequencing the respective genes, and the outcome of the continuous direct selection and the dual selection compared. It was found that continuous single selection selected five different mutations, three of which had not been observed in natural variants of TEM-1. All three of the unusual amino acid changes were found to diminish the MIC for amoxicillin, suggesting that they altered enzyme activity so that this antibiotic was no longer so readily hydrolysed. As expected from the MIC data for the mutants isolated by continuous single selection, only two mutant types were isolated by dual selection, both of which had been encountered previously in natural isolates. These mutations were also dominant among the isolates selected by continuous exposure to ceftazidime and also conferred higher MIC values for this antibiotic suggesting that these mutations create a better β-lactamase than did either of the rarer changes (Blazquez *et al.*, 2000).

3.2. Small concentration differences do matter

Recent work has pointed to the small differences in β-lactamase enzymes that might be sufficient to lead to selection. Negri and colleagues (Negri *et al.*, 2000) noted that many TEM variants possess multiple amino acid changes that could only arise by several rounds of mutagenesis and selection. Although mutators increase the mutation frequency by up to 10^3-fold (see below), they are unlikely to introduce several mutations into a single gene in any one generation. Thus, selection of each mutant is mandatory. It has been pointed out that TEM-12 differs from TEM-1 by a single amino acid change, which produces a small change in growth inhibition by cefotaxime: *E. coli* cells expressing TEM-1 are inhibited by 0.008 mg/L compared with 0.015 mg/L for those expressing TEM-12. TEM-12-producing strains have been selected by cefotaxime, suggesting that this small change in MIC is sufficient growth advantage to select for this mutation (Negri *et al.*, 2000). In a series of elegant experiments, it was demonstrated that cells expressing TEM-12 were selected over TEM-1 when the antibiotic concentrations were very low, but that TEM-1 was selected at higher cefotaxime concentrations. The authors concluded that this confirmed the idea of *selective windows*—a concentration range over which a particular mutation confers a specific growth advantage to the cells expressing it. The duration of the exposure of the cultures to the antibiotic, at concentrations in the selective window range, increased the selection for the TEM-12-bearing cells.

In these studies, it was noted that there was an unpredicted second, higher, concentration range in which the TEM-12 enzyme was more advantageous to cells than TEM-1. This proved to be due to the selection of an OmpF⁻ mutation that conferred the higher MIC when combined with TEM-12 rather than with TEM-1 (Negri *et al.*, 2000). The experiment was repeated in mice inoculated with a mixed population of cells expressing either TEM-1 or TEM-12, by varying the dose of cefotaxime. As predicted from the *in vitro* experiments, TEM-12 cells were selected over TEM-1 cells at low concentrations of cefotaxime but not at higher concentrations. The selective window appeared to be larger than in the *in vitro* experiments but the overall effect was similar (Negri *et al.*, 2000). These data show that apparently small differences in cell performance can be selected given the appropriate antibiotic concentration. Moreover, it demonstrates that spatial heterogeneity within the human body, giving rise to protected compartments with lowered antibiotic penetration, is likely to provide a selective environment for a range of resistance mutations.

4.   MUTATOR STRAINS

Mutator strains are those that have acquired high rates of mutagenesis due to a breakdown in the normal proofreading processes engaged during DNA replication and repair. Throughout evolution there has been a drive to maintain the fidelity of the genome sequence. Mutation is essential for evolution to take place, but the rate must be controlled to ensure the overall viability of the organism. Changes that occur naturally are minimised by repair systems that identify mismatched bases. In *E. coli*, four genetic loci (*mutS, mutL, mutH*, and *uvrD*) are associated with increased mutation rates due to loss of the mismatch repair system (Chopra *et al.*, 2003; Gross and Siegel, 1981; Matic *et al.*, 1997; Radman *et al.*, 2000). Similarly *mutD* (*dnaQ*) codes for the ε (proofreading) subunit of DNA polymerase III and ensures the fidelity of DNA replication and mutations in this locus alter the rates of mutation due to overload of the mismatch repair system (Chopra *et al.*, 2003; Oller *et al.*, 1993). Altering the metabolism of cells can also predispose them to exhibit high rates of mutagenesis. Cells that have undergone oxidative damage accumulate oxidised guanine bases, both as the free 8-oxo-dGTP and incorporated into DNA. Three gene products, MutT, MutM, and MutY eliminate these oxidised bases. A less obvious example is that of nucleotide diphosphate kinase (*ndk*), loss of which causes increases in the pools of dGTP and dCTP that in turn appears to favour A:T to G:C changes (Oller *et al.*, 1993). The mutation is synergistic with a *mutS* mutation since mismatch repair normally undoes the mutations caused by unbalanced pools of deoxynucleotide triphosphates. This study also found that not all types of mutator activity can be detected by the standard mechanism of measuring the rate of mutation to rifampicin resistance. Altered pools of dCTP/dGTP also caused duplications of repeat sequences, which would alter the reading frame of genes and cause loss of function.

It seems probable that mutation rates will rise due to changes in the environment once a pathogenic organism has entered a macrophage or another environment in which it is nutritionally disadvantaged. The impact of oxidative damage has been alluded to above. In starving colonies of *E. coli*, an increase in error-prone repair has been suggested to cause increased genetic heterogeneity (Taddei *et al.*, 1995, 1997). Recent work has shown that *Brucella* and *Salmonella* undergo modified gene expression when they enter macrophages that is consistent with them resisting the mutagenic effects of the natural electrophile methylglyoxal (MG) (Eriksson *et al.*, 2003; Eskra *et al.*, 2001; Kohler *et al.*, 2002). MG is produced by cells either in response to phosphate limitation or when the balance of carbon metabolism is perturbed leading to the accumulation of sugar phosphates (Booth *et al.*, 2003). Overproduction of this metabolite by cells is known to be mutagenic and is countered by repair mechanisms and by detoxification of MG by a glutathione-dependent glyoxalase pathway. *Salmonella* cells that have been engulfed by macrophages exhibit induction both of phosphate scavenging pathways and systems for protection against MG (Eriksson *et al.*, 2003). These data point to the potential for the intracellular environment to increase the rate of mutagenesis.

It has been known for almost 50 years that mutator bacteria exist within the natural populations, but they do so at a low frequency (approx. 1%) (LeClerc *et al.*, 1996). These observations have been repeated and extended at intervals, particularly during the investigation of the emergence of new pathogenic strains and the growth in interest in antibiotic-resistant isolates. Hypermutable *Pseudomonas aeruginosa* were isolated from a cystic fibrosis (CF) lung infection (Oliver *et al.*, 2000). From a range of perspectives, this is a very challenging environment for a bacterial cell—both natural and man-made challenges are prevalent. In contrast to isolates from acute clinical infections, those from the CF patients were found to be quite diverse in appearance and in physiology (Oliver *et al.*, 2000). Many of the CF isolates gave mucoid colonies, a property that is strongly associated with pathogenesis. It may affect adherence, resistance to antibiotics and to macrophages, quenching of oxygen radicals and hypochlorite. Mutations in the *mucA* gene, which forms part of the regulatory network governing *algD* that produces the immediate precursor of alginate, have been observed in *P. aeruginosa* isolates from CF patients. Loss of MucA leads to high levels of expression of AlgD and synthesis of alginate, which enhances survival of *P. aeruginosa* in the lung (Boucher *et al.*, 1997; Govan and Deretic, 1996).

Such considerations led to the hypothesis that mutators might be more frequent among CF patient isolates than among those from acute infections (Oliver *et al.*, 2000). Mutator isolates were obtained from 11 of 30 CF patients and 19.5% of all isolates from all the patients exhibited the mutator phenotype (scored by the acquisition of resistance to either rifampicin or streptomycin at a frequency 20 times that of the control strain PAO1; Oliver *et al.*, 2000). Each patient who had mutator clones exhibited unique *P. aeruginosa* lineages with no suggestion that there had been cross-infection. In contrast, it was observed that no mutators were found among 50 blood isolates and among 25 respiratory isolates from non-CF patients. The mutation rates were ~100 times greater in the mutators than in the non-mutator strains. In most cases, the origin of the mutator phenotype was a mutation in either *mutS*, which could be complemented by the cloned *mutS* gene or in the *mutY* gene (Oliver *et al.*, 2000). The patients from whom these strains were obtained had received several courses of different antibiotics and thus the isolates were screened for their MIC values for a range of antimicrobial agents. As expected, all of the CF isolates exhibited a tendency to be more resistant to antibiotics that non-CF isolates, but for almost all drugs, a higher frequency of resistance was observed among the CF isolates (Oliver *et al.*, 2000).

4.1.   Persistence of mutators

There is a strong prediction that mutators pose only a transient advantage while the selective pressure is strong, but should be at a disadvantage in the long run when selective pressure is relieved (Chopra *et al.*, 2003; Funchain *et al.*, 2000; Giraud *et al.*, 2001a). As a rule, significant frequencies of mutator strains have been found in clinical situations where the organism persists for extended periods and is probably subject to severe challenge from both host defences and antibiotic treatment. In the CF example, mutator bacteria survived for long periods in the patients, with similar isolates being obtained from one patient after a 4-year interval. The constantly changing environment of the lung as the patient's health deteriorates poses a continuing challenge to the colonising organisms, and consequently may act as a selective force in favour of mutators despite their potential to generate nonbeneficial mutations (Giraud *et al.*, 2001b). Recent work has identified mutators among fluoroquinolone resistant clones of *E. coli* isolated from the urinary tract (Lindgren *et al.*, 2003). The most resistant clones were found to have multiple mutations in *gyrA* and *parC* and were also found to exhibit mutation rates two orders of magnitude greater than the more sensitive clones that also possessed fewer mutations contributing to resistance (Lindgren *et al.*, 2003). Similarly, in a comparison of 603 commensal and pathogenic isolates of *E. coli* and *Shigella*, the highest rates of mutation were found to be associated with uropathogenic strains (Denamur *et al.*, 2002). The incidence of mutators did not differ significantly between commensals and pathogenic organisms, but was found to be enhanced among strains recovered from urinary tract infections. Strains recovered from pus were found to have the lowest rate of mutation among the whole collection of strains, but the reasons for this are unknown.

4.2.   Gram-positive bacteria

The role of mutator strains in the development of antibiotic resistance among Gram-positive organisms in the clinical setting is largely unknown. Analysis of the mutation rate of over 490 clinical *Staphylococcus aureus* isolates suggested that none were exceptional (O'Neill and Chopra, 2002). In contrast, *mutS* alleles of *S. aureus* have been created and they exhibit the expected increase in mutation rate and were used to select high level vancomycin-resistant clones that only arise from the acquisition of multiple mutations (Schaaff *et al.*, 2002). Among 200 clinical isolates of *Streptococcus pneumoniae*, ~8% had mutation rates equivalent to those seen in *mutS* mutants of this organism (Morosini *et al.*, 2003). However, these strains appeared to have normal *hexA* (*mutS*) and *hexB* (*mutL*) loci and the presence of the mutators did not appear to influence the level of resistance to penicillin. In laboratory experiments, the potential of mutators has again been demonstrated. Low concentrations of cefotaxime led to enrichment of a mixed population (*hexA*/wild type) of *S. pneumoniae* cells for the hypermutable strain due to the higher rates of acquisition of mutations in pbp2x (Thr550Ala) (Negri *et al.*, 2002). From these initial studies, it seems likely that mutators will play a role in the development of resistance among Gram positives in an appropriately selective environment.

5.   CASE STUDIES IN EVOLVING RESISTANCE IN PATIENTS

Two major studies have been published that describe the evolution of antibiotic resistance within the same patient over a period of time (Low *et al.*, 2001; Rasheed *et al.*, 1997). Each obtained serial isolates which when characterised by molecular methods were shown to have evolved from a common stock distinct from other *E. coli* strains. Both studies observed the periodic selection of different resistant mutants followed by their replacement with other related strains. Additionally, in both studies, the progressive increase in MIC for β-lactams was associated with changes in porin expression (and structure) and in the complement of β-lactamases. However, the two studies differ in the probable mechanism by which the switch of β-lactamase profile takes place.

5.1.   An infant with aplastic anaemia

Tenover and colleagues (Rasheed *et al.*, 1997) studied serial isolates, obtained over a 3-month period, from the blood of an infant with idiopathic aplastic anaemia. Extensive broad spectrum antibiotic treatment took place during this period to treat fever episodes, in parallel with treatment for the aplastic anaemia. The first confirmed *E. coli* isolate from blood was some ~10 weeks after initial admission. Eight days after the initial isolate, three distinct phenotypes were noted among the isolates from the blood on the same day suggesting that the population was diverging or that the infected locus contained an already diverged community. The most distinctive isolate had acquired high-level resistance to tetracycline, gentamicin, and sulfonamides. This strain was found to have a modified plasmid banding pattern consistent with the possible enlargement of a preexisting 120 kDa plasmid. The multiple drug resistance phenotypes could be transferred by conjugation, whereas the β-lactamases could not (Rasheed *et al.*, 1997). Midway through this period, the resistance spectrum of the isolates changed dramatically with the acquisition of high-level resistance to several oxyimino cephalosporins, for example, cefotaxime and ceftazidime. PCR analysis coupled with IEF/nitrocefin-based detection of β-lactamases showed that the new variants had acquired an SHV β-lactamase in addition to the TEM-type enzyme already present. Over the time course of the study, strains lacking the porin OmpF were also detected, although the final pair of isolates exhibited some OmpF protein. As indicated above, analysis of the genome by pulsed field gel electrophoresis (PFGE) and by arbitrarily-primed PCR showed that all of the strains originated from a single clone. Thus, there was considerable diversity selected in the patient over a relatively short time period

The nature of this patient's illness suggests that the major source of the infecting organisms was the intestine. Thus, there are several parallel processes that may be responsible for the evolution of the antibiotic resistance. The intestinal flora may have contained a group of related strains that were interacting to effect the transfer of drug resistant determinants, for example, the SHV β-lactamase. The OmpF$^-$ strains may have been resident the whole time and were then selected by changes in the antibiotic regime. Alternatively, mutations may have arisen during the course of treatment. It is not clear from the study if the strains were mutators. The reappearance of the porin-containing strains among the final blood isolates suggests the potential for coexistence in the gut of both OmpF-lacking and "normal" strains and thus both might have been the focus of evolution throughout the period of observation. Data from the author's group suggest that evolution of a related group of *E. coli* isolates in the gut is probably a frequent occurrence.

Studies in the author's laboratory have analysed the complexity of the faecal *E. coli* in the gut of an elderly patient exhibiting bacteraemia (J. Park, F. MacKenzie, I. M. Gould, and I. R. Booth, unpublished data). A remarkable diversity was observed below the molecular level. PFGE and single-gene RFLP patterns suggested that there were just two major clones, one of which was also isolated from the patient's blood. The minor clone, which was represented by 3/20 faecal isolates, exhibited a completely different PFGE pattern, a unique *ompC* gene sequence, had acquired sucrose metabolism and a very high MIC for amoxicillin. The pattern of plasmids in this strain was also quite distinct from the major clone, although by transformation experiments it was demonstrated that they also had at least one plasmid in common (J. Park, F. MacKenzie, I. M. Gould, and I. R. Booth, unpublished data). However, the major faecal clone, which was identical to the blood isolate, could be subdivided into two metabolic types with respect to the rate of metabolism of melibiose. In addition, the isolates exhibited distinct MIC values for amoxicillin and differed in the degree of inhibition of the β-lactamases by clavulanic acid. While not an exhaustive analysis, the data point to considerable diversity coexisting among the resident *E. coli* flora in a single patient (J. Park, F. MacKenzie, I. M. Gould, and I. R. Booth, unpublished data). Other studies have concluded that *E. coli* is the dominant carrier of antibiotic resistance in the faecal flora and have suggested that the main potential route of transmission is by conjugation (Osterblad *et al.*, 2000). In this study, comparisons were made between isolates from the community, from short-stay patients and from long-stay (geriatric) patients. In all three groups, *E. coli* was the main resistant organism, but it was notable that some of the hospital samples (short-stay) had increased levels of multidrug resistant Enterobacteriaceae that exhibited distinct profiles from the multidrug resistant *E. coli*.

5.2.   A long-standing *E. coli* infection of liver cysts

More complex scenarios were seen with a study of a patient with infected cysts (Low, 2002; Low *et al.*, 2001). A patient suffering Caroli syndrome, which is a congenital disease frequently associated with the formation of liver cysts, was treated for frequent episodes of bacteraemia associated with *E. coli* colonisation of the cysts. Via surgical examination, it was estimated that the patient had >100 cysts, many of which were infected. Cell densities in two cysts, sampled surgically, were found to be 2×10^5 and 3.5×10^7 viable cells per millilitre. Over a 2-year period, five *E. coli* samples were recovered from the blood and a further two from infected cysts during surgery. The isolates showed unique patterns of antibiotic resistance of increasing severity and the final isolates were resistant to meropenem, imipenem, ceftazidime, and cefotaxime as well as ciprofloxacin. Variable resistance was observed to trimethoprim, but the isolates were sensitive to gentamicin throughout (Low *et al.,* 2001). PFGE analysis suggested isolates that derived from a single infecting clone which then diversified. The clonal nature of the isolates was confirmed by the discovery that the *ompC* gene sequence had diverged significantly from that observed in *E. coli* K-12 and pathogenic *E. coli* (an observation subsequently extended to other clinical isolates; J. Park, F. MacKenzie, I. M. Gould, and I. R. Booth, unpublished data). Sequencing the *ompC* gene demonstrated that it had diverged more than 9% from the *E. coli* K-12 sequence and that this was more than double the rate of change of other genes in the same strains and higher than the observed rate of change in *E. coli* O157 and in two commensal *E. coli* obtained from the faeces of healthy volunteers (Low *et al.,* 2001). All seven isolates from the patient had the same basic *ompC* gene sequence (Low *et al.,* 2001). However, the strains progressively accumulated mutations in the *ompC* gene but did not at any stage express OmpF despite the presence of a functional structural gene. In the later isolates, the mutations rendered the OmpC protein unstable and consequently lowered levels of the protein were observed in the outer membrane.

A complex pattern of evolution of the β-lactamases was observed in these clinical isolates. Each isolate possessed the same dominant TEM-class enzyme (pI ~5), but possessed several other β-lactamase activities with more acidic (strains 1–4) or slightly more alkaline (strain 5) pI values. At least two different TEM-class proteins were detected in the first isolate by proteomic analysis of the periplasmic fractions. The first isolates had three large plasmids and this reduced to one by the final isolates. No attempt was made to study their transmission to other *E. coli.* The isolation of the *E. coli* in the individual liver cysts makes it unlikely that the variants arose by plasmid acquisition and would be consistent with at least two *bla*$_{TEM}$ genes in the first isolate and their divergence and selection during subsequent antibiotic regimes. In this regard, it was notable that while the first two isolates were non-mutators, the later isolates displayed a very strong mutator phenotype (J. Park, F. MacKenzie, I. M. Gould, and I. R. Booth, unpublished data). Not only did the isolates alter their plasmid-encoded β-lactamase pattern, but they also evolved increased expression of chromosomal AmpC due to mutations in the promoter and regulatory regions. This evolution was most marked over the period after the isolate had acquired mutator properties (J. Park, F. MacKenzie, I. M. Gould, and I. R. Booth, unpublished data).

The importance of the mutator status is also supported by the analysis of ciprofloxacin resistance (Figure 3). Ciprofloxacin had been used in treatment prior to the storage of the first blood culture for analysis. The MICs of the first

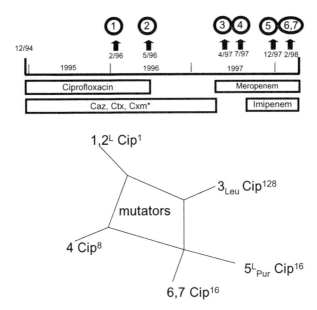

Figure 3. The evolution of drug resistance in a patient with infected liver cysts. The patient first presented at the end of 1994 and was monitored for 3 years with bacterial samples from the blood and cysts being acquired at intervals. The figure shows the time line for the infection, a summary of the antibiotic treatment (see Low *et al.*, 2001 for further details) and the times at which samples were obtained and analysed. The lower half of the diagram indicates a Split Tree Decomposition (STD) analysis of all the molecular data obtained from the patient samples. Isolates 3–7 were subsequently discovered to be mutators. Abbreviations: Cip, ciprofloxacin with MIC indicated as a superscript; 2^L and 5^L, the superscript indicates the isolate was obtained directly from a cyst; Leu and Pur (as subscripts) indicate the two isolates with auxotrophies. The STD analysis was performed by Dr Martin Maiden, University of Oxford.

two strains were in the sensitive range, but analysis of the *gyrA* gene sequence revealed one of the commonest mutations that gives rise to low-level resistance (D87Y) in the first two isolates and in all the subsequent strains. In the third isolate, the MIC rose to 128 mg/L but fell in subsequent isolates to 8 (isolate 4) and 16 (isolates 5, 6, and 7) mg/ml. Four of these strains were found to have additional mutations in *gyrA* (S83L) and *parC* (E84G). These mutations alone cannot explain the MIC values since isolate 1, 2, and 4 have the same mutations in *gyrA*, but have quite different MIC values and, similarly, isolates 3, 5, 6, and 7 have the same mutations in both *gyrA* and *parC* but have MIC values ranging from 128 to 8 mg/L. Mutations affecting efflux systems must have arisen in some of the strains, but to date none have been detected. The data illuminate two points. First, the multiple mutations within the components that can give rise to ciprofloxacin resistance point to the importance of the evolution of the mutator status in these strains. Second, some mutations are present in the first and in all subsequent isolates, but others are present only transiently in the isolated strains. This pattern of the mutations in sequential isolates points to the separate and independent evolution of the antibiotic resistance. This is graphically illustrated in Split decomposition analysis of all the sequence data, suggesting that the infected cysts provided unique environments in which different evolutionary paths were followed.

Analysis of the total number of changes that took place during the evolution from a core infecting population led to the conclusion that the strains evolved in parallel (Low *et al.*, 2001) (Figure 3). In this patient, we believe that separate cysts created unique environments in which mutation and selection took place at different rates and with separate outcomes. At intervals, the patient presented with a bacteraemia as a result of one, or more, of the cysts leaking their microbial load. The presence of the same mutations in some isolates but not in others that were isolated subsequently suggests that at different stages, some cysts became infected with organisms that originated in another cyst possibly during the bacteraemia. The mutator status of the later isolates from the liver cysts is interesting and is consistent with the prevalence of mutators in the antibiotic-resistant *P. aeruginosa* flora of CF patients and in uropathogenic *E. coli* (Denamur *et al.*, 2002; Oliver *et al.*, 2000). This is possibly the first notification of mutator organisms from a tissue within the body cavity since previous isolates have been from lung and bladder. It may also explain the observation that two of the isolates had demonstrable separate auxotrophies while others had a more general impairment of growth on minimal medium (Low *et al.*, 2001).

6.   CONCLUSIONS

The study of the molecular mechanisms of antibiotic resistance has been very informative. It provides a backcloth against which an understanding of the properties of both microorganisms and antibiotics can be understood. In turn this understanding has led to changes in clinical practice, though not as rapidly in all countries as many would desire. In addition, it is clear that the understanding of the pharmacodynamics of antibiotics can inform their design. Finally, it is appropriate to place the evolution of resistance in patients at the centre of our thinking. Basic insights into protein structure and mechanisms of action of enzymes and transporters can tell us the options that are at the bacterium's disposal to acquire resistance. However, it is through the analysis of the actual outcomes of failed therapy in patients that we begin to understand the checks and balances within biology that make some options preferred to others.

ACKNOWLEDGEMENTS

Special thanks to Alison Low and Jenni Park who undertook the seminal studies that have informed the views expressed here. Figure 1 is based upon an original drawing by Jenni Park. Thanks also to Fiona MacKenzie and Ian Gould for their moral, intellectual, and financial support in the development of the research programme on antibiotic resistance in patients. The generous support and encouragement of the Bacterial Physiology Research Group at Aberdeen is gratefully acknowledged. Some of the work described here was supported by Grampian University NHS Trust Clinical Microbiology Endowment Fund (A.S. Low) and by the Faculty of Medicine and Medical Sciences Studentship to Jenni Park. Thanks also to Hiroshi Nikaido, Jim Naismith, The Aberdeen Proteomics facility, and to Martin Maiden (University of Oxford). The author was supported during the completion of this work by a Wellcome Trust Research Leave Fellowship (040174).

REFERENCES

Blazquez, J., Morosini, M. I., Negri, M. C., and Baquero, F., 2000, Selection of naturally occurring extended-spectrum TEM beta-lactamase variants by fluctuating beta-lactam pressure. *Antimicrob. Agents Chemother.*, **44**, 2182–2184.

Booth, I. R., Ferguson, G. P., Miller, S., Gunasekara, B., and Kinghorn, S., 2003, Bacterial production of methylglyoxal: A survival strategy or death by misadventure? *Biochem. Soc. Trans.*, **31**, 1406–1408.

Boucher, J. C., Mudd, H., and Deretic, V., 1997, Mucoid Pseudomonas aeruginosa in cystic fibrosis: Characterization of muc mutations in clinical isolates and analysis of clearance in a mouse model of respiratory infection. *Infect. Immun.*, **65**, 3838–3846.

Canton, R., Coque, T. M., and Baquero, F., 2003, Multi-resistant Gram-negative bacilli: From epidemics to endemics. *Curr. Opin. Infect. Dis.*, **16**, 315–325.

Chopra, I., O'Neill, A. J., and Miller, K., 2003, The role of mutators in the emergence of antibiotic-resistant bacteria. *Drug Resist. Updates*, **6**, 137–145.

Denamur, E., Bonacorsi, S., Giraud, A., Duriez, P., Hilali, F., Amorin, C. *et al.*, 2002, High frequency of mutator strains among human uropathogenic Escherichia coli isolates. *J. Bacteriol.*, **184**, 605–609.

Dong, Y. Z., Zhao, X. L., Kreiswirth, B. N., and Drlica, K., 2000, Mutant prevention concentration as a measure of antibiotic potency: Studies with clinical isolates of Mycobacterium tuberculosis. *Antimicrob. Agents Chemother.*, **44**, 2581–2584.

Drlica, K., 2003, The mutant selection window and antimicrobial resistance. *J. Antimicrob. Chemother.*, **52**, 11–17.

Eriksson, S., Lucchini, S., Thompson, A., Rhen, M., and Hinton, J. C. D., 2003, Unravelling the biology of macrophage infection by gene expression profiling of intracellular Salmonella enterica. *Mol. Microbiol.*, **47**, 103–118.

Eskra, L., Canavessi, A., Carey, M., and Splitter, G., 2001, Brucella abortus genes identified following constitutive growth and macrophage infection. *Infect. Immun.*, **69**, 7736–7742.

Funchain, P., Yeung, A., Stewart, J. L., Lin, R., Slupska, M. M., and Miller, J. H., 2000, The consequences of growth of a mutator strain of Escherichia coli as measured by loss of function among multiple gene targets and loss of fitness. *Genetics*, **154**, 959–970.

Giraud, A., Matic, I., Tenaillon, O., Clara, A., Radman, M., Fons, M. *et al.*, 2001a, Costs and benefits of high mutation rates: Adaptive evolution of bacteria in the mouse gut. *Science*, **291**, 2606–2608.

Giraud, A., Radman, M., Matic, I., and Taddei, F., 2001b, The rise and fall of mutator bacteria. *Curr. Opin. Microbiol.*, **4**, 582–585.

Govan, J. R. W. and Deretic, V., 1996, Microbial pathogenesis in cystic fibrosis: Mucoid Pseudomonas aeruginosa and Burkholderia cepacia. *Microbiol. Rev.*, **60**, 539–574.

Gross, M. D. and Siegel, E. C., 1981, Incidence of mutator strains in Escherichia coli and coliforms in Nature. *Mutat. Res.*, **91**, 107–110.

Hyatt, J. M., McKinnon, P. S., Zimmer, G. S., and Schentag, J. J., 1995, The importance of pharmacokinetic-pharmacodynamic surrogate markers to outcome—Focus on antibacterial agents. *Clin. Pharmacokinet.*, **28**, 143–160.

Kohler, S., Foulongne, V., Ouahrani-Bettache, S., Bourg, G., Teyssier, J., Ramuz, M. *et al.*, 2002, The analysis of the intramacrophagic virulome of Brucella suis deciphers the environment encountered by the pathogen inside the macrophage host cell. *Proc. Nat. Acad. Sci. USA*, **99**, 15711–15716.

LeClerc, J. E., Li, B. G., Payne, W. L., and Cebula, T. A., 1996, High mutation frequencies among Escherichia coli and Salmonella pathogens. *Science*, **274**, 1208–1211.

Levin, B. R. and Antia, R., 2001, Why we don't get sick: The within-host population dynamics of bacterial infections. *Science*, **292**, 1112–1114.

Lindgren, P. K., Karlsson, A., and Hughes, D., 2003, Mutation rate and evolution of fluoroquinolone resistance in *Escherichia coli* isolates from patients with urinary tract infections. *Antimicrob. Agents Chemother.*, **46**, 3222–3232.

Livermore, D. M., 2003, Bacterial resistance: Origins, epidemiology, and impact. *Clin. Infect. Dis.*, **36**, S11–S23.

Low, A. S., MacKenzie, F. M., Gould, I. M., and Booth, I. R., 2001, Protected environments allow parallel evolution of a bacterial pathogen in a patient subjected to long-term antibiotic therapy. *Mol. Microbiol.*, **42**, 619–630.

Low, A. S., 2002, Characterization of antibiotic resistance mechanisms in clinical isolates of *Escherichia coli* from a single patient. In *Molecular and Cell Biology*. University of Aberdeen, p. 254.

Martinez, J. L. and Baquero, F., 2002, Interactions among strategies associated with bacterial infection: Pathogenicity, epidemicity, and antibiotic resistance. *Clin. Microbiol. Rev.*, **15**, 647–679.

Matic, I., Radman, M., Taddei, F., Picard, B., Doit, C., Bingen, E. *et al.*, 1997, Highly variable mutation rates in commensal and pathogenic Escherichia coli. *Science*, **277**, 1833–1834.

Morosini, M. I., Baquero, M. R., Sanchez-Romero, J. M., Negri, M. C., Galan, J. C., del Campo, R. *et al.*, 2003, Frequency of mutation to rifampin resistance in Streptococcus pneumoniae clinical strains: hexA and hexB polymorphisms do not account for hypermutation. *Antimicrob. Agents Chemother.*, **47**, 1464–1467.

Moxon, R. E., Rainey, P. B., Nowak, M. A., and Lenski, R. E., 1994, Adaptive evolution of highly mutable loci in pathogenic bacteria. *Curr. Biol.*, **4**, 24–33.

Negri, M. C., Lipsitch, M., Blazquez, J., Levin, B. R., and Baquero, F., 2000, Concentration-dependent selection of small phenotypic differences in TEM beta-lactamase-mediated antibiotic resistance. *Antimicrob. Agents Chemother.*, **44**, 2485–2491.

Negri, M. C., Morosini, M. I., Baquero, M. R., del Campo, R., Blazquez, J., and Baquero, F., 2002, Very low cefotaxime concentrations select for hypermutable Streptococcus pneumoniae populations. *Antimicrob. Agents Chemother.*, **46**, 528–530.

Oliver, A., Coanton, R., Campo, P., Bacquero, F., and Blazquez, J., 2000, High frequency of hypermutable *Pseudomonas aeruginosa* cystic fibrosis lung infection. *Science*, **288**, 1251–1253.

Oller, A. R., Fijalkowska, I. J., and Schaaper, R. M., 1993, The Escherichia-coli Galk2 papillation assay—Its specificity and application to 7 newly isolated mutator strains. *Mutat. Res.*, **292**, 175–185.

O'Neill, A. J. and Chopra, I., 2002, Insertional inactivation of mutS in Staphylococcus aureus reveals potential for elevated mutation frequencies, although the prevalence of mutators in clinical isolates is low. *J. Antimicrob. Chemother.*, **50**, 161–169.

Osterblad, M., Hakanen, A., Manninen, R., Leistevuo, T., Peltonen, R., Meurman, O. *et al.*, 2000, A between-species comparison of antimicrobial resistance in enterobacteria in fecal flora. *Antimicrob. Agents Chemother.*, **44**, 1479–1484.

Radman, M., Taddei, F., and Matic, I., 2000, DNA repair systems and bacterial evolution. *Cold Spring Harbor Symp. Quant. Biol.*, **65**, 11–19.

Rasheed, J. K., Jay, C., Metchock, B., Berkowitz, F., Weigel, L., Crellin, J. *et al.*, 1997, Evolution of extended-spectrum beta-lactam resistance (SHC-8) in a strain of *Escherichia coli* during multiple episodes of bacteremia. *Antimicrob. Agents Chemother.*, **41**, 647–653.

Ruiz, J., 2003, Mechanisms of resistance to quinolones: Target alterations, decreased accumulation and DNA gyrase protection. *J. Antimicrob. Chemother.*, **51**, 1109–1117.

Schaaff, F., Reipert, A., and Bierbaum, G., 2002, An elevated mutation frequency favours the development of vancomycin resistance in *Staphylococcus aureus. Antimicrob. Agents Chemother.*, **46**, 3540–3548.

Schentag, J. J., Gilliland, K. K., and Paladino, J. A., 2001, What have we learned from pharmacokinetic and pharmacodynamic theories? *Clin. Infect. Dis.*, **32**, S39–S46.

Sindelar, G., Zhao, X. L., Liew, A., Dong, Y. Z., Lu, T., Zhou, J. F. *et al.*, 2000, Mutant prevention concentration as a measure of fluoroquinolone potency against mycobacteria. *Antimicrob. Agents Chemother.*, **44**, 3337–3343.

Soriano, F. and Rodriguez-Cerrato, V., 2002, Pharmacodynamic and kinetic basis for the selection of pneumococcal resistance in the upper respiratory tract. *J. Antimicrob. Chemother.*, **50**, 51–58.

Taddei, F., Matic, I., and Radman, M., 1995, cAMP-dependent SOS induction and mutagenesis in resting bacterial populations. *PNAS*, **92**, 11736–11740.

Taddei, F., Halliday, J. A., Matic, I., and Radman, M., 1997, Genetic analysis of mutagenesis in aging Escherichia coli colonies. *Mol. Gen. Genet.*, **256**, 277–281.

Whittle, G., Shoemaker, N. B., and Salyers, A. A., 2002, The role of Bacteroides conjugative transposons in the dissemination of antibiotic resistance genes. *Cell. Mol. Life Sci.*, **59**, 2044–2054.

Wise, R., 2003, Maximizing efficacy and reducing the emergence of resistance. *J. Antimicrob. Chemother.*, **51**, 37–42.

Chapter 21

Impact of Pharmacodynamics on Dosing Schedules: Optimizing Efficacy, Reducing Resistance, and Detection of Emergence of Resistance

Johan W. Mouton
Department of Medical Microbiology and Infectious Diseases,
Canisius Wilhelmina Hospital, Nijmegen, The Netherlands

1. INTRODUCTION

One of the more important issues in the prudent use of antimicrobial agents is to use an antibiotic in such a manner that the effect of the antimicrobial is optimal. This involves not only choosing the best antibiotic for a specific indication, but also to administer the correct dose to ensure a maximal effect. At the same time, the dosing regimen given should also minimize the probability of emergence of resistance. In various studies, it has been shown, and will be further explored in this chapter, that using suboptimal doses increases the probability of emergence of resistant clones.

A second major issue is the early detection of resistance in the community and in the hospital. In several chapters in this book, emergence of resistance is evaluated based on the categorization of resistance as reported by the microbiology laboratory. The categorization in susceptible (S), intermediate susceptible (I), and resistant (R) has historically been based on both frequency distributions, some measure of activity, or both (Kahlmeter *et al.*, 2003; Mouton, 2002, 2003). Since these breakpoints were primarily designed to help

Antibiotic Policies: Theory and Practice. Edited by Gould and van der Meer
Kluwer Academic / Plenum Publishers, New York, 2005

clinicians and guide therapeutic decisions and not to detect emergence of resistance, resistance is identified when it might be too late. Breakpoints sometimes are too far off from the upper part of the wild-type distribution to notice mutations. Although these may have little clinical impact at the time, they may be a prebode for problems later on (Grohs *et al.*, 2003; Lim *et al.*, 2003). Thus, resistant clones may have emerged, but have not been detected simply because they are categorized as susceptible. Ideally, emergence of resistance, or in practical terms, an increase in minimum inhibitory concentration (MIC) is identified as early as possible.

In the first part of this chapter, dose–effect relationships are discussed and it is shown that these can be used to set clinical breakpoints that can be used in guiding therapy by the clinician. In the second part, a brief history of breakpoints is presented. It is shown that historically, breakpoints are not consistent and vary over time and place, resulting in a wide variety of definitions and percentages of resistance. It is discussed how the European Committee on Antimicrobial Susceptibility Testing (EUCAST) is approaching this challenge. In the third part, the relationship between dose and/or concentration and the emergence of resistance is explored. Although a relatively young area of research, data are accumulating fast and provide insight in how this can be applied to minimize emergence of resistance.

2.   DOSE–EFFECT RELATIONSHIPS

2.1.   The MIC

The major parameter that has been used for several decades as a measure of susceptibility of a microorganism to a drug is the MIC. However, in itself the MIC is not very informative: The MIC is determined at static concentrations, usually in a twofold series of dilutions and is read after 18–24 hr, thus at a specific point in time. The MIC has, however, been shown to be a reasonably reproducible measure of activity of an antimicrobial agent against a microorganism (EUCAST, 2003), although methods to determine the MIC vary slightly (Andrews, 2001; NCCLS, 2002).

Conversely, *in vivo* concentrations are not static but decline over time. Furthermore, it is not only the effect of an antimicrobial after 18–24 hr that one is interested in, but primarily the effect of a drug over time in relation to the concentration–time profile of the drug *in vivo*. As a consequence, some sort of relationship between the MIC as measured in a test tube has to be correlated to the *in vivo* effect as a function of the concentration–time profile and/or dosing regimen of the antimicrobial.

2.2. Concentration–time curves and the PK/PD index

The concentration–time curve of a drug possesses two major characteristics: the peak concentration (C_{max}) and the area under the concentration–time curve (AUC) (Figure 1). Both these pharmacokinetic parameters can be related to the MIC of a microorganism by determining the ratio between the two to obtain pharmacokinetic/pharmacodynamic (PK/PD) indices (Mouton *et al.*, 2002a). Thus the AUC/MIC ratio and the Peak/MIC ratio are obtained. A third important PK/PD characteristic of the drug with respect to the microorganism is the time the concentration of the antimicrobial remains above the MIC of the microorganism ($T_{>MIC}$). This latter value is usually expressed as a percentage of the dosing interval over 24 hr. By using different dosing regimens in animal models of infection and in *in vitro* pharmacokinetic models, by varying both the frequency and the dose of the drug it has been shown that there exists a relationship between a PK/PD index and efficacy. The efficacy is usually expressed as either the decrease or increase in the number of bacteria with respect to the initial inoculum (at the start of treatment) as a function of the PK/PD index. It has been shown that this efficacy measure correlates well with survival (Andes and Craig, 2002). In general, two patterns of activity are recognized: Antimicrobials showing a clear relationship between AUC and efficacy and those where efficacy is correlated to the time the concentration remains above the MIC, $T_{>MIC}$. An example is shown in Figure 2 for levofloxacin and ceftazidime. For levofloxacin there is a clear relationship between AUC and effect while there is virtually no relationship between $T_{>MIC}$ and efficacy, while for ceftazidime the $T_{>MIC}$ is correlated with effect.

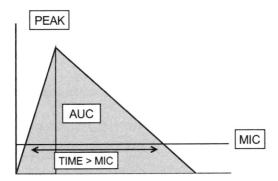

Figure 1. Diagram of a concentration–time curve showing the pharmacokinetic parameters Peak (or C_{max}) and AUC. The PK/PD indices are derived by relating the pharmacokinetic parameter to the MIC: AUC/MIC, C_{max}/MIC, and $T_{>MIC}$.

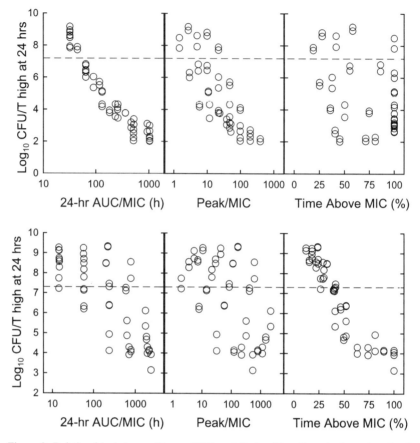

Figure 2. Relationship between $T_{>MIC}$, AUC, and Peak of levofloxacin (upper) and ceftazidime (lower) in a mouse model of infection with *S. pneumoniae* as obtained by various dosing regimens and efficacy expressed as cfu (colony forming units). The best relationship is obtained with the AUC for levofloxacin and $T_{>MIC}$ for ceftazidime. Reproduced from Andes and Craig, 2002.

Since the pharmacokinetic parameters describe the concentration–time profile of the drug, and the MIC is a measure of the activity of the drug, another way to look at the PK/PD index is normalizing the relationship between concentration–time profile and effect.

2.3.   The sigmoid dose–response curve

As can be observed from Figure 2, the relationship between PK/PD index and effect can be described by a sigmoid curve. The model most often used to describe this relationship is the E_{max} model with variable slope or Hill equation, where the effect $E = E_{max} \times EI_{50}{}^g/(EI_{50}{}^g + I^g)$. The EI_{50} is the PK/PD index necessary to obtain 50% of the maximum effect, I is the PK/PD index value, and g is the slope factor. Focusing on quinolones, this relationship has now been demonstrated in numerous studies. Observations *in vitro* (Lacy *et al.*, 1999; Lister, 2002; Lister and Sanders, 1999a, b; Madaras-Kelly *et al.*, 1996), animals (Andes and Craig, 1998; Bedos *et al.*, 1998a, b; Croisier *et al.*, 2002; Drusano *et al.*, 1993; Ernst *et al.*, 2002; Fernandez *et al.*, 1999; Mattoes *et al.*, 2001; Ng *et al.*, 1999; Onyeji *et al.*, 1999; Scaglione *et al.*, 1999) and clinical studies (Ambrose *et al.*, 2001; Forrest *et al.*, 1993, 1997; Highet *et al.*, 1999; Preston *et al.*, 1998) have shown a clear relationship between AUC/MIC ratio and effect and the results are fairly consistent. Importantly, the AUC/MIC values needed to obtain a certain outcome is similar for the various quinolones. It has to be emphasized here that this is true for the free fraction of the drug, that is, non-protein bound fraction, only (Cars, 1990; Liu *et al.*, 2002).

Figure 3 shows five characteristics of the sigmoid relationship in relation to the efficacy parameter often used to characterize the potency of the drug. Apart from the minimum (or no) effect and the maximum effect, these are the static effect, the 50% effect and the 90% of E_{max} values. The PK/PD index values correlating with these effects are the EI_s, EI_{50}, and EI_{90}, respectively. The static effect is where the number of bacteria remaining after treatment is equal to the initial inoculum. Both the static effect and the EI_{90} are applied as an effect measure—for various practical purposes the 100% effect is not suitable to use. The EI_{50} is, since it is in the steepest part of the curve, most sensitive to changes and is often employed to show differences between treatment effects.

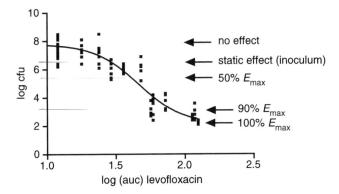

Figure 3. Diagram showing various characteristic effect levels of a sigmoid dose–response relationship, in this example levofloxacin based on data from Scaglione *et al.*, 1999.

2.4.   Determination of clinical breakpoints for susceptibility testing using the PK/PD index

The relationship between PK/PD index value and effects can be used to determine clinical breakpoints, presuming that these distinguish between bacteria that can likely be treated with a commonly used dosing regimen and those that can not. If the dosing regimen commonly used is known, the AUC and other pharmacokinetic parameters can easily be estimated for the average patient from pharmacokinetic studies available. From the PK/PD index–effect relationship, the EI_s, EI_{50}, and EI_{90} are identified and the only unknown parameter in the equation is (taking the EI_{90} as an example) the MIC; $EI_{90} =$ AUC/MIC or, to put it differently, MIC = AUC/EI_{90}. This MIC would be a reasonable estimate of the clinical breakpoint. Higher MICs than this breakpoint would result in lower AUC/MIC ratios and thus have a lower probability of successful treatment (resistant) while infections with microorganisms with lower MIC values would have a higher chance of successful treatment (susceptible).

The debate that ensues, is whether to apply the EI_s or the EI_{90} as a parameter to settle on the S breakpoint. Both have their advantages and disadvantages. The argument to use the EI_{90} is that, since the objective of treating patients is to treat them optimally, one should always aim for an EI_{90}. On the other hand, in non-immunocompromised patients and/or non severe infections, the EI_s is probably as good as the EI_{90}. Whichever of these two is used, what is clear and obvious is that this relationship can be used to determine breakpoints and that this also provides a method or system where the different quinolones are treated equally in the sense that their breakpoints are consistent with each other (Mouton, 2002; Turnidge, 1999). The same argument, of course also applies to other classes.

From these arguments, it is obvious that the breakpoint is dependent on the dosing regimen given. If a certain PK/PD index value is desired to obtain a certain effect, this is dependent on both the PK part and the MIC part of the ratio. Since for the same strain the MIC is a constant, the clinical breakpoint is entirely dependent on the dosing regimen. An example is provided in Figure 4 for amoxicillin/clavulanic acid. The per cent of $T_{>MIC}$ of amoxicillin over 24 hr is depicted here as a function of the MIC for four different dosing regimens (Mouton and Punt, 2001). The value needed for efficacy in this example is taken as 40%. Thus, drawing a horizontal line at the 40% level, the crossing point with the $T_{>MIC}$ function indicates the clinical breakpoint for that specific dosing regimen. It is clear from this figure, that the breakpoint value obtained is dependent of the dosing regimen and this is one of the reasons breakpoints differ between countries (there are other reasons, these are discussed below).

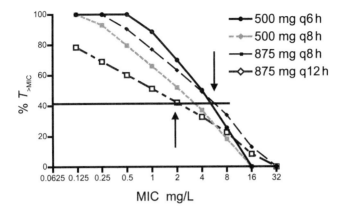

Figure 4. Diagram showing the relationship between $T_{>MIC}$ and MIC of amoxicillin for four different dosing regimens of amoxicillin–clavulanic acid to demonstrate that the clinical breakpoint is dependent on the dosing regimen. Assuming that 40% $T_{>MIC}$ is the time of the dosing regimen needed for effect, the breakpoint for the 875 mg q12 h is 2 mg/L while for the dosing regimen of 500 mg q6 h it is 8 mg/L. Based on Mouton and Punt, 2001.

2.5. Monte Carlo simulations

In the discussion above, the AUC/MIC or $T_{>MIC}$ was used as a reference value to determine tentative PK/PD breakpoints. However, the values used to calculate the breakpoint were mean values of the population. Thus, approximately half the population will maintain a PK/PD index lower than this value (for instance because of a higher than average clearance) and the other half will have a higher value. However, when PK/PD indices are being used as a value for the determination of breakpoints that are used to predict the probability of success of treatment as discussed above, this should be true not only for the population mean, but also for each individual within the population, also that part of the population with a higher elimination. An example is given in Figure 5. The figure shows the proportion of the population reaching a certain concentration of ceftazidime after a 1 g dose. It is apparent that there are individuals with a $T_{>MIC}$ of 50%, while others have, with the same dosing regimen, a $T_{>MIC}$ of more than 80%. It is, in particular, those individuals who have lower values than average that one should be concerned about, since if they have infection with a microorganism with an MIC at the breakpoint level, it may be reported as susceptible, while the PK/PD index value for those particular individuals is less than optimal. Thus, when using the relationship between PK/PD index and efficacy, the breakpoint chosen should take this interindividual variation into account. To that end, Drusano *et al.* suggested an

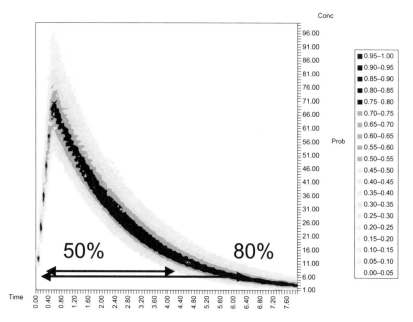

Figure 5. Simulation of ceftazidime after a 1 g dose using data from (Mouton *et al.*, 1990). The greyscale indicates the probability of presence of a certain concentration. Due to inter-individual variability, some individuals in the population will have a $T_{>MIC}$ 50%, while others will have a value of 80%. The population mean is in the middle of the black area. Reproduced from Mouton, 2003.

integrated approach of population pharmacokinetics and microbiological susceptibility information (Drusano *et al.*, 2000, 2001) by applying Monte Carlo simulations. This is a method which takes the variability in the input variables into consideration in the simulations (Bonate, 2001) and thereby generates slightly different pharmacokinetic parameters concordant to the variation in the population. Thus, PK/PD index values are generated not only for the population mean, but for every possible individual in the population. This is subsequently used to determine the breakpoints, not using the mean PK/PD index of the population but also taking into account the distribution. This approach has been used now by several authors (Ambrose and Grasela, 2000; Drusano *et al.*, 2001; Montgomery *et al.*, 2001; Mouton *et al.*, 2002b; Nicolau and Ambrose, 2001). An extensive discussion on this subject can be found in Mouton (2003); it is mentioned here because it may affect breakpoint values in some cases.

From the discussion above, two important conclusions have to be drawn. The first is, as discussed in the last section, that the clinical breakpoint value is dependent on the dosing regimen. The second one is that, when different agents within a class show more or less similar dose–effect relationships, the breakpoints of these agents relative to each other should be consistent.

3. BREAKPOINTS: A SHORT HISTORY AND OVERVIEW

The criteria used for categorization in "susceptible" and "resistant," and the meaning of S and R, have varied over time and place and are not always clear. Initially, the terms susceptible and resistant were largely based on frequency distributions, the distributions of the populations being well apart and largely confined to distinguishing susceptible from resistant *Staphylococcus aureus* strains, which correlated well with clinical success and failure. From that perspective the term breakpoint was first used in the Report of an International Collaborative Study on Antimicrobial Susceptibility Testing by Ericsson and Sherris in 1971 (Ericsson and Sherris, 1971). Since then, interpretation and use have diverged between various countries, persons and societies, with major differences in approach between Europe and the United States. This has led to breakpoint values diverging by as much as a factor 32 for some drugs. In Europe, most of the national breakpoint committees were, apart from creating reproducible antimicrobial susceptibility testing (AST) methods, primarily interested in providing breakpoints which could be used to predict clinical and bacteriological efficacy based on serum concentrations achieved in patients. In most cases, the duration of time that the free fraction remained above a certain value was considered to be the breakpoint and less attention was paid, except for notably the SRGA in Sweden and the BSAC in the United Kingdom, to the fact that the natural population distribution of MICs could be divided by using this approach too rigorously. Alternatively, the line in the United States tended to categorize bacteria based on frequency distributions and thus in some cases breakpoints that were higher than the highest concentration were achieved clinically. Table 1 shows an overview of the various approaches used in six European countries and the United States.

It was mentioned above that in particular the SRGA and the BSAC took the population distribution into account when setting their breakpoints. The reason for this was that the reproducibility and accuracy of MIC testing (or measuring zone diameters) in the routine medical microbiology laboratory does not allow a precise answer: In general, the accuracy of MIC measurement is within 1–2 twofold dilutions, while the wild-type population distribution spans a concentration range of 3–4 twofold dilutions. Thus, it is impossible to determine the exact MIC from a single measurement and whether a particular strain belongs to the upper part or the lower part of the population distribution. Indeed, the distribution of measurements of the same strain is very similar to that of the distribution

Table 1. Breakpoint systems used

Country	Committee	Method	References
France	CASFM	Formula[a] based on PK = $(C_{max}/3 + Ct\frac{1}{2} + C4h)/3 \times (1 - k)$	(Soussy *et al.*, 1994)
Great Britain	BSAC	Formula[b] based on PK = $C_{max} \times f \times s/(e \times t)$	(MacGowan and Wise, 2001)
Netherlands	CRG	70–80% $T_{>MIC}$ for non-protein bound fraction	(Mouton *et al.*, 2000)
Sweden	SRGA	Pharmacokinetic profile and frequency distribution, species dependent	(Swedish Reference Group of Antibiotics, 1997)
Norway	NWGA	67% $T_{>MIC}$	(Bergan *et al.*, 1997)
Germany	DIN	Pharmacokinetic profile, frequency distributions, efficacy	(Deutsches Institut fur Normung, 1990)
Europe	EUCAST	Clinical breakpoints: PK profile and efficacy Wild-type cut-offs: frequency distribution	(EUCAST, 2003)
US	NCCLS	Frequency distribution, pharmaco-kinetic profile	(NCCLS, 2002, 2003)

[a]C_{max}, maximum serum concentration; $Ct\frac{1}{2}$, concentration in serum after one half-life; $C4h$, minimum quantity obtained over 4 hr period that corresponds approximately to 10 bacterial generations; k, degree of protein binding.
[b]C_{max}, maximum serum concentration at steady state, usually 1 h post-dose; e, factor by which C_{max} should exceed MIC (usually 4); t, factor to allow for serum half-life; f, factor to allow for protein binding; s, shift factor to allow for reproducibility and frequency distributions (usually 1).

of strains within a species (Figure 6). Consequently, the MIC or zone diameter to indicate whether a strain is susceptible or resistant, the breakpoint, has be either at the lower or at the upper end of the distribution but not in the middle.

3.1.   Different breakpoints, different resistance rates

Since there are different breakpoints in use for some (most) drugs in various countries, it is obvious that the resistance rates in these countries differ, if only because of the definition used. It is therefore not very rational to compare resistance rates between countries if the breakpoints countries use differ and it is not known what the effect of these different breakpoints is on the resistance rates. An example of the various rates one might obtain using breakpoints from different countries is shown in Figure 7 for cefotaxime and *Escherichia coli*.

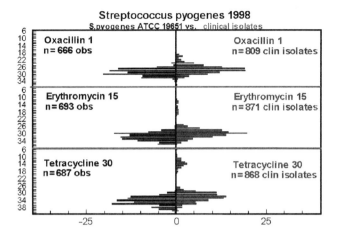

Figure 6. Distributions of zone diameters from repeated measurements of the quality control strain (left) and clinical isolates from the same species (right). The distributions are relatively similar, except for the resistant strains, which is particularly obvious for tetracycline. Reproduced from Mouton, 2002.

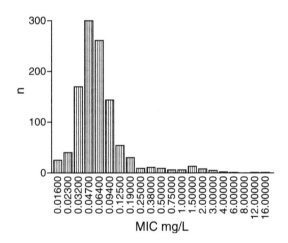

Figure 7. Distribution of cefotaxime MICs of *E. coli*. Based on data from de Man *et al.*, 1998.

The susceptibility data were taken from a survey of urinary isolates in the Netherlands (de Man *et al.*, 1998). Table 2 lists the current breakpoints for cefotaxime as used by various breakpoint committees (Kahlmeter *et al.*, 2003). Depending on the breakpoint used, the resistance rates range from 0.5% to 4%. For other drugs, the differences may be much larger.

Table 2. Similarities and differences in international breakpoint systems—current cefotaxime and ciprofloxacin breakpoints for *Enterobacteriaceae* in Europe and the United States. Reproduced from (Kahlmeter *et al.*, 2003)

Breakpoint committee (country)	Cefotaxime breakpoint (mg/L)		Ciprofloxacin breakpoint (mg/L)	
	Susceptible	Resistant	Susceptible	Resistant
BSAC (UK)	2	4	1	2
CA-SFM (France)	4	>32	1	>2
CRG (Netherlands)	4	>16	1	>2
DIN (Germany)	2	16	1	4
NCCLS (USA)	8	64	1	4
NWGA (Norway	1	32	0.12	4
SRGA (Sweden)	0.5	2	0.12	2

3.2.   Breakpoint harmonization in Europe

It is apparent that the difference in breakpoints within Europe leads to confusion and may lead to erroneous conclusions. Given the similarities in dosing within the various countries in Europe, it seems logical that breakpoints within Europe should be harmonized. The discussion above has also shown two important fundamentals in setting breakpoints that were not clearly distinguished or recognized in the past. The first is that, over the last decade, clear dose–effect relationships have been established for antimicrobials and microorganisms and that these relationships can be used to set breakpoints to help guide therapy. At the same time it is recognized that early detection of emergence of resistance is increasingly important, but that clinical breakpoints are not particularly suitable to detect that. The EUCAST has therefore concluded that these two objectives of using breakpoints serve different purposes and that therefore, apart from clinical breakpoints, a new type of breakpoints should be introduced (EUCAST, 2003). These are wild-type cut-off values and are set at the edge of the wild-type distribution. Microorganisms that have MICs higher than this value will be easily detected, even if the MICs are well below the clinical breakpoint. For instance, in a study by Grohs *et al.* (2003), the wild-type MIC for moxifloxacin is 0.125 mg/L while strains with single ParC mutations have MICs of 0.5 mg/L. Since the clinical breakpoint of moxifloxacin is above 0.5 mg/L, these strains, indicating emergence of resistance, would not be identified using clinical breakpoints. A wild-type cut-off value of 0.25 mg/L however would pick-up these strains, alert the laboratory that this strain is not wild-type and subsequently relevant action can be taken.

3.3. Use of PK/PD to set clinical breakpoints for quinolones

Based on PK/PD relationships, considering that the dosing regimens in various countries in Europe are largely similar, EUCAST has proposed clinical breakpoints for a variety of antimicrobials and is still in the process of completing those for all antimicrobials. The breakpoints for quinolones are shown in Table 4 and are based on the pharmacokinetic properties of the drug (Table 3; EUCAST, 2003). The breakpoint values are consistent in that the S breakpoint determined from the AUC/MIC ratio is similar for all quinolones. There is one notable exception, the S breakpoint for levofloxacin and *S. pneumoniae*. The S breakpoint here is 2 mg/L instead of 1 mg/L because a 1 mg/L breakpoint would divide the wild-type population. It has to be realized however that the downside of designating all *S. pneumoniae* strains with an MIC of ≤2 mg/L susceptible is, that the probability of a successful outcome is slightly less than one would conventionally presume. This may be overcome by using a higher dose.

From the discussions above it thus becomes apparent and evident that two different types of breakpoints are required. Clinical breakpoints based on PK/PD relationships confer a different meaning to resistance than early detection of microorganisms that do not belong to the natural bacterial population, but somehow have acquired resistance mechanisms. An example is shown in Figure 8 for ciprofloxacin and *E. coli*. The figure shows the MIC distributions of *E. coli* from a variety of sources (EUCAST, 2003) with arrows indicating the wild-type cut-off breakpoint (for epidemiological purposes and early detection of emergence of resistance) as well as clinical breakpoints (based on PK/PD and probability of cure).

It has to be emphasized that clinical breakpoints may vary over time and place because they are dose dependent, but that wild-type cut-off values do not, because the susceptibility of wild-type bacteria to a drug is a universal characteristic. It is thus that the wild-type cut-off is especially suited to compare rates of resistance between countries and over time.

Table 3. PK parameters of four fluoroquinolones based on dosing regimens generally used. Adapted from Mouton, 2002

Fluoro quinolone	Regimen	C_{max} (mg/L)	AUC (mg · hr/L)	Protein binding (%)
Ciprofloxacin	500 mg/12 hr	2.8	22.2	22
Levofloxacin	500 mg/24 hr	5.2	61.1	30
Ofloxacin	200 mg/12 hr	2.2	29.2	30
Moxifloxacin	400 mg/24 hr	4.5	48.0	40

Table 4. EUCAST breakpoints for quinolones (EUCAST, 2003)

Fluoroquinolone[a]	Species-related breakpoints (S≤/R>)											Non-species related breakpoints[g] S≤/R>
	Entero bacteriaceae[b]	Pseudo-monas	Acineto-bacter	Staphylo-coccus[c]	Entero-coccus	Strepto-coccus A,B,C,G	S. pneu-moniae[d]	H. influenzae M. catarrhalis[e]	N. gonorr-hoeae	N. menin-gitidis[f]	Anaerobe bactera	
Ciprofloxacin	0.5/1	0.5/1	1/1	1/1	—	—	0.125/2	0.5/0.5	0.03/0.06	0.03/0.06	—	0.5/1
Levofloxacin	1/2	1/2	1/2	1/2	—	1/2	2/2	1/1	IE	IE	—	1/2
Moxifloxacin	1/2	—	—	IE	—	IE	0.5/0.5	0.5/0.5	IE	IE	IE	1/2
Ofloxacin	0.5/1	—	—	1/1	—	—	0.125/4	0.5/0.5	0.12/0.25	IE	—	0.5/1

Notes: '—' = Susceptibility testing not recommended as the species is a poor target for therapy with the drug.

IE = There is insufficient evidence that the species in question is a good target for therapy with the drug.

[a]For breakpoints for other fluoroquinolones (e.g., pefloxacin and enoxacin)—refer to breakpoints determined by national breakpoint committees.

[b]*Salmonella* spp.—there is clinical evidence for ciprofloxacin to indicate a poor response in systemic infections caused by *Salmonella* spp. with low-level fluoroquinolone resistance (MIC > 0.064 mg/L).

[c]*Staphylococcus* spp.—breakpoints for ciprofloxacin and ofloxacin relate to high dose therapy.

[d]*Streptococcus pneumoniae*—wild-type *S. pneumoniae* are not considered susceptible to ciprofloxacin or ofloxacin and are therefore categorized as intermediate. For ofloxacin the I/R breakpoint was increased from 1.0 to 4.0 mg/L to avoid dividing the wild-type MIC distribution.

The breakpoints for levofloxacin relate to high dose therapy.

[e]*Haemophilus/Moraxella*—fluoroquinolone low-level resistance (ciprofloxacin MICs of 0.125–0.5 mg/L) may occur in *H. influenzae*. There is no evidence that low-level resistance is of clinical importance in respiratory tract infections with *H. influenzae*. An intermediate category was not defined since only few clinically resistant strains have been reported.

[f]*Neisseria meningitidis*—breakpoints apply to the use of ciprofloxacin in the prophylaxis of meningococcal disease.

[g]Non-species related breakpoints have been determined mainly on the basis of PK/PD data and are independent of MIC distributions of specific species. They are for use only for species that have not been given a species-specific breakpoint and not for those species where susceptibility testing is not recommended (marked with "—" or IE in the table).

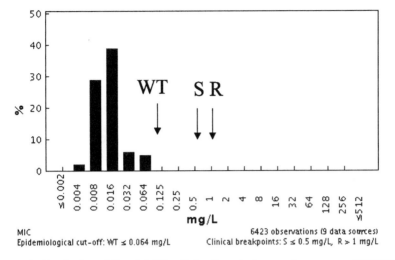

MIC   6423 observations (9 data sources)
Epidemiological cut-off: WT ≤ 0.064 mg/L   Clinical breakpoints: S ≤ 0.5 mg/L, R ≥ 1 mg/L

Figure 8. Distribution of *E. coli* MICs of ciprofloxacin from various sources (EUCAST, 2003). Arrows indicate the wild-type cut-off and clinical breakpoints, respectively.

4. PHARMACODYNAMIC RELATIONSHIPS AND EMERGENCE OF RESISTANCE

In the sections above, it was shown that there is a relationship between PK/PD index and efficacy and that this relationship can be used to optimize dosing regimens. The question that can be asked subsequently is, whether there is a relationship between PK/PD indices and emergence of resistance as well. To answer that question, two items need to be addressed. The first is whether the PK/PD index itself is the proper measure to use, more in particular whether the MIC part of the ratio should be applied to calculate the index or whether another pharmacodynamic measure of antibacterial activity is more appropriate. The second item that needs to be addressed is a measure of emergence of resistance. Both these items are discussed below.

4.1. Pharmacodynamic measures: PK/PD index, mutation prevention concentration and mutation selection window

The mutation prevention concentration (MPC) has been used recently as a measure for emergence of resistance. Although various definitions and method descriptions exist, the one used most often lately is comparable to an agar dilution MIC with a high inoculum. Agar plates are prepared with anti-microbial concentrations in twofold dilutions. The plates are inoculated with a high inoculum, typically 10^9 or more colony forming units (cfu). The MPC is the lowest concentration where no growth is observed.

The major shortcoming of the MPC is comparable to the use of the MIC as discussed above. The MPC, similar to the MIC, is determined at static concentrations and read after a certain period of time, while *in vivo* concentrations decline over time. In attempting to overcome this problem, to correlate the MPC to emergence of resistance, the mutant selection window is defined and it is hypothesized that when concentrations of an antimicrobial fall into this window there is a marked increase in risk of emergence of resistance. Concentrations should be above the MPC for the whole dosing interval or, at least, the duration of exposure at concentrations within this window should be as short as possible (Zhao and Drlica, 2002; Drlica, 2003) (Figure 9). Several authors have attempted to demonstrate the validity of the concept, mainly with the fluoroquinolones (Firsov *et al.*, 2003). Although some of these studies seem to favour the concept, it is difficult to prove and there are a number of theoretical and practical arguments against it. The two most important ones are again that both the MIC and the MPC are determined at static concentrations after a certain period of time and the window itself does not take into account the momentary effects of the concentrations, nor the effect of the concentrations

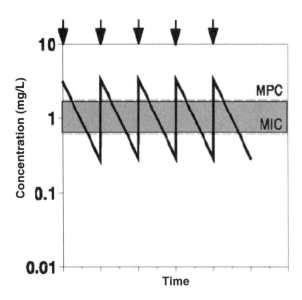

Figure 9. Diagram showing the mutation prevention window (shaded area).

over time. Another important argument is that the window is determined by concentrations determined in serum, while the concentration profile at the site of infection may be markedly different.

Another approach that has been taken is to correlate emergence of resistance to PK/PD indices, the effect parameter being the frequency of resistant clones after exposure. In this manner the total concentration–time profile is correlated to outcome in terms of emergence of resistance in a similar way to the efficacy of the drug. In a recent study by Firsov *et al.* (2003), it was demonstrated that there is a clear relationship between AUC/MIC ratio and frequency of emergence of resistance that followed a Bell-shaped curve when exposing four *S. aureus* strains to various dosing regimens of quinolones in an *in vitro* pharmacokinetic model (Figure 10). Indeed, the model fit to the data shown in the figure is analogous to the normal equation. At low AUC/MIC ratios the frequency of emergence of resistant clones is very low, as it is also low at relatively high values. There is a value of the AUC/MIC ratio that corresponds to the highest frequency of resistant clones after exposure, in this instance 43 hr. Similar relationships could be made for results obtained in other studies (Aeschlimann *et al.*, 1999; Firsov *et al.*, 2003; Peterson *et al.*, 1999). The question that arises is, how does the value

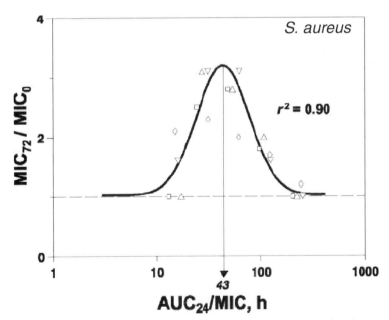

Figure 10. Bell-shaped relationship between AUC/MIC and frequency of resistance for quinolones. Reproduced from Firsov *et al.*, 2003.

of 43 hr compare to the PK/PD index value necessary for a maximum effect? In the paper mentioned this was not the subject of study, but other studies involving quinolones and Gram-positive microorganisms have found values between 30 and 50 hr. Thus, it may very well be that the PK/PD index value necessary to minimize emergence of resistance is higher than the value needed to obtain maximum efficacy. Experiments have also been performed *in vivo*. In an established abscess mixed model of infection Stearne *et al.* determined the frequency of emergence of resistant clones following exposure to ceftizoxime (Stearne *et al.*, 2002). The data presented allowed the determination of PK/PD index values for the various dosing regimens. Figure 11 shows the relationship between $T_{>MIC}$ and frequency of emergence of resistant clones. Again, a Bell-shaped curve is observed. Apparent from the figure is that the highest frequency of resistance is observed at values of 60–70% while the frequency is very low at $T_{>MIC}$ values above 87%. These values are higher than those where a maximum effect is reached. One study in humans has shown that the probability of emergence of resistance was far greater when AUC/MIC ratios were lower than 100 hr as compared to values above 100 hr (Figure 12) (Thomas *et al.*, 1998).

Taken together, these studies, although still few in number, show that there is a clear relationship between PK/PD index values and emergence of resistance and that suboptimal dosing leads to a higher probability of emergence of resistance. It is therefore important that dosing schedules of antimicrobials are derived which do take these relationships into account. This probably also applies to breakpoints, but this has, as yet, to be taken into consideration.

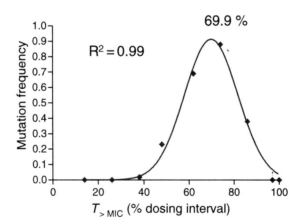

Figure 11. Bell-shaped relationship between $T_{>MIC}$ and frequency of resistance for ceftizoxime. Based on data from Stearne *et al.*, 2002.

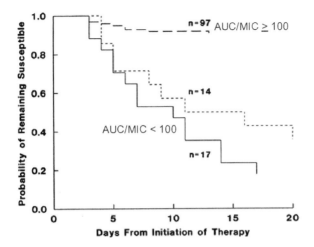

Figure 12. Relationship between AUC/MIC and probability of emergence of resistance (Thomas *et al.*, 1998).

5. CONCLUDING REMARKS

It is evident that increasing insight in dose–effect relationships of antimicrobials has led to a re-appreciation of the meaning of susceptible and resistance. In a world where resistance is an escalating problem there is both a need to have the ability for fast intervention as well as a requisite to treat patients in the best way possible. In retrospect, it is not surprising that the limitations of one breakpoint system encompassing both these key elements have lead to controversies, as they were anticipated in the report by Ericsson and Sherris in 1971(Ericsson and Sherris, 1971).

The approach that EUCAST is now taking, by formulating both clinical breakpoints and wild-type cut-off values (or epidemiological breakpoints), to distinguish between the two objectives can therefore also be regarded as a continuance of that earlier work. Although it may be difficult to reach global consensus on clinical breakpoints (actually impossible because of differences in dosing), global consensus should easily be reached for wild-type cut-off values. Comparisons between resistance rates will then be more rational, while early intervention will be more likely.

REFERENCES

Aeschlimann, J. R., Kaatz, G. W., and Rybak, M. J., 1999, *J. Antimicrob. Chemother.*, **44**, 343–349.

Ambrose, P. G. and Grasela, D. M., 2000, *Diagn. Microbiol. Infect. Dis.*, **38**, 151–157.

Ambrose, P. G., Grasela, D. M., Grasela, T. H., Passarell, J., Mayer, H. B., and Pierce, P. F., 2001, *Antimicrob. Agents. Chemother.*, **45**, 2793–2797.

Andes, D. and Craig, W. A., 2002, *Int. J. Antimicrob. Agents.*, **19**, 261–268.

Andes, D. R. and Craig, W. A., 1998, *Clin. Infect. Dis.*, **27**, 47–50.

Andrews, J. M., 2001, *J. Antimicrob. Chemother.*, **48** (Suppl. 1), 5–16.

Bedos, J. P., Azoulay-Dupuis, E., Moine, P., Muffat-Joly, M., Veber, B., Pocidalo, J. J. *et al.*, 1998a, *J. Pharmacol. Exp. Ther.*, **286**, 29–35.

Bedos, J. P., Rieux, V., Bauchet, J., Muffat-Joly, M., Carbon, C., and Azoulay-Dupuis, E., 1998b, *Antimicrob. Agents. Chemother.*, **42**, 862–867.

Bergan, T., Bruun, J. N., Digranes, A., Lingaas, E., Melby, K. K., and Sander, J., 1997, *Scand. J. Infect. Dis. Suppl.*, **103**, 1–36.

Bonate, P. L., 2001, *Clin. pharmacokinet.*, **40**, 15–22.

Cars, O., 1990, *Scand. J. Infect. Dis. Suppl.*, **74**, 23–33.

Croisier, D., Chavanet, P., Lequeu, C., Ahanou, A., Nierlich, A., Neuwirth, C. *et al.*, 2002, *J. Antimicrob. Chemother.*, **50**, 349–360.

de Man, P., Goessens, W. H., Verbrugh, H. A., and Mouton, J. W., 1998, In *European Congress Chemotherapy*. Hamburg, p. 85.

Deutsches Institut fur Normung, 1990, Deutsches Institut fur Normung, Beuth Verlag.

Drlica, K., 2003, *J. Antimicrob. Chemother.*, **52**, 11–17.

Drusano, G. L., D'Argenio, D. Z., Preston, S. L., Barone, C., Symonds, W., LaFon, S. *et al.*, 2000, *Antimicrob. Agents. Chemother.*, **44**, 1655–1659.

Drusano, G. L., Johnson, D. E., Rosen, M., and Standiford, H. C., 1993, *Antimicrob. Agents. Chemother.*, **37**, 483–490.

Drusano, G. L., Preston, S. L., Hardalo, C., Hare, R., Banfield, C., Andes, D. *et al.*, 2001, *Antimicrob. Agents. Chemother.*, **45**, 13–22.

Ericsson, H. M. and Sherris, J. C., 1971, *Acta Pathol. Microbiol. Scand. [B] Microbiol. Immunol.*, **217**, 1–90.

Ernst, E. J., Klepser, M. E., Petzold, C. R., and Doern, G. V., 2002, *Pharmacotherapy*, **22**, 463–470.

EUCAST, 2003.

Fernandez, J., Barrett, J. F., Licata, L., Amaratunga, D., and Frosco, M., 1999, *Antimicrob. Agents. Chemother.*, **43**, 667–671.

Firsov, A. A., Vostrov, S. N., Lubenko, I. Y., Drlica, K., Portnoy, Y. A., and Zinner, S. H., 2003, *Antimicrob. Agents. Chemother.*, **47**, 1604–1613.

Forrest, A., Chodosh, S., Amantea, M. A., Collins, D. A., and Schentag, J. J., 1997, *J. Antimicrob. Chemother.*, **40**(Suppl. A), 45–57.

Forrest, A., Nix, D. E., Ballow, C. H., Goss, T. F., Birmingham, M. C., and Schentag, J. J., 1993, *Antimicrob. Agents. Chemother.*, **37**, 1073–1081.

Grohs, P., Houssaye, S., Aubert, A., Gutmann, L., and Varon, E., 2003, *Antimicrob. Agents Chemother.*, **47**, 3542–3547.

Highet, V. S., Forrest, A., Ballow, C. H., and Schentag, J. J., 1999, *J. Antimicrob. Chemother.*, **43** (Suppl. A), 55–63.

Kahlmeter, G., Brown, D. F. J., Goldstein, F. W., MacGowan, A. P., Mouton, J. W., Österlund, A. *et al.*, 2003, *J. Antimicrob. Chemother.*

Lacy, M. K., Lu, W., Xu, X., Tessier, P. R., Nicolau, D. P., Quintiliani, R. *et al.*, 1999, *Antimicrob. Agents Chemother.*, **43**, 672–677.

Lim, S., Bast, D., McGeer, A., de Azavedo, J., and Low, D. E., 2003, *Emerg. Infect. Dis.*, **9**, 833–837.

Lister, P. D., 2002, *Antimicrob. Agents Chemother.*, **46**, 69–74.

Lister, P. D. and Sanders, C. C., 1999a, *J. Antimicrob. Chemother.*, **43**, 79–86.

Lister, P. D. and Sanders, C. C., 1999b, *Antimicrob. Agents Chemother.*, **43**, 1118–1123.

Liu, P., Muller, M., and Derendorf, H., 2002, *Int. J. Antimicrob. Agents.*

MacGowan, A. P. and Wise, R., 2001, *J. Antimicrob. Chemother.*, **48**(Suppl. 1), 17–28.

Madaras-Kelly, K. J., Ostergaard, B. E., Hovde, L. B., and Rotschafer, J. C., 1996, *Antimicrob. Agents. Chemother.*, **40**, 627–632.

Mattoes, H. M., Banevicius, M., Li, D., Turley, C., Xuan, D., Nightingale, C. H. *et al.*, 2001, *Antimicrob. Agents. Chemother.*, **45**, 2092–2097.

Montgomery, M. J., Beringer, P. M., Aminimanizani, A., Louie, S. G., Shapiro, B. J., Jelliffe, R. *et al.*, 2001, *Antimicrob. Agents. Chemother.*, **45**, 3468–3473.

Mouton, J. W., 2002, *Int. J. Antimicrob. Agents*, **19**, 323–331.

Mouton, J. W., 2003, *Infect. Dis. Clin. North Am.*, **17**, 579–598.

Mouton, J. W., Dudley, M. N., Cars, O., Derendorf, H., and Drusano, G. L., 2002a, *Int. J. Antimicrob. Agents*, **19**, 355–358.

Mouton, J. W., Horrevorts, A. M., Mulder, P. G., Prens, E. P. and Michel, M. F., 1990, *Antimicrob. Agents Chemother.*, **34**, 2307–2311.

Mouton, J. W. and Punt, N., 2001, *J. Antimicrob. Chemother.*, **47**, 500–501.

Mouton, J. W., Schmitt-Hoffmann, A., Shapiro, S., Nashed, N. and Punt, N. C., 2004, *Antimicrob. Agents Chemother.*, **48**, 1713–1718.

Mouton, J. W., van Klingeren, B., de Neeling, A. J., Degener, J. E., and Commissie Richtlijnen Gevoeligheidsbepalingen, 2000, *Nederlands Tijdschrift voor Medische Microbiologie*, **8**, 73–78.

NCCLS, 2002, *Performance Standards for Antimicrobial Susceptibility Testing; Twelfth International Supplement. M100-S12 (M7).* NCCLS, Wayne.

NCCLS, 2003, www.nccls.org.

Ng, W., Lutsar, I., Wubbel, L., Ghaffar, F., Jafri, H., McCracken, G. H. *et al.*, 1999, *J. Antimicrob. Chemother.*, **43**, 811–816.

Nicolau, D. P. and Ambrose, P. G., 2001, *Am. J. Med.*, **111**(Suppl. 9A), 13S–18S; discussion 36S–38S.

Onyeji, C. O., Bui, K. Q., Owens, R. C., Jr., Nicolau, D. P., Quintiliani, R., and Nightingale, C. H., 1999, *Int. J. Antimicrob. Agents*, **12**, 107–114.

Peterson, M. L., Hovde, L. B., Wright, D. H., Hoang, A. D., Raddatz, J. K., Boysen, P. J. *et al.*, 1999, *Antimicrob. Agents Chemother.*, **43**, 2251–2255.

Preston, S. L., Drusano, G. L., Berman, A. L., Fowler, C. L., Chow, A. T., Dornseif, B. *et al.*, 1998, *J. Am. Med. Association*, **279**, 125–129.

Scaglione, F., Mouton, J. W., Mattina, R. and Fraschini, F., 2003, *Antimicrob. Agents Chemother.*, **47**, 2749–2755.

Soussy, C. J., Cluzel, R., and Courvalin, P., 1994, *Eur. J. Clin. Microbiol. Infect. Dis.*, **13**, 238–246.

Stearne, L. E., Lemmens, N., Goessens, W. H. F., Mouton, J. W., and Gyssens, I. C., 2002, In *European Conference Clinical Microbiology and Infectious Diseases.* Milan.

Swedish Reference Group of Antibiotics, 1997, *Scandinavian J. Infect. Dis.*, **105S**, 5–31.

Thomas, J. K., Forrest, A., Bhavnani, S. M., Hyatt, J. M., Cheng, A., Ballow, C. H. *et al.*, 1998, *Antimicrob. Agents. Chemother.*, **42**, 521–527.

Turnidge, J., 1999, *Drugs*, **58**(Suppl. 2), 29–36.

Zhao, X., and Drlica, K., 2002, *J. Infect. Dis.*, **185**, 561–565.

Chapter 22

Types of Surveillance Data and Meaningful Indicators for Reporting Antimicrobial Resistance

Hervé M. Richet

Laboratoire de Bactériologie—Hygiène Hospitalière, Institut de Biologie des Hôpitaux de Nantes, Centre Hospitalier Universitaire de Nantes, 44093 Nantes Cedex 01, France

1.   INTRODUCTION

Available data indicate that many important human pathogens carry antimicrobial resistance characters and these pathogens are sometimes resistant to all or nearly all available antimicrobials causing a major public health threat worldwide in both community and healthcare settings. The control of this phenomenon is now considered a priority by organizations such as the World Health Organization (WHO). However, the control of this impending public health problem may be hampered in developed countries by the increasing complexity of the healthcare delivery system and by attempts to reduce the costs of healthcare. In contrast, lesser developed countries may lack the indispensable basic microbiology and infection control resources needed to analyze the situation and to set up control and prevention programs.

Conscious of the problem caused by the emergence of multidrug resistance, the European Union (EU) organized a conference in Denmark in 1998 entitled "The Microbial Threat" (The Copenhagen Recommendations, 1998). The recommendations elaborated by the EU experts participating in this meeting included the need to collect good quality data on resistant microorganisms essential to underpin effective interventions to counter the problem of resistance and developing guidelines for prescribing antimicrobial drugs. In addition, the EU experts emphasized that data collected have to be clinically and epidemiologically relevant.

Antibiotic Policies: Theory and Practice. Edited by Gould and van der Meer
Kluwer Academic / Plenum Publishers, New York, 2005

Earlier, a task force set up by the American Society for Microbiology (ASM) in 1995 had developed recommendations for dealing with the rise of antimicrobial resistance in bacterial pathogens (Cassell, 1995). Here are the main recommendations of the task force:

- Increase basic research and training
- Develop more consistent measures of antimicrobial resistance
- Develop more information on the impact of infectious diseases
- Surveillance measures to provide a long-term view of the emergence and development of antimicrobial resistance are urgently needed. Developing reliable baseline information about current levels of antimicrobial resistance in animal and human pathogens is considered critical to these efforts.

Both the EU experts and the ASM task force mention surveillance of antimicrobial resistance as a critical task to perform in order to contain and/or prevent antimicrobial resistance.

2.   DEFINITION OF SURVEILLANCE

According to the United States Centers for Disease Control and Prevention (CDC), surveillance is defined as "the ongoing, systematic collection, analysis, and interpretation of health data essential to the planning, implementation, and evaluation of public health practice, closely integrated with the timely dissemination of these data to those who need to know" (CDC, 1988a). The key words in this definitions are probably "planning, implementation, and evaluation of public health practice" and "timely dissemination of the data." These words are important because they mean that a surveillance program not used as an evaluation tool is useless and that surveillance programs should have an impact on the control and prevention of diseases under surveillance.

3.   EVALUATION OF THE ANTIMICROBIAL RESISTANCE SURVEILLANCE DATA PUBLISHED IN THE MEDICAL LITERATURE

The perception of antimicrobial resistance as a threat by a growing number of microbiologists, physicians, public health authorities, or scientific societies has led to the implementation of surveillance programs at the local, national, or international levels. However, the continuous emergence of new resistance phenotypes and the spread of multidrug-resistant organisms suggest that those surveillance programs as well as the efforts at controlling those phenomena are failing. Therefore, before addressing the issue of how to conduct surveillance of antimicrobial resistance it seems important to evaluate the characteristics of the existing surveillance programs by answering the following questions. What antimicrobial resistance surveillance activities are conducted? Are the data available/published on antimicrobial resistance predictive/representative of the global situation? Can the available/published data on antimicrobial resistance be used to intervene or to predict the evolution of this phenomenon?

3.1. Results of the review

To assess if the available surveillance data can serve such purpose, a review of all articles published in the medical literature from January 1, 2000 to October 31, 2000 and indexed in Medline under the key words "surveillance" and "antimicrobial resistance" was performed for a review article published in ASM news (Richet, 2001). The search retrieved 101 articles showing that surveillance data were collected in 32 individual countries in Africa, Asia, North and Latin America, Central and Western Europe, and the Pacific Region. Regarding the microorganisms under surveillance, a combination of multiple Gram-negative and Gram-positive bacteria was included in 28% of the surveillance programs, followed by *Streptococcus pneumoniae* (18%); *Enterococcus* spp. and *Escherichia coli* (9% each); *Salmonella* spp. and *Haemophilus* spp. (5.6% each); a combination of various Gram-negative bacilli, various *Enterobacteriaceae*, and *Staphylococcus aureus* (4.4% each); methicillin-resistant *S. aureus*, coagulase negative staphylococci, *Pseudomonas aeruginosa*, *Vibrio cholerae*, and *Moraxella catarrhalis* (3.3% each). The majority (81%) of the surveillance programs conducted surveillance of three or more different classes of antimicrobials. When only one class of antimicrobials was surveyed, β-lactams were the most common class of antimicrobial surveyed (13.3%), followed by fluoroquinolones (6.7%), cephalosporins (4.4%), glycopeptides (3.3%), aminoglycosides (3.3%), macrolides (3.3%), sulfonamides (2%), penicillins (1%), trimethoprim/sulfamethoxazole (1%).

Bacteremia was the most common infection surveyed and was included in 20% of the surveillance programs followed by respiratory tract infections (14.4%), diarrhea and urinary tract infection (11%), and meningitis (6.7%). In addition, surveillance of colonization was included in six surveillance programs.

The mean number of isolates included in the surveillance programs was 1,772, the median 502 ranging from 66 to 34,530 isolates. Data were collected during a mean of 37 months with a median duration of 24 months ranging from 1 to 156 months.

The setting of the surveillance was not mentioned in more than a third (36.7%) of the reviewed articles. Almost half (47.8%) of the surveillance activities were performed in healthcare facilities and 18% in the community. More specifically, hospital acquired infections were the object of surveillance in 24.4% of the programs, community acquired infections in 22% of the programs, adults in 15.6% of the programs and animals in 8% of the programs. The characteristics of the individuals/patients under surveillance were mentioned in only 22 of the articles, those characteristics included hospitalization in intensive care units (10%), HIV infection (4.4%), cancer (3.3%), newborns (2%), transplantation (1%), and immunosuppression (1%).

In 62% of the articles, the only results presented were the rates of resistance to the different antimicrobials tested. In the remaining articles, in addition to the resistance rates, the results of additional studies were shown. They included molecular typing of the isolates in 15.6% of the articles, the analysis of risk factors for infections caused by a resistant organism (12%), the genetic analysis of resistance characters (8%), the assessment of mortality (5.6%), the assessment of the relationship between antimicrobial use and antimicrobial resistance (4.4%), evaluation of antimicrobial therapy (2%), and evaluation of control measures (1%).

3.2.   Critical evaluation of the surveillance programs

To assess how the published data could be used to implement control and prevention strategies, we used as evaluation criteria the attributes of a surveillance program as defined by the CDC (1988b). Evaluation of surveillance systems should promote the best use of public health resources by ensuring that only important problems are under surveillance and that surveillance systems operate efficiently. Among the attributes of a "good surveillance system" are simplicity, flexibility, representativeness, timeliness, and usefulness.

3.2.1.   Simplicity

The simplicity of a surveillance system refers to both its structure and ease of operation. This includes the amount and type of information necessary to establish the diagnosis, the number and type of reporting sources, the methods of data transmission, and the number of organizations involved in receiving case reports.

Surveillance of antimicrobial resistance is most often a very simple process since the amount and type of information needed are laboratory data and nowadays, most laboratory data are computerized and supposedly easy to use for surveillance purpose. Most of the surveillance programs included in this review were very simple since they only included laboratory data and the results presented were restricted to the proportion of resistance to various antimicrobials. However, regarding surveillance of antimicrobial resistance, simplicity may be a limitation since, for instance, the sites of the infections and/or the specimens included were not mentioned in 33.3% of the articles and the patients' characteristics were not presented in 36.7% of the articles.

3.2.2. Flexibility

A flexible surveillance system can adapt to changing information, needs, or operating conditions and accommodate new diseases and health conditions. Flexibility is best judged retrospectively by observing how a system responded to a new demand. An example is the capacity of an *S. aureus* surveillance system to accommodate special surveillance for glycopeptide-intermediate or resistant *S. aureus*.

Flexibility is a very important attribute for antimicrobial resistance surveillance programs because they have to detect emerging resistance phenotypes or emerging pathogens. Unfortunately, no early warning system was described and the flexibility of the surveillance systems was never demonstrated.

3.2.3. Representativeness

A surveillance system that is representative, accurately describes the occurrence of a health event over time and its distribution in the population by place and person. Representativeness is assessed by comparing the characteristics of reported events to all such actual events. Judgment of the representativeness is possible based on knowledge of characteristics of the population (age, socioeconomic status, geographic location), natural history of the condition, mode of transmission, prevailing medical practices, multiple source of data, and quality of data. Regarding representativeness, there is both good and bad news. The good news is that the global nature of antimicrobial resistance is clearly addressed by the publication of antimicrobial resistance data collected in both developed and developing countries. Even if 18% of the data on antimicrobial resistance came from the United States, antimicrobial resistance data were also provided by lesser developed and even developing countries in Africa, Central and South America, Asia, and Central Europe. Since antimicrobial resistance is a global problem, these data are valuable because the extent of antimicrobial resistance in developing countries, where inappropriate antimicrobial usage may be more common, was until recently unknown. The spectrum of microorganisms under surveillance was also broad, including pathogens and opportunistic pathogens, although the pathogens were unevenly distributed with some such as *S. pneumoniae* probably being overrepresented. The bad news is that there was usually little information on the characteristics of the population, the natural history of the conditions such as the modes of transmission, the medical and laboratory practices, and the quality of the data, was rarely assessed. One problem with the representativeness of antimicrobial resistance surveillance activities is that many of those programs are sponsored by drug companies who will therefore influence the choice of the microorganisms and antimicrobials under surveillance. This may explain why *S. pneumoniae* was the leading microorganism surveyed and cephalosporins and fluoro-quinolones the leading antimicrobials.

3.2.4.   Timeliness

Timeliness reflects the speed or delay between steps in a surveillance system and may be evaluated in terms of availability of information for disease control, either for immediate control efforts or for long-term program planning. The timeliness of these antimicrobial resistance surveillance programs can be evaluated by looking at the interval between the end of data collection and their publication. The median interval between the end of the data collection and publication was 2 years ranging from 1 year to 5 years. The timeliness of these surveillance programs is probably not optimal since antimicrobial resistance is rapidly evolving and a 2 year delay is probably much too long. The published data probably do not reflect anymore the current state of antimicrobial resistance.

3.2.5.   Usefulness

A surveillance system is useful if it contributes to the prevention and control of adverse health events. A surveillance system can also be useful if it helps determine that an adverse event previously thought to be unimportant is actually important. The evaluation of the usefulness can be done by answering the following questions:

Does the system detect trends signaling changes in the occurrence of diseases?
The duration of surveillance, which was relatively short (median 24 months), as well as the fragmentary nature of the published data (different laboratories, different methods, different microorganisms, different antimicrobials, different settings when they are known, different populations) and the multiplicity of surveillance programs make the detection of trends rather difficult. What we can observe through the reviewed publications looks like a puzzle or a series of snapshots rather than a continuous process or a coherent ensemble.

Does the system provide estimates of the magnitude of morbidity and mortality related to the health problem under surveillance?
Regarding surveillance of antimicrobial resistance, the answer is mostly no since very few studies sought to assess the mortality and there was nothing

about morbidity or cost.It is obvious that the assessment of mortality, morbidity, or cost associated with antimicrobial resistance is a very difficult task requiring the collection of extensive clinical data including severity of disease scores and the use of sophisticated epidemiologic methods like matched case-control studies. Once again, most laboratories, even when they are computerized, receive and store very few clinical data and it is, for instance, usually impossible to assess, with the information present on the laboratory request form, if the infection is hospital or community acquired.

Does the system stimulate epidemiologic research likely to lead to control and prevention?

Control and prevention of antimicrobial resistance requires knowledge of risk factors leading to the emergence of new resistance genes or phenotypes, and knowledge of the modes of transmission of the resistance genes and organisms. The most common epidemiologic research performed was molecular typing of isolates in 16% of the articles, genetic analysis of the resistance genes in 8% of the articles, and assessment of the relationship between antimicrobial use and antimicrobial resistance in 4% of the articles.

Does the system identify risk factors associated with disease occurrence?

Risk factors for infections caused by resistant organisms were assessed in only 12% of the studies and the relationship between antimicrobial use and antimicrobial resistance was evaluated in only 4% of the publications. It is obviously difficult to implement efficient control strategies for a disease if the risk factors for this disease are not clearly identified. This is especially difficult for antimicrobial resistance because many factors related to the organism, the host, the therapies, procedures, or precautions may influence the risk of emergence and dissemination of antimicrobial resistance genes and/or antimicrobial resistant organisms.

Does the system permit assessment of the effects of control measures?

Only one article presented the results of an evaluation of control measures.

Does the system lead to improved clinical practice by the healthcare providers who are the constituents of the surveillance system?

Because the majority of antimicrobial therapies are prescribed before the susceptibility of the infecting strain is known, the production of epidemiologic data about antimicrobial resistance is essential. Therefore, the most useful benefit of the reviewed articles could be, if timeliness is not an issue, to improve the prescription of antimicrobials by making the prescription evidence based.

In conclusion, this review of the literature shows relatively little good news regarding our capacity to control the spread of antimicrobial resistance and this may be explained by the fact that the public health approach is rarely integrated into antimicrobial resistance surveillance activities, as shown by the fact that the majority of surveillance programs are laboratory based. Very few clinical data are collected and the data obtained by most surveillance programs cannot be used to implement control and/or prevention measures.

4.   HOW TO DESIGN A SURVEILLANCE PROGRAM

The major dilemma when setting up a surveillance program is to decide which bacteria, which antimicrobials, which denominators, which susceptibility testing method, which specimens, which patients, and which setting to use and which data to collect. Theses choices may be made easier by assessing the public health importance of these indicators. Therefore such choice should be based upon the expected number of cases, the severity of the health event measured by the mortality rate and the case-fatality ratio, the medical cost of the event, and its preventability. These surveillance programs should be useful, meaning that they should contribute to the prevention and control of antimicrobial resistance. To do that, microbiologist will have to collect not only laboratory data but the clinical data that are missing in the majority of the reviewed articles and are nevertheless needed to design prevention and control strategies. Criteria used to evaluate a surveillance system can also be used to determine which bacterial species and antimicrobials to survey in which population. In addition to the criteria presented in the previous chapter, which include simplicity, flexibility, representativeness, and timeliness, other criteria such as acceptability, sensitivity and predictive positive value can also be used.

Acceptability. Acceptability reflects the willingness of individuals and organization to participate in the surveillance system. Factors influencing acceptability include the public health importance of the health event, the recognition by the system of the individual's contribution, the responsiveness of the system to suggestions or comments, the time burden relative to available time, and also the legislative requirement for reporting.

Sensitivity. The sensitivity of a surveillance system can be considered on two levels. At the level of case reporting, the proportion of cases of a disease detected by the surveillance system can be evaluated for its ability to detect epidemics. This can be important with antimicrobial resistance.

Predictive positive value. The predictive positive value is the proportion of persons identified as having cases who actually do have the condition under surveillance.

5. PRACTICAL ASPECTS OF THE IMPLEMENTATION OF THE SURVEILLANCE PROGRAM

An antimicrobial resistance surveillance program should be based on the acquisition and reporting of reliable standardized, quality controlled, non-duplicative data. This objective may be achieved by obtaining appropriate specimens from the infected individual, by the successful identification and isolation of the causative organism, by the accurate determination of the antimicrobial susceptibility of the isolate and by transforming the data into useful information for action. The microorganisms, the antimicrobials, and the setting of the surveillance have to be chosen according to the criteria presented above but additional decisions should be taken about the choice of specimens, method and duration of the surveillance program, indicator, and denominator. In the meantime, if they do not exist, quality control programs should be implemented.

Case definition. Ideally, surveillance of antimicrobial resistance should involve the collection of both clinical and microbiological data. Therefore, the first step in setting up a surveillance program is to design a case definition. As usual, the case definition should serve as inclusion criteria and include information about time, place, and persons. Here are two examples of case definitions of infections and of the data to collect for key pathogens developed by WHO (2002).
Pneumonia:

(a) clinical: febrile illness with purulent productive cough, rapid breathing in children;
(b) laboratory: isolation of *S. pneumoniae* or *H. influenzae* from sputum or blood;
(c) appropriate specimens: sputum, blood (nasopharyngeal swabs may be used in children);
(d) optimal sampling location and surveillance type: primary health care facility.

Urinary tract infection:

(a) clinical: frequency and dysuria or fever in presence of indwelling catheter or other focus of infection;
(b) laboratory: isolation of *E. coli* from urine in significant numbers or blood;
(c) appropriate specimen: urine (midstream or catheter specimen);
(d) optimal sampling location: primary healthcare facility.

Mode of surveillance. Surveillance may be defined as passive or active. Passive surveillance is done when data are awaited and no attempts are made to seek reports actively from the primary data collector. Surveillance is said to be active when reports are sought from the primary data collector on a regular basis. Active surveillance is the method of choice.

Choice of specimens. A distinction should be made between specimens collected to diagnose an infection and those realized to detect colonization. The choice of the specimens to include depends on the objectives of the surveillance program. The types of specimens to include should be part of the case definition. Emerging resistance may present first in colonizing organisms.

Data to collect. Below is an example of minimum data set to collect for a specific organism according to WHO (2002). Recommended minimum data set for *S. aureus*:

(a) case-based data: unique identifier capable of cross linkage with laboratory data. Age or date of birth; gender; healthcare facility and care group; date of admission and of onset; presenting sign, symptoms; predisposing factors (surgery, trauma, indwelling devices);
(b) laboratory-based data: unique identifier capable of cross-linkage with clinical data. Specimen date and type; method of identification, resistance to specified antimicrobials.

Microbiological representativeness. Colonies used for susceptibility testing should be representative of the culture as a whole.

Expression of antimicrobial resistance data. Laboratories should use standards for reporting quantitative resistance data (e.g., minimal inhibitory concentrations or zone diameters) in ways that will detect decreased susceptibility. This is necessary because numerical antimicrobial test results reported non-quantitatively (e.g., as susceptible, intermediate, or resistant) may hide an emerging resistance character in microorganisms with a small decrease in susceptibility that may still be classified as susceptible.

Choice of the surveillance method. The measure of the incidence rate is the best way to conduct surveillance of antimicrobial resistance. Incidence quantifies the number of new cases of diseases in a population at risk during a specific time interval. The duration of surveillance should be based on the expected rate of the event under surveillance. Table 1 shows the estimate of sample size needed for documenting increasing antimicrobial frequencies. As an example, if 5% of isolates in a sample of 200 are resistant, an increase to 11% or more in a second sample will show a significant increase.

Numerator for surveillance. For reliable and accurate information, data should relate to a single episode of disease in each patient. Regarding surveillance of antimicrobial resistance, it is of the utmost importance to avoid

Table 1. Estimate of sample size needed for documenting increasing antimicrobial resistance (WHO, 2002)

% of resistance detected in original sample	% resistance (indicative of significant increase) detectable in a second sample at sample size of				
	100	200	400	600	1000
2	9	7	5	4	3
5	14	11	9	8	7
10	21	17	15	14	12
25	39	35	32	31	28
50	65	60	58	25	54

including duplicate results since a patient may have either consecutive cultures obtained from the same body site or cultures from different body sites yielding the same organism (e.g., urine and blood culture). Therefore, only the first positive culture from the patient for each disease episode should be reported for surveillance purposes. However, there is a need for a clear definition of what represents a "disease episode." Emerging resistance in an organism isolated previously from the same patient could count, according to a predefined case definition, as a new episode.

Denominator for surveillance. Whenever possible, rates should be expressed as incidence rates within a defined human population instead of using the number of isolates tested as denominators. This is important because the submission of microbiology specimens to the laboratory is inconsistent and varies broadly. Rates produced by using this method are of limited epidemiological relevance. In hospital settings, it is recommended to use the number of admissions and the number of days of hospitalization, which are particularly useful for inter- or intra-healthcare facility comparison.

Quality control program. Organizers of antimicrobial resistance programs should ensure that clinical laboratories providing data for the surveillance program have access to and routinely participate in pertinent training and proficiency testing programs with good performance and that they indicate in their reports the antimicrobial resistance testing methods they use (e.g., specific automated methods or manual techniques). Internal and external quality control should be performed.

6. CONCLUSIONS

Surveillance programs of antimicrobial resistance should stimulate epidemiologic research and improve control and prevention strategies. To control the development of antimicrobial resistance, investigators generally need to understand risk factors that lead to resistance, modes of transmission of resistance traits, and other features that contribute to the emergence of resistant pathogenic microorganisms.

Unfortunately, the results of the survey presented in this chapter may help to explain why it is proving so difficult to slow the expansion of antimicrobial resistance. The underlying reason could be that a public health approach is rarely integrated into antimicrobial resistance surveillance activities. For that to happen, more clinical microbiologists should be trained in epidemiology. In addition, the choice of microorganisms and antimicrobials to survey should be based on their relative public health importance, using criteria such as expected numbers of cases, severity of the infectious disease as measured by its mortality rate and case-fatality ratio, medical costs of such infections, and preventability. Moreover, when surveillance programs are being developed, they should be designed to be useful, meaning that they should contribute to the prevention and control of antimicrobial resistance. To do that, microbiologists will have to collect not only laboratory data but the clinical data that is so often missing from surveillance reports.

REFERENCES

Cassell, G. H., 1995, ASM task force urges broad program on antimicrobial resistance. *ASM News*, **61**, 116–120.

CDC, 1988a, *CDC Surveillance Update*. Atlanta, GA.

CDC, 1988b, Guideline for evaluating surveillance systems. *CDC, MMWR*, **37**.

Richet, H. 2001, Better antimicrobial resistance surveillance efforts are needed. *Am. Soc. Microbiol. News*, **67**, 1–6.

The Copenhagen Recommendations, 1998, Report from the Invitational EU Conference on the Microbial Threat. Ministry of Health, Ministry of Food, Agriculture and Fisheries, Denmark. Edited by the Statens Serum Institut, Copenhagen, Denmark (www.microbial.threat.DK)

WHO, 2002, Surveillance standards for antimicrobial resistance. Geneva, Switzerland.

Chapter 23

Data Mining to Discover Emerging Patterns of Antimicrobic Resistance

J. A. Poupard, R. C. Gagnon, and M. J. Stanhope
GlaxoSmithKline, Collegeville, PA, USA

1.   INTRODUCTION

Antimicrobial susceptibility testing results are typically presented as summary information in the form of percent susceptible-intermediate-resistant (SIR), as the minimum concentrations required to inhibit 50% or 90% of isolates ($MIC_{50/90}$s) or as MIC frequency distributions. However, extracting additional information from large databases, involving thousands of isolates tested against more than 20 antimicrobial agents containing 6–10 individual dilutions, involve data points numbering in the millions. In such cases, traditional methods of analysis are insufficient. One approach to dealing with these levels of complexity is by applying novel data mining procedures. Although it should be noted that there is no universal definition of the term "data mining," for the purposes of this chapter it is defined as:

A new discipline lying at the interface of statistics, database technology, pattern recognition and machine learning, and concerned with secondary analysis of large databases in order to find previously unsuspected relationships, which are of interest or value to their owners. (Hand, American Statistician, 1998 [Hand, 1998]).

Due to the complex nature of data mining in this context, a team effort is required involving experts from various fields, including specialists in computer/bioinformatics. Regardless of the number of disciplines involved, it is critical that someone specializing in microbiology or infectious diseases is

Antibiotic Policies: Theory and Practice. Edited by Gould and van der Meer
Kluwer Academic / Plenum Publishers, New York, 2005

included. This person should have full knowledge of the limitations, in design and execution, of the susceptibility testing procedures used to generate the data. In this capacity, the microbiologist evaluates the data generated and searches out inconsistencies in those data as well as in the quality control procedures associated with the primary database. Any inconsistencies that are revealed need to be pursued and resolved in order to assure validation of the primary database. Failure to do this thoroughly will undermine any further analysis conducted. As data mining of large multinational resistance databases is relatively novel, the complexity of these procedures is just now becoming apparent.

The goal of this chapter is to focus attention on methods for identifying new or unrecognized resistance patterns in large surveillance study databases. Application of these data for use in resistance modeling and infection control is addressed elsewhere in this book. Although methods applicable to these large databases certainly apply to information generated in individual hospital laboratories, the large surveillance databases pose a special problem because of the volume of data involved.

2.   BRIEF LITERATURE REVIEW

Although it may not have been called data mining, the extraction of information from large databases has been a hallmark of epidemiology studies for a long period of time (Kaslow and Moser, 2000). In the 1980s investigators started to apply data mining techniques in attempts to combine antimicrobial susceptibility testing information obtained from the clinical microbiology laboratory with hospital/medical center information systems to help identify specific hospital infection control problems. These studies attempted to characterize and rapidly identify outbreaks of infection, particularly in locations such as intensive care units. A series of papers associated with investigators at the University of Alabama focused on many of these and related issues (Brossette *et al.*, 1998, 2000; Moser *et al.*, 1999). In a 1998 paper, Brossette *et al.* described a concept they called data mining surveillance system (DMSS) to encourage the application of rules of association in identifying new patterns within large infection control and public health surveillance data. The DMSS was further refined in subsequent papers through its application to intensive care units and for infection control surveillance. More recently, Peterson and Brossette introduced the concept of virtual surveillance to encourage the application of data mining techniques on an ongoing basis (Peterson and Brossette, 2002).

Two significant data mining studies on antibiotic use and drug resistance in a hospital setting have been conducted (Lopez-Lazano *et al.*, 2000; Monnet *et al.*, 2001). These studies combined antimicrobial susceptibility testing data with information on antimicrobial use obtained from the hospital pharmacy to conduct time-series analysis employing data mining techniques based on the ARIMA (autoregressive integrated moving average) model proposed by Box and Jenkins (Box and Jenkins, 1976). They succeeded in demonstrating a temporal relationship between the use of two antibiotics in the hospital and the percentage of Gram-negative bacilli resistant or intermediate to those antibiotics. In somewhat similar studies, Brown *et al.* evaluated the use of binary cumulative sums (CUSUMs) and moving average (MA) control charts to identify clusters of nosocomial infections using changes in antimicrobial resistance of isolates (Brown *et al.*, 2002). In 2002 Poupard *et al.* summarized three methods for data mining of large multinational surveillance databases (Poupard *et al.*, 2002). It is becoming apparent that as hospital and third party payer databases expand, the use of novel applications of data mining from other fields will become increasingly applied to resistance surveillance databases.

3.   PRIMARY RESISTANCE SURVEILLANCE DATA

When planning surveillance studies, the basic database for analysis will need to contain line listings of MICs for the individual isolates. This is important, not only because breakpoints (SIR) change over time, but because the investigator may want to apply unique breakpoints that differ from the standard breakpoints; for example, in order to collect information on possible first-step mutations prior to an organism becoming resistant.

It should also be noted that as much patient demographic information and specific isolate information as possible is always preferred, regardless of the original intent of the planners. These kinds of information are often invaluable when resolving issues of unusual or interesting results from data mining procedures.

4.   THREE APPROACHES TO RESISTANCE INFORMATION DATA MINING

Three main approaches will be discussed for data mining of large databases containing drug susceptibility/resistance information to search for novel information or patterns of antimicrobial resistance: (1) the antibiotype method, (2) multivariate analysis, and (3) evolutionary genetics approaches. These methods were previously presented in a summary paper analyzing information generated by the Alexander Project, a 10-year multinational surveillance study of upper respiratory isolates (Poupard *et al.*, 2002).

4.1.   The antibiotype approach

This is a method that first converts the long string of MICs to any number of drugs into a series of 0s and 1s with 0s representing a susceptible result and 1s representing non-susceptible or resistant results; the basic antibiotype. It should be noted that the 1 can be based on selected, published breakpoints, or an artificial breakpoint such as an MIC result that the investigator determines to be a first-step mutation. The string of S (susceptible) and R (non-susceptible) results of an isolate tested against individual drugs is converted into a string of 0s and 1s. The use of the binary code is adaptable to all computer programs and enables the investigator to search the database for novel patterns. A string of all 0s indicates an isolate is susceptible (based on the chosen breakpoints) to all drugs tested which is often lost when one prepares summary tables based on percent SIR or on MIC$_{50/90}$s.

In order to perform more sophisticated analysis, the long string of numbers can be converted into a two- or three-digit number. This is done by grouping the string of 0s and 1s into subsets of three numbers. Each consecutive number in the set of three is assigned a value of 1, 2, or 4, with 0s remaining 0s. For example, $100 = 100; 010 = 020; 001 = 004; 111 = 124$, etc. Once converted in this way, each set of three numbers can then be transformed to a single number derived from the sum of the values for the non-susceptible results, so: $100 = 100 = 1$; $010 = 020 = 2$; $001 = 004 = 4$; $111 = 124 = 7$, etc. It is also possible to assign a two- or three-digit hyphenated code based on the number of non-susceptible results (1s) in the binary string. Each unique string with the same number of non-susceptible results would give a new second digit, and the process would be continued until each antibiotype has a unique number designation.

Use of this method permits the determination of the predominant antibiotype, as well as rare antibiotypes, and enables the evolution of these antibiotypes to be tracked over time or by designated locations. It also permits analysis of variability in a population of unique antibiotypes over time and can show the rise or decline in the all zero antibiotype within the population. Specific applications of this methodology to isolates of *Streptococcus pneumoniae* and *Haemophilus influenzae* can be found in the previously cited paper by Poupard *et al.* (2002).

4.2.   Multivariate analysis methods

Multivariate projection methods are applicable for obtaining a broad overview of large, complex (multidimensional) data. Projection methods are powerful tools for discovering patterns in such data and have been applied in many situations, for example, genetics (Nguyen and Rocke, 2002), cheminformatics (Janne *et al.*, 2001), and statistical process control (MacGregor and Kourti, 1995), among others. In this chapter, we describe their application to multinational surveillance data. These methods are not meant to supersede the traditional univariate approach to analysis of surveillance data. Rather, they are meant to enhance the univariate analysis and to enable a greater understanding of the underlying patterns of variability among the antibiotics, countries, dates of collection, isolate sources, etc. In addition to this greater understanding, identification of interesting isolates, which may have escaped detection using the univariate approach, is likely. Thus, multivariate processes extend the level of understanding of the data beyond that obtained from the standard univariate approaches and provide a framework for additional analysis.

We applied multivariate analysis to the 1998 Alexander Project collection of 8,952 *S. pneumoniae* isolates from 24 countries. Isolates were tested against 20 antibiotics and data were available for age, gender, and isolate source (Table 1). While it is possible, and generally of interest, to apply multivariate techniques to the entire dataset, such an application may give rise to results which are artifacts of the sampling, rather than reflecting true patterns of resistance. For example, preliminary examination showed that data were collected from any of six sources (throat, ear, sputum, blood, nasopharynx, sinus) and from patients of any age. However, these data were not well distributed across the countries of origin. Some countries had no ear isolates (e.g., Austria, France, Germany) whereas others had many (e.g., United States). Similarly, some countries have very few isolates from patients of 5 years of age or less

Table 1. Antibiotics (abbreviation used) analyzed from the 1998 Alexander Project

β-Lactams	Macrolides	Quinolones	Others
Penicillins	Erythromycin (Ery)	Ciprofloxacin (Cip)	Clindamycin (Cli)
Penicillin (Pen)	Clarithromycin (Cla)	Ofloxacin (Ofl)	Chloramphenicol (Chl)
Amoxicillin	Azithromycin (Azi)	Gemifloxacin (Gem)	Doxycycline (Dox)
(Amx)			Co-trimoxazole (Cot)
Amoxicillin/			
clavulanic acid			
(Aug)			
Cephalosporins and			
loracarbef			
Cefaclor (Fac)			
Loracarbef (Lor)			
Cefuroxime (Fur)			
Cefixime (Fix)			
Cefotaxime (Tax)			
Ceftriaxone (Axo)			
Cefprozil (Cpz)			

(e.g., Switzerland, Austria) and others had many (e.g., United States, Japan). As the prevalence of resistance has been shown to vary based on isolate source and age, particularly in patients of 5 years of age or younger (Sahm *et al.*, 2000; Thornsberry *et al.*, 1999), there was a risk that the uneven distribution of these data would produce misleading results. However, all the countries had large numbers of sputum isolates from patients older than 5 years of age and the analysis was therefore restricted to this subset, including 1,295 bacterial isolates.

For the 1998-subset data, 20 antibiotics were tested representing 20 dimensions (variables) against 1,295 bacterial isolates (observations). It is, of course, not possible to visually examine 20-dimensional data. The concept of multivariate projection is that high dimensional data are transformed into a lower dimensional space, allowing data to be examined visually while at the same time explaining the variation in the data. The percentage of the total information in the data that is represented by the lower dimensional space is determined by R^2, analogous to the familiar R^2 value from simple linear regression. Distributions of MIC data are generally not symmetric. For most modeling procedures, whether univariate or multivariate, transformation of the MIC data to achieve a close-to-symmetric distribution is common. For example, MIC distributions are sometimes summarized using the geometric mean, which is based on log MIC, a close-to-symmetric transformation of the MIC data. Such transformations are essential for data modeling, mainly to make the models more efficient (reliable) and to remove undue influence on the model from relatively few extreme values, in this case extreme MICs.

For our analysis, MIC data were transformed to achieve a close-to-symmetric distribution using a log transformation. Data were first modeled using principal components analysis (PCA). The mathematical complexities of PCA are discussed in many statistical texts; see, for example, Morrison (1990), or Eriksson *et al.* (2001). PCA was carried out on the log MICs using SIMCA (2000) software. The principal components were then summarized graphically using score and loading plots. In their lower dimensional space, score plots describe the coordinates of the observations (isolates) while loading plots describe the coordinates of the variables (antibiotics). In order to interpret the principal components, it is necessary to examine loading plots. The loading plot for each principal component describes the structure being revealed by the component. The largest component corresponds to the highest proportion of total R^2 explained and is called the principal component 1, or p[1]. In addition to plots of individual loadings, two-dimensional or three-dimensional plots of combinations of the loadings are also very informative. Variables contributing similar information, for example, those that are correlated, are grouped together in two-dimensional or three-dimensional loading plots. Variables that are negatively correlated are located on opposite sides of the plot. The further from the plot origin that a variable lies, the greater its impact on the PCA model.

For the 1998 MIC data, the 20 variables and 1,295 observations were represented by four principal components, accounting for 88% of the total information in the MICs. Thus, the 20-dimensional data are summarized and interpretable in only four dimensions. The loadings for the first four principal components for *S. pneumoniae* in 1998 are shown in Figure 1. As stated, the loadings reflect structure among the variables, in this case antibiotics.

- The first principal component, p[1], explained 61% of the information in the MIC data and described isolates with high MICs among all antibiotic classes except the quinolones (Figure 1a).
- The second component explained an additional 13% of the MIC information, separated macrolides, chloramphenicol, and doxycycline from the β-lactams and co-trimoxazole, but contained relatively little information about the quinolones (Figure 1b).

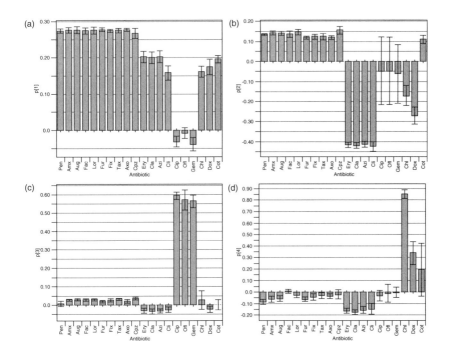

Figure 1. Loading plots for the first four principal components. (a) Antibiotic loadings for principal component 1 (p[1]); (b) antibiotic loadings for principal component 2 (p[2]); (c) antibiotic loadings for principal component 3 (p[3]); (d) antibiotic loadings for principal component 4 (p[4]).

- The third component (11% of information) uniquely described isolates with high quinolone MICs (Figure 1c).
- The fourth component (3% of information) picked up isolates with high chloramphenicol, doxycycline, and co-trimoxazole MICs, with relatively low macrolide MICs (Figure 1d).

The 1998 MIC data therefore can be described by four components, which, perhaps not surprisingly, closely follow the antibiotic classes.

Plots of two-dimensional loadings (antibiotics), combined with two-dimensional score plots (isolates), reveal unusual isolates, patterns among the isolates and relationships between the isolates and antibiotics (Figure 2). A two-dimensional plot of components p[1] vs p[2] revealed clustering by antibiotic class, with β-lactams clustered in the upper right-hand corner (Figure 2a). This cluster was due to the strong correlation among the β-lactam MICs (co-trimoxazole MICs were closely correlated with the β-lactams). The macrolides (plus clindamycin) were also clustered, but were separate from the β-lactams. Chloramphenicol and doxycycline were closely related to each other, but dissimilar from the other classes of antibiotics. The quinolones, being close to the origin, exerted no influence in the first two components. The two-dimensional score plot for the first two dimensions is shown in Figure 2b. The scores in dimension 1 are denoted t[1], and in dimension 2 are denoted t[2]. The score plot revealed distinct clusters among the isolates and was interpreted by relating the position of observations in the plot (which represent individual bacterial isolates) to the positions of variables in the p[1], p[2] loading plot. For example, the upper right quadrant of Figure 2b represents a cluster of isolates with high β-lactam and co-trimoxazole MICs (cluster 1). These isolates are associated with low macrolide MICs, and low to midrange doxycycline and chloramphenicol MICs, because these drugs are located in different regions of the loadings plot.

The median MICs from cluster 1 in Figure 2b were plotted in Figure 3a expressed as the number of dilutions from each antibiotic's respective MIC_{90}. As expected, the β-lactam median MICs were at their respective MIC_{90}s (i.e., 0 dilutions from the MIC_{90}) and macrolide, clindamycin MICs were between 9 and 11 dilutions below their MIC_{90}s. Cluster 2, at the bottom of the score plot Figure 2b, was associated with high macrolide and low β-lactam MICs; this was reflected by median MICs in that cluster, compared to the MIC_{90}s for each drug (Figure 3b). Clusters 3 and 4 were both high in the first component, the difference being the relative positions in the second component, and hence these clusters differed primarily by macrolide MICs (Figure 3c and 3d, respectively). Clearly the MICs were high for all drugs (except quinolones) in cluster 4, and in cluster 3 the increase in MICs among the macrolides was evident (compare with Figure 3a).

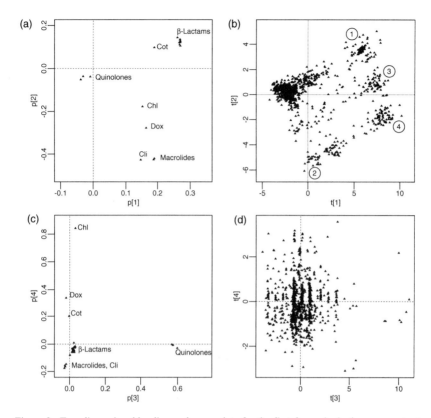

Figure 2. Two-dimensional loading and score plots for the first four principal components at the MIC level. (a) p[1] vs p[2] loading plots, showing clustering among antibiotics for the first two components; (b) t[1] vs t[2] score plots, showing major clusters among isolates for the first two components; (c) p[3] vs p[4] loading plots, showing clustering among antibiotics in the last two components; (d) t[3] vs t[4] score plots, showing major clusters among isolates in the last two components.

Figure 1c showed that the third principal component represented the quinolones and Figure 1d showed that the fourth component was dominated by chloramphenicol. Two-dimensional loading plots and corresponding two-dimensional score plots help to reveal how the isolates with high quinolone or chloramphenicol MICs cluster with the other isolates (Figure 2c and 2d). Isolates with high quinolone MICs are easily identified (on the far right of Figure 2d), and other outlying isolates, driven by high or low chloramphenicol MICs can be seen clearly. Note the vertical banding among the isolates of Figure 2d; these bands correspond to MICs of the quinolones, with the highest density bands being in the mid-MIC range.

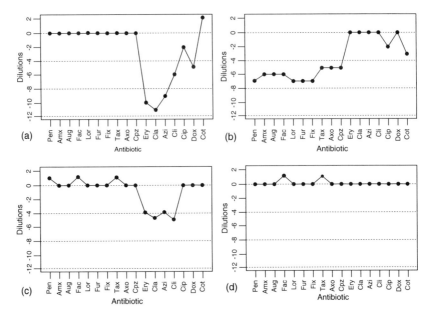

Figure 3. Median number of dilutions from each antibiotic's MIC_{90}, for major clusters in the first two principal components (p[1] and p[2]). (a) Cluster 1, median MICs for isolates in cluster 1 are at the MIC_{90} for β-lactams, but are between 9 and 11 dilutions below the MIC_{90} for the macrolides, between 3 and 5 dilutions below the MIC_{90} for clindamycin and doxycycline, and 2 dilutions below for chloramphenicol; (b) cluster 2, this is the inverse of cluster 1 with the β-lactams having low MICs relative to their MIC_{90}s, whereas macrolides and clindamycin are at their MIC_{90}s; (c) cluster 3, this shows MICs which are similar to cluster 1 except that macrolide, clindamycin MICs have increased compared with cluster 1; (d) cluster 4, this corresponds to high MICs for all non-quinolones, with MICs at the MIC_{90}s. Note that quinolones are not included in this figure as there was no information about the quinolones in the first two principal components.

The methodology described above can also be applied to the antibiotypes, and, as expected, the results were similar, though with some notable exceptions. First, the two-dimensional loadings plot for the first two components (Figure 4a) illustrates that antibiotic groupings were similar to the MIC groupings (Figure 2a), with the exception that amoxicillin and amoxicillin/clavulanic acid, at the antibiotype level, were distinct from the other β-lactams in p[1]. This distinction is due to the comparatively increased susceptibility of *S. pneumoniae* strains to amoxicillin vs other β-lactams when NCCLS breakpoints are used to determine the antibiotype (NCCLS, 2000). Figure 4b describes the distribution of isolates at the antibiotype level. The interpretation was similar to that for isolates at the MIC level (Figure 2b), although here there appeared to be a greater variety in the clustering pattern. The third and fourth components (Figure 4c) described (1) isolates with resistance to the quinolones ciprofloxacin and ofloxacin (there was no resistance to gemifloxacin in the isolates analyzed), and (2) the isolates with resistance to amoxicillin and amoxicillin/clavulanic acid (Figure 4c). Figure 4d shows the pattern of isolates in the third and fourth components. Isolates resistant to both amoxicillin and amoxicillin/clavulanic acid but not the two quinolones were clustered at the very top of the plot (cluster 1). There were two isolates with resistance to both amoxicillin and amoxicillin/clavulanic acid and one of the two quinolones, ciprofloxacin, or ofloxacin (cluster 2). A small group of six isolates were resistant to both quinolones, as well as to amoxicillin and

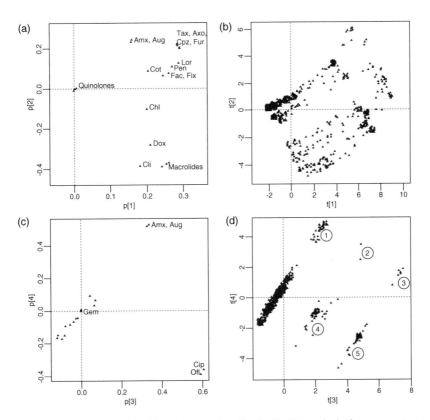

Figure 4. Two-dimensional loading and score plots for the first four principal components at the antibiotype level. (a) p[1] vs p[2] loading plots showing clustering among antibiotics in the first two components; (b) t[1] vs t[2] score plots, showing major clusters among isolates in the first two components; (c) p[3] vs p[4] loading plots, showing clustering among antibiotics in the last two components; (d) t[3] vs t[4] score plots, showing major clusters among isolates in the last two components.

amoxicillin/clavulanic acid (cluster 3). Cluster 4 included a group of isolates resistant to one of the two quinolones but not amoxicillin or amoxicillin/clavulanic acid. Finally, cluster 5 was a group of isolates resistant to both quinolones but not to amoxicillin and amoxicillin/clavulanic acid.

Other interactions among the antibiotics and isolates may be obtained by other two-dimensional and/or three-dimensional views of the components. The four components together explained over 80% of the information in the antibiotypes. In the above analysis, the country-to-country effect was not included. However, it is straightforward to include country as a variable. One way is to fit the same overall PCA model but use different colors or graph symbols in the score plots to depict the distribution of isolates for the different countries, and/or use graphical tools to display the countries individually. This works well and is a rather striking way to examine countries. Another approach is to use a related multivariate projection to model the relationship between countries and isolates; in this case the technique known as projection to latent structures (PLS) works well. The technical details of PLS have been well described (Eriksson *et al.*, 2001). With PLS, countries are considered predictor variables (or X variables) and antibiotics are considered response variables (or Y variables), with the isolates as observations. Both the X and Y, which are matrices with rows as isolates and columns as variables (countries in the X matrix and antibiotics in the Y matrix), are projected into lower dimensional space similar to a PCA projection, with the projections modified slightly to maximize the correlations between the X and Y variables. PLS models are easily fitted with the SIMCA software package.

The results of our PLS modeling have shown which countries are most (or least) highly associated with the antibiotic MIC or antibiotype patterns. As an example, we consider the 1998 antibiotype data previously modeled. In PLS we examine loadings for both X and Y variables to learn about their associations. To distinguish the PLS loadings from the PCA loadings, we used $w[i]$ and $c[i]$ for X and Y loadings, respectively, for component i. The loading plots for the first two components of the PLS projections are shown in Figure 5a (countries) and 5b (antibiotics). Figure 5b shows that the first Y component had high loadings for all antibiotics except the quinolones, and to a lesser extent amoxicillin, amoxicillin/clavulanate, clindamycin, and co-trimoxazole. Thus, this component picked up countries with resistance patterns dominated by most of the non-quinolone agents. X variables with high loadings are highly correlated with these Y loadings—hence from Figure 5a we see that France, Hong Kong, and Japan were the countries most highly associated with resistance to most non-quinolone agents. The next PLS component, captured in Figure 5c and 5d, was primarily driven by the United States, Japan, and the Slovak Republic, and picked up isolates that were differentiated based upon macrolide/β-lactam/doxycycline/co-trimoxazole resistance. Note that Figure

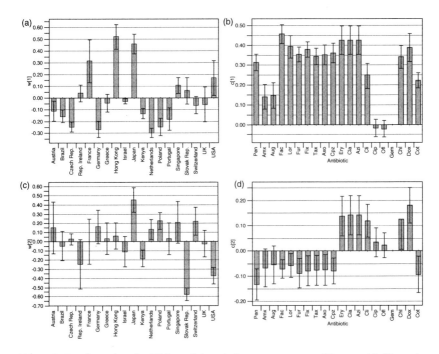

Figure 5. Loading plots for PLS models, relating country to antibiotype. (a) X matrix loading for component 1; (b) Y matrix loading for component 1; (c) X matrix loading for component 2; (d) Y matrix loading for component 2.

5b and 5d shows that there was no information about the quinolones in the first two PLS components.

As with PCA, it is of interest to plot two-dimensional plots. In Figure 6a we plotted w[1] and w[2] for the country loadings, and in Figure 6b c[1] and c[2] for the antibiotic loadings. Figure 6a shows which countries had the most extreme patterns of resistance (The Slovak Republic, United States, Japan, Hong Kong, and France) compared with the rest of the countries sampled. Figure 6b shows the projection into the Y space for the antibiotics, and spatially relating positions of antibiotics and countries in Figure 6a and 6b provides an understanding of the relationship between antibiotic resistance and country. Hong Kong, with high loadings in component 1 and loadings close to zero in component 2 had a relatively large number of isolates with resistance across all antibiotics. France was similar to Hong Kong, but a smaller component 1 loading suggests that this type of resistance was not as prevalent as in Hong Kong. Japan, high in component 1, but low in component 2 had isolates with resistance across all antibiotics, but also had a set of isolates with high

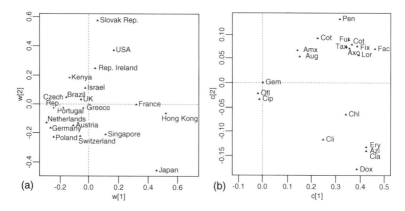

Figure 6. Two-dimensional loading plots for PLS models relating country to antibiotype. (a) X variable loadings for the first two components; (b) Y variable loadings for the first two components.

resistance among the macrolides, chloramphenicol, doxycycline, and clindamycin with low resistance among the β-lactams. Japan's second component of resistance was almost the opposite of the United States, which appeared to have isolates with resistance among the β-lactams but not among other antibiotics. The Slovak Republic had isolates generally associated with β-lactam resistance. The quinolones were not significant in the first or second component and hence no interpretations can be made regarding quinolone resistance and country for these two components. Note that this PLS model, with two components, explained only 30% of the information among the countries. The model points toward relationships between countries and antibiotics that are potentially of interest and worth further investigation.

The multivariate analysis clearly shows the broad patterns of antibiotic MICs and resistance for 1998 in *S. pneumoniae*. Based on PCA, β-lactam and macrolide resistance were responsible for the greatest variation among isolates, followed by quinolone resistance and resistance to chloramphenicol. At the MIC and antibiotype levels there were distinct clusters of isolates, which were largely determined by β-lactam and macrolide resistance. As the prevalence of quinolone resistance is still low, these agents were set apart from the other classes. At the antibiotype level, amoxicillin and amoxicillin/clavulanic acid were distinct from the population of other β-lactams, with less resistance to these two antibiotics.

4.2.1.   Multidimensional scaling

It is often of interest to genotype the isolates. Isolates that are similar genetically can be grouped together, and likewise, isolates dissimilar genetically, can be identified using a multitude of techniques. Multidimensional scaling (MDS) is one such technique, which is often applied to the analysis of genetic distances—see, for example, Agodi *et al.*, 1999. For a detailed description of this approach, see texts on multivariate statistics, such as Morrison, 1990. In this section we show an example from a set of 193 isolates sampled and genotyped from the 1998 and 1999 Alexander Project. In this example, the housekeeping gene *gki* was genotyped. After the nucleotide sequences were determined, the genetic divergence matrix for pairwise divergence among the isolates was computed (see Section 4.3 for details).

Similar to PCA and PLS, MDS approximates the data matrix, in this case the matrix of genetic divergence, in a lower dimensional space. Hence the 193-dimensional divergence matrix will be represented in a few, interpretable dimensions. We applied the PCA program from SIMCA to the divergence matrix; the two-dimensional loadings plot for p[1] and p[2] is shown in Figure 7. From this figure, along the p[1] axis we see isolates which are genetically highly divergent. In particular, a lone isolate collected in Italy in 1999 at the p[1] = −0.05 level (circled in Figure 7a) was very different from, for example, any of the isolates in the p[1] = 0.06–0.08 range. This was confirmed by plotting pairwise distances for the isolate at p[1] = −0.05 and two isolates at the other extreme. As an example we picked two isolates near p[1] = 0.08 and p[2] = 0 (one from the United Kingdom 1998, the other from Portugal 1999), and plotted the set of 193 distance pairs for these two, and for the former against the p[1] = −0.05 Italian 1999 isolate (Figure 8). In Figure 8a, the two isolates that

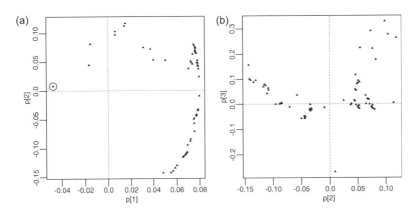

Figure 7. Multidimensional scaling of *gki* divergence matrix for 193 isolates from the Alexander project, 1998–9. (a) Two-dimensional plot for the first two components, identifying divergent isolates. The circled isolate is genetically divergent on p[1] and was collected in Italy in 1999; (b) Two-dimensional plot for the last two components, identifying divergent isolates and clusters of similar isolates.

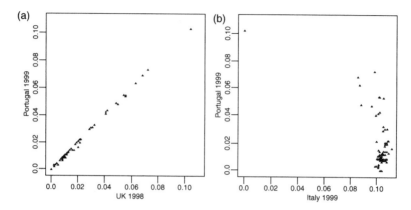

Figure 8. Comparison of pairwise distances for selected isolates. (a) Isolates close together in the first two components of the MDS model; (b) Isolates far apart in the first two components of the MDS model.

are close together on the plot are highly correlated; the Italy 1999 isolate is not correlated with these and the divergence from each of the other isolates is large (Figure 8b). Once genetically divergent isolates are identified, cross referencing with MIC and/or antibiotype data can be carried out, to determine the relationship of these divergent isolates to their MIC/resistance phenotype.

4.2.2.   Summary

The projection methods described above can handle extremely large numbers of observations and variables. For example, the Alexander Project 10-year dataset (1992–2001) has well over 35,000 isolates; despite this, analysis of the entire dataset is well within the computational boundaries of these projection methods. This may be of particular interest, for example, when investigating time trends over the 1992–2001 period. The many variables available in the Alexander Project data can also be modeled, such as age, gender, isolate source, and country. It is also possible to assess genetic divergence for these very large sets of data. Many other data mining methods exist for evaluation of such large datasets, and each method has its merits. Projection methods are robust methods that can handle extremely large data sets of predictor (X) and/or response (Y) matrices, and user-friendly software is readily available.

4.3.   Evolutionary genetic approaches

Technical developments associated with the polymerase chain reaction (PCR) and automated DNA sequencing technology, over the course of the last decade or so, have made it very easy to obtain large amounts of comparative sequence data from virtually any organism. Alongside this progression in laboratory technology, the evolutionary analysis of molecular sequence data has also advanced. This concomitant development of molecular biology technology and analytical perspective has resulted in the burgeoning field of molecular phylogenetics, which is now influencing all areas of biology. Despite this, the application of modern principles and techniques of molecular phylogenetics to the analysis of antibiotic resistance development and spread has not been widely undertaken. Nonetheless, we feel the application of such a perspective is ideally suited to extracting important information from large antimicrobial susceptibility databases, such as the Alexander Project. This is at least partly due to the fact that molecular phylogenetics falls within the broader realm of comparative biology and the nature of such databases provides numerous comparative possibilities. Databases such as the Alexander Project have a temporal perspective allowing comparisons between years, a geographical component allowing comparisons between collecting centers, as well as susceptibility data for numerous antibiotics for each isolate. This permits the correlation of isolate genetics with resistance over time and across geographical regions. The purpose of this section is to outline a few modern molecular phylogenetic perspectives on typical questions of antibiotic resistance development and spread in *S. pneumoniae*.

4.3.1. *S. pneumoniae* and questions of clonal spread

Analyses performed over recent years suggest that a small number of genotypes are responsible for >85% of fully penicillin-resistant pneumococci in the United States (MICs ≥2 mg/L) (Corso *et al.*, 1998; Gherardi *et al.*, 2000; Richter *et al.*, 2002). The majority of these studies used pulse field gel electrophoresis (PFGE), however, and in an organism such as *S. pneumoniae*, where nucleotide substitutions are relatively uncommon, PFGE is arguably too imprecise a method to gain an accurate picture of the species' genetic diversity. Furthermore, most of these studies examined only resistant isolates, whereas the inclusion of isolates with resistant as well as susceptible phenotypes is necessary to gain a more complete picture of the origins of resistance. An important additional level of specificity to isolate typing has been achieved with multiple locus sequence typing—MLST (Enright and Spratt, 1998; Maiden *et al.*, 1998; McGee *et al.*, 2001). However, the allele frequency data that arise from such studies do not lend themselves to forming an accurate picture of the cladistic history of the isolates.

Individuals from the same species diverge later than individuals from different species, which means that intraspecific molecular sequence data are typified by much lower levels of sequence variation. We have found this to be particularly the case in an organism like *S. pneumoniae*. This species exhibits very low levels of genetic diversity in comparisons of housekeeping gene sequences in clinical isolates sampled from globally distributed locations. In contrast, in comparisons involving the same housekeeping genes for the same country and year in the Alexander Project collection, *H. influenzae* has much higher genetic diversity than *S. pneumoniae*. For example, in the Alexander Project during collection of isolates for the United Kingdom in 1998, indexes of diversity for three different housekeeping gene sequences are 1.8–5.4 times higher for *H. influenzae* than for *S. pneumoniae* (Table 2).

The relative lack of genetic variation in *S. pneumoniae* means that it can be more difficult to accurately reconstruct the evolutionary history of isolates using traditional molecular phylogenetic procedures (which require at least moderate levels of sequence divergence between entities), particularly when the number of isolates is quite large. Furthermore, it is very difficult to root an *S. pneumoniae* phylogeny using another species. This is because most of the taxa widely recognized to be different species are much too distant to provide anything but a long branch attraction problem (a phylogenetic artifact which results in some sequences being artificially "dragged" to the base of the tree due to homoplasy with the outgroup), and other taxa which are currently classified as different species, may have no evolutionary basis for such a classification (Stanhope, unpublished data).

Methods have been developed over the course of the last decade, and are coming into increasing use over the last number of years, that employ phylogenetic statistical procedures for dealing with this problem of low sequence variation in the reconstruction of evolutionary history. One such method, known as statistical parsimony, was developed by Templeton in a series of important papers in the early to mid-1990s (Templeton, 1995; Templeton and Sing, 1993; Templeton *et al.*, 1987, 1992). In addition to being able to reconstruct accurate histories with low sequence divergence, the method also has a number of other important benefits: (1) the algorithm collapses the sequences into their

Table 2. Nei and Li's (1979) index of nucleotide diversity; based on total number of alleles, population frequency of each allele, and number of nucleotide substitutions per site between alleles; for *S. pneumoniae* ($N = 75$) and *H. influenzae* ($N = 73$) collected from the United Kingdom as part of the 1998 Alexander Project

Species	*gdh*	*gki*	*recP*
S. pneumoniae	0.01092	0.01479	0.00717
H. influenzae	0.02623	0.02633	0.03844

various haplotypes, estimates the maximum number of differences among these haplotypes, resulting from single substitutions, and then joins the haplotypes in a parsimony network with each step justified by a probability level of 95%; (2) the program TCS estimates haplotype outgroup probabilities for each haplotype, with the highest probability being judged the most likely ancestral haplotype for that set of sequences (Clement *et al.*, 2000); (3) it can be combined with a nested analysis procedure to partition the resulting network into a series of nested clades which can in turn be used to statistically test associations between genotype and phenotype, where "phenotype" could represent anything, such as geographic location, clinical setting, or antibiotic resistance phenotype (Templeton and Sing, 1993; Templeton *et al.*, 1987).

To illustrate this evolutionary approach using TCS, statistical parsimony networks were reconstructed from *gki* (840 bp) and *gdh* (1,245 bp) sequences for a set of isolates possessing a mixture of resistance phenotypes collected from Ohio, USA (Alexander Project collection, 2000) (Figure 9). For each of these two networks we tested the following null hypothesis: no association between genotype and isolates with penicillin (Pen) MICs >4 mg/L. For both these loci and networks the null hypothesis is rejected. Thus, we can conclude that, at least for this set of isolates, there was a significant correlation between genotype and isolates with penicillin (Pen) MIC >4 mg/L, indicating a significant degree of clonality to this type of resistance. By then examining the phenotypes of the constituent members of the various nests in the network, we can determine if there is a single clone or several clones that have convergently evolved this phenotype. In the present example, the vast majority of penicillin resistance of MIC >4 mg/L could be explained by two or three clones which have convergently evolved this phenotype. The fact that this test is not significant at the "0–step" clades (i.e., the individual haplotypes) indicates that both the relationships depicted in the network and the nested cladistical design associated with it were necessary to detect the genotype–phenotype association. In other words, any attempt to correlate haplotypes with resistance, while not taking into consideration the evolutionary history depicted in these networks, would have resulted in inaccurate conclusions.

Although in recent years there has been important work accomplished on the question of clonality and resistance in *S. pneumoniae*, it is also true that the issue can still benefit from the analytical procedures and perspective typical of modern molecular evolutionary biology. The statistical parsimony approach has various advantages, including the fact that it can be scaled up tremendously to include hundreds of isolate sequences. In contrast, traditional phylogenetic methods would be bogged down with such large numbers of highly similar sequences, as those typical of comparative data regarding *S. pneumoniae* housekeeping genes.

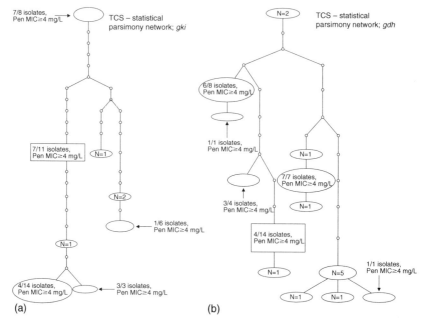

(a)   (b)

Figure 9. Statistical parsimony networks reconstructed from: (a) *gki* (840 bp) and (b) *gdh* (1,245 bp) sequences for a set of isolates ($N = 46$ for each locus) with a mixture of resistance phenotypes collected from Ohio, USA (Alexander Project, 2000). Small circles indicate nodes representing intermediate haplotype states not found in this sample of isolates. Each line represents a single mutational event. Larger ovals and two rectangles represent haplotypes that were sequenced in this sample of isolates, with the number of isolates of each indicated. The proportion of isolates of a particular haplotype possessing the penicillin MIC >4 mg/L phenotype are indicated (e.g., 4/14 isolates with penicillin MIC >4 mg/L). Haplotypes not represented by this resistance phenotype are simply labeled with the number of isolates with that sequence (e.g., $N = 5$).

4.3.2.   Horizontal transfer of resistance loci in *S. pneumoniae*

It has been known for many years that one of the principal mechanisms for *S. pneumoniae* resistance to β-lactam antibiotics is through specifically mutated penicillin-binding protein (*pbp*) genes. Unlike the housekeeping genes of *S. pneumoniae*, *pbp*s tend to be highly variable between isolates with a mosaic pattern in their homology comparisons to other species of streptococci. This pattern has provided evidence that these hyper-variable, resistance-conferring *pbp*s in *S. pneumoniae* have their origins in lateral gene transfer events involving other species, followed with intraspecific recombination events to create the mosaics. The frequency with which such lateral resistance transfer events take place is not well understood.

Arguments regarding lateral gene transfer events, whether it be regarding antibiotic resistance or not, are often based on basic interpretations of percent identity, or quite simply, BLAST (basic local alignment search tool) alignment scores. However, understanding whether a gene has been laterally transferred from another lineage is an evolutionary biology problem. A high profile example of the risks of failing to use such a perspective concerns claims, by the Human Genome Sequencing Consortium, that hundreds of human genes likely resulted from independent horizontal transfer events from bacteria at various points in the diversification of vertebrates (IHGSC, 2001). This conclusion was based on BLASTP alignment scores, or in other words, basic interpretations of sequence homology. However, subsequent, detailed phylogenetic analysis of this claim, did not find any support for horizontal gene transfer from bacteria to vertebrates (Stanhope *et al.*, 2001). Similarly, the best means to understand the history and dynamics of lateral transfer of *pbp*s in *S. pneumoniae* is to employ phylogenetic principles and techniques.

If two or more genes have the same evolutionary history then it is likely that there has been no lateral transfer involving those genes, and their history can be explained through a shared common ancestry and descent. In contrast, genes that have been laterally transferred will be discordant with the history depicted by genes that are not laterally transferred. Housekeeping genes are much less likely to be laterally transferred than are *pbp*s, and there is little *a priori* evidence to support their lateral transfer. Thus, the clearest phylogenetic evidence for lateral gene transfer of *pbp*s would be strongly supported by conflicting branching arrangements when comparing the phylogenies of *pbp* sequences and housekeeping genes for the same set of isolates. The phylogeny arising from the housekeeping genes can, therefore, be regarded as a "control phylogeny," or the best estimate of the true phylogeny. Nonetheless, in order to help verify their vertical inheritance, there are various methods available to detect the presence of recombinant sequences in any given sequence alignment (see, e.g., Posada, 2002). If one or more of such methods identify the presence of a recombinant, that particular sequence can be excluded from the set under analysis. These same methods can be used to distinguish between intragenic recombination of *pbp*s vs lateral transfer. In other words, it is possible that discordant phylogenies between housekeeping genes and *pbp*s could be the result of *pbp* lateral transfer, or intragenic recombination.

To illustrate this approach, we include an example of *pbp* sequences from a set of Alexander Project isolates from the United Kingdom collected in 1998. The control phylogeny in this case is based on an alignment of two concatenated housekeeping gene sequences, *gdh* and *gki* for a total of 2,085 bp; complete *gdh* and *gki* sequences were obtained for each isolate, these two sequences were joined together to make one long sequence for each isolate, and then the set of concatenated sequences were aligned. The low sequence divergence typical of *S. pneumoniae* housekeeping genes means that a single gene often does not have sufficient phylogenetic signal to reconstruct robust phylogenies using traditional methods (i.e., in comparison with methods such as TCS). However, low sequence divergence is not typical of *pbp*s, and thus in our experience it is possible to easily reconstruct reliable histories of at least *pbp*1a, *pbp*2b, and *pbp*2x genes. Traditional phylogenetic methods are best suited for comparisons between evolutionary histories—there is a wealth of theory, and statistics, for comparing the branching arrangements of bifurcating phylogenies, but at present such methods for comparing networks, such as those obtained from TCS, lag behind. Programs for computing phylogenies from molecular sequence are numerous, but the two most common and comprehensive are PHYLIP (Felsenstein, 1993) and PAUP* (Swofford, 2002).

In the present example the resulting housekeeping gene sequences showed no evidence for intragenic recombination and there was no evidence of lateral transfer for either housekeeping gene (no strongly supported conflicting nodes in individuals trees for each gene). Consequently, both loci were combined to yield the unrooted maximum likelihood tree depicted in Figure 10. Similarly, there was no evidence for intragenic recombination in the *pbp*2x sequences. Note that the composition of the italicized clade of isolates in the housekeeping tree includes two phenotypes: susceptible (isolates labeled 0) and those resistant to penicillin, cephalosporins, and co-trimoxazole (isolates labeled 17). Note that in the *pbp*2x tree, the "17" phenotype is either identical or very closely related to several other isolates with multidrug resistant phenotypes (labeled 23, 20, and 24), to which it was unrelated in the housekeeping tree. Furthermore, the "0" isolates from the housekeeping 0/17 clade no longer group with the multidrug resistant 17 isolates. These results indicate that there was the lateral transfer of a particular *pbp*2x allele into one or another of the "0" isolates of the 0/17 clone, and that this lateral transfer was correlated with a major shift in phenotype from all susceptible to multidrug resistant.

In general, our present collection of data suggests that *pbp* genes that are judged by our approach to be laterally transferred are often highly similar between a selection of multidrug resistant isolates. As these isolates share no close relationship on the basis of housekeeping genes, the most parsimonious explanation is lateral transfer rather than intragenic recombination. Whether the explanation is intragenic recombination or lateral transfer, the knowledge of the evolutionary history of the isolates themselves, based on their housekeeping gene sequences, can lead to important conclusions regarding the shifts in phenotype that have occurred coincidentally with the acquisition or alteration of the *pbp*s in question.

This comparative approach can be scaled up to include larger numbers of isolates from different locations and years allowing examination of (1) the relative frequency of lateral transfer of the resistance loci and whether this differs

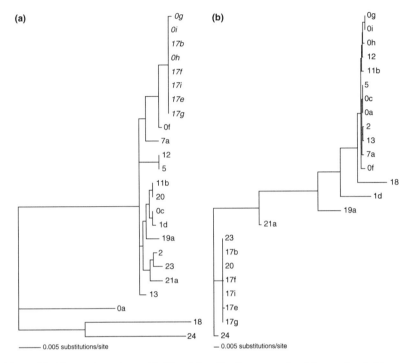

Figure 10. Unrooted maximum likelihood tree for *S. pneumoniae* (a) concatenated house-keeping genes *gdh* and *gki* and (b) *pbp*2x genes from isolates collected from the United Kingdom in 1998 (Alexander Project); branch lengths are drawn proportional to the amount of sequence change. The numbers and/or number–letter combinations refer to different isolates; the larger the number, the more antibiotics to which that isolate is resistant; 0 refers to isolates that are susceptible to all antibiotics tested for that particular year.

between years and countries; (2) which variants of these loci are associated with major shifts in resistance phenotype; (3) the sequence of lateral transfer events involving the resistance loci—that is, which resistance genes are acquired first, or all at once; (4) whether there are geographically specific *pbps* that are being laterally transferred amongst a set of isolates within a given region.

5.   CONCLUSIONS

Microbiologists are conditioned to approach a scientific subject with a hypothesis, a protocol outlining how to proceed, and a clear idea of what they are looking for. Data mining is a relatively new concept to microbiologists, and requires a change in mind set, as a key element of this approach is to apply methods developed for other fields like statistics and bioinformatics to a search for novel facts within the accumulated data. The papers cited in Section 2 of this chapter are included here to encourage workers interested in the subject of antimicrobial resistance to approach their subject within a new paradigm. The three general methods presented in some detail, antibiotypes, multivariate analysis, and evolutionary genetics, are techniques designed to stimulate the investigator rather than present any one set approach to the subject. Interested parties are encouraged to take the first step, namely, search for colleagues with expertise in other fields to become familiar with the rich data generated by antimicrobial surveillance and other research programs from the viewpoint of their individual specialities and search for novel aspects that will shed light on a problem that will only become more critical over the coming years. Data mining is one approach that may offer novel insights into our understanding of resistance, and ultimately may result in providing potential solutions to slow the rate of resistance against organisms of medical and environmental importance.

REFERENCES

Agodi, A., Campanile, F., Basile, G., Viglianisi, F., and Stefani, S., 1999, Phylogenetic analysis of macrorestriction fragments as a measure of genetic relatedness in *Staphylococcus aureus*: The epidemiological impact of methicillin resistance. *Eur. J. Epidemiol.*, **15**, 637–642.

Box, G. E. P. and Jenkins, G. M., 1976, *Time Series Analysis: Forecasting and Control*, 2nd edn. Holden Day, San Francisco, CA.

Brossette, S. E., Sprague, A. P., Hardin, J. M., Waites, K. B., Jones, W. T., and Moser, S. A., 1998, Association rules and data mining in hospital infection control and public health surveillance. *J. Am. Med. Inform. Assoc.*, **5**, 373–381.

Brossette, S. E., Sprague, A. P., Jones, W. T., and Moser, S. A., 2000, A data mining system for infection control surveillance. *Meth. Inf. Med.*, **39**, 303–310.

Brown, S. M., Benneyan, J. C., Theobald, D. A., Sands, K., Hahn, M. T., Potter-Bynoe, G. A. *et al.*, 2002, Binary cumulative sums and moving averages in nosocomial infection cluster detection. *Emerg. Infect. Dis.*, **8**, 1426–1432.

Clement, M., Posada, D., and Crandall, K. A., 2000, TCS: A computer program to estimate gene genealogies. *Mol. Ecol.*, **9**, 1657–1659.

Corso, A., Severina, E. P., Petruk, V. F., Mauriz, Y. R., and Tomasz, A., 1998, Molecular characterization of penicillin-resistant *Streptococcus pneumoniae* isolates causing respiratory disease in the United States. *Microb. Drug Resis.*, **4**, 325–337.

Enright, M. C. and Spratt, B. G., 1998, A multilocus sequence typing scheme for *Streptococcus pneumoniae*: Identification of clones associated with serious invasive disease. *Microbiology*, **144**, 3049–3060.

Eriksson, L., Johansson, E., Kettaneh-Wold, N., and Wold, S., 2001, *Multi- and Megavariate Data Analysis: Principles and Applications*. Umetrics Academy, Umea, Sweden.

Felsenstein, J., 1993, PHYLIP (Phylogeny Inference Package) version 3.6a2. Distributed by the author: http://evolution.genetics.washington.edu/phylip.html, Department of Genetics, University of Washington, Seattle.

Gherardi, G., Whitney, C. G., Facklam, R. R., and Beall, B., 2000, Major related sets of antibiotic-resistant pneumococci in the United States as determined by pulsed-field gel electrophoresis and pbp1a-pbp2b-pbp2x-dhf restriction profiles. *J. Infect. Dis.*, **181**, 216–229.

Hand, D. J., 1998, Data mining: Statistics and more? *Am. Statistician*, **52**, 112–118.

IHGSC, 2001, International Human Genome Sequencing Consortium: Initial sequencing and analysis of the human genome. *Nature*, **409**, 860–921.

Janne, K., Pettersen, J., Lindberg, N.-O., and Lundstedt, T., 2001, Hierarchical principal component analysis (PCA) and projection to latent structure (PLS) technique on spectroscopic data as a data pretreatment for calibration. *J. Chemometrics*, **15**, 203–213.

Kaslow, R. A. and Moser, S. A., 2000, Role of microbiology in epidemiology; before and beyond 2000. *Epidemiol. Rev.*, **22**, 131–135.

Lopez-Lazano, J. M., Monnet, D. L., Yague, A., Burgos, A., Gonzalo, N., Campillos, P. *et al.*, 2000, Modelling and forecasting antimicrobial resistance and its dynamic relationship to antimicrobial use: A time series analysis. *Int. J. Antimicrob. Agents*, **14**, 21–31.

MacGregor, J. F. and Kourti, T., 1995, Statistical process control of multivariate processes. *Control Eng. Practice*, **3**, 403–414.

Maiden, M. C., Bygraves, J. A., Feil, E., Morelli, G., Russell, J. E., Urwin, R. *et al.*, 1998, Multilocus sequence typing: A portable approach to the identification of clones within populations of pathogenic microorganisms. *Proc. Natl. Acad. Sci. USA*, **95**, 3140–3145.

McGee, L., McDougal, L., Zhou, J., Spratt, B. G., Tenover, F. C., George, R. *et al.*, 2001, Nomenclature of major antimicrobial-resistant clones of *Streptococcus pneumoniae* defined by the pneumococcal molecular epidemiology network. *J. Clin. Microbiol.*, **39**, 2565–2571.

Monnet, D. L., Lopez-Lazano, J. M., Campillos, P., Burgos, A., Yague, A., and Gonzalo, N., 2001, Making sense of antimicrobial use and resistance surveillance data: Application of ARIMA and transfer function models. *Clin. Microbiol. Infect.*, **7**, 29–36.

Morrison, D. F., 1990, *Multivariate Statistical Methods*, 3rd edn. McGraw-Hill, Hightstown, NJ.

Moser, S. A., Jones, W. T., and Brossette, S. E., 1999, Application of data mining to intensive care unit microbiologic data. *Emerg. Infect. Dis.*, **5**, 454–457.

NCCLS, 2000, *Performance Standards for Antimicrobial Susceptibility Testing; Tenth Informational Supplement*. National Committee for Clinical Laboratory Standards, Wayne, PA.

Nei, M. and Li, W. H., 1979, Mathematical model for studying genetic variation in terms of restriction endonucleases. *Proc. Natl. Acad. Sci. USA*, **76**, 5269–5273.

Nguyen, D. V. and Rocke, D. M., 2002, Multi class cancer classification via partial least squares with gene expression profiles. *Bioinformatics*, **18**, 1216–1226.

Peterson, L. R. and Brossette, S. E., 2002, Hunting health care-associated infections from the clinical microbiology laboratory: Passive, active and virtual surveillance. *J. Clin. Microbiol.*, **40**, 1–4.

Posada, D., 2002, Evaluation of methods for detecting recombination from DNA sequences: Empirical data. *Mol. Biol. Evol.*, **19**, 708–717.

Poupard, J., Brown, J., Gagnon, R., Stanhope, M. J., and Stewart, C., 2002, Methods for data mining from large multinational studies. *Antimicrob. Agents Chemother.*, **46**, 2409–2419.

Richter, S. S., Heilmann, K. P., Coffman, S. L., Huynh, H. K., Brueggemann, A. B., Pfaller, M. A. *et al.*, 2002, The molecular epidemiology of penicillin-resistant *Streptococcus pneumoniae* in the United States, 1994–2000. *Clin. Infect. Dis.*, **34**, 330–339.

Sahm, D. F., Jones, M. E., Hickey, M. L., Diakun, D. R., Mani, S. V., and Thornsberry, C., 2000, Resistance surveillance of *Streptococcus pneumoniae, Haemophilus influenzae* and *Moraxella catarrhalis* isolated in Asia and Europe, 1997–1998. *J. Antimicrob. Chemother.*, **45**, 457–466.

SIMCA, 2000, *8.0. Umetrics AB.* Umea, Sweden.

Stanhope, M. J., Lupas, A., Italia, M. J., Koretke, K. K., Volker, C., and Brown, J. R., 2001, Phylogenetic analyses do not support horizontal gene transfers from bacteria to vertebrates. *Nature*, **411**, 940–944.

Swofford, D. L., 2002, *PAUP* Version 4.0b10.* Sinauer Associates, Sunderland, MA.

Templeton, A. R., 1995, A cladistic analysis of phenotypic associations with haplotypes inferred from restriction endonuclease mapping or DNA sequencing. V. Analysis of case/control sampling designs: Alzheimer's disease and the apoprotein E locus. *Genetics*, **140**, 403–409.

Templeton, A. R., Boerwinkle, E., and Sing, C. F., 1987, A cladistic analysis of phenotypic associations with haplotypes inferred from restriction endonuclease mapping. I. Basic theory and an analysis of alcohol dehydrogenase activity in Drosophila. *Genetics*, **117**, 343–351.

Templeton, A. R., Crandall, K. A., and Sing, C. F., 1992, A cladistic analysis of phenotypic associations with haplotypes inferred from restriction endonuclease mapping and DNA sequence data. III. Cladogram estimation. *Genetics*, **132**, 619–633.

Templeton, A. R. and Sing, C. F., 1993, A cladistic analysis of phenotypic associations with haplotypes inferred from restriction endonuclease mapping. IV. Nested analyses with cladogram uncertainty and recombination. *Genetics*, **134**, 659–669.

Thornsberry, C., Ogilvie, P. T., Holley, H. P. Jr., and Sahm, D. F., 1999, Survey of susceptibilities of *Streptococcus pneumoniae, Haemophilus influenzae,* and *Moraxella catarrhalis* isolates to 26 antimicrobial agents: A prospective U.S. study. *Antimicrob. Agents Chemother.*, **43**, 2612–2623.

Chapter 24

Applications of Time-series Analysis to Antibiotic Resistance and Consumption Data

José-María López-Lozano[1], Dominique L. Monnet[2], Pilar Campillos Alonso[1], Alberto Cabrera Quintero[1], Nieves Gonzalo Jiménez[1], Alberto Yagüe Muñoz[1], Claudia Thomas[3], Arielle Beyaert[4], Mark Stevenson[5], and Thomas V. Riley[1,6]

[1]Departments of Microbiology, Preventive Medicine and Pharmacy, Hospital Vega Baja, Ctra Orihuela-Almoradí, s/n, 03314 Orihuela, Spain; [2]National Center for Antimicrobials and Infection Control, Statens Serum Institut, Artillerivej 5, 2300 Copenhagen S, Denmark; [3]School of Biomedical and Chemical Sciences (Microbiology), The University of Western Australia, Nedlands, Western Australia; [4]University of Murcia, Murcia, Spain; [5]School of Population Health, The University of Western Australia, Nedlands, Western Australia; [6]Division of Microbiology, The West Australian Centre for Pathology and Medical Research, Nedlands, Western Australia

1. PROBLEMS WHEN ATTEMPTING TO DEMONSTRATE A RELATIONSHIP BETWEEN ANTIBIOTIC CONSUMPTION AND BACTERIAL RESISTANCE

Data must be of the same type for both antibiotic consumption and bacterial resistance if one wants to study the relationship between these two parameters. These can be collected at the patient or at the collective level. Patient-level data allow the study of the effect of individual patient exposure to antibiotics on emergence and selection of bacterial resistance in this patient. However, these data are rarely available and do not take into account the possible conse-quences on resistance in other patients. Aggregated data, that is, data at the level of a hospital, a ward, or a primary care region, cannot take into account

Antibiotic Policies: Theory and Practice. Edited by Gould and van der Meer
Kluwer Academic / Plenum Publishers, New York, 2005

447

448   *José-María López-Lozano et al.*

misuse of antibiotics in individual patients and only represent the ecological pressure due to antibiotics; however, they often are the only available data in most hospitals. The type of aggregated data will determine the type of evidence in the demonstration of a relationship. If data are available for a specific period and for a large number of similar and independent settings, it will be possible to demonstrate a consistent association and/or a dose–effect relationship between antibiotic consumption and bacterial resistance (McGowan, 1987). Common problems with this approach are small sample size and possible selection bias. Multicenter studies, thus increasing statistical power and minimizing selection bias, have been implemented to circumvent these problems. However, even these multicenter studies often lack adequate sample size, include observations that are not independent, for example, hospitals or wards that exchange patients, cannot take into account the necessary delay between antibiotic use and bacterial resistance, and fail to deal with consumption of other antibiotic classes and other confounding factors. To illustrate this point, the data presented in Figure 1 suggest a relationship between the percentage of ceftazidime-resistant *Pseudomonas aeruginosa* and consumption of third-generation cephalosporins. However, this relationship clearly does not apply to two out of eight (25%) hospitals, where resistance is possibly explained by other factors such as consumption of other antibiotics, infection control practices, detection of resistance, etc. (McGowan, 1994; Monnet, 2000b).

When longitudinal data on antibiotic consumption and on bacterial resistance are available for a long period of time, it is possible to study concomitant variations, that is, changes in antibiotic consumption followed by changes in resistance in the same direction (McGowan, 1987). These variations are the most convincing proofs of causality when using aggregated data since they

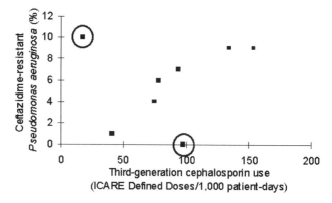

Figure 1. Percent ceftazidime-resistant/intermediate *P. aeruginosa* and third-generation cephalosporin use at eight US hospitals, CDC/NNIS Project ICARE Phase 1, 1994. Adapted from Monnet *et al.* (1998).

take into account the time sequence between the suspected cause, that is, antibiotic consumption, and the observed effect, that is, bacterial resistance. Such concomitant variations have been reported from various single hospitals or from single countries. However, when pooled data, covering one or several years, were used to analyze temporal associations observed between two time periods, these studies were not able to measure the delay necessary to observe an effect of antimicrobial use on resistance. Additionally, because data generally are available on a yearly basis, empirical attempts to take this delay into account consider a 1-year delay. For example, Goossens *et al.* (1986) found a correlation between the percentage of gentamicin-resistant Gram-negative bacilli and gentamicin use during the previous year. Interestingly, no correlation was observed when using gentamicin during the same year.

Our experience at Hospital Vega Baja (López-Lozano *et al.*, 2000b) shows that there is often a 1-year delay between a variation in antibiotic use and a consecutive variation in resistance, but not always. As shown in the example presented in Figure 2, we generally observed such variations of the percentage of ceftazidime-resistant Gram-negative bacilli following variations in ceftazidime use, except in 1996 when ceftazidime resistance increased following a decrease in ceftazidime use in 1995.

However, if we graphically represent the monthly percentage of ceftazidime-resistant Gram-negative bacilli and the monthly consumption of ceftazidime, we can observe variations at time periods shorter than 1 year (Figure 3). As shown below, there is a significant relationship between these monthly data.

Additionally, by simply smoothing the monthly series using a 5-month centered moving average transformation, the relationship becomes clear with ceftazidime consumption preceding ceftazidime resistance (Figure 4).

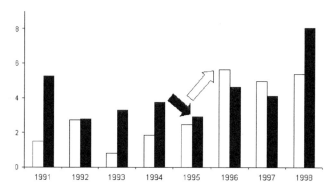

Figure 2. Yearly hospital ceftazidime use and percent ceftazidime-resistant/intermediate Gram-negative bacilli, Hospital Vega Baja, Orihuela, Spain, 1991–8. Adapted from Monnet *et al.* (2001). ■, ceftazidime use (DDD/1,000 patient-days); □, ceftazidime-resistant/ intermediate Gram-negative bacilli (%).

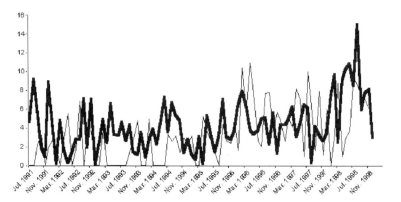

Figure 3. Monthly hospital ceftazidime use and percent ceftazidime-resistant/intermediate Gram-negative bacilli, Hospital Vega Baja, Orihuela, Spain, 1991–8. Adapted from López-Lozano *et al.* (2000). ▬, ceftazidime use (DDD/1,000 patient-days); ___, ceftazidime-resistant/intermediate gram-negative bacilli (%).

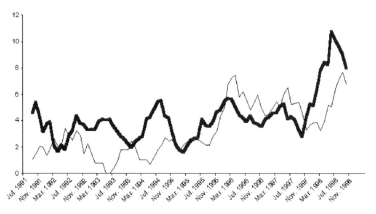

Figure 4. Five-period centered moving average of monthly hospital ceftazidime use and percent ceftazidime-resistant/intermediate Gram-negative bacilli, Hospital Vega Baja, Orihuela, Spain, 1991–8. Adapted from López-Lozano *et al.* (2000). ▬, ceftazidime use (DDD/1,000 patient-days); ___, ceftazidime-resistant/intermediate Gram-negative bacilli (%).

2.   DETERMINISTIC AND STOCHASTIC PROCESSES

To predict the trajectory of a missile following its launching—a classical example of a deterministic process—one can use physical laws if the direction and the velocity of the missile are known. To forecast weather—a very complex phenomenon, which depends on many complex, unstable, and sometimes unknown factors—one cannot use a mathematical model because not all causal factors are known. However, it is possible to construct a model to calculate the probability for a future value, for example, of temperature, to be between two specified confidence limits based on past values of temperature and of some well-known influencing factors such as atmospheric pressure or season. Such a model is called a probability model or a stochastic model. It is constructed to explain a stochastic process, which results from the interaction of several phenomena that act simultaneously to produce a variable outcome (Box and Jenkins, 1976).

Antibiotic resistance measured ecologically and over time can be considered a stochastic process since it is the result of the action of various and complex factors such as the level and the distribution of antibiotic consumption (variable), the transmission of resistant strains from one patient to another (conditioned by the level of infection control, which is also variable), the probability of spontaneous bacterial mutations (unpredictable, variable), and the number of isolates (conditioned by the number of microbiological samples and the frequency of sampling, also variable).

3. MODELING STOCHASTIC PROCESSES USING TIME-SERIES ANALYSIS

A time series corresponds to a group of observations taken sequentially over time and at equal intervals much shorter than the study period. Each time series is only one among many other possible realizations of the underlying stochastic process. Time series cannot be analyzed using classical regression techniques, that is, linear regression, because these techniques necessitate that the consecutive observations are independent from each other. Time-series analysis corresponds to a group of techniques aimed at adjusting a model to a time series for the purpose of predicting future behavior of the series based on its historical behavior and trying to explain its characteristics as well as the effect of some factors influencing the series. Unlike usual statistical methods, time-series analysis takes advantage of the relationships existing between consecutive observations, a phenomenon known as autocorrelation (Helfenstein, 1996). In 1976, Box and Jenkins provided a practical method for modeling time series by use of so-called auto regressive integrated moving average (ARIMA) models that analyze the temporal behavior of a variable as a function of its previous values, its trends and its abrupt changes in the near past (Box and Jenkins, 1976). Since then this method has been used in various fields where data are collected at repeated intervals for long periods of time, such as econometrics, engineering, weather forecast, and water resources research. More recently, it has been applied in medical specialties where such data are collected at the patient or population level, such as endocrinology, cardiology, environmental medicine, and the study of chronic diseases (Crabtree *et al.*, 1990; Helfenstein, 1996). Additionally, several modeling techniques of time-series analysis can be used to assess the relationships between a target (output) series and one or several explanatory (input) series. There are various types of multivariate time-series analysis models, for example, transfer function or polynomial distributed lag models. In transfer function models, delays or time lags between series are identified by the cross-correlation function (Helfenstein, 1996). In medicine, transfer function models have been used to study, for example, the relationship between weather parameters and mortality (Saez *et al.*, 1995) or influenza and mortality (Stroup *et al.*, 1988).

4. WHY USE TIME-SERIES ANALYSIS TO STUDY THE RELATIONSHIP BETWEEN ANTIMICROBIAL CONSUMPTION AND BACTERIAL RESISTANCE?

Time-series analysis is clearly relevant to the analysis of data produced by surveillance activity, which consist of repeated measurements at regular intervals during long periods of time. Such data, for example, the monthly percentage of antimicrobial-resistant microorganisms and the monthly number of defined daily doses (DDDs) of antimicrobial per 1,000 occupied bed-days (or patient-days), represent time series and can easily be produced by clinical microbiology laboratories and hospital pharmacies. Observation of these time series show short-term variations of both antibiotic resistance and antibiotic consumption, which could not be shown by using yearly data (Figures 2–4).

Additionally, our experience shows that these consecutive measurements of resistance levels on the one hand and of antibiotic use on the other hand, are highly autocorrelated, that is, show correlation from one period to another or several other periods as shown in the example in Figure 5. This autocorrelation should not be disregarded and classical statistical methods should not be used to analyze longitudinal antibiotic consumption and resistance data since these methods necessitate independent observations. Instead one should make use of methods that take this correlation into account such as time-series analysis.

The effect of antimicrobial use on resistance is not contemporaneous, that is to say, antimicrobial use needs some time in order to modify the ecological distribution of strains in a specific setting. Once a patient receives an antibiotic treatment his bacterial population might be modified acquiring resistance to this drug. Resistance measured as a group phenomenon is the sum of resistance expressed by an entire setting (animate and inanimate reservoirs) as a whole. Transmission of resistance is not immediate. Some time is necessary in order to pass from the initial carrier to others. This transmission is influenced by the

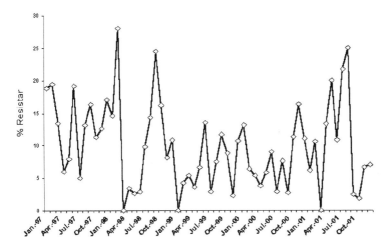

Figure 5. Example of auto-correlation in monthly resistance data.

hygiene control measures that can facilitate or restrict the diffusion of resistant strains from the initial patient to other patients via healthcare workers as well as the inanimate environment.

Because time-series analysis can assess the lagged relationship between two time series, it allows us to measure a non-contemporaneous relationship between two or more phenomena measured in a temporal basis (and at the same frequency).

A setting showing a certain resistance level can influence the levels for the same setting in the following time periods. This aspect of the resistance transmission is expressed as autocorrelation in the time series representing the temporal evolution of resistance.

Time-series analysis is based on the modeling of a series based on its past behavior as well as in the current or past evolution of one or more explaining series. In our case we will try to explain resistance as a function of: (1) its own past values, and (2) the current or past values of antimicrobial use and other possible influencing factors (compliance with hospital hygiene control measures).

5. EXAMPLES OF APPLICATION OF TIME-SERIES ANALYSIS

5.1. Study of the relationship between bacterial resistance and antimicrobial consumption

We studied the temporal relationship between hospital ceftazidime use and the percentage of ceftazidime-resistant or intermediate Gram-negative bacilli at

Hospital Vega Baja, a 400-bed general hospital in Southern Spain (López-Lozano *et al.*, 2000b). During the period July 1991–December 1998, the clinical microbiology laboratory isolated 6,244 non-duplicate (López-Lozano *et al.*, 2003) Gram-negative bacilli from hospital inpatients. The average observed monthly percentage of ceftazidime-resistant/intermediate Gram-negative bacilli was 3.3% (extremes: 0–10.9%) (Figure 3).

Based on these monthly data, we built an ARIMA model to predict the percentage of ceftazidime-resistant/intermediate Gram-negative bacilli. Following Box and Jenkins method, we found that time series had a stationary variance and mean, and therefore did not require stationarization. We identified an ARIMA model with two significant autoregressive terms of order (lag) 3 and 5 (months). The series residuals corresponded to white noise. The Akaike Information Criterion (AIC) was 659 and the determination coefficient (R^2) 0.38. Between July 1991 and December 1998, the average observed monthly hospital ceftazidime use was 4.4 DDD/1,000 patient-days (extremes: 0–15.0) (Figure 3). On the basis of these monthly data, we built a second ARIMA model to predict ceftazidime use. We identified an ARIMA model with two significant autoregressive terms of order (lag) 1 and 3 (months). We verified that the series residuals corresponded to white noise.

To investigate a possible relationship between the percentage of ceftazidime-resistant/intermediate gram-negative bacilli and hospital ceftazidime use, we built a transfer function model. The cross-correlation function (CCF) of the series of residuals obtained from the two previous ARIMA models showed only one significant correlation (parameter: 0.399, SE: 0.105) with a lag of 1 month between the hospital ceftazidime use series and the percentage of ceftazidime-resistant/intermediate Gram-negative bacilli series. We introduced a 1-month lag in the ceftazidime use series and estimated the parameters of an ARIMA(0, 0, 0) model with the lagged ceftazidime use series as the only dynamic predicting factor. Examination of the CCF of this last model residuals series with the ceftazidime use residuals series showed no other significant lag. Examination of the autocorrelation function (ACF) and the partial autocorrelation function (PACF) of the transfer function to determine the stochastic part of the model showed a good adjustment with two autoregressive terms of order 3 and 5. The series residuals corresponded to white noise. The parameters of the final transfer function model are presented in Table 1. The AIC was 416 and the $R^2 = 0.44$.

To predict the percentage of ceftazidime-resistant/intermediate Gram-negative bacilli by using this transfer function model and thus taking into account hospital ceftazidime use, we first needed to predict hospital ceftazidime use for the first 6 months of 1999. Then we predicted the percentage of ceftazidime-resistant/intermediate Gram-negative bacilli for the first 6 months of 1999 as 8.0, 4.7, 3.9, 4.5, 4.2, 3.4, and 4.2, respectively; which is slightly

Table 1. Transfer function model for percentage ceftazidime-resistant/intermediate Gram-negative bacilli taking into account hospital ceftazidime use, Hospital Vega Baja, Orihuela, Spain, 1991–8 ($R^2 = 0.44$) (López-Lozano *et al.*, 2000b)

Independent variable	Lag (months)	Coefficient (SE)	*T*-statistic	*P*-value
Constant	0	1.354 (0.760)	1.78	0.078
Ceftazidime use	1	0.420 (0.096)	4.34	<0.0001
AR	3	0.352 (0.096)	3.68	<0.001
AR	5	0.265 (0.098)	2.72	<0.01

different from what was obtained when predicting on the basis of the resistance series alone (data not shown). Finally, to give a graphical representation of the relationship between the series, we plotted both the smoothed resistance and use series using a 5-month moving average transformation (Figure 4). This figure showed an increasing trend in ceftazidime use but no trend for the percentage of ceftazidime-resistant/intermediate Gram-negative bacilli.

The model allows us to adjust an equation in which the present level of resistance would be a function of: (1) the level of resistance seen 3 months and 5 months before, and (2) ceftazidime use 1 month before in the same hospital. Note that the significant relationship is established with a delay of 1 month; it is not simultaneous. The interpretation of the parameter is: 1 DDD/1,000 patient-days, which is equivalent to 6.5 days of treatment approximately, would imply that resistance would increase 0.42% 1 month later, that is to say, it would change from $R = 5\%$ to $R = 5.42\%$.

5.2. To predict the short-term evolution of resistance

When a doctor is confronted with symptoms of a possible infection in a patient several issues arise: (1) if it really is an infection, (2) is possible to expect a satisfactory spontaneous progress, (3) if infection is due to a bacteria, a fungus, a virus, or another microorganism, and (4) in the event a bacterial origin is suspected and once the treatment is decided, the doctor must choose among the available antibiotics, the one that maximizes the possibilities of therapeutic success. Apart from clinical or pharmacologic considerations, the most influential factor in his decision is the possibility that the unknown microorganism responsible for the infection be resistant to the antibiotic he decides to use, (5) a basic factor is to determine the infecting microorganism. The problem that arises is that the microbiological analysis takes a while (one, two, three days) to show information about the germ that infects the patient, but meanwhile is often necessary to start the therapy, therefore an important factor is to predict the causal microorganism. In that sense, to know the usual local flora (hospital, community, etc.) will help us. (6) On the other hand, once it is established as to which is the microorganism that most probably caused the infection, we must choose among the available antibiotics. The knowledge about the spectrum of bacterial resistance in the setting in question (hospital, community, etc.) will help us in the choice of an antibiotic with minimal possibilities of the causal germ to be resistant. The doctor usually support his decision on guidelines made in other places, which recommend an antibiotic to use against this or that microorganism. Obviously, these recommendations should not be necessarily applied in our environment given that, as has been explained above, the stochastic nature of the phenomenon makes certain microorganisms show resistance phenotypic patterns completely different in different environments.

Time-series analysis let us make predictions based on previous values and related factors in terms of causality that we know, particularly antibiotic use. These predictions can be interesting in terms of practical application as an aid to their empiric antimicrobial treatment. Indeed, if we were able to decide every time as to what is the probability of a patient to have an infection due to this or that microorganism and if we were also able to estimate the probability of every microorganism to be sensitive or not to every available antibiotic, we could make recommendations about the empiric therapy, relevant to our local environment, that would minimize the probability of error in the therapeutic choice.

ViResiST (Spanish acronym for 'Resistance Surveillance using Time-series Analysis Techniques', www.viresist.org) is a project that focuses on antibiotic consumption and bacterial resistance in a short-term temporal dimension and at the local level (López-Lozano *et al.*, 2002b). It uses the following:

1. Microbiology data: Monthly hospital and community antibiotic susceptibility data for several years (usually from 1992) are exported into the ViResiST database. Duplicate and surveillance, that is, screening, isolates were excluded.
2. Pharmacy data: For the same period, monthly quantities of antimicrobials prescribed in the community and hospital are also exported into the ViResiST database. Data on use of individual antibiotics and antibiotic classes are stored as a number of DDD per 1,000 inhabitant-days for the community and per 1,000 patient-days for the hospital (ATC, 2002).

A Windows based interface allows an easy examination of the data as well as the selection and exportation of interesting time series (of monthly percentages of resistance and/or of monthly antibiotic consumption) in order to analyze them using time-series analysis techniques.

The first usefulness of the application program is a look up table with the results of the predictions of the ARIMA models fitted on each series of resistance. This table gathers the expected resistance for the current quarter as percentages of resistant strains of each microorganism to each antibiotic. In order to calculate these predictions we fit different ARIMA models using a semiautomatic method based on the software SCA (www.scausa.com). The application program also calculates the frequency of each microorganism in each type of sample, hospital department, and primary healthcare facility.

This information can be used by the clinician to improve his empiric therapy when he suspects that the patient has an infection caused by a certain microorganism and must decide among various antibiotics. Once the microorganism is known, the tables give the possibility of choosing the antibiotic with the lowest expected resistance. The expected resistance percentage and the frequency of microorganism in each type of sample can be used for the elaboration of empiric antimicrobial therapy guidelines based on local ecology at a certain hospital or health center.

5.3.   To evaluate interventions to control antibiotic resistance

Example. *Evaluation of a hospital-wide policy restricting third generation cephalosporin use to reduce* Clostridium difficile-*associated diarrhea: a time-series analysis (Thomas et al., 2003)*

The occurrence of *Clostridium difficile*-associated diarrhea (CDAD) has been observed closely at Sir Charles Gairdner Hospital (SCGH), a 560-bed urban teaching hospital in West Australia, since the early 1980s. Between 1983 and 1992, the incidence of CDAD increased from 23 to 50 cases per 100,000 patient-days at a time when consumption of third generation cephalosporin (3GC) antibiotics also increased (Riley *et al.*, 1994). Following the restriction of 3GC at the hospital in 1998, a significant decline in incidence of CDAD was observed (Thomas, 2002). The aim of this study was to describe the relationship between CDAD and 3GC consumption from 1993 to 2000, using time-series methodology, in order to evaluate the impact of the antibiotic policy change.

The study was conducted in SCGH from 1993–2000. Hospitalized patients that had a positive laboratory test, either direct faecal cytotoxin detection, or culture of *C. difficile*, were identified from the microbiology laboratory database from January 1993 to 2000. A period of 14 days or less between positive laboratory tests for *C. difficile* was considered to be part of the same episode.

Monthly use of 3GCs for the same period was obtained from the hospital's pharmacy as gram amounts dispensed and converted to the number of DDD per month (ATC, 2002).

Polynomial distributed lag (PDL) modeling was used to evaluate the relationship between 3GC use and the incidence of CDAD episodes in SCGH. PDL models for the monthly incidence rate of CDAD episodes, expressed as the number of *C. difficile* positive episodes/1,000 patient-days (the dependent series), and the monthly rate of 3GC use, expressed as DDDs/1,000 patient-days (the independent series), were constructed using Eviews 4.0 (Quantitative Micro Software, Irvine, California, USA). Diagnostic checking of the models consisted of ensuring all significant autocorrelation patterns were accounted for. The cumulative sum (CUSUM) and cumulative sum of squares (CUSUMQ) plots of the residuals were examined to locate possible structural changes in the model that were confirmed by applying Chow tests. The goodness of fit of the model was estimated using the determination coefficient (R^2).

Monthly incidence of CDAD and use of 3GC in SCGH, from January 1993 to December 2000 are illustrated in Figure 6. 3GC consumption fell from 28.95 DDD/1,000 patient-days (95% CI 28.63–29.26) prior to October 1998 to 3.29 DDD/1,000 patient-days (95% CI 3.12–3.46) after the policy was introduced. The average incidence of CDAD during the pre-intervention period was 0.61 episodes/1,000 patient-days (95% CI 0.56–0.65). During the post-intervention period the average incidence was 0.28 episodes/1000 patient days (95% CI 0.23–0.33).

The output for the final model of the relationship between 3GC antibiotics and CDAD is presented in Table 2. The incidence of CDAD observed in one month was related to (1) the incidence of CDAD four months previously and (2) the consumption of 3GC antibiotics during the same month. From the

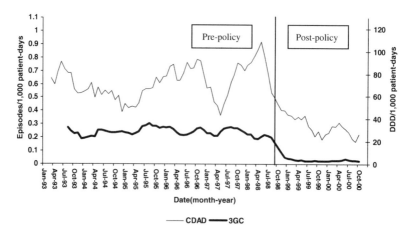

Figure 6. Five-period centered moving average of monthly incidence of *C. difficile*-associated diarrhoea episodes and hospital third-generation cephalosporin use, Sir Charles Gairdner Hospital, Perth, Australia, 1993–2000. Adapted from Thomas *et al.* (2003).

Table 2. Polynomial distributed lag model for monthly incidence of *Clostridium difficile*-associated diarrhoea (CDAD) episodes, Sir Charles Gairdner Hospital, Perth, Australia, March 1994–December 2000 ($R^2 = 0.62$). Adapted from Thomas *et al.* (2003)

Independent variable	Lag (months)	Coefficient	T-statistic	P-value
Past incidence of CDAD episodes	4	0.30	3.92	0.0002
Third-generation cephalosporin use	0	0.013	7.95	0.0000
Structural change (09/1997)	0	0.22	6.16	0.0000
Interaction between structural change (08/1998) and third-generation cephalosporin use	0	− 0.018	− 4.78	0.0000

coefficients in Table 2, if 3GC use increased by 1DDD/1000 patient-days in a particular month, an increase in 0.013 CDAD episodes/1000 patient-days could be expected in that same month.

Two structural changes were identified in the model. First, a shift in CDAD took place in September 1997 and, second, a substantial change in the influence of 3GC antibiotics occurred in August 1998. Prior to August 1998, 3GC use had a positive relationship with CDAD; after August 1998 this effect was canceled. That is, when consumption of 3GC fell, the impact on CDAD was so great that a relationship between 3GC and CDAD was no longer statistically detectable.

Time-series analysis was used to describe the temporal relationship between 3GC use and the incidence of CDAD in a public teaching hospital, in order to evaluate the impact of a restriction in the use of ceftriaxone at the hospital.

Using PDL modelling, 3GC consumption in one month was related to CDAD in the same month. Past incidence of CDAD was also related to current CDAD, however, the effect of a change in the incidence in 1 month was not produced until 4 months later. Explanations for this are not clear, however, a 4-month lag such as this relates to data from the preceding quarter and therefore may indicate some seasonal variation. For example, variation in hospital contamination, or the use of other antibiotics for community-acquired pneumonia during winter months.

The ceftriaxone policy implementation manifested as a structural change in the relationship between 3GC consumption and CDAD incidence from August 1998, 2 months prior to the policy introduction. An audit of 3GC was undertaken in the study hospital during July 1998 that may have influenced subsequent use of 3GC before the policy was introduced. During the period of the study, ceftriaxone was the only 3GC antibiotic in use from early 1997 after ceftazidime and cefotaxime were discontinued. However this change did not manifest as a structural change in the PDL model. An unexplained shift in CDAD was detected in September 1997 that may be attributed to unknown factors such as changes in consumption of other antimicrobials or infection control practices.

The results from this study indicate that in settings where 3GC consumption is high, reduction in the use of these antibiotics, -via policies aimed at prescription restriction, effectively reduces the occurrence of *C. difficile* infection in hospitals. Although others have reported similar observations for 3GCs (Ludlam *et al.*, 1999) this is the first study to evaluate such a policy by quantifying the relationship of antibiotics with CDAD using time-series models.

5.4.   Other applications

Time-series analysis techniques was used in order to study many other aspects of the evolution of resistance, the behavior of antimicrobial use, and its dynamic relationship: in order to evaluate the impact of antibiotic policies (Monnet *et al.*, 1999), modeling the relationship between the use of several antibiotics and the emergence of methicillin-resistant *Staphylococcus aureus* (MRSA) in various Scottish hospitals as well as its interrelationship (MacKenzie *et al.*, 2002a, 2003), modeling the interrelation between the emergence of MRSA in a hospital and in the surrounding community (MacKenzie *et al.*, 2002b), the relationship between the use of various antibiotics and the emergence of amikacin-resistant *P. aeruginosa* (López-Lozano *et al.*, 2002a), the relationship between the hospital hygiene conditions and the emergence of multiresistant microorganisms and *Candida albicans* in an intensive care unit (Allaouchiche *et al.*, 2002a, b) and the behavior and seasonality of antimicrobial use (Campillos *et al.*, 2002; López-Lozano *et al.*, 2002a).

6.   CONCLUSION

During the past 20 years, there have been numerous attempts to study the relationship between antimicrobial use and resistance using surveillance data. The method presented here represents another of those attempts. It proved helpful to demonstrate a temporal relationship between antimicrobial use and resistance and unlike most other methods, this method can take into account the use of several antimicrobials to explain specific types of resistance, quantify the effect of use on resistance, and estimate the delay between variations in use and subsequent variations in resistance. Additionally, it has allowed us to predict future levels of resistance based on past antimicrobial use and resistance data. Our observations show that ecologic systems such as that within the hospital tend to react to changes in antimicrobial use much faster than previously thought, that is, within a few months rather than several years. This finding has recently been confirmed by Corbella *et al.* who reported rapid variations in the percentage of imipenem-resistant *Acinetobacter baumannii* following changes in carbapenem use in a Spanish hospital (Corbella *et al.*, 2000), and by Lepper *et al.* who also reported changes in *P. aeruginosa* resistance to imipenem following changes in hospital imipenem use (Lepper *et al.*, 2002). The recent application of time-series analysis to antimicrobial use and resistance data from the primary healthcare sector in Denmark shows that this is probably also true outside hospitals (Monnet, 2000a). In conclusion, time-series analysis is a new tool that can help us make sense of antimicrobial use and resistance surveillance data, an area where modeling has proven difficult. Future developments must include confirmation of the usefulness of this method in other hospitals and in other countries.

REFERENCES

Allaouchiche, B., Monnet, D. L., López-Lozano, J. M. *et al.*, 2002a, Monthly incidence of ICU-acquired Candida sp. infections and its relationship to consumption of specific antibiotics in a French ICU: A time series analysis. 42nd Interscience Conference on Antimicrobial Agents and Chemotherapy, San Diego, CA, 27–30 September.

Allaouchiche, B., Monnet, D. L., López-Lozano, J. M. *et al.*, 2002b, Incidence of ICU-acquired infections due to antimicrobial-resistant microorganisms and relationship with antimicrobial use, infection control and workload in a French ICU. 42nd Interscience Conference on Antimicrobial Agents and Chemotherapy, San Diego, CA, 27–30 September.

Anatomical Therapeutic Chemical (ATC) classification index with Defined Daily Doses (DDDs), 2002, WHO Collaborating Centre for Drug Statistics Methodology, Oslo, Norway.

Box, G. E. P. and Jenkins, G. M., 1976, *Time Series Analysis: Forecasting and Control*, 2nd edition. Holden Day, San Francisco, CA.

Campillos, P., Tarazona, M. V., Burgos, A., Aznar, J., González, M., Alós, M. *et al.*, 2002, Distribution, trends and seasonal behaviour of antimicrobials used in five Spanish sanitary districts. 12th ECCMID, Milano (I), 24–27 April 2002, abstr. O231. *Clin. Microbiol. Infec.*, **8**(Suppl. 1).

Corbella, X., Montero, A., Pujol, M. *et al.*, 2000, Emergence and rapid spread of carbapenem resistance during a large and sustained hospital outbreak of multiresistant *Acinetobacter baumannii. Clin. Microbiol.*, **38**, 4086–4095.

Crabtree, B. F., Ray, S. C., Scmidt, P. M., O'Connor, P. J., and Schmidt, D. D., 1990, The individual over time: Time series applications in health care research. *J. Clin. Epidemiol.*, **43**, 241–260.

Goossens, H., Ghysels, G., Van Laethem, Y. *et al.*, 1986, Predicting gentamicin resistance from annual usage in hospital. *Lancet*, **2**, 804–805.

Helfenstein, U., 1996, Box-Jenkins modelling in medical research. *Stat. Meth. Med. Res.*, **5**, 3–22.

Lepper, P. M., Grusa, E., Reichl, H., Hogel, J., and Trautmann, M., 2002, Consumption of imipenem correlates with beta-lactam resistance in *Pseudomonas aeruginosa. Antimicrob. Agents Chemother.*, **46**(9), 2920–2925.

López-Lozano, J. M., Burgos, A., Gould, I. M., Campillos, P., MacKenzie, F. M., Monnet, D. L. *et al.*, 2002a, Trends and seasonal behaviour of community antimicrobial use in three

European countries. 42nd Interscience Conference on Antimicrobial Agents and Chemotherapy, San Diego, CA, 27–30 September. Abstract No. O-1004.

López-Lozano, J. M., Monnet, D. L., Burgos Sanjose, A., Gonzalo, N., Campillos, P., and Yagüe, A., 2000a, Modelling the temporal relationship between use of amikacin and other antimicrobials and amikacin resistance in *Pseudomonas aeruginosa* isolates: A time series analysis. 3rd European Congress of Chemotherapy, Madrid, 7–10 May 2000, abstr. T127. *Revista Española de Quimioterapia-Spanish J. Chemother.* **13**(Suppl. 2), 65.

López-Lozano, J. M., Monnet, D. L., Yagüe, A., Burgos, A., Gonzalo, N., Campillos, P. *et al.*, 2000b, Modelling and forecasting antimicrobial resistance and its dynamic relationship to antimicrobial use: A time series analysis. *Int. J. Antimicrob. Agents*, **14**, 21–31.

López-Lozano, J. M., Monnet, D. L., Yagüe, A., Campillos, P., Gonzalo, N., and Burgos, A., 2002b, Surveillance de la résistance bactérienne et modélisation de sa relation avec les consommations d'antibiotiques au moyen de l'analyse des séries chronologiques. *Bull. Soc. Fr. Microbiol.* **17**.

López-Lozano, J. M., Rodríguez, J. C., Sirvent, E., González, M., Royo, G., and Cabrera, A., 2003, Comparison of several criteria for rejecting multiple isolations on *Pseudomonas aeruginosa* antimicrobial sensitivity estimation. 13th European Congress of Clinical Microbiology and Infectious Diseases. Glasgow (UK). May. Abstract No. P728.

Ludlam, H., Brown, N., Sule, O., Redpath, C., Coni, N., and Owen, G., 1999, An antibiotic policy associated with reduced risk of *Clostridium difficile*-associated diarrhoea. *Age and Ageing*, **28**, 578–580.

MacKenzie, F. M., López-Lozano, J. M., Beyaert, A., Monnet, D. L., Camacho, M., Stuart, D. *et al.*, 2002a, Modelling the temporal relationship between hospital use of macrolides, third generation cephalosporins and fluoroquinolones, and the percentage of methicillin-resistant *Staphylococcus aureus* isolates: A time series analysis. 12th ECCMID, Milano (I), 24–27 April 2002, Abstract O231. Clin. Microbiol. Infect. **8**(Suppl. 1), 32.

MacKenzie, F. M., López-Lozano, J. M., Gould I. M, Wilson, R., Beyaert, A., Camacho, M. *et al.*, 2002b, Temporal relationship between prevalence of methicillin-resistant *Staphylococcus aureus* (MRSA) in one hospital and prevalence of MRSA in the surrounding community: A time series analysis. 42nd Interscience Conference on Antimicrobial Agents and Chemotherapy, San Diego (CA, USA), 27–30 September. Abstract No. K-103.

MacKenzie, F. M., López-Lozano, J. M., Monnet, D. L., Camacho, M., Wilson, R., Stuart, D. *et al.*, 2003, Relationship between methicillin-resistant *S. aureus* outbreaks in two Scottish hospitals. 13th ECCMID, Glasgow (UK), 10–12 May 2003, Abstract O338. *Clin. Microbiol. Infect.*, **8**(Suppl. 1), 32.

McGowan, J. E. Jr., 1987, Is antimicrobial resistance in hospitals related to antibiotic use? *Bull. N.Y. Acad. Med.*, **63**, 253–268.

McGowan, J. E. Jr., 1994, Do intensive hospital antibiotic control programs prevent the spread of antibiotic resistance? *Infect. Control. Hosp. Epidemiol.*, **15**, 478–483.

Monnet, D. L., 2000a, Antimicrobial use and resistance in Denmark. In: Abstracts of the 40th Interscience Conference on Antimicrobial Agents and Chemotherapy; 2000 September 17–20); Toronto, Canada. Washington, DC: American Society for Microbiology, 2000: p. 527.

Monnet, D. L., 2000b, Toward multinational antimicrobial resistance surveillance systems in Europe. *Int. J. Antimicrob. Agents*, **15**, 91–101.

Monnet, D. L., Archibald, L. K., Phillips, L., Tenover, F. C., McGowan, J. E. Jr., and Gaynes, R. P., 1998, Antimicrobial use and resistance in eight US hospitals: Complexities of analysis and modeling. Intensive Care Antimicrobial Resistance Epidemiology Project and National Nosocomial Infections Surveillance System Hospitals. *Infect. Control Hosp. Epidemiol.*, **19**(6), 388–394.

Monnet, D. L., Lopez-Lozano, J. M., Campillos, P., Burgos, A., Yague, A., and Gonzalo, N., 2001, Making sense of antimicrobial use and resistance surveillance data: Application of ARIMA and transfer function models. *Clin. Microbiol. Infect.*, 7(Suppl. 5), 29–36.

Monnet, D. L., Sørensen, T. L., Johansen, H. L., and López-Lozano, J. M., 1999, Impact of a Reimbursement Policy on Antimicrobial Prescriptions in the Primary Health Care Sector, Denmark, 1994–1998: A Time-Series Analysis. In: Abstracts of the 39th Interscience Conference on Antimicrobial Agents and Chemotherapy, San Francisco, CA. Washington, DC: American Society for Microbiology, 1999: p. 733.

Saez, M., Sunyer, J., Castellsagué, J., Murillo, C., and Antó, J. M., 1995, Relationship between weather temperature and mortality: A time series analysis approach in Barcelona. *Int. J. Epidemiol.*, **24**, 576–582.

Stroup, D. F., Thacker, S. B., and Herndon, J. L., 1988, Application of multiple time series analysis to the estimation of pneumonia and influenza mortality by age 1962–83. *Stat. Med.*, **7**, 1045–1059.

Riley, T. V., O'Neill, G. L., Bowman, R. A., and Golledge, C. L., 1994, *Clostridium difficile*-associated diarrhoea: Epidemiological data from Western Australia. *Epidemiol. Infect.*, **113**, 13–20.

Thomas C., 2003, The epidemiology and control of Clostridium difficile infection in a Western Australian Hospital. PhD thesis, The University of Western Australia.

Thomas, C., Stevenson, M., and Riley, T. V., 2002, *Clostridium difficile*-associated diarrhoea: Epidemiological data from Western Australia following a modified antibiotic policy. *Clin. Infect. Dis.*, **35**, 1457–1462.

Chapter 25

Biocide Use and Antibiotic Resistance

Jean-Yves Maillard
School of Pharmacy and Biomolecular Sciences, University of Brighton
Cockcroft Building, Moulsecoomb, Brighton BN2 4GJ, UK

1. INTRODUCTION

Bacterial resistance to antibiotics is a well-known phenomenon, which has been extensively described. Despite the early optimism in the 1960s that the advent of the antibiotic era would eradicate bacterial diseases, drug resistance in bacteria has increased at an alarming rate, so that bacterial resistance has been described for most if not all available antibiotics. Multidrug resistance in bacteria is indeed a major clinical problem (Hawkey, 2001). This situation is further exacerbated by the slow pace at which new molecules are being produced and by the rapid microbial adaptation to new antimicrobials. Several reasons have been put forward for the emergence of bacterial resistance to antibiotics, among which the overuse and sometimes misuse of antibiotics have been the most important. More recently, biocides (i.e., disinfectants, antiseptics, and preservatives) commonly used in hospital settings and elsewhere, for example, domiciliary and industrial environments, have been implicated in the emergence of antibiotic resistance in bacteria. This has raised concerns in the scientific community (Bloomfield, 2002; Levy, 2000; Russell, 1999a, 2000, 2002a; Russell and Maillard, 2000; Schweizer, 2001) and among institutions (Anon, 1997) and prompted many investigations into the possible linkage between biocide and antibiotic resistance in bacteria.

This chapter aims to give a brief description of disinfectant usage, evidence of resistance to these agents with possible linkage between biocide and

Antibiotic Policies: Theory and Practice. Edited by Gould and van der Meer
Kluwer Academic / Plenum Publishers, New York, 2005

antibiotic resistance, and the possible role, or not, of disinfectants in selecting for drug resistance.

2.   DISINFECTANTS: TYPES, ACTIONS, AND USAGES

2.1.   Use of disinfectant in the hospital environment

Biocides have been used for centuries, originally for the preservation of foodstuff and water, then for antisepsis and more recently for disinfection purposes (Hugo, 1999a). The number and diversity of chemicals used as biocides have increased tremendously within the last 50 years. Likewise, the usage of biocides has diversified over the years and more recently has found a commercial if not practical niche in the home environment (Bloomfield, 2002) with public awareness of hygiene-related issues (Favero, 2002; Maillard, 2002).

The first recorded use of biocides for a "clinical" purpose was probably that of the burning of juniper branches in the 14th century to combat the episodic scourge of the plague (Hugo, 1999a). The use of biocides for wound healing, and notably the use of mercuric chloride, have been practiced since the Middle Ages. However, the use of biocides in the hospital environment really progressed in the 19th century with the work of Semmelweis and notably of Lister with his study on antiseptic surgery (Hugo, 1999a; Rotter, 1998, 2001). Since then the number of chemical biocides used in the hospital environment has increased tremendously with the development of cationic biocides such as the bisbiguanide chlorhexidine, quaternary ammonium compounds (QACs), phenolic compounds, and the aldehydes and peroxygen compounds (Russell, 1999b).

In the hospital environment, biocides fulfil several functions (Rutala and Weber, 1999): (1) disinfection of equipment and surfaces, (2) antisepsis (e.g., handwashing), and (3) preservation of medical and pharmaceutical products. The activity of biocides and their use often depends upon their activity against the different types of microorganisms (Maillard, 2002), but also on the type of equipment or surfaces to disinfect. Therefore, high-level disinfection is used for equipment that comes into contact with sterile parts of the body, intermediate-level for those that come into contact with broken skin, and low-level for equipment and surfaces that come into contact with intact skin (Maillard, 2002; Rutala and Weber, 1999).

2.2.   Mechanisms of action of biocides

It is generally accepted that at "in-use" concentrations, biocides have multiple target sites on the bacterial cells, the inactivation/alteration/destruction of which lead to an inhibitory or lethal effect (Hugo, 1999b; Maillard, 2002). These agents can be divided according to their reactivity with the bacterial cells. Highly reactive biocides such as alkylating (e.g., glutaraldehyde, *ortho*-phthalaldehyde) and oxidising agents (e.g., hydrogen peroxide, peracetic acid, chlorine dioxide) are often used for high-level disinfection. Because of their nature and their interaction with the target cell (McDonnell and Russell, 1999; Russell, 1999b; Russell *et al.*, 1997), the emergence of bacterial resistance to these compounds is uncommon, but not impossible as observed with glutaraldehyde (see Section 3.1). Halogen-releasing agents such as chlorine and iodine-based products have a broad spectrum of activity. Some of these agents find an important role in clinical settings. Povidone-iodine is widely used for antisepsis and notably in surgical soap, whereas chlorine-releasing agents such as sodium hypochlorite and sodium dichloroisocyanurate (NaDCC) are used to disinfect blood spillages and other contaminated materials (Russell, 1999b). Less reactive biocides used for intermediate- to low-level disinfection encompass membrane-active agents such as bisbiguanides (e.g., chlorhexidine), QACs (e.g., benzalkonium chloride), and the phenolics (e.g., triclosan). These agents act primarily by disrupting the bacterial membranes before altering/inactivating other target sites within the bacterial cytoplasmic membrane and cytosol. They have also a broad spectrum of activity. Their efficacy is often limited (if any) against spores, some mycobacteria, and certain viruses, particularly non-enveloped ones (Maillard, 2001; Russell *et al.*, 1997). As for most biocides, the antimicrobial activity of these agents can be altered by several factors, such as concentration, pH, organic load, and temperature (McDonnell and Russell, 1999). Among these factors, concentration of biocides is of paramount importance (Russell and McDonnell, 2000), particularly for the emergence of bacterial resistance as described below. Indeed, although it is well recognised that biocides have multiple target sites on the microbial cells, some biocides, notably, the cationic and phenolic compounds, might have a "primary" target site, at a lower concentration (Maillard, 2002). This is particularly pertinent since these cationic and phenolic compounds have been implicated in a possible linkage between biocide and antibiotic resistance. Furthermore, there are sometimes similarities between biocide and antibiotic antimicrobial action. For example, triclosan and isoniazid both target the enoyl-acyl reductase carrier protein in mycobacteria whereas acridines, isothiazolones, chloracetamide, phenylethanol, β-lactams, fluoroquinolones, and novobiocin induce the formation of filaments in Gram-negative bacteria, although no cross-resistance between these biocides and antibiotics was observed (Ng *et al.*, 2002). Further useful information on the types and mechanisms of action of biocides can be found in the papers by Russell (1995, 1999b, 2002b), Russell and Russell (1995), Russell *et al.* (1997), Hugo (1999b), and Maillard (2002).

3.   BACTERIAL RESISTANCE TO BIOCIDES

Bacterial resistance to biocides has been reported since the 1950s, although the number of cases describing such resistance has increased steadily within the last two decades. This might reflect the increased use of biocides in the food, pharmaceutical industries and in the hospital and domiciliary environments. Biocides employed for intermediate- and low-level disinfection, antisepsis, and preservation have been particularly implicated in the emergence of bacterial resistance to biocides but also in the selection of bacterial strains with low-level antibiotic resistance. The mechanisms underlying such bacterial resistance are now better understood and it is emerging that bacterial cells have developed an arsenal of measures, which used together or not allow the cells to survive biocide exposure.

3.1.   Evidence of bacterial resistance to biocides

For the last 50 years, there have been multiple reports on bacterial resistance to various biocides. Generally, bacterial resistance to biocides has been well reported in the literature (Poole, 2002), particularly for these biocides that are used for low- and intermediate-level disinfection, or antisepsis, such as the cationic biocides, QACs, chlorhexidine, and phenolics. In most cases, low-level resistance has been reported, often based on MIC determination, although in some instances high-level resistance has been described, for example, with triclosan (Heath *et al.*, 1998, 2000a; Sasatsu *et al.*, 1993). Microbial contamination of cationic biocide formulations (QACs or chlorhexidine) by Gram-negative microorganisms has been reported as early as the 1950s and has been amply described since then (see Russell [2002c] for a comprehensive list of references). However, it has to be noted that in many instances, the emergence of resistance to these biocides often resulted from an improper usage/storage of these products often implying a decrease in the active "in-use" concentration (Russell, 2002c). Anderson *et al.* (1984) reported the contamination of a commercial iodophor solution with *Pseudomonas aeruginosa*, which was probably linked to the formation of a biofilm in the manufacturing plant. Kahan (1984) reported the contamination of a 0.05% chlorhexidine wound irrigation solution with *Burkholderia picketti*, which lead to septicemia in six patients. Likewise, contamination of chlorhexidine with *Burkholderia cepacia* (Speller *et al.*, 1971) and of a QAC by *Serratia marcescens* (Ehrenkranz *et al.*, 1980) has been reported. Contamination of such products often results from a poor appreciation of the importance of the concentration exponent of a particular agent (Russell and McDonnell, 2000). For example, the inappropriate use of disinfectants (Centers for Disease Control, 1974; Curie *et al.*, 1978; Sanford, 1970), the use of weak solutions (Prince and

Ayliffe, 1972), and the "topping-up" of containers lead to the observations of bacterial resistance to a biocide. The emergence of bacterial resistance in clinical practice is more of a concern. For example, Stickler (1974) reported the isolation of chlorhexidine-resistant *Proteus mirabilis* from patients with long-term indwelling catheterization after the extensive use of the biguanide to treat urinary tract infections. It has to be remembered that some microorganisms, notably spores, some mycobacteria, and Gram-negative bacteria such as *P. aeruginosa* are intrinsically resistant to some disinfectants, and therefore their presence where "in-use" or below "in-use" concentration of a biocide is used is not exceptional. Furthermore, the expectation to encounter such microorganisms should dictate the use of a more active biocidal formulation, even though, there have been some reports where resistant bacteria to high-level disinfectants have been isolated. For example, Griffiths *et al.* (1997) isolated two *Mycobacterium chelonae* resistant to the high-level disinfectant glutaraldehyde from endoscope washer disinfectors.

Acquired resistance is more of a concern, whereby a previously sensitive microorganism becomes resistant to a biocide. Staphylococci showing resistance to QACs have been isolated from healthcare facilities (Gillespie *et al.*, 1986) but also from other sources such as food preparation (Heir *et al.*, 1995). Biocide resistance in staphylococci has also been observed, particularly to cationic agents, acridines, and diamidines (reviewed by Russell, 2002a) and has been linked to the acquisition of multidrug resistance determinants (Section 4.2.4). Lear *et al.* (2000) isolated from biocide manufacturing sites, *Acinetobacter* and *Citrobacter* that showed high-level resistance to triclosan. Some staphylococci with low-level resistance to the bisphenol were also observed. Laboratory experiments also provide valuable information on the mechanism of the development of bacterial resistance to biocides. Tattawasart and colleagues (1999) were able to develop *Pseudomonas stutzeri* resistance to biguanides and other cationic compounds. Long-term and short-term exposure of *P. aeruginosa* to sub-inhibitory (residual) concentration of chlorhexidine showed some marked increase in resistance to the biguanide and to benzalkonium chloride (Thomas *et al.*, 2000). Winder *et al.* (2000) observed that *P. aeruginosa* was able to develop resistance to isothiazolones as a result of a constant exposure to sub-inhibitory concentrations. Likewise, McMurry *et al.* (1998a) demonstrated that resistance of *Escherichia coli* to triclosan and other compounds (Walsh *et al.*, 2003), and of *Mycobacterium smegmatis* (McMurry *et al.*, 1999) to the bisphenol can be increased, although in most cases low-level resistance was observed as measured by MIC determination. As mentioned above, the emergence of bacterial resistance to highly reactive biocides has also been described. Two isolates of *M. chelonae* have been isolated from endoscope washer disinfectors (Griffiths *et al.*, 1997; van Klingeren and Pullen, 1993). These two isolates were shown to be highly resistant to glutaraldehyde (Fraud *et al.*, 2001; Manzoor *et al.*, 1999; Walsh *et al.*, 2001). Resistance to hydrogen peroxide and peracetic acid has also been described, but in this case, microbial biofilms were responsible (see below).

Chapman (1998) and Chapman *et al.* (1998) described bacterial strains, which were virtually resistant to all known preservatives. Russell (2002a) pointed out that although the emergence of such strains would be expected, since preservatives have been used extensively for a long period of time, the number of cases concerning bacterial-resistance to "in-use" concentrations of biocides is low.

3.2.   Mechanisms of bacterial resistance to biocides

In order to survive biocide exposure, the main objective of the microorganism is to decrease the toxic concentration of chemicals and thus several "resistance" mechanisms can be activated. The main mechanisms are probably based on decreased uptake/penetration of the agents through "reinforcing" an impermeability barrier, and to some extent the decrease of the intracellular concentration of the agents through efflux systems and also degradation. A selected decrease in the toxic effect of the agents can also take place through phenotypic alteration and target modification.

Some of these mechanisms are intrinsic to microorganisms, whereas other can be acquired. Acquired resistance can arise from one or several processes: mutation or amplification of an endogenous chromosomal gene, and the acquisition of resistance determinants from extra-chromosomal elements (i.e., plasmids and transposons) (Poole, 2002). Phenotypic variations that might lead to biocide resistance are dealt with later (Section 3.2.5) (Chapman, 2003). In addition, one preliminary investigation on the adaptation of *Staphylococcus aureus* to chlorhexidine demonstrated the possible role of bacterial "alarmones" (extracellular induction components), which confer the Gram-positive bacteria a 3-fold increase in MIC. However, no evidence was found that such a mechanism was involved with increased level of triclosan resistance or in *E. coli* (Davies and Maillard, 2001).

3.2.1.   Impermeability

The most resistant microorganisms to disinfection are bacterial spores (Russell, 1990; Russell *et al.*, 1997). The particular structure of bacterial spores (i.e., the presence of several envelopes), their low water content, and the presence of small soluble proteins (SASPs) account for their high resistance to antimicrobials (Russell, 1990). Intrinsic resistance of mycobacteria to biocides involves mainly the outer cell layer, which acts as an impermeability barrier (McNeil and Brennan, 1991; Russell, 1996; Russell *et al.*, 1997). For example, Manzoor *et al.*

(1999) observed that the reduction in susceptibility of *M. chelonae* isolates to glutaraldehyde was associated with a change in cell wall polysaccharides. In Gram-negative bacteria, the outer envelope is responsible for resistance to both antibiotics and biocides. It has been suggested that Gram-negative bacterial susceptibility to biocides could be decreased with a change in cell permeability (McDonnell and Russell, 1999), for example, by changes to overall cell hydrophobicity (Tattawasart *et al.*, 1999), but particularly outer membrane ultrastructure (Tattawasart *et al.*, 2000a, b), protein composition (Brözel and Cloete, 1994; Gandhi *et al.*, 1993; Winder *et al.*, 2000), and fatty acid composition (Guérin-Méchin *et al.*, 1999, 2000; Jones *et al.*, 1989; Méchin *et al.*, 1999).

The importance of the outer membrane impermeability of Gram-negatives in decreasing biocide susceptibility has been exemplified by the use of permeabilizers such as ethylene diamine tetraacetic acid (EDTA), which induces a release of lipopolysaccharides (LPS) (Ayres *et al.*, 1998; McDonnell and Russell, 1999). Likewise, the use of permeabilizers in mycobacteria has been shown to reduce the intrinsic resistance of these microorganisms to biocides (Broadley *et al.*, 1995). The investigation of bacterial spheroplasts can also provide information on the role of the outer cell layer in bacterial resistance to biocides (Fraud *et al.*, 2003; Munton and Russell, 1970).

3.2.2. Multidrug efflux pumps

Multidrug efflux pumps have been shown to play a role in bacterial resistance to a wide range of antimicrobials (Nikaido, 1996; Paulsen *et al.*, 1996a). Multidrug efflux determinants have now been found to be widespread in Gram-negative and -positive bacteria. They can be separated into five main classes: (1) the small multidrug resistance (SMR) family, which is now described as part of the drug/metabolite transporter (DMT) superfamily, (2) the major facilitator superfamily (MFS), (3) the ATP-binding cassette (ABC) family, (4) the resistance-nodulation-division (RND) family, and (5) the multidrug and toxic compound extrusion (MATE) family (Borges-Walmsley and Walmsley, 2001; Brown *et al.*, 1999; McKeegan *et al.*, 2003; Poole, 2001, 2002; Putman *et al.*, 2000).

Bacterial resistance to QACs has been particularly studied and in many cases active-efflux systems have been involved (Heir *et al.*, 1995, 1999; Leelaporn *et al.*, 1994; Littlejohn *et al.*, 1992; Lomovskaya and Lewis, 1992; Sundheim *et al.*, 1998; Tennent *et al.*, 1989). In *S. aureus*, several genes have been characterised; *qacA* and *qacB*, which confer high-level resistance, and *qacC* and *qacD*, which confer low-level resistance (Littlejohn *et al.*, 1990; Rouche *et al.*, 1990), *smr* (Lyon and Skurray, 1987), *qacG* (Heir *et al.*, 1999), and *qacH* (Heir *et al.*, 1998). In addition, *qacC* and *qacD* are similar to the *ebr* gene encoding for resistance to ethidium bromide in *S. aureus*. As a result, resistance to QACs is often concurrent with resistance to ethidium bromide. In Gram-negative bacteria, similar efflux systems have been described. In *P. aeruginosa*, the MexAB-OprM, MexCD-OprJ, and MexEF-OprN (Schweizer, 1998), in *E. coli, emrA* (Lomovskaya and Lewis, 1992), *qacE* and *qacEΔ1* (Kazama *et al.*, 1998). The role of efflux pumps in antimicrobial resistance is discussed further in Section 4.2.2.

3.2.3.   Degradation

The degradation of biocides has been described in several investigations. Resistance to heavy metals (including mercury, copper, and silver) is usually associated with enzymatic reduction of the cation to the metal (Cloete, 2003). Likewise, the presence of aldehyde dehydrogenase in some strains has been responsible for the bacterial resistance to aldehyde and notably formaldehyde (Kummerle *et al.*, 1996). The mechanism of action of oxidising agents such as peroxygens is via the production of free radicals. Microorganisms have developed a system to prevent and repair radical-induced damage, particularly with the synthesis of proteins such as catalases, superoxide dismutase, and alkyl hydroperoxidases (Demple, 1996). In *E. coli*, these enzymes are encoded by multigene systems such as *soxRS* and *oxyR*, which can be induced by biocides (Dukan and Touati, 1996). In addition, such multigene systems can confer cross-resistance to other oxidants (Greenberg and Demple, 1989; Greenberg *et al.*, 1990). For example, induction by hydrogen peroxide confers resistance to hypochlorous acid and vice versa (Dukan and Touati, 1996). Degradation of the bisphenol triclosan has been described among environmental isolates (Hundt *et al.*, 2000), although there is no evidence that degradation of bisphenol takes place in clinical isolates.

3.2.4.   Alteration of target sites

The alteration of a specific target site is a well-established mechanism of bacterial resistance to antibiotics (Chopra *et al.*, 2002), but not to biocidal agents, with the exception of the bisphenol triclosan. As previously mentioned, biocides interact with the bacterial cell by targeting multiple nonspecific sites (Maillard, 2002). Triclosan has been found to interact with an enoyl-acyl reductase carrier protein in a range of microorganisms including *E. coli* (Heath *et al.*, 1998; McMurry *et al.*, 1998b), *P. aeruginosa* (Hoang and Schweizer 1999), *Haemophilus influenzae* (Marcinkeviciene *et al.*, 2001), *Bacillus subtilis* (Heath *et al.*, 2000b), *M. smegmatis* (McMurry *et al.*, 1999), *Mycobacterium tuberculosis* (Parikh *et al.*, 2000), and *S. aureus* (Heath *et al.*, 2000a; Slater-Radosti *et al.*, 2001). Indeed, crystallographic studies have shown that triclosan binds specifically with FabI (Heath *et al.*, 1999; Levy *et al.*, 1999, Roujeinikova *et al.*, 1999; Stewart *et al.*, 1999). Mutations in the

M. smegmatis inhA and *E. coli fabI* genes have been associated with resistance of these microorganisms to triclosan (Heath *et al.*, 2000a; McMurry *et al.*, 1999; Parikh *et al.*, 2000). However, while initially mutation in the *fabI* gene was described as responsible for triclosan resistance (McMurry *et al.*, 1998a), other mechanisms of resistance have been described since (Gilbert and McBain, 2003), notably involving efflux pumps (Schweizer, 2001).

3.2.5. Biofilms

Biofilms are responsible for infections in the hospital environment by growing on implants, catheters, and other medical devices (Costerton and Lashen, 1984; Costerton *et al.*, 1987; Gilbert *et al.*, 2003; Salzman and Rubin, 1995), but also by colonising various surfaces such as air conditioning systems and cooling towers. Biofilms are notoriously more resistant to biocide action that their planktonic cells (Allison *et al.*, 2000). The involvement of biofilms in the resistance to biocides is complex and involves several mechanisms that work together towards decreasing or inhibiting the detrimental effect of a biocide (Gilbert *et al.*, 2003). The emergence of bacterial resistance following phenotypic adaptation upon attachment to surfaces and within the biofilms is an important mechanism (Brown and Gilbert, 1993; Das *et al.*, 1998), although impairment of biocide penetration into the biofilm matrix, quenching by the exopolysaccharides, and enzymatic inactivation have a role to play (Gilbert *et al.*, 2003; Gilbert and Allison, 1999). For example, some biocides, such as iodine and povidone-iodine (Favero *et al.*, 1983), chlorine, and peroxygens (Huang *et al.*, 1995) have been shown to be inactivated to some extent by bacterial biofilm as a result of an interaction with the glycocalyx. Certain degradative enzymes, for example, formaldehyde dehydrogenase, can become concentrated within the glycocalyx and can lead to a decrease of biocidal activity, by hindering the penetration of biocide (Sondossi *et al.*, 1985). Such mechanisms might lead to the formation of a sublethal gradient-concentration, which in turn might induce further resistance mechanism such as efflux (Gilbert *et al.*, 2003).

4. EVIDENCE OF CROSS-RESISTANCE BETWEEN BIOCIDES AND ANTIBIOTICS

4.1. Evidence of bacterial resistance to biocides and antibiotics

Possible linkage between biocide and antibiotic resistance in bacteria has been discussed in the literature, particularly within the last few years

(Bloomfield, 2002; Levy, 2000; Russell, 2000, 2002a; Russell and Maillard, 2000; Russell et al., 1999; Schweizer, 2001).

In S. aureus, β-lactam resistance has been associated with QAC resistance (Akimitsu et al., 1999). Chuanchuen and colleagues (2001) observed that triclosan-resistant P. aeruginosa possessed elevated MICs to several antibiotics. Chlorhexidine-resistant P. stutzeri have also been found to be resistant to some antibiotics (Russell et al., 1998; Tattawasart et al., 1999). Cross-resistance in E. coli has also been observed between pine oil, triclosan, and multiple antibiotics (Cottell et al., 2003; Moken et al., 1997), and diverse biocides (e.g., QACs, amine oxide) and multiple antibiotics (Walsh et al., 2003), although in these studies, only a low-level resistance was observed. Triclosan exposure of triclosan-sensitive mutants of P. aeruginosa produced an important increase in resistance to the bisphenol and ciprofloxacin (Chuanchuen et al., 2001). However, other studies showed that the isolation of laboratory triclosan-resistant S. aureus was not necessarily correlated with an increase in antibiotic resistance (Slater-Radosti et al., 2001; Suller and Russell, 1999).

Methicillin-resistant S. aureus (MRSA) strains exhibiting low-level resistance to triclosan (MICs 2–4 μg/ml) and resistance to mupirocin (MIC > 512 μg/ml) have been isolated from patients treated with nasal mupirocin and daily triclosan baths (Cookson et al., 1991a). However, Suller and Russell (2000) did not find changes in triclosan MICs in MRSA strains after the acquisition of a plasmid encoding for mupirocin resistance. Likewise, chlorhexidine appeared to be as effective against MRSA strains as against their sensitive counterparts (Cookson et al., 1991b).

4.2.   Possible mechanisms involved in bacterial resistance to biocides and antibiotics

The main difference between antibiotic and biocide is their mechanism of action at their "in use" concentration. Whereas antibiotics have (generally) a unique target site within the bacterial cell (Chopra, 1998; Chopra et al., 2002), biocides are different in having multiple target sites, the inactivation/ alteration/ destruction of which lead to an inhibitory or lethal effect (Hugo, 1999b; Maillard, 2002). Microorganisms have developed several mechanisms to survive the lethal or static effects of antibiotics (Poole, 2002) and biocides (see Section 3.2). The main difference between these mechanisms is that the alteration of target sites and the degradation of drugs are far more common against antibiotics than against biocides. Nevertheless, some mechanisms bear some similarities, particularly, impermeability and efflux systems and can lead potentially to a linkage in resistance to both antibiotics and biocides. In addition, the induction of some mechanisms (e.g., efflux and change in permeability properties) and the acquisition of genetic elements play an important role in the dissemination of bacterial resistance to both antimicrobials.

4.2.1. Changes in membrane permeability

Changes in the outer membrane have been shown to confer resistance to both antibiotics and biocides in some investigations. For example, Tattawasart *et al.* (1999) showed that *P. stutzeri*, which developed cationic resistance, also presented an altered antibiotic susceptibility profile.

4.2.2. Induction of multidrug efflux systems

Multidrug efflux systems have been particularly involved in bacterial resistance to both biocides and antibiotics. The structure and function of these pumps have been reviewed recently by Borges-Walmsley and Walmsley (2001), Poole (2001), and McKeegan *et al.* (2003). In *P. aeruginosa*, the MexAB-OprM has been involved in bacterial resistance to several unrelated antibiotics (Schweizer, 1998), but also to some biocides. In *S. aureus*, several proteins involved in efflux mechanisms have been described: QacA, QacB, Smr (QacC, QacD, Ebr), SepA, QacEΔ1, QacG, and QacH (Borges-Walmsley and Walmsley, 2001; McKeegan *et al.*, 2003; Narui *et al.*, 2002). NorA is another efflux pump which has been linked to the low-level resistance to some antiseptics and to fluoroquinolones in *S. aureus* (Noguchi *et al.*, 2002). In *E. coli*, the Acr-AB-TolC multidrug efflux pump exhibits a broad-spectrum activity, targeting antibiotics and biocides such as QACs and phenolics (George and Levy, 1983; McMurry *et al.*, 1998b; Moken *et al.*, 1997; Zgurskaya and Nikaido, 2000).

Bacterial exposure to sub-effective concentration of some biocides, such as phenolics (Levy, 1992), has been shown to upregulate the expression of multidrug resistance operons and efflux pumps. In particular, the induction of the *mar* phenotype and its relevance to cross-resistance between triclosan and pine oil, and multiple antibiotic resistance has been well studied (Moken *et al.*, 1997). Overexpression of MarA leads to an increase in MIC triclosan and to antibiotics (McMurry *et al.*, 1998a). A similar phenomenon has been observed with the overexpression of SoxS and AcrAB efflux pumps (McMurry *et al.*, 1998a; Wang *et al.*, 2001).

4.2.3. Alteration of target sites

As mentioned previously, the mechanism of action of triclosan at a low concentration is (at the moment) unique among biocides, since the bisphenol has been shown to target, specifically, a bacterial enzyme. In *M. smegmatis*, mutation in *inhA* confers resistance to not only triclosan but also to isoniazid (McMurry *et al.*, 1999).

4.2.4.   Acquisition of genetic determinants

The presence of resistance determinants on mobile genetic elements favors the spread of resistance via horizontal gene transfer. The presence of genes encoding for resistance determinants have been detected on genetic mobile elements such as conjugative plasmids, transposons (Lyon and Skurray, 1987), and integrons (Paulsen *et al.*, 1993). For example, the *qacA/B* genes have been found on plasmids encoding β-lactamases and heavy metal resistance determinants. Likewise, the *smr* gene has been detected on large conjugative plasmids with multiple-resistance determinants (Bjorland *et al.*, 2001; Lyon and Skurray, 1987). In *S. aureus*, the plasmid pSK01 also carries trimethoprim and aminoglycosides resistance. Paulsen and colleagues (1993) found the *qacE* and *qacEΔ1* genes to be located on the 3′ conserved sequence of integrons in Gram-negative bacteria. These genes have been associated with multiple resistance to biocides and antibiotics in clinical isolates of Gram-negative bacteria (Kücken *et al.*, 2000). Lemaitre and colleagues (1998) showed that QACs resistance in listeriae resulted from the transfer of plasmids, which presented high frequency of transfer. Inorganic (Hg^{2+}) and organomercurial resistance can be carried on plasmids encoding for penicillinase in clinical isolates of *S. aureus* (Shalita *et al.*, 1980). Plasmids that carry genes encoding for antibiotic resistance but also metal ion resistance are not uncommon in Gram-negative bacteria (Silver *et al.*, 1989). Foster (1983) recognised that mercury resistance was inducible and transferable by conjugation and transduction.

Pearce *et al.* (1999) studied the effects of sub-inhibitory concentrations of several biocides in the transfer of antibiotic resistant genes in *S. aureus*. Povidone-iodine at a low concentration (0.005%) reduced the conjugation efficiency of the pWG613 plasmid, while other biocides (at sub-inhibitory concentrations), namely cetrimide and chlorhexidine, had no effect. However, both the biguanide and povidone-iodine effectively inhibited transduction, whereas cetrimide (0.0001%) increased the transduction efficiency.

4.2.5.   Biofilms

The multicomponent resistance response of a biofilm accounts for its resistance to both biocides and antibiotics (Gilbert *et al.*, 2003). It was previously mentioned that the concentration of degradative enzymes within the glycocalyx might play a role in biocidal resistance by enhancing the barrier activity through degradation of molecules. Enzymes such as β-lactamase have also been found within the glycocalyx (Giwercman *et al.*, 1991) and might play a role in the bacterial biofilm resistance to antibiotics (Stewart, 1996). Other mechanisms such as the ones described in Section 3.2.5, namely the presence of a "biofilm-associated, drug-resistant phenotype" (Ashby *et al.*, 1994), the induction of multidrug resistance operons, and efflux pumps, also play an important role (Maira-Litrán *et al.*, 2000).

5. DISINFECTANT USAGE AND ANTIBIOTIC RESISTANCE

Following the evidence of a possible linkage between biocide and antibiotic resistance in bacteria, several authors have claimed that the usage of biocides was deemed to cause the emergence of antibiotic resistance.

There are a number of cases whereby the use of a biocide was associated with emerging antibiotic resistance in bacteria. The use of silver sulphadiazine for the treatment of burn infection which was somewhat successful (Lowbury *et al.*, 1976) in decreasing patient mortality, was also linked to the emergence of plas-mid- mediated sulphonamide resistance (Bridges and Lowbury, 1977). Newsom *et al.* (1990) observed that although the use of chlorhexidine scrub-based preoperative showers effectively decreased skin flora, it might also promote the growth of staphylococci post-surgery, and notably those presenting high-resistance to methicillin. However, similar observations were made when antibiotic prophylaxis was used instead of biocide (Archer and Armstrong, 1983).

Triclosan has been particularly investigated since the number of products containing the bisphenols (and notably household products) has increased dramatically over the last few years. Although bacterial resistance to triclosan has raised concerns (Chuanchuen *et al.*, 2001; Heath and Rock, 2000; Lear *et al.*, 2000; Levy, 2001), emergence of bacterial resistance to other biocides commonly used in the hospital environment have been described, notably to chlorhexidine. Gram-negative bacteria with a significant increase in chlorhexidine resistance and to several antibiotics have been isolated from patients with long-term indwelling catheterization after the extensive use of the biguanide to treat urinary tract infections (Stickler 1974; Stickler and Chawla, 1988; Stickler *et al.*, 1983). However, no plasmid-linked association between antibiotic and chlorhexidine resistance was isolated. Tattawasart *et al.* (1999, 2000a, b) observed an increase in chlorhexidine insusceptibility in *P. stutzeri* and several antibiotics after repeated exposures to the biguanide. However, Thomas *et al.* (2000) noted that although exposure to low residual concentration of the biguanide resulted in an increased insusceptibility of *P. aeruginosa* to the biocide, there was no cross-resistance to any of the antibiotics tested. In staphylococci, chlorhexidine resistance has almost always been associated with antibiotic resistance. Reverdy *et al.* (1992) concluded that the spread of resistant staphylococci in the hospital environment was increased by the use of either antibiotics or biocides.

The selective pressure exerted by the use of cationic biocides has been deemed to play a role in the dissemination of *qac* genes and thus the widespread occurrence of multidrug efflux pumps (Heir *et al.*, 1998, 1999; Mitchell *et al.*, 1998; Paulsen *et al.*, 1996b, c; Sundheim *et al.*, 1998). However, Russell (2002c) pointed out that although a link between the introduction of cationic biocides and the spread of resistance determinants in staphylococci is conceivable, further evidence and investigations were needed to warrant this possibility.

Other biocides have also been involved in the emergence of antibiotic resistance in bacteria. Armstrong *et al.* (1981, 1982) isolated multiple-antibiotic resistant bacteria from drinking water. Disinfection and purification of water, and notably chlorination, have been suggested to increase the occurrence of antibiotic-resistant bacteria (Murray *et al.*, 1984).

It has been observed that clinical isolates of Gram-negative bacteria tend to be more resistant to biocides than their culture collection counterparts (Hammond *et al.*, 1987; Higgins *et al.*, 2001). This might be explained by the repeated exposures to the selective pressure provided by disinfection (Russell, 2002c). It was also proposed that the overuse of biocides might provide a selective environment for less susceptible microorganisms such as MRSA (Levy, 2000). However, at an "in-use" concentration of a biocide, antibiotic resistant microorganisms are not necessarily more resistant to biocides than their sensitive counterparts (Russell 1999b, 2000). It is pertinent to note that selective pressure from biocide exposure is not necessarily correlated with an increase in bacterial resistance to biocides and antibiotics. Lear *et al.* (2000, 2002) isolated only two microorganisms (*Acinetobacter johnsonii* and *Citrobacter freundii*) showing elevated resistance to triclosan from industrial sites where biocides are manufactured (apart from a multitude of intrinsically resistant pseudomonads). However, the authors were not able to correlate increased triclosan resistance with a change in antibiotic susceptibility profile (unpublished data). For further information on this subject, Russell (2002c) comprehensively reviewed the current evidence of biocide usage and the emergence of antibiotic resistance in hospital.

Disinfectants are used in general at a very high concentration, often exceeding 1,000 times that of their MIC, in order to achieve a rapid rate of kill. At a high concentration, a biocide will interact with multiple target sites on the bacterial cell, producing multiple damages, which ultimately will account for cell death. In addition, at these concentrations, it is unlikely that bacteria will become resistant through adaptation or other mechanisms. However, it has to be noted that some bacteria might be intrinsically resistant to a biocide, as already mentioned. Hence, one has to take into account the type of microorganism likely to be encountered and the level of disinfection that needs to be achieved.

Furthermore, the concentration of biocides and the conditions of use (i.e., presence of organic load, temperature, bioburden, etc.) are of paramount importance to achieve an overall lethal activity (Russell and McDonnell, 2000).

6. CONCLUSION

The introduction of biocides into the hospital environment and the selective pressure exerted by their extensive use might confer some level of, or select for, biocide-resistant bacteria, which might impact ultimately on antibiotic resistance. However, it is pertinent to remember that the selective pressure exerted by the extensive use, and sometimes misuse, of antibiotics themselves bears a far greater significance in the emergence of antibiotic resistant microorganisms (WHO, 2000). The development of antibiotic resistance caused by overuse is particularly well documented (French and Phillips, 1999; Gould and MacKenzie, 2002; Hossein *et al.*, 2002). However, there is an increasing body of evidence of shared bacterial resistance mechanisms between biocides and antibiotics. Although the historical use of biocides for hospital disinfection is unlikely to have caused the emergence of antibiotic resistance in hospitals, the improper use of biocides or the use of low sub-inhibitory concentration of a biocide might lead to the development of bacterial resistance to biocide (at last at the MIC level) and to some low-level resistance to antibiotics.

One of the main questions is the clinical relevance of such resistance. Several authors have expressed their doubts as to whether the bacterial resistance described in the literature (notably evidence from *in vitro* investigations) has relevance in practice. It has been pointed out that genes encoding for efflux pumps do not necessarily predominate in MRSA when compared to their sensitive counterparts (methicillin-sensitive *S. aureus*, MSSA) (Bamber and Neal 1999; Suller and Russell, 1999). Furthermore, biocides such as triclosan are still effective in killing hospital strains such as MRSA (Webster 1992; Webster *et al.*, 1994; Zafar *et al.*, 1995). It has been suggested that although a low-level biocide resistance might be produced by efflux systems, the concentration of cationic biocides used in practice is much higher. At these concentrations, where multiple target sites are attacked, it is unlikely that efflux mechanisms will operate efficiently (Favero, 2002; Russell and Maillard, 2000). Finally, some authors and institutions have advocated the rotation of biocide used for low-level disinfection (Murtough *et al.*, 2001).

When employed correctly, biocides have an important role to play in combating infection in the hospital environment (Larson *et al.*, 2000; Russell, 2002a). Therefore, it is important to ensure that biocides are used appropriately, which implies a compliance with disinfection and antisepsis (notably handwashing) regimens and possibly, staff training programs. Overall, more information is needed to establish whether the use of biocides in the clinical context induces the emergence of antibiotic resistance. Bacterial resistance to low (residual) concentration of biocide, notably to cationic and phenolic compounds, might need to be monitored.

Monitoring and understanding the mechanism(s) involved in the emergence of bacterial resistance to biocides are important, since they can indicate whether a disinfection failure results from an operator (no compliance with manufacturer instructions), a product, or the formation of bacterial biofilm. They might further indicate a possible risk for cross-resistance with antibiotics.

REFERENCES

Akimitsu, N., Hamamoto, H., Inoue, R., Shoji, M., Akamine, A., Takemori, K. *et al.*, 1999, Increase in resistance of methicillin-resistant *Staphylococcus aureus* to β-lactams caused by mutations conferring resistance to benzalkonium chloride, a disinfectant widely used in hospitals. *Antimicrob. Agents Chemother.*, **43**, 3042–3043.

Allison, D. G., McBain, A. J., and Gilbert, P., 2000, Biofilms: Problems of control. In D. G. Allison, P. Gilbert, H. M. Lappin-Scott, and M. Wilson (eds.), *Community Structure and Co-operation in Biofilms*. Cambridge University Press, Cambridge, pp. 309–327.

Anderson, R. L., Berkelman, R. L., Mackel, D. C., Davies, B. J., Holland, B. W., and Martone, W. J., 1984, Investigation into the survival of *Pseudomonas aeruginosa* in poloxamer-iodine. *Appl. Environ. Microbiol.*, **47**, 757–762.

Anonymons, 1997, Royal Pharmaceutical Society of Great Britain. Society's evidence on resistance to antimicrobial agents. *Pharm. J.*, **259**, 919–921.

Archer, G. L. and Armstrong, B. C., 1983, Alteration of staphylococcal flora in cardiac surgery patients receiving antibiotic prophylaxis. *J. Infect. Dis.*, **147**, 642–649.

Armstrong, J. L., Shigend, D. S., Calomaris, J. J., and Seidler, P. J., 1981, Antibiotic-resistant bacteria in drinking water. *Appl. Environ. Microbiol.*, **42**, 277–283.

Armstrong, G. L. Calomaris, J. J., and Seidler, P. J., 1982, Antibiotic resistant standard plate count bacteria during water treatment. *Appl. Environ. Microbiol.*, **44**, 308–316.

Ashby, M. J., Neale, J. E., Knott, S. J., and Critchley, I. A., 1994, Effect of antibiotics on non-growing cells and biofilms of *E. coli*. *J. Antimicrob. Chemother.*, **33**, 443–452.

Ayres, H. M., Payne, D. N., Furr, J. R., and Russell, A. D., 1998, Use of the Malthus-AT system to assess the efficacy of permeabilizing agents on the activity of antimicrobial agents against *Pseudomonas aeruginosa*. *Lett. Appl. Microbiol.*, **26**, 422–426.

Bamber, A. I. and Neal, T. J., 1999, An assessment of triclosan susceptibility in methicillin-resistant and methicillin-sensitive *Staphylococcus aureus*. *Lett. Appl. Microbiol.*, **26**, 422–426.

Bjorland, J., Sunde, M., and Waage, S., 2001, Plasmid-borne *smr* gene causes resistance to quaternary ammonium compounds in bovine *Staphylococcus aureus*. *J. Clin. Microbiol.*, **39**, 3999–4004.

Bloomfield, S. F., 2002, Significance of biocide usage and antimicrobial resistance in domiciliary environments. *J. Appl. Microbiol.*, **92**(Suppl.), 144–157.

Borges-Walmsley, M. I. and Walmsley, A. R., 2001, The structure and function of drug pumps. *Trends Microbiol.*, **9**, 71–79.

Bridges, K. and Lowbury, E. J. L., 1977, Drug resistance in relation to use of silver sulphadiazine cream in burns unit. *J. Clin. Pathol.*, **31**, 160–164.

Broadley, S. J., Jenkins, P. A., Furr, J. R., and Russell, A. D., 1995, Potentiation of the effects of chlorhexidine diacetate and cetylpyridinium chloride on mycobacteria by ethambutol. *J. Med. Microbiol.*, **43**, 458–460.

Brown, M. R. W. and Gilbert, P., 1993, Sensitivity of biofilms to antimicrobial agents. *J. Appl. Bacteriol.*, **74**(Suppl.), 87–97.

Brown, M. H., Paulsen, I. T., and Skurray, R. A., 1999, The multidrug efflux protein NorM is a prototype of a new family of transporters. *Mol. Microbiol.*, **31**, 393–395.

Brözel, V. S. and Cloete, T. E., 1994, Resistance of *Pseudomonas aeruginosa* to isothiazolone. *J. Appl. Bacteriol.*, **76**, 576–582.

Centers for Disease Control, 1974, *Disinfectant or Infectant: The Label Doesn't Always Say*. National Nosocomial Infections Study, Fourth Quarter 1973, pp. 18–23.

Chapman, J. S., 1998, Characterizing bacterial resistance to preservatives and disinfectants. *Int. Biodeterior. Biodegrad.*, **41**, 241–245.

Chapman, J. S., 2003, Disinfectant resistance mechanisms, cross-resistance, and co-resistance. *Int. Biodeterior. Biodegrad.*, **51**, 271–276.

Chapman, J. S., Diehl, M. A., and Fearnside, K. B., 1998, Preservative tolerance and resistance. *Int. J. Cosm. Sci.*, **20**, 31–39.

Chopra, I., 1998, Research and development of antibacterial agents. *Curr. Opin. Microbiol.*, **1**, 495–501.

Chopra, I., Hesse, L., and O'Neill, A. J., 2002, Exploiting current understanding of antibiotic action for discovery of new drugs. *J. Appl. Microbiol.*, **92**(Suppl.), 4–15.

Chuanchuen, R., Beinlich, K., Hoang, T. T., Becher, A., Karkhoff-Schweizer, R. R., and Schweizer, H. P., 2001, Cross-resistance between triclosan and antibiotics in *Pseudomonas aeruginosa* is mediated by multidrug efflux pumps: Exposure of a susceptible mutant strain to triclosan selects *nxfB* mutants overexpressing MexCD-OprJ. *Antimicrob. Agents. Chemother.*, **45**, 428–432.

Cloete, T. E., 2003, Resistance mechanisms of bacteria to antimicrobial compounds. *Int. Biodeterior. Biodegrad.*, **51**, 277–282.

Cookson, B. D., Farrely, H., Stapelton, P., Garvey, R. R. J., and Price, M. R., 1991a, Transferable resistance to triclosan in MRSA. *Lancet*, **i**, 1548–1549.

Cookson, B. D., Bolton, M. C., and Platt, J. H., 1991b, Chlorhexidine resistance in methicillin-resistant *Staphylococcus aureus* or just an elevated MIC? An in vitro and in vivo assessment. *Antimicrob. Agents Chemother.*, **35**, 1997–2002.

Costerton, J. W. and Lashen, E. S., 1984, Influence of biofilm on efficacy of biocides on corrosion-causing bacteria. *Mater. Perform.*, **23**, 13–17.

Costerton, J. W., Cheng, K. J., Geesey, G. G., Ladd, T. I., Nickel, J. C., and Dasgupta, M., 1987, Bacterial biofilms in nature and diseases. *Annu. Rev. Microbiol.*, **41**, 435–464.

Cottell, A., Hanlon, G. W., Denyer, S. P., and Maillard, J.-Y., 2003, *Bacterial Cross-Resistance to Antibiotics and Biocides: a Study of Triclosan-Resistant Mutants*. ASM abstract Q-278, Washington, DC (USA).

Curie, K., Speller, D. C. E., Simpson, R., Stephens, M. and Cooke, D. I., 1978, A hospital epidemic caused by a gentamicin-resistant *Klebsiella aerogenes*. *J. Hyg., Cambridge*, **80**, 115–123.

Das, J. R., Bhakoo, M., Jones, M. V., and Gilbert, P., 1998, Changes in biocide susceptibility of *Staphylococcus epidermidis* and *Escherichia coli* cells associated with rapid attachment to plastic surfaces. *J. Appl. Microbiol.*, **84**, 852–859.

Davies, A. J. and Maillard, J.-Y., 2001, Bacterial adaptation to biocides: The possible role of "alarmones." *J. Hosp. Infect.*, **49**, 300–302.

Demple, B., 1996, Redox signaling and gene control in the *Escherichia coli soxRS* oxidative stress regulon—a review. *Gene*, **179**, 53–57.

Dukan, S. and Touati, D., 1996, Hypochlorous acid stress in Escherichia coli: Resistance, DNA damage, and comparison with hydrogen peroxide stress. *J. Bacteriol.*, **178**, 6145–6150.

Ehrenkranz, N. J., Bolyard, E. A., Wiener, M., and Cleary, T. J., 1980, Antibiotic-sensitive *Serratia marcescens* infections complicating cardio-pulmonary operations: Contaminated disinfectants as a reservoir. *Lancet*, **ii**, 1289–1292.

Favero, M. S., 2002, Products containing biocides: Perceptions and realities. *J. Appl. Microbiol.*, **92**(Suppl.), 72–77.

Favero, M. S., Bond, W. W., Peterson, N. J., and Cook, E. H., 1983, Scanning electron microscopic observations of bacteria resistant to iodophor solutions. In *Proceedings of the International Symposium on Povidone*. University of Kentucky: Lexington, pp. 158–166.

Foster, T. J., 1983, Plasmid-determined resistance to antimicrobial drugs and toxic metal ions in bacteria. *Microbiol. Rev.*, **47**, 361–409.

Fraud, S., Maillard, J.-Y., and Russell, A. D., 2001, Comparison of the mycobactericidal activity of *ortho*-phthalaldehyde, glutaraldehyde and other dialdehydes by a quantitative suspension test. *J. Hosp. Infect.*, **48**, 214–221.

Fraud, S., Hann, A. C., Maillard, J.-Y., and Russell, A. D., 2003, Effects of ortho-phthalaldehyde, glutaraldehyde and chlorhexidine diacetate on *Mycobacterium chelonae* and *M. abscessus* strains with modified permeability. *J. Antimicrob. Chemother.*, **51**, 575–584.

French, G. L. and Phillips, I., 1999, Antimicrobial resistance in hospital flora and nosocomial infections. In C. G. Mayhall (ed.), *Hospital Epidemiology and Infection Control*, 2nd edn. Lippincot, Williams & Wilkens, Philadelphia, PA, pp. 1243–1264.

Gandhi, P. A., Sawant, A. D., Wilson, L. A., and Ahearn, D. G., 1993, Adaptation and growth of *Serratia marcescens* in contact lens disinfectant solution containing chlorhexidine gluconate. *Appl. Environ. Microbiol.*, **59**, 183–188.

George, A. M. and Levy, S. B., 1983, Amplifiable resistance to tetracycline, chloramphenicol, and other antibiotics in *Escherichia coli*. *Mol. Microbiol.*, **21**, 441–448.

Gilbert, P. and Allison, D. G., 1999, Biofilms and their resistance towards antimicrobial agents. In H. N. Newman and M. Wilson (eds.), *Dental Plaques Revisited: Oral Biofilms in Health and Diseases*. Bioline Press, Cardiff, pp. 125–143.

Gilbert, P. and McBain, A. J., 2003, Potential impact of increased use of biocides in consumer products on prevalence of antibiotic resistance. *Clin. Microbiol. Rev.*, **16**, 189–208.

Gilbert, P., McBain, A. J., and Rickard, A. H., 2003, Formation of microbial biofilm in hygienic situations: A problem of control. *Int. Biodeterior. Biodegrad.*, **51**, 245–248.

Gillespie, M. T., May, J. W., and Skurray, R. A., 1986, Plasmid-encoded resistance to acriflavine and quaternary ammonium compounds in methicillin-resistant *Staphylococcus aureus*. *FEMS Microbiol. Lett.*, **34**, 47–51.

Giwercman, B., Jensen, E. T., Hoiby, N., Kharazmi, A., and Costerton, J. W., 1991, Induction of β-lactamase production in *Pseudomonas aeruginosa* biofilms. *Antimicrob. Agents Chemother.*, **35**, 1008–1010.

Gould, I. M. and MacKenzie, F. M., 2002, Antibiotic exposure as a risk factor for emergence of resistance: The influence of concentration. *J. Appl. Microbiol.*, **92**(Suppl.), 78–84.

Greenberg, J. T. and Demple, B., 1989, A global response induced in *Escherichia coli* by redox cycling agents overlaps with that induced by peroxide stress. *J. Bacteriol.*, **171**, 3933–3939.

Greenberg, J. T., Monach, P., Chou, J. H., Josephy, P. D., and Demple, B., 1990, Positive control of a global antioxidant defense regulon activated by superoxide-generating agents in *Escherichia coli*. *Proc. Natl. Acad. Sci. USA*, **87**, 6181–6185.

Griffiths P. A., Babb, J. R., Bradley, C. R., and Fraise, A. P., 1997, Glutaraldehyde-resistant *Mycobacterium chelonae* from endoscope washer disinfectors. *J. Appl. Microbiol.*, **82**, 519–526.

Guérin-Méchin, L., Dubois-Brissonnet, F., Heyd, B., and Leveau, J.-Y., 1999, Specific variations of fatty acid composition of *Pseudomonas aeruginosa* ATCC 15442 induced by quaternary ammonium compounds and relation with resistance to bactericidal activity. *J. Appl. Microbiol.*, **87**, 735–742.

Guérin-Méchin, L., Dubois-Brissonnet, F., Heyd, B., and Leveau, J. Y., 2000, Quaternary ammonium compounds stresses induce specific variations in fatty acid composition of *Pseudomonas aeruginosa. Int. J. food. Microbiol.*, **55**, 157–159.

Hammond, S. A., Morgan, J. R., and Russell, A. D., 1987, Comparative sensitivity of hospital isolates of Gram-negative bacteria to antiseptics and disinfectants. *J. Hosp. Infect.*, **9**, 255–264.

Hawkey, P., 2001, The threat from multidrug resistance. Royal College of Pathologists: Education in Pathology. **113**, 23–24.

Heath, R. J. and Rock, C. O., 2000, A triclosan-resistant bacterial enzyme. *Nature*, **406**, 145.

Heath, R. J., Yu, Y. T., Shapiro, M. A., Olson, E., and Rock, C. O., 1998, Broad spectrum antimicrobial biocides target the FabI component of fatty acid synthesis. *J. Biol. Chem.*, **273**, 30316–30320.

Heath, R. J., Rubin, J. R., Holland, D. R., Zhang, E., Snow, M. E., and Rock, C. O., 1999, Mechanism of triclosan inhibition of bacterial fatty acid synthesis. *J. Biol. Chem.*, **274**, 11110–11114.

Heath, R. J., Li, J., Roland, G. E., and Rock, C. E., 2000a, Inhibition of the *Staphylococcus aureus* NADPH-dependent enoyl-acyl carrier protein reductase by triclosan and hexachlorophene. *J. Biol. Chem.*, **275**, 4654–4659.

Heath, R. J., Su, N., Murphy, C. K., and Rock, C. O., 2000b, The enoyl-[acyl-carrier-protein] reductases FabI and FabL from *Bacillus subtilis. J. Biol. Chem.*, **275**, 40128–40133.

Heir, E., Sundheim, G., and Holck, A. L., 1995, Resistance to quaternary ammonium compounds in *Staphylococcus* spp. isolated from the food industry and nucleotide sequence of the resistance plasmid pST827. *J. Appl. Bacteriol.*, **79**, 149–156.

Heir, E., Sundheim, G., and Holck, A. L., 1998, The *Staphylococcus qacH* gene product: A new member of the SMR family encoding multidrug resistance. *FEMS Microbiol. Lett.*, **163**, 49–56.

Heir, E., Sundheim, G., and Holck, A. L., 1999, The *qacG* gene on plasmid pST94 confers resistance to quaternary ammonium compounds in staphylococci isolated from the food industry. *J. Appl. Microbiol.*, **86**, 378–388.

Higgins, C. S., Murtough, S. M., Williamson, E., Hiom, S. J., Payne, D. J., Russell, A. D. *et al.*, 2001, Resistance to antibiotics and biocides among non-fermenting Gram-negative bacteria. *Clin. Microbiol. Infect.*, **7**, 308–315.

Hoang, T. T. and Schweizer, H. P., 1999, Characterization of *Pseudomonas aeruginosa* enoyl-acyl carrier protein reductase (FabI): A target for the antimicrobial triclosan and its role in acylated homoserine lactone synthesis. *J. Bacteriol.*, **181**, 5489–5497.

Hossein, I. K., Hill, D. W., Jenkins, L. E., and Magee, J. T., 2002, Clinical significance of the emergence of bacterial resistance in the hospital environment. *J. Appl. Microbiol.*, **92**(Suppl.), 90–97.

Huang, C. T., Yu, F. P., McFeters, G. A., and Stewart, P. S., 1995, Nonuniform spatial patterns of respiratory activity within biofilms during disinfection. *Appl. Environ. Microbiol.*, **61**, 2252–2256.

Hugo, W. B., 1999a, Historical introduction. In A. D. Russell, W. B. Hugo, and G. A. J. Ayliffe (eds.), *Principles and Practice of Disinfection, Preservation and Sterilization*, 3rd edn. Blackwell Science, Oxford, pp. 258–283.

Hugo, W. B., 1999b, Disinfection mechanisms. In A. D. Russell, W. B. Hugo, and G. A. J. Ayliffe (eds.), *Principles and Practice of Disinfection, Preservation and Sterilization*, 3rd edn. Blackwell Science, Oxford, pp. 1–4.

Hundt, K., Martin, D., Hammer, E., Jonas, U., Kindermann, M. K., and Schauer, F., 2000, Transformation of triclosan by *Trametes versicolor* and *Pycnoporus cinnabarinus. Appl. Environ. Microbiol.*, **66**, 4157–4160.

Jones, M. W., Herd, T. M., and Christie, H. J., 1989, Resistance of *Pseudomonas aeruginosa* to amphoteric and quaternary ammonium biocides. *Microbios*, **58**, 49–61.

Kahan, A., 1984, Is chlorhexidine an essential drug? *Lancet*, **ii**, 759–760.

Kazama, H., Hamashima, H., Sasatsu, M., and Arai, T., 1998, Distribution of the antiseptic-resistance genes *qacE* and *qacE∆1* in Gram-negative bacteria. *FEMS Microbiol. Lett.*, **159**, 173–178.

Kücken, D., Feucht, H. H., and Kaulfers, P. M., 2000, Association of *qacE* and *qacE∆1* with multiple resistance to antibiotics and antiseptics in clinical isolates of Gram-negative bacteria. *FEMS Microbiol. Lett.*, **183**, 95–98.

Kummerle, N., Feucht, H. H., and Kaulfers, P. M., 1996, Plasmid-mediated formaldehyde resistance in *Escherichia coli*: Characterization of resistance gene. *Antimicrob. Agents Chemother.*, **40**, 2276–2279.

Larson, E. L., Early, E., Cloonan, P., Sugrue, S., and Parides, M., 2000, An organizational climate intervention associated with increases handwashing and decreased nosocomial infections. *Behav. Med.*, **29**, 14–22.

Lear, J. C., Maillard, J.-Y., Goddard, P. A., Dettmar, P. W., and Russell, A. D., 2000, Isolation and study of chloroxylenol- and triclosan-resistant strains of bacteria. *J. Pharm. Pharmacol.*, **52**(Suppl.), 126.

Lear, J. C., Maillard, J.-Y., Dettmar, P. W., Goddard, P. A., and Russell, A. D., 2002, Chloroxylenol- and triclosan-tolerant bacteria from industrial sources. *J. Ind. Microbiol. Biot.*, **29**, 238–242.

Leelaporn, A., Paulsen, I. T., Tennent, J. M., Littlejohn, T. G., and Skurray, R. A., 1994, Multidrug resistance to antiseptics and disinfectants in coagulase-negative staphylococci. *J. Med. Microbiol.*, **40**, 214–220.

Lemaitre, J. P., Echannaoui, H., Michaut, G., Divies, C., and Rousset, A., 1998, Plasmid-mediated resistance to antimicrobial agents among listeriae. *J. Food. Prot.*, **61**, 1459–1464.

Levy, S. B., 1992, Active efflux mechanisms for antimicrobial resistance. *Antimicrob. Agents Chemother.*, **36**, 695–703.

Levy, S. B., 2000, Antibiotic and antiseptic resistance: Impact on public health. *Pediatr. Infect. Dis. J.*, **19**(Suppl.), 120–122.

Levy, S. B., 2001, Antibacterial household products: Cause for concern. *Emerg. Infect. Dis.*, **7**, 512–515.

Levy. C. W., Roujeinikova, A., Sedelnikova, S., Baker, P. J., Stuitje, A. R., Slabas, A. R. *et al.*, 1999, Molecular basis of triclosan activity. *Nature*, **398**, 383–384.

Littlejohn, T. G., Di Bernadino, D., Messerotti, L. J., Spiers, S. J., and Skurray, R. A., 1990, Structure and evolution of a family of genes encoding antiseptic and disinfectant resistance in *Staphylococcus aureus*. *Gene*, **101**, 59–66.

Littlejohn, T. G., Paulsen, I. P., Gillespie, M., Tennent, J. M., Midgely, M., Jones, I. G. *et al.*, 1992, Substrate specificity and energetics of antiseptic and disinfectant resistance in *Staphylococcus aureus*. *FEMS Microbiol. Lett.*, **95**, 259–266.

Lomovskaya, O. and Lewis, K., 1992, *emr*, an *Escherichia coli* locus for multidrug resistance. *Proc. Natl. Acad. Sci. USA*, **89**, 8938–8942.

Lowbury, E. J. L., Babb, J. R., Bridges, K., and Jackson, D. M., 1976, Topical chemoprophylaxis with silver sulphadiazine and silver nitrate chlorhexidine cream: Emergence of sulphonamide-resistant Gram-negative bacilli. *BMJ*, **i**, 493–496.

Lyon, B. R. and Skurray, R. A., 1987, Antimicrobial resistance of *Staphylococcus aureus*: Genetic basis. *Microbiol. Rev.*, **51**, 88–134.

Maillard, J.-Y., 2001, Virus susceptibility to biocides: An understanding. *Rev. Med. Microbiol.*, **12**, 63–74.

Maillard, J.-Y., 2002, Antibacterial mechanisms of action of biocides. *J. Appl. Microbiol.*, **92**(Suppl.), 16–27.

Maira-Litrán, T., Allison, D. G., and Gilbert, P., 2000, An evaluation of the potential of the multiple antibiotic resistance operon (*mar*) and the multidrug efflux pump *acrAB* to moderate resistance towards ciprofloxacin in *Escherichia coli* biofilms. *J. Antimicrob. Chemother.*, **45**, 789–795.

Manzoor, S. E., Lambert, P. A., Griffiths, P. A., Gill, M. J., and Fraise, A. P., 1999, Reduced glutaraldehyde susceptibility in *Mycobacterium chelonae* associated with altered cell wall polysaccharides. *J. Antimicrob. Chemother.*, **43**, 759–765.

Marcinkeviciene, J., Jiang, W., Kopcho, L. M., Locke, G., Luo, Y., and Copeland, R. A., 2001, Enoyl-ACP reductase (FabI) of *Haemophilus influenzae*: A steady-state kinetic mechanism and inhibition by triclosan and hexachlorophene. *Arch. Biochem. Biophys.* **390**, 101–108.

McDonnell, G. and Russell, A. D., 1999, Antiseptics and disinfectants: Activity, action and resistance. *Clin. Microbiol. Rev.*, **12**, 147–179.

McKeegan, K. S., Borges-Walmsley, M. I., and Walmsley, A. R., 2003, The structure and function of drug pumps: An update. *Trends Microbiol.*, **11**, 21–29.

McMurry, L. M., Oethinger, M., and Levy, S. B., 1998a, Overexpression of *marA*, *soxS*, or *acrAB* produces resistance to triclosan in laboratory and clinical strains of *Escherichia coli*. *FEMS Microbiol. Lett.*, **166**, 305–309.

McMurry, L. M., Oethinger, M., and Levy, S. B., 1998b, Triclosan targets lipid synthesis. *Nature*, **394**, 531–532.

McMurry, L. M., McDermott, P. F., and Levy, S. B., 1999, Genetic evidence that InhA of *Mycobacterium smegmatis* is a target for triclosan. *Antimicrob. Agents Chemother.*, **43**, 711–713.

McNeil, M. R. and Brennan, P. J., 1991, Structure, function and biogenesis of the cell envelope of mycobacteria in relation to bacterial physiology, pathogenesis and drug resistance; some thoughts and possibilities arising from recent structural information. *Res. Microbiol.*, **142**, 451–463.

Méchin, L., Dubois-Brissonnet, F., Heyd, B., and Leveau, J. Y., 1999, Adaptation of *Pseudomonas aeruginosa* ATCC 15442 to didecyldimethylammonium bromide induces changes in membrane fatty acid composition and in resistance of cells. *J. Appl. Microbiol.*, **86**, 859–866.

Mitchell, B. A., Brown, M. H., and Skurray, R. A., 1998, QacA multidrug efflux pump from *Staphylococcus aureus*: Comparative analysis of resistance to diamidines, biguanides and guanylhydrazones. *Antimicrob. Agents Chemother.*, **42**, 475–477.

Moken, M. C., McMurry, L. M., and Levy, S. B., 1997, Selection of multiple-antibiotic-resistant (Mar) mutants of *Escherichia coli* by using the disinfectant pine oil: Roles of the *mar* and *acrAB* loci. *Antimicrob. Agents Chemother.*, **41**, 2770–2772.

Munton, T. J. and Russell, A. D., 1970, Effect of glutaraldehyde on protoplasts of *Bacillus megaterium*. *J. Gen. Microbiol.*, **63**, 367–370.

Murray, G. E., Tobi, R. S., Jenkins, B., and Kusher, D. J., 1984, Effect of chlorination on antibiotic resistance profiles of sewage-related bacteria. *Appl. Environ. Microbiol.*, **48**, 73–33.

Murtough, S. M., Hiom, S. J., Palmer, M., and Russell, A. D., 2001, Biocide rotation in the healthcare setting: Is there a case for policy implementation? *J. Hosp. Infect.*, **48**, 1–6.

Narui, K., Noguchi, N., Wakasugi, K., and Sasatsu, M., 2002, Cloning and characterization of a novel chromosomal drug efflux gene in *Staphylococcus aureus*. *Biol. Pharm. Bull.*, **25**, 1533–1536.

Newsom, S. W. B., White, R., and Pascoe, J., 1990, Action of teicoplanin on perioperative skin staphylococci. In R. N. Gruneberg (ed.), *Teicoplanin-Further European Experience*. Royal Society of Medicine, London, pp. 1–18.

Ng, M. E. G.-L., Jones, S., Leong, S. H., and Russell, A. D., 2002, Biocide and antibiotics with apparently similar actions on bacteria: Is there the potential for cross-resistance? *J. Hosp. Infect.*, **51**, 147–178.

Nikaido, H., 1996, Multidrug efflux pumps of gram-negative bacteria. *J. Bacteriol.*, **178**, 5853–5859.

Noguchi, N., Tamura, M., Narui, K., Wakasugi, K., and Sasatsu, M., 2002, Frequency and genetic characterization of multidrug-resistant mutants of *Staphylococcus aureus* after selection with individual antiseptics and fluoroquinolones. *Biol. Pharm. Bull.*, **25**, 1129–1132.

Parikh, S. L., Xiao, G., and Tonge, P. J., 2000, Inhibition of InhA, the enoyl reductase from *Mycobacterium tuberculosis*, by triclosan and isoniazid. *Biochemistry*, **39**, 7645–7650.

Paulsen, I. T., Littlejohn, T. G., Radstrom, P., Sundstrom, L., Skold, O., Swedberg, G. *et al.*, 1993, The 3′ conserved segment of integrons contains a gene associated with multidrug resistance to antiseptics and disinfectants. *Antimicrob. Agents Chemother.*, **37**, 761–768.

Paulsen, I. T., Brown, M. H., and Skurray, R. A., 1996a, Proton-dependent multidrug efflux systems. *Microbiol. Rev.*, **60**, 575–608.

Paulsen, I. T., Skurray, R. A., Tam, R., Saier, M. H., Jr., Turner, R. J., Wiener, J. H. *et al.*, 1996b, The SMR family: A novel family of multidrug efflux proteins involved with the efflux of lipophilic drugs. *Mol. Microbiol.*, **19**, 1167–1175.

Paulsen, I. T, Brown, M. H., Litteljohn, T. G., Mitchell, B. A., and Skurray, R. A., 1996c, Multidrug resistance proteins QacA and QacB from *Staphylococcus aureus*. Membrane topology and identification of residues involved in substrate specificity. *Proc. Natl. Acad. Sci. USA*, **93**, 3630–3635.

Pearce, H., Messager, S., and Maillard, J.-Y., 1999, Effect of biocides commonly used in the hospital environment on the transfer of antibiotic-resistance genes in *Staphylococcus aureus*. *J. Hosp. Infect.*, **43**, 101–108.

Poole, K., 2001, Multidrug resistance in Gram-negative bacteria. *Curr. Opin. Microbiol.*, **4**, 500–508.

Poole, K., 2002, Mechanisms of bacterial biocide and antibiotic resistance. *J. Appl. Microbiol.*, **92**(Suppl.), 55–64.

Prince, J. and Ayliffe, G. A. J., 1972, In-use testing of disinfectants in hospitals. *J. Clin. Pathol.*, **25**, 586–589.

Putman, M., van Veen, H. W., and Konings, W. N., 2000, Molecular properties of bacterial multidrug transporters. *Microbiol. Mol. Biol. Rev.*, **64**, 672–693.

Reverdy, M. E., Bes, M., Nervi, C. Martra, A., and Fleurette, J., 1992, Activity of four antiseptics (acriflavine, benzalkonium chloride, chlorhexidine gluconate and hexamidine diisethionate) and of ethidium bromide on 392 strains representing 26 *Staphylococcus* species. *Med. Microbiol. Lett.*, **1**, 56–63.

Rotter, M. L., 1998, Semmelweis' sesquicentennial: A little-noted anniversary of handwashing. *Curr. Opin. Infect. Dis.*, **11**, 457–460.

Rotter, M. L., 2001, Argument for alcoholic hand disinfection. *J. Hosp. Infect.*, **48**(Suppl.), 4–8.

Rouche, D. A., Cram, D. S., Di Bernadino, D., Littlejohn, T. G., and Skurray, R. A., 1990, Efflux-mediated antiseptic gene *qacA* in *Staphylococcus aureus*: Common ancestry with tetracycline and sugar transport proteins. *Mol. Microbiol.*, **4**, 2051–2062.

Roujeinikova, A., Levy, C. W., Rowsell, S., Sedelnikova, S., Baker, P. J., Minshull, C. A. *et al.*, 1999, Crystallographic analysis of triclosan bound enoyl reductase. *J. Mol. Biol.*, **294**, 527–535.

Russell, A. D., 1990, Bacterial spores and chemical sporicidal agents. *Clin. Microbiol. Rev.*, **3**, 99–119.

Russell, A. D., 1995, Mechanisms of bacterial resistance to biocides. *Int. Biodeterior. Biodegrad.*, **36**, 247–265.

Russell, A. D., 1996, Activity of biocides against mycobacteria. *J. Appl. Bacteriol.*, **81**, 87–101.

Russell, A. D., 1999a, Bacterial resistance to disinfectants: Present knowledge and future problems. *J. Hosp. Infect.*, **43**(Suppl.), 57–68.

Russell, A. D., 1999b, Types of antimicrobial agents. In A. D. Russell, W. B. Hugo, and G. A. J. Ayliffe (eds.), *Principles and Practice of Disinfection, Preservation and Sterilization*, 3rd edn. Blackwell Science, Oxford, pp. 5–94.

Russell, A. D., 2000, Do biocides select for antibiotic resistance? *J. Pharm. Pharmacol.*, **52**, 227–233.

Russell, A. D., 2002a, Antibiotic and biocide resistance in bacteria: Comments and conclusion. *J. Appl. Microbiol.*, **92**(Suppl.), 171–173.

Russell, A. D., 2002b, Mechanisms of antimicrobial action of antiseptics and disinfectants: An increasingly important area of investigation. *J. Antimicrob. Chemother.*, **49**, 597–599.

Russell, A. D., 2002c, Introduction of biocides into clinical practice and the impact on antibiotic-resistant bacteria. *J. Appl. Microbiol.*, **92**(Suppl.), 121–135.

Russell, A. D. and Russell, N. J., 1995, Biocides: Activity, action and resistance. In P. A. Hunter, G. K. Darby, and N. J. Russell (eds.), *50 Years of Antimicrobials*. Society for General Microbiology Symposium 53. Cambridge University Press, UK.

Russell, A. D. and Maillard, J.-Y., 2000, Reaction and response: Is there a relationship between antibiotic resistance and resistance to antiseptics and disinfectants among hospital-acquired and community-acquired pathogens? *Am. J. Infect. Control.*, **28**, 204–206.

Russell, A. D. and McDonnell, G., 2000, Concentration: A major factor in studying biocidal action. *J. Hosp. Infect.*, **44**, 1–3.

Russell, A. D., Furr, J. R., and Maillard, J.-Y., 1997, Microbial susceptibility and resistance to biocides: An understanding. *ASM News*, **63**, 481–487.

Russell, A. D., Tattawasart, U, Maillard, J.-Y., and Furr, J. R., 1998, Possible link between bacterial resistance and use of antibiotics and biocides. *Antimicrob. Agents Chemother.*, **42**, 2151.

Russell, A. D., Suller, M. T., and Maillard, J.-Y., 1999, Do antiseptics and disinfectants select for antibiotic resistance? *J. Med. Microbiol.*, **48**, 613–615.

Rutala, W. A., and Weber, D. J., 1999, Infection control: The role of disinfection and sterilization. *J. Hosp. Infect.*, **43**(Suppl.), 43–55.

Salzman, M. B. and Rubin, L. G., 1995, Intravenous catheter-related infections. *Adv. Pediatr. Infect. Dis.*, **10**, 37–368.

Sanford, J. P., 1970, Disinfectants that don't. *Ann. Intern. Med.*, **72**, 282–283.

Sasatsu, M., Shimizu, K., Noguchi, N., and Kono, M., 1993, Triclosan-resistant *Staphylococcus aureus*. *Lancet*, **341**, 756.

Schweizer, H. P., 1998, Intrinsic resistance to inhibitors of fatty acid biosynthesis in *Pseudomonas aeruginosa* is due to efflux: Application of a novel technique for generation of unmarked chromosomal mutations for the study of efflux systems. *Antimicrob. Agents Chemother.*, **42**, 394–398.

Schweizer, H. P., 2001, Triclosan: A widely used biocide and its link to antibiotics. *FEMS Microbiol. Lett.*, **202**, 1–7.

Shalita, Z., Murphy, E., and Novick, R. P., 1980, Penecillinase plasmids of *Staphylococcus aureus*: Structural and evolutionary relationships. *Plasmid*, **3**, 291–311.

Silver, S., Nucifora, G., Chu, L., and Misra, T. K., 1989, Bacterial ATPases—primary pumps for exploring toxic cations and anions. *Trends Biochem. Sci.*, **14**, 76–80.

Slater-Radosti, C., van Aller, G., Greenwood, G., Nicholas, R., Keller, P. M., DeWolf, W. E., Jr. *et al.*, 2001, Biochemical and genetic characterization of the action of triclosan on *Staphylococcus aureus. J. Antimicrob. Chemother.*, **48**, 1–6.

Sondossi, M., Rossmore, H. W., and Wireman, J. W., 1985, Observation of resistance and cross-resistance to formaldehyde and a formaldehyde condensate biocide in *Pseudomonas aeruginosa. Int. Biodeterior. Biodegrad.*, **21**, 105–106.

Speller, D. C. E., Stephens, M. E., and Vinat, A., 1971, Hospital infection by *Pseudomonas cepacia. Lancet*, **i**, 798–799.

Stewart, P. S., 1996, Theoretical aspects of antibiotic diffusion into microbial biofilms. *Antimicrob. Agents Chemother.*, **40**, 2517–2522.

Stewart, M. J., Parikh, S., Xiao, G., Tonge, P. J., and Kisker, C., 1999, Structural basis and mechanism of enoyl reductase inhibition by triclosan. *J. Mol. Biol.*, **290**, 859–865.

Stickler, D. J., 1974, Chlorhexidine resistance in *Proteus mirabilis. J. Clin. Pathol.*, **27**, 284–287.

Stickler, D. J. and Chawla, J. C., 1988, Antiseptics and long-term bladder catheterization. *J. Hosp. Infect.*, **2**, 337–338.

Stickler, D. J., Thomas, B., Clayton, C. L., and Cawla, J. C., 1983, Studies on the genetic basis of chlorhexidine resistance. *Br. J. Clin. Pract.*, **25**(Suppl.), 23–28.

Suller, M. T. E. and Russell, A. D., 1999, Antibiotic and biocide resistance in methicillin-resistant *Staphylococcus aureus* and vancomycin-resistant enterococcus. *J. Hosp. Infect.*, **43**, 281–291.

Suller, M. T. E. and Russell, A. D., 2000, Triclosan and antibiotic resistance in *Staphylococcus aureus. J. Antimicrob. Chemother.*, **46**, 11–18.

Sundheim, G., Langsrud, S., Heir, E., and Holck, A. L., 1998, Bacterial resistance to disinfectants containing quaternary ammonium compounds. *Int. Biodeterior. Biodegrad.*, **41**, 235–239.

Tattawasart, U., Maillard, J.-Y., Furr, J. R., and Russell, A. D., 1999, Development of resistance to chlorhexidine diacetate and cetylpyridinium chloride in *Pseudomonas stutzeri* and changes in antibiotic susceptibility. *J. Hosp. Infect.*, **42**, 219–229.

Tattawasart, U., Hann, A. C., Maillard, J.-Y., Furr, J. R., and Russell, A. D., 2000a, Cytological changes in chlorhexidine-resistant isolates of *Pseudomonas stutzeri. J. Antimicrob. Chemother.*, **45**, 145–152.

Tattawasart, U., Maillard, J.-Y., Furr, J. R., and Russell, A. D., 2000b, Outer membrane changes in *Pseudomonas stutzeri* strains resistant to chlorhexidine diacetate and cetylpyridinium chloride. *Int. J. Antimicrob. Agents*, **16**, 233–238.

Tennent, J. M., Lyon, B. R., Midgley, M., Jones, I. G., Purewal, A. S., and Skurray, R. A., 1989, Physical and biochemical characterization of the *qacA* gene encoding antiseptic and disinfectant resistance in *Staphylococcus aureus. J. Gen. Microbiol.*, **135**, 1–10.

Thomas, L. Maillard, J.-Y., Lambert, R. J. W., and Russell, A. D., 2000, Development of resistance to chlorhexidine diacetate in *Pseudomonas aeruginosa* and the effect of "residual" concentration. *J. Hosp. Infect.*, **46**, 297–303.

Van Klingeren, B. and Pullen, W., 1993, Glutaraldehyde resistant mycobacteria from endoscope washers. *J. Hosp. Infect.*, **25**, 147–149.

Walsh, S. E., Maillard, J.-Y., Russell, A. D., and Hann, A. C., 2001, Possible mechanisms for the relative efficacies of *ortho*-phthalaldehyde and glutaraldehyde against glutaraldehyde-resistant *Mycobacterium chelonae. J. Appl. Microbiol.*, **91**, 80–92.

Walsh, S. E., Maillard, J.-Y., Russell, A. D., Charbonneau, D. L., Bartolo, R. G., and Catrenich, C., 2003, Development of bacterial resistance to several biocides and effects on antibiotic susceptibility. *J. Hosp. Infect.*, **55**, 98–107.

Wang, H., Dzink-Fox, J. L., Chen, M., and Levy, S. B., 2001, Genetic characterization of high-level fluoroquinolone resistant clinical *Escherichia coli* strains from China: Role of *acrA* mutations. *Antimicrob. Agents Chemother.*, **45**, 1515–1521.

Webster, J., 1992, Handwashing in a neonatal intensive care nursery: Product acceptability and effectiveness of chlorhexidine gluconate 4% and triclosan 1%. *J. Hosp. Infect.*, **21**, 137–141.

Webster, J., Faoagoli, J. L., and Cartwrigth, D., 1994, Elimination of methicillin-resistant *Staphylococcus aureus* from neonatal intensive care unit after hand washing with triclosan. *J. Paediatr. Child Health*, **30**, 59–64.

WHO, 2000, *Overcoming Antimicrobial Resistance*. World Health Organisation Report on Infectious Diseases. WHO, Geneva.

Winder, C. L., Al-Adham, I. S., Abdel Malek, S. M., Buultjens, T. E., Horrocks, A. J., and Collier, P. J., 2000, Outer membrane protein shifts in biocide-resistant *Pseudomonas aeruginosa* PAO1. *J. Appl. Microbiol.*, **89**, 289–295.

Zafar, A. B., Butler, R. C., Reese, D. J., Gaydos, L. A., and Mennonna, P. A., 1995, Use of 0.3% triclosan (Bacti-Stat) to eradicate an outbreak of methicillin-resistant *Staphylococcus aureus* in a neonatal nursery. *Am. J. Infect. Control.*, **23**, 200–208.

Zgurskaya, H. I. and Nikaido, H., 2000, Multidrug resistance mechanisms: Drug efflux across two membranes. *Mol. Microbiol.*, **37**, 219–225.

Chapter 26

Interventions to Improve Antibiotic Prescribing in the Community

Sandra L. Arnold
Le Bonheur Children's Medical Center, University of Tennessee,
50 N Dunlap St, Memphis TN 38103, USA

Antimicrobial resistant bacterial pathogens have become an increasing threat to the world's population. Many common pathogens have become resistant to usual antimicrobial therapy leading to escalating use of combinations of powerful, broad-spectrum antibiotics. Worldwide, community-acquired infections with drug resistant *Salmonellae, Neisseria gonorrheae,* and *Mycobacterium tuberculosis* are quite prevalent. Given the ease of travel in the global community, these pathogens among others, threaten the ability to treat infections, even in areas with relatively effective public health and disease control programs. Antibiotic resistance in other community-acquired pathogens has become more frequent in the last years of the twentieth century and threatens our ability to treat common community-acquired infections in both the developing and developed world.

Streptococcus pneumoniae is the most common community-acquired pathogen causing meningitis, bacteremia, pneumonia, and otitis media in young children. Population-based surveillance for invasive pneumococcal disease by the US Centers for Disease Control in selected regions has revealed an overall incidence of 21–24 cases per 100,000 population (Centers for Disease Control and Prevention, 2001). Antimicrobial resistance, especially penicillin resistance, among isolates of *S. pneumoniae* has increased throughout the world (where the incidence is substantially greater) since the late 1980s and early 1990s and threatens the ability to treat pneumococcal infections. Over the course of 8 years, the prevalence of invasive penicillin-resistant (intermediate

Antibiotic Policies: Theory and Practice. Edited by Gould and van der Meer
Kluwer Academic / Plenum Publishers, New York, 2005

and fully resistant) isolates increased from 2.8% to 6.8% in 10 Canadian pediatric hospitals (Scheifele *et al.*, 1996, 2000). Active surveillance by the US Centers for Disease Control for invasive *S. pneumoniae* infections has revealed an increase in the proportion of penicillin-resistant invasive isolates from 6.7% (Breiman *et al.*, 1994) to 27.5% in 8 years across the country, with some regions (Tennessee) reporting 54% of isolates resistant to at least one antibiotic (Whitney *et al.*, 2000). In Europe, 11.4% of invasive pneumococcal isolates from 1999 to 2001 were penicillin non-susceptible (EARSS, 2001). However, this was an average over all with penicillin non-susceptibility rates as high as 45% and 35% in France and Spain down to as low as 1–2% in the Netherlands.

Organisms other than *S. pneumoniae* causing infections in the community are also demonstrating clinically significant antibiotic resistance. While less of a problem in North America than Europe and Japan, macrolide resistant Group A streptococci have been isolated with increasing frequency (Cornaglia *et al.*, 1996; Maruyama *et al.*, 1979; Seppälä *et al.*, 1992). In addition to concerns about increasing resistance in respiratory tract pathogens, researchers have also documented increasing rates of antibiotic-resistant *Escherichia coli* urinary isolates. Several studies from different regions of the world have documented resistance to trimethoprim-sulfamethoxazole among community urinary iso-lates varying from 18% to 28% (Brown *et al.*, 2002; Ladhani and Gransden, 2003; Zhanel *et al.*, 2000).

The risk factors associated with colonization or infection with antibiotic-resistant *S. pneumoniae* have been well characterized (Arason *et al.*, 1996; Block *et al.*, 1995; Boken *et al.*, 1995; Duchin *et al.*, 1995; Jackson *et al.*, 1984; Nasrin *et al.*, 2002; Nava *et al.*, 1994; Pallares *et al.*, 1987; Radetsky *et al.*, 1981; Reichler *et al.*, 1992; Tan *et al.*, 1993; Zenni *et al.*, 1995). There is evidence that antibiotic exposure is associated with carriage of resistant *S. pneumoniae* at the level of the individual (Brook and Gober, 1996; Dagan *et al.*, 1998). Antibiotic use in the preceding 3–6 months has been associated with a 2–5-fold increase in the risk of nasopharyngeal colonization with antibi-otic-resistant *S. pneumoniae* (Arason *et al.*, 1996; Boken *et al.*, 1995; Duchin *et al.*, 1995; Nasrin *et al.*, 2002; Radetsky *et al.*, 1981; Reichler *et al.*, 1992; Zenni *et al.*, 1995). The risk of invasive disease with antibiotic-resistant pneumococci (compared with infection due to susceptible bacteria) is increased substantially following antibiotic exposure as well (Block *et al.*, 1995; Jackson *et al.*, 1984; Nava *et al.*, 1994; Pallares *et al.*, 1987; Tan *et al.*, 1993). In a study of the dynamics of pneumococcal carriage during antibiotic treatment (Dagan *et al.*, 1998), it was observed that 19/120 children had a new pneumococcal isolate colonizing the nasopharynx at 3–4 days into treatment, 16/19 of which were resistant to the antibiotic the child was taking.

The association between antibiotic use and resistance in bacteria has been demonstrated for many other important community-acquired pathogens.

Macrolide resistant Group A streptococcal infections in Europe and Japan have been linked to the popularity of azithromycin (Priest *et al.*, 2001; Seppälä *et al.*, 1995). Studies from Europe and Asia have documented substantial decreases in the rates of erythromycin resistance following decreases in use of macrolide agents (Fujita *et al.*, 1994; Seppälä *et al.*, 1997). In addition, antimicrobial resistance in urinary tract pathogens has also been associated with recent antibiotic use (Allen *et al.*, 1999; Brown *et al.*, 2002; Magee *et al.*, 1999; Steinke *et al.*, 2000; Zhanel *et al.*, 2000).

Thus, there is compelling evidence that exposure of the community and the individual to antibiotics enhances the risk of an individual's harboring a resistant organism. For *S. pneumoniae* carriage, it appears that frequent repeated exposure to antibiotics in children (who are most likely to be carriers of the organism) reduces the pool of circulating strains that are susceptible to antibiotics, allowing resistant strains to multiply and spread easily among children.

There is a major body of literature documenting the substantial misuse of antibiotics throughout the world for viral infections and other diseases, such as asthma, for which antibiotics are of no known benefit (Cronk *et al.*, 1954; Gordon *et al.*, 1974; Hardy and Traisman, 1956; Kaiser *et al.*, 1996; Lexomboon *et al.*, 1971; Stott and West, 1976; Taylor *et al.*, 1977; Townsend, 1960; Townsend and Radebaugh, 1962). Antibiotics were the second leading class of drugs prescribed in the United States according to the US National Ambulatory Medical Care Surveys (NAMCS) (McCaig and Hughes, 1995) between 1980 and 1992 with the majority of such prescriptions for respiratory tract infections. In another analysis of the 1992 NAMCS data (Nyquist *et al.*, 1998), antibiotic prescription rates for colds, upper respiratory tract infections, and bronchitis in children were 44%, 46%, and 75% respectively, with similar rates of inappropriate prescribing for adults (Gonzales *et al.*, 1997). In a pediatric practice in Memphis, Tennessee, 43% of children with a diagnosis of uncomplicated upper respiratory infection (URI) or asthma received a prescription for an antimicrobial (Arnold *et al.*, 1996). In a Kentucky Medicaid study, 60% of patients received an antibiotic prescription for the common cold (Mainous *et al.*, 1996).

In England and Scotland, the number of antibiotic prescriptions increased by 45.8% between 1980 and 1991. Rates of growth in antibiotic prescribing in Germany and France were 78% and 65%, respectively in the same period (Davey *et al.*, 1996; Taboulet, 1990). In Canada, an analysis of Saskatchewan Health Databases demonstrated that 85% of outpatient antibiotic prescriptions for respiratory tract infections in children under the age of 5 years in that province were inappropriate (Wang *et al.*, 1999).

In addition to using antibiotics for inappropriate indications, physicians are using more broad-spectrum antimicrobials, considered as second- or third-line agents to treat common infections. In the NAMCS survey of 1980–92, the authors noted an increase in the rates of use of amoxicillin and cephalosporins with a concomitant decrease in the use of penicillin (McCaig and Hughes, 1995). In another analysis of the NAMCS, the authors examined antibiotic treatment for sore throat in adults from 1989 to 1999 (Linder and Stafford, 2001). They not only found excessive antibiotic use (prescribed for 73% of visits for sore throat in a population expected to have a rate of streptococcal pharyngitis of 5–17%) but also that there was a significant decline in the use of recommended antibiotics (penicillin and erythromycin) and an overall rate of prescribing non-recommended antibiotics of 68%.

Changing physician behavior requires identifying and addressing barriers to change in practice. In a systematic review of interventions to improve physicians' practice and implement findings of medical research in practice, Oxman and colleagues found "no magic bullets" for improving the quality of healthcare (Oxman *et al.*, 1995). They found that simple interventions such as conferences or the mailing of unsolicited materials produced little or no change in behavior for a variety of areas of patient care. Complex (and expensive) interventions such as educational outreach visits or training of local opinion leaders were moderately effective in producing changes in physician behavior. Overall, conclusions could not be made regarding effectiveness of different types of interventions because the scope of identified practice deficiencies, types of physicians, and variations of interventions is too diverse. What these authors did recommend was the careful evaluation of focused interventions on specific practice areas after the elucidation of the root causes of suboptimal performance and the identification of barriers to change.

Research has enhanced the understanding of the underlying reasons for inappropriate antibiotic use for viral respiratory tract infections. The following factors appear to be significant determinants of antibiotic use in clinical situations where they are not indicated are: (1) physician–patient interaction; (2) physician characteristics; (3) physician time constraints; and (4) diagnostic uncertainty (Arnold *et al.*, 1999; Bauchner *et al.*, 1999; Hamm *et al.*, 1996; Hutchinson and Foley, 1999; Mainous *et al.*, 1997, 1998; McIsaac and Butler, 2000; Murray *et al.*, 2000; Palmer and Bauchner, 1997; Steinke *et al.*, 2000; Watson *et al.*, 1999).

Aspects of the physician–patient interaction have been studied extensively. Many physicians argue that they prescribe antibiotics for viral infections because they feel pressured to do so by patients or parents (Bauchner *et al.*, 1999; Palmer and Bauchner, 1997; Watson *et al.*, 1999); however, it appears that physicians frequently misjudge patients' or parents' intentions (Hamm *et al.*, 1996). Physicians with certain characteristics, such as longer duration of time in practice (Mainous *et al.*, 1998) or not being involved in medical teaching (Steinke *et al.*, 2000), appear to overuse antibiotics more frequently. Physicians who see more patients and presumably, spend less time with each patient (Arnold *et al.*, 1999; Hutchinson and Foley, 1999) prescribe more antibiotics than those who see fewer patients. The issue of diagnostic uncertainty (difficulty in distinguishing a benign, self-limited viral infection from a more serious infection requiring antibiotic therapy) has received less attention than these other areas in the medical literature and requires further study. It has been shown that physicians frequently overestimate the probability that a patient with a respiratory tract infection has a bacterial infection, particularly for pharyngitis (McIsaac and Butler, 2000), sinusitis (Murray *et al.*, 2000), and bronchitis (Mainous *et al.*, 1997; Murray *et al.*, 2000). These studies have mainly been performed in adult populations. Diagnostic uncertainty regarding the presence of invasive bacterial infections in young children with high fever (Bass *et al.*, 1993; Carroll *et al.*, 1983; Jaffe *et al.*, 1987; McGowan *et al.*, 1973; Teele *et al.*, 1975; Waskerwitz and Berkelhamer, 1981) is a distinct clinical problem in pediatrics which likely has an impact on the use of antibiotics in children with respiratory tract infection. When there is uncertainty in any potentially infectious condition, physicians tend to be cautious and prescribe an antibiotic if it could, at all, be beneficial.

Some of these identified barriers are amenable to change. However, given the multifactorial nature of the problem, it is unlikely that a single approach will work for all physicians in all regions. In addition, different patient populations and conditions may warrant a variety of interventions. This chapter will explore published studies of interventions to improve antibiotic prescribing in the ambulatory care setting. It will include reviews of studies targeting overall antibiotic use, use for specific conditions, and use of recommended and nonrecommended agents (usually defined locally). These studies predominantly address common infectious clinical syndromes seen in ambulatory care including urinary tract infections, respiratory tract infections (viral and bacterial), and infectious diarrhea in a large variety of ambulatory care settings around the world.

1. METHODS OF LITERATURE REVIEW

This review was initiated under the auspices of the Effective Practice and Organization of Care (EPOC) review group of the Cochrane Collaboration. The Cochrane Collaboration comprises a worldwide group of individuals dedicated to systematically reviewing the literature on topics relevant to modern healthcare and maintaining an up-to-date database of these high-quality systematic reviews. The focus of the EPOC review group is on "reviews of interventions designed to improve professional practice and the delivery of effective health services" (www.epoc.uottawa.ca/scope.htm).

The research relevant to this particular review was obtained by searching the specialized register maintained by the EPOC review group. This register was created to identify studies relevant to the EPOC scope. A search strategy was designed and compared to a gold standard of studies obtained by hand searching the *BMJ* and *Medical Care* journals as well as electronic searching of OVID for all relevant articles from the *Annals of Internal Medicine, BMJ, JAMA*, and *Lancet*. Compared to this gold standard, the EPOC search strategy was found to have a sensitivity of 92.4% and a precision of 18.5%. The register includes studies obtained by searching MEDLINE back to 1966, HealthSTAR back to 1975, EMBASE back to 1980, and CINAHL back to 1982. Monthly updates are obtained automatically from OVID. The Cochrane Controlled Clinical Trials register is searched every 3 months and the contents of several key journals are scanned regularly. The EPOC search strategies are available on the EPOC website (www.epoc.uottawa.ca/register.htm) and the register can be accessed via the trials search coordinator.

For the purpose of this study, additional articles were obtained by searching the bibliographies of retrieved articles and personal files. From all of these sources a total of 148 titles and abstracts were obtained searching through the end of 2002. On the first pass, many studies were excluded based on review of the title and abstract. These early exclusions occurred for the following reasons: systematic review of literature (used references as a source of studies), hospital-based study, duplicate publication in another language, not an antibiotic prescribing intervention, or we were unable to obtain the full study after contacting the authors or publishers (predominantly studies produced and published by INRUD—International Network for the Rational Use of Drugs).

Several types of studies were considered for inclusion in this review. Any randomized controlled trials (RCT) as well as quasi-randomized controlled trials were reviewed. A quasi-randomized controlled trial is one in which participants (or groups of participants) are prospectively assigned to one or more intervention groups using a quasi-random allocation method, for example, alternation, date of birth, patient or physician identifier number. In addition, controlled before and after (CBA) studies, where subjects or groups of subject were assigned to study group by some nonrandom method, were included. Finally, studies designed to detect a trend in behavior of a single group without controls were included. These are called interrupted time series (ITS) studies.

Studies of qualified physicians or other trained medical workers (including nurse practitioners) of all ages and level of experience, who prescribe antibiotics and provide primary care in community or academic ambulatory settings, as well as healthcare consumers were included. Studies including medical trainees only were excluded. Studies examining the prescribing of multiple drug classes were included provided that specific data on antibiotic prescribing could be extracted.

Interventions addressing professional practice or education, or financial or structural changes to the method of delivery of care, alone or in combination, were included.

1. Distribution of educational materials: distribution of published or printed recommendations for clinical care, including clinical practice guidelines, audio–visual materials, and electronic publications. The materials may have been delivered personally or through mass mailings.
2. Educational meetings: healthcare providers who have participated in conferences, lectures, workshops, or traineeships.
3. Local consensus processes: inclusion of participating providers in discussion to ensure that they agreed that the chosen clinical problem was important and the approach to managing the problem was appropriate.
4. Educational outreach visits: use of a trained person who met with providers in their practice settings to give information with the intent of changing the provider's practice. The information given may have included feedback on the performance of the provider(s).
5. Local opinion leaders: use of providers nominated by their colleagues as "educationally influential." The investigators must have explicitly stated that their colleagues identified the opinion leaders.
6. Patient mediated interventions: new clinical information (not previously available) collected directly from patients and given to the provider, for example, depression scores from an instrument.
7. Audit and feedback: any summary of clinical performance of healthcare over a specified period of time. The summary may include recommendations for clinical action. The information may have been obtained from medical records, computerized databases, or observations from patients.
8. Reminders: patient or encounter-specific information, provided verbally, on paper or on a computer screen, which is designed or intended to prompt a health professional to recall information. This would usually be encountered through their general education; in the medical records or through interactions with peers and so remind them to perform or avoid some action to aid individual patient care. Computer aided decision support and drugs dosage are included.
9. Marketing: use of personal interviewing, group discussion ("focus groups"), or a survey of targeted providers to identify barriers to change and subsequent design of an intervention that addresses identified barriers.
10. Mass media: (1) varied use of communication that reached great numbers of people including television, radio, newspapers, posters, leaflets, and booklets, alone or in conjunction with other interventions; (2) targeted at the population level.
11. Financial interventions: method of physician remuneration, patient-oriented approaches such as user fees, formularies.

Outcomes that were considered relevant for improving the manner in which ambulatory care physicians prescribe antibiotics were any one or more of: the decision to prescribe an antibiotic or the prescribing of a recommended choice, dose or duration of antibiotic. In addition, secondary outcomes of interest were considered. These included the incidence of colonization with or infection due to antibiotic-resistant organisms prior to and following changes in prescribing behavior, other adverse drug-related events (rash, gastrointestinal disturbances, allergic reactions) and the incidence of adverse events associated with reduced use or duration of antibiotics or use of narrow-spectrum antibiotics.

A total of 116 articles were retrieved from the EPOC specialized register. In addition, 32 articles were retrieved from personal files and review of bibliographies of retrieved articles and review articles for a total of 148 articles to be reviewed. Three studies were excluded as they were duplicates of studies published in another language (two studies) or as an abstract (one study). Two additional papers were long-term follow-up publications from earlier studies. These were considered as single studies along with the original publication. Thirty-six studies retrieved from the EPOC search were excluded prior to full review based on the title and abstract as they were neither intervention studies nor studies of medical healthcare workers, or were hospital-based studies or systematic reviews.

Sixty-eight further studies that were excluded after more detailed review were excluded based on failure to meet minimum methodological criteria, the most common reasons for which were: an ITS study without enough data points, a nonrandomized controlled study without baseline prescribing data, and unavailability of specific data for prescribing antibiotics (or data incompletely reported or mixed with other prescribing data). If the published report did not report appropriate baseline data or contained inextricable antibiotic prescribing data, the authors of the study were contacted in an attempt to obtain the data prior to excluding the study from review.

2.   PROBLEMS WITH INTERPRETATION OF PUBLISHED STUDIES

Among the 39 studies selected for review, there were 24 randomized (RCT) or quasi-randomized clinical trials (QRCT), 13 controlled before and after studies (CBA) and two interrupted time-series studies (ITS). The methodological quality of the included studies was highly variable. The most significant problem with the RCTs and the CBAs was the failure to account for clustering of patients within physicians or practices when the unit of randomization or assignment was physician, practice, or region. In many studies, groups of physicians (often in pre-existing practice groups or by geographic region) were assigned (randomly or not) to intervention and control groups to prevent contamination of subjects. Using such a cluster design requires larger sample sizes since the physicians within these groups cannot be considered completely independent as there will be similarities in the way that these physicians practice medicine and in the characteristics of the patients attending their practices (Diwan *et al.*, 1992). Most of these cluster randomized studies did not report sample size calculations. This is of concern as it is possible that many of these studies are underpowered to detect significant differences in prescribing between study groups.

In addition to issues regarding sample size, assignment to study group in clusters requires that the data be analyzed in a manner that adjusts for these clusters. Unit of analysis errors are considered to have occurred if the provider or groups of providers have been assigned to intervention groups but the outcomes are measured at the level of the individual patient or prescription. This is a significant problem as it ignores the correlation in the physician's prescribing behavior among all the patients. That is, a physician's decision to prescribe an antibiotic each time he or she sees a patient is not a completely independent event as the physician will tend to behave in a similar fashion from one patient to the next. In addition, physicians not only tend to manage patients with similar problems in similar ways but also tend to attract patients who have similar characteristics and treatment expectations. For this reason, data from physician intervention studies should be analyzed with methods adjusting for the clustering of patients under physicians and if necessary, the clustering of physicians in practices or other groups. Using simple statistical techniques such as paired *t*-tests, chi-square tests, or even simple least squares regression analysis ignores the effects of these correlations. This leads to inflated results, where the point estimates for the effect size are accurate but the standard errors are too small leading to erroneously small confidence intervals and *p*-values. This has the effect of making the intervention appear more effective than it is.

The choice of appropriate analysis depends upon the research design (Austin *et al.*, 2001; Moerbeek *et al.*, 2003). One may use regression with fixed effects which includes terms for the cluster (physician or physician group) and an interaction term between the cluster and the intervention. One may also use a summary measures method where the cluster is the unit of analysis and the outcome at the patient level is considered a repeated measure. The generalized estimating equation (GEE) has also been used which takes into account the correlation of the repeated measures within subjects (physicians' repeated decisions to prescribe or not to prescribe). Finally, multilevel or hierarchical regression is considered by many methodologists to be the most appropriate analytic method as it accounts for the multiple levels of data and leads to results that do not appear to overestimate the intervention effect. Multilevel regression is, however, complicated to perform and understand and thus the simpler methods described previously may be acceptable as long as the limitations are recognized. The studies for which appropriate statistical considerations of this problem were undertaken, generally utilized the GEE or multilevel regression in their cluster analyses, both in sample size calculations and in statistical analysis of the results.

In addition to problems with analysis there were other methodological issues with many of the studies to bear in mind while reviewing the summary of the results. In many of the randomized clinical trials the method of randomization was not reported and thus it is difficult to judge if the study is truly randomized. In addition, in the control groups in some of these studies, physicians, knew what the intervention was and might have altered their behavior during the period of the study blunting the apparent effect of the intervention. In order to prevent such contamination, many researchers utilized a cluster design so that physicians practicing in the same group or geographic region were all assigned to the same group. Another method used in a few studies was the application of the intervention to change performance in two different areas, with each group acting as a control for the other, to control for the Hawthorne effect (positive effect of being in a study).

In most of the studies, the methods reported for measuring the prescribing rates was objective, using pharmacy or insurance databases. However, there were a few studies where less objective measures were used. In these, the prescribing rates were frequently determined using chart review or interview of the patient. The reliability of measures was frequently not reported making it difficult to interpret the results of the studies. Studies using chart review or interview methods of data collection were performed primarily in developing countries.

Interrupted time-series studies were included if they reported enough data points for appropriate statistical inference. Only two ITS studies were included as the remainder did not report enough data points or could not define a point in time when the intervention occurred. There were significant problems with the statistical analysis of the two included ITS studies. Neither reported all available data points and neither used appropriate statistical methods to appropriately test for trends in the data over time. The data from these studies was, thus, reanalyzed using appropriate time-series methods.

A summary of the results from the reviewed studies follows. Studies are grouped according to intervention type. There are no studies from all the categories of potential interventions listed previously. In addition, results of some studies are reported more than once if the study compared two or more interventions that fall into different categories.

3.   DISCUSSION OF INCLUDED STUDIES

3.1.   Distribution of educational materials

There were four studies (Angunawela *et al.*, 1991; Avorn and Soumerai, 1983; Ray *et al.*, 1985; Schaffner *et al.*, 1983; Seppälä *et al.*, 1997) in which physicians received printed educational materials regarding appropriate antibiotic prescribing (see Table 1). One study (Ray *et al.*, 1985) reports long

Table 1. Distribution of educational materials

Study citation	Design	Analysis	Outcome measure	Results as reported in the study
Seppälä et al. (1997)	ITS	No time series analysis reported	Reduction in use of macrolide antibiotics	2.40 DDD per 1,000 popln in 1991 to 1.38 in 1992 ($p = 0.007$)
Avorn et al. (1983)	Cluster RCT	No cluster analysis	Reduction in use of cephalexin	Change in mean number of units prescribed -100 ($p = $ NS)
Angunawela et al. (1991)	Cluster RCT	No cluster analysis	Reduction in use of antibiotics for viral infections	Mean difference in proportion of patients receiving prescription of -7.4%
Schaffner et al. (1983)	CBA	No cluster analysis	Reduction in use of contraindicated antibiotics and cephalosporins	No difference (numbers not given)

term follow up data from a preceding study (Schaffner *et al.*, 1983). These materials consisted of information regarding treatment of specific infectious conditions and/or recommendations regarding the use of certain medications (including at least one antibiotic). No materials referring to the specific physician's practice (i.e., audit of practice pattern with feedback) were sent. These educational materials were unsolicited and sent by mail (Avorn *et al.*, 1983; Schaffner *et al.*, 1983; Anunawela *et al.*, 1991) or published in a national medical journal (Seppala *et al.*, 1997).

There were two RCTs (one cluster RCT in which groups of healthcare workers rather than individuals are randomized), one CBA, and one ITS involving distribution of educational materials. One study aimed to reduce use of the oral cephalosporin, cephalexin (among other non-antibiotic medications) in general practice and compared printed educational materials alone or combined with audit and feedback (see Section 3.2) to a no intervention control group (Avorn and Soumerai, 1983). The second study compared the effect of printed educational materials with or without group educational seminars to controls (three study groups) on reducing inappropriate use of antibiotics for acute viral respiratory tract infections (Angunawela *et al.*, 1991). In a CBA study from Tennessee, researchers compared a mailed educational brochure to an outreach visit by a pharmacist or physician (2 groups) and to controls to reduce the use of certain antibiotics (oral cephalosporins, clindamycin, tetracycline, and chloramphenicol) (Ray *et al.*, 1985; Schaffner *et al.*, 1983). Finally, Finnish medical authorities sought to reduce the use of erythromycin for Group A streptococcal infections with a published recommendation against their routine use (Seppälä *et al.*, 1997).

None of the first three studies showed any effect of printed educational materials alone, in changing prescribing behavior. This was despite the fact that the lack of adjustment for clustering likely resulted in an overestimation of the intervention effect in these studies. The Finnish study demonstrated a significant reduction in macrolide antibiotic use following publication of the recommendation which seemed to decay (not statistically significant) somewhat over time (by time-series analysis not presented in the published paper). Thus, written educational materials alone appear to have a limited impact overall with three of four studies demonstrating no change in prescribing behavior. The specific circumstances of the Finnish study, that is, there being an increase in resistance to macrolides in Group A streptococci may have substantially contributed to the effectiveness of the published recommendations in that particular study. Interventions tailored to circumstances may, therefore, be more likely to be effective.

3.2.   Audit and feedback with or without other educational materials

The process of audit and feedback refers to the collection and analysis of individual or group prescribing data followed by distribution of these materials back to prescribing physicians. Additional materials distributed accompanying the feedback material may include any one or more of the following: an explanation of the prescribing data, comparative data for other physicians or groups of physicians, and educational materials promoting more optimal prescribing practices. Audit and provision of feedback data may also be a part of the information provided at individual or group educational meetings or seminars. In this section, only those studies involving distribution of written audit material with or without accompanying comparative data and/or printed educational materials are considered. This type of intervention has been popular as it is relatively simple and inexpensive to perform. This means that even modest reductions in inappropriate prescribing may result in substantial cost savings for insurers although the impact on antibiotic resistance may be less impressive.

There were four studies, three RCTs (Hux *et al.*, 1999; Mainous *et al.*, 2000) (one cluster RCT [O'Connell *et al.*, 1999]) and one CBA trial (Rokstad *et al.*, 1995), in this category (Table 2). In the first RCT (Mainous *et al.*, 2000), authors examined the effect of audit and feedback alone and audit and feedback with patient educational materials compared with a no intervention control group in an intervention designed to reduce antibiotic prescribing for viral pediatric respiratory tract infection. The second study (Hux *et al.*, 1999), compared the effect of audit and feedback to controls on the prescribing of recommended first-line antibiotics for community-acquired infections to senior

Table 2. Audit and feedback

Study citation	Design	Analysis	Outcome measure	Results as reported in the study
Mainous *et al.* (2000)	Cluster RCT	No cluster analysis	Reduction in antibiotic use for viral URI	Overall increase in proportion of patients with colds having prescriptions, gain scores compared with controls significantly lower by Dunnett's T ($p < 0.05$)
Hux *et al.* (1999)	RCT	No cluster analysis	Increase in prescribing of first-line agents	Change of 2.5% in intervention compared with −1.7% in controls ($p < 0.01$)
O'Connell *et al.* (1999)	Cluster RCT	Acknowledged intracluster correlations for sample size calculations but did not use in analysis	Reduction in overall antibiotic use	No change in median prescribing rates per 100 medicare services
Rokstad *et al.* (1995)	CBA	No cluster analysis	Increase prescribing of first-line agents for UTI	Increase in mean number of prescriptions for trimethoprim of 32.5 (3.5–61.7) in intervention group and reduction in control group −31.8 (−54.7 to −3.5)

citizens. In the cluster RCT (O'Connell *et al.*, 1999), physicians were randomized by postal code regions to receive audit and feedback on the prescribing of oral antibiotics (and four other classes of medication) or no intervention. Finally, in the CBA study (Rokstad *et al.*, 1995), Norwegian general practitioners received audit and feedback material along with clinical practice guidelines for the management of urinary tract infection (UTI) in women (diagnostic, first-line antibiotics, duration of therapy), with a control group receiving similar information on the management of insomnia.

In these studies, audit and feedback with or without written educational materials or prescribing guidelines had little or no impact on antibiotic prescribing in general terms. Two studies demonstrated no significant change in prescribing behavior in response to the intervention (Mainous *et al.*, 2000; O'Connell *et al.*, 1999). The remaining two studies (Hux *et al.*, 1999; Rokstad *et al.*, 1995) showed a small but statistically significant increase in the prescribing of recommended first-line antibiotics. In none of these studies was appropriate statistical analysis to account for clustering of patients within physicians undertaken and thus, the results of the latter two studies might be rendered nonsignificant on reanalysis. The results from these studies indicate that there may be certain prescribing situations that are more amenable to interventions such as those studies seeking to increase prescribing of certain selected "first-line" agents. In contrast, interventions aimed at reducing antibiotic overuse (for viral infections, for example) may require more complex interventions.

3.3.   Educational group meetings or seminars

This category of study includes those in which the intervention consists of one or more group educational sessions. It excludes educational outreach visits which are generally one-on-one meetings between the prescribing physician and another individual (physician, pharmacist, drug detailer). Information reviewed at the meetings generally includes an oral and audiovisual review of recommendations for appropriate prescribing practices for particular conditions or particular antibiotics and/or review of a clinical practice guideline or other published recommendations. Some educational meetings include a review of practice audit material for the group.

There were seven cluster RCTs (Augunawela *et al.*, 1991; Bexell *et al.*, 1996; Lagerlov *et al.*, 2000; Lundborg *et al.*, 1999; Meyer *et al.*, 2001; Santoso, 1996; Veninga *et al.*, 2000), one cluster QRCT (Harris *et al.*, 1984), and two CBA studies (McNulty *et al.*, 2000; Perez-Cuevas *et al.*, 1996) reviewed in this section, Table 3. Three cluster RCTs, from Sweden (Lundborg *et al.*, 1999), the Netherlands (Veninga *et al.*, 2000), and Norway (Meyer *et al.*, 2001), used group educational meetings to promote prescribing of first-line agents for UTI. Controls in these studies received a similar intervention targeting asthma management. Three cluster RCTs, two from Africa (Bexell *et al.*, 1996; Meyer *et al.*, 2001) and one from Sri Lanka (Angunawela *et al.*, 1991) examined the effect of group educational sessions on reducing overall antibiotic use in community health clinics compared with controls receiving no education. The content of these interventions was based on rational prescribing guidelines promoted by the World Health Organization focusing on reducing antibiotic use for viral respiratory tract infection and acute diarrhea (De Vries *et al.*, 1994). Similarly, a study from Indonesia (Santoso, 1996), in which health centers were randomized to small or large group educational meetings or to no intervention, aimed to reduce inappropriate antibiotic use for acute diarrhea.

In the European UTI prescribing studies, there was a moderate increase in the use of recommended first-line agents in one study (Lundborg *et al.*, 1999). The other two studies showed modest reductions in the use of long courses of therapy but showed no significant change in the prescribing of first-line agents (Lagerlov *et al.*, 2000; Veninga *et al.*, 2000). The two African

Table 3. Educational meetings

Study citation	Design	Analysis	Outcome measure	Results as reported in the study
Angunawela *et al.* (1991)	Cluster RCT	No cluster analysis	Reduction in use of antibiotics for viral infections	Mean difference in proportion of patients receiving prescription of − 7.3%.
Lundborg *et al.* (1999)	Cluster RCT	Multilevel regression used for cluster analysis	Increase use of first-line agents and reduce duration of treatment for UTI, reduce use of antibiotics for asthma	Increase in proportion of UTI episodes with a first-line agent 18% in intervention, 1% in control ($p < 0.001$). Reduction in duration of antibiotics in both groups (7.51 to 7.41 days and 7.60 to 7.44 days). Increase in number of courses of antibiotics per asthma patient (0.26 to 0.32 and 0.27 to 0.26).
Veninga *et al.* (2000)	Cluster RCT	Use multilevel regression for cluster analysis	Increase use of first-line agents and reduce duration of treatment for UTI	No significant change in the proportion of patients receiving first-line agents (0.12, SE 0.12; relative effect size, 1%). Reduction in the number of DDD of antibiotics (-0.37 DDD, SE 0.02; relative effect size, 31%).
Harris *et al.* (1984)	QRCT	No cluster analysis	Reduction in the use of penicillins (among many other drugs)	No change in intervention group (16.5 prescriptions per 1,000 to 16.7/1,000) vs increase in control group (16.0/1,000 to 18.9/1,000).
Meyer *et al.* (2001)	Cluster RCT	No cluster analysis	Reduction in prescribing of antibiotics for URI and diarrhea	Reduction in proportion of patients with URI with antibiotic prescription 66.3 to 29.41 ($p < 0.05$) compared with 53.99 to 45.64 ($p =$ NS) in controls. For diarrhea change from 66.41 to 44.03 ($p < 0.05$) compared with 11.15 to 5.86 ($p =$ NS) in controls.
Santoso (1996)	Cluster RCT	No cluster analysis	Reduction in prescribing antibiotics for diarrhea	Reduction in proportion of patients receiving antibiotics for diarrhea in small groups (77.4 to 60.4, $p < 0.001$) and large groups (82.3 to 72.3, $p < 0.001$) but not for controls (82.6 to 79.3, $p =$ NS).
Lagerlov *et al.* (2000)	Cluster RCT	Use multilevel regression for cluster analysis	Increase in the number of UTIs treated with short course antibiotics (4 days)	Relative increase in proportion of UTIs treated with short course therapy, 13.1% ($p < 0.0001$) and relative reduction in long course therapy, 9.6% ($p = 0.0004$)
Bexell *et al.* (1996)	Cluster RCT	No cluster analysis	Reduction in overall antibiotic use	Proportion of patients prescribed antibiotics changed from 41.2% to 34.2% in intervention compared with 41.0% to 42.1% in controls.

Table 3. *Continued*

Study citation	Design	Analysis	Outcome measure	Results as reported in the study
McNulty *et al.* (2000)	CBA	No cluster analysis	Reduction in overall antibiotic use; increase in use of narrow-spectrum and reduction in use of broad-spectrum antibiotics	Reduction of 2,458 antibiotic prescriptions for intervention compared with reduction of 1,209 for control ($p = 0.09$); change in narrow-spectrum agent use of -139 for intervention compared with $-1,248$ in controls ($p = 0.003$); change in broad-spectrum use -1612 for intervention compared with $+561$ for controls ($p = 0.002$).
Perez-Cuevas *et al.* (1996)	CBA	No cluster analysis	Reduction in antibiotic use for rhinopharyngitis	Proportion of patients prescribed antibiotics reduced by 17.7% in intervention compared with 10.6% in controls ($p < 0.05$).

studies (Bexell *et al.*, 1996; Meyer *et al.*, 2001) produced a moderate reduction in overall antibiotic use at the intervention clinics while the study from Sri Lanka did not (Angunawela *et al.*, 1991). In the Indonesian study (Santoso, 1996), small group meetings resulted in a larger reduction in antibiotic prescribing for diarrhea than large group meetings; however, both type of meetings were only modestly effective.

The quasi-randomized cluster trial from Britain showed a minimal reduction in the use of penicillin with repeated small group meetings that included the use of audit material (Harris *et al.*, 1984). Antibiotic prescribing workshops in Northern England (McNulty *et al.*, 2000) were more successful in reducing the use of broad-spectrum antibiotics than in reducing overall antibiotic use for viral respiratory tract infections. Modest reductions in antibiotic use for viral URI were seen following group meetings targeting this behavior in Mexico (Perez-Cuevas *et al.*, 1996).

Overall, the results from studies of group educational meetings are modest. Greater effect sizes are seen when the recommendations of the study are for the use of certain proscribed "first-line" antibiotics and are much less impressive for those studies whose goal is to reduce overall inappropriate antibiotic use for viral respiratory tract infection or diarrhea. Only two studies appropriately adjusted for the clustering of patients within physicians and physicians within larger groups (Lagerlov *et al.*, 2000; Veninga *et al.*, 2000). Both of these studies demonstrated a significant reduction in the duration of antibiotic therapy for uncomplicated UTI.

3.4.   Educational outreach/academic detailing

Face-to-face detailing has been used by the pharmaceutical industry for many years to market its products to physicians. The studies in this section use similar methods of detailing physicians regarding appropriate prescribing. The detailers in these studies were either clinical or research pharmacists or medical doctors. In some studies the effectiveness of different types of detailers are compared. This method is attractive for appropriate prescribing education as it has been used very successfully by the pharmaceutical industry, possibly because of the one-on-one attention paid to the prescriber. Materials discussed in the sessions include the review of audit material from the physician's own practice.

In this section, there were eight studies (Avorn and Soumerai, 1983; De Santis *et al.*, 1994; Dolovich *et al.*, 1999; Font *et al.*, 1991; Ilett *et al.*, 2000; McConnell *et al.*, 1982; Peterson *et al.*, 1997; Ray *et al.*, 1985; long term follow-up data for Schaffner *et al.*, 1983) (see Table 4). The two cluster RCTs,

Table 4. Educational outreach visits

Study citation	Design	Analysis	Outcome measure	Results as reported in the study
Dolovich *et al.* (1999)	Cluster RCT	Controlled for location of pre-scriber in multivariable analysis	Increased prescribing of first-line agents (amoxicillin) for acute otitis media	Mean percent change in market share for amoxicillin in intervention region 0.63 compared with -0.72 in controls ($p = 0.15$).
De Santis *et al.* (1994)	Cluster RCT	No cluster analysis	Increase prescribing of penicillin and erythromycin for tonsillitis	Increase in proportion of prescriptions that are penicillin or erythromycin from pre to post intervention in both groups— 60.5% to 8 7.7% in intervention, 52.9% to 71.7% in controls.
Avorn and Soumerai (1983)	Cluster RCT	No cluster analysis	Reduction in use of cephalexin	Mean difference from controls between pre and post intervention periods -382 (mean number of units prescribed) ($p = 0.0006$).
Ilett *et al.* (2000)	RCT	No cluster analysis	Increase prescribing of first-line agents for UTI, bacterial tonsillitis, otitis media, bacterial bronchitis, mild pneumonia	Significant increase in prescribing of amoxicillin from median 293 prescriptions per physician to 594 ($p < 0.05$) and doxycycline from median 235 prescriptions per physician to 865 ($p < 0.05$) (within group comparison only). Both recommended first-line agents for otitis media and pneumonia, respectively.

Table 4. *Continued*

Study citation	Design	Analysis	Outcome measure	Results as reported in the study
McConnell *et al.* (1982)	RCT	No cluster analysis	Reduction in prescribing of tetracycline	Average number of tetracycline prescriptions per provider dropped from 12.6 to 1.8 ($p < 0.001$) in intervention group and 7.6 to 3.2 ($p < 0.001$) in controls.
Font *et al.* (1991)	RCT	No cluster analysis	Reduction in prescribing of antibiotics combined with symptomatic drugs, injectable cephalosporins and injectable penicillin/ streptomycin combination, increase in oral cephalosporins	Median number of prescriptions per month for combination agents fell from 86.53 to 75.89 ($p < 0.001$) for intervention group, 98.55 to 91.96 for controls ($p < 0.001$) between group comparison significantly different ($p < 0.01$); for oral cephalosporins 4.97 to 5.36 ($p < 0.05$), 6.43 to 5.98 (NS), between group comparison ($p < 0.01$); injectable cephalosporins no reduction; penicillin/streptomycin between group comparison NS.
Peterson *et al.* (1997)	CBA	No cluster analysis	Increase prescribing of first-line agents for UTI (amoxicillin clavulanic acid, cephalexin, trimethoprim)	Relative prescribing of first-line agents for UTI higher in intervention than control region compared with baseline ($p < 0.0001$)—ratio of recommended to non recommended antibiotics increased from 2.77 to 5.43 in intervention region and 4.29 to 4.92 in control region.
Schaffner *et al.* (1983)	CBA	No cluster analysis	Reduction in use of contraindicated antibiotics and cephalosporins	Relative reduction in number of prescriptions for contraindicated antibiotics of 67%, 85%, and 41% in pharmacist detailer, physician detailer, and control groups, respectively; for oral cephalosporins, corresponding changes were 35%, 50%, and 33% reduction.

randomized by geographic region, examined changes in the prescribing of first-line agents for acute otitis media (Dolovich *et al.*, 1999) and tonsillitis (De Santis *et al.*, 1994) in response to academic detailing by a pharmacist. Three of the physician randomized RCTs examined the effectiveness of academic detailing on reducing the use of particular antibiotics, oral cephalosporins (Avorn and Soumerai, 1983), tetracycline (McConnell *et al.*,

1982), and antibiotics combined with symptomatic medications, oral and injectable cephalosporins, and an injectable combination of penicillin and streptomycin (Font *et al.*, 1991). Another RCT used academic detailing to promote the use of certain first-line agents for a variety of community-acquired bacterial infections (UTI, bacterial tonsillitis, otitis media, bacterial bronchitis, and mild pneumonia) (Ilett *et al.*, 2000).

The two cluster RCTs combined mailed educational materials with academic detailing by a pharmacist (De Santis *et al.*, 1994) or a traditional pharmaceutical industry detailer (Dolovich *et al.*, 1999) which focused on the content of the previously mailed materials. One of these studies demonstrated a small but significant benefit from academic detailing (De Santis *et al.*, 1994) while the other study showed not benefit (Dolovich *et al.*, 1999). Among the physician randomized studies, two demonstrated a reduction in the use of specific classes of antibiotics (cephalexin [Avorn and Soumerai, 1983] and tetracyclines [McConnell *et al.*, 1982]) compared with controls and printed educational materials alone (Avorn and Soumerai, 1983). In the third study (Ilett *et al.*, 2000), very small changes in the prescribing of recommended first-line agents were achieved; however, the large number of outcomes measured (many antibiotics examined) raises concerns regarding multiple comparisons and the possibility that small, significant results may be due to chance alone.

In the three CBA studies, academic detailing was successful in reducing inappropriate antibiotic use as defined by the individual studies. One group successfully increased the number of prescriptions for first-line agents in UTIs (Peterson *et al.*, 1997). The remaining two studies aimed to reduce the use of certain antimicrobials considered to be contraindicated for use in general practice. Physicians prescribed fewer antibiotics in combination with symptomatic medication and fewer courses of injectable cephalosporins in Barcelona following academic detailing by a pharmacist (Font *et al.*, 1991). The Tennessee statewide intervention mentioned previously used pharmacist or physician detailers and reduced the use of certain agents considered to be contraindicated with a greater reduction seen with physician detailers than pharmacists (Schaffner *et al.*, 1983). This reduction was sustained after two years (Ray *et al.*, 1985).

The effect of academic detailing on improving antibiotic prescribing appears to be moderate in the best cases. Most of the studies utilized pharmacist detailers and thus, it is difficult to generalize about the differences between pharmacist and physician detailers. In the one study comparing the two types of detailing, physician detailing clearly had a more substantial effect on changing prescribing behavior (Ray *et al.*, 1985; Schaffner *et al.*, 1983). The one study utilizing traditional pharmaceutical representatives (Dolovich *et al.*, 1999) to provide education did not demonstrate a significant impact. The overall success of these studies may, in part, be due to the fact that the primary goal in most of these studies was to reduce prescribing of certain agents and/or promote prescribing of recommended first-line agents. None of these studies sought to reduce overall use of antibiotics for inappropriate indications which appears to be a much more difficult task.

3.5.   Financial/healthcare system changes

There are a variety of interventions which fall under the umbrella of financial interventions and/or changes in the healthcare system. These range from changes in formularies or benefits to the institution of user fees, or coinsurance for physician visits or medications. These are aimed at reducing utilization of specific drugs or physician services. Reforms to the local healthcare system usually involve changes in the manner in which healthcare delivery occurs such as the use of non-physician providers or changes in the organizational structure of health clinics or regions. It is inappropriate to compare interventions in this category head-to-head as they may involve disparate methods that defy comparison.

There are only two studies that fall into this category (MacCara *et al.*, 2001; Juncosa and Porta, 1997) (Table 5). One examines the effect of a formulary change (MacCara *et al.*, 2001) and the other the effect of primary care reform in a region of Spain (Junosa and Porta, 1997) Researchers in Nova Scotia, Canada, studied the effect of a change in the provincial drug formulary limiting the use of fluoroquinolones in the elderly to certain specific conditions (MacCara *et al.*, 2001). Using ITS data they reported a significant drop in the mean number of fluoroquinolone prescriptions between the two time periods; however, they did not use appropriate time-series analysis which compares rates of prescribing over time before and after the intervention. Time-series analysis detected the same significant drop in the level of prescribing between pre- and post-formulary changes; the slope for fluoroquinolone prescribing, however, increased between pre- and post-formulary change periods. Thus while there was an immediate and dramatic impact of the formulary change, time trends suggest that the level of prescribing over time might be expected to return to pre-formulary change levels.

In a CBA study from Spain, researchers examined the effect of nationwide primary care reform on drug prescribing in one county, using areas that had not yet undergone reform as controls (Juncosa and Porta, 1997). The primary care reform consisted, mainly, of changes in staffing (all working full time as opposed to part time) and reimbursement (per capita payment changed to fixed salary) as well as changes in the organization of services and integration of preventive medicine services into the traditional curative model. Overall, there was a greater reduction in all antibiotic prescriptions in the reform primary care network compared with the non-reformed network.

The study from Nova Scotia demonstrates the substantial initial impact that formulary changes may have on prescribing of particular medications

Table 5. Financial/healthcare system changes

Study citation	Design	Analysis	Outcome measure	Results as reported in the study
MacCara *et al.* (2001)	ITS (formulary change)	No time-series analysis reported	Number of prescriptions for fluoroquinolones	Fluoroquinolone use dropped by 80.2% following intervention
Juncosa and Porta (1997)	CBA (primary care reform)	No cluster analysis	Overall number of antibiotic prescriptions	56.6% relative reduction in prescribing in reform network compared with 32.5% relative reduction in non-reform network

(MacCara *et al.*, 2001). However, overall antibiotic use did not change significantly with substitution of macrolide antibiotics for fluoroquinolones. This likely represents a substantial cost saving to the drug plan but does little to control overall antibiotic use. This study also demonstrates the pitfalls in not using appropriate analytic techniques for time-series data which, when performed, indicated that levels of fluoroquinolone use were increasing after the change and could eventually reach pre-intervention levels. The Spanish study demonstrates that structural changes in the way in which healthcare is delivered can have an impact on physicians' practice without applying specific educational interventions (Juncosa and Porta, 1997). It would be interesting to see if the reduction in antibiotic use were maintained over time or simply represents the early enthusiasm for the restructured system.

3.6.   Reminders

Physician reminders, at the point of care, have been assessed by three physician randomized RCTs attempting to reduce antibiotic prescribing for two clinical syndromes, acute otitis media (Christakis *et al.*, 2001) and sore throat (two studies of a similar intervention) (McIsaac and Goel, 1998; McIsaac *et al.*, 2002) (Table 6). The principle behind this type of intervention is that if physicians are provided with information about specific treatments, at the time they are making prescribing decisions, inappropriate antibiotic use may be reduced.

In one study, an online prescription writer was used to present computer-based point-of-care evidence on the optimal duration of antibiotics for acute otitis media in children in an effort to reduce the duration of prescribing for this condition (Christakis *et al.*, 2001). The results indicated a significant increase in the proportion of prescriptions for otitis media that were of less than 10 days duration for intervention physicians compared with controls.

Table 6. Reminders

Study citation	Design	Analysis	Outcome measure	Results as reported in the study
Christakis et al. (2001)	RCT	No cluster analysis	Reduction in duration of therapy for acute otitis media to <10 days	Change in mean number of patients receiving <10 days of antibiotics 44.43% vs 10.48 % in controls ($p = 0.000$)
McIsaac et al. (1998)	RCT	No cluster analysis needed (only one patient per provider)	Reduction in antibiotic prescriptions for sore throat	Proportion of patients receiving prescription for sore throat 27.8% in intervention group, 35.7% in control group; odds ratio for an antibiotic prescription associated with intervention 0.44 (0.21–0.92)
McIsaac et al. (2002)	RCT	Cluster analysis performed	Reduction in antibiotic prescriptions for sore throat	Proportion of patients receiving antibiotics for sore throat 28.1% in intervention group, 27.9% for controls; odds ratio for antibiotic prescribing associated with intervention 0.57 (0.27–1.17)

In the two studies of reminders for sore throats (McIsaac and Goel, 1998; McIsaac et al., 2002) (both by the same authors), a paper-based decision support tool for diagnosis and management of sore throat was developed in an effort to reduce inappropriate antibiotic use for this condition. This decision support tool consisted of a scoring system for signs and symptoms of sore throat to help determine the likelihood that a particular patient suffers from streptococcal pharyngitis. Management recommendations were then based upon scores. In the first study, family physicians received a package containing information on the study, a score card for diagnosing streptococcal sore throat and a form to complete for one sore throat patient encounter. Physicians in the study group received a form that required them to calculate a score based on four items. The control form was identical except that the score items were listed and physicians were not asked to calculate a score. Since each physician contributed only one sore throat encounter, there was no need to adjust the analysis for clustering. The authors were able to demonstrate a reduction in the odds of prescribing an antibiotic for sore throat in the intervention group compared with controls.

In the second study (McIsaac et al., 2002), by the same authors, the same checklist and score were used. All physicians received a pocket card summarizing the score, eight patient encounter and consent forms and a one page survey of practice characteristics. Physicians in the intervention group received stickers to place on the patient encounter forms with the score items and a place to calculate the score. This resulted in the intervention physicians' receiving repeated prompts regarding the score and management recommendations. The control physicians completed similar encounter forms without stickers. In this study there was no reduction in the odds for prescribing an antibiotic associated with the repeated prompts compared with controls.

From this limited dataset, physician reminders appear to have some effect on altering prescribing behavior. In the otitis media study, intervention physicians prescribed fewer long courses of antibiotics. In addition, however, physicians had an opportunity to view information on the overall need for antibiotics in otitis media. There was no reduction in overall prescribing for otitis associated with this part of the intervention. While the sore throat score seemed to modestly reduce antibiotic use for this condition when used for a single patient, repeated prompts in the clinical setting did not have a beneficial effect, limiting the usefulness of this intervention. This may have been due to the effect of the control physicians' receiving the score (without using the repeated prompts) in the control group, leading to reduced prescribing overall. It may also be that physicians stopped interacting with the score after using it several times either out of fatigue or the sense that they could judge the probability of streptococcal infection well enough after using the score a few times.

3.7. Patient-based interventions

In several studies, patient-based interventions were evaluated in conjunction with other interventions. There were, however, five studies that examined the effect of a variety of patient-based interventions alone (Table 7). These studies evaluated the effect of patient educational materials (Mainous *et al.*, 2000), a patient information leaflet regarding antibiotics for acute bronchitis (Macfarlane *et al.*, 2002) and the use of delayed prescriptions for infections where patients desired antibiotics but physicians did not feel antibiotics were necessary (Arroll *et al.*, 2002; Dowell *et al.*, 2001; Little *et al.*, 2001).

In the one study examining the effect of patient educational materials alone (Mainous *et al.*, 2000), there was a modest effect seen in the group with this intervention compared with the control group. This effect was similar to that seen in the other intervention groups (audit and feedback alone, or combined with patient educational materials). In a study of prescribing for acute bronchitis in adults (Macfarlane *et al.*, 2002), patients who were not felt to need antibiotics by the physician but appeared to desire a prescription received a prescription along with either a flyer explaining why antibiotics were unnecessary for this condition or a blank leaflet. Patients in the intervention group were less likely to fill their prescriptions than those in the control group.

Three additional studies, utilizing the same concept, randomized patients to receive and fill a prescription immediately or to fill the prescription later

Table 7. Patient-mediated interventions

Study citation	Design	Analysis measure	Outcome	Results as reported in the study
MacFarlane *et al.* (2002)	RCT— patient randomized	No cluster analysis required	Reduction in number of patients filling prescription for antibiotic for acute bronchitis	Proportion of patients filling prescription 47% in intervention vs 62% in control; hazard ratio for taking antibiotic for intervention compared with control 0.66 (0.46–0.96)
Mainous *et al.* (2000)	RCT	No cluster analysis	Reduction in number of prescriptions for URI	Overall increase in proportion of patients with colds with prescriptions—gain score compared with controls significantly lower by Dunnett's T ($p < 0.05$)
Arroll *et al.* (2002)	RCT— patient randomized	No cluster analysis required	Reduction in the number of patients filling prescriptions for URI	48% of patients in delayed prescription group filled prescription vs 89% in controls; odds ratio for filling prescription for delayed prescription 0.12 (0.05–0.29)
Dowell *et al.* (2001)	RCT— patient randomized	No cluster analysis required	Reduction in the number of patients filling prescriptions for URI	45% of patients in delayed prescription group received a prescription vs 82% in the control group (information obtained from authors)
Little *et al.* (2001)	RCT— patient randomized	No cluster analysis required	Reduction in the number of patients filling prescriptions for otitis media	24% of patients in delayed prescription group filled a prescription vs 98% in the control group

(3–7 days depending on the study) if symptoms did not improve. In the two studies of adult patients with the common cold, patients were much less likely to fill the prescription in the delayed group (Arroll *et al.*, 2002; Dowell *et al.*, 2001). A similar effect of delayed prescriptions was seen in the study of prescribing for otitis media in children (Little *et al.*, 2001).

The attractiveness of this type of intervention is that physicians advise against the need for an antibiotic for viral infections while still feeling that they have satisfied perceived patient demands. Indeed, some of the physicians in these studies felt that they would continue to use this intervention with demanding patients. Ultimately, the goal is that the patient will not fill the prescription and recover from the illness with the knowledge that an antibiotic was not necessary. He or she may then be less inclined to request a prescription during the next illness or possibly even forgo a visit to the physician for this problem as he or she now knows that it is benign and self-limited. One could argue, however, that giving a patient a prescription when it is not necessary sends a mixed message and places the onus of the decision on the patient. Simply taking the time to explain why an antibiotic is not necessary may accomplish the same goal.

3.8. Multifaceted interventions

Many of the older studies presented in the preceding sections have indicated that it is very difficult to produce anything but modest changes in physicians' antibiotic prescribing practices especially if one is attempting to reduce overall use for viral infections as opposed to promoting use of first-line agents. As a result, seven of the most recent studies examined the effect of complex, multifaceted interventions on reducing antibiotic misuse (see Table 8). (Belongia *et al.*, 2001; Finkelstein *et al.*, 2001; Flottorp *et al.*, 2002; Gonzales *et al.*, 1999; Hennessy *et al.*, 2002; Perz *et al.*, 2002; Stewart *et al.*, 2000). These interventions include physician education in a variety of forums as well as education of the patient/parent and the general public as to the appropriate use of antibiotics. The public education message in these studies has focused on the individual and public health hazards of antibiotic overuse. The US Centers for Disease Control (CDC) was involved in the design and implementation of four (Belongia *et al.*, 2001; Finkelstein *et al.*, 2001; Hennessy *et al.*, 2002; Perz *et al.*, 2002) out of seven studies in this category.

In the four CDC sponsored studies, the interventions involved a combination of healthcare provider and consumer education aimed at reducing inappropriate antibiotic use for viral respiratory tract infections. Healthcare provider education was undertaken in the form of small group sessions, traditional CME lectures, hospital and clinic staff meetings and grand rounds, the exact combination depending on the particular study. The content of the educational message was based upon the Principles of Judicious Antimicrobial use for pediatric upper respiratory infections (Dowell *et al.*, 1998) drafted by the CDC, the American Academy of Pediatrics (AAP), and the American Academy of Family Physicians (AAFP). Patient and community education was primarily undertaken using printed educational materials produced by the CDC which were distributed at hospitals, during doctor visits, and at community centers and schools. Samples of this material can be found at http://www.cdc.gov/drugresistance/community/tools.htm. In addition, some studies used community meetings and local media to promote the judicious use of antibiotics.

Three of the CDC-sponsored studies in this category examined the effect of the intervention on rates of penicillin-resistant *S. pneumoniae* in the communities

Table 8. Multifaceted interventions

Study citation	Design	Analysis	Outcome measure	Results as reported in the study
Belongia et al. (2001)	CBA	No cluster analysis	Reduction in antibiotic use in children	11% reduction in intervention region vs 12% increase in control region ($p = 0.019$) for liquid antibiotics; 19% reduction in intervention region vs 8% reduction in control ($p = 0.042$) for solid antibiotics.
Finkelstein et al. (2001)	Cluster RCT	Used generalized estimating equation for cluster analysis	Reduction in antibiotic use in children	0.23 (0.08–0.39, $p < 0.01$) fewer antibiotic courses per person 3 to < 36 mos per year; 0.13 (0–0.27, $p = 0.06$) fewer antibiotic courses per person 36 to <72 mos per year
Hennessey et al. (2002)	CBA	No cluster analysis	Reduction in overall antibiotic use	First year: reduction from 1.24 to 0.81 antibiotic courses per person ($p < 0.01$) in intervention region vs 0.63 to 0.57 ($p = NS$) Second year: initial intervention region constant 0.85 to 0.81 antibiotic courses per person, 0.55 to 0.4, 0.55 to 0.41 ($p < 0.01$) in expanded intervention region (control from first year)
Perz et al. (2002)	CBA	Used binomial regression controlling for region	Reduction in antibiotic use in children	19% reduction in antibiotic prescriptions per 100 person years in intervention vs 8% reduction in controls (intervention attributable effect 11% (8–14%) reduction ($p < 0.001$)
Gonzales et al. (1999)	CBA	Cluster analysis performed using study site as a fixed effect in regression analysis	Reduction in antibiotic prescribing for acute bronchitis in adults	Proportion of cases with antibiotics prescribed reduced from 74% to 48% in full intervention site vs 82% to 77% in limited intervention site ($p = 0.02$ between site)
Stewart et al. (2000)	CBC	No cluster analysis	Increase prescribing of first-line agents	No change in prescribing first-line agents compared with control (adjusted odds

Table 8. *Continued*

Study citation	Design	Analysis	Outcome measure	Results as reported in the study
				ratio 1.02 [0.99–1.06]); increase in prescribing second-line agents in control region vs intervention region (odds ratio 1.40 [1.32–1.49])
Flottorp *et al.* (2002)	RCT	Multilevel regression use for cluster analysis	Reduction in prescribing antibiotics for sore throat	Proportion of cases of sore throat with antibiotics reduced from 48.1% to 43.8% in intervention vs 50.8% to 49.5% in control region ($p = 0.032$)

being studied (Belongia *et al.*, 2001; Hennessy *et al.*, 2002; Perz *et al.*, 2002). This very important secondary endpoint will be discussed in detail in the section entitled "Effect of interventions on antibiotic resistance."

Among the CDC-sponsored studies, there was one cluster randomized trial (Finkelstein *et al.*, 2001) and three CBA studies (Belongia *et al.*, 2001; Hennessy *et al.*, 2002; Perz *et al.*, 2002). All of these studies were successful in significantly reducing the inappropriate use of antibiotics for viral respiratory tract infections. Effect sizes ranged from moderate (12–16% relative reduction in antibiotic use) (Finkelstein *et al.*, 2001) to a more substantial relative reduction in antibiotic use of 31% (Hennessy *et al.*, 2002).

In a CBA study, Gonzales and colleagues (Gonzales *et al.*, 1999) applied a full intervention (consisting of physician education and patient materials in the office and sent to homes) to one site and compared the effect to an intervention limited to patient education materials at another site and two (no intervention) control sites. This study demonstrated a substantial absolute reduction in prescribing from baseline for the full intervention site compared with controls (24%) while the patient intervention alone had no significant effect. The remaining two studies demonstrated little (Stewart *et al.*, 2000) or no change (Flottorp *et al.*, 2002) in prescribing despite extensive interventions. The Norwegian study (Flottorp *et al.*, 2002) purported to use interventions that were tailored to locally identified barriers to change and included changes to the fee schedule for phone calls with patients in order to reduce the number of visits to physicians for sore throat and UTI. Despite this study's excellent design and execution, the authors could not demonstrate a reduction in the use of antibiotics for sore throat.

Overall the methodology of these studies was better than for studies in other categories. In addition, these combined interventions appeared to be more effective in reducing inappropriate antibiotic use than single interventions.

4.   EFFECT OF INTERVENTIONS ON
        ANTIBIOTIC RESISTANCE

Among the 39 studies reviewed, only 4 studies simultaneously assessed the effect of the interventions on bacterial antimicrobial resistance in the study communities (Belongia *et al.*, 2001; Hennessy *et al.*, 2002; Perz *et al.*, 2002; Seppälä *et al.*, 1997). Seppälä *et al.* (1997) examined the effect of a reduction in consumption of macrolides on the rate of isolation of macrolide-resistant Group A streptococci from throat swabs and pus specimens. The reduction in macrolide use in Finland was observed following a published recommendation against use of macrolides as first-line agents for Group A streptococcal infections. For the analysis of the trend in macrolide resistance, they used data from 1992 as the recommendations were issued in late 1991 and early 1992. This data could not be subjected to time-series analysis as there were an insufficient number of pre-intervention data points with which to perform such an analysis. Using logistic regression, it was determined that the odds of an isolate being erythromycin resistant in 1996 were half the odds of it being resistant in 1992 indicating a significant reduction in the rate of resistance (odds ratio 0.5 [95% CI 0.4–0.5]). This held true when the data were analyzed by region. Thus, these researchers were able to demonstrate that a reduction in consumption of one class of antibiotics could result in a reduction in resistance to that antibiotic class in the population. Of course, the reduction in macrolide consumption was accompanied by an increase in use of other antibiotics for Group A streptococcal infections; however, Group A streptococcal resistance to penicillin has never been documented in Finland and thus penicillin represents an effective alternative that, at least currently, is safe from the problems of antibiotic resistance.

The other three studies addressing the effect of interventions on antimicrobial resistance are three of the four studies sponsored by the CDC. In all three of these studies, the researchers examined the effect of multifaceted interventions on reducing overall antibiotic use in the study communities and on the rates of isolation of penicillin-resistant *S. pneumoniae* from individuals in the community. In the CBA community study from Wisconsin (Belongia *et al.*, 2001), the effect of the intervention on the rate of penicillin resistance (either intermediate or full resistance as defined by NCCLS criteria) in pneumococcal isolates was determined by performing nasopharyngeal swabs and cultures for pneumococcal carriage on children attending child care facilities in the study communities. The intervention resulted in a moderate reduction in antibiotic consumption following the intervention compared with the control regions. In a multivariate logistic regression model controlling for clustering of children within child care centers, nasopharyngeal carriage of a resistant pneumococcus was not associated with living in a particular community (odds ratio 0.46 [0.18–1.18]).

This indicates that over the period of the study (2 years) there was no significant reduction in the odds of harboring a penicillin-resistant pneumococcus due to the intervention.

In the multifaceted, community intervention trial in Alaska (Hennessy *et al.*, 2002), nasopharyngeal swabbing was performed on any community member (adult or child) who consented to the procedure. After the initial education intervention in region A, the proportion of pneumococcal isolates that were penicillin non-susceptible (intermediate or full resistance) decreased from 41% to 29% in region A but did not significantly change in the control regions B and C (24% to 22%). The intervention continued in region A and was expanded to regions B and C in the second year of the study. In the second year, the reduction in the rate of resistance observed in region A was not sustained and the proportion of pneumococci that were penicillin non-susceptible increased from 29% to 43%. In the expanded intervention regions, the proportion of pneumococci that were penicillin non-susceptible did not change following the intervention (22% vs 26%). Thus, the initial reduction in region A cannot be attributed to the reduction in antibiotic consumption observed following the intervention.

In the community intervention study in Tennessee (Perz *et al.*, 2002), the effect of the intervention on the rate of pneumococcal resistance was assessed by examining the rate of isolation of resistant (intermediate or full resistance) pneumococci from sterile site isolates (blood, cerebrospinal fluid, other usually sterile body fluids) from children under 15 years of age. While the intervention resulted in an overall reduction in antibiotic use, the proportion of invasive, sterile site isolates resistant to penicillin did not change over the 3 years of the study (60% year 1, 74% year 2, and 71% year 3).

The results of these three multifaceted community intervention studies indicate that despite the moderate success of the interventions in reducing overall antibiotic used in the relative short term (1–2 years), a similar effect on antibiotic resistance rates has not been demonstrated in the same time period.

5. EFFECT OF INTERVENTIONS ON PATIENT OUTCOMES

Two studies examining the effect of patient-based interventions on antibiotic use also examined the effect of withholding antibiotics on the clinical outcomes of enrolled patients. Both of these studies examined the effect of delaying antibiotic prescriptions, one for acute otitis media in children and one for the common cold in adults. Both studies documented reduced antibiotic use by those patients randomized to the delayed antibiotics groups of the respective studies.

In the study of delayed antibiotics in the common cold by Arroll and colleagues (Arroll *et al.*, 2002), patients completed daily symptom checklists until the tenth day after the initial medical visit. These checklists were collected and symptom scores were tabulated for the two groups (immediate and delayed prescription groups). Using the general linear model for repeated measures, it was found that patient temperatures did not differ between the two groups over the course of the study. In addition, the symptom scores (based on cough, nasal discharge, throat pain, headache, etc.—maximum score 15) for the two groups were essentially the same at time points throughout the 10 day follow-up.

In the otitis media study (Little *et al.*, 2001), parents were requested to complete a daily diary regarding presence of symptoms (earache, unwellness, sleep disturbance), perceived severity of pain, number of episodes of distress, use of paracetamol, and temperature measurements. Overall, parents of patients in the immediate antibiotic group reported fewer days of crying and sleep disturbance as well as less paracetamol use; however, there was no difference in mean pain scores, episodes of distress or absence from school.

It is not surprising that delaying (and ultimately preventing) antibiotic use in the common cold had no effect on resolution of symptoms given the viral etiology. In addition, the recruiting physicians could choose not to enroll anyone they felt required an antibiotic (presumably for bacterial rhinosinusitis, ear infection, or pneumonia detected clinically). The otitis study confirms that, in many cases, patients diagnosed with otitis media will improve spontaneously, despite the bacterial origin of their disease. There has been a great deal of controversy about this as it may be that acute otitis media is so loosely defined in many studies, including this one, that many of the children may not truly have bacterial infection but red painful ears due to viral infection. In that case, spontaneous resolution, would be the rule. The authors of this study describe it as pragmatic, however, pointing out that in the real world, many of these children are diagnosed with acute otitis media and treated with antibiotics. Delaying the collection of a prescription in this setting reduces antibiotic use and will not result in excess morbidity. One might, however, also argue that teaching appropriate diagnostic technique for acute otitis media, with antibiotics prescribed only to those children who meet strict diagnostic criteria, would have a similar effect.

6.   DISCUSSION OF RESULTS AND IMPLICATIONS FOR PRACTICE

Given the broad array of targeted behaviors, the variation in interventions (even within categories) and the differences in the clinical settings, it is difficult to generalize the results from these individual studies and arrive at broadly applicable recommendations for improving antibiotic prescribing in any community. However, several general observations may be cautiously made.

The simple, single intervention studies (mailed or published educational materials, audit and feedback) generally resulted in no or little change in prescribing behavior. Previous systematic reviews have concluded that passive methods of physician education such as traditional conferences and lectures as well as publication of guidelines have very limited impact (O'Brein *et al.*, 2003a–d; Oxman *et al.*, 1995). The most plausible explanation for this is that these interventions often fail to address the root causes of inappropriate prescribing. Simply drawing the physician's attention to the behavior (audit and feedback) or recommending an alternate behavior (educational materials) may not provide the physician with the tools to change a behavior that likely is quite ingrained and multifactorial in its origins. It bears mentioning that these low cost interventions may result in cost savings to governments and other insurers even if the results are marginal given the high cost of prescription medications. Small changes in prescribing are unlikely, however, to reduce the incidence of antibiotic resistant bacteria in a community.

The one exception to this lack of effect of simple interventions was the impressive change in macrolide use in Finland following publication of a guideline recommending against the use of this class of antibiotics for Group A streptococcal infection. This result is unexpected given the limited effect of published guidelines on physician behavior. The basis of the recommendation was patient safety as there were concerns regarding treatment failures due to the increasing rate of macrolide resistance. This emphasis on patient safety may account for the impressive impact of the written recommendations compared with other reports of this intervention. However, as time-series analysis demonstrated, the effect of this single intervention (occurring at one point in time, i.e., single publication) appeared to wane somewhat over time as indicated by the positive slope of the post-intervention prescribing rates. Thus, over the long term, the effect of this recommendation may wane as memory of the publication fades.

The more complex the intervention, the more likely it was to produce important changes in antibiotic prescribing behavior. Educational meetings produced modest improvements in prescribing. The study examining patient education materials alone or in combination with another intervention demonstrated the benefit of including patient-based education in the intervention (Mainous *et al.*, 2000). This observation provided the rationale and impetus for the community-based, multifaceted interventions undertaken by the CDC and other groups of researchers. Combinations of interventions in most instances, produced moderately large reductions in antibiotic use which was sustained in those studies with follow-up data, depending on the specific intervention and targeted behavior (Belongia *et al.*, 2001; Finkelstein *et al.*, 2001; Flottorp et al., 2002; Gonzales et al., 1999; Hennessy et al., 2002; Perz et al., 2002; Stewart et al., 2000). One notable exception was the Norwegian study (Flottorp et al., 2002) where researchers designed the intervention to specifically address previously identified barriers to change. There was no change in antibiotic use for sore throat in this study, despite the tailored interventions possibly due to the passive nature of the interventions or an inadequate duration of follow-up. This strengthens the impression that one cannot derive broad-based recommendations from these studies to apply to any clinical situation in any community.

It appears that interventions aimed at increasing the prescribing of certain recommended first-line antibiotics for specific infections are more likely to produce substantial changes in prescribing than those interventions targeting overall inappropriate antibiotic use. As discussed in the introduction, the root causes of antibiotic misuse in the community outpatient setting are manifold and may include physicians' succumbing to pressure from patients, lack of understanding by the physician as to the necessity for antibiotics in certain clinical conditions, diagnostic uncertainty as to the true nature of the patient's illness, and constraints on the physician's time to explain the nature of the illness and the reasons an antibiotic is not indicated. Convincing a physician or patient that a particular antibiotic (usually the most narrow spectrum agent for the condition) should be his or her first choice should be relatively simple as long as appropriate justification for the recommendation is made. It stands to reason, however, that completely eliminating prescribing for a particular indication, such as a viral URI, in a clinical situation in which the physician would usually prescribe an unnecessary antibiotic would be a more difficult behavioral change. While this generally holds true for most of the reviewed studies, promoting the prescribing of first-line agents was not as straightforward a task as might be predicted. One potential explanation for this is that physicians may consider these prescribing recommendations a limitation to their clinical freedom. In addition, physicians want to prescribe what they think are the best medications for the individual patient which often means a broad-spectrum agent to protect against potentially resistant organisms regardless of the ecological consequences.

Several of the trials addressed patient-based outcomes such as changes in antibiotic resistance patterns as a result of altered antibiotic use (Belongia et al., 2001; Hennessy et al., 2002; Perz et al., 2002; Seppälä et al., 1997) and illness outcomes following the withholding of antibiotics for certain conditions (Arroll et al., 2002; Little et al., 2001). Over the intervention periods (usually between 1 and 3 years) no substantial or persistent reductions in incidence of isolating resistant bacteria were observed in any of the studies except for the Finnish macrolide study (Seppälä et al., 1997), where changes were observed in macrolide resistance rates after approximately 2 years. In contrast, no sustained reduction in penicillin resistance was observed with overall reductions in antibiotic use in several communities (Belongia *et al.*, 2001; Hennessy *et al.*, 2002; Perz *et al.*, 2002). The reason for this has been suggested by a mathematical model of rates of change of antibiotic resistance among bacteria (Stewart *et al.*, 1998). The conclusions from the model suggest that the period time to observe reductions in the incidence of antibiotic-resistant organisms will be longer than the preceding increases. Thus, it may be many years before sustained reductions in antibiotic use produce reductions in penicillin-resistant pneumococci. In addition, larger reductions in antibiotic use may be necessary to produce more rapid changes in resistance patterns. Assessing the full effect of reductions in community-wide antibiotic use may be made complicated by the already observed reductions in invasive pneumococcal infections in immunized children and their contacts due to conjugate pneumococcal vaccines.

In the studies of the use of delayed antibiotic prescriptions for URIs and otitis media, significant patient morbidity was not observed (Arroll *et al.*, 2002; Dowell *et al.*, 2001; Little *et al.*, 2001). As the outcome of viral respiratory tract infections is not altered by antibiotics, these are not unexpected results for the studies of URIs. It is important, however, to have data that demonstrates this lack of morbidity for illnesses such as acute bronchitis and purulent rhinitis where the etiologic agent, while usually viral, is often thought to be bacterial by many practitioners. The demonstration that there is no benefit to immediate use of antibiotics may serve to convince many physicians and patients that antibiotics are not needed for these conditions. Delayed prescriptions for acute otitis media in children are frequently used in many European countries but have not gained popularity in North America. This pragmatic study (Little *et al.*, 2001) demonstrates that waiting a few days to use an antibiotic among children diagnosed with otitis media does not increase morbidity from this disease. It has been argued that studies like this do not validly assess the effect of antibiotics on acute otitis media as the diagnostic criteria are not strict enough, leading to the inclusion of many patients who did not truly have bacterial otitis media; however, this argument only serves to strengthen the conclusions from this study that antibiotics are not required for most cases of acute otitis media diagnosed in the primary care setting most likely because this condition is overdiagnosed. Withholding immediate antibiotic helps to weed out those children with URIs and red tympanic membranes from those with true bacterial middle ear disease, leading to more appropriate antibiotic use.

6.1.   Implications for practice

The selection of the most effective intervention to improve the prescribing of antibiotics appears to be condition and situation specific. In designing an intervention to change a particular prescribing behavior, the ultimate goal of the intervention must be defined and barriers to change identified and addressed by the intervention.

Small changes in the prescribing of narrow-spectrum, first-line agents may be achieved to varying degrees with the use of simple, single interventions such as guideline publication and distribution, educational meetings, and audit and feedback. These incremental changes may result in substantial cost savings as these interventions are relatively inexpensive. These interventions are unlikely to lead to a reduction in the incidence of antibiotic-resistant bacteria causing community-acquired infection.

It appears that more complex interventions produce more substantial reductions in antibiotic prescribing. These interventions are also very expensive and may not be cost-effective. Multifaceted interventions, however, are more likely to result in the sort of changes in prescribing that may eventually lead to reductions in the incidence of antibiotic-resistant bacteria. That this was not observed in the studies measuring this outcome is most likely a function of time. Long-term follow-up of the intervention communities may reveal changes in the antibiotic susceptibilities of community-acquired bacterial infections.

REFERENCES

Allen, U. D., MacDonald, N., Fuite, L., Chan, F., and Stephens, D., 1999, Risk factors for resistance to "first-line" antimicrobials among urinary tract isolates of *Escherichia Coli* in children. *CMAJ*, **160**, 1436–1440.

Angunawela, I., Diwan, V., and Tomson, G., 1991, Experimental evaluation of the effects of drug information on antibiotic prescribing: A study in outpatient care in an area of Sri lanka. *Int. J. Epidemiol.*, **20**, 558–564.

Arason, V. A., Kristinsson, K. G., Sigurdsson, J. A., Stefansdottir, G., Molstad, S., and Gudmundsson, S., 1996, Do antimicrobials increase the carriage rate of penicillin resistant pneumococci in children? *BMJ*, **313**, 387–391.

Arnold, S. R., Allen, U. D., Al-Zahrani, M., Tan, D. H., and Wang, E. E., 1999, Antibiotic prescribing by pediatricians for respiratory tract infection in children. *Clin. Infect. Dis.*, **29**, 312–317.

Arnold, K., Leggiadro, R., Breiman, R. *et al.*, 1996, Risk factors for carriage of drug-resistant *Streptococcus pneumoniae* among children in Memphis, Tennessee. *J. Pediatr.*, **128**, 757–764.

Arroll, B., Kenealy, T., and Kerse, N., 2002, Do delayed prescriptions reduce the use of antibiotics for the common cold? *J. Fam. Pract.*, **51**, 324–328.

Austin, P. C., Goel, V., and van Walraven, C., 2001, An introduction to multilevel regression. *Can. J. Pub. Health*, **92**, 150–154.

Avorn, J. and Soumerai, S. B., 1983, Improving drug-therapy decisions through educational outreach. A randomized controlled trial of academically based "detailing." *N. Engl. J. Med.*, **308**, 1457–1463.

Bass, J. W., Steele, R. W., Wittler, R. R. *et al.*, 1993, Antimicrobial treatment of occult bactermia: A multicenter cooperative study. *Pediatr. Infect. Dis. J.*, **12**, 466–473.

Bauchner, H., Pelton, S. I., and Klein, J. O., 1999, Parents, physicians and antibiotic use. *Pediatr.*, **103**, 395–401.

Belongia, E. A., Sullivan, B. J., Chyou, P-H., Madagame, E., Reed, K. D., and Schwartz, B., 2001, A community intervention trial to promote judicious antibiotic use and reduce penicillin-resistant *Streptococcus pneumoniae* carriage in children. *Pediatrics*, **108**, 575–583.

Bexell, A., Lwando, E., von Hofsten, B., Tembo, S., Eriksson, B., and Diwan, V. K., 1996, Improving drug use through continuing education: A randomized controlled trial in Zambia. *J. Clin. Epidemiol.*, **49**, 355–357.

Block, S. L., Harrison, C. J., Hedrick, J. A. *et al.*, 1995, Penicillin-resistant *Streptococcus pneumoniae* in acute otitis media: Risk factors, susceptibility patterns and antimicrobial management. *Pediatr. Infect. Dis. J.*, **14**, 751–759.

Boken, D. J., Chartrand, S. A., Goering, R. V., Kruger, R., and Harrison, C. J., 1995, Colonization with penicillin-resistant *Streptococcus pneumoniae* in child-care center. *Pediatr. Infect. Dis. J.*, **14**, 879–884.

Breiman, R. F., Butler, J. C., Tenover, F. C., Elliott, J. A., and Facklam, R. R., 1994, Emergence of drug-resistant pneumococcal infections in the United States. *JAMA*, **271**, 1831–1835.

Brook, I. and Gober, A. E., 1996, Prophylaxis with amoxicillin or sulfisoxazole for otitis media: Effect on the recovery of penicillin-resistant bacteria from children. *Clin. Infect. Dis.*, **22**, 143–145.

Brown, P. D., Freeman, A., and Foxman, B., 2002, Prevalence and predictors of trimethoprim-sulfamethoxazole resistance among uropathogenic *Escherichia coli* isolates in Michigan. *Clin. Infect. Dis.*, **34**, 1061–1066.

Carroll, W. L., Farrell, M. K., Singer, J. I., Jackson, M. A., Lobel, J. S., and Lewis, E. D., 1983, Treatment of occult bacteremia: A prospective randomized clinical trial. *Pediatrics*, **72**, 608–612.

Christakis, D. A., Zimmerman, F. J., Wright, J. A., Garrison, M. M., Rivara, F. P., Davis, R. L., 2001, A randomized controlled trial of point-of-care evidence to improve the antibiotic prescribing practices for otitis media in children. *Pediatrics*, **107**, e15.

Cornaglia, G., Ligozzi, M., Mazzariol, A., Valentini, M., Orefici, G., and Fontana, R., 1996, Rapid increase of resistance to erythromycin and clindamycin in *Streptococcus pyogenes* in Italy, 1993–1995. The Italian Surveillance Group for Antimicrobial Resistance. *Emerg. Infec. Dis.*, **2**, 339–342.

Cronk, G. A., Naumann, D. E., McDermott, K., Menter, P., and Swift, M. B., 1954, A controlled study of the effect of oral penicillin G in the treatment of non-specific upper respiratory infections. *Am. J. Med.*, **16**, 804–809.

Dagan, R., Leibovitz, E., Greenberg, D., Yagupsky, P., Fliss, D. M., and Leiberman, A., 1988, Dynamics of pneumococcal nasopharyngeal colonization during the first days of antibiotic treatment in pediatric patients. *Pediatr. Infect. Dis. J.*, **17**, 880–885.

Davey, P. G., Bax, R. P., Newey, J. *et al.*, 1996, Growth in the use of antibiotics in the community in England and Scotland in 1980–93. *BMJ*, **312**, 613.

De Santis, G., Harvey, K. J., Howard, D., Mashford, M. L., and Moulds, R. F., 1994, Improving the quality of antibiotic prescription patterns in general practice. The role of educational intervention. *Med. J. Aust.*, **160**, 502–505.

De Vries, T. P., Henning, R. H., Hogerzeil, H. V., and Fresle, D. A., 1994, *Guide to Good Prescribing*. WHO/DAP/95.1. World Health Organisation, Geneva.

Diwan, V. K., Eriksson, B., Sterky, G., and Tomson, G., 1992, Randomization by group in studying the effect of drug information in primary care. *Int. J. Epidemiol.*, **21**, 124–130.

Dolovich, L., Levine, M., Tarajos, R., and Duku, E., 1999, Promoting optimal antibiotic therapy for otitis media using commercially sponsored evidence-based detailing: A prospective controlled trial. *Drug Inform. J.*, **33**, 1067–1077.

Dowell, J., Pitkethly, M., Bain, J., and Martin, S., 2001, A randomised controlled trial of delayed antibiotic prescribing as a strategy for managing uncomplicated respiratory tract infection in primary care. *Br. J. Gen. Pract.*, **51**, 200–205.

Dowell, S. F., Marcy, S. M., Phillips, W. R., Gerber, M. A., and Schwartz, B., 1998, Principles of judicious use of antimicrobial agents for pediatric upper respiratory tract infections. *Pediatrics*, **101**, 163–184.

Duchin, J. S., Breiman, R. F., Diamond, A. *et al.*, 1995, High prevalence of multidrug-resistant *Streptococcus pneumoniae* among chidlren in a rural Kentucky community. *Pediatr. Infect. Dis. J.*, **14**, 745–750.

EARSS, 2001, System EARS. EARSS Annual Report 2001.

Finkelstein, J. A., Davis, R. L., Dowell, S. F. *et al.*, 2001, Reducing antibiotic use in children: A randomized trial in 12 practices. *Pediatrics*, **108**, 1–7.

Flottorp, S., Oxman, A. D., Havelsrud, K., Treweek, S., and Herrin, J., 2002, Cluster randomised controlled trial of tailored interventions to improve the management of urinary tract infections in women and sore throat. *BMJ*, **325**, 367.

Font, M., Madridejos, R., Catalan, A., Jimenez, J. J. M. A., and Huguet, M., 1991, Improving drug prescription in primary care: A controlled and randomized study of an educational method. *Med. Clin. (Barc.)*, **96**, 201–205.

Fujita, K., Murono, K., Yoshikawa, M., and Murai, T., 1994, Decline of erythromycin resistance of group A streptococci in Japan. *Pediatr. Infect. Dis. J.*, **13**, 1075–1078.

Gonzales, R., Steiner, J. F., Lum, A., and Barrett, P. H. J., 1999, Decreasing antibiotic use in ambulatory practice: Impact of a multidimensional intervention on the treatment of uncomplicated acute bronchitis in adults. *JAMA*, **281**, 1512–1519.

Gonzales, R., Steiner, J. F., and Sande, M., 1997, Antibiotic prescribing for adults with colds, upper respiratory tract infections, and bronchitis by ambulatory care physicians. *JAMA*, **278**, 901–904.

Gordon, M., Lovell, S., and Dugdale, A. E., 1974, The value of antibiotics in minor respiratory illness in children. A controlled trial. *Med. J. Aust.*, **1**, 304–306.

Hamm, R. M., Hicks, R. J., and Bemben, D. A. 1996, Antibiotics and respiratory infections: Are patients more satisfied when expectations are met? *J. Fam. Pract.*, **43**, 56–62.

Hardy, L. M. and Traisman, H. S., 1956, Antibiotics and chemotherapeutic agents in the treatment of uncomplicated respiratory infections in children. *J. Pediatr.*, **48**, 146–156.

Harris, C. M., Jarman, B., Woodman, E., White, P., and Fry, J. S., 1984, Prescribing—a suitable case for treatment. *J. R. Coll. Gen. Pract.*, **24**, 1–39.

Hennessy, T. W., Petersen, K. M., Bruden, D. *et al.*, 2002, Changes in antibiotic-prescribing practices and carriage of penicillin-resistant *Streptococcus pneumoniae*: A controlled intervention trial in rural Alaska. *Clin. Infect. Dis.*, **34**, 1543–1550.

Hutchinson, J. M. and Foley, R. N., 1999, Method of physician remuneration and rates of antibiotic prescription. *CMAJ*, **160**, 1013–1017.

Hux, J. E., Melady, M. P., and DeBoer, D., 1999, Confidential prescriber feedback and education to improve antibiotic use in primary care: A controlled trial. *CMAJ*, **161**, 388–392.

Ilett, K. F., Johnson, S., Greenhill, G. *et al.*, 2000, Modification of general practitioner prescribing of antibiotics by use of a therapeutics adviser (academic detailer). *Br. J. Clin. Pharmacol.*, **49**, 168–173.

Jackson, M. A., Shelton, S., Nelson, J. D., and McCracken, G. H., 1984, Relatively penicillin-resistant penumococcal infections in pediatric patients. *Pediatr. Infect. Dis. J.*, **3**, 129–132.

Jaffe, D. M., Tanz, R. R., Davis, A. T., Henretig, F., and Fleisher, G., 1987, Antibiotic administration to treat possible occult bacteremia in febrile children. *N. Engl. J. Med.*, **317**, 1175–1180.

Juncosa, S. and Porta M., 1997, Effects of primary health care reform on the prescription of antibiotics: A longitudinal study in a Spanish county. *European Journal of Public Health*, 7(1), 54–60.

Kaiser, L., Lew, D., and Stalder, H., 1996, Effects of antibiotic treatment in the subset of common-cold patients who have bacteria in nasopharyngeal secretions. *Lancet*, **347**, 1507–1510.

Ladhani, S. and Gransden, W., 2003, Increasing antibiotic resistance among urinary isolates. *Arch. Dis. Child*, **88**, 444–445.

Lagerlov, P., Loeb, M., Andrew, M., and Hjortdahl, P., 2000, Improving doctors' prescribing behavior through reflection on guidelines and prescription feedback: A randomised controlled study. *Qual. Health Care*, **9**, 159–165.

Lexomboon, U., Duangmani, C., Kusalasai, V., Sunakorn, P., Olson, L. C., and Noyes, H. E., 1971, Evaluation of orally administered antibiotics for treatment of upper respiratory infections in Thai children. *J. Pediatr.*, **78**, 772–778.

Linder, J. A. and Stafford, R. S., 2001, Antibiotic treatment of adults with sore throat by community primary care physicians: A national survey, 1989–1999. *JAMA*, **286**, 1181–1186.

Little, P., Gould, C., Williamson, I., Moore, M., Warner, G., and Dunleavey, J., 2001, pragmatic randomised controlled trial of two prescribing strategies for childhood acute otitis media. *BMJ*, **322**, 336–342.

Lundborg, C. S., Wahlstrom, R., Oke, T., Tomson, G., and Diwan, V. K., 1999, Influencing Prescribing for urinary tract infection and asthma in primary care in Sweden: A randomized controlled trial of an interactive educational intervention. *J. Clin. Epidemiol.*, **52**, 801–812.

MacCara, M. E., Sketris, I. S., Comeau, D. G., Weerasinghe, S. D., 2001, Impact of a limited fluoroquinolone reimbursement policy on antimicrobial prescription claims. *Annals of Pharmacotherapy*, **35**(7–8), 852–858.

Macfarlane, J., Holmes, W., Gard, P. *et al.*, 2002, Reducing antibiotic use for acute bronchitis in primary care: Blinded, randomised controlled trial of patient information leaflet * Commentary: More self reliance in patients and fewer antibiotics: Still room for improvement. *BMJ*, **324**, 91.

Magee, J. T., Pritchard, E. L., Fitzgerald, K. A., Dunstan, F. D. J., Howard, A. J., and Group WAS, 1999, Antibiotic prescribing and antibiotic resistance in community practice: Retrospective study, 1996–8. *BMJ*, **319**, 1239–1240.

Mainous, A. G. I., Hueston, W. J., and Clark, J. R., 1996, Antibiotics and upper respiratory infection. Do some folks think there is a cure for the common cold? *J. Fam. Pract.*, **42**, 357–361.

Mainous, A. G. I., Hueston, W. J., and Eberlein, C., 1997, Colour of respiratory discharge and antibiotic use. *Lancet*, **350**, 1077.

Mainous, A. G. I., Hueston, W. J., and Love, M. M., 1998, Antibiotics for colds in children; who are the high prescribers? *Arch. Pediatr. Adolesc. Med.*, **152**, 349–352.

Mainous III, A. G., Hueston, W. J., Love, M. M., Evans, M. E., and Finger, R., 2000, An evaluation of statewide strategies to reduce antibiotic overuse. *Fam. Med.*, **32**, 22–29.

Maruyama, S., Yoshioka, H., Fujita, K., Takimoto, M., and Satake, Y., 1979, Sensitivity of group A streptococci to antibiotics. Prevalence of resistance to erythromycin in Japan. *Am. J. Dis. Child*, **133**, 1143–1145.

McCaig, L. F. and Hughes, J. M., 1995, Trends in antimicrobial drug prescribing among office-based physicians in the United States. *JAMA*, **273**, 214–219.

McConnell, T. S., Cushing, A. H., Bankhurst, A. D., Healy, J. L., McIlvenna, P. A., and Skipper, B. J., 1982, Physician behavior modification using claims data: Tetracycline for upper respiratory infection. *West J. Med.*, **137**, 448–450.

McGowan, J. E., Bratton, L., Klein, J. O., and Finland, M., 1973, Bacteremia in febrile children seen in a "walk-in" pediatric clinic. *N. Engl. J. Med.*, **288**, 1309–1312.

McIsaac, W. J. and Butler, C. C., 2000, Does clinical error contribute to unnecessary antibiotic use? *Med. Dec. Making.*, **20**, 33–38.

McIsaac, W. J. and Goel, V., 1998, Effect of an explicit decision-support tool on decisions to prescribe antibiotics for sore throat. *Med. Decis. Making*, **18**, 220–228.

McIsaac, W. J., Goel, V., To, T., Permaul, J. A., and Low, D. E., 2002, Effect on antibiotic prescribing of repeated clinical prompts to use a sore throat score. *J. Fam. Pract.*, **51**, 339–344.

McNulty, C. A. M., Kane, A., Foy, C. J. W., Sykes, J., Saunders, P., and Cartwright, K. A. V., 2000, Primary care workshops can reduce and rationalize antibiotic prescribing. *J. Antimicrob. Chemother.*, **46**, 493–499.

Meyer, J. C., Summers, R. S., and Moller, H., 2001, Randomized, controlled trial of prescribing training in a South African province. *Med. Educ.*, **35**, 833–840.

Moerbeek, M., van Breukelen, G. J. P., and Bergerb, M. P. F., 2003, A comparison between traditional methods and multilevel regression for the analysis of multicenter intervention studies. *J. Clin. Epidemiol.*, **56**, 341–350.

Murray, S., Del Mar, C., and O'Rourke, P., 2000, Predictors of an antibiotic prescription by GPs for respiratory tract infections: A pilot. *Fam. Pract.*, **17**, 386–388.

Nasrin, D., Collignon, P. J., Roberts, L., Wilson, E. J., Pilotto, L. S., and Douglas, R. M., 2002, Effect of β lactam antibiotic use in children on pneumococcal resistance to penicillin: Prospective cohort study. *BMJ*, **324**, 1–4.

Nava, J. M., Bella, F., Garau, J. *et al.*, 1994, Predictive factors for invasive disease due to penicillin-resistant *Streptococcus pneumoniae*: A population-based study. *Clin. Infect. Dis.*, **19**, 884–890.

Nyquist, A-C., Gonzales, R., Steiner, J. F., and Sande, M., 1998, Antibiotic prescribing for children with colds, upper respiratory tract infections, and bronchitis. *JAMA*, **279**, 875–877.

O'Connell, D. L., Henry, D., and Tomlins, R., 1999, Randomised controlled trial of effect of feedback on general practitioners' prescribing in Australia. *BMJ*, **318**, 507–511.

Oxman, A. D., Thomson, M. A., Davis, D. A., and Haynes, B. R., 1995, No magic bullets: A systematic review of 102 trials of interventions to improve professional practice. *CMAJ*, **153**, 1423–1431.

Pallares, R., Gudiol, F., Linares, J. *et al.*, 1987, Risk factors and response to antibiotic therapy in adults with bacteremic pneumonia caused by penicillin-resistant pneumococci. *N. Engl. J. Med.*, **317**, 18–22.

Palmer, D. A. and Bauchner, H., 1997, Parents' and physicians' views on antibiotics. *Pediatrics*, **99**, e6.

Perez-Cuevas, R., Guiscafre, H., Munoz, O. *et al.*, 1996, Improving physician prescribing patterns to treat rhinopharyngitis. Intervention strategies in two health systems of Mexico. *Soc. Sci. Med.*, **42**, 1185–1194.

Perz, J. F., Craig, A. S., Coffey, C. S. *et al.*, 2002, Changes in antibiotic prescribing for children after a community-wide campaign. *JAMA*, **287**, 3103–3109.

Peterson, G. M., Stanton, L. A., Bergin, J. K., and Chapman, G.A., 1997, Improving the prescribing of antibiotics for urinary tract infection. *J. Clin. Pharm. Ther.*, **22**, 147–153.

Prevention CfDCa, 2001, Active Bacterial Core Surveillance: Division of Bacterial and Mycotic Diseases.

Priest, P., Yudkin, P., McNulty, C., Mant, D., and Wise, R., 2001, Antibacterial prescribing and antibacterial resistance in English general practice: Cross sectional study. *BMJ*, **323**, 1037–1041.

Radetsky, M. S., Istre, G. R., Johansen, T. L. *et al.*, 1981, Mulitply resistant pneumococcus causing meningitis: Its epidemiology within a day-care centre. *Lancet*, **2**, 771–773.

Ray, W. A., Schaffner, W., and Federspiel, C. F., 1985, Persistence of improvement in antibiotic prescribing in office practice, *JAMA*, **253**, 1774–1776.

Reichler, M. R., Allphin, A. A., Breiman, R. F. *et al.*, 1992, The spread of multiply resistant *Streptococcus pneumoniae* at a day care center in Ohio. *Clin. Infect. Dis.*, **166**, 1346–1353.

Rokstad, K., Straand, J., and Fugelli, P., 1995, Can drug treatment be improved by feedback on prescribing profiles combined with therapeutic recommendations? A prospective, controlled trial in general practice. *J. Clin. Epidemiol.*, **48**, 1061–1068.

Santoso, B., 1996, Small group intervention vs formal seminar for improving appropriate drug use. *Soc. Sci. Med.*, **42**, 1163–1168.

Schaffner, W., Ray, W. A., Federspiel, C. F., and Miller, W. O., 1983, Improving antibiotic prescribing in office practice. A controlled trial of three educational methods. *JAMA*, **250**, 1728–32.

Scheifele, D., Gold, R., Marchessault, V., and Talbot, J., 1996, Penicillin resistance among invasive pneumococcal isolates at 10 children's hospitals, 1991–1994. The LCDC/CPS Impact Group. The Immunization Monitoring Program, Active. *Can Commun. Dis. Rep.*, **22**, 157–159, 162–163.

Scheifele, D., Halperin, S., Pelletier, L., Talbot, J., IMPACT Mo, 2000, Invasive pneumococcal infection Canadian children, 1991–1998: Implications for new vaccine strategies. *Clin. Infect. Dis.*, **31**, 58–64.

Seppälä, H., Klaukka, T., Lehtonen, R., Nenonen, E., and Huovinen, P., 1995, Outpatient use of erythromycin: Link to increase erythromycin in group A streptococci. *Clin. Infect. Dis.*, **21**, 1378–1385.

Seppälä, H., Klaukka, T., Vuopio-Varkila, J. *et al.*,1997, The effect of changes in the consumption of macrolide antibiotics on erythromycin resistance in Group A streptococci in Finland. *N. Engl. J. Med.*, **337**, 441–446.

Seppälä, H., Klaukka, T., Vuopio-Varkila, J. *et al.*, 1997, The effect of changes in the consumption of macrolide antibiotics on erythromycin resistance in Group A streptococci in Finland. *N. Engl. J. Med.*, **337**, 441–446.

Seppälä, H., Nissinen, A., Jarvinen, H. *et al.*, 1992, Resistance to erythromycin in group A streptococci. *N. Engl. J. Med.*, **326**, 292–297.

Stalsby Lundborg, C., Wahlstrom, R. T. O., Tomson, G., and Diwan, V. K., 1999, Influencing prescribing for urinary tract infection and asthma in primary care in Sweden: A randomized controlled trial of an interactive educational intervention. *J. Clin. Epidemiol.*, **52**, 801–812.

Steinke, D. T., Bain, D. J. G., MacDonald, T. M., Davey, P. G., 2000, Practice factors that influence antibiotic prescribing in general practice in Tayside. *J. Antimicrob. Chemother.*, **46**, 509–512.

Stewart, F. M., Antia, R., Levin, B. R., Lipsitch, M., and Mittler, J. E., 1998, The population genetics of antibiotic resistance II: Analytic theory for sustained populations of bacteria in a community of hosts. *Theor. Popul. Biol.*, **53**, 152–165.

Stewart, J., Pilla, J., and Dunn, L., 2000, Pilot study for appropriate anti-infective community therapy. Effect of a guideline-based strategy to optimize use of antibiotics. *Can. Fam. Phys.*, **46**, 851–859.

Stott, N. C. and West, R. R., 1976, Randomised controlled trial of antibiotics in patients with cough and purulent sputum. *BMJ*, **2**, 556–559.

Taboulet, F., 1990, Presentation d'une methodologie premettant de mesurer en quantite et de comparer les consommations pharmaceutiques. *J. D'econ. Med.*, **8**, 409–437.

Tan, T. Q., Mason, E. O., and Kaplan, S. L., 1993, Penicillin-resistant systemic pneumococcal infections in children: A retrospective case-control study. *Pediatrics*, **92**, 761–767.
Taylor, B., Abbott, G. D., McKerr, M., and Fergusson, D. M., 1977, Amoxycillin and cotri-
moxazole in presumed viral respiratory infections of childhood: Placebo-controlled trial.
BMJ, **2**, 552–554.
Teele, D. W., Pelton, S. I., Grant, M. J. A. *et al.*, 1975, Bacteremia in febrile children under
2 years of age: Results of cultures of blood of 600 consecutive febrile children seen in
a "walk-in" clinic. *J. Pediatrics*, **87**, 227–230.
Thomson O'Brien, M. A., Freemantle, N., Oxman, A. D., Wolf, F., Davis, D. A., and
Herrin, J., 2003a, Continuing education meetings and workshops. Cochrane Database
of Systematic Reviews, **3**.
Thomson O'Brien, M. A., Oxman, A. D., Davis, D. A., Haynes, R. B., Freemantle, N., and
Harvey, E. L., 2003b, Audit and feedback versus alternative strategies. Cochrane
Database of Systematic Reviews, **3**.
Thomson O'Brien, M. A., Oxman, A. D., Davis, D. A., Haynes, R. B., Freemantle, N., and
Harvey, E. L., 2003c, Educational outreach visits. Cochrane Database of Systematic
Reviews, **3**.
Thomson O'Brien, M. A., Oxman, A. D., Haynes, R. B., Freemantle, N., and Harvey, E. L.,
2003d, Local opinion leaders. Cochrane Database of Systematic Reviews, **3**.
Townsend, E. H. J., 1960, Chemoprophylaxis during respiratory infection in private practice.
Am. J. Dis. Child, **99**, 566–573.
Townsend, E. J. J., and Radebaugh, J. F., 1962, Prevention of complications of respiratory ill-
nesses in a pediatric practice. *Sem. Pediatr. Infect. Dis.*, **266**, 683–689.
Veninga, C. C. M., Denig, P., Zwaagstra, R., and Haaijer-Ruskamp, F. M., 2000, Improving
drug treatment in general practice. *J. Clin. Epidemiol.*, **53**, 762–772.
Wang, E. E., Einarson, T. R., Kellner, J. D., and Conly, J. M., 1999, Antibiotic prescribing for
Canadian preschool children: Evidence of overprescribing for viral respiratory infections. *Clin. Infect. Dis.*, **29**, 155–160.
Waskerwitz, S. and Berkelhamer, J. E., 1981, Outpatient bacteremia: Clinical findings in
children under two years with initial temperatures of 39.5°C or higher. *J. Pediatr.*, **99**,
231–233.
Watson, R. L., Dowell, S. F., Jayaraman, M., Keyserling, H., Kolczak, M., and Schwartz, B.,
1999, Antimicrobial use for pediatric upper respiratory infections: Reported practice,
acutal practice and parent beliefs. *Pediatr.* **104**, 1251–1257.
Whitney, C. G., Farley, M. M., Hadler, J. *et al.*, 2000, Increasing prevalence of multidrug-
resistant *Streptococcus pneumoniae* in the United States. *N. Engl. J. Med.*, **343**,
1917–1924.
Zenni, M. K., Cheatham, S. H., Thompson, J. M. *et al.*, 1995, *Streptococcus pneumoniae*
colonization in the young child: Association with otitis media and resistance to penicillin. *J. Pediatr.*, **127**, 533–537.
Zhanel, G., G., Karlowsky, J. A., Harding, G. K. M. *et al.*, 2000, A Canadian national
surveillance study of urinary tract isolates from outpatients: Comparison of the activities
of trimethoprim-sulfamethoxazole, nitrofurantoin and ciptrofloxacin. *Antimicrob. Agents
Chemother.*, **44**, 1089–1092.

Chapter 27

Education of Patients and Professionals

Christine Bond
Professor of Primary Care: Pharmacy
Department of General Practice and Primary Care
University of Aberdeen, UK and
Consultant in Pharmaceutical Public Health, NHSGrampian, UK

1.   BACKGROUND

This chapter is about the rationale for, and mechanisms for, educating both patients and professionals. First therefore, it seems appropriate to justify the need for both of these.

Medical practice is now accepted to have moved on from a paternalistic model of professional practice in which the physician or other healthcare professional (HCP) told the patient what to do and expected their directions to be carried out without any discussion. There is consensus that patients and the public are partners in healthcare and should be part of decision-making both at individual patient and population level. However for this partnership to be a reality, for an informed consultation in which the patient has meaningful involvement, it is necessary for the patient to understand the options available and their wider ramifications, such as side effects and long-term outcomes. While the healthcare professional can be the sole educator of the patient, this has many obvious disadvantages and the NHS has a responsibility to educate the public prior to the consultation.

Having considered the need to educate patients we should do likewise for the professionals. It could be assumed that professionals are educated to the necessary standard to practice, by virtue of their qualification and professional

Antibiotic Policies: Theory and Practice. Edited by Gould and van der Meer
Kluwer Academic / Plenum Publishers, New York, 2005

accreditation. It might therefore be questioned why "Education of Patients and Professionals" should be a topic in its own right. However, the current mantra of "evidence based practice," the importance of Clinical Governance, and the ever increasing body of knowledge, means that there is an ongoing need to continually provide up to date information at postgraduate, as well as under-graduate level. In addition, practice is not just about having knowledge but being skilled in its application. This is another area of practice in relation to antibiotic usage which is particularly relevant and where training professionals is essential.

This chapter therefore addresses the education of both patients and profes-sionals, as two key strands of our strategy to improve the use of antibiotics.

2.   PATIENTS

2.1.   Current knowledge

The public are continually subjected to subliminal education via the media and lay networks. This can result in an increased awareness of health education messages, but does not necessarily mean the message is correctly understood. Disseminated in this way, information is unlikely to be objective or balanced, reviewed critically, or consistent. For example, recent media publicity report-ing an increased rate of hospital admissions for pneumonia since campaigns to reduce antibiotic prescribing (Price, 2003) could be misinterpreted and lead to increased antibiotic use with damaging consequences. In contrast the earlier high profile reporting of the Standing Medical Advisory Committee report (SMAC: Department of Health, 1999) ensured that the key message of the dangers of antibiotic resistance reached a wide audience with obvious benefit.

It is therefore important that the NHS takes the lead on educational cam-paigns. This should have the advantage of giving clear balanced information in a consistent way, which should empower the public and allow a true partnership with the healthcare professional. Antibiotics are no exception to this core prin-ciple. In response to the SMAC report, the UK government published a strategy and action plan (Department of Health, 2000) in which stated objectives included: "to promote optimal antimicrobial prescribing in clinical practice through . . . professional education" and "to encourage realistic public expecta-tions for antimicrobial prescribing . . . through a public information campaign and . . . patient involvement in prescribing decisions."

In theory, development of an educational programme should begin with a statement of learning objectives and an assessment of current knowledge. In practice and for a range of reasons, this is not always carried out, and much "education" about antibiotic use has already been disseminated to the public, in particular since the SMAC report was published. However, a small Scottish Study (Emslie and Bond, 2003) gave some indication of the public's awareness of issues around antibiotic use. Over half agreed that most infections cleared up by themselves, yet two thirds would expect antibiotics for a very sore throat and half for a persistent cough. This contradicted apparent knowledge, as three-quarters agreed that we should be more careful how we used antibiotics. To some extent this apparent contradiction reflects the difference in what people want for themselves and what they believe is right for the wider population.

2.2. Public expectations

There is much anecdotal evidence that patients expect to receive an antibiotic for many viral infections such as coughs, colds, ear infections and other minor ailments. This is particularly the case for conditions, which do not resolve within a few days. Again the Scottish Study carried out recently (Emslie and Bond, 2003) confirmed this and a fifth had consulted a doctor for their most recent respiratory tract infection, with three quarters of these going on to receive an antibiotic. The majority of these would expect them again for similar symptoms.

Such expectations and therefore attitudes translate into a pattern of consulting behaviour which is reinforced if the professional prescribes an antibiotic (Little *et al.*, 1997). Given that minor ailments, almost by definition, improve spontaneously, recovery is linked to the antibiotic as the causative agent rather than natural improvement with time, and there is an understandable belief that this was the reason for recovery. It is no wonder that the next time demand for such an antibiotic is even greater.

2.3. Methods of public education

The methods of disseminating health education messages are many and there is little published information on their relative efficacy. A key point is to ensure that professionals should all give a consistent message as reinforcement is a strong educational tool in its own right. It is also important in this context, as well as encouraging people not to use antibiotics and seek medical advice unnecessarily, to be clear about those circumstances when medical advice should be sought. The involvement of local champions and community initiatives are generally recognised as effective strategies.

2.3.1. Health education leaflets

Leaflets and supporting posters are no doubt the quickest and cheapest way of disseminating and targeting health information. Again there is little robust research evidence of benefit in terms of outcomes, but there is no doubt that the public pickup leaflets from self-selection stands. These should be placed in as wide a range of outlets as possible, including all health related locations (e.g., GP surgeries, clinics, hospital outpatient areas, pharmacies, dentists, opticians) as well as community buildings (e.g., community centres, libraries) schools and colleges, and the workplace.

Figure 1. Self help leaflets used in NHSGrampian, Scotland (reproduced with the permission of Health Promotions, NHSGrampian).

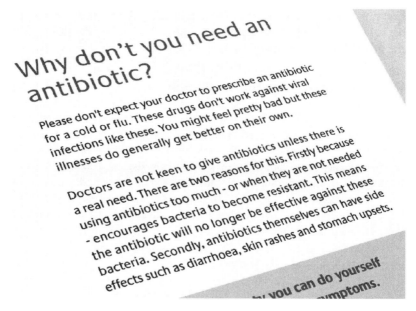

Figure 2. Selected text from self-help leaflets used in NHSGrampian, Scotland (reproduced with the permission of Health Promotions, NHSGrampian).

Health professionals should also use leaflets as part of the consultation (see later) in place of the prescription, perhaps augmented by further detailed, personalised printed materials. Leaflets should be eye catching and give a clear message with possibly more detailed information in smaller font. They should explain their purpose (Why), what behaviour is expected (What), how to get further help (How), who to consult (Who), and what to do (Actions to take). Figure 1 illustrates the range of leaflets recently produced in Grampian for flu, coughs, colds, and cystitis (reproduced with the permission of NHSGrampian). Figure 2 illustrates some of the text for the flu leaflet emphasising why antibiotics use should be restricted.

Leaflets can of course be developed nationally, locally, or at a practice level. It is important that messages are consistent, and a little preparatory research to explore what is currently available is important. In addition, examples from other countries, for example, Holland can be helpful (Van der Doef and Metz, 1996).

2.3.2.   Other mechanisms

There are of course other ways of disseminating information, such as publicity campaigns involving bus adverts, and posters with key messages on, for example, the back of bus seats, and on the underground! These methods are however costly and not evaluated.

As mentioned earlier the media including printed, audio, and visual material, is a powerful tool and if harnessed and controlled can be a useful ally, through either news and documentary programmes, medical slots or "soaps." Again the cost of these initiatives is expensive if pursued nationally, the extreme of which would be national prime time television. However, local options such as press, radio, and television can be more economical and if harnessed appropriately have the added benefit of local ownership.

Finally the Internet is increasingly being used as a source of information and this can only increase. Local health organisations should add key "messages for the day" to the public pages of their website and have dedicated pages to the topic. Again, these must be in lay language, clearly written and ideally in the same house style as the other current campaign materials. There should also be hot links to more detailed websites, but the quality of these should be checked out before the link is made, as endorsement of the content will be assumed.

2.4.   Behaviour change

There is much interest in both sociological and psychological theories explaining health seeking behaviour and understanding patients' *raison d'etre*. One of the most widely used models is the theory of behavioural change (Prochaska and di Clemente, 1983) which depicts six stages of subject change. These are pre-contemplation, contemplation, preparation, action, motivation,

Table 1. Cycle of behavioural change and appropriate education to manage response to illness

Stage of change and interpretation for self care behaviour	Mode of message most appropriate to stage of change	Professional role
Pre-contemplation—Subject unaware of any issues of antibiotic resistance and need to reduce overall use	"Soaps," news items, magazine articles, adverts	Professional advice, maintaining patient relationship
Contemplation—Increased awareness of problem but not concerned enough to change own behaviour	Documentaries, Dr columns/slots, magazine articles, Internet	Professional advice, suggesting other non-medical options
Preparation—Decides to change behaviour and finds out more about the issues and what he/she can do	Documentaries, leaflets, Internet	Professional advice, suggesting other non-medical options
Action—Is informed though appropriate advice on how to self care	Tailored leaflets, disease specific, patient specific	Appropriate professional support
Maintenance—Continued self empowerment and self care as appropriate.	Self care support, Internet books	Appropriate professional support
Sustained behaviour change	Self care support	Appropriate professional support

and sustained change. Although originally applied to addictive behaviour, they have recently been used to understand other lifestyle change such as weight reduction. They also can help us understand behaviour response to symptoms of illness. Educational methods can be tailored to the stage of change and help more subjects through the cycle to the "ideal" stage. Ways in which the educational methods described can be useful at various stages of the behavioural change cycle to alter response to illness are illustrated in Table 1.

3.   PROFESSIONALS

3.1.   Which professionals

Many professionals have a role in the management of minor ailments and can contribute to the appropriate use of antibiotics for these conditions. These professionals include doctors, nurses, and pharmacists, as well as dentists.

At the moment, only doctors, nurses, and dentists are the professionals allowed to prescribe antibiotics, with nurses being the recent addition (Department of Health, 2002), but it is anticipated that pharmacists may also soon have these prescribing rights (Department of Health, 2003). At the moment, most health-care professionals, whether or not they prescribe antibiotics can influence and modify patient behaviour through either giving advice on self-management or referring to the general practitioner (GP), thus reinforcing patients' behaviours. Therefore all the healthcare professionals, particularly including doctors, pharmacists, nurses, health visitors, district nurses, dentists must be educated with respect to:

- Knowledge of why and how antibiotic resistance develops
- Knowledge of different antibiotics and evidence of benefit in different conditions
- Knowledge of how to manage symptoms and use of laboratory tests to isolate causative organism and confirm best treatment
- Attitudes to the different perspectives of patients and population
- Beliefs about the risks of not treating with an antibiotic and litigation
- Behaviour with respect to managing a demanding patient
- Behaviour with respect to new strategies such as deferred prescriptions and laboratory testing (see above).

3.2. Undergraduate

All of the above learning outcomes must be delivered in the first instance in the undergraduate curricula of the professionals already identified. This should be done in an integrated way, so that, for example, information on drugs could be a part of pharmacology, resistance part of microbiology, and managing demanding patients part of communication skills. However, given the imperative to really highlight issues of antibiotic resistance and the central role of good professional practice in reversing that trend, there is also a case for targeted "antibiotic" sessions to bring all aspects of the topic together. This can be done through Problem Based Learning approaches and special study modules. There is also a potential for such sessions to be multidisciplinary and interactive, allowing students to experience the perspective of future professional colleagues and instil an ethos of common goals and team work at an early stage.

Of course to achieve the above, the individual curricula must include these learning objectives and there is still some way to go to achieve this, given the competing demands on limited time as knowledge advances. "Tomorrows Doctor," the General Medical Council's recommendation for undergraduate medical training, focuses on reducing the factual workload, emphasising the acquisition of generic and lifelong skills, and increased teaching in the community (General Medical Council, 1993). These principles are fully endorsed, but need examples from practice to make learning relevant and as a vehicle for delivery. Antibiotics particularly lend themselves to a wide range of relevant learning objectives such as critical appraisal of evidence, communication skills, relationships with patients and colleagues, risk assessment, and the patient-population dichotomy. Because management of antibiotic use can be used to deliver these generic issues, which cut across many specialities, it can fully justify a dedicated slot. Indeed the recent policy on antibiotics in Scotland (Scottish Executive Department of Health, 2002) had this as one of its main recommendations and in the update on Tomorrow's Doctors issued in 2002 this was partially recognised.

Graduates must know about and understand the principles of treatment including the following: The effective and safe use of medicines as a basis for prescribing, including side effects, harmful interactions, antibiotic resistance and genetic indicators of the appropriateness of drugs.

A recent initiative, often referred to as "The Scottish Doctor" is an agreed set of learning outcomes for graduates of the Scottish Medical Schools, developed under the auspices of the Scottish Deans Medical Curriculum group (2002). This identifies twelve learning domains under the three over-arching headings of "What the doctor is able to do," "How the doctor approaches his practice," and "The doctor as a professional." Again antibiotics are not specifically referred to but can be used as a topical exemplar in all three areas. The new core curriculum for pharmacy undergraduates (Royal Pharmaceutical Society of Great Britain, 2002) also identifies generic learning objectives for which antibiotics could be used as an exemplar, as do the curricula for nursing and dentistry.

3.3.   Postgraduate

With issues of Clinical Governance paramount, continuing postgraduate education has become a mandatory requirement and no longer a luxury. Changing and increasing knowledge requires all professionals to be up to date in order to deliver the best cost-effective and clinically effective care. Much research has been carried out to provide an "evidence base" for dissemination of knowledge (Grimshaw and Russell, 1993; Grimshaw *et al.*, 1995) and there is a need to balance locally and nationally driven initiatives. Outreach is seen as one of the key component of effective dissemination, but as with patients, reinforcement is also necessary to refresh and further update learning.

There is also some awareness that these methods and the hierarchy were developed on the basis of empirical work with doctors and that results could be different with other professional groupings. Recent research looking at disseminating information on the use of vaginal imidazoles in community pharmacy, showed no difference between four groups receiving the educational materials as mailed guidelines, guidelines and a continuing education evening, mailed guidelines and outreach, or a final group receiving all options. It is clear that knowledge alone in this group was already good (Watson *et al.*, 2002) but that other factors such as patient pressure affected actual outcomes. Therefore at postgraduate level, as well as undergraduate, managing the patient or customer is an important skill to acquire and is likely to be relevant across all the professions.

Continuing Education is now seen as one mechanism for delivery of education rather than an activity in its own right. Continuing Professional Development is the term currently applied to a reflective practitioner who identifies knowledge deficits, generates learning objectives, and carries out activities to meet these. Such activities could be attendance at a meeting, or identifying and reading relevant material. The theory is that because there is a need for the knowledge, which will be used in practice, that the learning is both more relevant and more effective. To some extent this has reduced the pressure on practitioners to read every new guideline but only to concentrate on those of direct use. However there are some directly relevant to antibiotic use (SIGN, 2003) and the reading of these by all professionals involved in antibiotic use should be strongly encouraged.

Hours of Continuing Personal Development are mandatory for ongoing annual registration with professional bodies. For example the Royal College of Nursing requires five days every three years, the Royal Pharmaceutical Society of Great Britain requires 30 hr per year and Royal College of General Practitioners 25 hr per year. However, Clinical Governance is now becoming even more intensive and intrusive and professional organisations have re-accreditation on the agenda.

Given the high profile of the advantages and disadvantages of antibiotic use in the medical literature, the SMAC report and others, it is likely that antibiotics will be an area identified by practitioners as part of their CPD, or Personal Development Plan. There is an onus on NHS organisations therefore to provide local training sessions and these should ideally be multidisciplinary, again to promote consistent messages, support mutual understanding of different *modus operandi*, and improve local delivery of care.

A final mention must be made in this section to formularies, which are both an educational tool and a guideline or protocol in their own right. Introduced largely in the early 1990s, they were devised as a means to focus practitioners on a small subset of the large numbers of preparations available for prescribing, with the belief that if thousands of drugs could be reduced to hundreds it became realistic to expect prescribers to be aware of the relative costs, benefits, and side effects of the drugs they were using. The very act of developing a practice formulary, whether it be from scratch or derived from an area formulary is educational, as it requires retrieval and appraisal of information on clinical and cost effectiveness before agreeing final items for inclusion. Having taken part in this process, adoption of the final recommendations should be relatively easy. Current and developing prescribing computer software and management screens facilitate implementation of this information into practice.

3.4.   Professionals as teachers for patients

Of course all professionals should have as part of their role a remit to educate their patients. The still emerging concept of a concordant consultation (Marinker, 1997) would again be a good way of involving patients in the sometimes challenging decision not to prescribe an antibiotic. "Concordance is a new approach to the prescribing and taking of medicines. It is an agreement reached after negotiation between a patient and a HCP that respects the beliefs and wishes of the patient in determining whether, when and how the medicine is to be taken. Although reciprocal, this is an alliance in which the HCP recognises the primacy of the patient's decision about taking the recommended medicine." To agree not to prescribe an antibiotic within a concordant framework depends to some extent on the patient being fully informed of the medical side of the decision, which includes diagnosis and management, as well as the HCP being fully aware of the patient agenda. Thus although a concordant consultation is one in which the professional and patient views have equal status, the onus is on the professional to both inform the patient and inform him/herself about the patient. In informing the patient, the professional is undertaking a serious educational role, which requires skills of analysis, synthesis, and delivery of health education relevant to the patient's needs and wishes and tailored to the stage of change. This can be time consuming and may not always be welcomed and the extent to which this is pursued fully in practice is a matter for professional judgement.

3.5.   Professionals as teachers of professionals

Pharmacists have for long been the self proclaimed experts in medicine, but while this has been recognised and utilised in the secondary care setting this has not been fully reflected in primary care.

This is now changing with many GP practices having on site pharmacists with a remit not to supply medicines but to advise on their prescribing. This advisory role can be delivered in a range of ways, either after global review of patient records and assessing appropriateness of drug against external validated indicators, or at individual patient review. Under the current regulations any change recommended has to be actioned by the GPs although the new supplementary prescribing regulations will alter this (Department of Health, 2003). These justified recommendations to the doctor and ensuing discussions for change, are educational in their own right.

In addition practice pharmacists can be given the task of providing "outreach" educational sessions on targeted areas, such as antibiotic use, including first and second line treatments for different infections and infection sites, as well as reminders of when not to prescribe and how to use local laboratory services. Trials of the use of community pharmacists in this way have demonstrated their potential effectiveness as teachers (Avorn and Soumerai, 1983; Watson *et al.*, 2001).

4.   A LOCAL EXAMPLE

For education to be successful, both professionals and patients must be given the same knowledge at the same time. For example, it is unlikely that educating professionals alone will be significant in changing practice if patients are still unaware of the need to change their behaviour and continue to inappropriately attend the GP and "demand" antibiotics. Even the best intentioned professional will be quickly worn down and enthusiasm to put into practice "negotiating" skills will soon wane, when faced with resistance or the time-consuming task of educating an unwilling and unreceptive patient himself. In addition, education at this point, when the patient has clearly come in with an agenda to get antibiotics, that is, is at the pre-contemplation stage, is unlikely to be successful.

Similarly, campaigns targeted at the patient alone, will likewise fail, if they attend their pharmacist for self-medication/symptomatic remedies, but are referred immediately to the GP who because he thinks the patient expects it, prescribes an antibiotic (Little *et al.*, 1997)

To coordinate patient and professional education together probably needs to be done at a locality level. An example of such a campaign, carried out in my local NHS area illustrates that this can be done, using some of the educational techniques already described in the preceding sections.

The campaign, started in 1998, was in response to the then recently published SMAC report and local trends for slightly higher than average use of antibiotics per 1,000 population than for Scotland. Taking on board the principles outlined above, eye-catching leaflets were designed by the local health promotion department, promoting the message that for colds and flu antibiotics were "not the answer." These were displayed primarily in doctors' surgeries and pharmacies. More detailed "serious" looking informative typed A4 sheets were also prepared with self-help guidance including symptoms, symptomatic relief, and when to call the doctor. It was envisaged that these sheets could be used as part of the consultation, proffered instead of a prescription and ultimately computer generated and individualised. Exact wording of both of these leaflets had to be developed with care, because on the one hand, there is a need to discourage unnecessary GP attendance, but on the other hand issues of liability might arise if a patient wrongly did not seek medical help and serious morbidity or mortality resulted.

At the same time, information was sent to GPs and pharmacists reminding them of best practice in using antibiotics, sending copies of the SMAC report, copies of the patient leaflets, information on local prescribing trends, and their exact practice position with respect to these—that is, practice feedback and audit, to coin one of the well used terms in dissemination strategies.

In addition advertising slots were purchased on buses and local radio and the press were invited to cover the campaign.

Preliminary evaluation at the end of the first year showed pharmacists and GPs had valued the initiative, believed it had helped and requested more leaflets for other common conditions. These have already been mentioned in Section 2.3.1. A subsequent more in-depth evaluation of patient knowledge and behaviour, while confounded because of a lack of a control group and limited sampling, indicated that many patients did appreciate the issues around antibiotic resistance, although were not always ready to translate this knowledge into their own behaviour when experiencing an infective illness. This second evaluation also demonstrated the need to involve the wider healthcare team in the campaign (Emslie and Bond, 2003). With the increased role of nurses in the management of minor illnesses and as prescribers and as educators, this was particularly pertinent and nurses and health visitors were included in campaigns in subsequent years.

Examination of current prescribing trends indicates that the prescribing of antibiotics in Grampian has decreased faster than in other Health Board areas in Scotland. Although absolute differences are small and there has been no control group, the indicators are that the campaign has had modest success in reversing the previous trends.

5.   SUMMARY

Education is an important tool in changing professional and patient behaviour. However, both of these are complex and linked, and subject to a wide range of external factors. Evaluation of health education campaigns are difficult because of the cost of robust study designs, which eliminate confounders. Nonetheless evaluation of campaigns targeted at professional behaviour change have indicated success and if we combine them with concurrent patient initiatives this is likely to be symbiotic and valuable.

REFERENCES

Avorn, J. and Soumerai, S. N., 1983, Improving drug therapy decisions through educational outreach: A randomised controlled trial of academically based 'detailing'. *N. Engl. J. Med.*, **308**, 1457–1463.

Department of Health, 1999, The path of least resistance. www.doh.gov.uk/smac1.htm

Department of Health, 2000, UK antimicrobial resistance strategy and action plan June, Published by the NHS Executive. www.doh.gov.uk/arbstrat.htm

Department of Health, 2002, Extending independent nurse prescribing within the NHS in England—A guide for implementation.

Department of Health, 2003, Supplementary prescribing by nurses and pharmacists within the NHS in England—A guide for implementation.

Emslie, M. J. and Bond, C. M., 2003, Public knowledge, attitudes, and behaviour to antibiotics: A survey of patients in general practice. *Eur. J. Gen. Pract.*, **9**, 85–91.

General Medical Council, 1993, Tomorrow's Doctors. www.gmc-uk.org/med_ed/ tomdoc.htm

Grimshaw, J., Freemantle, N., Wallace, S., Russell, I., Hurwitz, B., Watt, I. *et al.*, 1995, Developing and implementing clinical practice guidelines. *Qual. Health Care*, **4**(1), 55–64.

Grimshaw, J. M. and Russell, I. T, 1993, Effect of clinical guidelines on medical practice: A systematic review of rigourous evaluations. *Lancet*, **342**, 1317–1322.

Marinker, M., 1997, From compliance to concordance: Achieving shared goals in medicine taking Report of the Royal Pharmaceutical of Great Britain.

Little, P., Gould, C., Williamson, I., Warner, G., Gantley, M., and Kinmonth, A. L., 1997, Reattendance and complications in a randomised trial of prescribing strategies for sore throat: The medicalising effect of prescribing antibiotics. *BMJ*, Aug 1997, **315**, 350–352.

Price, D., Honeybourne, D., Little, P. *et al.*, 2003, Community acquired pneumonia mortality: A potential link to antibiotic prescribing trends in general practice. Respiratory Medicine.

Prochaska, J. O. and Di Clemente, C. C., 1983, Stages and processes of self-change of smoking: Toward an Integrative model of change. *J. Consult.Clin. Psychol.*, **51**(3), 390–395.

Royal Pharmaceutical Society of Great Britain, 2002, Accreditation of UK pharmacy degree courses.

Scottish Deans' Medical Curriculum Group, 2002, The Scottish Doctor; Undergraduate learning outcomes and their assessment: A foundation for competent and reflective practitioners.

Scottish Executive, 2002, Antimicrobial Resistance Strategy and Scottish Action Plan www.scotland.gov.uk/library5/health/arsap.pdf

SIGN, 2003, Diagnosis and management of childhood otitis media in primary care SIGN guideline 66.

Van der Doef, E. and Metz, R. G., 1996, *What Should I Do*? RTFB, Southampton, UK.

Watson, M. C., Bond, C. M., Grimshaw, J. M., Mollison, J., Ludbrook, A., and Walker, A. E., 2002, Educational strategies to promote evidence based community pharmacy practice: A cluster randomised trial (RCT). *Fam. Prac.*, **19**, 529–536.

Watson, M. C., Gunnell, D., Peters, T., Brookes, S., and Sharp, D., 2001, Guidelines and educational visits from community pharmacists to improve prescribing in general practice: A randomised controlled trial. *J. Health Serv. Res. Policy*, **6**(4), 207–213.

Chapter 28

The Influence of National Policies on Antibiotic Prescribing

Moyssis Lelekis and Panos Gargalianos
*The General Hospital of Athens "G. Gennimatas", 154 Messogeion Avenue
Athens 11527, Greece*

1.   INTRODUCTION

Resistance to antimicrobial agents has emerged as a major problem both in the community and the hospital. As S. B. Levy said "the warnings were there long ago, but too few people heeded them. Thus an emerging problem has grown to a crisis" (Levy, 2001). From the earliest signs of resistance, there was a reluctance among providers and manufacturers to link the problem to the misuse of antibiotics. It is only lately that antibiotic use and especially misuse has been recognized as a major cause of bacterial resistance, although the direct quantitative relationship between antibiotic use and resistance is still lacking (Levy, 2001). Inevitably the control of antibiotic use, by implementing antibiotic policies has been recognized as a tool for controlling and reversing antimicrobial resistance. In fact the principle aim of a policy is to bring about a change in prescribing, which will lead to reduced resistance, decreased costs, and improved quality (judicious, safe, and appropriate) of antibiotic prescribing (Nathwani, 1999).

Since the problem of resistance is a global one, affecting most countries, there is a trend towards involvement of international organizations (e.g., the European Union [Jenkins, 1998]) in the battle against resistance. However, despite the existing discussion concerning the implementation of uniform international antibiotic policies, there are some limitations to this approach. First, local resistance problems dictate different solutions and different

Antibiotic Policies: Theory and Practice. Edited by Gould and van der Meer
Kluwer Academic / Plenum Publishers, New York, 2005

prescribing practices even within a single country. Second, huge economic and social differences between different countries necessitate different approaches at different levels (Keuleyan and Gould, 2001).

Nevertheless the minimum standards for a European Antibiotic Policy have been proposed (Keuleyan and Gould, 2001):

- Establishment of a European antimicrobial resistance surveillance system to provide data about antibiotic usage, resistance, and linked clinical information
- Establishment of European guidelines for good clinical practice
- Support of educational programs for practitioners and consumers
- Establishment of an infection prevention and control policy
- Support of research in the field
- Supply of funds for developing countries, for example, from WHO and the World Bank.

2.   NATIONAL ANTIBIOTIC POLICIES

Very often there is a confusion concerning relevant terms. For example the terms policy and guideline are often used interchangeably, despite the fact that they mean different things (Gould, 2002). Policy should refer to local, regional, or national antibiotic stewardship programs as a whole, while guideline should refer to specific treatment or prophylaxis recommendations for individual diseases, syndromes, etc. (Gould, 2002).

A National Policy should address all relevant issues for antibiotic use, both in the community and the hospital, including veterinary and agricultural use. A national expert committee should be established in each country for that purpose (Keuleyan and Gould, 2001).

Some important issues to be included in the policy are (Keuleyan and Gould, 2001):

- Existing laws should be enforced to prevent non-prescription, over the counter sale of antibiotics
- Guidelines for antibiotic treatment and prophylaxis should be prepared and adapted institutionally at a local level
- Consumption of antibiotics should be monitored to estimate the national consumption of antibiotics
- A national antimicrobial resistance surveillance system should be established and coordinated with international systems
- A national control of infections program should be implemented
- Educational programs should be elaborated for both healthcare workers and the public

- Collaboration with international organizations (WHO, APUA, ESGAP, etc.) should be established
- Appropriate funding should be made available by government or any other organization.

3.   THE PRACTICE OF NATIONAL ANTIBIOTIC POLICIES

The mid-1990s are considered an important time period, since at this time, antibiotic stewardship programs were implemented in many countries (Chahwakilian, 2000). However, in practice the above-mentioned components of a policy are only rarely or not at all found altogether in the policies implemented in various countries. At the European level, the publication by Cars *et al.* (2001) presents the differences in antibiotic consumption that exist among European countries. Even though the data refer only to non-hospital use of antibiotics, it is reasonable to consider them a reflection of antibiotic policies implemented in the various European countries.

3.1.   Southern Europe

In Spain the problem of antimicrobial resistance, mainly among community-acquired bacterial pathogens, resulted in the generation of a Task Force by the Ministry of Health to address the problem (Baquero *et al.*, 1996). Members of this panel were academic experts in clinical microbiology, infectious diseases (in both children and adults), internal medicine, clinical epidemiology, veterinary microbiology, and public health administration. The proposals of this task force were included in the document "Antibiotic resistance in Spain: what can be done" whose objective was to provide a comprehensive framework to support the necessary actions. The recommendations of this task force included all aspects and interested parts of the problem and is a paradigm of a global approach to the issue of antibiotic policies as a way to fight antimicrobial resistance:

1. Data on Antibiotic resistance:
   (a) Continuous collection of data on resistance in human isolates, isolates from animals, and food of animal origin.
   (b) Detection of antibiotics and antibiotic resistance in the environment.
   (c) Continuous collection of data on human consumption of antibiotics.
   (d) Regular estimation of direct uncontrolled procurement of antibiotics by the public.

2. Physicians (the prescribers):
   (a) Periodic collection of pharmacotherapeutic antimicrobial profiles of office-based physicians.
   (b) Periodic updating of physicians concerning consensus recommendations on clinical and laboratory diagnosis of the most common community-acquired infections and possible treatment protocols.
   (c) Assurance of easy access for community-based clinicians to microbiological laboratories and data on local resistance patterns of the more frequent pathogens.
   (d) Promotion of education on rational use of antibiotics.
   (e) Optimization of the medical consultation time spent per patient in community clinics.
   (f) Establishment of committees on antibiotic policies at the national, regional, and local levels.
3. Patients and consumers:
   (a) Dissemination of information about antibiotic resistance.
   (b) Dissemination of information to the food industry on the dangers of uncontrolled use of antibiotics in feed or water for therapeutic or prophylactic purposes.
4. Pharmacists (the dispensers):
   (a) Promotion of pharmacists as agents for the rational use of antimicrobials.
   (b) Continuous inspection to prevent the over the counter delivery of antibiotics.
5. Veterinary medicine:
   (a) Standardization of veterinarians prescription forms for the therapeutic or prophylactic use of antimicrobials in animals.
   (b) Encouragement of the rational use of antimicrobials in veterinary medicine and as growth promoters including specific education on antibiotic use in veterinary schools.
6. Pharmaceutical industry:
   (a) Support to the industry for research and development of new antibiotics to overcome resistance.
   (b) Provision of incentives for the companies ready to cooperate in various ways for the control of resistance.
7. The Health administration:
   (a) Consideration of the potential impact on antibiotic resistance should be incorporated into the evaluation of new antimicrobials.
   (b) Provision of better information in package inserts.
   (c) Establishment of programs for rational use of antibiotics in humans.
   (d) Reevaluation of products used as feed additives and of antibiotics used in veterinary medicine.

(e) Organization of local or regional committees for identifying outbreaks of resistant organisms and for analyzing and instituting control procedures.

(f) Evaluation of the role of vaccines in the control of antimicrobial resistance.

Indirect information on the efficacy of the above extensive measures is available through a recent publication concerning antibiotic consumption in Spain from 1985 to 2000 (Bengoa *et al.*, 2002). The data analysis states clearly, that consumption had three periods. During the first one (1985–1989), there was a mild decreasing trend, while during the second lasting until 1995, there was a generalised increase in all antibiotic classes. During the third period, beginning in 1996, there is a sustained and generalized decline of antibiotic consumption. This last significant decrease is attributed by the authors to the campaign launched by public administration, to promote the rational use of these agents (Bengoa *et al.*, 2002).

In Greece, in the 1980s, extremely high resistance rates were reported among important nosocomial pathogens, which correlated well with very high antibiotic consumption in hospitals (ESGAR, 1987; Giamarellou *et al.*, 1986). As a result, an antibiotic restriction policy was established in Greek hospitals in late 1980s concerning the newer and more potent antibiotics. In fact, a restrictive order form was introduced for third generation cephalosporins, carbapenems, gly-copeptides, newer quinolones, and aztreonam. Even though this policy has never been subject to regular audit, sporadic reports showed that it worked, at least in the beginning (Giamarellou and Antoniadou, 1997; Lelekis *et al.*, 1993). Since it was apparent that the compliance to this policy had a downwards route, the policy was reinforced in 2003, with a regulatory order by the Ministry of Health.

In the year 2000, the National Committee for Nosocomial Infections published a guide for antimicrobial chemotherapy and prophylaxis for the hospitalized patient. The guide was distributed to all hospital doctors. In late 2002, an audit performed in two tertiary care hospitals and three provincial secondary care ones, revealed that 76% of physicians were aware of the existence of this guide and 61% had a copy. However, only 49% of those who had it, used it on a regular basis for their everyday practice (Lelekis *et al.*, 2003). It is worth noting that hospital formularies do not exist in Greek hospitals and a significant number of generics are used, (five for ciprofloxacin and four for cefuroxime in many hospitals).

In the community, according to state regulations, antibiotics should be dispensed only with a physician's prescription. However there is no control whatsoever to detect and punish the over the counter sale of antibiotics which continues in significant rate (Contopoulos-Ioannidis *et al.*, 2001). By regulatory order oral third generation cephalosporins and newer quinolones are restricted in the community. However the exceptions in this order for quinolones allows a significant consumption of these agents (ESAC, 2003). A campaign for the public has been initiated since year 2001. It includes television spots and booklets discouraging the over the counter sale of antibiotics and their use in cases of viral infections. Moreover a guide for the use of antibiotics in the community is being prepared and will soon be distributed to ambulatory care physicians. Besides interventions for antibiotic prescribing, there has been a change in the willingness of microbiologists to refer antibiotic resistance rates for the most important pathogens to a centralized resistance surveillance system. Thus, it is now easier to have data in a more generalized level (Vatopoulos *et al.*, 1999).

In Italy, a nationwide survey was conducted in year 2000 to quantify the prevalence in Italian hospitals of policies aimed to reduce the emergence and dissemination of resistant strains (Moro *et al.*, 2003). The overall response rate was 80% (428/535). Of the respondents, 9.6% claimed to have implemented a surveillance system of antimicrobial resistance, 90% had a hospital formulary in place, 50% had a pharmacy committee, and 18% had an antibiotic policy subcommittee that met at least once a year in 1999. Restriction policies were implemented on 41% based on written justifications for antibiotics and 8% provided susceptibility results for first line antibiotics only. In only 11% of cases, antibiotic consumption was monitored, by using the defined daily dose (DDD) as a unit of measure. Moreover 58.6% claimed to have defined clinical practice guidelines for hand washing, 36.4% for isolation procedures and 1.6% for the control of methicillin-resistant *Staphylococcus aureus* (MRSA) infections. The overall conclusion of this survey was that there is a lot more to be done in Italy concerning antibiotic policies.

3.2.   Northern Europe

In Ireland the National Disease Surveillance Centre was asked by the government to make recommendations about combating antimicrobial resistance. The committee prepared a report known as SARI (Strategy for the Control of Antimicrobial Resistance in Ireland) (National Disease Surveillance Centre, 2001). There are five major constituents of this strategy:

1. The surveillance of antimicrobial resistance both in the community and the hospitals.
2. The monitoring of the supply and use of antimicrobials. Especially in this field it is recommended that:
   (a) The tight legislative controls that exist in the area of antibiotic prescribing should be maintained and enforced.
   (b) A system for the collection and analysis of antimicrobial use and prescribing in hospitals and the community should be established.

(c) A basic set of data agreed by the committee should be collected, that is, the origin of the prescription (hospital or community), the agent and dose prescribed, the indication and the length of treatment.

3. The development of guidance in relation to the appropriate use of antimicrobials:

(a) Expert opinion on infectious diseases should be available 365 days a year to all medical practitioners both in the community and the hospitals.

(b) National guidelines for appropriate use of antibiotics should be drawn up and introduced in practice both in the community and the hospitals.

(c) A process for reduction of the inappropriate use of antibiotics should be defined, different for the community and the hospitals.

(d) Interventions aimed at changing clinical practice should be supported, encouraged, and reinforced by a process of regular audit.

(e) A priority should be given to improvement of vaccine uptake, especially for influenza and pneumococcus.

4. Education:

(a) Education should commence at undergraduate level.

(b) Education should be directed at, all clinical professional groups providing patient care, the pharmaceutical industry, and the general public.

5. The development of principles in relation to infection control in the hospital and community setting.

Furthermore the committee recommended the prioritization of funding the campaign against resistance and defined the areas of future research in the area of antimicrobial prescribing and the development of new treatment modalities and new antimicrobial agents.

In the UK almost a decade ago, a survey was performed among consultant microbiologists and pharmacists nationwide on methods used to control antibiotic usage (Working Party Report, 1994). A written policy for surgical prophylaxis was available in 51% of hospitals, 62% had a written policy for therapy, and 79% an antibiotic formulary. Compliance was monitored in approximately 40% and steps were taken in half to control non-compliance. A restricted list was operated in 77% of hospitals and 90% of respondents believed formularies to be beneficial.

After that, the British Society for Antimicrobial Chemotherapy (BSAC), who commissioned the survey, recommended that the following minimum control measures should be implemented in all UK hospitals (Working Party Report, 1994).

1. Formulary and policies should be updated frequently with appropriate funding for both staff and printing costs.

2. There should be widespread consultation before inception and effective enforcement thereafter via good educational programs.

3. A broadly representative committee should be set up to consider timely introduction of new antibiotics with the authority to ensure their availability or non availability. Quality of care should take preference over financial considerations such as antibiotic cost.
4. Agents not included in the formulary should be only available for formal clinical trials or after discussion with the requesting consultant.
5. Microbiology and pharmacy departments should have adequate facilities to ensure that educational programs can be carried out.
6. Programs for assessment and appropriate adoption of automatic stop dates, antibiotic prescription forms, and utilization coordinators should be introduced.
7. There should be compulsory notification of pharmaceutical promotional activities to the formulary committee in order that permission for these activities be granted and that a member of the committee may be present if appropriate.
8. Laboratories should regularly make local sensitivity patterns widely known and routinely should only report on those agents, which appear on their formulary and policy.

Despite apparent use of control measures like these, there were reports of increasing antibiotic prescribing both in the hospitals and the community (Gould, 1996). The experience of Grampian is very characteristic. In the northeastern part of Scotland despite the implementation of a very strict antibiotic policy, a significant increase in antibiotic consumption was recorded between 1992–3 and 1996–7 (Gould and Jappy, 2000). The results from this experience highlight the current difficulty in controlling prescribing budgets, the increasing use of antibiotics and the consequent increase of antimicrobial-resistant organisms.

In 1998 the Government's response to the House of Lords Select Committee on Science and Technology's report "Resistance to antibiotics and other antimicrobial agents," indicated its intention to implement a comprehensive strategy to tackle the problem of resistance. Key elements of this strategy are (Standing Medical Advisory Committee, 1998):

1. Surveillance to provide the data on resistant organisms, illness due to them, and antimicrobial usage necessary to inform action.
2. Prudent antimicrobial use to reduce the pressure for resistance by reducing unnecessary and inappropriate exposure of microorganisms to antimicrobial agents in clinical practice, veterinary practice, animal husbandry, agriculture and horticulture. Prudent antimicrobial use in humans in particular should be promoted through
   (a) Professional education in all levels of seniority

(b) Prescribing support (national, local guidelines etc)
(c) Organizational support.
3. Infection control to reduce the spread of infection in general and of resistant microorganisms in particular.

All these need to be supported by the provision of tailored information, education, communication, research, the necessary infrastructure, organization support, and where necessary, legislation or regulation.

The first result of this strategy was a public education campaign advising patients not to pressure their doctors to give them antibiotics for colds and flu and recommended 3 days treatment for simple urinary tract infections (Gould, 2001). With the evidence of reduced expectations by patients, there has been downturn in community prescribing but this had already started before 1998.

Lastly, a favourable impact on antibiotic prescribing is expected after the last reforms of the National Health Service, with its drive to improve quality and ensure better education of and performance by doctors. This includes the introduction of Clinical Governance, which intends to make doctors responsible for the quality of their antibiotic prescribing and empower their employers to ensure that this quality is achieved (Gould, 2001).

Even though significant differences exist between various regions, antibiotic use in the Netherlands is the lowest in Europe (Bruinsma *et al.*, 2002; Cars *et al.*, 2001). Antibiotic use in this country was always restrictive. Earlier studies revealed that most Dutch hospitals had formularies, which contained recommendations for antibiotic prescribing (Stobberingh *et al.*, 1993). Furthermore, most hospitals have, in general, adopted a conservative approach when devising antibiotic policies and there has been a tendency to limit the use of newer, more broad-spectrum agents (Janknegt *et al.*, 1994). It is worth noting that until 1996, Dutch governmental legislation obliged hospitals to have a drugs and therapeutics committee. The development and implementation of formulary agreements was an objective explicitly stated. After 1996 the law changed and policies and methods for rational pharmacotherapy can now be designed and implemented on the institution's own view (Fijn *et al.*, 1999). In a survey performed among all Dutch hospitals in 1998 the participation was 99%. From this survey it was apparent that the presence of a drugs and therapeutics committee and antibiotic policies in general hospitals appears independent of hospital characteristics. However, formulary agreements and treatment guidelines were less likely to be present in hospitals with only one pharmacist employed. More than half of the hospitals claimed to have restrictive formulary agreements. This was more likely the case for large hospitals and hospitals in the eastern and southern provinces (Fijn *et al.*, 1999).

In order to further promote the responsible use of existing antibiotics, a foundation was established in the Netherlands. The SWAB (Stichting Werkgroep

Antibioticabeleid—Foundation Antibiotics Policy Work Group) has, as a primary goal, to optimize the use of antibiotics in the Netherlands, in order to diminish the development of resistance (van Kasteren *et al.*, 1998a). One of the SWAB projects was to develop national guidelines for the use of antibiotics in hospitals. The guidelines are prepared by a committee of experts and reviewed by external consultants: infectious diseases specialists, medical microbiologists, and pharmacists. The revised version was then submitted for publication. The aim of SWAB is to make prevention of antibiotic resistance a major factor in the choice of the antibiotic. Up to now SWAB guidelines have been published for pneumonia (van Kasteren *et al.*, 1998b), for antimicrobial therapy in adults hospitalized with bronchitis (van Kasteren *et al.*, 1998c), for antimicrobial therapy of adults with sepsis in hospitals (van Kasteren *et al.*, 1999), perioperative prophylaxis (van Kasteren *et al.*, 2000), and selective decontamination of intensive care patients on mechanical ventilation (Bonten *et al.*, 2001). Interesting is the fact that a multicenter audit in Dutch hospitals of adherence to guidelines for surgical chemoprophylaxis, revealed that there was a willingness to adhere to guidelines (van Kasteren *et al.*, 2003).

In Belgium, a country with significantly higher antibiotic usage, the nosocomial usage of antimicrobial agents increased by 7% in volume and 21% in value between 1991 and 1993. Moreover there was a trend toward an increased usage of broad-spectrum and newer drugs (Struelens and Peetermans, 1999). The introduction of a normative reimbursement system for antimicrobial prophylaxis of surgical procedures in Belgian hospitals in 1997, has had a dramatic impact on antibiotic consumption (Struelens and Peetermans, 1999). Following the recommendations of the European Conference "The Microbial Threat," a Committee for the Coordination of Antibiotic Policy (BCCAP [*Commission de Coordination de la Politique Antibiotique/Commissie voor Coordinatie van de Antibiotica Beleid*]) was created in 1999, jointly by the Ministry of Social Affairs, Public Health and Environment and the Ministry of Agriculture (Nagler *et al.*, 1999). All interested parties are represented in the Committee. Its main tasks are:

- collection and organization of all available information on antibiotic use and resistance
- publication of reports on antibiotic use and resistance
- information and increase of public awareness on antibiotic resistance and the risks associated with the irrational use of antibiotics
- making recommendations on relevant points, such as detection of resistance, cross-resistance mechanisms, use and consumption of antibiotics in both man and animal, etc.
- making recommendations for research on antibiotic resistance and on the transfer of resistance among bacteria and among ecosystems.

Given the volume of antibiotic use in ambulatory care, a campaign was launched by the committee (Bauraind *et al.*, 2001a). The aims of the campaign were:

- to inform the public about antibiotic resistance and to warn it about the medical and general health issues related to the inappropriate use of antibiotics
- to foster the patient–physician and patient–pharmacist dialog about the appropriate use of antibiotics.

The materials used for this campaign were, booklets, folders and posters targeting patients, TV-spots and radio-spots targeting the public, direct press and media communications targeting general public and MDs, Pharm and Web sites for general public and MDs. The campaign was launched in November 2000 and it lasted until March 2001. Its evaluation (Bauraind *et al.*, 2001b) revealed that:

- it improved the awareness of the public and reduced requests for antibiotics
- it reduced antibiotic prescriptions only transiently
- media and especially TV are the most powerful tools for such a campaign for both the public and GPs.

After these, the final conclusion was that this campaign should be repeated and further improved (Bauraind *et al.*, 2001b).

In France, in an effort to control ambulatory care costs, regulatory practice guidelines, known as "références médicales opposables" (RMOs) or regulatory medical references, were introduced by law in 1993 (Durieux *et al.*, 2000). According to the law, physicians who do not comply with RMOs can be fined. In terms of antibiotic prescribing, the aim of this strategy was not to decrease the number of prescriptions, but to reduce the total cost of antibiotics and reduce especially the prescription of broad-spectrum, expensive agents (Chahwakilian, 2000). Moreover they were not planned to promote good antibiotic prescribing practice. The RMO policy was questioned in 1997, when the reform of the French Health System changed the rules (Colin *et al.*, 1997). According to the reform, French physicians having private practice could be collectively fined at the end of each year if they overspent the budget prescribed by the French parliament. On the contrary they could receive a bonus if they stayed within this budget. This regulation resulted in protest on the part of physicians as being unethical and it created an intense conflict. In a survey done in 1998 among French family physicians, it was quite obvious that despite financial penalties French physicians' knowledge of RMOs was poor (Durieux *et al.*, 2000).

The introduction of RMOs did not decrease the overall volume of outpatient antibiotic use and had only a modest economic impact. However, the prescription

patterns changed with this policy, leading to a decrease in the use of fluoroquinolones and oral cephalosporins and to a substantial increase in macrolide use
(Choutet, 2001).

Thus, in 1999 a national plan of actions for the control of antimicrobial
resistance was initiated in France and represents the most complete effort in
this direction. It includes (Chahwakilian, 2000):

1. Surveillance of resistance and antibiotic consumption for both the community and the hospital and also evaluation of the practices.
2. Infection control programs in the hospitals and the day care facilities.
3. Actions for the promotion of good antibiotic prescribing with the application of the existing recommendations, with the use of rapid diagnostic tests,
   and with the development of visits free of charge.
4. Reinforcement of the control of the advertisement of antibiotics.
5. Promotion of research, especially in the field of optimization of antibiotic
   prescribing and intervention studies on this subject.

In a comparison of France and Germany in terms of outpatient antibiotic
use, many interesting points were made, some of which refer to antibiotic policies implemented in these two countries (Harbarth *et al.*, 2002).

Analyses of national sales data show that the use of oral antibiotics in
France is almost three times higher than in Germany. This can be attributed to
several reasons:

First, antibiotic practices vary tremendously between France and Germany.
A pan-European survey revealed that in Germany many more patients do not
get a prescription for an antibiotic at the first consultation visit for respiratory
tract infections, even in cases of suspected pneumonia (Harbarth *et al.*, 2002).
This lower rate of prescriptions can be explained at least partly by a higher
recourse to diagnostic investigations and a watchful waiting approach very
common in Germany (Woodhead *et al.*, 1996).

Second, there are differences in terms of cultural factors. Expectancy of
French people for an antibiotic prescription for respiratory tract symptoms is
much higher than that of Germans (Bouvenot, 1999; Pradier *et al.*, 1999). This
allows German doctors to follow a much more conservative and watchful-waiting
approach in cases of non-life threatening situations. It is worth noting that in a
pan-European survey the demand index for an antibiotic prescription was 2.2
for French people surpassed only by Turkish patients (index 2.4—Branthwaite
and Pechere, 1996).

Third, there are differences in regulatory practices. Antibiotic prescriptions
are affected by reimbursement policies and the structure of the pharmaceutical

market. Historically the French drug economy has been regulated by product price control and has been considered as a low price, high quantity system, whereas the German one was a high price, low quantity system. As a result

(a) Generics have played a minimal role in French pharmaceutical market in contrast to Germany (<5% vs 39%). This feature contributes to the trend in France for using newer antibiotics (Garratini and Tediosi, 2000).

(b) Until recently, French pharmacies were better remunerated if they dispensed large volumes of expensive drugs such as oral broad-spectrum cephalosporins (Garratini and Tediosi, 2000). In Germany, on the contrary, pharmacy remuneration is calculated by applying regressive percentages to different price bands; the lower the price, the higher the pharmacy's share (Huttin, 1996).

(c) The French pricing system forces pharmaceutical companies to implement aggressive promotional efforts and marketing campaigns to compensate for low prices. Thus German and French physicians are exposed to different marketing information and pressure for prescribing antibiotics (Le Pen, 1997).

A very important difference is that health authorities in Germany have more regulatory power and thus a broader impact on drug use. In 1993 the introduction of capped physician budgets and a system of reference pricing in Germany, led to a switch in prescribing preferences and an incentive for German physicians to avoid expensive drugs priced above the reference price (Danzon and Chao, 2000). As a result, from 1994 to 1997 the volume of antibiotics prescribed decreased from 334 to 305 million DDDs (Giulani *et al.*, 1998).

The consumption of antibacterials has remained relatively stable in Scandinavia and is low compared with most other countries (Bergan, 2001). However, among these countries there are differences concerning the volume and the distribution of different classes of antimicrobials (Bergan, 2001). The consumption is highest in Iceland and Finland and lowest in Denmark and Norway.

In Denmark there has been a long standing tradition for maintaining registries for monitoring of a variety of healthcare parameters. The Danish Medicines Agency is legally obliged to monitor the consumption of all medicinal products in Denmark. This is done by regular reporting from all pharmacies, including pharmacies located in hospitals to the agency (Sørensen and Monnet, 2000). For every antimicrobial dispensed, the patient's central personal registration (CPR) number, the date, the place (pharmacy, hospital pharmacy, institute, etc.) the reimbursement (if any) and the license number of the prescribing doctor are automatically recorded and the data are transferred to the Danish Medicines Agency.

In Denmark there is no over the counter sale of antimicrobials. The emergence of resistant *Staphylococcus aureus* in the mid-1960s had many favourable results due to the actions taken (Sørensen and Monnet, 2000). At that time the post nursing school education as a "hygiene nurse" was established. Furthermore, a formal collaboration started in every county between hygiene nurses, clinical microbiologists, hospital pharmacists, hospital doctors, and general practitioners in close collaboration with the national center for hospital hygiene, in order to focus on reduction of the number of nosocomial infections, reduce resistance, and optimize the use of antimicrobials. Subsequently a number of interventions were taken at all levels of the Danish healthcare system ranging from medical audit projects in small groups of general practitioners to national legislation.

The low level of resistance of the major pathogens allows for conservative recommendations for the treatment of infectious diseases both in the hospital and the community. These recommendations are printed and sent free of charge to every doctor in Denmark (Sørensen and Monnet, 2000). Another way to reduce the use of antibiotics at the general practitioner level is by encouraging the use of rapid diagnostic tests. Fast negative results prevent inappropriate prescriptions of antibiotics. Doctors are reimbursed for the test itself and are paid for performing it (Sørensen and Monnet, 2000). At a national level a powerful tool for controlling antibiotic use in Denmark is the reimbursement policy. Changes in reimbursement policy resulted in a decrease in antibiotic use (Sørensen and Monnet, 2000). In Danish Hospitals antimicrobials are not subsidized. Hospital pharmacies buy antimicrobials from pharmaceutical companies after negotiation about the price through a union of hospital pharmacies and sell them to the medical wards. Aside this economic incentive, most hospitals have implemented local policies for the promotion of prudent use of antibiotics (Monnet and Sørensen, 1999).

In Sweden in response to the increasing sales of antibiotics and the spread of penicillin non-susceptible pneumococcus, a national project named STRAMA (Swedish Strategic Program for The Rational Use of Antimicrobial Agents and Surveillance of Resistance) was initiated in 1994 (Mölstad and Cars, 1999). Since the immediate threat was the emergence of resistant pneumococcus, the project focused initially on treatment of RTIs, antibiotic usage in pre-school children and surveillance of resistance in pneumococci.

At the national level, STRAMA was formed by clinical specialists from the Swedish Reference Group for Antibiotics (SRGA) representing infectious diseases, microbiology, general practice, ENT and pediatricians, as well as representatives from scientific and administrative authorities. The primary objective was to stimulate the formation of local STRAMA groups in each county. National STRAMA coordinates surveillance programmes and follows antibiotic consumption and the incidence of resistance nationwide. It also makes recommendations on identified problems. The main goal was to contain inappropriate use of antibiotics and antimicrobial resistance both in the community and the hospital (Mölstad and Cars, 1999).

The regional STRAMA groups that were formed in the counties of Sweden had, as a main objective to evaluate antibiotic utilisation and patterns of resistance in their geographical area. They influence healthcare workers to improve diagnostic procedures and change the prescribing patterns of antibiotics. They also decide whether proposed recommendations should be implemented in their county.

In 1995 the national STRAMA group prepared antibiotic treatment policies for respiratory tract infections caused by resistant pneumococci and in the following years on the treatment of urinary tract infections, chronic bronchitis, and skin and wound infections. Guidelines were also issued for the use of macrolides, vancomycin, and fluoroquinolones. Furthermore, STRAMA produced a brochure with patient information on respiratory tract infections, antibiotics and resistance, which was distributed to all Swedish medical health centres and surgeries. The folder gave the physician a possibility to offer written, non-commercial information to patients with respiratory tract infections.

Another activity was the production of guidelines for the management of resistant pneumococci in day care centres and family day care. A return visit free of charge was proposed to encourage expectancy, instead of an antibiotic prescription to patients with respiratory tract infections, who had no obvious signs and symptoms of bacterial infection.

During the period that the STRAMA project has been running, antibiotic consumption has decreased continuously. Between 1993 and 1997 it was reduced by 22% and the reduction was most pronounced for children 0–6 years old. Since large differences were recorded between counties, a further decrease may be possible. The national frequency of resistant pneumococci did not increase during the 1990s, while the increasing incidence in southern counties seems to have been curtailed (Mölstad and Cars, 1999).

The Finnish experience is a paradigm of successful intervention for lowering antimicrobial resistance in the community (Seppälä *et al.*, 1997). In 1991, a recommendation was made to decrease the use of macrolides in infections caused typically by Group A streptococcus. As a result, the usage of macrolides was nearly halved in all parts of the country and the resistance rate of streptococcus to macrolides decreased significantly (from 19% to 9%).

Compared to other Nordic countries Finland has noticeably higher antibiotic consumption. Only Iceland has the same high level of use as Finland. This high level of antibiotic usage in Finland could at least partly be attributed to the prescription of antibiotic courses of long duration (Rautakorpi *et al.*, 1997). In late 1990s a countrywide programme called MIKSTRA was launched, having as its main goal to develop an optimal antibiotic policy for outpatient infections (Rautakorpi *et al.*, 2001). This joint research and development program was planned for a 5-year duration and included new evidence-based recommendations for diagnosis and treatment of the most common outpatient infections with a careful follow up, cost–benefit analysis of the new recommendations, educational initiatives for both health professionals and the public, as well as measurement of the changes in the attitudes of the health professionals and the public. In particular the infection-specific, evidence-based guidelines were issued by the Finnish Medical Society Duodecim with the support of the Finnish Ministry of Social Affairs and Health. Six such guidelines were published in 1999–2000 in a medical journal and on the Internet (Rautakorpi *et al.*, 2001). The targeted infections were otitis media, throat infection, acute sinusitis, acute bronchitis, skin infections, and urinary tract infections. Changes in bacterial resistance of the most common community pathogens were to be surveyed by the Finnish Study Group for Antimicrobial Resistance.

The Icelandic experience is an example of a successful nationwide policy, which was carried out without being overseen by an official body (Kristinsson, 1999). In fact, the intervention was carried out by key opinion leaders in clinical microbiology, infectious diseases, general practice, and pediatrics. It included use of television, radio, and newspaper for conveying the message. Furthermore, articles in the *Icelandic Medical Journal*, medical meetings, and conferences were used for the same purpose. The intervention focused on the overuse of antibiotics in children. Physicians were urged to avoid prescribing antibiotics for viral illnesses and to give antibiotics less frequently for otitis media, sinusitis, and bronchitis in children. Moreover, new prescribing guidelines were prepared for general practitioners and the issue of prudent use of antibiotics was a frequent topic for discussions both in the journal of the Icelandic Medical Association and at numerous meetings.

The general feeling of Icelandic physicians is that the campaign changed the parents' attitude towards antibiotics. From 1991 until 1997 that this campaign lasted, a decrease in antibiotic use was recorded. The overall decrease was only 10%, but it was 35% for the use in children (Kristinsson *et al.*, 1998). Interestingly and most importantly, the incidence of penicillin–non susceptible pneumococci showed a significant decline during the same period (Kristinsson *et al.*, 1998).

3.3.   Central–Eastern Europe

The significant political and social changes that have taken place in Eastern Europe have caused dramatic changes in healthcare as well. These changes had an impact on antibiotic policies and antibiotic use. In a survey of national and local antibiotic policies in central and eastern Europe, interesting information was obtained about the situation 10 years after the decentralization of state drug policies (Krcméry *et al.*, 2000):

Ten countries were studied in this survey: Poland, Czech Republic, Slovak Republic, Hungary, Germany, Austria, Slovenia, Croatia, Bulgaria, and Russia. The survey revealed that Czech Republic and Slovak Republic have the most strict antibiotic policies, followed by Poland, Russia, and Croatia. In particular a National Antibiotic Committee at a ministerial level exists in Russia, Czech Republic, and Slovak Republic. Only Czech Republic, Slovak Republic, and Russia have a National Antibiotic Formulary. These same countries have hospital-central guidelines for antibiotic use. In Czech Republic, Slovak Republic, and Croatia there are restrictions in the community prescribing of antibiotics by GPs. Furthermore in all countries except Germany and Austria there are restrictions in the use of some antibiotics in the hospital. In most hospitals third and fourth generation cephalosporins, carbapenems, and glycopeptides are restricted and can be used only after consulting an infectious disease physician or pharmacologist.

The republic of Moldova is one of the new countries born after the breakdown of Soviet Union and can be used as a representative of these new countries. The establishment of market economy and the existence of 1,500 registered pharmacies in a population of 4.35 million citizens led to a flood of medicines on the market in Moldova. Antibiotics are sold mainly over the counter and even on the streets (Cebotarenco, 2001). Doctors are extremely underpaid and can gain money by selling or prescribing antibiotics. Only one sixth or one seventh of antibiotics are recommended by a doctor. In 1995 a non-government, medical organization, the Association DRUGS, was established in Moldova. The purpose of this organization was to provide to every interested person unbiased, up to date information on the safety and efficacy of drugs. Since its beginning the Association DRUGS has organized decades of seminars and training sessions on various aspects of rational drug use in the hospitals. Moreover a drug bulletin was published and distributed free of charge to physicians and pharmacists. Through the Association DRUGS a network was created for rational drug use in Moldova which included authorities from the Ministry of Health, the National Institute of Pharmacy, Medical and Pharmaceutical University of Moldova, pharmacists, physicians, journalists, and non-government organizations.

Given the chronic under funding that covers only 5–7% of the hospitals needs, the strategy to fight antimicrobial resistance included only training programs for physicians in an attempt to change antibiotic prescribing. Such a program was developed in 1998–9 with the support of APUA and the United States Pharmacopoeia in collaboration with the Association DRUGS. The program had five components (Cebotarenco, 2001).

1. Questionnaire assessment of knowledge, attitude, and practice survey (KAPS) of middle and lower lever pediatricians.

2. "Improving antibiotic use" training.
3. Distribution of the manual of antibacterial therapy in Russian to the participants of the training issued by US Pharmacopoeia.
4. Distribution of the APUA Newsletter among paediatricians and publishing the result of the KAPS survey in the informational bulletin of the Association DRUGS.
5. Using mass media to attract the public's attention to the topics of antibiotic rational use and antibiotic resistance.

4.   CONCLUSIONS

Increasing antimicrobial resistance presents a major threat to public health and has caused great concern worldwide. The problem calls for action at a local and global level. Since antibiotic misuse is considered the major condition for the emergence of resistance, many countries have implemented different policies and interventions for controlling antibiotic use most of the time depending on existing resources and infrastructure. Some are simple, based only on education, others are more extensive taking into consideration all aspects involved in antibiotic prescribing. The results of an antibiotic policy should be judged according to its impact on antimicrobial resistance and quality of prescribing and not solely on cost saving. In these terms some national policies have been successful, others not and for some it is still too early to make any judgment, since they are still in the early phase of implementation. However, the effects of many national policies may not be optimal from a global perspective if countries fail to take account of the cross-border effect of their actions (Smith and Coast, 2002). Therefore, besides national policies, international cooperation is a factor that can play a significant role in the battle against antimicrobial resistance. Specifically for Europe, its expansion with the acceptance of new members, gives the challenging opportunity to collectively tackle antimicrobial resistance. This can be done either by implementing uniform European antibiotic policies, or if it is not feasible, by promoting the close cooperation among the European countries on various relative issues.

REFERENCES

Baquero, F. and the Task Force of The General Direction for Health Planning of the Spanish Ministry of Health, 1996, Antibiotic Resistance in Spain: What can be done? *Clin. Infect. Dis.*, **23**, 819–823.
Bauraind, I., Goossens, H., Tulkens, P. M., De Meyere, M., De Mol, P., and Verbist, L., 2001a, A public campaign for a more rational use of antibiotics. 11th European

Conference on Clinical Microbiology and Infectious Diseases (ECCMID), Istanbul, Turkey, Poster No 410.

Bauraind, I., Van den Bremt I, Bogaert, M., Goossens, H., Mouchet, P., Trefois, P. *et al.* 2001b, 41st Interscience Conference on Antimicrobial Agents and Chemotherapy (ICAAC), Chicago, II, 2001, Oral Session LB No 023.

Bengoa, E. L., Sanz, M. M., and de Abajo Iglesias F. J., 2002, Evolución del consumo de antibióticos en España, 1985–2000. *Med. Clin.(Barc)*, **118**, 561–568.

Bergan, T., 2001, Antibiotic usage in Nordic countries. *Int. J. Antimicrob. Agents*, **18**, 279–282.

Bonten, M. J., Kullberg, B. J., Filius, P. M., and De Stichting Werkgroep Antibioticabeleid, 2001, Optimizing antibiotics policy in the Netherlands. VI. SWAB advice: No selective decontamination of intensive care patients on mechanical ventilation. *Ned. Tijdschr. Geneeskd.*, **145**, 353–357.

Bouvenot, G., 1999, French National Institute for observation of prescriptions and consumption of medicines. Prescription and consumption of antibiotics in ambulatory care. *Bull. Acad. Natl. Med.*, **183**, 601–613.

Branthwaite, A. and Pechere, J. C., 1996, Pan-European survey of patients' attitudes to antibiotics and antibiotic use. *J. Int. Med. Res.*, **24**, 229–238.

Bruinsma, N., Filius, P. M. G., De Smet, P. A. G. M., Degener, J., Endtz, Ph., Van den Bogaard, A. E. *et al.*, 2002, Antibiotic usage and resistance in different regions of the Dutch community. *Microb. Drug Resist.*, **8**, 209–214.

Cars, O., Mölstad, S., and Melander, A., 2001, Variation in antibiotic use in the European Union. *Lancet*, **357**, 1851–1853.

Cebotarenco, N., 2001, Moldova: Nongovernmental organizations and antibiotic programmes. In J. L. Avorn, J. F. Barrett, P. G. Davey, S. A. McEwen, T. F. O'Brien, and S. B. Levy (eds.), *Alliance for the Prudent Use of Antibiotics*. Antibiotic Resistance: Synthesis of recommendations by expert policy groups. WHO/CDS/DRS/ 2001.10, pp. 127–129.

Chahwakilian, P., 2000, Politiques d'antibiothérapie à l' échelon national dans divers pays. *Rev. Pneumol. Clin. Suppl*, **1**, 1S9–1S12.

Choutet, P., 2001, Impact of opposable medical references prescription guidelines for antibiotic prescriptions in ambulatory medicine. *Therapie*, **56**, 139–142.

Colin, C., Geffroy, L., Maisonneuve, H., Menard J., Guiraud-Chaumeil, B., Fourquet, F. *et al.*, 1997, Country profile : France. *Lancet*, **349**, 791–797.

Contopoulos-Ioannidis, D. G., Koliofoti I., Koutroumpa, I. C., Giannakakis, I., and Ioannidis, J. P. A., 2001, Pathways for inappropriate dispensing of antibiotics for rhinosinusitis: A randomized trial. *Clin. Infect. Dis.*, **33**, 76–82.

Danzon, P. M. and Chao, L. W., 2000, Cross-national price differences for pharmaceuticals: How large and why? *J. Health Econ.*, **19**, 159–195.

Durieux, P., Gaillac, B., Giraudeau, B., Doumenc, M., and Ravaud, P., 2000, Despite financial penalties, French physicians' knowledge of regulatory practice guidelines is poor. *Arch. Fam. Med.*, **9**, 414–418.

Ekdahl, K., Hanson, H. B., Mölstad, S., Södeström, Walder, M., and Persson, K., 1998, Limiting the spread of penicillin-rsistant Streptococcus pneumoniae: experiences from the South Swedish Pneumococcal Intervention Project. *Microb. Drug Resist.*, **4**: 99–105.

ESAC (European Surveillance of Antibiotic Consumption), 2003. In www. ua.ac.be/ESAC.

ESGAR (European Study Group on Antibiotic Resistance), 1987, In vitro susceptibility to aminoglycoside antibiotics in blood and urine isolates consecutively collected in twenty-nine European laboratories. *Eur. J. Clin. Microbiol.*, **6**, 378–385.

Fijn, R., de Jong-van de Berg, L. T. W., and Brouwers, J. R. B. J., 1999, Rational Pharmacotherapy in The Netherlands: Formulary management in Dutch hospitals. *Pharm. World Sci.*, **21**, 74–79.

Garratini, L. and Tediosi, F., 2000, A comparative analysis of generics markets in five European countries. *Health Policy*, **51**, 149–162.

Giamarellou, H. and Antoniadou, A., 1997, The effect of monitoring of antibiotic use on decreasing antibiotic resistance in the hospital. In Antibiotic Resistance: Origins, Evolution, Selection and Spread. *Ciba Found Symp.*, **207**, 76–86.

Giamarellou, H., Touliatou, K., Koratzanis, G., Petrikkos, G., Kanellakopoulou, K., Lelekis, M. *et al.*, 1986, Nosocomial consequences of antibiotic usage. *Scan. J. Infect. Dis.*, **49**(Suppl.), 182–188.

Giulani, G., Selke, G., and Garattini, L., 1998, The German experience in reference pricing. *Health Policy*, **44**, 73–85.

Gould, I. M., 1996, Hospital antibiotic use and its control—the UK experience. In M. Wolff, A. C. Cremieux, C. Carbon, F. Fachon (eds.), Du bon usage des antibiotiques á l' hôpital. Paris: Arnette Blackwell, 1996:107–119.

Gould, I. M., 2001, United Kingdom: Tackling antibiotic resistance in the UK and Europe. In J. L. Avorn, J. F. Barrett, P. G. Davey, S. A. McEwen, T. F. O'Brien, and S. B. Levy (eds.), *Alliance for the Prudent Use of Antibiotics*. Antibiotic Resistance: Synthesis of recommendations by expert policy groups. WHO/CDS/DRS/ 2001.10 :149–152.

Gould, I. M., 2002, Antibiotic policies and control of resistance. *Curr. Opin. Infect. Dis.*, **15**, 395–400.

Gould, I. M. and Jappy, B., 2000, Trends in hospital antimicrobial prescribing after 9 years of stewardship. *J.Antimicrob. Chemother.*, **45**, 913–917.

Harbarth, S., Albrich, W., Brun-Buisson, C., 2002, Outpatient antibiotic use and prevalence of antibiotic-resistant pneumococci in France and Germany: A sociocultural perspective. *Emerg. Infect. Dis.*, **8**, 1460–1467.

Huttin, C, 1996, A critical review of the remuneration systems for pharmacists. *Health Policy*, **36**, 53–68.

Janknegt, R., Monkelbaan, J. F., Stobberingh, E. E., and Wijnands, W. J. A., 1994, Antibiotic policies in Dutch hospitals for the treatment of patients with serious infection. *J. Antimicrob. Chemother.*, **34**, 1059–1069.

Jenkins, T., 1998, Advice of the economical and social committee on, Antibiotic resistance: A threat for public health. *Off. J. Eur. Community*, **C407(02)**, 7–17.

Keuleyan, E., Gould, I. M., 2001, Key issues in developing antibiotic policies: from an institutional level to Europe-wide. European Study Group on Antibiotic Policy (ESGAP), Subgroup III. *Clin. Microbiol. Infect.*, 7(Suppl. 6), 16–21.

Krcméry, V., Jeljaszewicz, J., Grzesiowski, P., Hryniewicz, W., Metodiev, K., Stratchounski, L. *et al.*, 2000, National and local antibiotic policies in Central and Eastern Europe. *J. Chemother.*, **12**, 471–474.

Kristinsson, K. G., 1999, Modification of prescribers' behavior: The Icelandic approach. *Clin. Microbiol. Infect.*, **5**, 4S43–4S47.

Kristinsson, K. G., Hjalmarsdottir, M. A., and Gudnason, T., 1998, Continued decline in the incidence of penicillin non-susceptible pneumococci in Iceland, [abstract C-22] In 38th ICAAC, San Diego, American Society for Microbiology, 74.

Lelekis, M., Pachios, G., Goneos, J., Perdios, J., Gargalianos, P., and Kosmidis, J., 1993, Study of antibiotic use in Greek hospitals. In *Recent Advances in Chemotherapy*. American Society for Microbiology 1993, p.107–108.

Lelekis, M., Chini, M., Vasilopoulos, J., Gika, M., Tsogas, N., Katsimpris, K. *et al.* 2003, Study of the acceptance and the use by nosocomial doctors of the "Guide for antimicrobial chemotherapy and prophylaxis for the hospitalized patient". 6th Hellenic Congress of Infectious Diseases, Athens, Abstract No. 46.

Le Pen, C., 1997, Pharmaceutical economy and the economic assessment of drugs in France. *Soc. Sci. Med.*, **45**, 635–643

Levy, S. B., 2001, Antibiotic resistance: Consequences of inaction. *Clin. Infect. Dis.*, **33** (Suppl. 3), S124–S129

Mölstad, S. and Cars O., 1999, Major change in the use of antibiotics following a national programme: Swedish Strategic programme for the rational use of Antimicrobial agents and surveillance of resistance (STRAMA). *Scan. J. Infect. Dis.*, **31**, 191–195

Monnet, D. L. and Sørensen, T. L., 1999, Interpreting the effectiveness of a national antibiotic policy and comparing antimicrobial use between countries. *J. Hosp. Infect.*, **43**, 239–248.

Moro, M. L., Petrosillo, N., and Gandin, C., 2003, Antibiotic policies in Italian hospitals: Still a lot to achieve. *Microb. Drug Resist.*, **9**, 219–222.

Nagler, J., Peetermans, W., Goosens, H., and Struelens, M., 1999, National Committee for the coordination of antibiotic policy in Belgium: The next step in the fight against antimicrobial resistance. *Acta Clin. Belg.*, **54**, 319–320.

Nathwani, D., 1999, How do you measure the impact of an antibiotic policy? *J. Hosp. Infect.*, **43**(Suppl.): S265–S268.

National Disease Surveillance Centre, 2001, A strategy for the control of antimicrobial resistance in Ireland (SARI). Report of the subgroup of the Scientific Advisory Committee of the National Disease Surveillance Centre, Dublin: NDSC, 2001.

Pradier, C., Rotily, M., Cavailler, P., Haas, H., Pesce, A., Dellamonica, P. *et al.*, 1999, Factors related to the prescription of antibiotics for young children with viral pharyngitis by general practitioners and pediatricians in southeastern France. *Eur. J. Clin. Microbiol. Infect. Dis.*, **18**, 510–514.

Rautakorpi, U.-M., Huovinen, P., and Klaukka, T., 1997, Is the duration of antibiotic treatment too long? *Suom. Lääkäril.*, **52**, 3228–3230.

Rautakorpi, U.-M., Klaukka, T., Honkanen, P., Mäkelä, M., Nikkarinen, T., Palva, E. *et al.*, on behalf of the MIKSTRA Collaborative Study Group, 2001, Antibiotic use by indication: A basis for active antibiotic policy in the community. *Scand. J. Infect. Dis.*, **33**: 920–926.

Seppälä, H., Klaukka, T., Vuopio-Varkila, J., Muotiala A, Helenius H, Lager K *et al.*, 1997, The effects of changes in the consumption of macrolide antibiotics on erythromycin resistance in group A streptococci in Finland. Finnish Study Group for Antimicrobial Resistance. *New Engl. J. Med.*, **337**, 441–446.

Smith, R. D. and Coast, J., 2002, Antimicrobial resistance: A global response. *Bull. World Health Organ.*, **80**, 126–132.

Sørensen, T. L. and Monnet, D. L., 2000, Control of antibiotic use in the community: The Danish experience. *Infect. Control Hosp. Epidemiol.*, **21**, 387–389.

Standing Medical Advisory Committee, 1998, The path of least resistance. Department of Health, London.

Stobberingh, E. E., Janknegt, R., and Wijnands, G., 1993, Antibiotic guidelines and antibiotic utilization in Dutch hospitals. *J. Antimicrob. Chemother.*, **32**, 153–161.

Struelens, M. J. and Peetermans, W. E., 1999, The antimicrobial resistance crisis in hospitals calls for multidisciplinary mobilization. *Acta Clin. Belg.*, **54**, 2–6.

van Kasteren, M. E., Wijnands W. J., Stobberingh, E. E., Janknegt, R., Verbrugh, H. A., and van der Meer, J. W., 1998a, Optimizing antibiotics use in the Netherlands. I. The Netherlands Antibiotics Policy Foundation (SWAB). *Ned. Tijdschr. Geneeskd.*, **142**, 949–951.

van Kasteren, M. E., Wijnands W. J., Stobberingh, E. E., Janknegt, R., and van der Meer, J. W., 1998b, Optimization of the antibiotics policy in the Netherlands.II. SWAB guidelines for the antimicrobial therapy of pneumonia in patients at home and as nosocomial infections. The Netherlands Antibiotic Policy Foundation. *Ned. Tijdschr. Geneeskd.*, **142**, 952–956.

van Kasteren, M. E., Wijnands, W. J., Stobberingh, E. E., Janknegt, R., and van der Meer, J. W.,1998c, Optimizing of antibiotics policy in the Netherlands. III. SWAB guidelines for antimicrobial therapy in adults hospitalized with bronchitis. Foundation Antibiotics Policy Work Group. *Ned. Tijdschr. Geneeskd.*, **142**, 2512–2515.

van Kasteren, M. E., Stobberingh, E. E., Janknegt, R., Wijnands, W. J., and van der Meer, J. W., 1999, Optimizing the antibiotics policy in the Netherlands. IV. SWAB- guidelines for antimicrobial therapy of adults with sepsis in hospitals. Foundation Antibiotics Policy Work Group. *Ned. Tijdschr. Geneeskd.*, **143**, 611–617.

van Kasteren, M. E., Gyssens, I. C., Kullberg, B. J., Bruining, H. A., Stobberingh, E. E., and Goris, R. J., 2000, Optimizing antibiotics policy in the Netherlands. V. SWAB guidelines for perioperative antibiotic prophylaxis. Foundation Antibiotics Policy Team. *Ned. Tijdschr. Geneeskd.*, **144**, 2049–2055.

van Kasteren, M. E., Kullberg, B. J., de Boer, A. S., Mintjes-de Groot, J., and Gyssens, I. C., 2003, Adherence to local hospital guidelines for surgical antimicrobial prophylaxis: A multicenter audit in Dutch hospitals. *J. Antimicrob. Chemother.*, **51**, 1389–1396.

Vatopoulos, A., Kalapothaki V., and Legakis N. J., 1999, An electronic network for the surveillance of antimicrobial resistance in bacterial nosocomial isolates in Greece. The Greek Network for the Surveillance of Antimicrobial Resistance. *Bull. World Health Organ.*, **77**, 595–601.

Woodhead, M., Gialdoni-Grassi, G., Huchon, G. J., Leophonte, P., Manresa, F., and Schaberg, T., 1996, Use of investigations in lower respiratory tract infection in the community: A European survey. *Eur. Respir. J.*, **9**, 1596–1600.

Working Party Report, 1994, Hospital antibiotic control measures in the UK. *J. Antimicrob. Chemother.*, **34**, 21–42.

Chapter 29

Antibiotic Use in the Community

Sigvard Mölstad[1] and Otto Cars[2]
[1]Unit of Research and Development in Primary Care, Jönköping Sweden;
[2]Department of Infectious Diseases, University Hospital, Uppsala, Sweden

Antibiotic use and misuse is one of the key drivers of antibiotic resistance and monitoring of antibiotic consumption is an essential tool in the strategy to contain this problem. Although publicly available data on antimicrobial use has increased in recent years, there is still a lack of comprehensive data on antibiotic consumption for most countries. In many parts of the world national surveillance systems are lacking and from large areas of the world, no reliable data on antibiotic use is available. In many developing countries the lack of access to essential pharmaceuticals, over the counter sales, and suboptimal dosage are major problems.

There are few population-based studies that address antibiotic use in the community. Varying time periods, data collection system, and units of measurements make it difficult to compare trends in antibacterial usage between countries. The objective of this chapter is not to make a systematic review of the literature, but to comment on some of the recently published papers where comparable data are presented.

1.   UNITS OF MEASUREMENTS

There is currently no consensus on the unit of measurement to compare the consumption of antibiotics between hospitals or communities. An increasing number of publications, however, use the system defined by the WHO Collaborating Centre for Drug Statistics Methodology (1999) where antibiotic

Antibiotic Policies: Theory and Practice. Edited by Gould and van der Meer
Kluwer Academic / Plenum Publishers, New York, 2005

use is expressed in defined daily dose (DDD) according to the Anatomic Therapeutic Chemical (ATC) classification system. Since both the classification of drugs within the ATC system and DDD may change over time, it is important to consider such changes when analysing time trends.

Defined daily dose is a unit based on the average adult dose used for the main indication of the drug, standardised by WHO. To make comparisons between countries or geographical areas possible, the number of DDDs per 1,000 inhabitants and day (DID) can be calculated. This measure will be influenced by several factors, for example, the dosage and duration of treatment, which may vary between countries even for the same indication. Since the DDD is based on the average dose for adults it may underestimate the number of treatment courses in children. To be able to adjust for demographic differences between and within countries, data should therefore be available for different age groups.

Another measure of antibiotic use is the number of antibiotic prescriptions per 1,000 inhabitants and year. This measure is probably more appropriate when evaluating antibiotic use in children. A third unit of measurement is the number of antibiotic prescriptions per office visits and/or per diagnosis. Data on indications for prescriptions are essential to assess the compliance with treatment guidelines. Such data are also valuable to evaluate if a change in prescribing over time in an area is the result of changes in the physicians prescribing pattern or in the consultation pattern of the population. In most developed countries, approximately 80–90% of antibiotics used in humans are for ambulatory care (outpatients) and 10–20% for hospital care (inpatients). In comparisons between countries and regions it should be borne in mind that the allocation of antibiotic use (community vs hospital) may vary, for example, for nursing homes, day care centres, and prescriptions by dentists.

2.   DDD/1,000 INHABITANTS AND DAY (DID)

2.1.   Europe

2.1.1.   European Union

Data for non-hospital national sales for 1993 and 1997 was obtained for all 15 Members States of the European Union. For 13 countries (Austria, Belgium, Germany, Finland, France, Greece, Italy, Ireland, Luxembourg, Netherlands, Spain, Portugal, and the United Kingdom) from the Institute for Medical Statistics (IMS) and for Sweden and Denmark from the National Corporation of Swedish Pharmacies and the Medical Product Agency, respectively (Cars *et al.*, 2001). IMS collects data from different sources, including wholesalers, manufacturers, pharmacies, physicians, and hospitals, and calculates national estimates. Data from Sweden and Denmark included all outpatient sales from community pharmacies. Between 1993 and 1997, large increases were noted in Italy (34%) and Luxembourg (12%), and a reduction was seen in five countries: Sweden had the largest (21%) and Greece the smallest (4%). Total sales of antibiotics in 1997 varied 4-fold between the countries: France (36.5 DID), Spain (32.4 DID), Portugal (28.8 DID), and Belgium (26.7 DID) had the highest sales, whereas the Netherlands (8.9 DID), Denmark (11.3 DID), Sweden (13.5 DID), and Germany (13.6 DID) had the lowest. In Figure 1, these data were compared with previously published data for France, Australia, United States, Canada, United Kingdom, and West Germany (McManus *et al.*, 1997). There were also profound variations in use of different classes of antibiotics. In 11 of the 15 countries, the most commonly used antibiotics were broad-spectrum penicillin, which varied between 56% (Spain) and 20% (Germany) of total sales. In Finland, the most common drug was a tetracycline (28%), in Austria a macrolide (26%), and in Denmark and Sweden narrow-spectrum penicillins (40% and 36%, respectively).

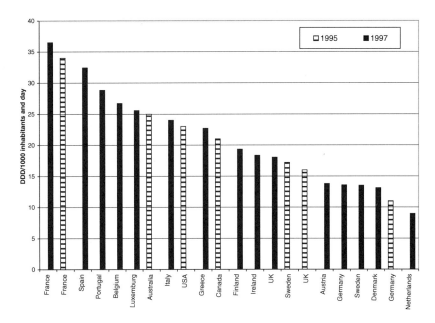

Figure 1. Antibiotic consumption in DDD/1,000 inhabitants and day (DID) in 1997 in 15 EU countries (France, Spain, Portugal, Belgium, Luxemburg, Italy, Greece, Finland, Ireland, United Kingdom, Austria, Germany, Sweden, Denmark, and the Netherlands) (Cars *et al.*, 2001) and for France, Australia, United States, Canada, United Kingdom, and West Germany in 1994 (Intercontinental Medical Service, Melbourne; McManus *et al.*, 1997).

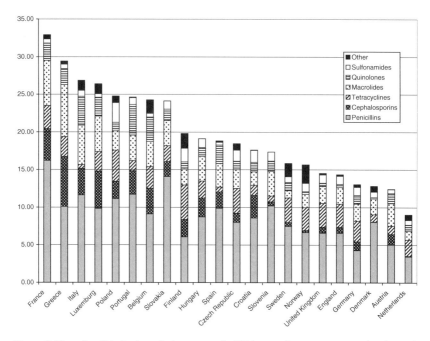

Figure 2. Use of antibiotics in ambulatory care in 2001 according to a retrospective data collection by ESAC.

The European Surveillance of Antimicrobial Consumption (ESAC) is a recently created international network of national surveillance systems funded by the European Commission aiming to collect comparable and reliable data on antibiotic consumption in European Countries. The first report showed retrospective data for 1997–2001 from ambulatory care from 21 countries in Europe (www.ua.ac.be/ESAC). Use in Europe remained at a high average level in the participating countries of about 20 DID for the study period. In 2001, the use varied between 9.2 DID (Netherlands) and 32.9 (France). Finland, Sweden, Norway, Denmark, the Netherlands, and Latvia were low consumers using relatively narrow-spectrum penicillins and tetracyclines more extensively and less cephalosporins and quinolones (Figure 2). Southern European countries (Portugal, Italy, Greece, and France) were high consumers using broad-spectrum penicillins, and exceptionally high proportions of cephalosporins, macrolides, and quinolones.

2.1.2.   Nordic countries

In the Nordic countries (Denmark, Finland, Iceland, Norway, and Sweden) all drugs are distributed through a national network of pharmacies, which

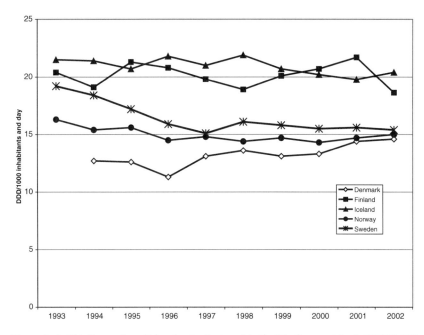

Figure 3. Antibiotic use (hospital and outpatient use) in the Nordic countries in DDD/1,000 inhabitants and day (DID), between 1993 and 2002. For Denmark, 1994–6, only outpatient care was included.

enables a precise follow-up of the amounts of different antibiotics dispensed. Iceland and Finland have the highest use and Denmark the lowest use of antibiotics. The use of antibiotics has remained steady during the last decade, but a decrease has been noted in Sweden between 1993–7 (Swedres, 2002; Figure 3).

2.1.3. Eastern Europe

In 1998, information on used amount of antibiotics was gathered from different sources in the respective countries, including pharmaceutical manufacturers' offices, pharmaceutical market research companies, and Ministries of Health. The pattern of total antibiotic use (ambulatory and hospital use) in the studied countries varied markedly, from 25.8 DID in Slovakia, 22.5 in Poland, 21.1 in Hungary, 11.2 in Russia, and 8 in Belarus (Stratchounski *et al.*, 2001). Russia and Belarus had a lower level of consumption and used lesser quantities of penicillins, cephalosporins, macrolides, and quinolones but greater amounts of aminoglycosides and chloramphenicol.

2.1.4.   Australia

Data on sales of oral antibiotics to retail and hospital markets was obtained from a commercial market research organisation, Intercontinental Medical Service, Melbourne. Antibiotic use remained steady between 1990 and 1995, with an estimated 24.7 DID in 1990 and 24.8 DID in 1995 (McManus *et al.*, 1997). Using the same data source, a comparison with France, United States, Canada, United Kingdom, and West Germany was reported. In 1995, France had the highest use, approximately 34 DID, Australia 25, United States 23, Canada 21, United Kingdom 16, and West Germany 11 DID (Figure 1).

3.   PRESCRIPTIONS/1,000 INHABITANTS AND YEAR

3.1.   European Union

Information on the number of prescriptions was obtained for Austria, Belgium, Finland, France, Germany, Greece, Italy, the Netherlands, Portugal, Spain, and the UK from IMS (Mölstad *et al.*, 2002). For Denmark and Sweden information was obtained from the Danish Medicines Agency and the National Corporation of Swedish Pharmacies. Between 1994 and 1997, the number of prescriptions per 1,000 inhabitants increased in France and Greece while Portugal, Spain, and Sweden reported a decrease. In 1997, Greece (1,350), Spain (1,320), and Belgium (1,070) had the highest numbers of antibiotic pre-scriptions per 1,000 inhabitants while the Netherlands (390), Sweden (460), and Austria (480) had the lowest. The most common antibiotics were extended-spectrum penicillins in 6 out of 13 countries, macrolides in Austria, Finland, and Germany, narrow-spectrum penicillins in Denmark and Sweden, and cephalosporins in Greece.

3.2.   United States

In a sample survey 2,500–5,000 office-based physicians reported data on office visits between 1980 and 1992 for all ages, including information on antimicrobial drug prescribing (McCaig and Hughes, 1995). From 1980 to 1992, no major change in the frequency of prescriptions for all anti-microbial drugs was registered. However, a relative increase in the number of prescriptions/1,000 inhabitants and year was found for the more expensive, broad-spectrum antimicrobial drugs, such as cephalosporins and decreasing rates were observed for less expensive drugs with narrower-spectrum, such as penicillins. No trend was found for trimethoprim-sulfamethoxazole, erythromycin, or tetracyclines.

In a sample survey between 1992–2000, 1,100–1,900 physicians and 200–300 outpatient departments provided information on antibiotic prescriptions. Between 1992 and 2000, the number of antimicrobial prescriptions decreased from 151 million to 126 million (McCaig *et al.*, 2002). During the same period the number of physician visits increased slightly from 908 millions to 1.0 billion. Antimicrobial prescriptions declined by 23%, from 599 to 461/1,000 inhabitants and year. The prescribing per visit declined by 25%, from 166 to 125 antimicrobial drug prescriptions/1,000 visits and year. The decline in prescribing rates varied between different age groups and was 32% for patients <15 years of age, 9% in the age group 15–24 years, and 17% for patients 25–44 years of age. No decline was observed in the age group >45.

There was a decrease in the use of amoxicillin/ampicillin, cephalosporins, and erythromycin (−43%, −28%, −76%, respectively). Decreasing trends were also found for other penicillins, tetracyclines, and trimethoprim-sulfamethoxazole. There was increase in the use of azithromycin /clarithromycin, quinolones, and amoxicillin/clavulanate (+388%, +78%, and +72%, respectively). The most often used antibiotics in year 2000 were amoxicillin/ ampicillin followed by cephalosporins, azithromycin/clarithromycin, quinolones, and amoxicillin/clavulanate.

3.3. Canada

Data reported by IMS Health, Canada, showed that prescribing of antibiotics increased steadily in Canada between 1990 and 1995 to reach 27.3 million (www.imshealthcanada.com). By the end of 1999, the number had decreased between 1990 and 1995 by close to 2 million to 25.5 million. The most commonly dispensed antibiotic, amoxicillin, decreased each year, while second-line antibiotics increased by 23% in 1999. Approximately 800 prescriptions per 1,000 inhabitants were dispensed in 1999, but with regional variation between 750 and 1,190 prescriptions per 1,000 inhabitants and year were dispensed.

The Drug Information Network is a prescription database, which reflects drug use for the population in the Canadian province Manitoba. Total use of antibiotics declined by 13.7% between 1995 and 1998 for non-institutionalised population (Carrie *et al.*, 2000). In 1995, 899 prescriptions per 1,000 inhabitants were issued, a figure that decreased to 778 in 1997. In 1997, penicillins were the most commonly used class of antibiotics, representing almost 50% of total antibiotic prescriptions, followed by macrolides (8.7%), sulfonamides (12.5%), cephalosporins (8.7%), tetracyclines (5.4%), and fluoroquinolones

(4%). The most commonly prescribed antibiotics showed a decrease: amoxicillin −17%, erythromycin −29%, trimethoprim-sulfamethoxazole −18.7%, narrow-spectrum penicillins −19.2% and tetracycline −18.9%. Antibiotics used more frequently included ciprofloxacin, +21.9%, cefuroxime +30.7% and azithromycin/clarithromycin +29.5%.

3.4.   Australia

Data on dispensed antibiotics were obtained from a database, monitored by the Drug Utilization Sub-Committee (McManus *et al.*, 1997). The database contains information on all subsidised prescriptions together with an estimate of non-subsidised prescriptions from a sample of about 250 community pharmacies. In 1995, amoxicillin, though declining, remained the most dispensed antibiotic followed by amoxicillin + clavulanate, cefaclor, doxycycline, and cefalexin. The use of quinolones was low (2.2%), most likely because of prescribing restrictions. The overall antibiotic pattern in Australia was similar to that in the United Kingdom.

4.   INDICATIONS FOR ANTIBIOTIC PRESCRIPTIONS

Available population-based studies which describe antibacterial use by indication use physician surveys or prescription databases as the data collection mechanism. In most studies the most common indication for an antimicrobial prescription is a respiratory tract infection (RTI) (60–70%) followed by urinary tract infections (UTI) (10–15%), and skin and soft tissue infections (10%). In children, more than 90% of infectious episodes are respiratory tract infections (including otitis media) and children also receive antimicrobial treatment more often than adults. In the elderly urinary tract infections becomes more prevalent as a diagnosis and increasing utilisation of antimicrobials are observed in the veteran population.

4.1.   Europe

In the comparative studies from countries in Europe there are no national data on indications for prescribing, dose, or duration of treatment. However there are some studies on diagnosis for prescriptions in outpatient care.

In the West Midlands General Practice Research Database, United Kingdom, there was between 1993 and 1997 an overall decrease in prescribing from 963 to 807 prescriptions per 1,000 patients and year (Frischer *et al.*,

2001). Non-specific lower respiratory tract infections and throat infections accounted for the main decreases in antibiotic prescribing.

In Norway, the most commonly recorded diagnosis was urinary tract infections, followed by acute bronchitis, ear infections, and non-specific upper respiratory tract infections in 1989 (Straand, 1998). The most prescribed antibiotics were narrow-spectrum penicillins (29%), followed by tetracyclines (24%), trimethoprim-sulfamethoxazole (17%), and erythromycin (12%). Narrow-spectrum penicillin was used in a majority of ear infections, tonsillitis, non-specific upper respiratory tract infections and sinusitis. Tetracyclines were most often prescribed for acute bronchitis and pneumonia and trimethoprim-sulfamethoxazole for urinary tract infections.

In Sweden, a 1-week survey on 7,700 visits for infectious diseases in five counties was conducted in the year 2000 (Stalsby *et al.*, 2002). Respiratory tract infections accounted for 70% of the diagnoses, of which 54% were prescribed an antibiotic, of which narrow-spectrum penicillin accounted for 62% of prescriptions, followed by tetracycline (14%). Most cases of acute otitis media and acute tonsillitis were prescribed narrow-spectrum penicillin. In acute bronchitis, 50% of cases were prescribed an antibiotic, of which tetracycline was the most common followed by narrow-spectrum penicillin and amoxicillin. In urinary tract infections, trimethoprim, pivmecillinam, and fluoroquinolones accounted for one third each. About 50% of skin and soft tissue infections were treated with isoxazolylpenicillins whereas cefalosporins were used in 12%.

In Finland, indications for antibiotic prescribing has been studied in a point prevalence study conducted in 30 Health Centres (Rautakorpi *et al.*, 2001). A total of 7,800 visits for infections were recorded. The most common cause for a visit was a respriratory tract infection (74%), followed by skin/wound infections and urinary tract infections (both 6%). Of the otitis media, 53% were treated with amoxicillin, 16% with a macrolide, and 16% with co-trimoxazole. Patients with acute bronchitis received antibiotic treatment in 70% of cases, mostly macrolides (39%) and doxycycline (36%).

4.2. United States

In a sample survey 2,500–5,000 office-based physicians reported data on office visits, including information on antimicrobial drug prescribing between 1980 and 1992 (McCaig and Hughes, 1995). During the years, an increasing trend in the visit rate to office-based physicians for otitis media was observed, while the visit rate for sinusitis among adults was found to be higher in 1992 than in each of the other study years. The five leading diagnoses for which oral antibiotics were prescribed were otitis media, upper respiratory tract infection, bronchitis, pharyngitis, and sinusitis.

A sample of community-based physicians in the National Ambulatory Medical Care Survey was used to collect data on 60,252 visits in 1991–2, 62,169 visits in 1994–5, and 37,467 visits in 1998–9 (Steinman *et al.*, 2003). The estimated annual national number of prescriptions decreased from 230 million prescriptions in 1991–2 to 190 million prescriptions in 1998–9. Antibiotics were less frequently used in 1998–9 to treat acute respiratory tract infections, such as the common cold and pharyngitis. However, use of broad-spectrum agents increased from 24% to 40% of antibiotic prescriptions in adults and from 23% to 40% in children. For common cold and unspecified upper respiratory tract infection adult use of broad-spectrum antibiotics more than doubled. For adults, use of azithromycin and clarithromycin also increased from 1% to 16% of prescription. During the period, the use of fluoro-quinolones increased from 17% to 35% of antibiotics prescribed for urinary tract infections and from less than 1% to 13% of antibiotics to treat common cold and unspecified upper respiratory tract infection. Among children, use of broad-spectrum antibiotics increased for otitis media and more than doubled for common cold and unspecified upper respiratory tract infection. In 1998–9, nonpneumonic acute respiratory tract infections accounted for 54% of adult and 77% of paediatric prescriptions for broad-spectrum antibiotics. In addition, nonpneumonic acute respiratory infection accounted for at least two thirds of prescriptions for azithromycin/clarithromycin, amoxicillin + clavu-lanate, and second and third generation cephalosporins. Also substantial regional differences in antibiotic use were noted.

In a sample survey, 2,500–3,500 office-based physicians reported data on 6,500–13,600 paediatric visits during 2-year periods from 1989 to 1990 through 1999 to 2000 (McCaig *et al.*, 2002), population and visit-based antimicrobial prescribing rates were calculated for children and adolescents younger than 15 years. Respiratory tract infections (otitis media, pharyngitis, bronchitis, sinusitis, and upper respiratory tract infection) represented 75% of all antimicro-bial prescriptions. The average number of prescriptions per 1,000 children and adolescents younger than 15 years decreased from 838 in 1989–91 to 503 in 1999–2000. The visit-based rate decreased from 330 antimicrobial prescriptions per 1,000 office visits to 234. For the five major diagnoses of respiratory tract infections, the population-based prescribing rate decreased from 674 to 379 per 1,000 children and adolescents younger than 15 years, and the visit-based from 715 to 613 prescriptions per 1,000 office visits. Both population-based and visit-based prescribing rates decreased for pharyngitis and upper respiratory tract infection; however, for otitis media and bronchitis, declines were only observed in the population-based rate. This indicates that fewer visits were made because of bronchitis and otitis media, but those who consulted were prescribed antibi-otics as often as before. Prescribing rates for sinusitis remained stable.

In the same report, similar trends were found for children <5 years old, as for those <15 years old (McCaig *et al.* 2002). The overall prescribing decreased by 40%, from 1,422 antimicrobial prescriptions per 1,000 children younger than five years in 1989–90 to 851 in 1999–2000. A similar decreasing trend was also found for the five respiratory tract infections combined (otitis media, pharyngitis, bronchitis, sinusitis, and upper respiratory tract infection) between 1989–90 and 1999–2000, representing a decrease of 43%, from 1,184 to 678 antimicrobial prescriptions per 1,000 children. Antibiotic prescribing decreased between 1989–90 and 1999–2000 for all five included respiratory tract infections; for otitis media (42% decrease, from 722 to 418), pharyngitis (51% decrease from 224 to 109), bronchitis (71% decrease from 112 to 32), and upper respiratory tract infections (40% decrease from 120 to 72). In this study, it was not possible to link diagnosis to a particular drug.

4.3.   Canada

Information on indications was retrieved from The Drug Information Network, a prescription database, which reflects drug use for the Manitoba population. For new cases of upper respiratory tract infection or pharyngitis, an antibiotic was recorded for 57% of urban patient encounters and for 73% of rural patient encounters (Carrie *et al.*, 2000). For sinusitis the most prescribed antibiotics was doxycycline (21%), amoxicillin–clavulanate (18%), and cefaclor (15%). For bronchitis, the most prescribed antibiotics was amoxicillin (18%), followed closely by roxithromycin (16.5%) and cefaclor. In urinary tract infections TMP-SMZ (28.5%) was most commonly prescribed, followed by cephalexin (18.9%), and amoxicillin–clavulanate (17.2%).

4.4.   Australia

Diagnoses for which patients were prescribed antibiotics were obtained from a survey, based on a sample of 420 general practitioners, stratified in line with the total population by age, location, and practice size (McManus *et al.*, 1997). In 1995, for sinusitis, the most prescribed antibiotics were tetracycline (21%), amoxicillin–clavulanate (18%), and cefaclor (15%). For otitis media, the most prescribed antibiotics were cefaclor (36%), amoxicillin (21%), and amoxicillin–clavulanate (21%) and for bronchitis, amoxicillin (18%) was followed by roxithromycin (17%) and cefaclor (15%). In urinary tract infections, trimethoprim-sulfamethoxazole (29%) was most commonly prescribed followed by cephalexin (19%), and amoxicillin–clavulanate (17%).

5.   POSSIBLE CAUSES FOR OBSERVED VARIATIONS IN ANTIBIOTIC USE

The large variations in antibiotic use in DID or numbers of prescriptions per 1,000 inhabitants and year comparing countries are difficult to explain by medical reasons alone. There is no comprehensive review on guidelines for the treatment of infectious diseases comparing different countries. But differences in antibiotic use may to some extent reflect differences in national guidelines or recommendations, for example, choice of antibiotics, dosage, and length of treatment. In addition, there are different recommendation on antibiotic use for identical indications. For example, antibiotics are recommended for uncomplicated otitis media in children in most countries but not for children older than 6 months in the Netherlands. In addition, different dosage may explain some of the differences in DID for β-lactam antibiotics, since higher dosage may be needed in areas with high prevalence of resistance. But this does not explain the difference in number of prescriptions per 1000 inhabitants, comparing countries, or the large regional differences within countries. In addition, studies show that the number of DIDs, numbers of antibiotic prescriptions, and choice of antibiotic class by general practitioners may differ not only between different countries but also between individual physicians within a country (Huchon *et al.*, 1996; Melander *et al.*, 2000; Örtqvist, 1995; Veninga *et al.*, 2000).

A large part of the differences between countries probably represent irrational use of antibiotics. Several non-medical factors may influence the use of antibiotics, for example, differences in healthcare systems (Basky, 1999), antibiotic dosage regimens, patient expectations and attitudes towards taking drugs (Branthwaite and Pechere, 1996; van Duijn *et al.*, 2003), cultural differences (Deschepper *et al.*, 2002), influence of the pharmaceutical industry or over-the counter sales. It has been suggested that the differences in antibiotic usage in Western Europe might be primarily explained by differences in health systems and that a high number of physicians in a country is associated with a high utilisation of antibiotic drugs (Huchon *et al.*, 1996; Veninga *et al.*, 2000). Antibiotics can also, in some countries in Europe, be obtained at pharmacies without a doctor's prescription, but data on the quantities and qualities of such sales are scarce (Bremon *et al.*, 2000; Contopoulos-Ioannidis *et al.*, 2001).

The available studies on antibiotic use in the community indicate that prescribing increased during the 1980s. Highest use in DID was reported from France, Greece, Belgium, and Australia. Highest rate of prescribing was reported from Greece, Spain, and Belgium. In a few countries (United States, Sweden, United Kingdom, and Canada) a decrease in prescribing of antibiotics has been reported after 1995.

Narrow-spectrum penicillin is still the preferred antibiotic in Scandinavian countries and amoxicillin in most other countries. But in most countries there has been an evident decline in the use of penicillins and a marked increase in the prescribing of broad-spectrum antibiotics such as macrolides, cephalosporins and quinolones. This general trend, descibed in a review (Carrie and Zhanel, 1999) and also from the Netherlands (Kuyvenhoven *et al.*, 2003), should be of concern to the medical community. Respiratory tract infections are the most common reason for an antibiotic prescription, the majority of which have viral etiology.

Better data are needed to compare antibiotic use in different countries and areas within countries with outcome variables such as resistance, morbidity, and mortality. International cooperation is needed to compare healthcare systems, national guidelines and indications for antibiotic prescribing in different age groups, as well as dosage and treatment duration. The quality of antimicrobial prescribing in any country, however, cannot be determined by these international comparisons alone. Each country must determine the appropriateness of its antimicrobial use and work within its constraints to optimise antimicrobial use.

6. CONCLUSIONS

Sufficient data on antibiotic use in the community is lacking in most countries. Few countries can show antibiotic use data over time, to analyse trends in prescribing. All countries have an important role to ensure that validated national and regional, data on antibiotic prescribing and sales are made publicly available. The collection of DID data in the ESAC project is an excellent start, but to evaluate if antibiotic prescribing is rational, data on number of prescriptions, indications, dose and duration of treatment, as well as different age groups is essential. Such data are needed in order to evaluate the impact of antibiotic prescribing on resistance, morbidity, complications, and mortality.

REFERENCES

Basky, G., 1999, Fee for service doctors dispense more antibiotics in Canada. *BMJ*, **18**, 1232.

Branthwaite, A. and Pechere, J. C., 1996, Pan-European survey of patients' attitudes to antibiotics and antibiotic use. *J. Int. Med. Res.*, **24**, 229–238.

Bremon, A. R., Ruiz-Touvaar, M., Gorricho, B. P., de Torres, P. D., and Rodrigues, R. L., 2000, Non-hospital consumption of antibiotics in Spain: 1987–1997. *J. Antimicrob. Chemother.*, **45**(3), 395–400.

Carrie, A., Colleen, M., and Zhanel, G., 2000, Antibiotics use in a Canadian Province, 1995–1998. *Ann. Pharmacother.*, **34**, 459–64.

Carrie, A. G. and Zhanel, G. G., 1999, Antibacterial use in community practice: Assessing qantity, indications and appropriateness, and relationship to the development of antibacterial resistance. *Drugs*, 6, 871–881.

Cars, O., Mölstad, S., and Melander, A., 2001, Variation in antibiotic use in the European Union. *Lancet*, 357, 1851–1853.

Contopoulos-Ioannidis, D. G., Koliofoti, I. D., Koutroumpa, I. C., Giannakakis, I. A., and Ioannidis, J. P., 2001, Pathways for inappropriate dispensing of antibiotics for rhinisinusitis: A randomized trial. *Clin. Infect. Dis.*, 33, 76–82.

Deschepper, R., Vander Stichele, R., and Haaijer-Ruskamp, F., 2002, Cross-cultural differences in lay attitudes and utilisation of antibiotics in a Belgian and a Dutch city. *Patient Educ. Couns.*, 48, 161–169.

Frischer, M., Heatlie, H., Norwood, J., Bashford, J., Millson, D., and Chapman, 2001, Trends in antibiotic prescribing and associated indications in primary care from 1993 to 1997. *J. Public Health Med.* 23, 69–73.

Huchon, G. K., Gialdroni-Grassi, G., Leophonte, P., Manresa, F., Schaberg, T., and Woodhead, M., 1996, Initial antibiotic therapy for lower respiratory tract infection in the community: A European survey. *Eur. Respir. J.*, 9, 1590–1595.

Kuyvenhoven, M., van Balen, F., and Verheij, T., 2003, Outpatient antibiotic prescriptions from 1992 to 2001 in The Netherlands. *J. Antimicrob. Chemother.*, Sept. 1.

McCaig, L., Besser, R., and Hughes, J., 2002, Trends in antimicrobial prescribing rates for children and adolescents. *JAMA*, 287, 3096–3102.

McCaig, L., Besser, R., and Hughes, J., 2003, Antimicrobial drug prescriptions in ambulatory care settings, United States, 1992–2000. *Emerg. Infect. Dis.*, 4, 432–437.

McCaig, L. and Hughes, J., 1995, Trends in antimicrobial drug prescribing among office-based physicians in the United States. *JAMA* 273(3), 214–219.

McManus, P., Hammond, L., Whicker, S., Primrose, J., Mant, A., and Fairall, S., 1997, Antibiotic use in the Australian community, 1990–1995. *Med. J. Australia.* 167, 124–127.

Melander, E., Ekdahl, K., Jönsson, G., and Mölstad, S., 2000, The frequency of penicillin-resistant pneumococci in children is correlated to community-level utilisation of antibiotics. *Pediatric. Infect. Dis. J.*, 19, 1172–1177.

Mölstad, S., Stålsby-Lundborg, C., Karlsson, A. K., and Cars, O., 2002, Antibiotic prescription rates vary markedly between 13 European countries. *Scand. J. Infect. Dis.*, 34, 366–371.

Örtqvist, A., 1995, Antibiotic treatment of community-acquired pneumonia in clinical practice: A European perspective. *J. Antimicrob. Chemother.*, 35, 205–212.

Rautakorpi, U. M., Klaukka, T., Honkanen, P., Mäkelä, M., Nikarinen, T., Palva, E. *et al.*, 2001, Antibiotic use in the community: A basis for active antibiotic policy in the community. *Scand. J. Infect. Dis.*, 33, 920–926.

Stalsby, Lundborg, C., Olsson, E., Molstad, S., 2002, Antibiotic prescribing in outpatients; a 1-week diagnosis-prescribing study in 5 counties in Sweden. *Scand. J. Infect. Dis.*, 34, 442–448.

Steinman, M., Gonzales, R., Linder, J., Landefeld, S., 2003, Changing use of antibiotics in community-based outpatient practice, 1991–1999. *Ann. Intern. Med.*, 138, 525–533.

Straand, J., Skinlo, Rokstad, K., and Sandvik, H., 1998, Prescribing systemic antibiotics in general practice; A report from the More and Romsdal prescription study. *Scand. J. Prim. Health Care*, 16, 121–127.

Stratchounski, L., Bedenkov, A., Hryniewisz, W., Krcmery, V., Ludwig, E., and Semenov, V., 2001, The usage of antibiotics in Russia and some countries in Eastern Europe. *Int. J. Antimicrob. Agents*, 18, 283–286.

Swedres, 2002, Available at www.strama.se.

van Duijn, H., Kuyvenhoven, M., Tudor Jones, R., Butler, C., Coenen, S., and Van Royen, P., 2003, Patients' views on respiratory tract symptoms and antibiotics. *Brit. J. of. Gen. Pract.*, **53**, 491–492.

Veninga, C. C. M., Stålsby-Lundborg, C., Lagerløv, P., Hummers-Pradier, E., Denig, P., Haaijer-Ruskamp, F. M., and the Drug Education Project Group, 2000, Treatment of uncomplicated Urinary tract infections: Exploring differences in adherence to guidelines between three European countries. *Ann. Pharmacother.*, **34**, 19–26.

WHO Collaborating Centre for Drug Statistics Methodology, 1999, *ATC Index with DDDs*. Oslo.

Chapter 30

Antibiotic Use in the Community:
The French Experience

Agnès Sommet and Didier Guillemot
Institut Pasteur, Paris, France

1. INTRODUCTION

Antibiotic use is the principal driving force for the emergence and spread of bacterial resistance in the community. This relationship has been particularly well-documented for penicillin G-resistant *Streptococcus pneumoniae* (Bronzwaer *et al.*, 2002; Melander *et al.*, 2000; Nasrin *et al.*, 2002) and probably also holds true for *Escherichia coli* resistance to fluoroquinolones and for other bacterial species, such as *Shigella*, *Salmonella*, and *Campylobacter*. Both rates and extents of bacterial resistance to antibiotics in France are among the highest in the world. Moreover, in the French community, population antibiotic exposure is also one of the highest. This situation worries both the scientific community and public health authorities and has been designated a national public health priority.

2. ANTIBIOTIC USE IN FRANCE

2.1. Data sources

Historically, the first description of the evolution of antibiotic use in the community in France was obtained from the Enquête Décennale sur la Santé et les Soins Médicaux (EDSSM, Decennial Inquiry on Health and Medical

Antibiotic Policies: Theory and Practice. Edited by Gould and van der Meer
Kluwer Academic / Plenum Publishers, New York, 2005

Care). The survey is a source of information on health and outpatient care that has been conducted every 10 years since 1960. It uses a one-stage probability sampling procedure based on the last population census. The sampling unit is the household (all persons living in each house sampled are included), and the survey covers a 3-month period. All household members are asked to note every medical event, physician consultation, diagnosis as stated by the practitioner, and drug purchase that occurred during the 3-month period. Investigators visit households five times during this period, to check the accuracy of each individual's information. Results of this inquiry, based on broad representative samples of the French population (>20,000 inhabitants) with a response rate exceeding 90%, are representative of an entire year, excluding the summer months. The first evaluations of antibiotic consumption using the EDSSM database, are relatively recent. These analyses intended to compare trends in antibiomicrobial drug use in the French community between 1981 and 1992. By extrapolation, the annual estimation was 0.7 antibiotic purchases per person and year in 1981 (Guillemot *et al.*, 1998). The main antibiotic classes bought were β-lactams and macrolides. The overall annual estimate of antibiotic use increased to 1.2 antibiotic purchases per person and year in 1991. Penicillins remained the main antibiotic class used, while cephalosporins came in second, rising to 20.8% annual rate use. Between 1980 and 1991, the overall annual rate of antibiotic use increased by an average of 3.9% each year for children and 3.0% for adults. In 1980, respiratory tract infections with presumed viral aetiology (acute nasopharyngitis, acute tracheitis, acute bronchitis, influenza) were already the leading indications for antimicrobial use. These clinical situations were also associated with the strongest augmentation of antibiotic use between 1980 and 1992. The frequency of these diagnoses treated with antibiotics, rose by 115% for children and by 86% for adults during this period. Because antibiotic use in such circumstances cannot be justified, unnecessary antibiotic use can be estimated at >50%.

Antibiotic use in the French population can also be estimated from different independent databases. The mandatory annual reporting by pharmaceutical companies of their antibiotics sales in France, also represent a major source of information. These data have been available since 1988. The French Health Products Safety Agency (AFSSAPS, Agence Française de Sécurité Sanitaire et des Produits de Santé) currently analyses these data by counting the number of boxes of antibiotics sold. The results were recently reported by the European Surveillance of Antimicrobial Consumption (ESAC). Since the recorded information enables accurate identification of each galenic formulation and packaging form of each drug, the results can be expressed as defined daily dose (DDD) per person and year. The estimations of French antibiotic sales rose from 12.8 DDD per person and year in 1997 to 13.3 DDD per person and year in 2001. This increase is in keeping with European comparisons obtained by analysing Information Medical Services (IMS) databases as recently published (Cars *et al.*, 2001). According to that analysis, antibiotic sales in France were estimated at 36.5 DDD per 1,000 inhabitants per day (DID) in 1997, that is, 13.3 DDD per person and year. Pertinently, in 2001, outpatient care accounted for 90% of the overall antibiotic sales, as opposed to 10% for hospital use. β-Lactams represent circa 60% of total use and the very high consumption of broad-spectrum penicillins, cephalosporins, and macrolides can be noted for every year (Elseviers *et al.*, 2003). The strength of the AFSSAPS database is its exhaustiveness and its potential availability since 1988. But it only provides national and annual data, without more precise information on regional trends or monthly or weekly rates.

France, like other European countries, is not lacking in databases. In addition to those cited above, other sources are available (Table 1). For example, the Centre for Research and Documentation on Health Economics (CREDES, Centre de Resource et de Documentation en Economie de la Santé) conducts an annual survey devoted to the surveillance of healthcare expenditures in France. The recent analysis of these data clearly showed that children are the predominant consumers of antibiotics. Children under 6 years old are around four times more exposed than adults (Sommet *et al.*, submitted). Moreover, the French national health insurance (CNAMTS, Caisse Nationnale d'Assurance Maladie des Travailleurs Salariés, CANAM, Caisse Nationnale d'Assurance Maladie des Professions Indépendantes, MSA, Mutualité Sociale Agricole) can provide quasi-exhaustive information on antibiotic deliveries. These data are currently being analysed to determine the impact of the French national campaign "to preserve the efficacy of antibiotics."

2.2.   Units of measurement

The DDD system was developed by an independent scientific committee participating in the WHO Collaborative Centre for Drug Statistics Methodology to measure and compare drug use at an international level (WHO, 2001). DDD is recognised worldwide and is very useful to compare countries. Nevertheless, since DDD is based on a theoretical daily dose, it does not necessarily reflect the recommended or prescribed daily dose. In France, antibiotic prescriptions follow recommendations established by the French Society of Infectious Diseases in collaboration with the French National Drug Agency. French daily recommended doses are often higher than DDD. For example, the DDD for amoxicillin is 1 g per person per day, whereas the French recommended dose for pneumonia is 3 g per person per day (Quatrième Conférence de Consensus en Thérapeutique Anti-Infectieuse, 1992). For this reason, application of the DDD system could overestimate the frequency of antibiotic use. Thus, the frequency of new antibiotic prescriptions for a defined period of time (generally 1 year) and expressed as the number of new antibiotic courses per person and year, cannot be extrapolated from results of antibiotic consumption expressed as DDD. As Wessling *et al.* showed the number of real antibiotic users and the number estimated by the DDD system differed by 4–28% (Wessling and Boethius, 1990). Nevertheless, new prescription frequencies for a defined period of time is an important indicator to monitor therapeutic practices and should be studied within the framework of a public health plan.

Another reason for the difference noted between DDD and prescribed dose is that DDD is defined as a dose for an adult. What does antibiotic use expressed in DDD mean in two different geographic areas where the numbers of children differ? This difference leads to major difficulties in comparison, for example, of rural and urban zones, to affirm differences in terms of antibiotic exposure. Therefore, to take these variations into account, France will probably use both DDD and frequencies of new antibiotic prescriptions as tools for antibiotic consumption surveillance.

3.   DETERMINANTS OF ANTIBIOTIC CONSUMPTION

It is a very difficult task to explain or to speculate as to why antibiotic use in the French community is so high. Despite clear guidelines on antibiotic use for presumed viral respiratory tract infection (PVRTI) over the last 10 years, the proportion of patients with PVRTI for whom antibiotics were prescribed remained high. Furthermore, antibiotic use tended to increase between 1984 and 1995, to treat acute media otitis and bronchitis, but remained almost stable for rhinopharyngitis or tonsillitis (Observatoire National des Prescriptions et Consommations des Médicaments, 1998)

In contrast to certain northern European countries, in France, acute otitis media is considered as a pathology requiring antibiotic therapy. No element explaining the increased prescription of antibiotics for bronchitis has been clearly identified. The frequency of antibiotic prescriptions in sore throats reached 90% in 1984. Obviously, no increase would be expected but no spontaneous decrease occurred despite a viral cause of circa two thirds of these infections. Although the frequency of rhinopharyngitis infections treated by antibiotics rose, the proportion of these infections treated with antibiotics was not really modified, suggesting either an enhanced incidence of these infections or more consultations for them. At the present time, approximately 80 million antibiotic prescriptions per year are written for the 60 million inhabitants of continental France. One of the hypotheses to explain the important increase of antibiotic use could be the increased incidence of these diseases, especially

PVRTI in children (Holmes *et al.*, 1996). In France as in others countries, epidemiologically documented data concerning these community infections are lacking. Thus, reasons for the enhanced antibiotic exposure of community populations is far from being clearly documented.

3.1. Sociocultural and historical factors

A reason for the late collective awakening to the potential impact of population exposure to antibiotic may be the traditional approach of microbiology in France where infectious agents are studied at bacterial strain and at patient levels, rather than on an epidemiological scale.

Another fact is that in France, regardless of the pharmaceutical class, drug consumption is very high. In 2002, national expenditure on health was estimated at 139 billion euros, representing 9.5% of GND with the main part attributed to drug consumption (286 billion euros) (Fenina and Geffroy, 2002). In particular, antibiotics, psychotropic agents and drug for venous insufficiency are overprescribed, in comparison to other European countries (Bouvenot, 1999; Garattini, 1998; Pelissolo *et al.*, 1996).

Because antibiotic exposure is also high in other south-western European nations—Spain, Portugal, and Belgium (Cars *et al.*, 2001)—this distribution could reflect the influence of sociocultural factors.

Another factor is the collective concept of antibiotics. These drugs are considered by a major part of the French population to be very powerful drugs with few and minor side effects. Patients often believe that antibiotics are the answer for many infections (Hamm *et al.*, 1996). Recently, before the onset of the national campaign for the optimisation of antibiotic use in the community, the CNAMTS promoted a survey on the public's perception of antibiotics: 39% of the questioned people thought that antibiotics were effective against viral infections. Their expectations might have a major impact on overprescribing, regardless of doctors' opinions (Macfarlane *et al.*, 1997).

3.2. Medical density

It is well known that patient pressure influences doctors' decisions to prescribe, even when antibiotics are clinically unnecessary (Butler *et al.*, 1998). In an area where medical density is high (303 general practitioners/100,000 inhabitants in France, 424 in Spain, 554 in Italy) (WHO Estimation of Health Personnel, 2003), patient pressure may be stronger when patients who did not receive an antibiotic are dissatisfied and could very easily consult another general practitioner.

3.3.   Pharmaceutical firms pressure

The main objective of pharmaceutical firms, like other producers, is to generate profits. Antimicrobial agents represent the fourth most common therapeutic class for pharmaceutic expenditures, after cardiovascular, neurocerebral, and digestive system treatments (Lancry *et al.*, 2002). Because antibiotics account for a large part of industry profit in France, production and promotion is, historically, one of the highest among European countries (European Federation of Pharmaceutical Industries and Associations, 2003). In fact pharmaceutical firms may have generated a culture of use by pressuring physicians to prescribe. This situation is now probably behind us.

4.   COLLECTIVE AWAKENING AND PROGRESSIVE MOBILIZATION OF FRENCH PUBLIC HEALTH AUTHORITIES

Over the past 10 years, several actions have been promoted by French public health authorities to optimise antibiotic use and thereby control bacterial resistance to them. In 1996, National Guidelines for the Good Use of Antibiotics in Hospitals were published by the French National Agency for Medical Practice Evaluation (Agence Nationale pour le Développement de l'Evaluation Médicale, 1996). Hospital drug committees had to target antibiotic use and implement the in-hospital monitoring of antibiotics. The use of patient-identified prescriptions for antibiotics was recommended to be followed by systematic re-evaluation of these prescriptions 3 days after their initiation.

The French National Report on the Use of Antibiotics in the Community was issued in 1998 (Observatoire National des Prescriptions et Consommations des Médicaments, 1998). That evaluation promoted by the French Health Products Safety Agency (AFSSAPS) suggested for the first time that 50% of community antibiotic prescriptions were unnecessary. In 1999, the French Institute for Public Health Surveillance (InVS, Institut de Veille Sanitaire) published the National Report on the Control of Bacterial Resistance in which a scheme was defined for the surveillance and control of antimicrobial resistance (Groupe de travail sur la Maîtrise de la résistance aux antibiotiques, 1999). In addition, that report identified the need to promote and support research in different domains (microbiology, pharmacoepidemiology, socio-economy, public health [Guillemot and Courvalin, 2001]).

Also in that year, antibiotic use was investigated for the first time in all French healthcare centres (Direction des Etudes Médico-Economiques et de l'Information Scientifique, 2002). Sixty-six per cent of the centres reported that their institution had established guidelines on antibiotic use and this was

Table 1. Characteristics of different French data sources on antibiotic consumption

Data source	Data provider	Strengths	Weaknesses
Sales data, (Outpatient and hospital care)	Pharmaceutical firms (annual drug sales declared by firms to AFSSAPS)	Exhaustiveness Available for each year since 1988	Annual data Do not precisely reflect the use
EDSSM (outpatient care)	CREDES, INSEE	Available since 1960 Reflects real acquisition	Based on a sample of the population, during a 3-month period
		Information on symptoms (not clinical diagnoses) are available	Only every 10 years Only at a national level
Healthcare and Health insurance survey (ambulatory care)	CREDES	Available for each year between 1988 and 1998, and every 2 years since 1998	Based on a sample of the population, during two 1- month periods each year. Only every 2 years
		Reflects real use	Only at a national level
		Information on symptoms (not clinical diagnoses) are available	
EPAS	CNAMTS	Prospective data collection Individual follow-up of antibiotic use	Based on a sample of the population Does not precisely reflect the exact deliveries. No clinical information
Health insurances information system	CNAMTS, CANAM, MSA	Exhaustiveness	
		Detailed and precise information on galenic formulation and packaging form of each drug Individual follow-up of antibiotic use may be possible in the future	No clinical information No information on posology

Notes: AFSSAPS: French Health Products Safety Agency; EDSSM: Decennial Inquiry on Health and Medical Care; CREDES: Centre for Research and Documentation on Health Economics; EPAS: Permanent Sample of Health Insured Individuals; CNAMTS: French National Employees' Health Insurance; CANAM: French National Independent Workers' Health Insurance; MSA: French National Farmers' Health Insurance; INSEE: National Institute for Statistics and Economic Studies.

more likely for those with an on-site drug committee, a specialist in infectious diseases or in hospitals with more than 200 beds. Only 21% of the centres had patient-identified antibiotic delivery, while 43% had mixed records for patient-identified and global delivery of antibiotics. Prescription monitoring was fully computerised in about 15% of the centres.

More recently a childhood infection-prevention programme was promoted with an information campaign tailored to physicians and parents in the south of France (Carbon *et al.*, 2002; Pradier *et al.*, 2003). One of the main innovations of that experimental programme was its use of peer counselling of prescribers, which has been incorporated into the National Programme to Preserve Antimicrobial Efficacy, launched in October 2002.

That national programme, which is scheduled to run from 2002 to 2005 (Cremieux *et al.*, 2001) has seven objectives: (1) to improve the public's and physicians' information on the importance of appropriate antibiotic use and the risks of overprescribing; (2) to perform a rapid diagnostic test for streptococcal tonsillitis, available for all health professionals; (3) to improve appropriate antibiotic use in hospital care; (4) to improve information exchanges between the community and hospital and in particular in the counselling provided by the specialist for the prescription of antibiotics in the community; (5) to improve training of prescribers; (6) to improve surveillance of both antibiotic consumption and bacterial resistance of the main pathogens both in the community and in the hospital; and (7) to improve the national coordination of these different actions. To achieve these objectives, a national committee representing professionals and the different scientific and public health agencies has been asked to monitor the implementation of the plan.

The last piece of the French puzzle will probably be put in place over the next few months. The parliament will have to define the national public health policy for the next 5 years. Antibiotic resistance will probably be one of the objectives having top priority with the aim of reducing antibiotic use in the community by 25% in order to decrease the rate of penicillin non-susceptible pneumococci to less than 30%.

Because the national plan to preserve the efficacy of antibiotics started in October 2002, it is still too early to evaluate its impact on antibiotic consumption and bacterial resistance. A retro-prospective analysis will take into account immediate historical and future data (2000–5) from different sources available in France (Table 1).

5.   CONCLUSION

The continental French population has one of the highest antibiotic exposures and one of the highest rates of antimicrobial resistance among European countries and perhaps among all countries in the world. However, several years ago, stopping the spread of antimicrobial resistance became a public health priority. Now, public health authorities, scientists, and health professionals are highly motivated and mobilised to lower antibiotic use, while maintaining optimal clinical efficacy of antimicrobials.

REFERENCES

Agence Nationale pour le Développement de l'Evaluation Médicale, 1996, *Bon usage des antibiotiques à l'hôpital. Recommandations pour maîtriser le développement de la résistance bactérienne.* Service des Etudes, Paris.

Bouvenot, G., 1999, Prescriptions and consumption of venotonic drugs in France (a propos of the report of the French National Institute for prescriptions and consumption of drugs). *Bull. Acad. Natl. Med.*, **183**, 865–875, discussion 875–878.

Bronzwaer, S. L., Cars, O., Buchholz, U. *et al.*, 2002, A European study on the relationship between antimicrobial use and antimicrobial resistance. *Emerg. Infect. Dis.* **8**, 278–282.

Butler, C. C., Rollnick, S., Pill, R., Maggs-Rapport, F., and Stott, N., 1998, Understanding the culture of prescribing: Qualitative study of general practitioners' and patients' perceptions of antibiotics for sore throats. *BMJ*, **317**, 637–642.

Carbon, C., Guillemot, D., Paicheler, G., Feroni, I., Lamar, L., and Aubry-Damon, H., 2002, *Rapport d'évaluation 2002 de la campagne menée par le Groupe d'Etude et de Prévention des Infections de l'Enfant pour une utilisation prudente des antibiotiques chez l'enfant. Alpes Maritimes, 1999–2005.* Institut de Veille Sanitaire, Saint Maurice, France.

Cars, O., Molstad, S., and Melander, A., 2001, Variation in antibiotic use in the European Union. *Lancet*, **357**, 1851–1853.

Cremieux, A., Reveillaud, O., and Schlemmer, B., 2001, *Plan national pour préserver l'efficacité des antibiotiques.* Ministère de la Santé, Paris. http://www.sante.gouv.fr/htm/actu/antibio.

Direction des Etudes Médico-Economiques et de l'Information Scientifique, Agence Française de Sécurité Sanitaire des Produits de Santé, 2002, *Etat des lieux en 1999–2000 dans les établissements de santé de l'existence de recommandations pour la prescription et des modalités de dispensation des antibiotiques.* Paris, France.

Elseviers, M., Ferech, M., Vanderstichele, R., and Goossens, H., 2003, Consumption of ambulatory care in Europe. In *European Conference of Clinical Microbiolgy and Infectious Diseases*. Glasgow.

European Federation of Pharmaceutical Industries and Associations, 2003, The Pharmaceutical Industry in Figures. Brussels, Belgium, http://www.efpia.org.

Fenina, A. and Geffroy, Y., 2002, Comptes nationaux de la Santé 2002. Ministère de la Santé, Paris. http://www.sante.gouv.fr.

Garattini, S., 1998, The drug market in four European countries. *Pharmacoeconomics*, **14**(Suppl. 1), 69–79.

Groupe de travail sur la Maîtrise de la résistance aux antibiotiques, 1999, *Propositions au Secrétaire d'Etat à la Santé et à l'Action Sociale pour un plan national d'actions pour la maîtrise de la résistance aux antibiotiques.* Réseau National de Santé Publique, Saint-Maurice, France.

Guillemot, D. and Courvalin P., 2001, Better control of antibiotic resistance. *Clin. Infect. Dis.*, **33**, 542–547.

Guillemot, D., Maison, P., Carbon, C. *et al.*, 1998, Trends in antimicrobial drug use in the community-France, 1981–1992. *J. Infect. Dis.*, **177**, 492–497.

Hamm, R. M., Hicks, R. J., and Bemben, D. A. 1996. Antibiotics and respiratory infections: Are patients more satisfied when expectations are met? *J. Fam. Pract.*, **43**, 56–62.

Holmes, S. J., Morrow, A. L., and Pickering, L. K., 1996, Child-care practices: Effects of social change on the epidemiology of infectious diseases and antibiotic resistance. *Epidemiol. Rev.*, **18**, 10–28.

Lancry, P. J., Pigeon, M., and Criquillion, B., 2002, MEDICAM: Les médicaments remboursés par le Régime Général d'Assurance Maladie au cours des années 2000 et 2001. Caisse Nationale d'Assurance Maladie des Travailleurs Sociaux, Paris.

Macfarlane, J., Holmes, W., Macfarlane, R., and Britten, N., 1997. Influence of patients' expectations on antibiotic management of acute lower respiratory tract illness in general practice: questionnaire study. *BMJ*, **315**, 1211–1214.

Melander, E., Ekdahl, K., Jonsson, G., and Molstad, S., 2000, Frequency of penicillin-resistant pneumococci in children is correlated to community utilization of antibiotics. *Pediatr. Infect. Dis. J.*, **19**, 1172–1177.

Nasrin, D., Collignon, P. J., Roberts, L., Wilson, E. J., Pilotto, L. S., and Douglas, R. M., 2002, Effect of beta lactam antibiotic use in children on pneumococcal resistance to penicillin: Prospective cohort study. *BMJ*, **324**, 28–30.

Observatoire National des Prescriptions et Consommations des Médicaments, 1998, Etude de la prescription et de la consommation des antibiotiques en ambulatoire. Agence du Médicament, Direction des Etudes et de l'Information Pharmaco-Economiques, Paris.

Pelissolo, A., Boyer, P., Lepine, J. P., and Bisserbe, J. C., 1996, Epidemiology of the use of anxiolytic and hypnotic drugs in France and in the world. *Encephale*, **22**, 187–196.

Pradier, C., Dunais, B., Ricort-Patuano, C. *et al.*, 2003, Campagne "Antibios quand il faut" mise en place dans le département des Alpes Maritimes. *Méd. Mal. Infect.*, **33**, 9–14.

Quatrième Conférence de Consensus en Thérapeutique Anti-Infectieuse, 1992, Les Infections des Voies Respiratoires. *Méd. Mal. Infect.*, **22**, 38–201

Sommet, A., Sermet, C., Boëlle, P. Y., Tafflet, M., Bernède, C., and Guillemot, D. Trends in antibiotic use in the French community, from 1992 to 2000. submitted.

Wessling, A. and Boethius, G., 1990. Measurement of drug use in a defined population. Evaluation of the defined daily dose (DDD) methodology. *Eur. J. Clin. Pharmacol.*, **39**, 207–210.

WHO Collaborating Center for Drug Statistics Methodology, 2001, Guidelines for ATC classification and DDD assignment. Oslo, Norway, http://www.whocc.no/atcddd.

WHO Estimation of Health Personnel, 2003, WWO Statistical Information System, http://www3.who.int/whosis/health_personnel.

Chapter 31

Antibiotic Policies in Developing Countries

Aníbal Sosa
The Alliance for the Prudent Use of Antibiotics, 75 Kneeland Street,
Boston MA 02111, USA

1.   BACKGROUND

As reported in 1998–9, infectious diseases continue to be a leading cause of death, accounting for a quarter to a third of the estimated 54 million deaths annually worldwide. Lower respiratory infections, diarrhoeal disease, tuberculosis (TB), and malaria were the leading causes of global infectious disease burden. While a decrease in diarrhoeal disease is projected for 2020, malaria is increasing and TB and HIV are growing far faster than projected (Gordon *et al.*, 2000). For the new millennium, HIV/AIDS, TB, and malaria are the leading infectious disease killers (WHO, 1999). Along with bioterrorism and homeland security, antimicrobial resistance has become one of the most important public health issues faced by the industrialised world, particularly G8 country members. Antimicrobial resistance typically develops when antibiotics are prescribed in inadequate amounts and/or for a condition that does not warrant their usage, such as for viral infections. It is estimated that annual cost of infections caused by antibiotic-resistant bacteria is US$4–5 billion (Harrison and Lederberg, 1998). In the United States, estimates exist for the overall economic impact of antimicrobial resistance in healthcare settings. These estimates need to be taken with caution since available data is imprecise and incomplete (McGowan, 2001).

Antibiotic Policies: Theory and Practice. Edited by Gould and van der Meer
Kluwer Academic/Plenum Publishers, New York, 2005

2.   ANTIBIOTIC USE AND COST TRENDS

Most information available comes from industrialised countries with minimal data from the developing world. In the United States, antimicrobial prescription rates by office-based physicians remained unchanged from 1980 through 1992; however, prescriptions for children increased by 48% (McCaig and Hughes, 1995). A large portion of these prescriptions was for the treatment of colds, upper respiratory tract infections, and bronchitis—conditions where there is no proven benefit of antibiotic therapy. Trends in antimicrobial pre-scribing at visits to office-based physicians, hospital outpatient departments, and hospital emergency departments from 1992 to 2000 in the United States declined by 25% (McCaig *et al.*, 2003). Amoxicillin and the cephalosporins were most prescribed (annual drug prescription rate per 1,000 population) in outpatient settings between 1980 and 1992 (McCaig and Hughes, 1995; Steinman *et al.*, 2003). Even though antibiotic use in ambulatory patients is decreasing, use of broader-spectrum antimicrobial drugs increased from 24% to 48% in adults and from 23% to 40% in children (Figures 1 and 2). The cost of broad-spectrum antibiotics exceeded $50 for a typical 7-day adult course (Steinman *et al.*, 2003).

The overall profile of antibiotic use in Australia by drug class was similar to that in the United Kingdom with the majority of antibiotics being prescribed in outpatient and community settings (Beilby *et al.*, 2002). Antibiotic use in Australia remained steady between 1990 and 1995, with an estimated 24.7 DDDs/1,000 population/day dispensed through community pharmacies in 1990 and 24.8 DDDs/1,000 population/day in 1995. Amoxicillin, although declining in use, remained the most dispensed antibiotic (McManus *et al.*, 1997). Each year Australia imports about 700 tons of antibiotics; one third of this is destined for human use and the remaining two thirds for use in animals. Australia is one of the world's highest users per capita of oral antibiotics (Geue, 2000), with the majority of these prescriptions originating from hospi-tals (South *et al.*, 2003).

For 1987–92, drug expenditure in Thailand increased at around 23% per annum, higher than the growth rate of health expenditure (14%) and GNP (8%). In 1993 drug consumption cost 50,000 million Baht (US$1,164.63) or 840 Baht (US$19.56) per capita (Riewpaiboon *et al.*, 1997).

After more than a decade of increasing antibiotic consumption in Chile, physicians became alarmed by the associated economic costs as well as rising antibiotic resistance due to indiscriminate use of these drugs (Salvatierra-Gonzalez and Benguigui, 2000). A study compared antibiotic consumption in Chile over a 10-year period from 1988 to 1998. Antibiotic consumption was measured as defined daily doses (DDD)/1,000 inhabitants/day (WHO, 1996a). The results showed a marked increase in consumption for most antibiotics studied (Table 1) (Bavestrello and Cabello, 2000).

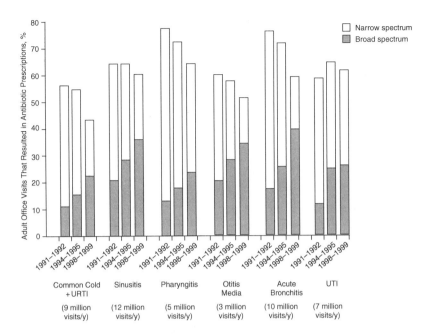

Figure 1. Antibiotic prescribing among adults between 1991–2 and 1998–9.

Notes: Overall use of antibiotics decreased in adult visits for the common cold and unspecified upper respiratory tract infections (URTIs) ($P = 0.011$), for pharyngitis ($P = 0.02$), and for acute bronchitis ($P < 0.001$). Among adults receiving an antibiotic, broad-spectrum agents made up an increased proportion of antibiotic prescriptions for each condition shown (for pharyngitis, $P = 0.002$; for all other conditions, $P < 0.001$). (Results are shown at the level of the patient visit: Broad spectrum indicates visits involving at least one broad-spectrum antibiotic; narrow spectrum indicates visits involving only narrow-spectrum agents.) The mean number of visits occurring annually during the study period is shown for each condition. UTI = urinary tract infection.

Source: Reproduced with permission of American College of Physicians.

The analysis of 10 years of antibiotic consumption showed that while many antibiotics were being sold, the most dramatic increases were seen in the sales of amoxicillin (498%), amoxicillin–clavulanic acid (16,460%), cephalosporins (309%), and fluoroquinolones (473%). The cost to the Chilean population for these drugs totalled US$45.8 million in 1997 (Table 2). Three months later, a second antibiotic consumption study evaluated the impact of the enforced measure. A 3-month period in 1999 was compared with the same 3-month period of 1998 (Table 3). Researchers found that ampicillin consumption decreased by 53%, amoxicillin by 36%, and erythromycin by 30%. Sales of antibiotics dropped by US$6.5 million, with no adverse patient outcomes reported (Table 4).

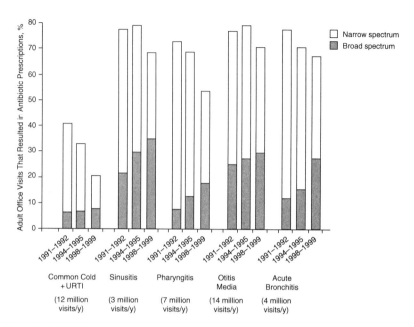

Figure 2. Antibiotic prescribing among children between 1991–2 and 1998–9.

Notes: Overall use of antibiotics decreased in paediatric visits for the common cold and unspecified URTIs ($P < 0.001$) and for pharyngitis ($P = 0.002$). Among children receiving an antibiotic, broad-spectrum agents made up an increased proportion of antibiotic prescriptions for each condition shown (for P values, see text). (Results are shown at the level of the patient visit: Broad spectrum indicates visits involving at least one broad-spectrum antibiotic; narrow spectrum indicates visits involving only narrow-spectrum agents.) The mean number of visits occurring annually during the study period is shown for each condition.

Source: Reproduced with permission of American College of Physicians.

Table 1. Comparison of antibiotic consumption in Chile over 10 years

Antibiotic	1988[a]	1997[a]	% change
Cloxacillin	0.39	0.417	7
Ampicillin	0.54	0.613	14
Amoxicillin	0.87	5.204	498
Amoxicillin–Clavulanic acid	0.0025	0.414	16,460
Chloramphenicol	0.097	0.079	−18
TMP-SMX	0.965	1.163	20
Cephalosporins	0.064	0.262	309
Fluoroquinolones	0.049	0.281	473

[a]DDD/1,000 inhabitants/day.

Table 2. Antibiotic sales in 1988, 1996, and 1997

Antibiotic group	1988[a]	1996[a]	1997[a]
Macrolides	1,599,000	10,790,383	14,745,019
BS[b] penicillin	3,062,000	11,606,151	13,869,812
NS[c] penicillin	2,997,000	3,814,038	4,141,114
Cephalosporins	1,601,000	3,692,174	4,275,996
Fluoroquinolones	709,000	3,512,470	4,203,548
TMP-SMX	1,968,000	2,199,671	2,316,532
Tetracyclines	772,000	1,682,546	2,014,394
Chloramphenicol	319,000	306,255	308,233
Total	13,027,000	37,603,688	45,874,648

[a]In US dollars.
[b]BS: Broad spectrum.
[c]NS: Narrow spectrum.

Table 3. Recent antibiotic consumption in Chile during a 1-year period

Antibiotic	Fourth quarter		% change
	1998[a]	1999[a]	
Clarithromycin	0.366	0.446	21.8
Azithromycin	0.342	0.353	3.2
Erythromycin	0.536	0.370	−30.9
Amoxicillin	4.639	2.943	−36.5
Ampicillin	0.391	0.182	−53.4
Amoxicillin–Clavulanic Acid	0.459	0.443	−3.5
Ciprofloxacin	0.317	0.386	21.8
Cloxacillin	0.424	0.338	−20.3
Fenoximetilpenicillin	0.184	0.111	39.7
TMP-SMX	0.913	0.633	−30.7
Tetracycline	0.199	0.340	−31.8

[a]DDD/1,000 inhabitants/day.

Table 4. Antibiotic sales in 1998 and 1999

Antibiotic group	1998[a]	1999[a]
Macrolides	14,763,740	12,448,627
BS penicillin	13,747,242	11,227,954
NS penicillin	4,121,100	3,275,517
Cephalosporins	4,448,543	4,193,849
Fluoroquinolones	4,075,818	4,115,261
TMP-SMX	2,316,260	1,767,612
Total	43,514,701	37,030,819

[a]In US dollars.

3. IMPACT ON HEALTH BUDGETS OF ANTIBIOTIC USE

Infectious diseases increase demands on national health budgets that already utilise some 7–14% of GDP in developed countries, up to 5% in the better-off developing countries, but currently less than 2% in least developed states. In many developing countries where many competing needs exist, healthcare budget represent only 2–5% of their total annual budget. This creates a very vulnerable public health infrastructure with small support for sustainability. In addition, when unexpected health crisis or natural disasters arrive in these countries, such as infectious outbreaks, emerging infectious diseases, earthquake, flooding, fire, etc., public health officials are faced with the difficult task of cutting existing budgets and allocating monies to solve crisis leaving scarce resources to carry on existing efforts.

Public health officials, policy makers, and clinicians must come together and develop a plan to minimise antimicrobial resistance and its economic impact. Strategies have to take into account all stakeholders affected by new policy measures such as physicians, patients, the healthcare businesses, the drug industry, and the public as a "society" (McGowan, 2001).

In outpatient settings, healthcare practitioners often feel compelled to empirically treat patients with acute infections without waiting for the results of cultures and susceptibility tests. In hospital settings, clinicians can often rely on bacteriology data and if available also on antimicrobial resistance surveillance data from routine screening of specimens. Quite often, medical, veterinarian, and dentist surgeons administer prophylactic antibiotics pre-, during, and/or post-surgery for much longer period of time required or recommended by guidelines issued by professional associations.

Differences between infectious disease and intensive care physicians prescribing attitudes have been observed (Sintchenko *et al.*, 2001). Fear of malpractice suits also play a role in the decision-making process. When selecting empiric therapy for infectious diseases, one must consider *in vitro* susceptibility patterns and prevalence resistance rates along with other factors, such as tissue penetration, expected efficacy, adverse effects, cost, cost-effectiveness, and selection of resistant strains.

Other factors contributing to the development and pervasiveness of antibiotic resistance include: unregulated drug approval, quality control and marketing; lack of patient resources/access to quality healthcare; patient non-compliance and self-medication; physician misuse of antibiotics; and lack of reliable information sources for physicians such as standard treatment guidelines (STGs) and laboratory facilities to confirm diagnoses.

Substandard and counterfeit drugs are also problematic, as are quality control deficiencies in pharmaceuticals and pharmaceutical companies. Many countries in Latin America, Africa, and Asia do not have agencies or regulatory mechanisms for the approval, quality control, and marketing of medicines; and rely on the good faith of pharmaceuticals that submit data approved by other country agencies (Food and Drug Administration [FDA]).

Poverty and lack of healthcare resources exacerbate the resistance problem in the region. Patients with limited incomes generally cannot afford to see a health-care worker or receive laboratory tests to determine the aetiology of the disease (if the tests are available). These patients can sometimes only afford poor quality or substandard or counterfeit drugs, or a drug course shorter than that which would be optimally effective. Counterfeit drugs often contain the wrong ingredient, contain no active ingredient, or contain only weak and inadequate amounts of the active ingredient (WHO, 2000). One study showed that patients who self-medicated were more likely to use an inadequate drug or dose, and to follow treatment for less than 5 days (Bojalil and Calva, 1994). Perhaps the single most significant problem in many countries is the widespread availability of antibiotics without a prescription. In these countries antibiotics may be purchased from pharmacies, street vendors, convenience stores, outdoor markets, fairs, etc. One study examining the impact of unregulated antibiotic sales in Bolivia tested healthy children from urban areas and found that 97% carried *Escherichia coli* insensitive to ampicillin (Bartoloni *et al.*, 1998). A survey of drugstore sales of antibiotics in Mexico revealed that only 57% of antibiotic purchases were made with a prescription, and that the person selling the drug gave instructions on its proper use in only 15% of observed transactions (Calva, 1996).

In 2001, Prof. Leonid S. Stratchounski, Director of the Institute of Antimicrobial Chemotherapy Smolensk, Russia lead a study to inventory the content of antibiotics for systemic use (ASU) in home medicine cabinets (HMCs) of the non-medical population in Russian cities and to find out for which indications people report they would use ASU on their own (Stratchounski *et al.*, 2002). One thousand two hundred families in twelve cities participated in the study. Two thousand five hundred forty five packages of antibiotics were identified.

The number of different antimicrobials (by international nonpatent names) in HMCs was 65. The average number of antimicrobials per household was 2.6 (from 1 to 11 ASU in a single HMC). Families with two and more antibiotics represented 72.2%.

Their findings seems to indicate that (1) antimicrobials are widely prevalent among inhabitants in Russia; (2) the most "popular" antimicrobials (co-trimoxazole, chloramphenicol, tetracycline, sulfonamides) may cause serious adverse drug reactions; (3) antibiotics are often used imprudently; and (4) the population does not have enough knowledge regarding antibiotics in general.

Physicians, who are overworked, underinformed, or feeling other pressures to overprescribe are also contributing to the spread of resistance. Other problems, such as choice of broad-spectrum antibiotics over appropriate narrow-spectrum alternatives, or prescribing antibiotics in incorrect doses and/or treatment durations, can occur even when clinical presentations necessitate antibiotic prescription. In a study by Paredes *et al.*, of 40 physicians in Lima, Peru, who were questioned on the proper use of antibiotics to treat diarrhoea (Paredes, 1996), 36 correctly reported that the majority of diarrhoeal disease is of viral origin and antibiotics are not indicated. Yet 35 of these 36 unnecessarily prescribed antibiotics for this condition. Although the physicians were clearly informed of appropriate prescribing practices, other factors including control of disease, patient demand, patient's family's demands, and practitioner self-confidence, persuaded them to misprescribe. In other cases, inappropriate prescribing has been attributed to insufficient training in infectious diseases and antibiotic treatment, insufficient use of microbiological information, and the difficulty of empiric drug choice (Nathwani and Davey, 1999).

In May of 2001, APUA conducted a survey on physician antibiotic prescribing practices and knowledge in seven countries in Latin America. Respondents prescribed antibiotics in the presence of acute respiratory infection (ARI)/pneumonia as a first (27%) priority reason. Second (14%) priority reason was the presence of UTI. Physicians reported using penicillin most often (29%), followed by ampicillin (18%), and cephalosporins (12%). Reasons for using these particular antibiotics were that they are the drugs of choice for the most prevalent clinical cases (ARI) in the countries surveyed, and also that they are accessible, less expensive, and wide spectrum (Sosa and Travers, 2001).

A lack of proper diagnostic facilities and laboratories is another serious issue, so that many physicians must rely on empirical treatment of a disease rather than evidence-based treatment (WHO, 2000). When available, it is helpful to use bacterial studies in order to confirm diagnosis and make the best treatment decision. Empirical treatment is the norm in most situations in the region. This survey, however, showed inconsistent and unclear knowledge on empiric treatment of the most common illnesses, pointing to a need for wider dissemination of STGs.

In many developing countries with other pressing health priorities, antimicrobial resistance is an even greater problem than in the United States, yet drug regulation and use policies are very limited or nonexistent. In the past few years, industrialised countries have at some point issued important antibiotic policy changes. Australia, Canada, United Kingdom, and the United States have implemented measures such as guidelines for practitioners, national campaigns for primary care providers, consumer awareness campaigns, applied research and surveillance, guidelines for hospital and veterinary use of antibiotics, national medicine weeks, grassroots projects involving both practitioners and consumers, creation of national prescribing services, etc. (Hemming and Harvey, 1999). A few countries in South America, such as Chile, Uruguay, and Venezuela have begun setting up regulatory measures towards the introduction of a national antibiotic policy (Bavestrello and Cabello, 2000).

The European Society for Clinical Microbiology and Infectious Diseases (ESCMID) have created a Study Group on Antibiotic Policies to investigate and document policy issues surrounding control measures, key infectious diseases, prophylactic use of antibiotics, resistance rates, antibiotic consumption, etc. (ESCMID, 2002).

Some countries in the developing world have enacted regulation to restrict the sale and dispensing of antibiotics without a prescription. Often, it does not get enforced or monitored, or there are unsanctioned dealers or illegal distributors who can operate without any retribution (Becker *et al.*, 2002) and antibiotics are still sold over-the-counter (OTC) (personal communication, Dr Celia Carlos, Philippines). Pharmacists also substitute antibiotic prescriptions from one class to another and not just for generic products (Unknown, 2001). These unsanctioned providers are commonly not trained to diagnose key infectious diseases or correctly prescribe appropriate doses (Hemming and Harvey, 1999).

4. ACCESS TO ESSENTIAL ANTIBIOTICS AT ALL LEVELS OF CARE

Essential antibiotics are those antibiotics needed to treat the most prevalent infectious diseases that affect that region. Essential antibiotics need to be identified, listed, and secured at all levels of care at all times. This can be determined by collecting data on morbidity and mortality due to infectious diseases and if possible by analysing surveillance data on key pathogens and their susceptibility patterns. In addition, essential antibiotics can be classified according to spectrum of action, administration, cost, availability of generics, side effects, and ability to tolerate existing storage and transportation conditions.

Access to essential antibiotics can be influenced by many factors. In countries where there is not a universal healthcare system, individual income level plays a major role in deciding what and when to purchase essential antibiotics. In most cases, only the privileged middle and upper class can obtain prescribed or unprescribed antibiotics. Studies conducted by Asturias and collaborators in Guatemala ascertained that middle and upper class families are more subject to issues of antimicrobial resistance due to an unrestricted access to antibiotics (Asturias *et al.*, 2003). These individuals not only buy antibiotics when they please, but also get inadequate dosages and do not adhere to therapy guidelines. There is not an understanding of the differences between improvement and cure of an infectious disease. In addition, as occurs in many other countries, most antibiotic use is unjustified due to the viral etiology of most frequent infectious clinical episodes.

The poor on the other hand will depend on their own ability to purchase drugs or government assistance programmes such as social security dispensing systems that might be able to secure access to cheaper drugs such as properly stored and transported generics.

Developing countries in some instances choose not to respect product patents as they are the only effective means of making available pharmaceuticals necessary to save lives and protect the health of their citizens (Brock, 2001). In other instances, countries have the right to grant compulsory licenses to permit the manufacture of generic equivalents of patented drugs in situations of national emergency or when negotiations have failed to ensure a fair price which would enable parties to satisfy their objectives of recouping investment costs and of maximally preventing avoidable deaths such as the case of HIV/AIDS (Schüklenk and Ashcroft, 2001). In November 2001, during the Fourth Ministerial Conference in Doha, Qatar, a declaration—the DOHA Declaration—was issued on international intellectual property laws. It helped deflect some of the effects of international patent protection policy. In the sections dealing with TRIPS (trade-related aspects of intellectual property rights) governments agreed to allow certain countries to ignore worldwide patent protection for certain drugs if the drugs are important to protecting public health (TRIPS, 2001).

Availability of antibiotics in developing countries is also influenced by other intrinsic factors such as counterfeited, expired, inactive, or substandard antibiotics (Hemming and Harvey, 1999). It is also common to find non-physician healthcare workers, located in remote locations, who lack training to treat minor illnesses. Sometimes, these untrained practitioners are responsible for misuse of antibiotics (Okeke *et al.*, 1999). In addition, the use of natural healers, or "curanderos" is another reason that antibiotics are often used indiscriminately without the supervision of a trained pharmacist or practitioner.

4.1. Countries in the developing world with and without antibiotic policies

A quick electronic survey was emailed to APUA country chapter leaders in developing countries in Latin America, Africa, Asia, and Middle East. Thirteen responses were obtained and are presented in Table 5.

Developed countries are facing a large illegal distribution of prescription drugs by online unlicensed pharmacies, or the "swap meet" pharmacy where medications are sold for cash with no questions asked. Such illegal distribution

Table 5. Sale, dispensing, and prescribing of antibiotics by illegal distributors and unsanctioned dealers of antibiotics

Country	ATB Policy	ATB Policy Enforcement	Antibiotic Prescription	Unsanctioned, Illegal dispensing or OTC of Law ATB
Argentina	No	No	Yes	Yes
Brazil	Yes	No	Yes	Yes
Chile	Yes	Yes	Yes	No
Colombia	Yes	No	Yes	Yes
Costa Rica	Yes	No	Yes	Yes
Dominican Republic	No	No	Yes	Yes
Ecuador	No	No	No	No
Guatemala	No	No	No	Yes
Lebanon	No	No	No	No
Nepal	Yes	Yes	Yes	Yes (OTC)
Philippines	Yes	Yes	Yes	Yes (OTC)
Taiwan	No	No	Yes	Yes
Venezuela	Yes	No	Yes	Yes (OTC)

Note: Over-the-counter (OTC) sale is prevalent in the majority of developing countries in Latin America, and the Caribbean, Asia, and Africa.

of prescription medication is a booming business in the United States, Canada, and Europe (Unknown, 2002). In response, the US FDA has delineated plans to help curtail illegal marketing of prescription drugs on the Internet.

Unsanctioned dealers can often be found in developing countries. These are either individuals or groups operating in a nonexistent drug regulatory system. They are not considered illegal because they are not defined as such by any legal process. They are visibly tolerated and the community relies on them to obtain essential drugs. It is assumed that these drugs are not counterfeited, expired, or inactive, but there is no proof of a quality assurance system in place that guarantees the quality and purity of drugs sold.

In many parts of Africa, where antibiotics are commonly available from unsanctioned providers, it will be worth educating the general population about the consequences of irrational antibiotic resistance (Okeke and Sosa, 2003). Unsanctioned providers often reach out to people with limited access to orthodox healthcare, and are commonly not trained to diagnose infections or correctly prescribe appropriate doses. They serve as an unofficial outlet for many antibiotics, often capsules and tablets of cheaper antimicrobials, but are not limited to these. For example, Becker *et al.* (2002) recently described the inappropriate distribution of injectable antibiotics, including second- and third-line drugs such as oxacillin and third-generation cephalosporins, medicines that should be conserved for managing resistant infections.

The use of subtherapeutic doses creates a situation where highly resistant strains can be selected sequentially and this is a condition that prevails in many cases when antibiotics are used without proper prescription or in patient non-compliance. In addition, poor quality drugs can lead to treatment failure and consequently, excess mortality and morbidity (Prazuck *et al.*, 2002). Poor quality drugs can provide subinhibitory selective pressure, of which neither the patient nor the prescriber are aware. Reports of substandard antibiotics have come from many countries and a significant proportion of these are in Africa. These reports describe preparations containing anything between 0–80% of stated label claim. Some of them contain such low concentrations that they can only be considered counterfeit, that is, deliberately manufactured with low or no active drug content. Others may have complied with pharmacopoeia standards at some time but have, in the course of distribution and display, been degraded by heat and humidity. Where possible, patients must be advised of the wisdom of obtaining medicines at reputable outlets, where they have been properly stored and where expiration and lot information is available (Okeke and Sosa, 2003). Prazuck *et al.* found similar findings in Northern Myanmar. Their findings suggest that public health policies based on national treatment guidelines should rigorously include the monitoring of quality control of available antimicrobial products. They found that only 3 of the 21 antimicrobials purchased to treat STD displayed the official "registered" label, 3 were expired, 6 did not have an expiration date, and 1 product did not contain the active principle and did not show any *in vitro* activity against bacteria. Seven of twenty-one products did not contain the stated dosage. In the absence of such measures, specific treatment strategies are likely to fail and to generate drug resistance (Prazuck *et al.*, 2002).

Counterfeiting is a problem in India and the Philippines. India exports counterfeited drugs to the Middle East, southern Africa, and Europe causing serious threats to public health. Of the Philippine pharmaceutical market, 30% is estimated to be counterfeited drugs. Intellectual property protection in many developing and transition countries is inadequate which in turn makes it very difficult to implement the data exclusivity protection (Scrip, 2003).

Excessive use of antimicrobials in developing countries brings serious consequences for infections responsible for high infant mortality. These issues are very complex and must be addressed effectively with an aggressive approach that attacks the root causes (Gundersen, 1992).

4.2.   Impact of donations of antimicrobials agents on country AMR policies

In general, drug donations to developing countries abide by international guidelines published in 1996 by WHO (WHO, 1996). Developed countries such as the United States, Australia, and Germany have specific bodies charged with the task of ensuring certain criteria, and in particular making sure that donations match countries needs. Sometimes, drugs destined for a specific region or country, surface in another one by illegal practices of traders and wholesalers. This violates existing international regulations.

In 1999, the Medicines Crossing Borders Project issued a step-by-step guide called *Good Drug Donation Practices* (WHO, 2002); its main purpose is "to ensure the quality of donations once ready for shipment, or to evaluate a donation once made."

Anti-infectives are the drugs most often donated. Donations of antimicrobials as well as other drugs must be offered for as long as they are needed. For example, the provision of anti-retrovirals for HIV and AIDS often lacks continuity and sustainability, jeopardising patient's compliance and adherence. In turn, HIV resistance develops quite rapidly and patients are left with fewer options. Country health officials must have protocols for accepting drug donations to minimise inappropriate use and waste that will cost them money to dispose.

4.2.1. Reality-check of antimicrobial donations to developing countries

1. Donations do not always follow international guidelines; families living abroad often send antimicrobial agents for "just in case" infections usually defined by the presence of fever and/or purulent fluid (sputum in most instances).
2. Patients receive unnecessary or unacceptable antimicrobials.
3. In general, doses of antimicrobials donated by family members are incomplete.
4. Emergence of resistance is not easy to monitor; in some instances antimicrobials received are fairly new in the market and not yet available in the country.
5. There is an increasing incidence of antibiotic-resistant pathogens in both hospitals and the community without a corresponding increase in new antimicrobial drugs.

In September 2002, The US Federal Interagency Task Force on Antimicrobial Resistance held a consultants meeting in San Diego, California (CDC, 2002). Their summary report, *Input for A Public Health Action Plan to Combat Antimicrobial Resistance Part II: Global Issues*, issued important recommendations in an attempt to curb antimicrobial resistance due to donated antimicrobials.

4.3.   The decision-making process in prescribing antibiotic therapy

In the decision-making process for prescribing antibiotic therapy it is important to review scientific literature and guidelines issued from specialty societies. It is also important to consider other factors such as adverse drug reactions, allergies, pharmaco-economics, pharmacodynamics, pharmacokinetics, previous antibiotic therapy, treatment failures, and compliance issues. Primary care physicians are being increasingly challenged to take good care of our patients despite demands to keep costs down, get the patient well, and have prescribing habits profiled. In countries where government agencies subsidise drugs, prescribers are required to prescribe only what is listed in their formulary.

Irrational prescribing and over-prescribing of drugs is prevalent in many countries. Most prescribing in developing countries is performed by paramedical workers who only have minimal or no training at all (Holloway and Gautam, 2001). When studying antibiotic dispensing by drug retailers in Kathmandu, Nepal, Wachter *et al.* described the prescribers' lack of adequate understanding of the disease processes in question (Wachter *et al.*, 1999).

The government of Nepal spends 42% of the national health budget on drugs, while international donors spend nearly three times that amount. Drug distribution and use in Nepal is largely unregulated and the pharmaceutical industry is a major and growing market. Of the drugs, 80% are purchased outside of the government-supplied health system, mostly through private retail shops and pharmacies. When antibiotics are prescribed, they are often not used correctly. Of all drugs prescribed in Nepal, 72% of scripts are not in compliance with standard norms and 38% of patients have misunderstood dosage and administration requirements.

4.4.   Need for AMR surveillance data for key pathogens

The goal of surveillance of antimicrobial resistance is to provide the information necessary to secure an approach to the management of communicable diseases that minimises morbidity and mortality while also containing the emergence of pathogens resistant to antimicrobials. Surveillance of antimicrobial resistance must involve the collection and collation of both clinical and microbiological data. Surveillance of prevalent bacterial pathogens and susceptibility profiles can be an expensive endeavour that few local, regional, or national institutions can afford; however, availability of AMR surveillance data can be useful in prescribing the most effective available antibiotic. It is important that each country has in place minimal requirements for an effective surveillance system.

Significant differences in susceptibility profiles within one pathogen, such as clinical isolates of *Pseudomonas aeruginosa* in two countries in South America (Venezuela and Paraguay), generate important implications. There is a need to monitor antimicrobial resistance at the local level that serve as an alert for clinicians and public health officials (Rodriguez *et al.*, 2002). Emergence of multidrug resistant pathogens calls for a proactive approach. Sentinel surveillance is useful in high-risk areas.

In 1993, Nepal instituted an AMR intervention that began by having an external contractor (ICDDR, B) assessing the microbiology infrastructure and capabilities to perform antibiotic susceptibility testing (AST). They have been collecting bacterial isolates, identifying them and performing internal and external quality assurance and AST to determine if they are resistant to relevant antimicrobials (Sosa and Stelling, 2003). These antimicrobial resistance surveillance efforts have focused on five priority pathogens: *Shigella* sp, *Vibrio cholerae*, *Neisseria gonorrhoeae*, *Streptococcus pneumoniae*, and *Haemophilus influenzae*.

4.4.1. Translating AMR surveillance data into action

AMR surveillance data can be used to make decisions about treatment at the clinical level and decisions about drug policy at the national level. For example, resistance of a pathogen to an antimicrobial often results in treatment failure so the detection of this resistance should lead to a recommendation to avoid using this antimicrobial for therapy. Sudha *et al.* have come up with proposed criteria for choosing empiric antimicrobial therapy based on the sensitivity index (SI) value in relation to prevalence of resistance. SI can be defined as the ratio of per cent susceptible to per cent nonsusceptible strains of microorganism in a particular region during a specific time period (Sudha *et al.*, 2003). This method could be useful for countries in the developing world; however, caution must be exercised since this approach needs to be validated.

Another approach could be the determination of Resistance Alert Thresholds (RAT). This approach, though controversial, remains elusive since RAT values need to be assigned arbitrarily using resistance prevalence data for key pathogens (Tapsall, 2001).

4.5. Effectiveness and cost-effectiveness of measures to contain antimicrobial resistance

In 2002, Wilton *et al.* reviewed 43 studies, mainly hospital-based, from the United States. Some of these studies had community level interventions. These studies covered policies on: restricting the use of antimicrobials (five studies, suggesting that restriction policies can alter prescriber behaviour, although with limited evidence of subsequent effect on AMR); prescriber education, feedback, and use of guidelines (six studies, with no clear conclusion); combination therapies (seven studies, showing the potential to lower drug-specific resistance, although for an indeterminate time period); vaccination (three studies showing cost/effectiveness). Important conclusions are derived from this analysis, and particularly for developing countries (Wilton *et al.*, 2002).

Public health officials, policy-makers, and clinicians can customise local antibiotic policies that have the potential to significantly impact AMR and reduce its burden in terms of cost, morbidity, and mortality. Drug and Therapeutic Committees (DTCs) are usually charged with this task.

High-risk hospital areas are hard hit by nosocomial infections and multidrug-resistant (MDR) pathogens. Emergence of MDR *Acinetobacter anitratus* species in neonatal and paediatric intensive care units in Durban, South Africa, caused 56.5% mortality. Most of the infants had received broad-spectrum antibiotics considered a risk factor for acquiring MDR (Jeena *et al.*, 2001). A variation of a restrictive antibiotic policy is the use of a directive policy-driven letter targeted to a specific antibiotic. This type of intervention sends instructions to prescribers to discourage the use of an antibiotic. Changes in prescribing practices can be observed; however, if reason for change is not fully described, unexpected, untoward results (increase in morbidity and cost) can also be encountered and can be counterproductive (Beilby *et al.*, 2002).

Drug and Therapeutic Committees offer practical approaches for promoting rational use of drugs. Key responsibilities include formulary management, STGs, essential drug list (formulary list), indicators of rational use of drugs, and interventions to change inappropriate use of drugs. They are a meeting ground between clinicians, pharmacists, and financial managers to negotiate the balance between cost and quality.

Drugs and Therapeutic Committees are charged with the responsibility of designing STGs to assist prescribers in choosing the most appropriate antimicrobials. These guidelines can address syndromes or be disease-specific. They can provide first-, second-, and third-line antibiotics according to a country's list of essential drugs described in the national formulary.

A study by Sosa *et al.* of physician prescribing practices in seven countries in Latin America and the Caribbean focused on physician knowledge, attitude, and behaviour as a way to quickly begin understanding and working on the resistance problem. The study illustrated that (1) there exists an acute need to raise physician awareness regarding resistance patterns in key pathogens; (2) there is a general lack of information for physicians to appropriately treat ARIs and acute diarrhoeal diseases empirically; (3) there is a significant need for the collection and dissemination of more local AB resistance data to improve prescribing patterns and patient outcomes; (4) reference laboratories are either lacking and/or physicians are unaware of their existence (62.9% stated that they do not have access to such a lab); and (5) physician training on antibiotics in Latin America had no significant effect on appropriate antibiotic usage (Sosa and Travers, 2001). We have often experienced that AMR surveillance data are not appropriately disseminated among prescribers and decision-makers.

The classical example of combination antimicrobial therapy is the use of combined antimicrobial drugs for the treatment of TB. Combination therapy is also the primary method of treating febrile neutropenia, infections involving multiple organisms, and for obtaining synergism and preventing or limiting resistance development. It is also used in other critical infections such as HIV/AIDS, malaria, sepsis by Gram-negative bacteria, nosocomial infections, and cystic fibrosis. Use of Direct-Observed-Therapy Short Course (DOTS)-Plus to treat multidrug resistance TB has proven to be more effective than DOTS alone. DOTS-Plus uses second-line antituberculosis drugs considered "reserve or second line." Second-line agents that would be used under DOTS-plus are more expensive, more difficult to administer and often poorly tolerated. Although controversial, it does cause fewer deaths (Sterling *et al.*, 2003).

Drug combinations have a downside in that they increase the potential for drug related side effects, the likelihood of colonisation by resistant organisms, and the possibility of antagonism between the agents (Cunha, 2003). When using initial empiric combination therapy for hospital-acquired infections, patient outcomes could be be improved (Kollef, 2001). In general, single, narrow-spectrum antibiotic therapy is more appropriate in community-acquired infections.

Several infectious microorganisms including *S. pneumoniae* and *H. influenzae* are serious pathogens in children under the age of 5, causing pneumonia and meningitis, respectively. These bacteria possess virulence mechanisms able to cause high morbidity and mortality. In addition, both have acquired resistance to many antimicrobial agents. In developing countries, immunisation can reduce the use of antimicrobials and hence the emergence of antimicrobial resistance (O'Dempsey *et al.*, 1996).

The widespread use of vaccination for *S. pneumoniae* and *H. influenzae* could reduce the need for antibiotics in infants and toddlers through:

- Reduction in the need to cover potential invasive pathogens in acute febrile episodes
- Reduction in respiratory tract morbidity
- Potential reduction in non-responsive episodes
- Reduction of carriage and spread of antibiotic-resistant organisms.

No effective reduction in antibiotic use will occur, however, without additional appropriate measures

* Education of medical teams
* Education of the community
* Providing legal protection to those withholding antibiotics by redefining the "reasonable physician."

An *S. pneumoniae* surveillance network was started in 1994 in six Latin American countries (SIREVA-Vigía project) coordinated by the Pan-American Health Organization. This regional initiative was designed to obtain information about the *S. pneumoniae* serotype distribution in order to determine the ideal composition for a conjugate vaccine that could be useful for the region. Additionally, the project was aimed at monitoring the rates of *S. pneumoniae* antimicrobial resistance (Gamboa *et al.*, 2002). This programme has identified high levels of antibiotic resistance in Latin America that call for the development of rational antibiotic use guidelines.

4.6.   Consumer awareness

Educational campaigns can effectively reduce antibiotic use and ultimately, antibiotic resistance rates. An intervention conducted in the United States resulted in a decrease in antibiotic prescribing for bronchitis, from 74% to 48%, compared to a 2% decrease at control/limited intervention sites (Gonzales *et al.*, 1999). Successful integrated approaches have been carried out in India, Indonesia, and Pakistan (Tawfik, 2000) combining a Verbal Case Review (VCR) and a tool called Information Sharing, Feedback, Contracting, and Ongoing Monitoring (INFECTOM). The approach successfully educated practitioners in those countries about standard protocols, and compared their knowledge with actual practices. The results indicate that ongoing monitoring helps to encourage consistent improvement in prescribing practices. To influence parents, another approach could be to field test an antibiotic use and resistance curriculum for elementary and high schools.

4.7.   Training of healthcare workers and
other individuals

Justifying training of non-medical providers (healthcare workers, for instance) could be controversial; however, the reality is that most physicians in a developing country are located in the capital where most of the population also resides. The presence of unskilled practitioners in remote rural areas calls for training them in approaches similar to one used by WHO Integrated Management of Children Illnesses (IMCI) which calls for non-physician health workers (HWs) to evaluate every sick child presenting to a first-level health facility (HF) for respiratory infections, diarrhoea, malaria, measles, and malnutrition, regardless of the child's presenting complaint(s). To find out effectiveness of the IMCI approach, a study was conducted to evaluate the level of performance achieved by IMCI-trained HWs at the end of training and the level of performance maintained during the first 3 months post-training with monthly or bimonthly clinical supervision. They found that HWs achieved reasonably high performance levels managing ill children with mild and moderate disease classifications but performed at a much lower level when managing severely ill children at the end of training (Odhacha, 1998). In 2001, WHO conducted an evaluation of the cost, effectiveness, and impact of IMCI in Tanzania, Uganda, Bangladesh, and Perú. Results from Tanzania showed that baseline data on mortality, family practices, and care seeking for sick children demonstrated that IMCI is addressing the most common major health problems of children in the study illnesses: malaria, pneumonia, and diarrhoea. The prevalence of anaemia, though, was very high, with almost 10% of children between the ages of 6 and 12 months having life-threatening anaemia (WHO, 2001).

5. EMERGENCE OF NEW RESISTANT PATHOGENS

The emergences of resistant pathogens challenges physicians to effectively and judiciously use antimicrobial agents. Pathogens such as vancomycin- and methicillin-resistant *Staphylococcus aureus*, vancomycin-resistant enterocci, and MDR *P. aeruginosa* and *Acinetobacter baumannii* are continuously changing their susceptibility profile due to the selective pressure exercised by the indiscriminate use of antimicrobials. Global warning systems (GAARD, INSPEAR have emerged in response to these pathogens. Preparedness by clinicians, epidemiologists, and microbiologists will allow them to team up to confront potential outbreaks.

5.1. Development of new antimicrobials

Only two completely new classes of antimicrobials have been developed and marketed since 1970 (streptogramins, Synercid®, and oxazolidinones, Zyvox®). Daptomycin is a novel lipopeptide antibiotic under development,

which demonstrated bactericidal activity *in vitro* against Gram-positive
bacteria. It is active *in vitro* against organisms resistant to currently approved
antibiotics, including methicillin-resistant *S. aureus*, vancomycin-resistant
enterococci, and penicillin-resistant *S. pneumoniae*. Other new antimicrobials
are variations of existing drugs and therefore are more vulnerable to resistance.
On average, research and development (R&D) of a single new antimicrobial
takes 10–20 years and costs US$500 million dollars to bring onto the market.
According to the Tufts Center for the Study of Drug Development (CSDD)
(http://csdd.tufts.edu), the average cost of developing a new drug is $802 mil-
lion dollars, more than double since the 1980s (Tufts-CSDD, 2001). By May
of 2003, Tufts CSDD released a new total cost of $897 to develop a new
prescription drug, including cost of post-approval research. However, money
spent on R&D of drugs targeted at ARI, diarrhoeal diseases, malaria, and TB
combined was under that minimum amount in 1999 (WHO, 2000).

5.2.   Customising interventions according to
level of care

Some countries have placed antibiotics according to level of care. This type
of policy needs to be accompanied by an STG and instructions for the clinical
management of severe diseases that requires the proper diagnosis and readi-
ness to transfer patient to the next level of care whenever possible. Countries
with different levels of healthcare access needs to make sure that one or more
antibiotic management policies will be in place with the appropriate training
of providers, at different levels of healthcare access.

In tertiary and secondary care there should be in place: antibiotic restriction formulary, infection control committees (ICC), DTCs, ongoing training
and education of healthcare practitioners, surveillance in high-risk areas
(ICUs).

In primary care there should be in place: STGs for key infectious diseases,
training and education of healthcare practitioners including health educators
and HCWs in rural settings.

At the ambulatory level there should be in place controls on: OTC sale and
dispensing of antibiotics, unsanctioned dealers, illegal distributors, access to
antibiotics without a prescription in neighbouring countries, online Internet
sales, and training of dispensers.

Since antimicrobial resistance is a global problem, harmonisation of
national policies is important but so far it has been an elusive goal (Gould
et al., 2000).

REFERENCES

Asturias, E., Dueger, E., Torres, O., Grazioso, C., and Figueroa, J., 2003, *Risk Factors for Antibiotic Resistance of Streptococcus pneumoniae Among Guatemalan Children from Different Socio-economic Strata and Health Care Delivery Systems: Preliminary Surveillance for a Program for the Judicious Use of Antibiotics in the Outpatient Setting.* National Congress of Pediatrics, Guatemala City, Guatemala.

Bartoloni, A., Cutts, F., Leoni, S., Austin, C., Mantella, A., and Guglielmetti, P., 1998, Patterns of antimicrobial use and antimicrobial resistance among healthy children in Bolivia. *Trop. Med. Int. Health*, **3**, 116–123.

Bavestrello, L. and Cabello, A., 2000, How Chile tackled overuse of antimicrobials. *WHO Essential Drugs Monitor*, Issue No 28–29,13–14.

Becker, J., Drucker, E., Enyong, P., and Marx, P., 2002, Availability of injectable antibiotics in a town market in southwest Cameroon. *Lancet Infect. Dis.*, **2**, 325–326.

Beilby, J., Marley, J., Walker, D., Chamberlain, N., Burke, M., and Group, F. S., 2002, Effect of changes in antibiotic prescribing on patient outcomes in a community setting: A natural experiment in Australia. *Clin. Infect. Dis.*, **34**, 55–64.

Bojalil, R., and Calva, J., 1994, Antibiotic misuse in diarrhea. A household survey in a Mexican community. *J. Clin. Epidemiol.*, **47**(2), 147–156.

Brock, D., 2001, Some questions about the moral responsibilities of drug companies. *Dev. World Bioethics*, **1**, 33–37.

Calva, J., 1996, Antibiotic use in a periurban community in Mexico: a household and drugstore survey. *Soc. Sci. Med.*, **42**(8), 1121–1128.

CDC 2002, Summary Report of a Public Health Meeting: Input for a Public Health Action Plan to Combat Anitmicrobial Resistance. Part II: Global Issues, The U.S. Federal Interagency Task Force on Antimicrobial Resistance.

Cunha, B., 2003, Penicillin resistance in pneumococcal pneumonia. *Postgrad. Med.*, **113**(1) 42–44, 47–48.

ESCMID, 2002, Study Group on Antibiotic policies, European Society of Clinical Microbiology and Infectious Diseases.

Fidler, D. P., 1998, Legal Issues Associated with Antimicrobial Drug Resistance. *Emerging Infect. Dis.*, **4**(2).

Gamboa, L., Camou, T., Hortal, M., and Castaneda, E., 2002, Dissemination of Streptococcus pneumoniae Clone Colombia5–19 in Latin America. *J. Clin. Microbiol.*, **40**(11), 3942–3950.

Geue, A., 2000, Containing antibiotic resistance: Australia's contribution to the global strategy. *Communicable Diseases Intelligence* **24**(10).

Gonzales, R., Steiner, J., Lum, A., and Barrett, P., 1999, Decreasing antibiotic use in ambulatory practice: impact of a multidimensional intervention on the treatment of uncomplicated acute bronchitis in adults. *J. Am. Med. Assoc.*, **281**(16), 1512–1519.

Gordon, D., Noah, D., Fidas, G., and Gannon, J., 2000, *The Global Infectious Disease Threat and Its Implications for the United States.* Central Intelligent Agency, Washington, DC.

Gould, I. M., 2002, Antibiotic policies and control of resistance. *Curr. Opinion Infect. Dis.*, **15**, 395–400.

Gundersen, S., 1992, Resistance problems in developing countries—use and misuse of anti-infective agents. *Tidsskr Nor Laegeforen*, **112**(21), 2741–2746.

Harrison, P. and Lederberg, J., 1998, Antimicrobial Resistance: Issues and Options. Workshop Report. Institute of Medicine, National Academy of Sciences, Washington p. 115.

Hemming, M., and Harvey, K., 1999, International Strategies on Antibiotics, Health Action International News.

Holloway, K., and Gautam, B., 2001, Consequences of over-prescribing on the dispensing process in rural Nepal. *Trop. Med. Int. Health*, **6**(2), 151–154.

Jeena, P., Thompson, E., Nchabeleng, M., and Sturm, A., 2001, Emergence of multi-drug-resistant Acinetobacter anitratus species in neonatal and paediatric intensive care units in a developing country: Concern about antimicrobial policies. *Ann Trop Paediatr.* **21**(3), 245–251.

Kollef, M., (2001). *Antibiotic Resistance Among Gram-Positive Bacteria in the Hospital Setting: What Can We Do About It?* Medscape Infectious Diseases.

McCaig, L., Besser, R., and Hughes, J., 2003, Antimicrobial drug prescription in ambulatory care settings, United States, 1992–2000. *Emerg. Infect. Dis.,* **9**(4), 432–437.

McCaig, L. F. and Hughes, J. M., 1995, Trends in antimicrobial drug prescribing among office-based physicians in the United States. *JAMA* **273**(3), 214–219.

McGowan, J. E. (2001). Economic impact of antimicrobial resistance. *Emerg. Infect. Dis.,* **7**(2).

McManus, P., Hammond, M., Whicker, S., Primrose, J., Mant, A., and Fairall, S., 1997, Antibiotic use in the Australian community, 1990–1995. *MJA*, **167**(3), 124–127.

Nathwani, D. and Davey, P., 1999, Antibiotic-prescribing—are there lessons for physicians? *Q. J. Med.*, **92**(5), 287–292.

O'Dempsey, T., McArdle, T., Lloyd-Evans, N., Baldeh, I., Lawrence, B., Secka, O., and Greenwood, B., 1996, Pneumococcal disease among children in a rural area of west Africa. *Pediatr. Infect. Dis. J.,* **15**(5), 431–437.

Odhacha, A., 1998, Health worker performance after training in integrated management of childhood illness—Western province, Kenya, 1996–1997. *MMWR,* **47**(46), 998–1001.

Okeke, I., Lamikanra, A., and Edelman, R., 1999, Socioeconomic and behavioral factors leading to acquired bacterial resistance to antibiotics in developing countries. *Emerg. Infect. Dis.* **5**(1).

Okeke, I., and Sosa, A., 2003, Antibiotic resistance in africa—discerning the enemy and plotting a defense. *Africa Health* **25**(3), 10–15.

Paredes, P., 1996, Factors influencing physicians' prescribing behaviour in the treatment of childhood diarrhoea: Knowledge may not be the clue. *Soc. Sci. Med.*, **42**(8), 1141–1153.

Prazuck, T., Falconi, I., Morineau, G., Bricard-Pacaud, V., Lecomte, A., and Ballereau, F., 2002, Quality control of antibiotics before the implementation of an STD program in Northern Myanmar. *Sex Transm. Dis.* **29**(11), 624–627.

Riewpaiboon, A., Sunderland, V., Passmore, P., Sirinavin, S., Jirasmitha, S., Tangcharoensathien, V. *et al.*, 1997, Evaluation of cost and appropriateness of antibiotic prescribing for community-acquired pneumonia using a novel computer program. In *State of the Art and Future Directions.* Chiang Mai, Thailand and School of Pharmacy, Curtin University of Technology, Western Australia.

Rodriguez, A. J., Samaniego, D. R., Soskin, A., Rodriguez, C., Canese, J., de Canese, J., *et al.*, 2002, Comparative study of antimicrobial resistance of Pseudomonas aeruginosa strains isolated from patients of Caracas and Asuncion in a 4-year-period. *Chemotherapy* **48**(4), 164–167.

Salvatierra-Gonzalez, R., and Benguigui, Y., (eds.), 2000, *Resistencia Antimicrobiana en la Americas: Magnitud del problema y su contencion.* Washington, DC, OPS/HCP/HCT/163/2000.

Schüklenk, U. and Ashcroft, R., 2001, Affordable access to essential medication in developing countries: Conflicts between ethical and economic imperatives. *J. Med. Philos.,* **27**(2), 179–195.

Scrip (2003). *USTR Looks at Counterfeits and Data Exclusivity.* PJB Publications Ltd, Surrey. p. 18.

Sintchenko, V., Iredell, J., Gilbert, G., and Coiera, E., 2001, What do physicians think about evidence-based antibiotic use in critical care? A survey of Australian intensivists and infectious disease practitioner. *Int. Med. J.* **31**, 462–469.

Sosa, A. and Stelling, J., 2003, *Using AMR Surveillance Data to Select Antimicrobial Agents for the Treatment of Key Bacterial Infectious Diseases in Nepal.* The Alliance for the Prudent Use of Antibiotics (APUA), Boston, MA.

Sosa, A. and Travers, K., 2001, *PAHO/APUA Report on Physician Antibiotic Prescribing Practices And Knowledge in Seven Countries in Latin America.* The Alliance for the Prudent Use of Antibiotics (APUA), Boston, MA.

South, M., Royle, J., and Starr, M., 2003, A simple intervention to improve hospital antibiotic prescribing. *Med. J. Aust.* **178**, 207–209.

Steinman, M., Gonzales, R., Linder, J., and Landefeld, C., 2003, Changing use of antibiotics in community-based outpatient practice, 1991–1999. *Ann. Intern. Med.* **138**(7), 525–533.

Sterling, T. R., Lehmann, H. P., and Frieden, T. R., 2003, Impact of DOTS compared with DOTS-plus on multidrug resistant tuberculosis and tuberculosis deaths: Decision analysis. *BMJ* **326**(7389), 574.

Stratchounski, L., Andreeva, I., Ratchina, S., Galkin, D., Petrotchenkova, N., Demin, A. et al., 2002, The Inventory of Antibiotics in Home Medicine Cabinets—The Results of Russian Nation-wide Cross-sectional Observational Study, 42nd ICAAC Abstracts, American Society for Microbiology, San Diego, CA.

Sudha, V., Prasad, A., Soni, G., Khare, S., and Singh, S., 2003, Simple Criteria for Choosing Antimicrobials for Empiric Therapy of Typhoid Based on Existing Susceptibility Patterns of S. typhi Strains—a Model Proposed. 4th International Symposium on Antimicrobial Agents and Resistance (ISAAR 2003), Seoul, Korea.

Tapsall, J., 2001, Antimicrobial Resistance—Susceptibility Surveillance Linked to Public Health Action. Presentation at the Antibiotic Resistance Surveillance Workshop. Joint Expert Technical Advisory Committee on Antibiotic Resistance (JETACAR). Melbourne, Australia.

Tawfik, Y., 2000, Promising interventions for improving private practitioner's practices in child survival. *Global Health Link: The Newsletter of the Global Health Council* **105**, 19–20.

TRIPS, 2001, *Declaration on the TRIPS Agreement and Public Health.* The Fourth WTO Ministerial Conference, Doha, Qatar, World Trade Organization.

Tufts-CSDD, 2001, Tufts Center for the Study of Drug Development Pegs Cost of a New Prescription Medicine at $802 Million.

Tufts-CSDD, 2003, Total Cost to Develop a New Prescription Drug, Including Cost of Post-Approval Research, is $897 Million.

Unknown, 2001, The Microbial Threat. Visby Conference 2001. Visby, EU2001.SE.

Unknown, 2002, Illegal prescription drug distribution is big business. *Infect. Dis. News.*

Wachter, D., Joshi, M., and Rimal, B., 1999, Antibiotic dispensing by drug retailers in Kathmandu, Nepal. *Trop. Med. Int. Health* **4**(11), 782–788.

WHO, 1996a, ATC. Classification Index. Geneva, WHO Collaborating Centre for Drug Statistics Methodology.

WHO, 1996b, Guidelines for Drug Donations. Geneva, World Health Organization.

WHO, 1999, Removing Obstacles to Healthy Development. Geneva.

WHO, 2000, Overcoming Microbial Resistance. Infectious Disease Report. Geneva, World Health Organization-http://www.who.int/infectious-disease-report/2000/htm.

WHO, 2001, The Multi-Country Evaluation of IMCI Effectiveness, Cost and Impact (MCE). Department of Child and Adolescent Health and Development. Geneva.

WHO, 2002, The medicines Crossing Borders Project—improving the quality of donations." *Essential Drugs Monitor* **31**, 35.

Wilton, P., Smith, R., Coast, J., and Millar, M., 2002, Strategies to contain the emergence of antimicrobial resistance: A systematic review of effectiveness and cost-effectiveness. *J. Health Serv. Res. Policy*, **7**(2), 111–117.

Chapter 32

Antimicrobial Resistance and its Containment in Developing Countries

Denis K. Byarugaba

Department of Veterinary Microbiology and Parasitology, Makerere University, P.O. Box 7062, Kampala, Uganda

1. INTRODUCTION

Antimicrobial resistance (AR) is a complex problem and a big challenge to public and animal health globally. In the last several years, the frequency and spectrum of antimicrobial-resistant infections have increased in both animals and man in developed nations and the developing countries (DCs) (Andrade *et al.*, 2003; Ashkenazi *et al.*, 1995; Bennish *et al.*, 1992; Bogaerts *et al.*, 1997; Daniels *et al.*, 1999; Filius and Gyssens, 2002; Flournoy *et al.*, 2000; Gupta *et al.*, 2001; Russo and Johnson, 2003). This increasing frequency is attributed to a combination of microbial characteristics, selective pressures of antimicrobial use, and societal and technologic changes that enhance the development and transmission of drug-resistant organisms (Aarestrup *et al.*, 2001; Cohen, 1997; Ellner, 1987; Lipsitch and Samore, 2002; McEwen and Fedorka-Cray, 2002; Rubin and Samore, 2002; Vidaver, 2002). AR is a natural consequence of infectious agents' adaptation to exposure to antimicrobials used in medicine, food animals, crop production, and the widespread use of disinfectants in farm and household chores (Allen *et al.*, 1999; Harris *et al.*, 2002; Zaidi *et al.*, 2003). The increased use of antimicrobial agents has increased selective pressure on organisms. As the opportunity for organisms to be exposed to the agents is extended, so is their opportunity to acquire mechanisms of resistance. It is now accepted that antimicrobial use is the single most important factor responsible

Antibiotic Policies: Theory and Practice. Edited by Gould and van der Meer
Kluwer Academic / Plenum Publishers, New York, 2005

for increased AR (Aarestrup, 2000, 1999; Lang *et al.*, 2002; Lipsitch and Samore, 2002; Lopez-Lozano *et al.*, 2000; Rubin and Samore, 2002).

Antimicrobial use is influenced by the interplay of the knowledge, expectations, and interactions of prescribers, users, and drug companies playing within the economic field and the political environment that influence the animal and public health systems and policies (Nyamogoba and Obala, 2002). While in the developed nations AR is mainly associated with overuse of the antimicrobial agents, it is paradoxical that underuse and under-dosing are the primary factors responsible for the problem in DCs (WHO, 2000). Poverty, hunger, malnutrition, coupled with other natural catastrophies in most DCs play an important role in the development and spread of AR. Unending wars and civil strife enhance the spread of the diseases of poverty such as cholera and typhoid as recently in Liberia (Moszynski, 2003) and breed good ground for development of resistance.

The recent AIDS epidemic is one single disease that has devastated societies and communities in DCs in the recent times. It has greatly enlarged the population of immunocompromised patients and left them completely defenceless and at great risk of numerous infections. Over 90% of these patients (over 30 million) are said to be found in DCs with about 22 million being found in Africa (Clark and O'Brien, 2003). This pandemic is ripping apart the social and economic fabric of this part of the world. Many AIDS patients use antimicrobial drugs more frequently to guard against or treat infections, thus increasing the selection pressure for resistant organisms. Besides, many come down with several chronic opportunistic infections, such as tuberculosis, that require hospitalisation for longer periods increasing further the risks for acquiring highly resistant pathogens found in hospital settings and spreading them (Leegaard *et al.*, 2002; Phongsamart *et al.*, 2002). Without making significant headway with AIDS prevention and mitigation in DCs, controlling AR will still remain an unreachable goal. This chapter discusses the current situation of AR in DCs, the factors contributing to the development and spread of the problem, the insurmountable task of controlling the problem amidst other major priorities and possible opportunities for tackling it.

2.   EPIDEMIOLOGY OF ANTIMICROBIAL RESISTANCE

The epidemiology of AR is complex and influenced by many factors. Understanding the problem requires that the rates of changes in the susceptibility to specific antibiotics of organisms isolated from both man and animals be available. This enables a scientific risk assessment to be made and baselines for intervention to be established (Vose *et al.*, 2001). In a bacterial population, the three conditions responsible for the rapid spread of AR are the presence of resistance determinants, vertical or horizontal spread of these resistance determinants, and selection pressure (Edelstein, 2002; Franklin *et al.*, 2002; O'Brien, 2002).

The genetic basis and the phenotypic expression of resistance is extremely complex. This is because there are different classes of antimicrobials each with a different molecular target and a single bacterial species may exhibit more than one resistance mechanism against a single class of antimicrobes (Smith and Lewin, 1993). Some bacterial species develop (or acquire) resistance more readily than others when exposed to apparently similar selective pressures (Ellner *et al.*, 1987). Whether particular organisms become resistant to a particular antimicrobial agent may depend on many factors including the basic physiology of the bacteria, the characteristics of genetic mutations that occur, the prevalence of resistance genes that might be acquired, or the quantity and quality of exposure to the antimicrobial agent (Parry, 1989). For example *Escherichia coli* and enterococci (*Enterococcus faecium* and *Enterococcus faecalis*) are used internationally as Gram-negative and positive indicator bacteria, respectively, because of their ability to harbour several resistance determinants (Aarestrup, 1999; Caprioli *et al.*, 2000; Catry *et al.*, 2003).

Rapid spread of genes resistant to antimicrobial agents can occur in a bacterial population and from one ecosystem to another. Particular antibiotic resistance genes first described in human specific bacteria have been found in animal specific species of microorganisms and vice versa, suggesting those bacterial populations can share and exchange these genes (Aarestrup, 2000). The development of resistance in one bacterial population can, and does, spread to other populations over time.

The selection pressure determines the rate and extent of the emergence of AR. The prevalence and persistence of AR is the result of the complex interaction between antimicrobial drugs, microorganisms, the host, and the environment (O'Brien, 2002). It alters the populations through the elimination of susceptible organisms and the survival of resistant ones. The use of antimicrobials is the main cause of selection pressure(Aarestrup, 2000; Drlica, 2003; Sexton, 2000). Pathogenic organisms are clearly the target population of antimicrobial drugs, on which, by consequence a selection pressure can be exerted. Antimicrobial drugs also exert selection pressure on commensal bacteria (Blake *et al.*, 2003). Commensal bacteria are present on the skin, in the upper respiratory tract, and especially in the digestive tract. The levels of AR in faecal commensal bacteria can reflect the selection pressure exerted by the use of antimicrobial agents in a certain environment (van den, 2001).

Monitoring AR gives very useful data. However, in pathogenic bacteria, it may be less accurate because the resistance patterns of pathogenic strains isolated from autopsy or following therapy failure can be altered by the preceding antimicrobial treatment (Caprioli *et al.*, 2000). Susceptibility patterns of indicator bacteria derived from healthy individuals are suggested as a good predictor of the resistance situation in the bacterial population as a whole (van den, 2001). The difference in the epidemiology of AR between the developed nations and DCs is the environmental factors especially to do with hygiene and disease control and management.

3.   USE OF ANTIMICROBIAL AGENTS

The relationship between antibiotic usage and AR for many types of pathogens is largely mediated by indirect effects or population level selection (Lipsitch and Samore, 2002). When resistant and susceptible organisms compete to colonise or infect hosts, increasing use of the antibiotic will result in increase in frequency of organisms resistant to that drug in the population, even if the risk for treated patients is modest (Lipsitch and Samore, 2002). Antimicrobial use and patient to patient transmission are not independent pathways for promotion of AR, rather they are inextricably linked. Understanding in detail, for each pathogen, the mechanism by which antimicrobial use selects for AR in treated patients and in the population is of more than academic importance (Smith, 2000). For practitioners, these mechanisms matter for making well-informed decisions about the design of treatment protocols, the choice of antibiotics and doses for particular indications. For policy makers, these issues have direct bearing on the design of campaigns to encourage more rational antibiotic use and on the priorities in regulating the use of antimicrobial agents for human and animal use.

3.1.   Use of antimicrobials in food animals and their impact on public health

The issue of antimicrobial use in food-producing animals has been controversial for more than three decades. Recent scientific evidence has highlighted concerns over the human health impact of resistant bacteria acquired from animals via food. Assessments examining the human health impact of antimicrobials used in food-producing animals have demonstrated quantitatively that resistance development in food-producing animals does impact on human health (Barber, 2001; Padungton and Kaneene, 2003; White *et al.*, 2002).

The use of antimicrobials in veterinary practice as therapeutic and prophylactic agents in addition to use as antimicrobial growth promoters greatly influences the prevalence of resistance in animal bacteria and poses a great risk for the emergence of antibiotic resistance in human pathogens (Aarestrup,

1999; Barber, 2001). Bacterial food borne illnesses have a major public health impact and are a growing problem worldwide. Animals serve as reservoirs for many of these food borne pathogens. Epidemiological information already indicates that food of animal origin is the source of a majority of food borne bacterial infections caused by *Campylobacter, Yersinia, E. coli* 0157, non-typhoid *Salmonella*, and other pathogens (Aarestrup and Engberg, 2001; Aarestrup *et al.*, 2001; Wegener *et al.*, 1997). Antimicrobial use increases the frequency of AR in zoonotic pathogens (Aarestrup, 1999). Human to human transmission of zoonotic pathogens is rare, although it may occur in settings where humans are immunocompromised or where the gut community has been disturbed by heavy medical antibiotic use. Therefore, the incidence of AR in zoonotic agents of humans is directly related to the prevalence of AR bacteria in food animals (Barber *et al.*, 2003; White *et al.*, 2002). Antimicrobial agents have been used for many years for either treatment of infections, growth promotion in farm animals, or for prophylactic use in intensive animal rearing. The worldwide use of antibiotics for animal health purposes was estimated at 27,000 tonnes of which 25% was used in Europe (Degener, 2002). Table 1 shows a comparison between the use of antimicrobial agents between a DC (Kenya) and a developed nation ranked as a low ratio user (Denmark). Clearly there is more use of antibiotics in Denmark than Kenya.

The use of antimicrobials in food-producing animals can lead to emergence of resistant bacterial strains, allergic reactions in animals and humans, and other adverse effects depending on the compound (Aarestrup *et al.*, 2001; Tollefson and Flynn, 2002). Bacteria from animals may reach the human population by many routes. Drug-resistant bacteria of animal origin, such as *E. coli* and *Enterococcus* spp. can colonise the intestine of humans (Wegener *et al.*, 1997). Heavily exposed individuals like slaughterhouse workers, food handlers, and farmers feeding antimicrobials to animals have a higher frequency

Table 1. Comparison of consumption of antibiotics (kg) in food animals between a developing country (Kenya) and a developed nation (in brackets, Denmark)

Antimicrobial class	Year		
	1996	1998	1999
Tetracycline	15,889 (12,900)	7,782 (12,100)	3,324 (16,200)
β-lactams	572 (13,000)	1,921 (21000)	1195 (21,300)
Aminoglycosides	752 (7,100)	2,421 (7,800)	843 (7,500)
Macrolides	165 (7,600)	8 (7100)	000 (8,700)
Sulfonamides	499 (2,100)	934 (100)	6,604 (100)
Quinolones	8 (NA)	177 (400)	252 (150)

Sources: Mitema *et al.* (2001), Anonymous (1999), NA—Not available.

of resistant organisms. Besides transfer of drug resistance there is another important problem of drug residues in animal products at slaughter. While this problem has been addressed in developed nations by defining Maximum Residue Levels (MRL) for most drugs for the maximum acceptable level of antibiotics in animal products, very little information is available in DCs and there is almost nothing done to enforce some of these international standards. Only isolated studies have been done to collect data on which rational decisions for intervention can be based.

4.   THREAT OF ANTIMICROBIAL RESISTANCE

Antimicrobial resistance is a natural biological phenomenon. It has become a significant public health problem because it has been amplified many times as a natural consequence of bacterial adaptation to exposure to antimicrobials used in medicine, food animals, crop production, and the widespread use of disinfectants in farm and household products (Bloomfield, 2002; McEwen and Ferdorka-Cray, 2002; Vidaver, 2002). There is an increasing decline in effectiveness of the existing antimicrobial agents. Infections have thus become more difficult and expensive to treat and epidemics harder to control (Cohen, 1994). While the use of drugs combined with improvements in sanitation, housing, and nutrition, and the advent of wide-spread immunisation programmes have led to a dramatic drop in deaths from diseases that were previously widespread, untreatable, and frequently fatal in developed nations. The situation in DCs remains as grim or worse than that of the previous generations of the industrialised nations. Preventable infectious diseases are still present, threatening the life and people's livelihoods, causing deaths and suffering of the underprivileged populations due to poverty, hunger, and inadequate resources. Previously treatable infectious diseases such as tuberculosis, malaria, acute respiratory diseases, and diarrhoea now cause the highest morbidity and mortality in DCs. Table 2 shows examples of some of the most important pathogens in DCs in which the resistance problem has developed.

4.1.   Resistance in malaria

Malaria is one of the most serious public health problems and a leading cause of morbidity and mortality in many tropical countries. Malaria kills more than 1 million people a year, and 3,000 deaths a day (WHO, 2002). Hundreds of millions of people most of them children and pregnant women in sub-Saharan Africa suffer acute attacks of malaria-induced fever, often several times a year. It is a complex disease that varies widely in epidemiology and clinical manifestation in different parts of the world. This variability is the result of factors such

Table 2. Resistance to important bacterial pathogens in developing countries

Disease	Agent	Resistance
Malaria	*Plasmodium falciparum*	Chloroquine
Pneumonia	*Streptococcus pneumoniae*	Penicillin
Dysentery	*Shigella dysenteriae*	Multiresistant
Typhoid	*Salmonella typhi*	Multiresistant
Gonorrhoea	*Neisseria gonorrhoeae*	Pen and tetracycline
Tuberculosis	*Mycobacterium tuberculosis*	Rifampicin and INH
HIV/AIDS	HIV-1 non-B subtype	NRTIs, PIs, and NNRTIs
Nosocomial	*Staphylococcus aureus*	Methicillin resistant.

Notes: NRTI, nucleoside reverse transcriptase inhibitor; NNRTI, non-nucleoside reverse transcriptase inhibitor; Pen, penicillin; PI, protease inhibitor; INH, isoniazid.

as the species of malaria parasites that occur in a given area, their susceptibility to commonly used or available anti-malarial drugs, the distribution and efficiency of mosquito vectors, climate and other environmental conditions, and the behaviour and level of acquired immunity of the exposed human populations. Drug resistance has been implicated in the spread of malaria to new areas and re-emergence of malaria in areas where the disease had been eradicated (Bloland, 1993). Drug resistance also plays a significant role in the occurrence and severity of epidemics in some parts of the world. Population movement has introduced resistant parasites to areas previously free of drug resistance. The economics of developing new pharmaceuticals for tropical diseases, including malaria, are such that there is a great disparity between the public health importance of the disease and the amount of resources invested in developing new cures (Ridley, 1997). This disparity comes at a time when malaria parasites have demonstrated resistance to almost every anti-malarial drug currently available, significantly increasing the cost and complexity of achieving parasitological cure.

For decades chloroquine (CQ) was the mainstay of anti-malarial therapy but the emergence of *Plasmodium falciparum* resistance has challenged the control efforts (Campbell, 1991). In East Africa, the level of resistance has risen steadily over the last 20 years, with recent studies indicating that CQ fails to clear parasites in upto 50–80% of patients (Bayouni *et al.*, 1989; Branding-Bennett *et al.*, 1988; Fowler *et al.*, 1993; Premji *et al.*, 1993; Watkins *et al.*, 1988; Wolday *et al.*, 1995). Table 3 shows clinical failure rates for CQ and sulfadoxine-pyrimethamine (SP) in Uganda.

Despite these rising levels of resistance, CQ has remained the first-line anti-malarial in many African countries because of its low cost and observed clinical efficacy (Barat, 1998; Hoffman *et al.*, 1998). The spread of CQ resistance has been temporarily associated with increased malaria-related morbidity and

Table 3. *In vivo* clinical failure rates of *P. falciparum* to chloroquine (CQ) and sulfadoxine-pyrimethamine (SP) in Uganda (1996–9) in children under <5 years, during 14 days of follow up

District	Year	Geographic location	Clinical failure rates			
			CQ		SP	
			n	%	n	%
Bundibugyo	1996	Rural	60	33	38	5
Kabarole	1996	Road side	60	58	29	3
Jinja	1996	Urban	52	12	35	7
Tororo	1996	Road side	33	60	—	—
Tororo	1997	Rural	78	22	—	—
Mbarara	1997	Hospital	53	81	64	25
Kampala	1998–9	Urban	142	62	45	11

Sources: Kamya *et al.*, 2002, Kilian *et al.* (1997).

mortality in Africa, highlighting the urgent need to change anti-malarial drug policy (Trape, 2001) in some countries (Kamya *et al.*, 2002). Because of the current healthcare infrastructure and the influence of the private sector, approaches to malaria therapy in sub-Saharan Africa will favour increased access to drugs over restricted access which may lead to short-term reductions in malaria morbidity and mortality but is also likely to further increase resistance. Mathematical models for the transmission dynamics of drug sensitive and resistant strains have been developed and can be a useful tool to help to understand the factors that influence the spread of drug resistance and can therefore help in the design of rational strategies for the control of drug resistance (Koella and Antia, 2003). However, significant advances against anti-malarial drug resistance is probably unlikely to be achieved without major improvements in health infrastructure leading to higher quality services that are more readily available.

4.2.   Resistance in tuberculosis

Tuberculosis is an enormous global health problem. The burden of the disease in DCs continues to grow largely fuelled by the HIV/AIDS pandemic and poor public health infrastructure (Yun *et al.*, 2003). The incidence of tuberculosis in Africa is estimated at 259 persons per 100,000 population per year compared to about 50 per 100,000 population per year in Europe and America (Dye *et al.*, 1999). Resistance of *Mycobacterium tuberculosis* to antibiotics is a man-made amplification of spontaneous mutations in the genes of the tubercle bacilli. Treatment with a single drug due to irregular drug supply, inappropriate prescription, or poor adherence to treatment suppresses the growth of susceptible strains to that drug but permits the multiplication of drug-resistant strains. Dramatic outbreaks of multidrug-resistant tuberculosis (MDR-TB) in HIV-infected patients have recently focused international attention on the emergence of strains of *M. tuberculosis* resistant to antimycobacterial drugs (Yun *et al.*, 2003). MDR-TB, defined as resistance to the two most important drugs, isoniazid (INH) and rifampicin (RMP) is a real threat to tuberculosis control. The prevalence of drug resistance, including MDR in different countries, has been extensively reviewed by WHO (1997) and a comparison between some developed nations and DCs is given in Table 4.

The prevalence of primary resistance to any drug ranged from 2% to 41%. Primary MDR-TB was found in every country surveyed except Kenya (surprisingly) (WHO, 1997). The prevalence of acquired drug resistance ranged from 5.3% to 100% and was much higher than that of primary drug resistance. Patients infected with strains resistant to multiple drugs are extremely difficult to cure, and the necessary treatment is much more toxic and expensive. Drug resistance is therefore a potential threat to the standard international method of TB control: the DOTS strategy ("Directly Observed

Table 4. Comparison of prevalence of anti-tuberculosis primary drug resistance between developed nations and DCs (acquired drug resistance figures shown in brackets)

Country	Patients tested	Overall resistance (%)	Resistance to 1 drug (%)	MDR (%)
USA	13,511 (833)	12.3 (23.6)	8.2 (12.5)	1.6 (7.1)
England	2,742 (148)	6.9 (32.4)	4.6 (12.2)	1.1 (16.9)
France	1,491 (195)	8.2 (21.5)	5.6 (12.3)	0.5 (4.1)
Brazil	2,095 (793)	8.6 (14.4)	6.4 (7.3)	0.9 (5.4)
Argentina	606 (288)	12.5 (41.3)	6.6 (12.2)	4.6 (22.2)
Swaziland	334 (44)	11.7 (20.5)	6.6 (9.1)	0.9 (9.1)
Zimbabwe	676 (36)	3.3 (13.9)	1.3 (5.6)	1.9 (8.3)
Botswana	407 (114)	3.7 (14.9)	3.4 (7.0)	0.2 (6.1)
Lesotho	330 (53)	8.8 (34.0)	6.1 (20.8)	0.9 (5.7)
SierraLeone	463 (172)	28.1 (52.9)	16.6 (16.3)	1.1 (12.8)
Kenya	445 (46)	6.3 (37.0)	5.4 (30.4)	0.0 (0.0)

Definitions: Primary resistance is the presence of resistant strains of *M. tuberculosis* in a patient with no history of such prior treatment. Acquired resistance is that which is found in a patient who has received at least 1 month of prior anti-tuberculosis drug treatment. Multidrug resistance (MDR) is defined as resistance to at least INH and RMP, the two most potent drugs and the mainstay of anti-tuberculosis treatment. Overall resistance refers to the proportion of isolates that are resistant to one or more drugs in either primary or acquired resistance.

Source: WHO, 1997.

Treatment, Short-course"). Although preventive therapy decreases the risk of developing active tuberculosis in HIV-infected persons, only a few studies have reported decrease in mortality (Pape *et al.*, 1993). While the availability of effective short-course treatment has been beneficial, tuberculosis still remains one of the greatest global health problems. The number of tuberculosis cases continues to increase in sub-Saharan Africa in contrast to decreasing numbers of cases in developed nations (Johnson and Ellner, 2000). This is principally due to poverty, civil strife, economic turmoil, and bad governance. In DCs, the majority of deaths are not due to MDR-TB but rather due to lack of effective and rationally delivered therapy for drug susceptible tuberculosis.

4.3.   Resistance in HIV/AIDS

The AIDS epidemic is one single disease that has devastated societies and communities in DCs in the recent times. It has greatly enlarged the population of immunocompromised patients and left them completely defenceless and at great risk of numerous infections. About 22 million people around the world have died from AIDS, and about 40 million more are currently infected with the HIV virus (Clark and O'Brien, 2003). About 83% of AIDS deaths and 71% of HIV infections have occurred in war-ravaged, poverty-stricken DCs in sub-Saharan Africa (Mason and Katzenstein, 2000). This pandemic is ripping apart the social and economic fabric of this part of the world. Only by giving those infected with HIV effective treatments, will people be prevented from dying of AIDS in the future.

Many AIDS patients use antimicrobial drugs more frequently to guard against or treat infections, thus increasing the selection pressure for resistant organisms (Tumbarello *et al.*, 2002; Zar *et al.*, 2003). Besides, many contract several chronic opportunistic infections, such as tuberculosis, that require hospitalisation for longer periods increasing further the risks for acquiring highly resistant pathogens found in hospital settings and spreading them (Yun *et al.*, 2003). Despite the fact that many of the anti-retroviral (ARV) drugs are simply unaffordable to many patients in many DCs, resistance has been reported. Most patients receive cheaper, lower cost drugs instead of highly active antiretroviral therapy (HAART) which is more effective but more expensive and unaffordable to many. Although they may get some beneficial response more drug resistance has been found in these patients, especially against lamivudine, than those on HAART (Weidle *et al.*, 2002). Resistance to at least one drug has been reported in 61% of patients studied in Uganda even in patients who had never received treatment (Weidle *et al.*, 2002) implying infection with resistant virus. The high cost of ARV drugs is a very high risk to development of resistance in poor resource countries.

4.4.   Resistance in respiratory infections

Acute respiratory infections (ARI) are a leading cause of childhood mortality, causing 25–33% of all deaths in children in DCs (Berman, 1991). Bacterial ARI are associated with higher case-fatality ratios than infections caused by viruses. *Streptococcus pneumoniae* is the most common cause of bacterial ARI (Huerbner *et al.*, 2000; Kristinsson, 1997). ARI are often treated empirically with antibiotics. Drug-resistance trends are not well documented in most DCs due to limited laboratory capacity (O'Dempse *et al.*, 1996; Forgie, 1992). It is clear, however, that the prevalence of strains resistant to penicillin-related compounds and to co-trimoxazole is increasing (Appellbaum, 1992).

Epidemiological studies have demonstrated that recent antibiotic use is strongly associated with carriage of resistant pneumococci both at the community and individual levels. Among individuals who develop invasive pneumococcal disease, recent antibiotic use poses an increased risk of infection with a resistant strain (Kristinsson, 1997). The biological mechanisms behind the association between recent antibiotic use and carriage of resistant strains are not completely understood. A key factor influencing the emergence and spread of resistant pneumococci is unnecessary antibiotic use for viral respiratory illnesses in humans. This is due to misdiagnosis of conditions because both viral and bacterial agents can cause symptoms of ARI, as well as physician and patient pressures to prescribe antibiotics.

However, while antibiotic overuse is a problem in some developing nations, in DCs, poor access to adequate healthcare is still a primary problem and children requiring antibiotic therapy do not receive it (Berman, 1991).

4.5.   Resistance in enteric bacteria

Important enteric pathogens are becoming increasingly resistant to the major antibiotics that are needed for optimal treatment of patients. Some of the important pathogens include *Vibrio cholerae, Shigella* spp, *Salmonella* spp, and *Campylobacter jejuni* (Hofer *et al.*, 1999; Kariuki, 2001). They cause quite different clinical syndromes; their ecology, epidemiology, and modes of transmission are distinct; and they are widely separated genetically. The fact that these different organisms are becoming increasingly antibiotic-resistant underlines the pervasiveness of the pressures that lead to the emergence and spread of resistance. The prevalence of one of them, *Salmonella typhi* in Egypt is summarised in Table 5.

In DCs, laboratory services, appropriate transport of specimens, and access to healthcare services remain problematic (Hofer *et al.*, 1999; Keusch, 1988).

Table 5. Prevalence of drug resistance to *Salmonella typhi* in Egypt over a 5-year period (1988–1992)

Year	Percent resistance (Number of isolates in brackets)					
	Amp	Chl	TMP-SMX	EM	Tet	MDR
1988	24 (58)	25 (28)	24 (28)	75 (2)	11 (28)	24 (60)
1989	45 (51)	30 (10)	45 (38)	69 (51)	41 (41)	45 (51)
1990	65 (48)	59 (48)	29 (48)	92 (48)	NA	60 (48)
1991	65 (114)	61 (93)	55 (111)	90 (93)	56 (111)	61 (127)
1992	66 (114)	28 (114)	53 (114)	100 (110)	49 (114)	53 (114)

Notes: Amp—Ampicillin; Chl—Chloramphenicol; TMP-SMX—Trimethoprim-Sulfamethoxazole; EM—Erythromycin; Tet—Tetracycline; MDR—Multidrug resistance; NA—Not available.

Source: Wasfy *et al.* (2002).

Data may be subject to bias, as information often comes from hospital-based surveillance and therefore may reflect the more severe infections. Similarly, the data from epidemics may not reflect the situation during non-epidemic periods.

4.6.   Resistance in nosocomial infections

Drug resistance in hospital-acquired infections poses serious constraints on the options available for treatment (Jones, 1992; Kariuki *et al.*, 1993). Nosocomial infections caused by bacteria such as methicillin-resistant *Staphylococcus aureus* (MRSA) pose a serious therapeutic problem worldwide (Pulimood *et al.*, 1996), although information on its frequency in most African countries is not available (Kesah *et al.*, 2003). MRSA is a well-known etiologic agent of very serious infections and more so in HIV-infected patients (Tumbarello *et al.*, 2002). The endemicity of this pathogen in many countries today and the resistance of such isolates to most anti-staphylococcal antibiotics represent a grave threat to public health. In Africa, data on MRSA, particularly antibiotic susceptibilities, are extremely limited. Table 6 shows available data on the prevalence of MRSA in some DCs. The rate of MRSA is relatively high in Nigeria, Kenya, and Cameroon (21–30%), and below 10% in Tunisia and Algeria. All MRSA isolates were sensitive to vancomycin, with MICs ≤ 4 mg/L (Kesah *et al.*, 2003). While the resistance levels may not be so worrying, there is a need to maintain surveillance and control of MRSA infections in Africa where alternative treatment may not be readily available or affordable.

Table 6. Prevalence of MRSA in sub-Saharan Africa 1996–7

Country	Total number of isolates	Number of MRSA	% Resistance
Cote D'Ivore	155	26	16.8
Morocco	167	24	14.4
Tunisia	186	15	8.1
Algeria	208	10	6.7
Senegal	168	21	12.5
Nigeria	142	42	29.6
Cameroon	127	27	21.3
Kenya	137	38	27.7

Source: Kesah *et al.* (2003).

4.7. Resistance in animal diseases

In human medicine, AR and especially the multiple resistance of *S. aureus*, pneumococci, enterococci, and *Enterobacteriaceae* isolated both from nosocomial and non-hospital-related infections (Schwartz *et al.*, 2001) have been found to cause therapy failure and higher morbidity and mortality rates. Major economic losses and animal welfare problems could arise in veterinary medicine, as well, if AR evolves towards a comparable critical level. Resistance of animal pathogens in veterinary medicine are already alarming and are comparable to the situation in humans in DCs. Table 7 shows resistance patterns of clinical isolates from bovine mastitis in Uganda. While the data from clinical isolates may be biased because of previous antibiotic therapy, they nevertheless give an indication of the extent of the problem.

It is now generally accepted that AR in veterinary medicine forms a potential public health hazard. Indeed, the commensal gastrointestinal flora of healthy animals are a reservoir of resistance genes (van den, 2000) that can colonise the flora of humans through the food chain or by direct contact. If, moreover, the underlying resistance genes are horizontally transferred into human pathogenic bacteria, this can result in treatment failure as a consequence of AR (Wegener *et al.*, 1997). The importance of this indirect path of resistance transfer is less clear than for the direct transfer of resistant zoonotic organisms. Because livestock animals are carriers of food-borne pathogens such as *Salmonella* and *Campylobacter* spp, these species also undergo similar selection pressure due to the use of antimicrobial drugs (Aarestrup and Engberg, 2001). As a result, treatment failure in humans arises as a consequence of the intake of these selected resistant organisms either through the food chain and/or through direct contact (Kruse, 1999), irrespective of the horizontal transfer of resistance genes.

Table 7. Resistance patterns of bacteria isolates from bovine clinical mastitis in Uganda

Antibiotic	Per cent resistance to clinically important isolates					
	S. aureus	*Strepto-coccus* spp	*E. coli*	*Klebsiella* spp	*Pseudomonas aeruginosa*	*Coryne-bacterium pyogenes*
Gentamicin	26	0	9	21	30	0
Kanamycin	19	0	22	22	70	0
Chloramphenicol	23	21	35	32	80	20
Erythromycin	47	41	62	8	92	60
Neomycin	43	51	42	3	NA	NA
Streptomycin	39	40	63	66	90	100
Tetracycline	100	100	100	67	70	80
Cloxacillin	96	90	100	100	100	100
Ampicillin	100	76	100	100	97	60
Penicillin	95	90	NA	NA	NA	NA

Source: Byarugaba *et al.* (2001).

5.   IMPLICATIONS OF HIV/AIDS ON THE ANTIMICROBIAL RESISTANCE PROBLEM

Many AIDS patients use antimicrobial drugs more frequently to guard against or treat infections, thus increasing the selection pressure for resistant organisms (Badri *et al.*, 2001). Besides, many come down with several chronic opportunistic infections, such as tuberculosis, that require hospitalisation for longer periods, increasing further the risks for acquiring highly resistant pathogens found in hospital settings and spreading them (Leegaard *et al.*, 2002). The impact of HIV/AIDS depends on the dual infection with other secondary infections such as tuberculosis, meningitis, pneumonia, and other infections (Leegaard *et al.*, 2002; Phongsamart *et al.*, 2002). HIV-infected individuals are at least 30-fold more likely to develop reactivation tuberculosis as HIV infection progresses and are more susceptible to exogenous re-infection and rapid progression to active tuberculosis when re-infected with *M. tuberculosis* (Puerto Alonso *et al.*, 2002).

The interaction of tuberculosis and HIV/AIDS is complex (Phongsamart *et al.*, 2002; Puerto Alonso *et al.*, 2002). The addition of several medications for treatment of tuberculosis for HIV patients already on ARV drugs makes adherence sometimes difficult. Complications often make it worse. These are important risk factors to development of resistance. Implementing and sustaining preventive therapy in HIV-infected persons in Africa bears formidable challenges that require comprehensive efforts for counselling, monitoring of side effects, and supervision of treatment (Mason and Katzenstein, 2000). Given the lack of good infrastructure in most DCs this becomes even more difficult.

The increased use of antibiotics in HIV/AIDS infected persons due to their immunocompromised status breeds good ground for development of AR (Tumbarello *et al.*, 2002). Studies have been published on the use of trimethoprim-sulfamethoxazole (TMP-SMX) (co-trimoxazole) for prophylaxis in HIV/AIDS patients (Badri *et al.*, 2001). In developed countries it is used for prophylaxis against *Pneumocystis carinii* pneumonia in HIV-infected individuals but has gained importance in Africa as well for prophylaxis against other infections such as toxoplasmosis, isosporiasis, *S. pneumonia*, and non-typhi *Salmonella* species which are frequent causes of morbidity and mortality in HIV-infected Africans. In DCs where ARV drugs are not affordable this prophylaxis, because of its low cost, may benefit the HIV-infected patients and has been reported to decrease septicaemia and enteritis significantly (Mason and Katzenstein, 2000). However there is also a danger of developing resistance to this drug as has been reported for many isolates, thus further compromising the benefits of these prophylaxis strategies (Zar *et al.*, 2003).

Evidence from prevention of malaria during pregnancy suggests that parasitological response to treatment among individuals infected with the HIV may also be poor. HIV-seropositive women require more frequent treatment with SP (sulfadoxine-pyrimethamine) during pregnancy in order to have the same risk of placental malaria as is seen among HIV-seronegative women (Parise *et al.*, 1998). Parasitological response to treatment of acute malaria among HIV-seropositive individuals has not been evaluated though.

6. ECONOMIC IMPLICATIONS OF ANTIMICROBIAL RESISTANCE

Antimicrobial resistance is not only a medical problem but also an economic one (Paladino *et al.*, 2002). Resistant organisms cause infections that are more difficult to treat, requiring drugs that are often less readily available, more expensive, and more toxic (Carmeli *et al.*, 1999; Howard *et al.*, 2001). In some cases, certain strains of microbes have become resistant to all available antimicrobial agents (Russo and Johnson, 2003). Resistant Gram-negative and Gram-positive bacteria have been associated with increased direct medical costs ranging from several thousand dollars to tens of thousands of dollars per patient (Paladino *et al.*, 2002). With increasing frequency and levels of AR, drug therapy must be viewed in an economic sense. Several factors impact on cost-effective antimicrobial therapy, including drug cost, drug efficacy and duration of treatment, dose regimen, diagnostic strategies, microbial resistance, and patient compliance.

Comparison has been made of the impacts of infections due to antimicrobial-resistant bacteria with those of infections due to antimicrobial-susceptible strains of the same bacteria. Data shows that for both nosocomial and community-acquired infections, the mortality, the likelihood of hospitalisation and the length of hospital stay were usually at least twice as great for patients infected with drug-resistant strains as for those infected with drug-susceptible strains of the same bacteria (Holmberg *et al.*, 1987). Tuberculosis, treatment costs have been estimated at US$20 for regular treatment, while the cost of treating MDR-TB rises to US$2,000 (WHO, 2001). For HIV/AIDS, it has been indicated that resistance to one protease inhibitor results in resistance to the entire family of drugs thus implying higher costs of treatment of the insensitive strains. This applies to all other major killers such as ARIs, diarrhoeal diseases, malaria, and other STDs which are very prevalent in DCs. Poor outcomes could be attributed both to the expected effects of ineffective antimicrobial therapy and to the unexpected occurrence of drug-resistant infections complicated by prior antimicrobial therapy for other medical problems. Although the adverse economic and health effects of drug-resistant bacterial infections can only be roughly quantified, AR is an important health problem and an economic burden to society (Cosgrove and Carmeli, 2003).

Unfortunately the costs for production of new drugs and introduction onto the market are enormous, estimated at a minimum of US$300 million. This partly explains the reason why since 1970, there have been few classes of antimicrobial agents developed. Their development also takes a period of 10–20 years. Although DCs have the enormous potential of hotspot virgin tropical forests that harbour a lot of plant resources that may provide solutions to many of the current resistance problems, their exploitation requires similar huge financial resources and time to develop them to market level.

Besides the direct costs, there are also biological costs associated with development of AR (Gillespie and McHugh, 1997; Nyamogoba and Obala, 2002) and more importantly, costs related to loss of life and hours spent without productive work during long hospitalisation with resistant disease agents. Table 8 shows the extent of the deaths resulting from major killer diseases in the DCs.

Together, HIV/AIDS, tuberculosis, and malaria claimed 5.7 million lives in 2002 and caused debilitating illness in many millions more (Goeman *et al.*, 1991; WHO, 2002). These are the lives of infants, young children, and young mothers and fathers and the economically most important group in their prime productive years. Resistance to these disease agents puts a heavy burden on the already strained public health.

Table 8. Occurrence of the major killer diseases in the world and proportion of their occurrence in developing countries

Disease	Deaths per year (in million)	New cases per year (in million)	Percentage in developing countries (%)
HIV/AIDS	3	5.3	92
Tuberculosis	1.9	8.8	84
Malaria	>1	300	Nearly 100

Source: WHO (2002).

7. FACTORS CONTRIBUTING TO DEVELOPMENT AND SPREAD OF RESISTANCE

The development and spread of AR is a complex problem driven by numerous interconnected factors, many of which are linked to the use of antimicrobials both in animals, plants and man. It is now accepted that antimicrobial use is the single most important factor responsible for increased AR (McGarock, 2002; Mitema *et al.*, 2001; Rubin and Samore, 2002). Antimicrobial use is influenced by interplay of the knowledge, expectations and interactions of prescribers and patients, economic incentives, characteristics of the health systems, the regulatory environment, and availability of resources. Organisms themselves possess characteristics that enable them to be resistant and these may be enhanced by other environmental factors (Allen *et al.*, 1999; Harris *et al.*, 2002; Tumbarello *et al.*, 2000; Zaidi *et al.*, 2003). Thus extreme poverty in most DCs leads to poor sanitation, hunger, starvation and malnutrition, poor access to drugs, and poor healthcare delivery, all of which may precipitate AR.

7.1. Poverty

Most DCs have inadequate healthcare systems due to limited resources. A health sector situational analysis in Uganda, one of the poorest countries, indicated that with a population of 26 million and a growth rate of 2.5% and fertility rate of 6.9, 49% of this population live below the poverty line, surviving on less than $1 per day. According to disease burden studies, 75% of the life years of Ugandans are lost to premature death due to preventable diseases with malaria being responsible for 15.4%, acute respiratory tract infections 10.5% and diarrhoea 8.4%. HIV/AIDS also, still, claims thousands of lives despite the significant decreases in the prevalence rates in the country. The geographical access to healthcare is limited to about 49% of the population, that is, population living within 5 km of a health unit (Health Facilities Inventory, 1992). Rural communities are particularly affected because health facilities are mostly located in towns and along main roads. Even among these facilities, many do not provide the full range of essential primary healthcare services. This poor health status prevails in most DCs and AR is seen as a secondary problem and not an immediate priority.

7.2.   Malnutrition and hunger

The current prevalence of malnutrition among African children under 5 years has been estimated to be 30% and an estimated 4–5 million children were thought to be infected with HIV at the beginning of this century (Brabin *et al.*, 2001; WHO, 2002). Among refugee children in the former Zaire, those who were malnourished (low weight for height) had significantly poorer parasitological response to both CQ and SP treatment (Wolday *et al.*, 1995). If it is proven that malnutrition or HIV infection plays a significant role in facilitating the development or intensification of anti-malarial drug resistance, the prevalence of these illnesses could pose a tremendous threat to existing and future anti-malarial drugs. Some characteristics of recrudescent or drug resistant infections appear to provide a survival advantage or to facilitate the spread of resistance conferring genes in a population which is further aggravated by low nutritional and immune status.

7.3.   Civil wars and unrest

Civil wars in many DCs have been responsible for breakdown of many health services and have forced many people into refugee camps, or internally displaced people camps. In such concentration camps there is often very poor hygiene and sanitation, good ground for spread of infection, and high selection and spread of resistant organisms. The perpetual wars also limit the enforcement of laws governing manufacture, market authorisation, distribution, and monitoring the sale of drugs, thus predisposing the population to poor and counterfeit drugs.

7.4.   Natural catastrophies

Climatic and weather patterns influence the occurrence, incidence, and distribution of infectious diseases such as cholera and malaria (Shears, 2001). Heavy rains and floods overwhelm the already dilapidated drainage systems, spreading organisms, some of them very resistant, while dry weather encourages spread of airborne pathogens (Byarugaba *et al.*, 2001).

7.5.   Human population growth

The fast growth of globalisation accompanied by enormous growth of trade and travel has increased the speed and facility with which both infectious diseases and resistant microorganisms can spread between continents. The ability of DCs to cope with this globalisation and the attendant spread of resistant organisms all over the world is severely stretched (WHO, 2000). The increased growth has also meant increased demand for resources and services. The rapid population increases and urbanisation without corresponding increase in resources and services has resulted in overcrowding and poor hygiene and sanitation which greatly facilitate the spread of such diseases as typhoid, tuberculosis, respiratory infections, and pneumonia.

7.6.   Societal factors

Societal factors are one of the major drivers of inappropriate antimicrobial use. These include self-medication, non-compliance, misinformation, poor immune status, wrong conceptions and beliefs, advertising pressures, and treatment expectations (Byarugaba *et al.*, 2001; Lipsitch and Samore, 2002; Parimi *et al.*, 2002). The most important underlying factor in DCs, however, is poverty (Hossain *et al.*, 1982). Lack of resources, compounds with other factors as it relates to ignorance, lack of purchasing power, lack of education, and lack of access to proper health facilities and diagnostic facilities (Lindtjorn, 1987). Poverty also means people do not get sufficient food to keep their bodies healthy and fit to fight off infections. They live in poor hygiene, are exposed to water-borne infections due to drinking contaminated water which exposes them to diseases of poverty, such as cholera, typhoid, and other infections. Often economic hardships in DCs lead to premature cessation of treatment or sharing of a single treatment course by a whole family (Byarugaba *et al.*, 2001).

7.7.   Service providers

In many DCs healthcare providers are influenced by financial gains from their practice as they are operated on cost–benefit basis (Byarugaba *et al.*, 2001; Indalo, 1997). This financial interest makes them prescribe antimicrobials even when they are not clinically indicated. Additional profit is sometimes gained by recommending newer and more expensive antimicrobials in preference to older cheaper agents and combination therapies where it is uncalled for, such as in malaria (Anakbonggo and Birungi, 1997). Pharmaceutical companies have been known to pay commissions to prescribers who use more of their products. Besides, the aggressive advertisement methods of many drug companies in electronic and print media leave the prescribers and dispensers vulnerable and the laws regarding such advertising are never implemented due to lack of commitment and resources on the part of the law enforcement departments (Lipsky, 1997). Other factors relating to healthcare providers include, lack of proper training, unprofessional conduct, lack of diagnostic facilities, wrong selection of antimicrobial agents, patient pressures for specific antimicrobials, and fear of treatment failure (Byarugaba *et al.*, 2001).

7.8.   Health institutions

Hospitals are a critical component of the AR problem worldwide (Allen *et al.*, 1999; Goldmann *et al.*, 1996; Harris *et al.*, 2002; Tumbarello *et al.*, 2002; Zaidi *et al.*, 2003). The combination of highly susceptible patients, intensive and prolonged antimicrobial use and cross-infection have resulted in highly resistant bacterial pathogens. Many of the health facilities are overcrowded and have limited capacity compared to the inflow of patients, as they were built many years ago for a smaller population.

Hospitals are also the eventual site of treatment for many patients with severe infections due to resistant pathogens acquired in the community. Recent introduction of free medical care to the population in Uganda saw most health-care units flooded with patients, completely overwhelming the wards. Other small health centers and clinics encounter similar problems. The situation is even worse for those located in rural settings far away from town centers where drug supplies are limited and laboratory facilities are unavailable (Byarugaba *et al.*, 2001).

7.9.   Policies

Globalisation has come with many structural adjustments to keep pace with developments in the rest of the world. The underlying principles of many national drug policies aim to contribute to the attainment of good standards of health by the population, through ensuring the availability, accessibility, and affordability, at all times, of essential drugs of appropriate quality, safety, and efficacy, and by promoting their rational use. However, many of these laws only exist on paper or are poorly communicated to the stakeholders and thus implementation is still difficult due to gross under funding of the health sector. Besides, some result out of inadequate consultation.

Structural adjustment programmes originating from development partners have also had an impact on healthcare delivery systems. Under the current decentralisation in many DCs, districts have the obligation to deliver a package of health services to the population, while the ministries only provide them with the technical support and supervision. Many of the districts do not have sufficient resources and personnel to implement some of the policies and to provide adequate health services. The policies of privatisation and liberalisation have also affected the procurement and supply of drugs. With inadequate supervision from regulatory agencies, these have resulted in sale of expired and counterfeit drugs (Byarugaba *et al*, 2001).

7.10. Use of antimicrobial drugs in food animal production

As already mentioned in Section 3.1, the use of antimicrobials in veterinary practice as therapeutic and prophylactic agents in addition to use as antimicrobial growth promoters greatly influences the prevalence of resistance in animal bacteria and poses a great risk for the emergence of antibiotic resistance in human pathogens (Barber, 2001; McDermott *et al.*, 2002; Shah *et al.*, 1993). There is already high resistance against many antibiotics in livestock (Catry *et al.*, 2003b). The extent to which antibiotic use in animals contributes to the AR in humans is still under debate. However, a wealth of epidemiological information already indicates that food of animal origin is the source of a majority of food borne bacterial infections caused by *Campylobacter, Yersinia, E. coli* (Schroeder *et al.*, 2002a, b), non-typhoid *Salmonella* (Angulo *et al.*, 2000), and other pathogens. Direct evidence indicates that antimicrobial use in animals selects for antimicrobial-resistant bacteria that may be transferred to humans through food or direct contact with animals (Aarestrup, 1999; Aarestrup *et al.*, 2001). In DCs the deliberate use of antimicrobial drugs as growth promoters is still on a limited scale. However, many antimicrobial drugs are used in animals for prophylaxis and therapy (Nakavuma *et al.*, 1994; Mitema *et al.*, 2001). Several guidelines are available for appropriate use of drugs in animals (Anthony *et al.*, 2001; Nicholls *et al.*, 2001), but very little is being done in DCs.

8. STRATEGIES FOR CONTAINMENT OF RESISTANCE IN DEVELOPING COUNTRIES

The problem of AR is a global one and will require a concerted effort and cooperation among nations. Several suggestions have been made (Burke, 2002; Coast and Smith, 2003; Kettler, 2000; Schwartz *et al.*, 1997; Lees and Shojaee, 2002; Wise, 2002) although there is absence of good published evidence concerning cost-effectiveness in reducing emerging resistance (Wilton *et al.*, 2002). Containment will depend on coordinated interventions that simultaneously target the behaviour of providers and patients, and change important features of the environment in which they interact, as well as managerial and policy issues which have been well articulated (Barrett, 2000; Dagan *et al.*, 2001; de Man *et al.*, 2000; Gruson *et al.*, 2000; Hooton and Levy, 2001; Murthy, 2001; Poole, 2001; Samaranayake and Johnson, 1999; Schwartz *et al.*, 1997; Schwarz *et al.*, 2001; Sehgal, 1999; Semjen, 2000; Sirinavin *et al.*, 1998; Slots, 1999; Smith, 1999; Tan *et al.*, 2002; Weinstein, 2001; Wise, 2002).

However, responsible use of the available antimicrobial agents is of paramount importance despite the challenges DCs face (Byarugaba *et al.*, 2003). Use of antimicrobial agents in food producing animals should follow prudent principles to minimise the selection and spread of resistant zoonotic bacteria like *Salmonella* spp and *Campylobacter* spp. Emphasis should be placed on disease preventive measures such as good farm management and hygiene to reduce bacterial load, rather than use of antimicrobials (WHO, 2000). The goal of any programme to monitor the quantities of antimicrobials used in animals is to generate objective quantitative information to evaluate usage patterns by animal species, antimicrobial class, and type of use. These data are essential for risk analysis and planning and can be helpful in interpreting resistance surveillance data and evaluating the effectiveness of prudent use efforts and mitigation strategies. The total consumption of antimicrobials for human usage, food animal, and other uses is a key factor in any consideration of this issue. Data on antimicrobial consumption is scanty and has been reported in only a few countries (Anonymous, 1999; Grave, 1999; Greko, 2000) and less so in DCs (Mitema *et al.*, 2001).

The global coalition against poverty and the concerted efforts to combat AIDS are very important strategies within which AR can be contained. Through massive education, sustained prevention efforts will prevent transmission of infections and liberate resources that can be used to combat AR. Pursuance of the millennium development goals and other global partnerships against important development challenges are very critical to containment of AR not only among the DCs but also among the developed nations. Without a bold, concerted action, not only will millions die in Africa, but the entire world will suffer. To allow sub-Saharan Africa to become socially and economically devastated will have a major impact on the economies of every country of the world and impact on AR. The African Comprehensive HIV/AIDS Partnership (ACHAP) is one answer to the problem.

The public–private partnership has provided a new sense of optimism for fighting many devastating pandemics such as AIDS and has given lots of people hope. ACHAP offers all interested parties a multifaceted paradigm that addresses not only the need for ARV medications, but also the other social and medical facets of the HIV/AIDS problem facing sub-Saharan Africa including indirectly the problems of AR.

9. CONCLUSIONS

The problem of AR in DCs is mainly due to poverty and the factors related to it. The problem of resistance is not quite appreciated by most stakeholders as there are more priorities to address such as provision of basic healthcare or sanitation. These overshadow the problems of AR. Containment of AR in DCs therefore will heavily rely on formation of global partnership with developed countries and organisations to formulate and implement integrated mechanisms to sensitise service providers, the policy-makers, and the users to understand the problem and the consequences of lack of control. Prevention and control will require prudent use of existing agents, discovery of new antimicrobial agents, new vaccines and enhanced public health efforts to reduce transmission. AIDS is a very important player in the development and spread of AR. Without making significant headway with AIDS prevention and mitigation and eradication of poverty, the control of AR will remain a secondary issue in DCs.

REFERENCES

Aarestrup, F. M., 1999, Association between the consumption of antimicrobial agents in animal husbandry and the occurrence of resistant bacteria among food animals. *Int. J. Antimicrob. Agents*, **12**, 279–285.

Aarestrup, F. M., 2000, Occurrence, selection and spread of resistance to antimicrobial agents used for growth promotion for food animals in Denmark. *APMIS Suppl.*, **101**, 1–48.

Aarestrup, F. M. and Engberg, J., 2001, Antimicrobial resistance of thermophilic *Campylobacter*. *Vet. Res.*, **32**, 311–321.

Aarestrup, F. M. and Seyfarth, A. M., 2000, Effect of intervention on the occurrence of antimicrobial resistance. *Acta Vet. Scand. Suppl.*, **93**, 99–102.

Aarestrup, F. M., Seyfarth, A. M., Emborg, H. D., Pedersen, K., Hendriksen, R. S., and Bager, F., 2001, Effect of abolishment of the use of antimicrobial agents for growth promotion on occurrence of antimicrobial resistance in fecal enterococci from food animals in Denmark. *Antimicrob. Agents Chemother.*, **45**, 2054–2059.

Allen, U. D., MacDonald, N., Fuite, L., Chan, F., and Stephens, D., 1999, Risk factors for resistance to "first-line" antimicrobials among urinary tract isolates of *Escherichia coli* in children. *CMAJ*, **160**, 1436–1440.

Anakbonggo, W. W. and Birungi, H., 1997, Prescibing audit with feedback intervention in six regional hospitals and Mulago referral/teaching hospital, Uganda. *The first International Conference on Improving Use of Medicines*, Chiang, Thailand, 1–4 April.

Andrade, S. S., Jones, R. N., Gales, A. C., and Sader, H. S., 2003, Increasing prevalence of antimicrobial resistance among *Pseudomonas aeruginosa* isolates in Latin American medical centres: 5 year report of the SENTRY Antimicrobial Surveillance Program (1997–2001). *J. Antimicrob. Chemother.*, **52**, 140–141.

Angulo, F. J., Johnson, K. R., Tauxe, R. V., and Cohen, M. L., 2000, Origins and consequences of antimicrobial-resistant nontyphoidal *Salmonella*: Implications for the use of fluoroquinolones in food animals. *Microb. Drug. Resist.*, **6**(1), 77–83.

Anonymous. DANMAP 99-Consumption of antimicrobial agents and occurrence of anti-microbial resistance in bacteria from food animals, food and humans in Denmark. ISSN 1600-2032.

Anthony, F., Acar, J., Franklin, A., Gupta, R., Nicholls, T., Tamura, Y. *et al.*, 2001, Antimicrobial resistance: Responsible and prudent use of antimicrobial agents in veterinary medicine. *Rev. Sci. Tech.* **20**, 829–839.

Appelbaum, P. C., 1992, Antimicrobial resistance in *Streptococcus pneumoniae*: An overview. *Clin. Infect. Dis.*, **15**, 77–83.

Ashkenazi, S., May-Zahav, M., Sulkes, J., Zilberberg, R., and Samra, Z., 1995, Increasing antimicrobial resistance of *Shigella* isolates in Israel during the period 1984 to 1992. *Antimicrob. Agents Chemother.*, **39**, 819–823.

Badri, M., Ehrlich, R., Wood, R., and Maartens, G., 2001, Initiating co-trimoxazole prophylaxis in HIV-infected patients in Africa: An evaluation of the provisional WHO/UNAIDS recommendations. *AIDS*, **15**, 1143–1148.

Barber, D. A., Miller, G. Y., and McNamara, P. E., 2003, Models of antimicrobial resistance and foodborne illness: Examining assumptions and practical applications. *J. Food Prot.*, **66**, 700–709.

Barrett, D. C., 2000, Cost-effective antimicrobial drug selection for the management and control of respiratory disease in European cattle. *Vet. Rec.*, **146**, 545–550.

Bayoumi, R., Babiker, H., Ibrahim. S. *et al.*, 1989, Chloroquine resistant *Plasmodium falciparum* in Eastern Sudan. *Acta Tropica*, **46**, 157–165.

Bell, J. M., Turnidge, J. D., and Jones, R. N., 2002, Antimicrobial resistance trends in community-acquired respiratory tract pathogens in the Western Pacific Region and South Africa: Report from the SENTRY antimicrobial surveillance program, (1998–1999) including an in vitro evaluation of BMS284756. *Int. J. Antimicrob. Agents*, **19**, 125–132.

Bennish, M. L., Salam, M. A., Hossain, M. A., Myaux, J., Khan, E. H., Chakraborty, J. *et al.*, 1992, Antimicrobial resistance of *Shigella* isolates in Bangladesh, 1983–1990: Increasing frequency of strains multiply resistant to ampicillin, trimethoprim-sulfamethoxazole, and nalidixic acid. *Clin. Infect. Dis.*, **14**, 1055–1060.

Berman, S., 1991, Epidemiology of acute respiratory infections in children of developing countries. *Rev. Infect. Dis.*, **13**(Suppl. 6), S454–S462.

Blake, D. P., Hillman, K., Fenlon, D. R., and Low, J. C., 2003, Transfer of antibiotic resistance between commensal and pathogenic members of the Enterobacteriaceae under ileal conditions. *J. Appl. Microbiol.*, **95**, 428–436.

Bloland, P. B., 2001, Drug resistance in malaria. WHO/CDS/CSR/DRS/2001.4

Bloomfield, S. F., 2002, Significance of biocide usage and antimicrobial resistance in domiciliary environments. *Symp. Ser. Soc. Appl. Microbiol.*, **31**, 144S–157S.

Bogaerts, J., Verhaegen, J., Munyabikali, J. P., Mukantabana, B., Lemmens, P., Vandeven, J. *et al.*, 1997. Antimicrobial resistance and serotypes of *Shigella* isolates in Kigali, Rwanda (1983 to 1993): Increasing frequency of multiple resistance. *Diagn. Microbiol. Infect. Dis.*, **28**, 165–171.

Brabin, B. J., Premji, Z., and Verhoeff, F., 2001, An analysis of anemia and child mortality. *J Nutr.*, **131**(2S–2), 636S–648S.

Brandling-Bennett, A. D., Oloo, A. J., Watkins, W. M., Boriga, D. A., Kariuki, D. M., and Collins, W. E., 1988, Chloroquine treatment of falciparum malaria in an area of Kenya of intermediate chloroquine resistance. *Trans. R. Soc. Trop. Med. Hyg.*, **82**(6), 833–837.

Burke, J. P., 2002, Rational approaches to combating resistance. *Int. J. Clin. Pract. Suppl.*, 29–36.

Byarugaba, D. K., Olilia, D., Azuba, R. M., Kaddu-Mulindwa, D. H., Tumwikirize, W., Ezati, E. *et al.*, 2001, Development of sustainable strategies for the management of antimicrobial resistance in man and animals at district and national level in Uganda. Feasibility Report, May 2001, Makerere University.

Byarugaba, D. K., 2003, Prudent use and containment of antimicrobial resistance in developing countries. In: OIE International Standards on Antimicrobial Resistance. World Organisation of Animal Health, Paris, France.

Campbell, C. C., 1991, Challenges facing antimalarial therapy in Africa. *J. Infect. Dis.*, **16**, 1207–1211.

Caprioli, A., Busani, L., Martel, J. L., and Helmuth, R., 2000, Monitoring of antibiotic resistance in bacteria of animal origin: Epidemiological and microbiological methodologies. *Int. J. Antimicrob. Agents*, **14**, 295–301.

Carmeli, Y., Troillet, N., Karchmer, A. W., and Samore, M. H., 1999, Health and economic outcomes of antibiotic resistance in *Pseudomonas aeruginosa*. *Arch. Intern. Med.*, **159**, 1127–1132.

Catry, B., Laevens, H., Devriese, L. A., Opsomer, G., and De Kruif, A., 2003, Antimicrobial resistance in livestock. *J. Vet. Pharmacol. Ther.*, **26**, 81–93.

Clark, P. A. and O'Brien, K., 2003, Fighting AIDS in Sub-Saharan Africa: Is a public- private partnership a viable paradigm? *Med Sci Monit.*, 9(9), 28–39.

Coast, J. and Smith, R. D., 2003, Solving the problem of antimicrobial resistance: Is a global approach necessary? *Drug Discov. Today*, **8**, 1–2.

Cohen, M. L., 1997, Epidemiological factors influencing the emergence of antimicrobial resistance. *Ciba Found. Symp.*, **207**, 223–231.

Cosgrove, S. E. and Carmeli, Y., 2003, The impact of antimicrobial resistance on health and economic outcomes, *Clin. Infect. Dis.*, **36**, 1433–1437.

Dagan, R., Klugman, K. P., Craig, W. A., and Baquero, F., 2001, Evidence to support the rationale that bacterial eradication in respiratory tract infection is an important aim of antimicrobial therapy. *J. Antimicrob. Chemother.*, **47**, 129–140.

Daniels, I. R. and Zaman, S. R., 1999, Increasing prevalence of antimicrobial resistance among uropathogens. *JAMA*, **282**, 325–326.

de Man, P., Verhoeven, B. A., Verbrugh, H. A., Vos, M. C., van den Anker, J. N., 2000, An antibiotic policy to prevent emergence of resistant bacilli. *Lancet*, **355**, 973–978.

Degener, J. E., 2002, European study on the relationship between antimicrobial use and antimicrobial resistance. *Emerg. Infect. Dis.*, **8**, 278–282.

Doern, G. V., 2001, Antimicrobial use and the emergence of antimicrobial resistance with *Streptococcus pneumoniae* in the United States. *Clin. Infect. Dis.*, **33**(Suppl. 3), S187–S192.

Drlica, K., 2003, The mutant selection window and antimicrobial resistance. *J. Antimicrob. Chemother.*, **52**, 11–17.

Dye, C., Scheele, S., Dolin, P., Pathania, V., Raviglione, M. C., 1999, Consensus statement. Global burden of tuberculosis: estimated incidence, prevalence and mortality by country. WHO Global Surveillance and Monitoring Project. *JAMA*, **282**, 677–686.

Ebright, J. R., Moore, E. C., Sanborn, W. R., Schaberg, D., Kyle, J., and Ishida, K., 1984, Epidemic Shiga bacillus dysentery in Central Africa. *Am. J. Trop. Med. Hyg.*, **33**, 1192–1197.

Edelstein, P. H., 2002, Predicting the emergence of antimicrobial resistance. *Clin. Infect. Dis.*, **34**, 1418.

Ellner, P. D., Fink, D. J., Neu, H. C., and Parry, M. F., 1987, Epidemiologic factors affecting antimicrobial resistance of common bacterial isolates. *J. Clin. Microbiol.*, **25**, 1668–1674.

Felmingham, D., Reinert, R. R., Hirakata, Y., and Rodloff, A., 2002, Increasing prevalence of antimicrobial resistance among isolates of *Streptococcus pneumoniae* from the PROTEKT surveillance study, and compatative in vitro activity of the ketolide, telithromycin. *J. Antimicrob. Chemother.*, **50** (Suppl. S1), 25–37.

Filius, P. M. and Gyssens, I. C., 2002, Impact of increasing antimicrobial resistance on wound management. *Am. J. Clin. Dermatol.*, **3**, 1–7.

Flournoy, D. J., Reinert, R. L., Bell-Dixon, C., and Gentry, C. A., 2000, Increasing antimicrobial resistance in gram-negative bacilli isolated from patients in intensive care units. *Am. J. Infect. Control*, **28**, 244–250.

Forgie, I. M., Campbell, H., Lloyd-Evans, N., Leinonen, M., O'Neill, K. P., Saikku, P. *et al.*, 1992, Etiology of acute lower respiratory tract infections in children in a rural community in The Gambia. *Pediatr. Infect. Dis. J.*, **11**, 466–473.

Fowler, V. G. Jr., Lemnge, M., Irare, S. G., Malecela, E., Mhina, J., Mtui, S. *et al.*, 1993, Efficacy of chloroquine on *Plasmodium falciparum* transmitted at Amani, eastern Usambara Mountains, north-east Tanzania: An area where malaria has recently become endemic. *J. Trop. Med. Hyg.*, **96**(6), 337–345.

Franklin, G. A., Moore, K. B., Snyder, J. W., Polk, H. C., Jr., and Cheadle, W. G., 2002, Emergence of resistant microbes in critical care units is transient, despite an unrestricted formulary and multiple antibiotic trials. *Surg. Infect. (Larchmt.)*, **3**, 135–144.

Gaynes, R., 1997, The impact of antimicrobial use on the emergence of antimicrobial-resistant bacteria in hospitals. *Infect. Dis. Clin. North Am.*, **11**, 757–765

Gesner, M., Desiderio, D., Kim, M., Kaul, A., Lawrence, R., Chandwani, S. *et al.*, 1994, *Streptococcus pneumoniae* in human immunodeficiency virus type 1-infected children. *Pediatr. Infect. Dis. J.*, **13**, 697–703.

Gilks, C. F., Brindle, R. J., Otieno, L. S., Simani, P. M., Newnham, R. S., Bhatt, S. M. *et al.*, 1990, Life-threatening bacteraemia in HIV-1 seropositive adults admitted to hospital in Nairobi, Kenya. *Lancet*, **336**, 545–549.

Gillespie, S. H. and McHugh, T. D., 1997, The biological cost of antimicrobial resistance. *Trends Microbiol.*, **5**, 337–339.

Goeman, J., Meheus, A., and Piot, P., 1990, The epidemiology of sexually transmitted diseases in Africa and Latin America. *Semin. Dermatol.*, **9**, 105–108.

Goldmann, D. A., Weinstein, R. A., Wenzel, R. P., Tablan, O. C., Duma, R. J., Gaynes, R. P. *et al.*, 1996, Strategies to Prevent and Control the Emergence and Spread of Antimicrobial-Resistant Microorganisms in Hospitals. A challenge to hospital leadership. *JAMA*, **275**, 234–240.

Gove, S., 1997, Integrated management of childhood illness by outpatient health workers: Technical basis and overview. *Bull World Health Org.*, **75**(Suppl. 1), 7–24.

Gruson, D., Hilbert, G., Vargas, F., Valentino, R., Bebear, C., Allery, A. *et al.*, 2000, Rotation and restricted use of antibiotics in a medical intensive care unit. Impact on the incidence of ventilator-associated pneumonia caused by antibiotic-resistant gram-negative bacteria. *Am. J. Respir. Crit Care Med.*, **162**, 837–843.

Gupta, K., Hooton, T. M., and Stamm, W. E., 2001, Increasing antimicrobial resistance and the management of uncomplicated community-acquired urinary tract infections. *Ann. Intern. Med.*, **135**, 41–50.

Harris, A. D., Perencevich, E., Roghmann, M. C., Morris, G., Kaye, K. S., and Johnson, J. A., 2002, Risk factors for piperacillin-tazobactam-resistant *Pseudomonas aeruginosa* among hospitalized patients. *Antimicrob. Agents Chemother.*, **46**, 854–858.

Hasan, R. and Babar, S. I., 2002, Nosocomial and ventilator-associated pneumonias: Developing country perspective. *Curr. Opin. Pulm. Med.*, **8**, 188–194.

Health Facilities Inventory, 1992, Ministry of Health, Uganda, Kampala.

Hima-Lerible, H., Menard, D., and Talarmin, A., 2003, Antimicrobial resistance among uropathogens that cause community-acquired urinary tract infections in Bangui, Central African Republic. *J. Antimicrob. Chemother.*, **51**, 192–194.

Holmberg, S. D., Solomon, S. L., and Blake, P. A., 1987, Health and economic impacts of antimicrobial resistance. *Rev. Infect. Dis.*, **9**, 1065–78.

Hofer, E., Quintaes, B. R., dos Reis, E. M., Rodrigues, D. P., Seki, L. M., Feitosa, I. S. *et al.*, 1999, The emergence of multiple antimicrobial resistance in *Vibrio cholerae* isolated from gastroenteritis patients in Ceara, Brazil. *Rev. Soc. Bras. Med. Trop.*, **32**, 151–156.

Hooton, T. M. and Levy, S. B., 2001, Antimicrobial resistance: A plan of action for community practice. *Am. Fam. Physician*, **63**, 1087–1098.

Hossain, M. M., Glass, R. I., and Khan, M. R., 1982, Antibiotic use in a rural community in Bangladesh. *Int. J. Epidemiol.*, **11**, 402–405.

Huebner, R. E., Wasas, A. D., and Klugman, K. P., 2000, Trends in antimicrobial resistance and serotype distribution of blood and cerebrospinal fluid isolates of *Streptococcus pneumoniae* in South Africa, 1991–1998. *Int. J. Infect. Dis.*, **4**, 214–218.

Indalo, A. A., 1997, Antibiotic sale behaviour in Nairobi: A contributing factor to antimicrobial drug resistance. *East. Afr. Med., J.*, **74**, 171–173.

Johnson, J. L. and Ellner, J. J., 2000, Adult Tuberculosis overview: African versus Western perspectives. *Curr. Opin. Pulm.*, **6**, 180–186.

Jones, R. N., 1992, The current and future impact of antimicrobial resistance among nosocomial bacterial pathogens. *Diagn. Microbiol. Infect. Dis.*, **15**, 3S–10S.

Jones, R. N., 1999, The impact of antimicrobial resistance: Changing epidemiology of community-acquired respiratory-tract infections. *Am. J. Health Syst. Pharm.*, **56**, S4–S11.

Kamya, M. R., Bakyaita, N. N., Talisuna., A. O., Were, W. M., and Staede, S. G., 2002, Increasing antimalarial drug resistance in Uganda and revision of the national drug policy. *Trop. Med. Int. Health*, 7(12), 1031–1041.

Kariuki, S. and Hart, C. A., 2001, Global aspects of antimicrobial-resistant enteric bacteria. *Curr. Opin. Infect. Dis.*, **14**, 579–586.

Kariuki, S., Olsvik, O., Mitema, E., Gathuma, J., and Mirza, N., 1993, Acquired tetracycline resistance genes in nosocomial *Salmonella typhimurium* infection in a Kenyan hospital. *East Afr. Med. J.*, **70**, 255–258.

Kesah, C., Ben Redjeb, S., Odugbemi, T. O., Boye, C. S., Dosso, M., Ndinya Achola, J. O. *et al.*, 2003, Prevalence of methicillin-resistant *Staphylococcus aureus* in eight African hospitals and Malta. *Clin. Microbiol. Infect.*, 9(2):153–156.

Kettler, H., 2000, *Narrowing the Gap Between Provision and Need for Medicines in Developing Countries*. The Office of Health Economics, London.

Keusch, G. T., 1988, Antimicrobial therapy for enteric infections and typhoid fever: state of the art. *Rev. Infect. Dis.*, **10**(Suppl. 1), S199–S205.

Kilian, A. H. D., Prislin, I., Kabagambe, G. *et al.*, 1997, In vivo resistance of *Plasmodium falciparum* to chloroquine sulphadoxine-pyrimethamine and co-trimoxazole in Kabarole and Bundibugyo Districts, Uganda. *Am. J. Tropic. Med. Hyg.*, **38**, 237–243.

Koella, J. C. and Antia, R., 2003, Epidemiological models for the spread of anti-malarial resistance. *Malar. J.*, **19**, 2(1), 3.

Koornhof, H. J., Wasas, A., and Klugman, K., 1992, Antimicrobial resistance in *Streptococcus pneumoniae*: A South African perspective. *Clin. Infect. Dis.*, **15**, 84–94.

Koulla-Shiro6S., Benbachir, M., Rahal, K., and Borg, M., 2003, Prevalence of methicillin-resistant *Staphylococcus aureus* in eight African hospitals and Malta. *Clin. Microbiol. Infect.* **9**, 153–156.

Kristinsson, K. G., 1997, Effect of antimicrobial use and other risk factors on antimicrobial resistance in pneumococci. *Microb. Drug. Resist.*, **3**, 117–123.

Kruse, H., 1999, Indirect transfer of antibiotic resistance genes to man, *Acta Vet. Scand. Suppl.*, **92**, 59–65.

Lang, A., De Fina, G., Meyer, R., Aschbacher, R., Rizza, F., Mayr, O. *et al.*, 2001, Comparison of antimicrobial use and resistance of bacterial isolates in a haematology ward and an intensive care unit. *Eur. J. Clin. Microbiol. Infect. Dis.*, **20**, 657–660.

Leegaard, T. M., Caugant, D. A., Froholm, L. O., Hoiby, E. A., Ronning, E. J., Sandven, P. *et al.*, 2002, Do HIV-seropositive patients become colonised with drug-resistant micro-organisms? *Eur. J. Clin. Microbiol. Infect. Dis.*, **21**, 856–863.

Lees, P. and Shojaee, A. F., 2002, Rational dosing of antimicrobial drugs: Animals versus humans. *Int. J. Antimicrob. Agents*, **19**, 269–284.

Lind, I., Arborio, M., Bentzon, M. W., Buisson, Y., Guibourdenche, M., Reimann, K. *et al.*, 1991, The epidemiology of *Neisseria gonorrhoeae* isolates in Dakar, Senegal 1982–1986: Antimicrobial resistance, auxotypes and plasmid profiles. *Genitourin. Med.*, **67**, 107–113.

Lind, I., 1990, Epidemiology of antibiotic resistant *Neisseria gonorrhoeae* in industrialized and developing countries. *Scand. J. Infect. Dis. Suppl.*, **69**, 77–82.

Lindtjorn, B., 1987, Essential drugs list in a rural hospital. Does it have any influence on drug prescription? *Trop. Doct.*, **17**, 151–155.

Lipsitch, M. and Samore, M. H., 2002, Antimicrobial use and antimicrobial resistance: A population perspective. *Emerg. Infect. Dis.*, **8**, 347–354.

Lipsky, M. S. and Taylor, C. A., 1997, The opinions and experiences of family physicians regarding direct-to-consumer advertising. *J. Fam. Pract.*, **45**, 495–499.

Lopez-Lozano, J. M., Monnet, D. L., Yague, A., Burgos, A., Gonzalo, N., Campillos, P. *et al.*, 2000, Modelling and forecasting antimicrobial resistance and its dynamic relationship to antimicrobial use: A time series analysis. *Int. J. Antimicrob. Agents*, **14**, 21–31.

Manfredi, R., Nanetti, A., Valentini, R., and Chiodo, F., 2001, Acinetobacter infections in patients with human immunodeficiency virus infection: Microbiological and clinical epidemiology. *Chemotherapy*, **47**, 19–28.

Manie, T., Brozel, V. S., Veith, W. J., Gouws, P. A., 1999, Antimicrobial resistance of bacterial flora associated with bovine products in South Africa. *J. Food Prot.*, **62**, 615–618.

Manie, T., Khan, S., Brozel, V. S., Veith, W. J., and Gouws, P. A., 1998, Antimicrobial resistance of bacteria isolated from slaughtered and retail chickens in South Africa. *Lett. Appl. Microbiol.*, **26**, 253–258.

Mason, P. R. and Katzenstein, D. A., 2000, HIV Treatment in Developing Countries. *Curr. Infect. Dis. Rep.*, **2**, 365–370.

McDermott, P. F., Zhao, S., Wagner, D. D., Simjee, S., Walker, R. D., and White, D. G., 2002, The food safety perspective of antibiotic resistance. *Anim. Biotechnol.*, **13**, 71–84.

McEwen, S. A. and Fedorka-Cray, P. J., 2002, Antimicrobial use and resistance in animals. *Clin. Infect. Dis.*, **34**(Suppl. 3), S93–S106.

McGarock, H., 2002, Unjustified antibiotic prescribing in the community: A major determinant of bacterial antimicrobial resistance. *Pharmacoepidemiol. Drug Saf.*, **11**, 407–408.

Mitema, E. S., Kikuvi, G. M., Wegener, H. C., and Stohr, K., 2001, An assessment of antimicrobial consumption in food producing animals in Kenya. *J. Vet. Pharmacol. Ther.*, **24**, 385–390.

Murthy, R., 2001, Implementation of strategies to control antimicrobial resistance. *Chest*, **119**, 405S–411S.

Nakavuma, J., Musisi, L. N., Byarugaba, D. K., and Kitimbo, F. X., 1994, Microbiological Diagnosis and Drug Resistance patterns of infectious causes of mastitis. Uganda *J. Agri. Sci.*, **2**, 22–28.

Nicholls, T., Acar, J., Anthony, F., Franklin, A., Gupta, R., Tamura, Y. *et al.*, 2001, Antimicrobial resistance: Monitoring the quantities of antimicrobials used in animal husbandry. *Rev. Sci. Tech.*, **20**, 841–847.

Nyamogoba, H. and Obala, A. A., 2002, Nosocomial infections in developing countries: Cost effective control and prevention. *East Afr. Med. J.*, **79**, 435–441.

O'Dempsey, T. J. D. *et al.*, 1996, A study of risk-factors for pneumococcal disease among children in a rural area of West Africa. *Int. J. Epidemiol.*, **25**, 885–893.

O'Brien, T. F., 2002, Emergence, spread, and environmental effect of antimicrobial resistance: How use of an antimicrobial anywhere can increase resistance to any antimicrobial anywhere else. *Clin. Infect. Dis.*, **34**(Suppl. 3), S78–S84.

Padungton, P. and Kaneene, J. B., 2003, *Campylobacter* spp in human, chickens, pigs and their antimicrobial resistance. *J. Vet. Med. Sci.*, **65**, 161–170.

Palmer, D. L., 1980, Epidemiology of antibiotic resistance. *J. Med.*, **11**, 255–262.

Pape, J. W., Jean, S. S. H. J. L., Hafner, A., and Johnson, W. D. Jr., 1993, Effect of isoniazid on incidence of active tuberculosis and progression of HIV infection. *Lancet*, **242**, 268–272.

Parimi, N., Pinto Pereira, L. M., and Prabhakar, P., 2002, The general public's perceptions and use of antimicrobials in Trinidad and Tobago. *Rev. Panam. Salud Publica*, **12**, 11–18.

Parise, M. E., Ayisi, J. G., Nahlen, B. L., Schultz, L. J., Roberts, J. M., Misore, A., Muga, R., Oloo, A. J., and Steketee, R. W., 1998, Efficacy of sulfadoxine-pyrimethamine for prevention of placental malaria in an area of Kenya with a high prevalence of malaria and Human Immunodeficiency Virus infection. *Am. J. Trop. Med. Hyg.*, **59**(5), 813–22.

Parry, M. F., 1989, Epidemiology and mechanisms of antimicrobial resistance. *Am. J. Infect. Control*, **17**, 286–294.

Phongsamart, W., Chokephaibulkit, K., Chaiprasert, A., Vanprapa, N., Chearskul, S., and Lolekha, R., 2002, *Mycobacterium avium* complex in HIV-infected Thai children. *J. Med. Assoc. Thai.*, **85**(Suppl. 2), S682–S689.

Piot, P. and Tezzo, R., 1990, The epidemiology of HIV and other sexually transmitted infections in the developing world. *Scand. J. Infect. Dis. Suppl.*, **69**, 89–97.

Poole, K., 2001, Overcoming antimicrobial resistance by targeting resistance mechanisms. *J. Pharm. Pharmacol.*, **53**, 283–294.

Premji, Z., Minjas, J. N., and Shiff, C. J., 1993, Chloroquine resistant *Plasmodium falciparum*. Coastal Tanzania. A challenge to the continued strategy of village based chemotherapy for malaria control. *Trop. Med. and Parasitol.*, **45**, 47–48.

Puerto Alonso, J. L., Garcia-Martos, P., Marin, C. P., Saldarreaga, M. A., Vega Elias, J. L. *et al.*, 2002, Evaluation of an applied program for the prevention and control of tuberculosis. *Enferm. Infecc. Microbiol. Clin.*, **20**, 150–153.

Pulimood, T. B., Lalitha, M. K., Jesudason, M. V., Pandian, R., Selwyn, J., and John, T. J., 1996, The spectrum of antimicrobial resistance among methicillin resistant *Staphylococcus aureus* (MRSA) in a tertiary care centre in India. *Indian J. Med. Res.*, **103**, 212–215.

Rowe, A. K., Deming, M. S., Schwartz, B., Wasas, A., Rolka, D., Rolka, H. *et al.*, 2000, Antimicrobial resistance of nasopharyngeal isolates of *Streptococcus pneumoniae* and *Haemophilus influenzae* from children in the Central African Republic. *Pediatr. Infect. Dis. J.*, **19**, 438–444.

Rubin, M. A. and Samore, M. H., 2002, Antimicrobial use and resistance. *Curr. Infect. Dis. Rep.*, **4**, 491–497.

Russo, T. A. and Johnson, J. R., 2003, Medical and economic impact of extraintestinal infections due to *Escherichia coli*: Focus on an increasingly important endemic problem. *Microbes. Infect.*, **5**, 449–456.

Saez-Llorens, X., Castrejon de Wong, M. M., Castano, E., De Suman, O., De Moros, D., and De, A. I., 2000, Impact of an antibiotic restriction policy on hospital expenditures and bacterial susceptibilities: A lesson from a pediatric institution in a developing country. *Pediatr. Infect. Dis. J.*, **19**, 200–206.

Samaranayake, L. P. and Johnson, N. W., 1999, Guidelines for the use of antimicrobial agents to minimise development of resistance. *Int. Dent. J.*, **49**, 189–195.

Saurina, G., Quale, J. M., Manikal, V. M., Oydna, E., and Landman, D., 2000, Antimicrobial resistance in Enterobacteriaceae in Brooklyn, NY: Epidemiology and relation to antibiotic usage patterns. *J. Antimicrob. Chemother.*, **45**, 895–898.

Schroeder, C. M., Meng, J., Zhao, S., DebRoy, C., Torcolini, J., Zhao, C. *et al.*, 2002a, Antimicrobial resistance of *Escherichia coli* O26, O103, O111, O128, and O145 from animals and humans. *Emerg. Infect. Dis.*, **8**, 1409–1414.

Schroeder, C. M., Zhao, C., DebRoy, C., Torcolini, J., Zhao, S., White, D. G. *et al.*, 2002b, Antimicrobial resistance of *Escherichia coli* O157 isolated from humans, cattle, swine, and food. *Appl. Environ. Microbiol*, **68**, 576–581.

Schwartz, B., Bell, D. M., and Hughes, J. M., 1997, Preventing the emergence of antimicrobial resistance. A call for action by clinicians, public health officials, and patients. *JAMA*, **278**, 944–945.

Schwarz, S., Kehrenberg, C., and Walsh, T. R., 2001, Use of antimicrobial agents in veterinary medicine and food animal production. *Int. J. Antimicrob. Agents*, **17**, 431–437.

Segal-Maurer, S., Urban, C., Rahal, J. J., Jr. 1996. Current perspectives on multidrug-resistant bacteria. Epidemiology and control. *Infect. Dis. Clin. North Am.*, **10**, 939–957.

Sehgal, R., 1999, Combating antimicrobial resistance in India. *JAMA*, **281**, 1081–1082.

Semjen, G., 2000, The effects of intervention on antimicrobial resistance. *Acta Vet. Scand. Suppl.*, **93**, 105–110.

Sexton, J. D., Deloron, P., Bugilimfura, L., Ntilivamunda, A., and Neill, M., 1988, Parasitologic and clinical efficacy of 25 and 50 mg/kg of chloroquine for treatment of *Plasmodium falciparum* malaria in Rwandan children. *Am J. Trop. Med Hyg.*, **38**, 237–243.

Sexton, D. J., 2000, The impact of antimicrobial resistance on empiric antibiotic selection and antimicrobial use in clinical practice. *J. Med. Liban.*, **48**, 215–220.

Shah, P. M., Schafer, V., and Knothe, H., 1993, Medical and veterinary use of antimicrobial agents: Implications for public health. A clinician's view on antimicrobial resistance. *Vet. Microbiol.*, **35**, 269–274.

Shears, P., 2001, Antibiotic resistance in the tropics. Epidemiology and surveillance of antimicrobial resistance in the tropics. *Trans. R. Soc. Trop. Med. Hyg.*, **95**, 127–130.

Sirinavin, S., Suvanakoot, P., Sathapatayavongs, B., and Malatham, K., 1998, Effect of antibiotic order form guiding rational use of expensive drugs on cost containment. *Southeast Asian J. Trop. Med. Public Health*, **29**, 636–642.

Slots, J., 1999, Addressing antimicrobial resistance threats. *Aust. Endod. J.*, **25**, 12–14.

Smith, J. T. and Lewin, C. S., 1993, Mechanisms of antimicrobial resistance and implications for epidemiology. *Vet. Microbiol.*, **35**, 233–242.

Smith, D. W., 2000, Decreased antimicrobial resistance following changes in antibiotic use. *Surg. Infect. (Larchmt.)*, **1**, 73–78.

Smith, R. D., 1999, Antimicrobial resistance: the importance of developing long-term policy. *Bull. World Health Organ.*, **77**, 862.

Struelens, M. J., 1998, The epidemiology of antimicrobial resistance in hospital acquired infections: problems and possible solutions. *BMJ.*, **317**, 652–654.

Tan, L., Nielsen, N. H., Young, D. C., and Trizna, Z., 2002, Use of antimicrobial agents in consumer products. *Arch. Dermatol.*, **138**, 1082–1086.

Thirumoorthy, T., 1990, The epidemiology of sexually transmitted diseases in Southeast Asia and the western Pacific. *Semin. Dermatol.*, **9**, 102–104.

Tollefson, L. and Flynn, W. T., 2002, Impact of antimicrobial resistance on regulatory policies in veterinary medicine: status report. *AAPS. PharmSci.*, **4**, 37.

Trape, J. F., 2001, The public health impact of chloroquine resistance in Africa. *Am J. Trop. Med. Hyg.*, **64**, 12–17.

Tumbarello, M., de Gaetano, D. K., Tacconelli, E., Citton, R., Spanu, T., Leone, F. *et al.*, 2002, Risk factors and predictors of mortality of methicillin-resistant *Staphylococcus aureus* (MRSA) bacteraemia in HIV-infected patients. *J. Antimicrob. Chemother.*, **50**, 375–382.

Urassa, W., Lyamuya, E., and Mhalu, F., 1997, Recent trends on bacterial resistance to antibiotics. *East Afr. Med. J.*, **74**, 129–133.

van den, B. A., 2001, Antimicrobial resistance, pre-harvest food safety and veterinary responsibility. *Vet. Q.*, **23**, 99.

Vidaver, A. K., 2002, Uses of antimicrobials in plant agriculture. *Clin. Infect. Dis.*, **34**(Suppl. 3), S107–S110.

Vose, D., Acar, J., Anthony, F., Franklin, A., Gupta, R., Nicholls, T. *et al.*, 2001, Antimicrobial resistance: Risk analysis methodology for the potential impact on public health of antimicrobial resistant bacteria of animal origin. *Rev. Sci. Tech.*, **20**, 811–827.

Wasfy, M. O., Frenck, R., Ismail, T. F., Mansour, H., Malone, J. L., Mahoney, F. J., 2002, Trends of multiple-drug resistance among Salmonella serotype Typhi isolates during a 14-year period in Egypt. *Clin. Infect. Dis.*, **35**, 1265–1268.

Watkins, W. M., Percy, M., Crampton, J. M., Ward, S., Koech, D., and Howells, R. E., 1988, The changing response of *Plasmodium falciparum* to antimalarial drugs in east Africa. *Trans. R. Soc. Trop. Med. Hyg.*, **82**(1), 21–6.

Wegener, H. C., Bager, F., and Aarestrup, F. M., 1997, Surveillance of antimicrobial resistance in humans, food stuffs and livestock in Denmark. *Euro. Surveill.*, **2**, 17–19.

Weidle, P. J., Malamba, S., Mwebaze, R., Sozi, C., Rukundo, G., Downing, R. *et al.*, 2002, Assessment of a pilot antiretroviral drug therapy programme in Uganda: Patients' response, survival, and drug resistance. *Lancet*, **6**, **360**(9326), 34–40.

Weinstein, R. A., 2001, Controlling antimicrobial resistance in hospitals: Infection control and use of antibiotics. *Emerg. Infect. Dis.*, **7**, 188–192.

White, D. G., Zhao, S., Simjee, S., Wagner, D. D., and McDermott, P. F., 2002, Antimicrobial resistance of foodborne pathogens. *Microbes. Infect.*, **4**, 405–412.

WHO, 1997. WHO/IUATLD Global Working Group on Antituberculosis Drug Resistance Surveillance. Guidelines for surveillance of drug resistance in tuberculosis. World Health Organization, Geneva, Switzerland, 1997. WHO/TB/96.216.

WHO, 2000. Overcoming antimicrobial resistance, WHO Report on Infectious Diseases 2000. Geneva, WHO/CDS/2000.2. 2000.

WHO, 2002. Scaling up response to infectious diseases: A way out of poverty. Report on Infectious Diseases, WHO/CDS/2002.7.

Wilton, P., Smith, R., Coast, J., and Millar, M., 2002, Strategies to contain the emergence of antimicrobial resistance: A systematic review of effectiveness and cost-effectiveness. *J. Health Serv. Res. Policy*, **7**, 111–117.

Wise, R., 2002, Antimicrobial resistance: Priorities for action. *J. Antimicrob. Chemother.*, **49**, 585–586.

Wolday, D., Kibreab, T., Bukenya, D., and Hodes, R., 1995, Sensitivity of *Plasmodium falciparum in vivo* to chloroquine and pyrimethamine-sulfadoxine in Rwandan patients in a refugee camp in Zaire. *Trans. Royal Soc. Trop. Med. Hyg.*, **89**, 654–656.

Yun, H. J., Whalen, C. C., Okwera, A., Mugerwa, R. D., and Ellner, J. J., 2003, HIV disease progression and effects of tuberculosis preventive therapy in HIV-infected adults. *Ann Epidemiol.*, **13**(8), 577–578.

Zaidi, M. B., Zamora, E., Diaz, P., Tollefson, L., Fedorka-Cray, P. J., and Headrick, M. L., 2003, Risk Factors for Fecal Quinolone-Resistant *Escherichia coli* in Mexican Children. *Antimicrob. Agents Chemother.*, **47**, 1999–2001.

Zar, H. J., Hanslo, D., and Hussey, G., 2003, The impact of HIV infection and trimethoprim-sulphamethoxazole prophylaxis on bacterial isolates from children with community-acquired pneumonia in South Africa. *J. Trop. Pediatr.*, **49**, 78–83.

Zellweger, C. and Tauber, M. G., 2002, Antibiotic resistance: Infections of the upper respiratory tract and bronchi. When are antibiotics necessary? *Ther. Umsch.*, **59**, 21–29.

Chapter 33

Antibiotic Use in Animals—Policies and Control Measures Around Europe

Pascal Sanders
AFSSA Site de Fougeres, Laboratoire d'études et de recherches sur les Médicaments Vétérinaires et les disinfectants, LNR residua d'antibiotiques, La Hautes Marche—Javené, 35302 Fougeres, France

Antibiotics are among the most used drugs in veterinary medicine. They have been used in animals bred for consumption since the 1950s and have been used also in companion animals with the development of veterinary medicine for dogs and cats. Many of these drugs are closely related to antibiotics that are used for the treatment of human infections. In the European Union, few antibiotics are still authorised as growth promoters. The development of antimicrobial policies and control measures in Europe is different between countries but is driven by international recommendations translated in European recommendations and regulations. First, the relevant authorities are in charge of the benefit/risk assessment of antibiotics before marketing authorisation and have to support the development of an antimicrobial resistance programme allowing the post-marketing antimicrobial resistance surveillance and encourage research in this area. They also have to control the distribution of antimicrobial agents and to provide data on consumption. Second, the industry has to provide data for marketing authorisation, to market only officially licensed and approved drugs and to provide the authorities with the information necessary to evaluate the amount of antimicrobial agents marketed. They have to contribute to the dissemination of information in compliance with the granted authorisation to authorised professionals involved in the prescription and distribution of the products. Third, training of antibiotic users, involving all the relevant professional organisations

Antibiotic Policies: Theory and Practice. Edited by Gould and van der Meer
Kluwer Academic / Plenum Publishers, New York, 2005

including regulatory authorities, industry, veterinary and agricultural schools, research institutes, and professionals associations should focus on information on disease prevention and management strategies to reduce the need for antimicrobial use.

1. INTRODUCTION

The development of antibiotic resistance is a major challenge for medicine (Anonymous, 2001). The increasing emergence and spread of antibiotic resistance is the result of decades of usage of antibiotics in human and animals with a misperception of the ecological impact of this usage on the bacterial flora. Even though the phenomenon of antimicrobial resistance was identified in the 1960s, the development of a global mobilisation of international communities against this threat was only effective in the 1990s. In industrialised countries, most of the problems of antibiotic resistance are generated within hospitals because of the relatively intensive use of antibiotics (Raveh *et al.*, 2001). Even if the nature of the relationship between antibiotic consumption and antibiotic resistance is complex (Fekete, 1995) from an epidemiological point of view, it is widely accepted that they are linked. The development of antibiotic resistance in hospitals is also driven by the epidemic spread of resistance clones that often prove highly adaptable and virulent. The development of antibiotic resistance in bacteria which are agents of disease in the community, was observed over the last 20 years, notably with the development of antibiotic resistance in *Streptococcus pneumoniae* and in food borne agents (Acar and Rostel, 2001). Even though in the case of *S. pneumoniae*, the phenomenon is driven by the overprescription of antibiotics in the treatment of upper respiratory tract infections in children by physicians (Guillemot *et al.*, 1998), the development of resistance in food borne pathogens is related to veterinary usage (Threlfall *et al.*, 2003).

In the European Union, few antibiotics are authorised as growth promoters to increase the feed efficacy and daily growth of animals. The conditions of use are defined by a European directive. In the 1990s, the selection of vancomycin-resistant enterococci by avoparcin and streptogramin resistance by virginiamycin has been extensively studied by several teams in the European Union (Aarestrup, 1999). The transfer of these strains or of resistance genes from animal to humans in contact with animals and to consumers has been analysed and discussed.

Many of these drugs are closely related to antibiotics that are used for the treatment of human infections. The benefit of antibiotic use in food producing animals, as veterinary medicine, is the control of infectious diseases, mainly in young animals (Raynaud *et al.*, 1988). Veterinary practitioners prescribe antimicrobials to treat a disease developing in the flock, mainly intestinal or respiratory diseases. For several industrialised animal productions (poultry, pigs, veal calves, sheep, fish), due to the importance of the population size in the same environment, group treatments are most often prescribed by oral route to treat diseased animals and animals in contact with them. For other types of production (calves, sows, horses), individual treatments are prescribed. Antibiotics can be used prophylactically by veterinarians if the previous knowledge of the epidemiology in the herd dictates. To prevent human risk related to consumption of food produced by animals and to decrease the incidence of food borne diseases, the risk associated with antibiotic treatment has to be assessed (Lathers, 2002c; Vose, 2001). Moreover, animals and food produced by animals are subjected to international trade regulations which impose an international coordination for the development of sanitary regulations. A large part of the antibiotic policy, defined as an antibiotic stewardship programme (Gould, 2002, Pedersen *et al.*, 1999) has been defined internationally (WHO, 2000) and applied at the European level on the basis of national, regional, and local experience.

2.   MARKET AUTHORISATION

2.1.   Feed additives

After the Swann report (1969), the Directive 70/524 concerning additives in feeding-stuffs (Table 1) defined the marketing authorisation rules at the European level for all the member states. In the late 1990s, the use of several compounds such as avoparcin (01/04/97), ardacin (01/04/98), virginiamycin, tylosin phosphate, spiramycin, and bacitracin zinc (01/07/99) were suspended by the European commission. At the time of writing, authorised growth promoters are flavophospholipol, sodium monensin, salinomycin, and avilamycin, and the European commission proposed to stop their usage in 2006.

European regulations about feed additives define the level of incorporation in feed and the period of usage in animals. The presence of additives in feed is labelled. The use of additives in feed differ between animal species and depends on the type of production. It is routinely used in different kinds of industrial production in Europe and excluded from others according to technical requirements. Some drugs used for their coccidiostatic effect have an antimicrobial effect, mainly on Gram-positive bacteria, such as ionophores. The Commission Directive 2001/79/EC fixes guidelines for the assessment of additives in animal nutrition. In the preamble of the directive, it is recognised that the increasing

Table 1. European regulations and guidelines for antimicrobials marketing authorisation

	Veterinary medicine	Feed additives
European regulations	Council Directive 81/851/EEC of 28 September 1981 on the approximation of the laws of the member states relating to veterinary medicinal products. Official Journal L317 of 6 Nov 1981	Council Directive 70/524/EEC of 23 November 1970 concerning additives and feeding-stuffs. Official Journal L270, 14/12/1970, p. 0001–0017
	Council Directive 81/852/EEC of 28 September 1981 on the approximation of the laws of the member states relating to analytical, pharmaco-toxicological, and clinical standards and protocols in respect of the testing of veterinary medicinal products. Official Journal L317 of 6 November 1981, amended by Council Directive 87/20/EEC, 92/18/EEC, and 93/40/EEC	Council Directive 96/51/EC of 23 July 1996 amending Directive 70/524/EEC concerning additives in feedingstuffs Official Journal L235, 17/09/1996, pp. 0039–0058
	Council regulation 2377/90 Regulation (EEC) No. 2377/90 of 26 June 1990 laying down a community procedure for the establishment of maximum residue limits for veterinary medicinal products in foodstuffs of animal origin. Official Journal L67 of 7/03/1997.	
Guidelines	EMEA/CVMP/627/01—Guideline for the demonstration of efficacy for veterinary medicinal products containing antimicrobial substances.	Commission Directive 2001/79/EC of 17 September 2001 amending Council Directive 87/153/EEC fixing guidelines for the assessment of additives in animal nutrition. Official Journal of the European Communities 2001, 6/10/2001m L267/1–25.
	EMEA/CVMP/244/01—Guideline on pre-authorisation studies to assess the potential for resistance resulting from the use of antimicrobial veterinary medicinal products.	
	EMEA/CVMP/234/01—Revised guideline on safety evaluation of antimicrobial substances regarding the effects on human gut flora.	

prevalence of antibiotic-resistant bacteria is of major concern to public health and that the resistance caused by the use of antibiotics as feed additives contributes to the overall levels of resistance. The guidelines for additives thus require the dossier to include assessments of (1) the risk of selection and/or transfer of antibiotic resistance, (2) the risk of any increased persistence and shedding of enteropathogens in order to ensure the safety of use of those additives, (3) the risk for the consumer which could result from the consumption of food containing residues of the additive and of the environmental impact of feed additives. These guidelines amend previous guidelines established in the light of advances in scientific and technical knowledge. If the active substance has an antimicrobial activity at feed concentration level, the minimum inhibitory concentration (MIC) should be determined in appropriate pathogenic and non-pathogenic, endogenous, and exogenous bacteria. The ability of the additive to select cross-resistance to relevant antibiotics and to select resistant bacterial strains under field conditions in the target species has to be studied by relevant tests. The effect of the additive has to be determined on a number of opportunistic pathogens present in the digestive tract (e.g., Enterobacteriaceae, Enterococci, and Clostridia) and on the shedding or excretion of relevant zoonotic microorganisms (e.g., *Salmonella* spp, *Campylobacter* spp.). Moreover, the directive 2001/79/EC encourages the development of field studies to monitor antimicrobial resistance and the data provided by these monitoring programmes are required in the evaluation dossier.

2.2. Veterinary drugs

The process of harmonisation of risk–benefit assessment for veterinary drugs began at the beginning of the 1980s (Table 1). Three evaluation procedures are defined: national, mutual recognition, and centralised for new innovative products. European guidelines are provided by the European Medicine Evaluation Agency (EMEA) and define the type of pre-clinical and clinical studies necessary to define therapeutical indications of an antimicrobial and the dosage regimen in the different animal species. On the basis of international recommendations, EMEA experts have reviewed the problem of development of antimicrobial resistance after veterinary usage and reinforced risk evaluation (Anonymous, 1999). Several guidelines have been modified to provide new data about risk of selection of resistant bacteria during treatment and the mechanism of resistance and to promote the development of a dosage regimen based on the pharmacokinetic/pharmacodynamic approach for new products (Toutain *et al.*, 2002).

For a new veterinary medicine, information is requested about the drug, its mechanism of action, the activity spectrum determined by MICs against a wide range of Gram-negative and Gram-positive bacteria including target bacteria (*Escherichia coli, Pasteurella multocida, Mannheima haemolytica, Staphylococcus* spp., *Streptococcus* spp., etc), zoonotic bacteria (*Salmonella* spp., *Campylobacter* spp.), and commensal bacteria (*E. coli*, Enterococci), and also bacteria representative of the human and animal intestine flora. This pharmacodynamic information is completed with the evaluation of cross-resistance, data on the mechanisms of resistance, their genetics, and the risk of spread. The pharmacokinetics of the drug in animals is described with particular insight on the distribution in the intestinal lumen. To regularly re-evaluate the drug, the monitoring of antimicrobial resistance for veterinary pathogens targeted by the drug is requested from pharmaceutical companies which have to encourage and support monitoring programmes in member states.

The risk assessment of veterinary drugs residue was defined by the council regulation, 2377/90. The process makes it possible to determine maximum residue limits in tissue and animal products (milk, eggs, honey) before clinical trials in the European Union. The effect of antimicrobial residues on the human intestinal flora is evaluated by the definition of a microbiologically acceptable daily intake on the basis of *in vitro* or *in vivo* data. For compounds recently approved, the microbiological acceptable daily intake (ADI), defined on the basis of antimicrobial effect evaluation is, the basis for the maximum residue limit determination.

3.   MONITORING AND SURVEILLANCE

3.1.   Antimicrobial resistance

The development of harmonised antimicrobial resistance and surveillance programmes in animals and animal derived food was one of the first tasks in the development of an antibiotic policy to provide information for risk assessment and evaluate the effect of an antibiotic policy (WHO, 2000). A review of the national monitoring programmes (Gnanou and Sanders, 2000) described as state of the art, was the basis for an analysis of the needs to improve the European system. Recommendations established by the members of a European concerted action (Wray and Gnanou, 2000) suggested developing a monitoring programme of veterinary pathogens, zoonotic bacteria, and indicator bacteria (Caprioli *et al.*, 2000). Almost all of these recommendations were rediscussed by an international expert group and proposed by OIE at the international level (Franklin *et al.*, 2001). Monitoring the antimicrobial resistance of zoonotic bacteria will be requested in the future zoonosis directive (Common Position [EC] N°13/2003) and will be a part of surveillance networks implemented in the European Union for communicable diseases (Common Decision of 17 July 2003). In the meantime, several European countries have improved the existing system.

For example, the monitoring programme of antimicrobial resistance of veterinary pathogens has been extended from cattle (Martel *et al.*, 1995) to other major species (pig and poultry) in France (Jouy *et al.*, 2002). Several countries such as France (Sanders *et al.*, 2002), Germany (Guerra *et al.*, 2003), Spain (Brinas *et al.*, 2003; Moreno *et al.*, 2000; Teshager *et al.*, 2000) and the United Kingdom (Goodyear, 2002) have implemented monitoring programmes of antimicrobial resistance for *E. coli* and *Enterococcus faecium* collected in faeces or the caecal content of slaughtered animals based on the same epidemiological approach permitting international comparisons (Bywater *et al.*, 2003). For salmonella, the cooperation between national scientific teams from the veterinary and human sectors has permitted comparisons between isolates from different origins and was the basis for molecular epidemiological studies (Baggesen *et al.*, 2000; Threlfall *et al.*, 2003). These studies made it possible to distinguish different ways of dissemination according to serotypes and to analyse the complexity of animal and human relationships.

In 2003, a concerted action ARBAO II began to provide an external quality assurance system for antibiotyping and a web resource to collect national data about antimicrobial resistance in parallel with the global salmonella survey supported by WHO (http://www.vetinst.dk). This concerted action is an important step in the process of standardisation and harmonisation of bacterial antimicrobial susceptibility testing methods in the veterinary field. The interlaboratory study, based on proficiency testing of selected bacterial species (*Salmonella, E. coli, Streptococcus, Staphylococcus, Pasteurella, Mannheimia, Campylobacter*) will contribute to the comparison among the national surveillance programmes as recommended by OIE (White *et al.*, 2001).

3.2. Antibiotic use

The use of antibiotics as feed additives depends on the type of production of the herds and is based on zootechnical choices made by the producer. In the European Union, antibiotics used as veterinary drugs are the only prescribed medicines. To be used in food producing animals, antibiotics should first be listed in directive 2377/90 which defines the list of products with maximum residue limits after risk assessment by the Committee of Veterinary Medicinal Products of the EMEA. The prescription of antibiotics is under the responsibility of veterinarians. Depending on the country, the drug is delivered by a pharmacist and/or a veterinarian. In several countries, drugs listed in animal health programmes established by veterinarians and approved by authorities can also be sold by producer organisations after authorisation by authorities and under veterinarian control.

The monitoring of antimicrobial usage in food animals for the protection of human health was recommended by the scientific community and international bodies (Nicholls *et al.*, 2001). To develop effective antibiotic policies and follow their effect, it is necessary to know which drugs are administered to food producing animals and to determine, why, when, and how they are used. However, no study of drug consumption fails to come up against complex problems associated with data acquisition and processing (Chauvin *et al.*, 2001; Grave *et al.*, 1999b). Information about drug usage should directly be collected from pharmaceutical companies. This approach was been implemented by several countries, mainly Nordic countries (Aarestrup, 1999; Grave, 1999b; Wierup, 2001b) where veterinary drug regulations allow them to collect sales. The United Kingdom (VMD, 2003) and the Netherlands (Anonymous, 2002b) also collect this data and France has begun recently. The total consumption or sales can be expressed in terms of active substance weight, in kilograms or tons at national level. This methodology is limited by the fact that many veterinary drugs are approved and sold for different animal species. A single active substance may consequently be used in several species. Only new products (approved for one species) or products specifically designed for one animal type (e.g., fish, poultry) or usage (intra-mammary treatment) can be directly compared with the potential consumer population or usage. Moreover, it is difficult to accurately determine the population at risk, even though some statistics exist. A more detailed analysis, relating consumption to the metabolic body weight of consumers (animals or humans) could be carried out. These calculations could be based on approximation both of the consumer population and of the average body weight. This approach is also limited by the rapid growth of food producing animals with large differences in body weights and growth rates between species (Chauvin *et al.*, 2001). Another approach is to define a defined daily dose (DDD) as in human medicine. This approach was used to study the drug usage for mastitis in cattle in Sweden and Norway (Grave *et al.*, 1999a). The main problem encountered when using the DDD in veterinary medicine is due to dosage variations between different animal species. Besides, few DDD have been defined to date in veterinary medicine (Grave *et al.*, 1999b).

Another approach is to study drug consumption at the population level. Epidemiological studies in herds made it possible to collect information about drug usage in herds under investigation and analyse trends in antibiotic resistance in bacterial populations. The effect of different antibiotic use and management systems can also be investigated. This epidemiological approach is very effective but has a high cost and is labour intensive. The relationship between antibiotic exposure and antibiotic resistance in bacteria should also be investigated by collecting information about animal treatment at the time of sampling for the monitoring of antimicrobial resistance in intestinal flora.

If the registration of treatment in the flock is effective, this approach provides data to investigate the link between drug use and resistance. On the basis of their experience, the Danish have developed a new surveillance system, called VetStat, which provides detailed data on the consumption pattern of all prescription medicines at herd level according to species, age, group, and diagnostic group (Stege *et al.*, 2003). It is based on the registration of drug usage at farm level identified in a national database where the number of animals registered by species and age group is recorded. The veterinary prescriber must provide information about the identity and quantity of medicine prescribed, the target animals, age group, and the disease. The drug usage is described by the disease, the ATC classification system and the defined animal daily dosage (ADD) which is a technical unit describing mean daily maintenance dosage defined for each package species age group. Standardised animal body weights by species and age group are used to allow the calculation of the mean percentage of animals treated per day within the period of interest. Data may be grouped according to herd identity, period, species, age group, veterinarian, practice, ATC-Code, disease groups, or geographic region. Farm level statistics are directly available via the Internet to the farmers, veterinarians, and the veterinary authority as a tool for comparing drug usage between herds and at national level. An important application will be pharmaco-epidemiological research combining data concerning medication and vaccine strategies with antimicrobial resistance.

In human medicine, the variation of the antibiotic dosage regimen prescribed and the level of exposure of the treated subject has been identified as major factors for the selection of antibiotic resistant strains (Guillemot *et al.*, 1998; Gyssens, 2001). Collecting information about the variability of the dosage regimen prescribed and the dosage regimen really applied on the animals should be a way of investigating risk factors related to veterinary practice. These data could be collected at the level of veterinarians by way of postal survey (Chauvin *et al.*, 2002). This approach makes it possible to describe the prescription patterns in different pathologies. Large variations in the duration of treatment have been recorded (Chauvin *et al.*, 2002a).

Analysing drug consumption requires developing new information tools in many countries to collect robust information about drug prescription at the herd level. An international agreement concerning the handling of topics, such as dosage regimen variations (daily dose, length of treatment), differentiation between short- and long-acting formulations, drug potency, liability of dosage, and other variations in the therapeutic course should be sought. Common definitions should be reached to obtain comparable figures for international research (Chauvin *et al.*, 2001). These tools are now in development in few European countries but need to be discussed and compared to select the best tools according to the epidemiological objectives (Chauvin *et al.*, 2002b).

4.   RISK ANALYSIS

4.1.   Principles

Risk analysis is a process composed of hazard identification, risk assessment, risk management, and risk communication (Figure 1). It encompasses assessing and managing the risk together with the appropriate communication between risk assessors, stakeholders, and risk managers. Typically, a policy framework is established by risk managers to describe the types of risk that need to be addressed. A strategy for assessing the risk is formulated in consultation with technical experts and risk assessors. The policy framework also provides an explanation of the type of risk management options that can be considered under the legislative and regulatory framework of the country. Finally, the policy framework should explain the risk decision-making process, including methods for evaluating and quantifying risks and the level of risk deemed to be acceptable (European Commission, 2000; Vose *et al.*, 2001). The risk assessment is a step-by-step approach with regular exchange with risk managers. A preliminary qualitative assessment is recommended by the OIE

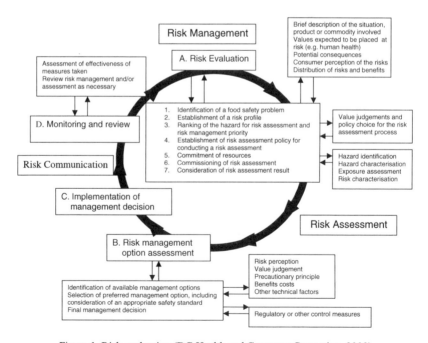

Figure 1. Risk evaluation (DG Health and Consumer Protection, 2000).

ad hoc group (Vose *et al.*, 2001) to advise risk managers on the feasibility of quantitatively assessing the risk and on the identified risk management strategies. The report is made public. On the basis of this report, managers will determine whether the risk is sufficiently severe to warrant further action. If the risk is considered sufficiently important and if it is feasible, risk managers may then instruct risk assessors to fully assess the risk (qualitatively and/or quantitatively) and the reduced level of risk that would exist after each identified risk reduction option. Through several iterations, the risk assessment and the risk reduction options are refined. The aspect of risk communication is particularly helpful in ensuring transparency of the risk analysis as a whole and the efficient collection of data. On the basis of the result of the risk assessment, risk managers determine the appropriate actions to take in order to manage the risk in the most efficient manner and make public their decision. Risk managers have to implement their decision and organise the follow-up of these regulatory and other measures in order to evaluate the impact of these decisions. The data collected by the follow-up must be assessed in order to allow possible amendments of the risk analysis policy, the assessment strategy, the outcome of the scientific assessment and the regulatory and other actions that have been taken.

Several reports about antimicrobial resistance in bacteria from animal origin have been established in different member states and at the level of the European Commission. In the case of several antimicrobials used as feed additives, the risk decision of the risk manager (DG Health and Consumer Protection) was based on a review of the opinion of the scientific committee on animal nutrition, member states reports and now, the opinion of the European Food Safety Authority (EFSA).

As the antimicrobial resistance risk is combined with a microbiological risk, the progress of risk assessment is linked with the development of knowledge and tools for qualitative, semi-quantitative, and quantitative risk assessment. The risk assessment process is subdivided into four components: risk release, exposure, consequence, and estimation (Figure 1). The risk release is the description of the biological pathways necessary for the use of an antimicrobial in animals to generate resistant strains or resistant determinants into a particular environment, and estimating the probability of that complete process occurring either qualitatively or quantitatively. The fact that data on new compounds (veterinary drugs, additives) has to be provided by the pharmaceutical industry to describe the impact of treatment on the intestinal flora and environment will contribute to our knowledge. Moreover, the knowledge of the epidemiology of some zoonotic bacteria is continuously improved and used to reduce their release by different ways according to the principles of the zoonoses directive (Wegener *et al.*, 2003). The development of the monitoring and surveillance programmes for several antimicrobial resistant bacteria, combined with epidemiological studies in herds and food industry, contributes to the description of the biological pathways necessary for exposure of animals and humans to the hazards released from a given source and allows estimation of the probability of the exposure occurring either qualitatively or quantitatively. These are the objectives of exposure assessment. Several epidemiological studies have been recently published (Helms *et al.*, 2002, 2003) to describe the relationship between specified exposures to a biological agent and the consequences of this exposure. For some zoonotic infections, the presence of antibiotic resistance increases the risk of failure of an antibiotic treatment as well as the morbidity and mortality, notably for people with other diseases (Helms *et al.*, 2003).

4.2.   Feed additives

4.2.1.   Benefits

Antibiotic usage in animals are analysed as a risk–benefit for animals, producers, and consumers. By common understanding, modern animal agriculture is practiced with the intent of returning a profit to the livestock producer. Two economic considerations influence the selection of programmes used by any livestock producer (Gustafson and Bowen, 1997). The first is that many diseases commonly affect groups of animals rather than individuals. It is more economical to prevent a disease than to rely on a treatment. This type of use can prevent death losses as well as the loss of performance that often accompanies both clinical and subclinical infections. These prevention efforts can include a variety of practices, such as immunisation, adequate nutrition, appropriate husbandry practices, and effective sanitation and isolation. In many cases, it will also be appropriate and beneficial to use prophylactic antibiotics during crucial periods of the animal life.

The second economic benefit gained from the use of antibiotics in animal agriculture is derived from the enhancement of some production parameters. They represent an extremely important tool in the efficient production of pork, beef, poultry meat, and other animal products. When used at low (subtherapeutic) levels in feeds, antibiotics improve growth rate and efficiency of feed utilisation, reduce mortality, and morbidity, and improve reproductive performance (Cromwell, 2002). Using feed additives for the past 50 years improved the consumer access to food from animal origin with a regular price decrease. In the same period, animal genetics, animal husbandry, and feed technology have continuously improved and modified the management of herds. In Europe, the risk associated with the selection of resistant zoonotic bacteria was evaluated in the late 1960s (Swann, 1969) and contributed to the ban of usage of broad-spectrum antibiotic as additives at subtherapeutic levels in feed. According to the European directive, only narrow-spectrum drugs active against Gram-positive bacteria are approved as potential growth promoters. Most of them are not used or have a limited usage in human medicine at the first time of approval.

4.2.2. Risk

The work done in Europe by several teams gave good evidence that approved drugs could select for resistant bacteria. For example, the presence of vancomycin-resistant enterococci selected by avoparcin use was demonstrated in the intestinal flora of treated animals, in people in contact with the animals, in the environment, and on the food produced by the animals (Aarestrup, 1999; Klein *et al.*, 1998; Willems *et al.*, 2000). The spread of genetic elements carrying resistance genes from animals to consumers was described (Jensen *et al.*, 1998). The final steps in risk analysis which are the passage of the resistant genes to bacteria responsible for human infection have not been assessed. But, using a theoretical approach, it was demonstrated that the animal usage of antibiotics reduced the time needed for emergence of resistant pathogens in human medicine (Lipsitch *et al.*, 2002; Smith *et al.*, 2002).

The available data demonstrate the selection of resistant bacteria in the animal intestinal flora of treated animals and the spread of these bacteria from animals to humans by contact or by food. The data are, however, insufficient to reasonably assess the potential risk issue at the human level. Indeed, prevalence of resistant enterococci is low in Europe and the worst scenario, which is the emergence of a totally resistant *Staphyloccocus aureus* after transfer and combination of genetic elements, has not occurred. Moreover, pathogenic *E. faecium* for humans differ from those of animal origin and harbour specific virulence genes (Willems *et al.*, 2000). The risk manager has considered that the risk is potentially of such severity that one cannot wait for sufficient data before taking action. According to the OIE ad hoc group recommendations and the precautionary principle, it is reasonable for risk managers to take a temporary risk avoidance action that minimises any exposure to the risk. The five points to consider before taking this action have been reached. First, a risk assessment was done on the basis of available data. The risk avoidance action which was a suspension of market authorisation, provided a reduction of antimicrobial resistance in targeted bacteria. This effect was previously observed in Denmark (Monnet, 2000). The action did not limit trade but showed a negative impact of the production cost in comparison with countries outside Europe. The benefit of use of feed antimicrobial additives was detrimental for the quality perception by the consumer. Even if the risk for human health was not accurately evaluated, the market authorisation suspension was a benefit in term of safety perception by the consumer. The decision was taken in conjunction with a commitment to acquire the necessary data to help assess the severity of the risk. The process remained transparent by the publication of opinions on the web.

4.3. Veterinary drugs

4.3.1. Benefits

Antibiotics are also used to prevent and treat diseases in animals. In the case of bacterial infection, the efficacy of an antimicrobial treatment has to be demonstrated by pre-clinical studies and clinical studies. In experimental infections, antimicrobials reduce mortality and morbidity. The clinical cure of animals is the result of antibiotic activity and immune response. The clinical efficacy shown by clinical signs does not necessarily reflect a bacteriological cure defined as the eradication of bacteria from the site of infection. Few veterinary drugs are evaluated on the basis of their clinical and bacteriological efficacy (e.g., intramammary treatment of cow mastitis). For most of them, evaluation is based on clinical criteria such as the improvement of clinical signs (e.g., temperature, cough, or diarrhoea), mortality, and relapse a few weeks after the end of the treatment. Zootechnical parameters, such as feed efficacy and growth rate, are also compared. For ethical reasons, the pre-clinical and clinical trials are performed in comparison with previously approved drugs. Data provided by these trials cannot assess whether the ultimate goal, which is a total bacterial cure, is achieved.

4.3.2. Risks

If bacterial eradication does not occur, fewer bacteria are present at the site of infection and the ratio between resistant and susceptible bacteria can be changed allowing a more resistant population to grow and become predominant. The consequences are negative for animal health. The risk for the animal owner is the further development of disease with a higher cost of treatment. The second risk is the transfer of antibiotic-resistant pathogenic bacteria to people in contact with animals (animal owners and relatives, technical staff, veterinary practitioner). The third risk, linked with the exposure of intestinal or commensal flora to antibiotics is the selection in treated animals of antibiotic-resistant zoonotic bacteria (*Salmonella* spp., *Campylobacter* spp., verotoxigenic *E. coli, Yersinia* spp., etc.). These bacteria could infect people in contact with the animals and could reach consumers via the food chain. The fourth risk is the diffusion of these bacteria in the environment and their spread in soil and water. The fifth and sixth risks are those related to the selection of antibiotic-resistant genes in bacteria from the intestinal and commensal flora of animals, with the risk of transfer of these genes to pathogenic bacteria (Blake *et al.*, 2003). All these bacteriological risks are the subject of many discussions and some epidemiological studies of salmonella infection demonstrate their occurrence (Helms *et al.*, 2002). For some zoonotic infections, the presence of antibiotic resistance increases the risk of failure of the antibiotic treatment as well as morbidity and mortality, notably for people with other diseases (Helms *et al.*, 2003).

The risk evaluation related to antimicrobial treatment is complex due to the combination of different events and the ecology principles driving this risk (Vose *et al.*, 2001). It should be stated that the development of a risk assessment framework is on-going in Europe, with the development of knowledge and research in different member states. Most of this research on risk assessment is focused on susceptible bacteria (Rocourt *et al.*, 2003; Rosenquist, 2003; Wegener *et al.*, 2003) more than resistant bacteria.

For veterinary medicines, the risk evaluation is very developed and was recently strengthened. The use of antimicrobials by veterinarians contributes to the health of animals. The selection of resistant bacteria depends largely on the conditions of this use. Bad conditions of use, such as the inappropriate choice of antibiotics, dosage, or duration of treatment, will favour the selection and preservation of resistant bacteria. The difficulty is to encourage good practices of use (adequate prescription) which have a limited effect on the selection of resistant bacteria but have a beneficial clinical effect, and to distinguish these from bad practices, that is, useless or inadequate prescription.

4.4.   Communication

Communication is a critical step in the risk evaluation process (Figure 1). Communication about antimicrobial resistance selected by non-human use needs to be developed. First, the misperception about modern animal husbandry by the general population is well known. Few people in European cities have a personal experience of animal husbandry. This is due to the decreased importance of agriculture in Europe. Moreover, with the development of the bio-security concept in animal production, a large part of animal production is now secured against pathogens by use of a closed environment with limited access. This hides major industrial production, such as poultry and pigs from the general population. Moreover, the development of misperceptions about animal production by consumers has created a need for some people to return to "natural" food. This need has contributed to the development of the concept of organic food which limits the usage of drugs and pesticides and was reinforced by the communication around this product even if the advantages are not scientifically demonstrated (Anonymous, 2003). This need was also supported by the development of environmental concerns by European consumers.

Moreover, perceptions about antimicrobial resistance due to animal production have been affected by previous experiences such as hormone administration to calves, and BSE and *Listeria* outbreaks. The efforts by animal producers, veterinarians, food industry, and administrations to manage these different hazards in the past by implementation of regulations and development of hazard analysis and critical control point (HACCP) principles are not appreciated by consumers because their effect is countered by each new outbreak highlighted by the media, even if the global frequency of outbreaks has been reduced (Fife-Schaw and Rowe, 1996; Kirk *et al.*, 2002).

To develop communication around antimicrobial resistance in bacteria of animal origin, it is necessary to develop the understanding of physicians about drug usage in veterinary medicine and the control of zoonotic agents in the food chain. Animal producers, in collaboration with veterinarians, have to develop their means of communication on the modern animal husbandry and food industry. The development of quality assurance in animal production contributes to the reinforcement of consumer confidence. By developing quality and traceability-based contracts with producers, food producers, and distributors contribute to the development of safety, but they have to regularly verify the compliance with the contract by the way of analytical self-controls. The source of emergent antimicrobial resistant bacteria was thus investigated and identified during an outbreak (Desenclos *et al.*, 2002; Helms *et al.*, 2002). Through the rapid alert surveillance system for food, the production originating from a source (herd, food workshop) was traced and where possible, recalled from the market. Quality assurance provides a powerful tool to control outbreaks and limit health impact but needs a good knowledge of crisis communication by the food-chain manager to manage any new event and limit its economic impact.

5.   WAYS OF PROGRESS

5.1.   Research

The problem of antimicrobial resistance is one of the major topics investigated by research projects in the area of food safety. Epidemiological studies of food-borne disease regularly bring new knowledge on the factors influencing the transfer of bacteria from animal to man. These works allow us to test new means of risk reduction following the HACCP principles. The current studies on the effect of the antimicrobial resistance on the mortality and the morbidity of pathogenic bacteria have to develop. The results of these studies allow the managers to grade the risks and to adapt the efforts of control. The pharmacological studies of antimicrobials in animals have to integrate the effects on the pathogenic bacteria and the commensal flora. The animal individual variability has to be assessed by population analysis. Finally, collection of information about usage practices by development of pharmaco-epidemiology is necessary to distinguish correct usage practices from the wrong ones. The current development of the antimicrobial usage surveillance network will allow us to develop this area as epidemiological studies in farms focusing on the relationship between drug usage and antimicrobial resistance.

5.2.   Training

The relationship between antimicrobial use and the selection of antimicrobial resistance is complex. Veterinarians, as clinicians, are well trained to identify diseases and diagnose etiological agents. During their initial training, microbiology and epidemiology are taught to give them a scientific background of epizootic and zoonotic transmission. As future drug prescribers, the basis of pharmacology was taught by pharmacologists and the uses of antimicrobial drugs to prevent and treat bacterial diseases were explained, during the course on animal diseases, by clinicians. The development of antimicrobial resistance is a subject which involves bacteriologists, pharmacologists, clinicians, and epidemiologists. Unfortunately, this problem was not well developed in the recent past in veterinary schools and universities. More recently, the development of resistance has been taught by bacteriologists who emphasise the genetics and transfer of resistance between bacteria rather than by pharmacologists who should describe the relationship with dosage regimens (Lees and Aliabadi, 2000; Toutain *et al.*, 2002), clinicians who are the key players in the choice of treatment (Lathers, 2002a), and epidemiologists who should describe the spread of antimicrobial resistance bacteria at the level of animal population and the relationship with drug usage with a global perspective (Lathers, 2002b; Lipsitch *et al.*, 2002).

In the European Union, over the last 10 years, the subject has begun to be a research area for scientific teams other than microbiologists. Some concepts developed in human medicine have to be discussed and adapted to veterinary medicine, before being taught to veterinary students and veterinary practitioners. To reduce antimicrobial resistance, multiple and often conflicting recommendations have been made in human medicine (Highet *et al.*, 1999; Lipsitch and Samore, 2002; Schentag, 2001). For example, strategies to minimise the burden of resistance in hospitals, have included reduction of all antimicrobial classes, increased use of prophylactic antimicrobials to reduce colonisation, rotation of different antibiotic classes in a temporal sequence, and simultaneous use of different antimicrobials for different patients. On the basis of these varying recommendations, it is difficult to export them to veterinary medicine without the development of experience and expertise.

For example, the use of a pharmacokinetic/pharmacodynamic approach to define the dosage regimen of antibiotics has been the subject of research by several teams (Aliabadi and Lees, 2002; Lauritzen *et al.*, 2003). The relationship between pharmacokinetic/pharmacodynamic surrogates and selection of resistance, notably for new fluoroquinolones products, has been discussed in human medicine and a threshold ratio for several surrogates was proposed (Highet *et al.*, 1999). The concept has been introduced and discussed in veterinary medicine and different studies in animals were performed or are on-going. Unfortunately, at the same time, this concept was used in veterinary medicine by pharmaceutical companies more for product promotion than for prudent use.

Indeed, the development of an adapted antimicrobial policy needs to distinguish the different impact of antimicrobial treatments at the individual level and at the herd level. This effect will be different according to the genetics of resistance, the pharmacology of the antimicrobial family, and use in the different animal species. The experience developed in hospital or community human medicine has to be evaluated in animal production. While prudent use of antimicrobials is encouraged by the European veterinary professional organisation FVE (Choraine, 2000) and the pharmaceutical industry, it remains as general as the international guidelines provided by international organisations. The technical support describing a rational use for each animal production is poorly developed. In several countries, an antimicrobial formulary has been promoted but its effect has not been measured (van Kasteren *et al.*, 1998). Few epidemiological studies make it possible to evaluate the effect of different antimicrobials on the level of resistance in herds and on animal products. Therefore experts and teachers have difficulty in proposing technical solutions to practitioners. In Nordic countries, an antibiotic policy based on the objective of reducing antimicrobial use as additives was followed by a transitory increase of antibiotic use as veterinary drugs (Wierup, 2001a, b). Another approach, applied in Germany, is a strict control of veterinary prescription and drug delivery in animal husbandry.

5.3.   Information

The problem of antimicrobial resistance in non-human use is recognised and the subject of recurrent discussions between animal producers, veterinarians, physicians, and consumers. The recent development in the European Union of national surveillance programmes for antimicrobial resistance and drug consumption as well as the development of knowledge about the transmission of zoonotic agents has been driven by European policy and recommendations.

Several qualitative risk analyses have been produced and are available to the public via the Internet and several teams are dealing with the development of a quantitative risk assessment. While this information is shared between experts, it is relatively difficult to train and explain the different concepts, to animal producers and veterinarians due to the multifaceted nature of the risks. This is done during scientific and professional meetings and using professional journals. It is necessary to overcome the contradictions between human and veterinary medicine and present the problem with a global ecological perspective. The progress in the exchange of experience over the last few years, in the writing of international recommendations and the development of common investigations during zoonotic outbreaks, is the sign of an effective cooperation between veterinarians and physicians on this subject in several countries.

6. CONCLUSION

The development of an antibiotic policy for the containment of antimicrobial resistance in the animal sector in the European Union follows international recommendations. The suspension of authorisations for several antimicrobials used as growth promoters has contributed to the reduction of the prevalence of resistance in indicator bacteria, demonstrated by different national programmes of surveillance (Monnet *et al.*, 2000; Sanders *et al.*, 2002). The last authorised antibiotics (avilamycin, salinomycin, and flavophospholipol) do not belong to any antimicrobial family used in veterinary or human medicine. The development of national surveillance and monitoring programmes follows the scientific recommendations. These programmes, organised at a national level, share a common experimental external quality assurance system and the collected data should be summarised via the Internet in the coming few months. This project is the second step of a European collaboration in this field supported by the European commission. The development of a network of excellence focused on zoonosis, sharing the competence of a major veterinary and human health institute (Med-Vet-Net) will probably contribute to the improvement of global surveillance of zoonosis. Antimicrobials used in veterinary medicine are only prescribed by veterinarians in the European Union. Before approval, drugs are assessed for the definition of maximum residue limits and environmental impact and efficacy. The guidelines of the EMEA have been modified to improve the assessment of antimicrobial risk and are taken into account during the evaluation of feed additives by the expert committees of member states and the new European Food Safety Authority (EFSA). Veterinarians and animal producers are aware of antimicrobial resistance. The major problem is now to develop specific recommendations about antimicrobial use for each animal species and type of production. This step requires a better knowledge of

668

Pascal Sanders

antimicrobial usage, current practices, and their effect on antimicrobial resistance (Chauvin *et al.*, 2001). It also requires to identify the worst usage and develop communications about good usage, allowing a low risk of selection (Anthony *et al.*, 2001). The development of pharmaco-epidemiology in veterinary medicine is necessary to propose, in the future, adapted antimicrobial management for each of animal production. In the past, antimicrobials were developed on the basis of evaluation of the effect on a few subjects, extrapolated to the general population. In the future, the main change will be to develop a global approach of adapted individual therapy (Vinks, 2002).

REFERENCES

Aarestrup, F. M., 1999, Association between the consumption of antimicrobial agents in animal husbandry and the occurrence of resistant bacteria among food animals. *Int. J. Antimicrob. Agents*, **12**, 279–285.

Acar, J. and Rostel, B., 2001, Antimicrobial resistance: An overview. *Rev. Sci. Tech. Off. Int. Epiz.*, **20**, 797–810.

Aliabadi, F. S. and Lees, P., 2002, Pharmacokinetics and pharmacokinetic/pharmacodynamic integration of marbofloxacin in calf serum, exudate and transudate. *J. Vet. Pharmacol. Ther.*, **25**, 161–174.

Anonymous, 1999, Antibiotic resistance in European Union associated with therapeutic use of veterinary medicines. Report and Quantitative Risk Assessment by the Committee of Veterinary Medicinal Products. EMEA/CVMP/342/99.

Anonymous, 2001, WHO global strategy for containment of antimicrobial resistance. WHO/CDS/CSR/DRS/2001.2.

Anonymous, 2002a, FAAIR Scientific Advisory Panel. Policy recommendations. *Clin. Infect. Dis.*, **34**(Suppl. 3), S76–S77.

Anonymous, 2002b, FIDIN Antibiotics Report 2001. Put Together by the FIDIN Working Group for Antibiotic Management, June 2002, The Hague. Tijdschr Diergeneeskd. **127**, 495–6.

Anonymous, 2003, Rapport sur l'évaluation nutritionnelle et sanitaire des aliments issus de l'agriculture biologique.

Anthony, F., Acar, J., Franklin, A., Gupta, R., Nicholls, T., Tamura, Y. *et al.*, 2001, Antimicrobial resistance: Responsible and prudent use of antimicrobial agents in veterinary medicine. *Rev. Sci. Tech. Off. Int. Epiz.*, **20**, 829–839.

Baggesen, D. L., Sandvang, D., and Aarestrup, F. M., 2000, Characterization of Salmonella enterica serovar typhimurium DT104 isolated from Denmark and comparison with isolates from Europe and the United States. *J. Clin. Microbiol.*, **38**, 1581–1586.

Blake, D. P., Hillman, K., Fenlon, D. R., and Low, J. C., 2003, Transfer of antibiotic resistance between commensal and pathogenic members of the Enterobacteriaceae under ileal conditions. *J. Appl. Microbiol.*, **95**, 428–436.

Brinas, L., Moreno, M. A., Teshager, T., Zarazaga, M., Saenz, Y., Porrero, C. *et al.*, 2003, Beta-lactamase characterization in Escherichia coli isolates with diminished susceptibility or resistance to extended-spectrum cephalosporins recovered from sick animals in Spain. *Microb. Drug Resist.*, **9**, 201–209.

Bywater, B. J., Deluyker, H., Deroover, E., De Jong, A., Marion, H., McConville, M. *et al.*, 2003, European Antimicrobial Susceptibility Surveillance in Animals (EASSA) programme: Results on antimicrobial susceptibility of enteric bacteria from pigs. *J. Vet. Pharmacol. Therap.*, **26**(Suppl. 1), 165.

Caprioli, A., Busani, L., Martel, J. L., and Helmuth, R., 2000, Monitoring of antibiotic resistance in bacteria of animal origin: epidemiological and microbiological methodologies. *Int. J. Antimicrob. Agents*, **14**, 295–301.

Chauvin, C., Madec, F., Guillemot, D., and Sanders, P., 2001, The crucial question of standardisation when measuring drug consumption. *Vet. Res.*, **32**, 533–543.

Chauvin, C., Beloeil, P. A., Orand, J. P., Sanders, P., and Madec, F., 2002a, A survey of group-level antibiotic prescriptions in pig production in France. *Prev. Vet. Med.*, **55**, 109–120.

Chauvin, C., Madec, F., Guittet, M., and Sanders, P., 2002b, Pharmaco-epidemiology and -economics should be developed more extensively in veterinary medicine. *J. Vet. Pharmacol. Ther.*, **25**, 455–459.

Choraine, P., 2000, Medicines availability—the millennium muddle. *Vet J.*, **160**, 1–2.

Commission Decision, 2003, Of 17 July amending Decision 2000/96/EC as regards the operation of dedicated surveillance networks. *Off. J. EU*, **L185**, 55–58.

Commission Position (EC), 2003, N°13/2003 adopted by the Council on February 2003 with a view to adopting Directive 2003/.../EC of the European Parliament and of the Council of ... on the monitoring of zoonoses agents, amending Council Decision 90/424/EEC and repealing Council Directive 92/117/EEC. *Off. J. EU*, **C90E**, 9–24.

Cromwell, G. L., 2002, Why and how antibiotics are used in swine production. *Anim. Biotechnol.*, **13**, 7–27.

Desenclos, J. C., Vaillant, V., and De Valk, H., 2002, Food-borne infections: Do we need further data collection for public health action and research. *Rev. Epidemiol. Sante Publique.* **50**, 67–79.

European Commission, 2000, Opinion of the scientific committee on veterinary measures relating to public health on food-borne zoonoses.

Fekete, T., 1995, Bacterial genetics, antibiotic usage, and public policy: The crucial interplay in emerging resistance. *Perspect. Biol. Med.*, **38**, 363.

Fife-Schaw, C. and Rowe, G., 1996, Public perceptions of everyday food hazards: A psychometric study. *Risk Anal.*, **16**(4), 487–500.

Franklin, A., Acar, J., Anthony, F., Gupta, R., Nicholls, T., Tamura, Y. *et al.*, 2001, Antimicrobial resistance: Harmonization of national antimicrobial resistance monitoring and surveillance programmes in animals and in animal-derived food. *Rev. Sci. Tech. Off. Int. Epiz.*, **20**, 859–870.

Gnanou, J. C. and Sanders, P., 2000, Antibiotic resistance in bacteria of animal origin: methods in use to monitor resistance in EU countries. *Int. J. Antimicrob. Agents*, **15**, 311–322.

Goodyear, K. L., 2002, DEFRA Antimicrobial Resistance Coordination Group. Veterinary surveillance for antimicrobial resistance. *J. Antimicrob. Chemother.*, **50**, 612–614.

Gould, I. M., 2002, Antibiotic policies and control of resistance. *Curr. Opin. Infect. Dis.*, **15**, 395–400.

Grave, K., Greko, C., Nilsson, L., Odensvisk, K., Mork, T., and Ronning, M., 1999a, The usage of veterinary antibacterial drugs for mastitis in cattle in Norway and Sweden during 1990–1997. *Prev. Vet. Med.*, **42**, 45–55.

Grave, K., Lingaas, E., Nilsson, L., Bangen, M., and Ronning, M., 1999b, Surveillance of the overall consumption of antibacterial drugs in humans, domestic animals and farmed fish in Norway in 1992 and 1996. *J. Antimicrob. Chemotherap.*, **43**, 243–252.

Guerra, B., Junker, E., Schroeter, A., Malorny, B., Lehmann, S., and Helmuth, R., 2003, Phenotypic and genotypic characterization of antimicrobial resistance in German

Escherichia coli isolates from cattle, swine and poultry. *J. Antimicrob. Chemother.*, **52**, 489–492.

Guillemot, D., Carbon, C., Balkau, B., Geslin, P., Lecoeur, H., Vauzelle-Kervroeden, F. *et al.*, 1998, Low dosage and long treatment duration of beta-lactam risk factors for carriage of prenicillin-resistant Streptococcus pneumoniae. *J. Am. Med. Assoc.*, **279**, 365–370.

Gustafson, R. H. and Bowen, R. E., 1997, Antibiotic use in animal agriculture. *J. Appl. Microbiol.*, **83**, 531–541.

Gyssens, I. C., 2001, Quality measures of antimicrobial drug use. *Int. J. Antimicrob. Agents*, **17**, 9–19.

Helms, M., Vastrup, P., Gerner-Smidt, P., and Molbak, K., 2002, Excess mortality associated with antimicrobial drug resistant *Salmonella* Typhimurium. *Emerg. Infect. Dis.*, **8**, 490–495.

Helms, M., Vastrup, P., Gerner-Smidt, P., and Molbak, K., 2003, Short and long term mortality associated with foodborne gastrointestinal infections: Registry based study. *BMJ*, **326**, 357–361.

Highet, V. S., Forrest, A., Ballow, C. H., and Schentag, J. J., 1999, Antibiotic dosing issues in lower respiratory tract infection: Population-derived area under inhibitory curve is predictive of efficacy. *J. Antimicrob. Chemother.*, **43**(Suppl. A), 55–63.

Jensen, L. B., Ahrens, P., Dons, L., Jones, R. N., Hammerum, A., and Aarestrup, F. M., 1998, Molecular analysis of Tn1546 in *Enterococcus faecium* isolated from animals and humans. *J. Clin. Microbiol.*, **36**, 437–442.

Jouy, E., Meunier, D., Martel, J. L., Kobisch, M., Coudert, M., and Sanders, P., 2002, Méthodologie du réseau national de surveillance de la résistance aux antibiotiques chez les principales bactéries pathogènes des animaux de rente (RESAPATH). *Bull. Acad. Vét. de France*, **155**, 259–266.

Kirk, S. F., Greenwood, D., Cade, J. E., and Pearman, A. D., 2002, Public perception of a range of potential food risks in the United Kingdom. Appetite. **38**, 189–197.

Klein, G., Pack, A., and Reuter, G., 1998, Antibiotic resistance patterns of enterococci and occurrence of vancomycin-resistant enterococci in raw minced beef and pork in Germany. *Appl. Environ. Microbiol.*, **64**, 1825–1830.

Lathers, C. M., 2002a, Clinical pharmacology of antimicrobial use in humans and animals. *J. Clin. Pharmacol.*, **42**, 587–600.

Lathers, C. M., 2002b, Educational issues in clinical pharmacology: Who are our audiences and what are their specialized needs? One specialized need: "Understanding the role of veterinary medicine in public health". *J. Clin. Pharmacol.*, **42**, 718–730.

Lathers, C. M., 2002c, Risk assessment in regulatory policy making for human and veterinary public health. *J. Clin. Pharmacol.*, **42**, 846–866.

Lauritzen, B., Lykkesfeldt, J., and Friis, C., 2003, Evaluation of a single dose versus a divided dose regimen of danofloxacin in treatment of *Actinobacillus pleuropneumoniae* infection in pigs. *Res. Vet. Sci.*, **74**, 271–277.

Lees, P., and Aliabadi, F. S., 2000, Rationalising dosage regimens of antimicrobial drugs: A pharmacological perspective. *J. Med. Microbiol.*, **49**, 943–945.

Lipsitch, M., Singer, R. S., and Levin, B. R., 2002, Antibiotics in agriculture: When is it time to close the barn door. *PNAS*, **99**, 5752–5754.

Lipsitch, M. and Samore, M., 2002, Antimicrobial use and antimicrobial resistance: a population perspective. *Emerging Infect. Dis.*, **8**, 347–354.

Martel, J. L., Chaslus-Dancla, E., Coudert, M., Poumarat, F., and Lafont, J. P., 1995, Survey of antimicrobial resistance in bacterial isolates from diseased cattle in France. *Microb. Drug Resist.*, **1**, 273–283.

Molbak, K., Baggesen, D. L., Aarestrup, F. M., Ebbesen, J. M., Engberg, J., Frydendahl, K. *et al.*, 1999, An outbreak of multidrug-resistant, quinolone-resistant *Salmonella enterica* serotype Typhimurium DT104. *N. Engl. J. Med.*, **4**, 1420–1425.

Moreno, M. A., Dominguez, L., Teshager, T., Herrero, I. A., and Porrero, M. C., 2000, Antibiotic resistance monitoring: The Spanish programme. The VAV Network. Red de Vigilancia de Resistencias Antibioticas en Bacterias de Origen Veterinario. *Int. J. Antimicrob. Agents*, **14**, 285–290.

Monnet, D. L., Hemborg, H. D., Andersen, S. R., Scholler, C., Sorensen, T. L., and Bager, F., 2000, Surveillance of antimicrobial resistance in Denmark. *Euro Surveill.*, **5**, 129–132.

Nicholls, T., Acar, J., Anthony, F., Franklin, A., Gupta, R., Tamura, Y. *et al.*, 2001, Antimicrobial resistance: Monitoring the quantities of antimicrobials used in animal husbandry. *Rev. Sci. Tech. Off. Int.*, **20**, 841–847.

Pedersen, K. B., Aarestrup, F. M., Jensen, N. E., Bager, F., Jensen, L. B., Jorsal, S. E. *et al.*, 1999, The need for a veterinary antibiotic policy. *Vet Rec.*, **145**, 50–53.

Powers, W. J., 2003, Keeping science in environmental regulations: The role of the animal scientist. *J. Dairy Sci.*, **86**, 1045–1051.

Raveh, D., Levy, Y., Schlesinger, Y., Greenberg, A., Rudensky, B., and Yinnon, A. M., 2001, Longitudinal surveillance of antibiotic use in the hospital. *QJM*, **94**, 141–152.

Raynaud, J. P., Gorse, P., and Ruckebush, Y., 1988, Interventions thérapeutiques, métaphylactiques et prophylactiques en élevage intensif—Problèmes et réalités politiques. *Revue Méd. Vét.* **139**, 205–225.

Rocourt, J., BenEmbarek, P., Toyofuku, H., and Schlundt, J., 2003, Quantitative risk assessment of Listeria monocytogenes in ready-to-eat foods: The FAO/WHO approach. *FEMS Immunol. Med. Microbiol.*, **35**, 263–267.

Rosenquist, H., Nielsen, N. L., Sommer, H. M., Norrung, B., and Christensen, B. B., 2003, Quantitative risk assessment of human campylobacteriosis associated with thermophilic Campylobacter species in chickens. *Int. J. Food Microbiol.*, **83**, 87–103.

Sanders, P., Gicquel, M., Humbert, F., Perrin-Guyomard, A., and Salvat, G., 2002, Plan de surveillance de la résistance aux antibiotiques chez les bactéries indicatrices isolées de la flore intestinale des porcs et de la volaille 1999–2001. *Bull. Acad. Vét. De France*, **155**, 267–277.

Schentag, J. J., 2001, Antimicrobial management strategies for Gram-positive bacterial resistance in the intensive care unit. *Crit. Care Med.*, **29**(Suppl. 1), N100–N107.

Smith, D. L., Harris, A. D., Johnson, J. A., Silbergeld, E. K., Morris, J. G. Jr., 2002, Animal antibiotic use has an early but important impact on the emergence of antibiotic resistance in human commensal bacteria, *Proc. Natl. Acad. Sci. USA*, **99**, 6434–6439.

Smith, D. L., Johnson, J. A., Harris, A. D., Furuno, J. P., Perencevich E. N., and Morris, J. G. Jr., 2003, Assessing risks for a pre-emergent pathogen: Virginiamycin use and the emergence of streptogramin resistance in Enterococcus faecium. *Lancet Infect. Dis.*, **3**, 241–249.

Stege, H., Bager, F., Jacobsen, E., and Thougaard, A., 2003, VETSTAT—The Danish system for surveillance of the veterinary use of drugs for production animals. *Prev. Vet. Med.*, **57**, 105–115.

Swann, M., 1969, Report of the Joint Committee on the Use of Antibiotics in Animal Husbandry and Veterinary Medecine. Her Majesty's Stationary Office, London.

Teshager, T., Herrero, I. A., Porrero, M. C., Garde, J., Moreno, M. A., and Dominguez, L., 2000, Surveillance of antimicrobial resistance in Escherichia coli strains isolated from pigs at Spanish slaughterhouses. *Int. J. Antimicrob. Agents*, **15**, 137–142.

Threlfall, E, J., Teale, C. J., Davies, R. H., Ward, L. R., Skinner, J. A. Graham, A. *et al.*, 2003, A comparison of antimicrobial susceptibilities in nontyphoidal salmonellas from humans and food animals in England and Wales in 2000. *Microb. Drug Resist.*, **9**, 183–189.

Toutain, P. L., del Castillo, J. R., and Bousquet-Melou, A., 2002, The pharmacokinetic-pharmacodynamic approach to a rational dosage regimen for antibiotics. *Res Vet. Sci.*, **73**, 105–114.

van Kasteren, M. E., Wijnands, W. J., Stobberingh, E. E., Janknegt, R., Verbrugh, H. A., and van der Meer, J. W., 1998, Optimizing antibiotics use policy in the Netherlands. I. The Netherlands Antibiotics Policy Foundation (SWAB). *Ned. Tijdschr. Geneeskd.*, **142**, 949–951.

Vink, A. A., 2002, The application of population pharmacokinetic modeling to individualized antibiotic therapy. *Int. J. Antimicrob. Agents*, **19**, 313–322.

VMD, 2003, Sales of antimicrobial products authorised for use as veterinary medicines, growth promoters, coccidiostats and antiprotozoals, in the UK in 2001. VMD Report.

Vose, D., Acar, J., Anthony, F., Franklin, A., Gupta, R., Nicholls, T. *et al.*, 2001, Antimicrobial resistance: Risk analysis methodology for the potential impact on public health of antimicrobial resistant bacteria of animal origin. *Rev. Sci. Tech. Off. Int. Epiz.*, **20**, 811–827.

Wegener, H. C., Hald, T., Wong, D. L., Madsen, M., Korsgaard, H., Bager, F. *et al.*, 2003, Salmonella control programs in Denmark. *Emerg. Infect. Dis.*, **9**, 774–780.

White, D. G., Acar, J., Anthony, F., Franklin, A., Gupta, R., Nicholls, T. *et al.*, 2001, Antimicrobial resistance: Standardization and harmonization of laboratory methodologies for the detection and quantification of antimicrobial resistance. *Rev. Sci. Tech. Off. Int. Epiz.*, **20**, 849–858.

Wierup, M., 2001a, The experience of reducing antibiotics used in animal production in the Nordic countries. *Int. J. Antimicrob. Agents*, **18**, 287–290.

Wierup, M., 2001b, The Swedish experience of the 1986 year ban of antimicrobial growth promoters, with special reference to animal health, disease prevention, productivity, and usage of antimicrobials. *Microb. Drug Resist.*, **7**, 183–190.

Willems, R. J., Homan, W., Top, J., van Santen-Verheuvel, M., Tribe, D., Manzioros, X. *et al.*, 2001, Variant esp gene as a marker of a distinct genetic lineage of vancomycin-resistant *Enterococcus faecium* spreading in hospitals. *Lancet*, **357**, 853–855.

Willems, R. J. L., Top, J., van den Braak, N., van Belkum, A., Endtz, H., Mevius, D. *et al.*, 2000, Host specificity of vancomycin-resistant *Enterococcus faecium*. *J. Infect. Dis.*, **182**, 816–823.

World Health Organization (WHO), 2000, WHO global principles for the containment of antimicrobial resistance in animals intended for food. Report of a WHO Consultation with the Participation of the Food and Agriculture Organization of the United Nations and the Office International des Epizooties. Geneva, Switzerland, 5–9 June 2000. WHO, Geneva.

Wray, C. and Gnanou, J. C., 2000, Antibiotic resistance monitoring in bacteria of animal origin: Analysis of national monitoring programmes. *Int. J. Antimicrob. Agents*, **14**, 291–294.

Chapter 34

The Pharmaceutical Company Approach to Antibiotic Policies

Anthony R. White
GlaxoSmithKline, New Frontiers Science Park, Third Avenue, Harlow,
Essex CM19 5AW, UK

1.   INTRODUCTION

The word "policy" is defined as "prudent conduct, sagacity; course or general plan of action (to be) adopted by government, party or person." In this sense, antibiotic policies range from local hospital or community policies to national treatment guidelines for specific infections, regulatory policy, and regional and global strategies. All have the potential to impact on the pharmaceutical industries' (Pharma) integrated activities of discovering, developing, and marketing antibacterials. This article reviews the relationship between antibacterial policies and the role of Pharma.

For almost 70 years, Pharma has played a major role in discovering, developing, and making available antibacterials for human and veterinary use (Garrod *et al.*, 1981). Today, more than in recent decades, there are high demands on Pharma to deliver to patients new, effective antibacterials to replace those agents for which clinical utility is compromised by emerging bacterial resistance. This "crisis," with the daunting prospect that one of the most beneficial discoveries and developments in the history of mankind could become extinct within a mere few human generations, has understandably also prompted a wide range of recommendations, guidelines, and policies with the intent of preserving and protracting the utility of our current antibacterial armamentarium. However, in some cases, these policies themselves are

Antibiotic Policies: Theory and Practice. Edited by Gould and van der Meer
Kluwer Academic/Plenum Publishers, New York, 2005

673

reducing the chance of success of Pharma delivering the antibacterial agents for the future.

In order to succeed, Pharma will need to adapt to the new environment, but there is a very real and immediate danger that the extent of change in the environment is presenting significant hurdles that will prevent or delay the discovery and delivery of new agents. A greater understanding of the collective impact of policies on Pharma's ability and willingness to conduct sustainable research and development (R&D) of antibacterials is needed along with a cooperative partnership approach. There is a need for governments, non-governmental organizations (NGOs), policy-makers, regulators, and infectious disease experts to collaborate with Pharma to establish ways to maximize the chance of bringing new antibacterials into clinical use and sustain their utility. Unfortunately, individual Pharma companies continue to exit from antibacterial discovery and development and the pool of major companies is likely to soon be occupied by a handful or less, severely restricting the ability of Pharma to meet the medical needs of infectious diseases, particularly those which require flexibility and responsiveness, such as epidemics and those arising from bioterrorism. Once R&D efforts are terminated it will require substantial new investment, both in monetary and intellectual terms, as well as time to re-initiate effective programmes. Without continued, sustainable, and successful antibacterial R&D, we will certainly be facing the prospect of "antibacterial extinction."

2.   THE PAST

There is no doubt that *research-based* Pharma has for many years been instrumental in providing the wide range of clinically important agents which are available today. Garrod *et al.* (1981) described three eras of antimicrobial chemotherapy:

- *alkaloids*: from 1619 and use of cinchona bark to treat malaria
- *synthetic compounds*: from 1909 and the discovery of the arsenical salvarsan by Ehrlich, and including Prontosil/sulfonamides in 1935
- *antibiotics*: from 1929 with the discovery of penicillin.

Pharma has been predominant in discovering or developing antibacterials in the two most recent of these three eras. While not all antibacterials were discovered by Pharma, many of the early members of today's antibacterial classes were brought to clinical use through the development and scale-up activities of Pharma, in some cases through collaboration by a number of companies, notably in the development of penicillin G in the 1940s. The majority of discoveries and developments that have produced the major classes of antibacterials, as well as the many significant developments within these classes, have been the result of the Pharma industries' innovation and expertise.

It is probable that without the efforts of research-based Pharma and the associated integrated competitive and commercial aspects, many of today's agents would not yet have been discovered, developed, or available to physicians. The now somewhat famous declaration in 1969 by the US Surgeon General, William Stewart, testifying before US Congress that it was time to, "Close the book on infectious diseases," illustrates the degree of success that was perceived to have been achieved at that time.

3.   THE PRESENT

3.1.   Pharma and the "crisis"

Although there is a sector of the Pharma industry that focuses on the sale of generic versions of agents discovered and developed by the research-based companies, these activities mainly require only manufacturing and distribution expertise. Now, as in the past, it is the research-based companies to which we need to continue to look to for the provision of agents for the future.

The discovery, development, and appropriate use of new agents is a key theme in the major antibacterial strategies and policies to combat resistance which have been reviewed by Carbon *et al.* (2002). Against this background it may be considered surprising or even alarming that the Interscience Conference on Antimicrobial Agents and Chemotherapy for 2003 (ICAAC, 2003) includes a symposium entitled *Why is Big Pharma Getting Out of Anti-Infective Drug Discovery?* A review of abstracts on *new* antibacterial agents or targets/methods presented at ICAAC 2002 indicates some 47 companies involved in this field including 10 of the "top 12" companies as determined by sales (Table 1).

Of the total, 32 companies presented data from established classes (β-lactams/ inhibitors, lipopeptides/glycopeptides, fluoroquinolones, oxazolidines, protein synthesis inhibitors) and 8 of these companies were from the top 12 companies by sales. Twenty-one companies presented data on targets or methods, thirteen exclusively, and seven on novel agents or immunomodulators (Table 1). This analysis represents only a snapshot of those companies presenting at ICAAC and not the total discovery programmes currently being undertaken, some on areas not yet disclosed, others since terminated.

Despite the wide range of presentations by Pharma at ICAAC 2002, there has undoubtedly been a consolidation in antibacterial R&D in the large Pharma companies, and a dearth of novel agents emerging from the pipelines.

Table 1. Companies involved in the development of new antibacterial agents presented at ICAAC 2002

Type of agent	Companies involved
Carbapenems	Meiji Seika Kaisha Ltd, Sankyo Company Ltd
Cephalosporins	LG Chem Investment Ltd, Takeda Chemical Industries Ltd, Basilea Pharmaceutica AG, *Shiongi & Co Ltd, Johnson & Johnson*
Oxapenems	Amura Ltd
Lipo- and glycopeptides	Cubist Pharmaceuticals Inc, *Eli Lilly & Co, Wyeth Research*, Biosearch Italia SpA, *Aventis*, Theravance Inc
Fluoroquinolones	Wakanuga Pharmaceutical Co Ltd, *Abbott Laboratories*, Wockhardt Research Centre, *Bayer AG*, Dong Wha Pharm Co Ltd
Oxazolidines	Ranbaxy Research Laboratories, AstraZeneca, Dong-A Pharm, ImaGene, Morphochem AG, *Johnson & Johnson*, Versicor Inc, *Pharmacia Corporation*, Dr Reddy's Laboratories Ltd
Protein synthesis inhibitors (macrolides and ketolides)	*Aventis, Johnson & Johnson*, Enanta Pharmaceuticals Inc, Versicor Inc, Novartis Pharmaceuticals Corporation, British Biotech Pharmaceuticals Ltd, Optimer Pharmaceuticals Inc, *Bayer AG*
Other new agents and immunomodulators	*F Hoffman La Roche AG* (diaminopyrimidine), Meiji Seika Kaisha Ltd (caprazamycin), *Bayer AG* (dipeptide), Versicor (LpxC inhibitor), Genesoft (hetero-aromatic polyamides), Xechem Inc (rapamycin), Biocryst Pharmaceuticals Inc (purine nucleoside phosphorylase inhibitor)
Targets and discovery methods	AstraZeneca, *Aventis*, Quorex Pharmaceuticals, Daiichi Pharmaceutical Co Ltd, Proteomic Systems Inc, Arrow Therapeutics, Genome Therapeutics Corporation, GPC Biotech AG, *Eli Lilly & Co*, British Biotech Pharmaceuticals Ltd, NewBiotics Inc, PanTherix Ltd, Schering-Plough Research Institute, *Wyeth Research, GlaxoSmithKline Pharmaceuticals*, Influx Inc, Genelabs Technologies Inc, *Bayer AG*, ImaGene, Pantheco A/S, *Pharmacia Corporation*

Note: Names in italics are major pharmaceutical companies that feature in the top 12 (by sales).

Increasing discovery activity in smaller companies may offset this, although the capabilities of smaller companies to develop antibacterials to market may be limited in the face of increasing development hurdles. The future model of discovery and development may well have to rely on collaborations and cooperations between small/discovery and large development/supply Pharma.

Mergers and acquisitions have resulted in the key players becoming significantly fewer and each withdrawal from antibacterial development has an increasingly significant impact on the potential for new antibacterials becoming available. The top 12 leaders in the field of antibacterials in terms of sales in 2002 (GlaxoSmithKline Pharmaceuticals, Pfizer/Pharmacia, Bayer AG, Abbott Laboratories, Aventis, Johnson & Johnson, Roche-Chugai, Bristol-Myers Squibb, Wyeth Research, Shionogi Seiyoku, Eli Lilly & Co, and Merck & Co) have evolved from some 20 to 30 companies with antibacterial R&D heritages. Some are the result of none or few amalgamations while others are many-fold.

Of the top 12 companies, less than half are believed to be currently active in antibacterial R&D. Within the remaining companies the discovery effort has been directed towards either defined niche disease areas, or large "block-buster" commercial areas, with a consequent reduction in diversity across the industry. Individual specializations within Pharma, such as was the case when companies almost exclusively worked on penicillins (Beecham Research Laboratories), cephalosporins (Glaxo), or quinolones (Bayer AG), for example, are becoming less as the companies look to satisfy commercially attractive target product profiles from whichever molecular classes that are available through their R&D efforts or in-licensing. The emphasis on development of line extensions of existing molecules has also increased as a means to improve patient benefits such as convenience and efficacy and importantly for Pharma, to maintain development and commercial activity to offset patent loss and bridge the gap to the introduction of new agents. Critically, of the companies which have exited or severely reduced their antibacterial R&D (e.g., Aventis, Bristol-Myers Squibb, Eli Lilly & Co, Roche-Chugai, and Wyeth Research) (IDSA, 2003), these decisions have been taken relatively recently at a time when the need for new effective antibacterials is arguably greater, but also at a time when policies and regulations for the development and use of antibacterials have proliferated.

The crucial issue will be the ability of companies, big or small, alone or in collaboration, to bring novel agents to the market and whether the agents being discovered and developed today will satisfy both the medical and the commercial needs to provide a sustainable future for antibacterial chemotherapy. Continued investment and commitment of big or small Pharma into research for new antibacterials to meet current and future clinical needs is crucial and yet is at a crisis and being impacted by antibacterial policies which are increasingly being developed and implemented.

The Infectious Disease Society of America (IDSA) newsletter, March 12, entitled *The Future of Antimicrobial Drug Availability: An Impending Crisis* (IDSA, 2003), highlights the current issues facing the development of antimicrobial drugs. Infectious diseases are the second leading cause of death and the leading cause of disability-adjusted life years worldwide. Antibacterials are key tools in treating many globally important infectious diseases, including meningitis, pneumonia, diarrhoeal illness, skin and bone infections, tuberculosis, sexually transmitted infections (gonorrhoea, syphilis, and chlamydia),

HIV/AIDS-related bacterial infection, and diseases that may be spread as a result of bioterrorist acts. Withdrawal from antibacterial research will have a major impact on global health. A letter to the journal of the American Society for Microbiology (Appelbaum, 2003) states that, "A crisis has developed and we are doing nothing about it," highlighting that Pharma companies are reducing or closing their antibacterial research and that pipelines are, "practically empty." Appelbaum suggests that, unlike drugs for long-term treatments, there is a lack of funding for research into antibacterials, one of the reasons for this being the overly stringent approval criteria imposed by the Federal Drug Agency (FDA), making it practically impossible or prohibitively expensive to bring new antibacterials to market (Appelbaum, 2003).

For those who have worked in the antibacterial field from the hey-day of discoveries in the 1960s and 1970s it is very apparent that the number of new chemical agents being progressed to man and subsequently marketed is much reduced. IDSA reported that of 89 new medicines reaching the US market in 2002 none was an antibacterial, with only 7 new antibacterials being approved since 1998 (IDSA, 2003). A review of company reports for the 11 major Pharma companies listed only 4 new antibacterials in the drug pipeline out of 290 agents listed (1.38%) (IDSA, 2003) and that on average the time to develop new antibacterials was 7–10 years from discovery to first approval. Although Pharma has continued to explore ways to increase efficiencies and decrease development times, these have largely remained stable due to the increased numbers of patients required for new drug applications (NDAs), as well as increased complexity and costs in meeting regulatory approval requirements (Kaitin and DiMasi, 2000). An important consideration for Pharma companies involved in R&D and commercialization of therapies for a range of disease areas, is that other therapeutic areas are increasingly more profitable than antibacterials. Today, there are fewer large companies specializing in antibacterials as a main R&D area. Investment decisions are made across and between therapeutic areas and Pharma has an obligation to maintain shareholder value and commercial viability or else no new drugs would become available in any therapy area.

The reasons for the "crisis" in antibacterial R&D, manifested by fewer new agents or classes reaching clinical use, are complex and multifactorial but undoubtedly the impact of the changing environment of regulations, guidelines, and policies on the development and use of antibacterials is an important contributory factor.

3.2.   Antibacterial resistance and policies

Stated simply, antibacterials are developed and used to overcome bacterial infection. The progressive discovery and introduction of new classes and agents extended the range of bacteria and associated infections that could be successfully treated. Antibacterial resistance, which has emerged from the beginning of antibacterial chemotherapy, has provided a key medical reason for the development of new agents, along with the wish to improve efficacy, safety and tolerability, and dosing convenience. Anti-infectives are unique (with the exception of some increasing evidence in oncology) in that the target of their action can change and provide the rationale, medical need, and commercial incentive for the development of new agents. The need for new agents or classes to combat antibacterial resistance is arguably greater today than it has been for some decades.

Concerns about emerging resistance, increasing loss of utility of existing antibacterials and associated negative impacts on health have given rise to a wide range of strategies (strategic plans, action plans), policies (rules and directives), and guidelines or guidance. Some are advisory and others mandatory but all constitute broad policy with respect to the development and use of antibacterials. The sources of policy activities range from global NGOs, such as the World Health Organization (WHO), regionally based bodies such as the European Union (EU) and European Medicines Evaluation Agency (EMEA), country-based public health and regulatory bodies such as the Centres for Disease Control (CDC), and FDA, and physician associations (IDSA), to the local, and institutional level within countries. A comprehensive summary of the more influential recommendations for the control of antimicrobial resistance has been prepared by Carbon *et al.* (2002). Some 17 separate national recommendations, from eight countries are listed (Australia, Belgium, Canada [2], France [2], Finland, Sweden [2], United Kingdom [3], United States [5]). In addition there are five international recommendations: two from the EU (Copenhagen recommendations, 1998; Community Strategy against Antimicrobial Resistance, 2001), one from ICAAC 1999 (Summit on Antimicrobial Resistance), one from Toronto (Toronto Declaration, 2000), and one from the WHO Global Strategy for Containment of Antimicrobial Resistance (2001). All these recommendations date from 1994 to 2001 with the majority being published between 1998 and 2000. Undoubtedly, there are more or updated recommendations in the pipeline and continuing activities of bodies addressing the topic, such as the USA CDC Task Force (IDSA, 2003; Shlaes and Ryan, 2003).

Common themes, which emerge from the range of recommendations related to antibacterials and antibacterial resistance, are the need for:

- surveillance of resistance and antibacterial usage
- optimizing antibiotic use (reduce inappropriate use), guidelines, and policies
- education of professionals and patients into judicious/prudent use
- prevention through infection control, interventions, immunization
- focused development of new agents, diagnostics, and strategies

- regulatory/label guidance, prescribing, and advert restriction
- audit of evaluation of intervention and compliance.

The objectives of most strategies and policies are to better control antibacterial use with the intent of reducing or preventing antibacterial resistance. This, accompanied by parallel activities to monitor and understand resistance (such as its consequences and relationship to antibiotic use), improve infection control, and development of new agents (antibacterials and vaccines) and therapeutic approaches, is hoped to provide a sustainable solution in combating bacterial infection.

Carbon *et al.* (2002) state that these recommendations rightly focus on controlling resistance in the community through surveillance of resistance and usage, and education of prescribers and the public in judicious antibiotic use. Nevertheless, the authors point to the need for more research to fill substantial knowledge gaps, notably the reversibility or containment of resistance with the optimization of antibiotic usage has yet to be definitely established (Carbon *et al.*, 2002; Davey *et al.*, 2003). It is suggested by Carbon *et al.* (2002) that for now, antimicrobial management programmes should focus on ensuring the most appropriate use of antimicrobials rather than simply on limiting choices. Interventions must also be audited for effectiveness/cost (Christiansen *et al.*, 2002). It is also suggested that the programmes in the recommendations can not succeed if they do not imply the active contribution of health professionals, politicians, consumers, and Pharma, and that the creation of optimal conditions for this type of cooperation is of critical importance (Carbon *et al.*, 2002; Chahwakilian, 2000). The key role of Pharma in these strategies is to deliver new antibacterials, although the cumulative impact of the parallel recommendations may present significant barriers to achieving this aim (Figure 1).

3.2.1. Surveillance of resistance and usage

Surveillance programmes of bacterial susceptibility have become commonplace in the field of antibacterials over the last decade and feature in most recommendations and policies relating to combating resistance. The principles of surveillance, key studies, data, and issues have been extensively reviewed (Bax *et al.*, 2001; Felmingham *et al.*, 2002; Hunter and Reeves, 2002). Good quality surveillance methodology and interpretation are essential to the understanding of resistance emergence and its control. The main functions for surveillance were described by Felmingham *et al.* (2002) as:

- quantification of resistance: resistance prevalence/distribution/changes over time
- guidance for antibiotic use: individual patient level/guidelines/policies

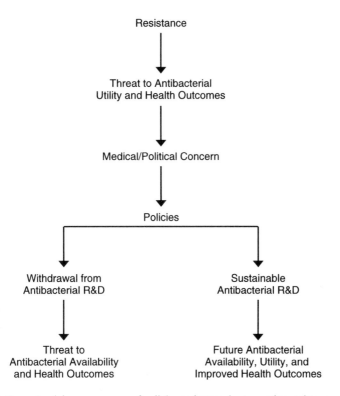

Figure 1. Two potential consequences of policies and strategies to combat resistance.

- research/education: epidemiology of resistance/link with usage
- industry: R&D/licensing/marketing/post-marketing
- resistance control: design of strategies/impact of strategies/interventions.

Pharma is both a generator and user of surveillance. Additionally, all the functions for surveillance described by Felmingham *et al.* (2002) have the potential to impact Pharma's development strategies and potential success in delivering new antibacterials. The relationship between resistance, antibiotic usage patterns, and clinical outcomes is an important factor in determining policies for antibacterial use. Scientifically robust methodologies and interpretations are critical but have not yet been fully defined or attained for many surveillance studies (Bax *et al.*, 2001).

Pharma itself has been actively involved in the initiation and support of a number of key surveillance studies over many years (Felmingham *et al.*, 2002).

These studies are mostly aligned with the companies' discovery and marketing interests but nevertheless have provided good quality data in many cases where data were lacking and even pre-date the recent calls in strategy and policy documents for antibacterial surveillance. Indeed, global studies, such as the Alexander Project (Felmingham *et al.*, 2002), initiated in 1992 and funded by SmithKlineBeecham (now GlaxoSmithKline) pioneered the standardization and quality control of methodology across countries and laboratories in the surveillance of bacterial resistance in community-acquired lower respiratory tract infection and provided a model for a number of subsequent global and national studies.

Pharma involvement and funding in surveillance may be a continuing necessary and desirable role. Pharma need data from which to establish and drive the need for R&D activities into new antibacterials. Surveillance data are also needed for product labelling, such as the resistance prevalences required, for example, by EMEA. Susceptibility data from surveillance are important to support product differentiation and benefits in the market place, and may also be required as part of a Phase IV regulatory commitment for a new agent. Academia and policy-makers also need such data to assess the antibacterial environment and the appropriate policy and practice interventions and outcomes. All parties need good quality, scientifically valid data. However, there may be differences in needs regarding the scope of studies, such as the organisms that are included, the agents tested, or the reporting of data (minimum inhibitory concentrations vs susceptible/intermediate/resistance), but the demand for quality and accuracy of interpretation should be equal.

A key issue is the source and sustainability of funding of surveillance, along with ownership, interpretation, and dissemination of data. Global or regional, longitudinal surveillance studies are extremely expensive and resource intensive. Individual Pharma companies have initiated and solely sponsored studies to support the development and marketing of new agents (Bax *et al.*, 2001; Felmingham *et al.*, 2002; Hunter and Reeves, 2002). This creates a situation where at a particular time a number of Pharma companies may compete for collecting centres for key isolates and be funding a number of laboratories to screen similar isolates and agents while at other times or for other pathogens or agents, there may be little Pharma support, or data available from any source. Surveillance funded by government, societies, or other non-Pharma sources may not meet Pharma's specific needs. Conversely, Pharma's studies may not meet broader needs (Bax *et al.*, 2001; Lewis, 2002), or be perceived as "distorted" source of data because they are funded by Pharma (Lewis, 2002). Pharma support for large single-sponsored surveillance studies is dependent on successful product registrations and marketing. Similarly, Pharma support for broad regional and national studies, of value to strategists and policy-makers in these areas will become limited in an environment of reduced investment and commercial success. Collaborative surveillance between and within Pharma and academia or regulators may become increasingly necessary but is difficult to implement in terms of mutually acceptable objectives, scope, methods, timelines, funding agreement, management, and publication/presentation rights. Bax *et al.* (2001) stated that consortium funding will be necessary for large schemes to be successful, and that there is no "ideal" surveillance system.

New approaches to the funding, conduct, and availability of surveillance data while recognizing individual Pharma confidentiality and commercial interests, include the purchasing of surveillance data from commercial surveillance companies or entering into collaborative surveillance studies with specific protocols. Sources of purchased data, such as from Focus Technologies (formerly MRL/TSN) (Bax *et al.*, 2001; Hunter and Reeves, 2002), may provide solutions to many requirements although the extent, quality, and utility of the data is dependent on the database and may not be as well defined or specific for a purpose as a pre-defined protocol. Examples of collaborative pre-defined protocols are the British Society for Antimicrobial Chemotherapy (BSAC) surveillance projects for respiratory tract pathogens (Reynolds and BSAC Extended Working Party on Respiratory Resistance Surveillance, 2001) and bacteraemia pathogens in which a number of Pharma sponsors support a core protocol and panel of antibacterials, along with the ability to have research agents tested under confidentiality. The results from the core protocol, which has been developed by experts in academia, specialist testing laboratories, and Pharma are made available via the website to the BSAC membership (www.bsacsurv.org). This approach, combining support from a number of companies, with the wide and timely dissemination of core data resolves some of the issues and conflicts relating to quality, interpretation, and dissemination associated with single-company studies and also is hoped to increase sustainability. The current utility of this programme is, however, restricted due to the data and methods being specific to the United Kingdom.

In conjunction with the generation and use of susceptibility surveillance data, there is need for a clear understanding of the limitations, validity, and consequences of studies to correlate antibacterial usage with resistance. The relationship between antibacterial usage and resistance is complex, and while there is a clear association between use and resistance, cause and effect has not been substantially defined (Carbon *et al.*, 2002; Davey *et al.*, 2003). There are a number of confounding variables that are important to consider in the interpretation of simple cause and effect (Davey *et al.*, 2003). Reversibility or containment of resistance with optimization (or reduction) of antibacterial usage has yet to be definitely established (Carbon *et al.*, 2002). Consequently, policies designed to curtail or prevent antibacterial resistance based on reduction or change in antibacterial use need to be based on sound science and importantly, the objective of maintaining or improving patient outcomes. Bax *et al.*

(2001) concluded that for surveillance:

1. decisions need to be made regarding the critical populations to be monitored for both local and national comparisons of resistance rates, that is,
   what level of resistance warrants intervention, and what should be the
   nature of the intervention;
2. quantitative data are more valuable than qualitative data and all testing must
   be shown to yield results that are capable of valid comparison;
3. molecular techniques should be used to define strains and detect mechanisms of resistance (see Poupard, Chapter 23);
4. there is a place for both narrow focus and broader surveillance studies: both
   require funding, but broader studies, including a wide range of organisms
   and compounds, require consortium funding.

There is a need for the interested parties such as academia, policy-makers, regulators, and Pharma to consider mutually beneficial surveillance programmes,
with agreed methodologies and interpretations persuant to their purpose. This
approach may not only provide efficiencies in utility, support, and sustainability but also a better understanding of relationship between antibacterial use
and resistance and the role and impact of policies and interventions.

3.3. Policies, guidelines, and education on antibacterial use

In addition to the key influential recommendations described by Carbon
et al. (2002) there are many other national and local guidelines, guidances, policies, and educational campaigns which impact antibacterial use. Most such
policies either describe the need for implementation of "judicious" or "prudent"
use of antibacterials, or use terms such as "misuse" and "overuse" as causes of
resistance and imply that a reduction in antibacterial prescribing *per se* will
have a beneficial and sustainable effect on resistance. There is a recognition that
while prescribing guidelines and other prescribing support systems should help
control bacterial resistance in the community, their actual effect on resistance
patterns is largely unknown (Finch and Low, 2002) and the reversibility or containment of resistance with the optimization of antibiotic usage has yet to be
definitely established (Carbon *et al.*, 2002; Davey *et al.*, 2003).

Davey *et al.* (2003) described the control of antibiotic prescribing as a
crucial part of any strategy to limit the development of resistance as advocated
by the UK House of Lords recommendations and the Copenhagen recommendations (Carbon *et al.*, 2002). These authors recognized the difficulty in proving that antibiotic policies help to resolve a problem of resistance that has
already developed and also recognized the greater difficulty in proving that

policies prevent the development of resistance. Most policies (and guidelines) are concerned with which antibiotic is prescribed and Davey *et al.* (2003) state that it is more likely that development of resistance will be contained by policies which also try to limit unnecessary prescribing of antibiotics (i.e., prescribing for non-bacterial or self-limiting illness).

The general advantages of an antibiotic policy are described by Davey *et al.* (2003) as:

- promotion of awareness of benefits, risks, and cost of prescribing
- facilitation of educational and training programmes
- reduction of aggressive marketing by Pharma
- encouragement of rational choice between drugs based on analysis of pharmacology, clinical effectiveness, safety, and cost;

and specifically to antimicrobials as:

- the promotion of education on local pathogen epidemiology and susceptibility along with awareness of infection control.

The general benefits are described as:

- improved efficacy of prescribing (sensitivity and specificity)
- improved clinical outcome
- reduced medical liability;

and specifically to antimicrobials as:

- limited emergence and spread of resistance.

Davey *et al.* (2003) suggest that antibacterial policies can definitely improve the quality of prescribing and may be used to limit cost and that limiting superinfections and antibiotic resistance should be viewed as additional benefit (and probably not a realistic primary aim). Similarly, Ball *et al.* (2002) stated that when prescribing is necessary, quality prescribing may not only combat resistance and optimize patient outcomes but also provide cost benefits. In a review of the impact of guidelines in the management of community-acquired pneumonia, several studies were identified which showed that an increase in the proportion of patients who receive prompt, appropriate therapy was associated with improvements in outcome measurements such as mortality and length of hospital stay (Nathwani *et al.*, 2001).

Gould undertook a review of the role of antibiotic policies in the control of antibiotic resistance, noting that the pragmatic and essential approach to the control of antibiotic resistance is to control antibiotic use (Gould, 1999).

He also noted that policies can be efficacious in reducing costs and levels of antibiotic use, but the subject of debate is whether antibiotic control measures can reduce current levels of resistance rather than just halting it. There are examples where control of antibiotic use has reduced the incidence of outbreak and resistant organisms in hospitals and similar examples where reduction of prescribing in the community has led to a degree of control of resistance. There are a number of confounding factors which make it difficult to attribute cause and effect entirely to reduction in antibiotic prescribing (Davey *et al.*, 2003; Gould, 1999). Mathematical modelling seeks to take these factors into account and, while also casting doubts on whether reduction alone will prevent or reduce resistance (Austin *et al.*, 1999; Gould, 1999), may help with future understanding of measures which contribute to or reduce antibacterial resistance, and as such should be encouraged. Use of this approach to define policies has been limited (if used at all) (Davey *et al.*, 2003). Nevertheless, policy decisions should be supported by evidence, and not just by concerns about future development of resistance or the desire to reduce costs (Davey *et al.*, 2003).

Importantly, appropriate use of antibacterials needs a clear definition. It is recognized that recommendations and guidelines should emphasize that the role of antibacterials is for treating bacterial infection, when known or suspected (Ball *et al.*, 2002; Davey *et al.*, 2003). Use of antibacterials outside of this scenario is unnecessary, contributes to overuse and is an avoidable risk factor in the development of resistance (Ball *et al.*, 2002; Davey *et al.*, 2003). However, when use of antibacterials is warranted (necessary) to treat bacterial infection, such use may also be inappropriate or suboptimal in terms of choice of antibacterial, dose, or duration. For example, Schlemmer (2001) concluded that antibiotic misuse does have an impact on promoting antibiotic resistance, and that antibacterial choice, dosage, dosing regimen, or duration of therapy must also, therefore, be considered. Carbon *et al.* (2002) suggested that, for now, antimicrobial management programmes should focus on ensuring the most appropriate use of antimicrobials rather than simply on limiting use or choices. A Consensus Group (Ball *et al.*, 2002) has recently identified principles for appropriate prescribing of antibacterials in lower respiratory tract infection with which to underlay prescribing and guideline formulation. These principles include:

- identification of bacterial infection by optimized diagnosis
- severity assessment where relevant
- recognition and incorporation of ambient resistance data
- targeting bacterial eradication (or maximal reduction in bacterial load)
- use of pharmacodynamic (PD) indices to optimize choice and dosage
- objective assessment of true (overall) costs of resistance and related treatment failure.

Guidelines, policies, and educational campaigns should seek to better define, support, and implement appropriate use based on principles such as those from the Consensus Group (Ball *et al.*, 2002). "Prudent" use or reduction in use alone will not adequately define or ensure optimum or appropriate use of antibacterials. The definition and implementation of appropriate prescribing based on evidence-based optimization of quality antibacterial use (right drug/dose/duration based on PD indices and local resistance patterns), rather than reduction in quantity alone may contribute towards the aim of slowing or preventing the emergence of antibacterial resistance, and also offer associated cost and efficiency benefits. Importantly, this approach may also maintain an environment that supports the use of new effective antibacterials and a commercial rationale for investment and development by Pharma. An ill-defined environment, focusing on the need to reduce quantity with little differentiation between optimal and suboptimal agents is not conducive to the development and marketing of agents with improved activity and efficacy benefits.

3.4. Discovery, development, and commercialization in the face of policies

Increasing development costs and hurdles and shrinking market sizes, means that Pharma has to get more value out of the R&D investment by improving productivity, reducing development timelines, reducing risk, and increasing the value of agents progressed to market. In the case of antibacterials, the situation is compounded by increasingly demanding regulatory policies affecting the activities required to secure product registration, the product label and clinical use (Shlaes and Ryan, 2003). The extent and success of antibacterial research has declined (IDSA, 2003) and is partly a reflection of its relatively low commercial attractiveness compared to other therapeutic areas. The relative return on investment (ROI) or net present value (NPV) of antibacterials is a reflection of the market size and chance of success in capturing a commercially acceptable part of that market. This important "market share" will be a reflection of the properties and utility of the agents along with the success of marketing activities. Differentiation of comparative benefits, such as efficacy against resistant organisms, broader spectrum of pathogens for a given indication, superior efficacy over commonly used treatments, improved dosage regimens and compliance, or better safety characteristics are important features in marketing an antibacterial and in the acceptance and uptake by policy-makers and in the demonstration of cost–benefits to formularies. Policies or guidelines which decrease the ability of Pharma to demonstrate these benefits, reduce the chance of gaining differentiated label indications (such as the inclusion of resistant organisms), or restrict usage to limited indications (such as only to resistant organisms) will negatively impact NPVs and ROIs for new agents. Development of an antibacterial agent for restricted use only may not be a commercially attractive proposition for Pharma companies with interests in other therapeutic areas.

In order for Pharma to succeed and maintain investment, hurdles in productivity in discovery and development need to be overcome and the key issue of ROI based on product benefits and use needs to be addressed.

3.4.1. Antibacterial discovery and genomics

Although policies themselves do not directly impact upon the process of discovery of new antibacterials, they do have a cumulative negative impact on the chance of success of developing and marketing new differentiated antibacterials and consequently, the willingness of companies to initiate or continue investing in antibacterial discovery research. The urgent need for new classes of antibacterials has, however, increased the pressure in Pharma to maximize efficiencies in identifying new molecular targets for antibacterials and compounds active against those targets. Genomics and high-throughput screening (HTS) offer important potential advances in this area.

On 25 April 1953 Francis Crick and James D. Watson published their paper "Molecular structure of nucleic acids" in *Nature* (Watson and Crick, 1953), describing the double helix structure of DNA. This signalled the launch of the modern era of molecular biology and provided the foundation of our understanding of molecular medicine, transforming much of scientific research including how we approach the discovery of new drugs. On 15 April 2003, nearly 50 years after the publication in *Nature*, Carl B. Feldbaum of the Biotechnology Industry Organization (BIO) announced the "finished" version of the human genome sequence and the completion of the Human Genome Project providing an accurate map of the 3.1 billion units of DNA. At the same time, it was reported that the genome of the newly identified SARS (Severe Acute Respiratory Syndrome) virus had been sequenced in only 6 days. The *Bacillus anthracis*, Ames strain used in postal terrorist attacks in the United States in 2002 was reported as sequenced on 1 May 2003.

The role of genomics in antimicrobial discovery should be considered against the background of Pharma's efforts over many years in fine-tuning the existing classes of antibacterials to improve their spectra, efficacy, and safety through semisynthetic and wholly synthetic chemistry approaches. Linezolid, an oxazolidinone, represented the first novel compound class introduced to the market in more than 25 years (Diekema and Jones, 2000). Traditional chemical modification to produce new members of existing classes has resulted in significant improvements in β-lactams, macrolides, and fluoroquinolones (e.g., Rasmussen and Projan, 2003), but has been a relatively slow and labour-intensive process with limited additional success in recent years. A team of bench chemists in the 1970s might produce up to 30 novel compounds each per annum for biological screening against a panel of target organisms, resulting in a total, perhaps of hundreds of compounds being screened. For example, methicillin, discovered in 1960 had a Beecham Research number of BRL 1241, and was followed by many valuable semisynthetic penicillins over the next decade, notably ampicillin (in 1961), carbenicillin (1967), and amoxicillin, designated BRL 2333 (1972), indicating the extent of synthesis and screening of this highly successful semisynthetic range of compounds, at approximately 100 per year. Unfortunately, this approach of chemical modification based on knowledge of structure–activity relationships combined with improvements in throughputs of screening has failed to increase the number and diversity of antibacterials in recent times.

Mills (2003) describes genomics-related technologies as they are currently applied to the discovery of small molecule antibacterial therapies. With respect to microbial genomics, the DNA sequence of *Haemophilus influenzae* was published in 1995 (the first free-living whole-organism sequence), followed by *Mycoplasma genitalium* and many other sequenced complete bacterial genomes, of which 79 are publicly available, including those of more than 40 human pathogens.

The process of antibacterial discovery through genomics revolves around the "minimum gene set for cellular life hypothesis" (Koonin, 2000; Mills, 2003) and a number of defined stages in the process.

1. Inventory of essential genes:
   (a) genome-scale transposon mutagenesis to identify all non-essential genes (inferring essential genes);
   (b) expression of anti-sense RNA to probe suspected essential genes, and knocking out conserved genes of unknown function to identify novel "broad-spectrum" essential genes.
2. Identification of target genes:
   (a) select target gene either as narrow- or broad-spectrum by specific conservation profile based on comparative genomic analysis;
   (b) prove essential for *in vitro* growth, for example, by gene knockout in relevant bacteria;
   (c) clone and sequence gene and optimally express protein product;
   (d) purify and develop assay, and screen and identify target inhibitors "hits";
   (e) characterize hits in terms of potency, mechanism of action, spectrum, and selectivity.

For broad-spectrum targets a key issue is to study isozymes from several genetically diverse bacterial species along with further screening against a panel of microbes for cellular activity, particularly to address the potential role of intake cell membrane barriers or efflux mechanisms (MDRs) on antibacterial activity. Lead compounds can be further optimized in terms of potency, spectrum, and selectivity through studies on the mode of action (MOA). Optimization of pharmacokinetic (PK) and PD properties through distribution, metabolism, and PK (DMPK) studies and drug delivery/formulation are increasingly important areas in the development process. In particular, knowledge of the role of PK/PD in relation to appropriate dosing and the potential for resistance development is evolving and is being used increasingly by Pharma.

The proof of principle of this genomics-driven, target-based approach, starting with a conserved gene and leading to antimicrobial compound is the discovery of BB-3497, a peptide deformylase inhibitor with Gram-positive and Gram-negative activity. Optimization by HTS is exemplified by methionyl tRNA synthetase. A target identified through genomics and genetics and to which whole-cell screening of target-specific inhibitors has been undertaken is LpxC, a metallo-enzyme essential for growth of Gram-negative bacteria. High-density DNA microarrays is a technology undergoing validation which has the potential to generate a vast amount of functional information on coordinated gene expression under various growth conditions.

There are a number of new antibacterials currently in late-stage development, although all of these are from existing established classes (e.g., penem, fluoroquinolone, glycopeptide, lipopeptide, glycylcycline) (Rasmussen and Projan, 2003). It is hoped that use of genomics to identify novel targets with biological and clinical relevance and HTS to increase the chance of "hits" may result, in time, in a range of novel antibacterials for development (Mills, 2003; Rasmussen and Projan, 2003). Many, if not all, of the bacterial targets have now been identified, but HTS has yet to deliver the range of novel antibacterial compounds that hit these targets. Encouragingly, the most advanced HTS plants now have the capacity to screen 300,000 compounds per day. Worryingly, there are increasing hurdles in the development and commercialization of antibacterial agents that might adversely impact the willingness of companies to initiate or continue investing in antibacterial discovery research.

3.5.   Antibacterial development, labelling, and benefits

New antibacterials are unlikely to be recommended as first-line clinical agents in policies designed to control antibacterial use unless they offer benefits, often assessed in relation to cost over existing branded or generic agents. Efficacy against organisms resistant to other antibacterial choices is a potential benefit but if it is the sole indication it may be of limited commercial attractiveness. There are a number of significant hurdles in being able to demonstrate and use differential benefits. These relate to governmental/ regulatory policy with respect to clinical development, product labelling, and promotion.

Shlaes and Ryan (2003) reviewed the strategic and regulatory considerations in the clinical development of anti-infectives. A new and evolving set of challenges exists for Pharma in the clinical development of antibacterials with a number of requirements from existing or proposed regulatory guidelines posing significant hurdles and complexity. The key points of concern to Pharma are listed below.

1. Removal of *in vitro* activity from product label (FDA)—limits potential for description of agents' full spectrum of activity (such as against rare or bioterrorism pathogens), differentiation and promotion.
2. Post-marketing surveillance of resistance and label updates (as required (FDA and national EU agencies))—presents "unlevel playing field" with established/older products.
3. Requirement for *in vitro* percentage susceptibility of organisms from EU countries (Committee for Proprietary Medicinal Products [CPMP])— requires product-specific broad surveillance (organisms and geography).
4. Powering of clinical trials to 10% non-inferiority powered at 90% level (FDA, CPMP)—substantially increased patient numbers and cost.
5. Placebo-controlled superiority trials requested for acute otitis media and acute exacerbations of chronic bronchitis (FDA, CPMP)—increased patient numbers and increased risk of not achieving relevant outcomes.
6. Paediatric rule (FDA); the requirement to progress paediatric indications/ registration—increased or earlier resource/cost for paediatric studies.
7. Specific indications, for example, pharyngitis, otitis media, rather than upper respiratory tract infection (FDA, CPMP)—larger number of Phase III randomized controlled comparator studies required to achieve a broad label.
8. Bacteriological study populations and endpoints requirements (FDA, CPMP)—bacteriological endpoints required for breakpoints, labelling, and differentiation.
9. National comparators and resistance in EU (CPMP)—additional studies needed to address national comparators.

The regulatory guidelines are increasingly focusing on the need for demonstrations of non-inferior clinical efficacy in randomized controlled trials in the specific indication. In the United States, inclusion in the product label of only those organisms isolated from the indication under study and for which clinical cure was achieved is proposed, irrespective of the *in vitro* activity spectrum. Label indications are a key point to control antibiotic use (Schlemmer, 2001) and any information given to the prescriber that is able to help select for proper indications is potentially an important tool for good antimicrobial practice (Schlemmer, 2001). However, Schlemmer (2001) also states that there is a critical need for trials to select for those patient populations who really need antibiotics, or to look for new endpoints able to differentiate between drugs, rather than only demonstrating equivalence. This approach may ensure that clinical evidence supports the claim for beneficial activity against a particular organism in a particular indication, or population, but inevitably will restrict the breadth of label.

The commercial implications of limited labels based on proof of superior benefits (and not just equivalence), compared with broader labelling based on equivalence will be critical to Pharma. The value of studying new endpoints for differentiation will also be dependent on acceptance of the endpoints by regulators, and their incorporation into product labelling, enabling promotion and communication. The differential benefits would also need to translate into guidelines and formulary inclusion. It has been argued by Monnet and Sorensen (2001), that antimicrobials that represent real innovations are readily accepted by hospital prescribers and naturally gain market shares. The hurdle is in being able to demonstrate the innovative benefits to the satisfaction of the regulators and decision-makers, and in there being sufficient commercial return from use in a narrow-defined patient group.

A particular issue is that there will be significant hurdles in the ability of Pharma to demonstrate clinical efficacy against emerging resistant organisms, which may only rarely be isolated in clinical trials. PK/PD data which allow predictions of bacteriological efficacy to be made and tested, for example, in *in vivo* infection models of simulated human PK, or in small PD studies in patients, have been proposed by a number of bodies, including the EU CPMP (EMEA). However, the use of such data has yet to be fully accepted and adopted by the FDA as a surrogate for clinical efficacy and breakpoint determination. PK/PD data are of particular interest when trying to define the best dosage and dosing regimens for new compounds. As the bacteriological endpoint correctly defines the outcome in an infectious process it would serve to assess the PK/PD relationship of a drug (Schlemmer, 2001). Using PK/PD data to predict bacterial eradication or clinical outcome should be considered as the only way to select for optimal therapeutic regimens regarding antibiotic choice and dosing regimen as well as determining the optimal therapy duration (Ball *et al.*, 2002; Schlemmer, 2001). A greater role of predictive PK/PD data and modelling under certain circumstances and for specific target populations, has been proposed by CPMP (EMEA) and is supported in general by Pharma.

Global Pharma need to conduct clinical development programmes that satisfy the highest requirements of both FDA and EMEA. In Europe, there is a need to include comparative agents approved and relevant to individual EU countries, or for which a consensus justification can be argued, for technical approval (EMEA). Phase III clinical trials are designed primarily to meet the requirements of the regulators. However, these traditional comparative non-inferiority trials fail to provide important evidence of potential benefits over existing therapies. With clinical success for antibacterials in the 85% or more level, demonstrations of superiority require vast patient populations. Pharma rely on other data to indicate superiority and benefit, ranging from potency (minimum inhibitory concentrations, etc.), per cent susceptibility based on breakpoint, *in vitro* and *in vivo* models, PK/PD, and defined clinical studies to demonstrate bacteriological and health-outcome benefits. Much of the clinical work undertaken to secure regulatory approval fails to best demonstrate the role and clinical benefits of new agents (Bax *et al.*, 1999) and yet is often used as the evidence base for reimbursement, guidelines, and policies. There has also been a proposal to update and harmonize antibacterial product labels between European countries, particularly for "old" antibiotics and their datasheets (summary of product characteristics) at the time of generic introductions because lack of relevant product information promotes inappropriate use (European Conference on Antibiotic Use in Europe, 15–17 November, 2001; Schlemmer, 2001). The burden of cost to provide new data to demonstrate the benefits of marketed and new agents would lie with the Pharma R&D companies. Failure to produce data to support current claims could result in removal of differentiation from established branded products, while provision of data may strengthen the labels of generic products.

Bax *et al.* at the Whitley Park Symposium in 1999 reviewed the limitations of the current clinical evaluation process and stated that the lack of development in how to define precisely both drug value and appropriate use will seriously hamper the drug industry's ability to develop important new medicines discovered by new technologies (Bax *et al.*, 1999). The authors suggested that what is needed is further and much more rapid development of conventional means of drug evaluation, such as the clinical trial, as well as visionary use of new methods in epidemiology, and new use of electronic data merged in a way that identifies drug effects on both a population and individual basis.

4. THE FUTURE

4.1. Collaboration, collaboration, collaboration

The role of R&D-based Pharma is to discover, develop, and commercialize antibacterials. There is a danger that the wide range of policies which have been introduced under the auspices of "combating resistance" will themselves hamper Pharma's ability to deliver a key strategy, if not the most important weapon in the battle against bacterial infection and resistance, the introduction of novel antibacterials. Policies and improvements in the quality of antibacterial prescribing have an important role to play, as do diagnostics, vaccines, infection control, and other interventions. Pharma is clearly an important partner along with government, NGOs, regulators, infectious diseases experts, practioners, and patients in the battle against bacterial infection. The "conflicts of interest between the prescriber, the regulator, and the profit maker" was the subject of a supplement to *Clinical Microbiology and Infection*. The editor (Gould, 2001a), recognized that the calls for restrained use of antibacterials to combat and control resistance counters the natural instincts of doctors to do the best for their patients and of Pharma, which of necessity, needs to profit to exist and continue.

It is, however, unfortunate that Pharma can be regarded first as "profit makers" and can even be excluded from contributing to debates on the subject of antibacterial use (European Conference on Antibiotic Use in Europe, 15–17 November 2001). There needs to be a distinction between companies which are solely suppliers of generic products and those who have the objective, dedication, and expertise to discover and introduce new antibacterials. These are undoubtedly profit makers, being subject to commercial and shareholder expectations, but perhaps should be considered foremost as "providers" with respect to new antibacterial solutions. Gould describes an international partnership of medical societies and industry, in association with regulators as a model for the future in the battle against antibacterial resistance (Gould, 2001b). Schlemmer (2001) highlighted that while antibacterial policies were the keystone in promotion of good antimicrobial practice, more accurate and relevant information on antibiotics is urgently needed and that it is the responsibility of Pharma and regulators to move together towards an improvement in antibiotic evaluation. The objective would be to give prescribers more critical product information and to help the experts create better guidelines.

WHO included Pharma via the International Federation of Pharmaceutical Manufacturers Association (IFPMA) in consultation on their *Containment Strategy and Model Prescribing Guidelines* (www.who.int/emc/amr.html) and the European Federation of Pharmaceutical Industries and Associations (EFPIA) have represented Pharma in EU and EMEA discussions and consultations on clinical trial guidelines and labelling. In the United States a multiagency task force of FDA, CDC and the National Institute of Health (NIH) issued a document entitled *Public Health Action Plan to Combat Antimicrobial Resistance* in 2001 with a priority to create an interagency antimicrobial resistance product development working group coordinated by the FDA, US Department of Agriculture (USDA), and CDC (Shlaes and

Ryan, 2003). A second priority was to, "Investigate and act upon potential approaches for stimulating and speeding the entire antimicrobial resistance product development process, from drug discovery through licensing," including exploring the economics of the situation and incentives. While Monnet and Sorensen (2001) argue that much investment from Pharma is into marketing and not R&D and that the so-called "R&D Scare Card" is not justified, it is clear that Pharma companies are increasingly taking the commercial decision not to invest in R&D for antibacterials, but in other areas. According to Shlaes and Ryan, incentives for increasing Pharma R&D into priority infectious diseases include the need for more information on the market and disease areas (based on market identification, disease-specific bioinformatic systems, development of surrogate endpoints), more predictability (based on market assessment, international regulatory harmonization, and reinforcement of intellectual property rights), and more cost–risk sharing (market creation, patent extension, orphan drug legislation) (Shlaes and Ryan, 2003).

A number of bodies, such as WHO, IDSA, and EU task forces have begun processes to assess the situation and identify some areas of action. These include:

- summarizing the value of antibacterials
- identifying the status of clinically relevant resistance and priority pathogens
- evaluating the state of antibacterial research/regulatory submissions/industry involvement and outlook
- maintaining databases of funding agencies, their research interests, new opportunities for funding
- documenting the vulnerabilities related to the manufacture of antibacterials/shortages/negative impacts on public health: gap analysis
- undertaking the identification, prioritization, and tracking of global public health needs (a global agenda)
- reviewing/documenting government activities to foster development of new antibacterials
- reviewing recommendations for "incentivizing"
- modifying regulatory approval processes such as priority review, fast track, waiver of user fees, orphan drug status, modified/smaller clinical trials, and/or reduced number of efficacy studies per indication (and use of surrogates such as PK/PD)
- harmonizing clinical trial methodologies and implementing coordinated action by regulatory agencies in registration of new products
- increasing education and training opportunities.

Research-based Pharma needs to be included as a partner in these collaborative initiatives in order to help shape the future in a way which will be conducive to

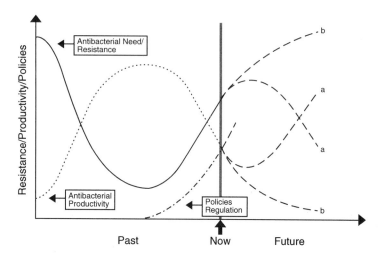

Figure 2. From the past to the future: Hypothetical potential scenarios evolving from the relationship between antibacterial need/resistance, antibacterial productivity and policy. (a) Sustainable balance of antibacterial productivity, resistance, and regulation/policy; oscillations may decrease over time with better understanding of the relationships and balanced management. (b) Crisis of imbalance of increased policy/regulation, decreased incentives and antibacterial productivity and increasing/uncontrolled resistance.

continued investment into R&D for new antibacterials while improving their use and sustainability. This will require a full consideration of the contributing factors, actions, and impacts in order to find a way to maintain a critical balance (Figure 2). A new model may need to be defined which explores alternative roles of small companies, government, and NGO bodies along with big Pharma in the conduct and funding of antibacterial discovery, development, and distribution.

5.   CONCLUSIONS

Policies are important and necessary in controlling antibacterial use in an attempt to preserve the utility of existing and future antibacterials. They should aim to minimize unnecessary and inappropriate use, as avoidable risk factors in the development of resistance, while maintaining or improving patient outcomes. The effect of antibacterial policies on the containment or reduction of resistance is largely unknown and should focus not on quantity of prescribing alone but on quality to ensure appropriate prescribing in terms of antibacterial choice, dosage, dosing regimen, or duration. Policies that relate to antibacterial use have the potential to reduce the antibacterial market size and commercial attractiveness relative to other therapeutic areas. Policies that relate to development activities and labelling have the potential to increase development costs, restrict use, and limit the ability to differentiate benefits. Collectively policies are negatively impacting big Pharma's willingness to continue investment in this field. The combined and cumulative effects of the various policy items related to surveillance, use, clinical trial guidelines, product labelling, and promotion require a broad, collaborative, and global view of the impact on Pharma R&D. Collaboration is required to identify agreed and mutually acceptable ways to streamline development, implement appropriate prescribing, demonstrate and promote product benefits, and maintain financial incentives.

Antibacterial discovery, development, and use are related within a dynamic environment that includes bacteria, their resistances, and policies. As with any dynamic environment, change will alter the balance (Figure 3). Policies are contributing to a change and a negative impact on the potential and willingness of Pharma to develop new antibacterials. While change may require adaptation for survival and success, careful consideration and collaboration is needed by all parties to help to ensure that insurmountable hurdles are not created leading to the inability of Pharma to survive and succeed in antibacterial R&D. Failure of Pharma to deliver new antibacterials would lead to "antibacterial extinction."

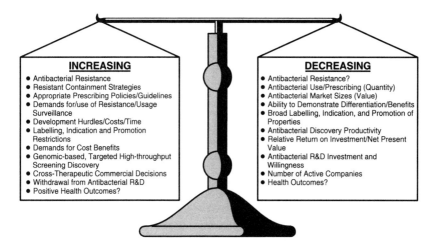

Figure 3. A question of balance: increasing and decreasing factors in the environment of antibacterial resistance, usage, discovery, and development.

REFERENCES

Appelbaum, P. C., 2003, Bacterial resistance concerns. *ASM News*, **69**(4), 161.

Austin, D. J., Kristinsson, K. G., and Anderson, R. M., 1999, The relationship between the volume of antimicrobial consumption in human communities and the frequency of resistance. *Proc. Natl. Acad. Sci. USA*, **96**, 1152–1156.

Ball, P., Baquero, F., Cars, O., File, T., Garau, J., Klugman, K. *et al.*, 2002, Antibiotic therapy of community respiratory tract infections: Strategies for optimal outcomes and minimized resistance emergence. *J. Antimicrob. Chemother.*, **49**, 31–40.

Bax, R., Bywater, R., Cornaglia, G., Goossens, H., Hunter, P., Isham, V. *et al.*, 2001, Surveillance of antimicrobial resistance—what, how and whither? *Clin. Microbiol. Infect.*, **7**, 316–325.

Bax, R., Gabbay, F., Phillips, I., and the Witley Park Study Group, 1999, Antibiotic clinical trials—the Witley Park Symposium. *Clin. Microbiol. Infect.*, **5**, 774–788.

Carbon, C., Cars, O., and Christiansen, K., 2002, Moving from recommendation to implementation and audit: Part 1. Current recommendations and programs: a critical commentary. *Clin. Microbiol. Infect.*, **8**(Suppl. 2), 92–106.

Chahwakilian, P., 2000, Politiques d'antibiothérapie a l'échelon national dans divers pays. *Rev. Pneumol. Clin.*, **1S**, 9–12.

Christiansen, K., Carbon, C., and Cars, O., 2002, Moving from recommendation to implementation and audit: Part 2. Review of interventions and audit. *Clin. Microbiol. Infect.*, **8**(Suppl. 2), 107–128.

Davey, P., Nathwani, D., and Rubinstein, E., 2003, Antibiotic policies. In R. Finch, D. Greenwood, S. Norrby, and R. Whitley (eds.), *Antibiotic and Chemotherapy*. Churchill Livingstone, London.

Diekema, D. I. and Jones, R. N., 2000, Oxazolidinones: A review. *Drugs*, **59**, 7–16.

European Conference on Antibiotic Use in Europe. 15–17 November 2001, Brussels, Belgium.

Felmingham, D., Feldman, C., Hryniewicz, W., Klugman, K., Kohno, S., Low, D. E. *et al.*, 2002, Surveillance of resistance in bacteria causing community-acquired respiratory tract infections. *Clin. Microbiol. Infect.*, **8**(Suppl. 2), 12–42.

Finch, R. G. and Low, D. E., 2002, A critical assessment of published guidelines and other decision-support systems for the antibiotic treatment of community-acquired respiratory tract infections. *Clin. Microbiol. Infect.*, **8**(Suppl. 2), 69–91.

Garrod, L. P., O'Grady, F., and Lambert, H. P., 1981, The evolution of anti-microbic drugs. In L. P. Garrod, H. P. Lambert, and F. O'Grady (eds.), *Antibiotic and Chemotherapy*. Churchill Livingstone, London.

Gould, I. M., 1999, A review of the role of antibiotic policies in the control of antibiotic resistance. *J. Antimicrob. Chemother.*, **43**, 459–465.

Gould, I. M., 2001a, Introduction. *Clin. Microbiol. Infect.*, **7**(Suppl. 6), 1.

Gould, I. M., 2001b, Minimum antibiotic stewardship measures. *Clin. Microbiol. Infect.*, **7**(Suppl. 6), 22–26.

Hunter, P. A. and Reeves, D. S., 2002, The current status of surveillance of resistance to antimicrobial agents: Report on a meeting. *J. Antimicrob. Chemother.*, **49**, 17–23.

ICAAC, 2002, Abstracts of the 42nd Interscience Conference on Antimicrobial Agents and Chemotherapy. September 27–30, 2002, American Society for Microbiology, San Diego, CA.

ICAAC, 2003, Abstracts of the 43rd Interscience Conference on Antimicrobial Agents and Chemotherapy. September 14–17, 2003, American Society for Microbiology, Chicago, IL.

IDSA, 2003, The future of antimicrobial drug availability: An impending crisis. March 12.

Kaitin, K. I. and DiMasi, J. A., 2000, Measuring the pace of new drug development in the user fee era. *Drug Inf. J.*, **34**, 673–680.

Koonin, E. V., 2000, How many genes can make a cell: the minimal-gene-set concept. *Annu. Rev. Genomics Hum. Genet.*, **1**, 99–116.

Lewis, D., 2002, Antimicrobial resistance surveillance: Methods will depend on objectives. *J. Antimicrob. Chemother.*, **49**, 3–5.

Mills, S. D., 2003, The role of genomics in antimicrobial discovery. *J. Antimicrob. Chemother.*, **51**, 749–752.

Monnet, D. L. and Sorensen, T. L., 2001, The patient, the doctor, the regulator and the profit maker: Conflicts and possible solutions. *Clin. Microbiol. Infect.*, **7**(Suppl. 6), 27–30.

Nathwani, D., Rubinstein, E., Barlow, G., and Davey, P., 2001, Do guidelines for community-acquired pneumonia improve the cost-effectiveness of hospital care? *Clin. Infect. Dis.*, **32**, 728–741.

Rasmussen, B. A. and Projan, S. J., 2003, Antibacterial drug discovery: What's in the pipeline? In R. Finch, D. Greenwood, S. Norrby, and R. Whitley (eds.), *Antibiotic and Chemotherapy*. Churchill Livingstone, London.

Reynolds, R., and BSAC Extended Working Party on Respiratory Resistance Surveillance, 2001, BSAC Respiratory Resistance Surveillance Programme: First results of the winter 1999–2000 collection. *Int. J. Antimicrob. Agents*, **17**(Suppl. 1), S143. Abstract P127.088.

Schlemmer, B., 2001, Impact of registration procedures on antibiotic policies. *Clin. Microbiol. Infect.*, **7**(Suppl. 6), 5–8.

Shlaes, D. M. and Ryan, J. L., 2003, Strategic and regulatory considerations in the clinical development of anti-infectives. In R. Finch, D. Greenwood, S. Norrby, and R. Whitley (eds.), *Antibiotic and Chemotherapy*. Churchill Livingstone, London.

Watson, J. D. and Crick, F. H., 1953, Molecular structure of nucleic acids: A structure for deoxyribose nucleic acid. *Nature*, **171**, 737–738.

Chapter 35

Antibiotic Use—Ecological Issues and Required Actions

Ian M. Gould
*Department of Medical Microbiology, Aberdeen Royal Infirmary,
Foresterhill, Aberdeen, Scotland AB25 2ZN, UK*

Antibiotics (and their synthetic derivatives) are used extensively and often for completely unnecessary and inappropriate reasons (Gould, 2002a). Estimates are varied but total use worldwide may well be over 100, 000 tons per annum (Kümmerer, 2003). This can be split into three main areas of use: agriculture, community, and hospital.

I will discuss each of these in turn while accepting that there is, inevitably, some overlap as in the environmental contamination that occurs from all these areas through excretion of active agents in urine and faeces into sewage systems. I will try to relate use to resistance in each area and also discuss other factors of relevance to the impact of antibiotic use on resistance. Finally I will consider the major issues that need to be addressed urgently (including better use of the diagnostic laboratory) if we are to save these incredibly valuable and irreplaceable agents for future generations.

1. AGRICULTURE

Estimates in the United States of America are that up to 80% of antibiotic use by weight is in agriculture—mainly for rearing animals and also for crop spraying such as the use of streptomycin and tetracycline on fruit trees to prevent fruit rot. Within Europe, use of antibiotics for growth promotion, which are related to those used in humans was banned by the European Commission (EC) in 1999 (Wegener, 2003; WHO report, 2003). Despite some appeals from manufacturers, this ban has so far been upheld and has led to significant reductions in antibiotic use in agriculture (Figure 1). Experience from Scandinavia, which had imposed a voluntary ban previously, suggested that there would be little detriment to farming profits and no untoward effects in animals, although therapeutic antibiotic use has increased slightly. Antibiotic use for growth promotion remains, however, a significant issue in most non-European countries and yet has questionable benefits. Of course, veterinary antibiotic use remains significant, even within the EC, for prophylaxis and therapy and is especially high where there is intensive animal breeding as in chicken and pig farming (Taylor, 1999). One example of the problem serves to illustrate this: a few chickens in a battery of 20,000 birds become ill. Without treatment the presumed bacterial respiratory infection will spread rapidly with major bird and economic loss. The veterinary practitioner therefore prescribes enrofloxacin, a close relative of ciprofloxacin. It is, however, inconvenient to administer this solely to those birds that are sick so the antibiotic has to be administered en mass to all the birds, in their food. Little wonder that ciprofloxacin resistance is a significant problem in Campylobacter, the main food borne pathogen from chickens, but only since the introduction of enrofloxacin and related drugs into the industry (Piddock, 1995).

I have estimated that, per capita animals slaughtered annually, antibiotic use measured by weight in UK agriculture is of a similar magnitude to annual

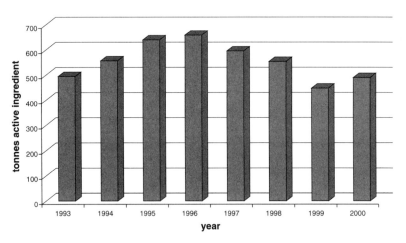

Figure 1. Sales of antimicrobial therapeutic products and growth promoters in food and non-food animals in the UK.

human use per capita in UK medicine. Is this significant, if not high use, of importance for human medicine? Should we be more stringent in regulating agricultural use of antibiotics? Certainly no one is suggesting that animals should not have access to proper veterinary treatment. Many of the antibiotics are, however, very similar to many of those used in human medicine and similar mechanisms of resistance are undoubtedly selected for in both environments. But is there significant spread of resistant bacteria or their genetic elements into human medicine? The answer to this is yes, periodically, but generally it is probably not an overwhelming problem. Campylobacter with quinolone resistance is selected relatively easily as it requires only a single mutation to raise the minimum inhibitory concentration (MIC) above the breakpoint due to the relative natural insenstivity of Campylobacter to quinolones. There is, however, no well-defined plasmid mediated resistance to quinolones so there is, as yet, no obvious hazard that this resistance will spread outside of Campylobacter. Furthermore, it is rare for this illness to require antibiotic therapy and most strains remain susceptible to macrolides. Periodically there are clonal outbreaks of multiresistant *Salmonella* spp., that do cause clinical problems, including ciprofloxacin resistant strains and extended-spectrum β-lactamase producers (ESBL) but, as yet, they are not a major cause of excess morbidity or mortality (Threlfall *et al.*, 2003).

Perhaps of more immediate concern are the possibilities for transfer of resistance, through the food chain, by commensal bacteria such as enterococci and *Escherichia coli*. While it is likely that modern slaughter house hygiene practices have made a major beneficial impact on contamination of carcasses by such faecal organism, resistance in them is widespread and concerns have been raised, but with little evidence to date. Prior to it being banned, avoparcin a glycopeptide closely related to vancomycin, was commonly used in pig rearing and there are many well-documented cases of farmers and pig meat consumers being colonised with vancomycin-resistant enterococci (VRE) of animal origin (Chavers *et al.*, 2003; Stobberingh *et al.,* 1999). Since use of avoparcin was stopped, the prevalence of VRE in the pig (and human) population has declined significantly. In some farm animals, where antibiotic use is intensive, levels of resistance can be higher than in human medicine. For example, ampicillin resistance in *E. coli* has been recorded as high as 70%, which is well above the commonly quoted figure of 40–50% in humans.

On balance, then, antibiotic use in agriculture can undoubtedly cause resistance problems in human medicine but up until now, these do not appear to be insurmountable. There are causes for concern, however, particularly in *Salmonella* spp. which can cause serious invasive disease requiring urgent and appropriate antibiotic therapy. There is no room for complacency.

2.   COMMUNITY USE

This varies several fold in intensity between different countries, whether measured per capita or by population density (see Chapter 29 by Mölslad and Cars). Both have significance for the epidemiology of antibiotic resistance, with high relative use by both measurements being associated with more resistance (Bruinsma *et al.*, 2003). Measurements of use are often open to questions over their accuracy because there is probably a significant amount of black market antibiotic available in many countries, much of it of dubious quality, with questionable amounts of active ingredients. Non-prescription, over-the-counter use is difficult to measure but is a significant problem, even in some European countries, where it is illegal. There are also difficulties in measuring the effects of all this antibiotic use—namely the levels of antibiotic resistance. Even in wealthy countries there are few unbiased data so that problems of resistance may be overstated. Few studies are funded to go out and survey selected groups. Most data is taken from clinical samples submitted for analysis. These tend to be biased to those from more difficult or seriously ill patients who will probably have received prior antibiotic therapy. We know that previous antibiotic exposure, especially in the few months prior, increases the likelihood of carrying resistant organisms several fold (Steinke *et al.*, 2001).

There is no doubt of the strong association between antibiotic use in the community and its inevitable consequence, namely antibiotic resistance. The relationship has been established at all levels from individual patients, through families to patient groups, towns, cities, regions, countries, and continents. Relationships are often complicated and not always clear cut. Even social class and deprivation have a relationship although this may be complicated by levels of access to and type of medical care (Howard *et al.*, 2001). Penicillin resistance, for instance, has never developed in *Streptococcus pyogenes*, despite over 50 years of intense exposure. Clearly, the genetic make up of the organism and/or the mechanisms of resistance open to it do not suit the production of β-lactamase or a change in its penicillin binding proteins. And yet the same organism has become resistant to macrolides such as erythromycin and the closely related *Streptococcus pneumoniae* to both macrolides and pencillin (PRP), although it took 30 or 40 years of exposure for these resistance mechanisms to become well enough distributed to cause major clinical concern. While, it may largely be down to genetics of resistance mechanisms, other factors also play a role such as clonal spread, intensity of antibiotic use, human behaviour, and pharmacodynamic/kinetic issues (see Chapter 21 by Mouton).

Clonal spread has clearly had a major role in the worldwide spread of penicillin (and multiresistant) pneumococci, with close links between resistance to β-lactams, macrolides, tetracyclines, chloramphenicol, co-trimoxazole, and now even quinolones. Use of any one drug can maintain selection pressure for all linked resistances, so it is sometimes difficult to envisage antibiotic rotation policies having much effect on such resistance. Restriction of key agents does seem to work in some situations, although the role of the natural evolution and decline and fall of epidemic clones has probably also played a role in the control of some outbreaks. Even then, resistance often declines slowly in the community and rarely declines to zero, such that levels can be anticipated to rise again quickly if antibiotic use increases. Up until now, plasmid-mediated resistance and its spread is not an issue in pneumococci, which are not tolerant of plasmids. The increased use of day-care centres for child care has played a significant role in some outbreaks of multiresistant pneumococci, because of increased opportunity for spread and oropharyngeal colonisation of a very susceptible population. Carriage rates of up to 50% are not uncommon in such populations and often antibiotic use is very intensive.

Some antibiotics may be more likely to maintain outbreaks/cause resistance than others. In the extensive experience gained in Iceland, penicillin resistance seemed to be more related to high co-trimoxazole and macrolide use than actual penicillin use and others have described macrolide and cephalosporin use as more likely to select for penicillin resistance than penicillin use itself, by a factor of up to 3 or 4 times (Kristinsson, 1999). This may be because these agents are less bactericidal than penicillin and its derivatives (such as amoxicillin) and thus less likely to eradicate the pathogen, allowing selection of resistant variants which might otherwise have been in a minority and suppressed because they were not at a selective advantage. Other issues to be considered in this context are the dose and duration of the antibiotic. For similar reasons, long term, low dose antibiotic treatment is more likely to select resistant strains in individual patients. Canet and Garau (2002), have calculated that for each day of pencillin use over 7 days in the preceding 6 months, the odds of a child carrying a PRP increased by 4%.

Macrolide use is of particular interest. The question has been raised that resistance only becomes a clinical problem where a threshold of use is exceeded in that community. Whether this threshold exists for other antibiotic groups or, indeed, what this threshold is for macrolides, remains to be established but it presumably varies depending upon the community's epidemiology. For macrolides it has also been suggested that the newer, long-acting agents are more likely to select for resistance than the older, shorter half-life agents. This may be because of prolonged low residual tissue concentrations which select surviving resistant variants with, what are essentially, only bacteriostatic agents (Lonks *et al.*, 2002). Different mechanisms of resistance to the same antibiotic can also play their role in the epidemiology of resistance. There are two main mechanisms of resistance to macrolides—a mutation in the methylase gene (*erm B*) giving rise to high level resistance and cross resistance to lincosamides and streptogramins and a low level resistance due to efflux (MefA) with no cross-resistance. While the clinical significance of the latter has been questioned there is good evidence that it can cause treatment failures.

3.   HOSPITAL

Although total use of antibiotics in hospitals is much less than in the community the intensity of use, magnified by cross infection, ensures a multitude of resistant bacteria in today's hospitals. Within hospitals, intensive care units (ICUs) have levels of antibiotic use many fold higher than general wards and it is no surprise that these units have been termed the genesis units for selection, maintenance, and spread of antibiotic resistant bacteria.

While published data suggests that only between 20% and 40% of patients admitted to acute hospitals will receive antibiotics at some point during their stay, by other measures, antibiotic use in some hospitals can be said to have reached saturation levels. Recent data from ESAC (http://www.ua.ac.be/main. asp?c=*ESAC) and ARPAC (www.abdn.ac.uk/arpac/) suggests that many hospitals have reached the notable total consumption figure of 100 defined daily doses (DDD) per 100 patient days. This is because many patients will receive several agents simultaneously often in daily doses, greater than DDD. Commonly used antibiotics in hospitals include β-lactams, in particular second and third generation cephalosporins and co-amoxiclav, macrolides, and quinolones. Data from the SCOPE (Polk *et al.*, 2001) and ARPAC surveys suggests that quinolones are much more commonly used in North American hospitals than European hospitals, which favour β-lactams. This may well be because of different preferences in the treatment of community-acquired pneumonia for which β-lactams (often in combination with macrolides) are generally still favoured in Europe (BTS, 2001), while North American guidelines tend to favour monotherapy with a quinolone (Bartlett *et al.*, 2000; Mandell *et al.*, 2000). What is certain is that both regimens are being increasingly used with disastrous consequences for the microbial ecology of hospitals (MacKenzie *et al.*, 2002).

Methicillin-resistant *Staphylococcus aureus* (MRSA) in particular, is an epidemic organism of huge proportions in hospitals on both continents and numerous publications now link this epidemic to quinolone, macrolide, and β-lactam (cephalosporin) use, both at a patient and ecological level (Graffunder and Venezia, 2002). In the United Kingdom, the British Thoracic Society guidelines, infact, only recommend combination therapy for severe pneumonia. In my own hospital, however and no doubt in many others, it is extremely difficult to stop the use of combination therapy for the majority of patients admitted with possible respiratory infection. Part of the reason for this seems to be difficulties in ensuring experienced medical assessment of severity of illness in acute admission units. Furthermore, although recent statements of good practice recommend daily review of antibiotic therapy (Scottish Infections Standards and Strategies Group, 2003) (Figure 2), the average duration of this usually unnecessarily broad-spectrum antibiotic therapy is 3 days (Kumarasamy *et al.*, 2003). Furthermore, a recent review questioned the evidence base for such treatment, suggesting that monotherapy with a cephalosporin or, indeed, even a penicillin was still the treatment of choice for most severe pneumonias (Oosterheert *et al.*, 2003). This should certainly be the case in Scotland where penicillin resistant pneumococci and atypical pneumonia are equally rare (Gould, 2002b) (www.show.scot.nhs.uk/scieh/).

All antibiotics, no doubt, are associated with problems of resistance sooner or later but it does seem that cephalosporins get unusually bad press (Dancer, 2001). Maybe this is deserved or maybe it just reflects their huge use. For MRSA and VRE certainly, cephalosporins do seem to be much more selective than penicillins. Indeed co-amoxiclav seems to have an MRSA protective effect, possibly due to some affinity of both amoxicillin and clavulanic acid for the modified penicillin binding proteins of MRSA.

What is now beyond doubt is that antibiotic use in most hospitals is too high and must be reduced to retain the efficacy of key agents, particularly while there are a dearth of new agents in development. While prevention of selection of resistant strains by combination therapy, pharmacodynamic dosing schedules, and the use of bactericidal agents do have their place, bacteria are too adaptable for these strategies to be the complete answer. Selective digestive decontamination (SDD) and antibiotic cycling (Kollef, 2001) remain unproven and are unlikely to be widely adopted, the former because of the increased total antibiotic consumption in ICUs adopting SDD (see Chapter 15 by Hanberger *et al.*) and the latter because there are too few options to make it viable. In addition, mathematical models favour random use of different agents over cycling as a way of preventing development of resistance (Lipsitch and

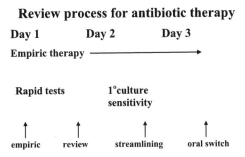

Figure 2. Review process for antibiotic therapy.

Samore, 2002; Lipsitch *et al.*, 2000). Another major drawback of many of these strategies is that they take no account of the commensal flora, concentrating solely on the targeted pathogen. Clearly this is naive, as we all have more commensal bacteria than human cells and the inevitable consequence of any antibiotic administration (appropriate or inappropriate) is that this commensal flora will be damaged with predictable effects such as thrush, diarrhoea, and induction or selection of resistance. Whether the recent trends to encourage more use of oral antibiotics in hospitals (for good reasons) will alter this equation of resistance development is unknown (Lelekis and Gould, 2001). Most drugs used for oral switch or step down are poorly absorbed, leaving huge concentrations in the gastrointestinal tract to upset the bowel flora.

4.   WHAT CAN BE DONE NOW ABOUT ANTIBIOTIC RESISTANCE?

As a microbiologist I can see the need for yet more research into the microbial and genetic aspects of resistance and the subtleties of its relationship to antibiotic use but as a practising clinician with an overview of how antibiotics are misused in both the hospital and community, I am sure that we know enough about such relationships to inform our prescribing in a useful way right now. The really urgent agenda is to change prescribers' habits. Futhermore, while continued development of new antibiotics is desirable, it is really just feeding the "addiction" of doctors to antibiotic prescribing. It is not addressing the underlying problems although it has served us well in the past (but for how much longer is it possible?).

What is likely to pay the most dividends in the shortest time is to research into and act on the determinants of antibiotic abuse—the factors that can influence the prescriber and patient alike—such as cultural, social, educational, economic, and regulatory differences between countries. A lot is already known about these issues (Eng *et al.*, 2003; Harbarth, 2002) and how they can influence the antibiotic prescribing process. In addition, more emphasis needs to be put on public health issues of antibiotic resistance, infection control, and use of diagnostic tests.

One example of this approach is that GPs in some countries are addressing the problem by successfully reducing the number of inappropriate antibiotic prescriptions, at least in children where many of the new "evidence based" guidelines are targeted. Where this has been most successful it has often been linked in with a multifaceted educational approach, targeting patients and the general public through multimedia campaigns, not to demand unnecessary antibiotics from their GP (Belongia and Schwartz, 1998; Bengoa *et al.*, 2002; McCaig *et al.*, 2003; Perz *et al.*, 2002) (Table 1). Patient demand (and use of

Table 1. CDC's national campaign

Current campaign activities include

1. Developing and distributing *educational materials* promoting appropriate antibiotic use.
2. *Funding states* to develop, implement, and evaluate local campaigns.
3. Developing and pilot testing a *medical school curriculum* promoting appropriate use of antibiotics.
4. Developing and testing Health Plan Employer Data and Information Set *(HEDIS) measures* for appropriate antibiotic use.
5. Funding a *national advertising campaign* promoting the appropriate use of antibiotics.

Source: www.cdc.gov/drugresistance/community.

often illegal, over-the-counter antibiotics) is known to be highly variable, even within Europe. Although data from hospitals is lacking, there is little to suggest such widespread successes in hospitals where approaches have tended to be targeted to single institutions.

In a study of outpatient antibiotic use and prevalence of antibiotic resistant pneumococci in France (13%) and Germany (7%), the sociocultural differences seemed paramount (Harbarth, 2002). There were markedly different rates of antibiotic prescribing for respiratory tract infection (RTIs). These were not to do with microbiologically defined need but complex reasons such as patient demand associated with health belief differences, social determinants such as different child care practices, regulatory issues, and economic factors (retail prices for pharmaceutical products are much lower in France than Germany). Finally, Japan is well known to have high levels of antibiotic resistance because doctors and hospitals rely on profits from drug sales as a source of income.

The last part of the equation may be the most difficult to solve—our relationship as prescribers and consumers with the Pharmaceutical Industry (Pharma). I have previously called for a partnership, based on discussion and resolution of our obvious conflicts (Gould, 2001). We rely on Pharma for new agents but the economics of drug development necessitate aggressive promotion to recoup investment, shortening the useful life of many valuable agents by encouraging high use (Wazana, 2000). A related issue is the poor supply of key agents such as injectable trimethoprim and tetracycline which are no longer available, presumably as production is no longer economically justifiable, yet we find these agents valuable for treating MRSA infections. Penicillin G has previously been in short supply also (Strausbaugh *et al.*, 2001; Harbarth, 2000).

One other area I would like to discuss, no doubt as I am involved with it every day, is the (lack of) use of the diagnostic microbiology laboratory which can play a major role in improving antibiotic prescribing. I make no apology for this indulgence as I believe that improved use of diagnostics is a very easy way, available to prescribers right now, to improve their antibiotic prescribing.

5.   HOW CAN THE DIAGNOSTIC LABORATORY HELP?

The issues are several fold, including access to the laboratory, speed and accuracy of analysis, and communications of results. It is fashionable to emphasise the importance of molecular diagnostics in this context, particularly in view of the speed of these tests and there is no doubt that rapid results can improve patient outcome and quality of antibiotic use. Nevertheless, molecular tests have not, so far, become well integrated into clinical diagnostic microbiology and the near future does not hold a great deal of promise.

In fact, much more can be done using conventional technology both in the community and hospital. Ease of access with same day delivery and processing may seem a luxury, at least in primary care, but has been shown to be feasible and reduce antibiotic prescribing by at least 50% (Shackley *et al.*, 1997) (Figure 3). Even with conventional tests, most specimens in primary and secondary care can yield useful preliminary results after overnight incubation.

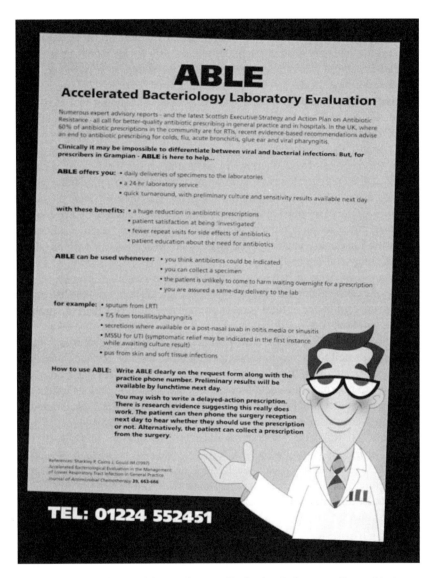

Figure 3. Accelerated Bacteriology Laboratory Evaluation for better quality antibiotic prescribing in general practice and in hospitals.

Source: JAC (1997: 39, 663–666).

In primary care most patients can wait overnight for results to see if an antibiotic is indicated. Results can be communicated electronically, or by fax, to the GP and the patient can then telephone the surgery to see whether, or not, to activate a delayed prescription. For emergency specimens in hospitals, gram stains and antigen tests should be available within minutes. The rate limiting step is transport of the specimens to the laboratory. Ninety per cent of positive blood cultures will be positive within 24 hr (MacKenzie *et al.*, 2003) and these and all intensive care specimen results should be telephoned to the duty doctor, or better still, the patients visited by a medical microbiologist to ensure appropriate (often streamlined) therapy. In particular, appropriate therapy can be started if the empiric choice is inactive, or a broad-spectrum, empiric choice can be downgraded based on real or anticipated susceptibilities of the isolate (Figure 2). Clinicians need much more education on the issues of streamlining and step down as their natural inclination is to ignore negative laboratory results and keep the patient on the original empiric choice if they are doing well (Cunney and Smyth, 2000).

In my own hospital, one of the most common reasons for an inadequate choice of empiric antibiotic therapy is MRSA bacteriaemia when the patient has been given a quinolone or third generation cephalosporin. If MRSA is a problem in an institution, then empiric therapy should take account of this. In my own hospital all our MRSA are susceptible to gentamicin so this agent plus a broad-spectrum penicillin is recommended as empiric therapy for life-threatening sepsis. A crucial role of the laboratory is to analyse its historical data by origin of specimen to inform the local antibiotic policy about the best options for empiric therapy. Computer programmes are now available for such purposes, based on time-series analysis of previous year's data (www. viresist.org/).

Quality control in the laboratory is, of course, essential and probably the best scheme in Europe is run by UKNEQAS (www.ukneqas.org.uk/). It now offers 24 cultures per year, specifically designed to highlight problems with detection of antibiotic resistances of clinical relevance. Current challenges for labs are detection of ESBLs, carbapenemases, glycopeptide intermediate staphylococci, and the general issue of susceptibility of non-fermenting Gram-negative rods which really have to be tested by MIC methodology. Standardisation of laboratory methods for susceptibility testing is another topical issue, predominantly so that results can be used for surveillance and EUCAST is making great strides forward on a European platform (www.escmid.org).

6.   THE FUTURE

The issues are complex and worthy of research. Most importantly, ways must be found of improving the pipeline of new drugs because development of these agents is the most tried and tested way of solving our current problems. For one reason or another, the pharmaceutical industry, which has been responsible for discovering and developing the great majority of agents in current clinical use, does not appear to consider the area worthy of much investment at the moment. Moreover, the feasibility of being able to develop many truly novel agents in the foreseeable future was questioned at a recent European meeting. These are issues that need to be urgently addressed. Also, we need to develop a better understanding of how bacteria develop resistance, the importance of mutator strains, integrons, clonality, etc., but all this will take a lot of time and money.

In the meantime, antibiotic resistance continues to get much worse on the back of huge levels of continued antibiotic misuse. We must, therefore, urgently implement what we know to reduce inappropriate prescribing and at least, stem the tide of resistance. Whether we can actually reverse current resistance problems is more problematic although I believe the answer is often yes. This is certainly the case in hospitals, often in combination with better infection control to stem the spread of clones. In the community, resistance may well decline more slowly due to the absence of a dilutional effect seen in hospitals from patient discharge. Clearly, clonal spread also is important in the community and other issues than antibiotic prescribing have to be taken into the equation such as vaccination and access to day-care centres and nursing homes. Society needs to be more informed about the issues and allowed to make value judgements about the risks and benefits of antibiotic prescribing. Do the low levels of prescribing seen in Holland represent an ideal for us all or can they be reduced even further or not as much? What are the long-term disadvantages of heavy antibiotic use and resistance? Both are much higher in Spain and Greece than in Northern Europe. Are the population, as a whole better or worse off for this? We need the answers to these many questions as soon as possible. In the meantime, it is the responsibility of everyone and especially doctors, nurses, pharmacist, and patients to prescribe and consume antibiotics wisely.

REFERENCES

Bartlett, J. G., Dowell, S. F., Mandell, L. A., File, T. M., Musher, D. M., and Fine, M. J., 2000, Practice guidelines for the management of community-acquired pneumonia in adults. Infectious Diseases Society of America. *Clin. Infect. Dis.*, **31**, 347–382.

Belongia, E. A. and Schwartz, B., 1998, Strategies for promoting judicious use of antibiotics by doctors and patients. *BMJ*, **317**, 668–671.

Bengoa, E. L., Francisco, M. M. S., and De Abajo Iglesias, J., 2002, Evolución del consumo de antibióticos en España, 1985–2000. *Med. Clin. (Barc.)*, **118**, 561–568.

Bruinsma, N., Hutchison, J. M., van den Bogaard, A. E., Giamarellou, H., Degener, J., and Stobberingh, E. E., 2003, Influence of population density on antibiotic resistance. *J. Antimicrob. Chemother.*, **51**, 385–390.

BTS Guidelines for the Management of Community Acquired Pneumonia in Adults, 2001, *J. British Thorac. Soc.*, **56**, suppl. iv, 1–64.

Canet, J. and Garau, J., 2002, Importance of dose and duration of β-lactam therapy in nasopharyngeal colonization with resistant pneumococci. *J. Antimicrob. Chemother.*, **50**, 39–43.

Chavers, L. S., Moser, S. A., Benjamin, W. H., Banks, S. E., Steinhauer, J. R., Smith, A. M. *et al.*, 2003, Vancomycin-resistant enterococci: 15 years and counting. *J. Hospital Infect.*, **53**, 159–171.

Cunney, R. J. and Smyth, E. G., 2000, The impact of laboratory reporting practice on antibiotic utilisation. *Int. J. Antimicrob. Agents*, **14**, 13–19.

Dancer, S. J., 2001, The problem with cephalosporins. *J. Antimicrob. Chemother.*, **48**, 463–478.

Eng, J. V., Marcus, R., Hadler, J. L., Imhoff, B., Vugia, D. J., Cieslak, P. R. *et al.*, 2003, *Emerg. Infect. Dis.*, **9**, 1128–1135.

Gould, I. M., 2001, Conflicts of interest between the prescriber, the regulator and the profit maker. *Clin. Microbiol. Infect.*, **7**, suppl. 6, 1.

Gould, I. M., 2002a, Antibiotic policies and control of resistance. *Curr. Opin. Infect. Dis.*, **15**, 395–400.

Gould, I. M., 2002b, BTS guidelines on CAP. *Thorax*, **57**, 657–658.

Graffunder, E. M. and Venezia, R. A., 2002, Risk factors associated with nosocomial methicillin-resistant *Staphyloccous aureus* (MRSA) infection including previous use of antimicrobials. *J. Antimicrob. Chemother.*, **49**, 999–1005.

Harbarth, S., 2000, Antibiotic policy and penicillin-G shortage. *Lancet*, **355**, 1650.

Harbarth, S., 2002. Outpatient antibiotic use and prevalence of antibiotic-resistant pneumococci in France and Germany: A sociocultural perspective. *Emerg. Infect. Dis.*, **8**, 1460.

Howard, A. J., Magee, J. T., Fitzgerald, K. A., and Dunstan, F. D. J., 2001, Factors associated with antibiotic resistance in coliform organisms from community urinary tract infection in Wales. *J. Antimicrob. Chemother.*, **47**, 305–313.

Kristinnson, K. G., 1999, Modification of prescribers' behavior: The Icelandic approach. *Clin. Microbiol. Infect.*, **5**, 4S43–4S47.

Kümmerer, K., 2003, Significance of antibiotics in the environment. *J. Antimicrob. Chemother.*, **52**, 5–7.

Kumarasamy, Y., Cadwgan, T., Gillanders, I. A., Jappy, B., Laing, R., and Gould, I. M., 2003, Optimizing antibiotic therapy—the Aberdeen experience. *Clin. Microbiol. Infect.*, **9**, 406–411.

Lelekis, M. and Gould, I. M., 2001 Sequential antibiotic therapy for cost containment in the hospital setting: Why not? *J. Hosp. Infect.*, **48**, 249–257.

Lipsitch, M., Bergstrom, C. T., and Levin, B. R., 2000, The epidemiology of antibiotic resistance in hospitals: Pardoxes and prescriptions. *PNAS*, **97**, 1938–1943.

Lipsitch, M. and Samore, M. H., 2002, Antimicrobial use and antimicrobial resistance: A population perspective. *Emerg. Infect. Dis.*, **8**, 347–354.

Lonks, J. R., Garau, J., and Medeiros, A. A., 2002, Implications of antimicrobial resistance in the empirical treatment of community-acquired respiratory tract infections: The case of macrolides. *J. Antimicrob. Chemother.*, **50**, 87–91.

Kollef, M. H., 2001, Is there a role for antibiotic cycling in the intensive care unit? *Crit. Care Med.*, **29**, 135–142.

McCaig, L. F., Besser, R. E., and Hughes, J. M., 2003, Antimicrobial drug prescriptions in ambulatory care settings, United States, 1992–2000. *Emerg. Infect. Dis.*, **9**, 432–436.

MacKenzie, F. M., Lopez-Lozano, J. M., Beyaert, A., Camacho, M., Stuart, D., Wilson, R. *et al.*, 2002, Modelling the temporal relationship between hospital use of macrolides, third generation cephalosporins and fluroquinolones, and the percentage of methicillin-resistant *Staphylococcus aureus* isolates: A time series analysis. 12th ECCMID, Milano (1), 24–27 April 2002, abstract 0231. *Clin. Microbiol. Infect.*, **8**, 32.

MacKenzie, A. R., Robertson, L., Jappy, B., Laing, R. B. S., and Gould, I. M., 2003, Audit of an antibiotic policy and microbiological investigations for treating bacteraemia in a large teaching hospital. *Int. J. Antimicrob. Agents*, **22**, 618–621.

Mandell, L. A., Marrie, T. J., Grossman, R. F., Chow, A. W., and Hyland, R. H., 2000, Summary of Canadian guidelines for the initial management of community-acquired pneumonia: An evidence-based update by the Canadian Infectious Disease Society and the Canadian Thoracic Society. *Can. Resp. J.*, **7**, 371–382.

Oosterheert, J. J., Bonten, M. J. M., Hak, E., Schneider, M. M. E., and Hoepelman, I. M., 2003, How good is the evidence for the recommended empirical antimicrobial treatment of patients hospitalized because of community-acquired pneumonia? A systematic review. *J. Antimicrob. Chemother.*, **52**, 555–563.

Perz, J. F., Craig, A. S., Coffey, C. S., Jorgensen, D. M., Mitchel, E., Hall, S. *et al.*, 2002, Changes in antibiotic prescribing for children after a community-wide campaign, *JAMA*, **287**, 3103–3109.

Piddock, L. J. V., 1995, Quinolone resistance and *Campylobacter* spp. **36**, 891–898.

Polk, R., Johnson, C., Clarke, J., Edmond, M., and Wenzel, R., 2001, Trends in fluoro-quinolone (FQ) prescribing in 35 U.S. hospitals and resistance for *P. aeruginosa*: A SCOPE-MMIT report (abstract). In 41st ICAAC Abstracts Addendum, December 16–19, 2001. American Society for Microbiology, 1, Washington, DC.

Shackley, P., Cairns, J., and Gould, I. M., 1997, Accelerated bacteriological evaluation in the management of lower respiratory tract infection in general practice. *J. Antimicrob. Chemother.*, **39**, 663–666.

Scottish Infections Standards and Strategies (SISS) Group. 2003, Good practice guidance for antibiotic prescribing in hospital. *J. R. Coll. Physicians*, **33**, 281–284.

Steinke, D. T., Seaton, R. A., Phillips, G., MacDonald, T. M., and Davey, P. G., 2001, Prior trimethoprim use and trimethoprim-resistant urinary tract infection: A nested case-control study with multivariate analysis for other risk factors. *J. Antimicrob. Chemother.*, **47**, 781–787.

Stobberingh, E., van den Bogaard, A., London, N., Driessen, C., Top, J., and Willems, R., 1999, Enterococci with glycopeptide resistance in turkeys, turkey farmers, turkey slaugh-terers, and (sub) urban residents in the South of the Netherlands: Evidence for transmis-sion of vancomycin resistance from animals to humans? *Antimicrob. Agents Chemother.*, **43**, 2215–2221.

Strausbaugh, L. J., Jernigan, D. B., Liedtke, L. A., and The Infectious Diseases Society of America Emerging Infections Network, 2001, National shortages of antimicrobial agents: Results of 2 survey form the Infectious Diseases Society of America Emerging Infections Network. *Clin. Infect. Dis.*, **33**, 1495–501.

Taylor, D. J., 1999, Antimicrobial use in animals and its consequences for human health, *Clin. Microbial. Infect.*, **5**, 119–124.

Threkfall, E. J., Fisher, I. S. T., Berghold, C., Gerner-Smidt., Tschäpe, H., Cormican, M. *et al.*, 2003, Trends in antimicrobial drug resistance in *Salmonella enterica* serotypes Typhi and Paratyphi A isolated in Europe, 1999–2001. **22**, 487–491.

Wazana, A., 2000, Physicians and the pharmaceutical industry. **283**, 373–380.

Wegener, H., 2003, Ending the use of antimicrobial growth promoters is making a difference. **69**, 443–448.

WHO report, 2003, Will antibiotic misuse now stop? **1**, 85.

Author Index

Aarestrup, F.M., 617–19, 621, 637, 650, 656, 661
Abad, X.F., 360
Abi-Said, D., 316
Abramson, M.A., 286–7
Acar, J., 668
Achong, M.R., 212–13, 217
Adler, J.L., 209, 218
Aeschlimann, J.R., 403
Afif, W., 289
Agodi, A., 435
Akimitsu, N., 474
Aksaray, S., 274
Albrich, W.C., 271
Al-Eidan, F.A., 207, 209, 211
Aliabadi, F.S., 665–6
Allaouchiche, B., 271, 460
Allen, B., 241
Allen, U.D., 493, 617, 633, 636
Allerberger, F., 273
Allison, D.G., 473
Alonso, P.C., 447
Alp, E., 351
Ambrose, P.G., 391, 394
Anakbonggo, W.W., 635
Andersen, B., 131
Andersen, C.T., 263
Anderson, R.L., 468
Anderson, R.M., 90

Andes, D., 87, 98, 389–91
Andrade, S.S., 617
Andrews, J.M., 388
Angulo, F.J., 637
Angunawela, I., 500–1, 504–6
Ansari, F., 219
Ansari, S.A., 361
Anthony, F., 637, 668
Antia, R., 367, 624
Antoniadou, A., 549
Appelbaum, P.C., 627, 678
Arason, V.A., 76, 92, 492
Archer, G.L., 477
Archibald, L., 76, 261, 269–70
Ariano, R.E., 239
Armstrong, B.C., 477
Armstrong, J.L., 478
Armstrong-Evans, M., 354
Arnold, K., 493
Arnold, M.S., 289, 299
Arnold, S. L., 491
Arnold, S. R., 494
Arroll, B., 513–14, 520, 522–3
Asano, Y., 345
Ashby, M.J., 477
Ashcroft, R., 602
Ashkenazi, S., 617
Asmundsdottir, L.R., 316
Asturias, E., 601

717

Aucken, H.M., 281
Austin, D.J., 76, 90, 686
Austin, P.C., 499
Avorn, J., 173, 228–9, 500–1,
    507, 509
Aygun, G., 354, 360
Ayliffe, G.A.J., 287, 291, 297, 355, 469
Ayres, H.M., 471

Babar, S.I., 642
Badri, M., 630–1
Bager, F., 111
Baggesen, D.L., 655
Bagshawe, K.D., 291
Bailey, T.C., 273
Balfour, H.H., Jr, 340
Ball, P., 685–7, 692
Ballemans, C.A.J.M., 355
Ballow, C.H., 92–3
Bamber, A.I., 479
Bamberger, D.M., 164, 211, 217–18
Baquero, F., 367, 547
Baran, J., 316
Barchiesi, F., 316
Barenfanger, J., 234–5, 237–8
Barnes, A.J., 317
Barrett, D.C., 638
Barrett, F.F., 209
Barrett, S.P., 282, 291, 299
Barriere, S.L., 231, 238
Bartlett, J.G., 706
Bartlett, R.C., 211–12, 218
Bartoloni, A., 599
Basky, G., 578
Bass, J.W., 495
Bassett, D.C.J., 480
Bauchner, H., 494
Baughman, R.P., 272
Bauraind, I., 555
Bavestrello, L., 594, 601
Bax, R., 680–3, 693
Bayoumi, R., 623
Becker, J., 601, 603
Bedard, M., 239
Bedos, J.P., 391
Beilby, J., 594, 608
Bell, J.M., 640
Belliveau, P.P., 161
Belongia, E.A., 515–18, 521–3, 708
Bendall, M.J., 208–9

Bengoa, E.L., 549, 708
Benguigui, Y., 594
Bennett, J.E., 324
Bennish, M.L., 617
Beović, B., 251–2, 258–9
Berg, R.W.A., 354, 360
Bergan, T., 397, 557
Bergmans, D.C., 262
Bergogne-Berezin, E., 273
Berkelhamer, J.E., 495
Berman, J.R., 227, 231–2, 240
Berman, S., 627
Bexell, A., 504–6
Beyaert, A., 447
Birungi, H., 635
Bjorland, J., 476
Bjornsson, T.D., 72
Blake, D.P., 619, 663
Blazquez, J., 371–3
Blignaut, E., 316
Block, C., 361, 492
Bloland, P.B., 623
Bloom, B.S., 296
Bloomfield, S.F., 465–6, 474, 622
Blot, S.I., 282
Bodasing, N., 340
Boeckh, M., 321
Boethius, G., 586
Bogaerts, J., 617
Bohme, A., 321
Boivin, G., 333
Bojalil, R., 599
Boken, D.J., 492
Boldt, J., 188
Bolyard, E.A., 354–5
Bonacorsi, G., 332
Bonanni, P., 332
Bonapace, C.R., 90, 94–5
Bonasso, J., 242
Bonate, P.L., 394
Bond, C.M., 531–3
Bonhoeffer, S., 91, 179
Bonten, M.J.M., 270–2, 353, 554
Boogaerts, M., 323
Booth, I.R., 367, 375, 379–80
Booth, J.C., 339
Borderon, 263
Borges-Walmsley, M.I., 471, 475
Bosch, X., 237
Bosso, J.A., 90

Boucher, J.C., 375
Bourdain, N., 263
Bouvenot, G., 556, 587
Bow, E.J., 322
Bowen, R.E., 660
Box, G.E.P., 423, 451
Boyce, J.M., 274, 291, 296, 298–9, 351, 353–4, 360
Brabin, B.J., 634
Brachman, P.S., 284, 299
Bradley, J.M., 291
Bradley, S.J., 209, 211–12
Brady, M.T., 353
Brandling-Bennett, A.D., 623
Branthwaite, A., 556, 578
Breiman, R.F., 492
Bremon, A.R., 578
Brennan, P.J., 470
Briceland, L.L., 203
Bridges, C.B., 337
Bridges, K., 477
Brinas, L., 655
Broadley, S.J., 481
Brock, D., 602
Bronzwaer, S.L., 591
Brook, I., 492
Brossette, S.E., 422
Brown, E.M., 159, 169
Brown, M.H., 471
Brown, M.R.W., 473
Brown, P.D., 492–3
Brown, S. M., 423
Brözel, V.S., 471
Bruinsma, N., 553, 704
Burd, M., 354
Burgess, D.S., 90, 94, 97
Burke, J.D., 231
Burke, J.F., 186
Burke, J.P., 240, 271, 637
Burstin, H.R., 131
Burton, M.E., 203, 206, 209, 211, 213, 217
Burwen, D.R., 262
Busse, W.W., 332
Busuttil, R.W., 323
Butler, C.C., 494–5, 587
Byarugaba, D.K., 630, 634–8
Byers, K.E., 353–4
Byl, B., 199, 202, 204, 208–10, 212, 217
Bywater, B.J., 655

Cabana, M.D., 172, 193–4
Cabello, A, 594
Cairns, C., 241
Calfee, D.P., 354
Calva, J., 599
Campbell, C.C., 623
Campillos, P., 460
Canet, J., 705
Canton, R., 367
Canuto, M.M., 315
Capellà, D., 275
Capitano, B., 298
Caprioli, A., 619–20, 654
Carbon, C., 590, 675, 679–80, 683–4
Carman, B., 338
Carman, W.F., 331
Carmeli, Y., 92, 261, 631–2
Carrie, A.G., 573, 577, 579
Carroll, W.L., 495
Cars, O., 105–7, 197, 204, 264, 391, 547, 553, 558–9, 567–8, 585, 587, 704
Casewell, M.W., 294, 297, 299
Cassell, G.H., 410
Catry, B., 619, 637
Cebotarenco, N., 561
Celik, Y., 283, 298, 300
Černelč, V., 259
Chadwick, P.R., 332
Chaffee, B.W., 68
Chahwakilian, P., 555–6, 680
Chaix, C., 288, 298–9
Chao, L.W., 557
Chapman, J.S., 470
Chauvin, C., 656–7, 668
Chavers, L.S., 703
Chawla, J.C., 241, 477
Chawla, P., 246
Chen, D.K., 92–3
Chen, Y.C., 316
Chopra, I., 367, 369, 374, 376–7, 472, 474
Choraine, P., 666
Choutet, P., 556
Chow, J.W., 92
Christakis, D.A., 511–12
Christiansen, K., 680
Chryssanthou, E., 315
Chuanchuen, R., 474, 477
Čižman, M., 251, 253–4
Clark, P.A., 618, 626
Classen, D.C., 189, 194, 211, 215

Clement, M., 439
Cloete, T.E., 471–2
Coast, J., 284–5, 562, 637
Coello, R., 283, 287, 297
Cohen, M.L., 617, 622
Cohen, S.H., 292
Colin, C., 555
Colombo, A.L., 317
Conlon, C., 241
Contopoulos-Ioannidis, D.G., 549, 578
Conus, P., 206
Cook, A.A., 231–2, 242
Cooke, D.M., 209, 219
Cookson, B.D., 474
Corbella, X., 460–1
Cornaglia, G., 492
Corso, A., 437
Cosgrove, 632
Costerton, J.W., 473
Cottell, A., 474
Couch, R.B., 339
Courvalin, P., 588
Cox, R.A., 291
Crabtree, B.F., 451–2
Cradle, R.M., 234–5, 238
Craig, W.A., 87, 98, 389–91
Crane, V.S., 231–2
Cremieux, A., 590
Crick, F.H., 688
Croisier, D., 391
Cromwell, G.L., 660
Cronk, G.A., 493
Crowcroft, N.S., 115
Cruse, P.J., 186
Cunha, B.A., 213, 609
Cunney, R.A., 203
Cunney, R.J., 712
Curie, 468

Dacre, J., 291
Dagan, R., 492, 638
Dahl, S.L., 164, 211
Dahms, R.A., 92
Dancer, S.J., 288, 294, 298, 707
Danforth, D., 358
Daniels, I.R., 617
Danzon, P.M., 557
Das, J.R., 473
Davey, P.G., 207, 292, 493, 600, 680, 683–6
Davies, A.J., 481

De Champs, C., 208–9, 212
De Jonge, E., 271
De Man, P., 204, 208–9, 212, 397, 638
De Santis, G., 507–9
De Vries, T.P., 504
Dean, B., 240
Degener, J.E., 621
Dellinger, E.P., 218
Dembry, L.M., 353
Demple, B., 472
Denamur, E., 382
Denning, D.W., 313, 315, 321
Deretic, V., 375
Deschepper, R., 578
Desenclos, J.C., 664
Destache, C.J., 203, 206, 209, 211, 213
Dharan, S., 355, 358
Di Clemente, C.C., 535
Diaz, E., 275
Dickerson, L.M., 237, 240
Diekema, D.I., 262, 268–9, 285–6, 688
Dienstag, J.L., 333, 336
Diggory, P., 340
Dijk, Y.V., 354, 360
DiMasi, J.A., 678
Diwan, V.K., 499
Doern, G.V.
Doern, G.Y., 203, 641
Dollman, C.M., 133
Dolovich, L., 507–9
Dong, Y.Z., 370–1
Donskey, 92
Dowell, J., 513–14, 523
Dowell, S.F., 515
Drake, S., 339
Drew, R.H., 284
Drlica, K., 367, 370–1, 402, 619
Drori-Zeides, T., 211–13, 218
Drummond, M., 299
Drusano, G.L., 87, 391, 393–4
Duchin, J.S., 492
Duckworth, G.J., 291
Duffy, K.R., 299
Dufour, P., 299
Dukan, S., 472
Dunagan, W.C., 199, 202–3, 212–13, 217
Dunn, L.J., 291
Dupont, B., 316
Durbin, W.A.J., 211–12, 214–15
Durieux, P., 555

Dybul, M., 337
Dye, C., 624
Dykewicz, C.A., 321

Ebright, J.R, 641
Eccles, M., 117
Echols, R.M., 212, 215
Eckert, G.M., 68
Edelstein, P.H., 619
Edmond, M.B., 262, 266, 284
Eggimann, P., 273, 323
Ehrenkranz, N.J., 204, 206–9, 211, 468
Ekdahl, K., 563
Elhanan, G., 203–4
Elizaga, J., 338
Ellner, J.J., 617, 619
Ellner, P.D., 626
Elseviers, M., 585
Emery, V.C., 345
Emmerson, A.M., 282–3
Emslie, M.J., 532–3
Ena, J., 91
Eng, J.V., 708
Engberg, J., 621
Engelenburg, F.A.C., 360
Engelhart, S., 354, 360
Engemann, J.J., 287, 299
Enright, M.C., 437
Enzweiler, K.A., 90, 92, 94, 96
Ericsson, H.M., 395, 405
Eriksson, L., 426, 432
Eriksson, S., 375
Erlandsson, C.-M., 262
Ernst, E.J., 391
Eron, L.J., 198, 207–9
Eskra, L., 375
Evans, C., 27
Evans, M.E., 213, 215
Evans, M.R., 351, 353
Evans, R.S., 180, 199, 203–5, 207, 209,
    212–13, 215, 229, 272
Everitt, D.E., 202, 204, 211–12, 218

Farr, B.M., 285, 288–9
Favero, M.S., 466, 473, 479
Fedorka-Cray, P.J., 617, 622
Feinstone, S.M., 353
Fekete, T., 650
Feldbaum, C.B., 688
Felmingham, D., 680–2

Felsenstein, J., 442
Fenina, A., 587
Fernandez, J., 391
Feucht, C.L., 232, 234–6
Fialkow, P.J., 484
Fife-Schaw, C., 664
Fijn, R., 553
Filius, P.M., 617
Finch, R.G., 169, 684
Finkelstein, J.A., 515–17, 521
Finkelstein, R., 202, 207
Firsov, A.A., 402–3
Fischer, J.E., 263
Fitzpatrick, F., 291
Fletcher, C.V., 240
Flottorp, S., 515, 517, 521–2
Flournoy, D.J., 617
Fluckiger, U., 204, 210, 212, 217–18
Fluit, A.C., 270
Flynn, W.T., 621
Foley, R.N., 494
Fonseca, S.N., 263
Font, M., 507–9
Foord, R., 186
Forgie, I.M., 627
Forrest, A., 391
Foster, T.J., 476
Fowler, V.G., Jr., 202, 205, 209, 211–13,
    218, 623
Fox, R., 340
Fraise, A.P., 287, 298
Francioli, P., 206
Frank, M.O., 209, 211, 295
Franklin, A., 654
Franklin, G.A., 619
Fraser, G.L., 209, 211–13, 231–3, 235, 239,
    284, 297
Fraser, V.J., 297
Fraud, S., 470
French, G.L., 479
Fridkin, S.K., 125–6, 136, 266, 270, 273
Frieden, T.R., 91, 207, 213
Friedrich, L.V., 76, 90–1
Frighetto, L., 241
Frimodt-moller, N., 294
Frischer, M., 574
Fry, D.E., 283–4
Fujita, K., 493
Funchain, P., 376
Fürst, J., 259

Gagnon, R.C., 421
Gamboa, L., 610
Gandhi, P.A., 471
Garattini, S., 587
Garau, J., 705
Garcia-Rodriguez, J.A., 276
Garey, K.W., 231
Gargalianos, P., 545
Garratini, L., 557
Garrod, L.P., 673
Gastmeier, P., 107, 263
Gautam, B., 606
Gaynes, R., 642
Gebel, J., 361
Geffroy, Y., 587
Geissler, A., 295, 298
Geldner, G., 297
Gentry, C.A., 231–2, 234–6, 238–9
George, A.M., 475
Gerding, D.N., 272
Gern, J.E., 332
Gesner, M., 642
Getchell-White, S.I., 353, 360
Geue, A., 594
Gherardi, G., 437
Giamarellou, H., 549
Gilbert, P., 473, 476
Gilks, C.F., 642
Gill, O.N., 286
Gillespie, M.T., 469
Gillespie, S.H., 632
Giraud, A., 376
Girmenia, C., 325
Girotti, M.J., 202
Girou, E., 288–9, 298
Giulani, G., 557
Giwercman, B., 476
Gnanou, J.C., 654
Gober, A.E., 492
Goel, V., 511–12
Goeman, J., 632
Goldmann, D.A., 228–30, 636
Gómez, J., 211
Gonzales, 493, 515–17, 522, 610
Goodyear, K.L., 655
Goossens, H., 449
Gopal Rao, G., 289
Gordon, D., 593
Gordon, M., 493
Gorecki, P., 202

Gortner, L., 263
Gottschau, A., 283
Gould, I.M., 205, 229, 241, 257, 479, 546,
    552–3, 612, 651, 686, 694, 707–8, 710
Govan, J.R.W., 375
Gove, S., 642
Grabenstein, J.D., 242
Graffunder, E.M., 706
Gransden, W., 492
Grasela, D.M., 394
Grasela, T.H., 240
Grasela, T.H., Jr., 204, 212
Grave, K., 638, 656
Greenberg, J.T., 472
Greenberg, S.B., 332
Greenough, W.B., 283
Greko, C., 638
Griffith, C.J., 359
Griffiths, L.R., 202
Griffiths, P.A., 469
Grilli, R., 169
Grimshaw, J., 118, 538
Grimshaw, J.M., 171
Grohs, P., 388, 398
Groll, A.H., 314
Gross, M.D., 374
Gross, P.A., 218
Gross, R., 231–3, 235
Grundmann, H., 289
Gruson, D., 262, 271, 295, 638
Gubareva, L.V., 333, 336
Guérin-Méchin, L., 471
Guerra, B., 5
Guglielmo, B.J., 240
Guillemot, D., 205–6, 584, 588, 650, 657
Gulick, R.M., 332, 342
Gums, J.G., 231, 233, 235
Gundersen, S., 604
Gupta, K., 617
Gustafson, R.H., 660
Gutierrez, F., 213
Gwaltney, J.M. Jr., 353
Gyssens, I.C., 19, 172, 194, 199–207, 211–17,
    228, 257, 617, 657

Hails, J., 354
Haley, R.W., 282–4, 299
Hall, C.B., 353
Halls, G.A., 204–6
Hamilton, D.C., 210, 212, 216

Hamilton-Miller, J.M., 241
Hamm, R.M., 494, 587
Hammond, S.A., 478
Hanberger, H., 261–3, 266, 268, 272, 707
Hand, D.J., 421
Harbarth, S., 77, 205, 289, 708–10
Hardy, L.M., 493
Hargreaves, J., 353
Harris, A.D., 77, 101, 617, 633, 636
Harris, C.M., 504–6
Harrison, P., 593
Hart, C.A., 643
Hart, G.D., 345
Harvey, K., 601–2
Hasan, R., 642
Hawkey, P., 465
Hay, A.J., 333, 336–7, 341
Hayden, F.G., 333, 336–8, 341
Hayden, M.K., 351
Hayman, J.N., 231–2
Heath, R.J., 468, 472–3, 477, 483
Heininger, A., 203, 212, 215
Heir, E., 469, 471, 478
Helfenstein, U., 451–2
Helms, M., 660, 663–4
Hemming, M., 601–2
Hemsell, D.L., 188
Hendley, J.O., 353
Hendrickx, E., 101
Hennnessy, T.W., 515–19, 522–3
Herbrecht, R., 320
Herold, B.C., 299
Herwaldt, L., 286, 288, 292
Higgins, C.S., 478
Highet, V.S., 391, 665–6
Hill, A.B., 43
Hill, D.A., 294
Hima-Lerible, H., 642
Himmelberg, C.J., 164
Hiramatsu, K., 299
Hirschman, S.Z., 231
Hoang, T.T., 472
Hoffer, E., 627
Hoffman, 623
Holland, J.J., 268, 335
Hollenbeak, C.S., 283
Holloway, K., 606
Holmberg, S.D., 285, 632
Holmes, A.H., 228
Holmes, L., 227

Holmes, S.J., 587
Hooton, T.M., 638
Hori, S., 294
Hornberg, C., 287
Hossain, M.M., 635
Hossein, I.K., 479
Howard, A.J., 631, 704
Howe, R.A., 282
Huang, C.T., 473
Huchon, G.K., 578
Huebner, R.E., 643
Hughes, J., 572–3, 575
Hughes, J.M., 493–4, 594
Hughes, W.T., 321
Hugo, W.B., 466–7, 474
Humphreys, H., 354
Hundt, K., 472
Hunter, P.A., 680, 682–3
Husain, S., 317
Hutchinson, J.M., 494
Huttin, C., 557
Hux, J.E, 502–3
Hyatt, J.M., 370

Ibrahim, E.H., 272
Ibrahim, K.H., 229, 231
Ilett, K.F., 507
Indalo, A.A., 635
Ives, N.J., 314

Jackson, M.A., 492
Jacqua-Stewart, M.J., 299
Jaffe, D.M., 495
Janknegt, R., 260, 284–5, 294, 553
Janne, K., 425
Janoir, C., 92
Jappy, B., 205, 229, 552
Jarvis, W.R., 228, 282–5, 288, 299
Jayaratne, P., 294
Jeena, P., 608
Jenkins, G.M., 423, 545
Jenney, A., 231–3
Jensen, L.B., 661
Jernigan, J.A., 287–8, 292–3
Jewell, M., 298
Jiménez, N.G., 447
Johnson, C.K., 90, 94, 96
Johnson, J.L., 626
Johnson, J.R., 617, 631
Johnson, N.W., 638

Johnston, J., 202
Jones, M.W., 471
Jones, R.N., 90, 94, 97, 628, 688
Jönsson, B., 283, 299
Jouy, E., 655

Kahan, A., 468
Kahlmeter, G., 387, 397–8
Kaiser, L., 493
Kaitin, K.I., 678
Kamya, M.R., 624
Kaneene, J.B., 620
Kaplan, J.E., 344
Karanfil, L.V., 353–4
Karchmer, T.B., 289, 297, 299
Kariuki, S., 627–8
Kashuba, A.D., 205
Kaslow, R.A., 422
Katzenstein, D.A., 626, 631
Kaufman, D., 323
Kawai, H., 353
Kazama, H., 472
Keady, D., 284
Keegan, 475
Kelkar, P.S., 190
Kesah, C., 628–9
Kettler, H., 637
Keuleyan, E., 257, 546
Keusch, G.T., 627
Keye, J.D., 311
Khan, M.M., 283, 298, 300
Khare, M., 284
Kibbler, C.C., 291
Kicklighter, S.D., 323
Kiivet, R.A., 262–3
Kilian, A.H.D., 624
Kim, T., 287
Kirk, S.F., 664
Kish, M.A., 169
Klein, G., 661
Knox, K.L., 176, 228
Koella, J.C., 624
Kohler, S., 375
Kollef, M.H., 261, 263, 271, 284, 297, 609, 714
Koonin, E.V., 689
Koornhof, H.J., 643
Korin, J., 300
Koulla-Shiro, S., 643
Kourti, T., 425
Kovacicova, G., 317
Kowalsky, S.F., 215

Krcmery, V., 252, 317, 560
Kristinsson, K.G., 560, 627, 705
Kruse, 629
Kücken, D., 476
Kumarasamy, Y., 707
Kümmerer, K., 701
Kummerle, N., 472
Kunin, C.M., 199, 207, 212–13
Kunori, T., 293
Kunova, A., 316
Kuti, J.L., 241
Kuyvehoven, M., 579

Lacey, R.W., 282
Lacy, M.K., 391
Ladhani, S., 492
Lagerlov, P., 504–6
Lancry, P.J., 588
Landman, D., 209, 295
Lang, A., 618
Lankford, M.G., 290
Larson, E.L., 479
Lashen, E.S., 473
Lathers, C.M., 651, 665
Lauritzen, B., 666
Law, M.R., 286
Lawson, D.H., 8
Lawson, W., 227, 243
Lawton, R.M., 169
Layton, M.C., 296
Le Pen, C., 557
Lear, J.C., 469, 477
LeClerc, J.E., 375
Lederberg, J., 593
Lee, J., 231–3, 235–6, 238
Lee, J.C., 484
Leegaard, T.M., 618, 630
Leelaporn, A., 471
Lees, P., 637, 665–6
Lelekis, M., 241, 545, 549, 708
Lemaitre, J.P., 476
Lemmen, S.W., 209, 211–12, 262
Lepper, P.M., 461
Lesar, T.S., 240
Lesch, C.A., 84
Levin, B.R., 76, 90, 367, 384
Levin, S.A., 101
Levy, C.W., 472
Levy, S.B., 101, 465, 472, 474–5, 477–8, 638
Lewin, C.S., 619
Lewis, D., 682

Lewis, D.L., 355, 471
Lewis, K., 472
Lexomboon, U., 493
Li, J.T.C., 190
Li, S.C., 203, 209, 213
Li, W.H., 438
Lim, S., 388
Limaye, A.P., 333, 336
Lin, S.J., 238, 313
Linares, L., 282
Lind, I., 644
Linder, J.A., 494
Lindgren, B., 283, 299
Lindgren, P.K., 376
Lindtjorn, B., 635
Lingnau, W., 273
Lipsitch, M., 76, 90, 617–18, 620, 635, 661, 665, 707–8
Lipsky, B.A., 229, 636
Lipsy, R.J., 215, 217
Lister, P.D., 391, 466
Little, P., 513–14, 520, 522–3, 533, 541
Littlejohn, T.G., 471
Liu, P., 391
Livermore, D.M., 367
Lloyd-Evans, N., 353
Lok, A.S., 339
Lomovskaya, O., 471–2
Lonks, J.R., 705
Lopez-Lozano, J.M., 71, 90, 92, 117, 159, 422, 449–50, 456, 460, 618
Love, S.S., 351
Low, A.S., 369, 377, 379–80, 382
Low, D.E., 169, 684
Lowbury, E.J.L., 477
Lucas, G.M., 75
Lucent, J.C., 298
Ludlam, H., 460
Lundborg, C.S., 504–5
Lynch, T.J., 239
Lynch, W., 283
Lyon, B.R., 471, 476

Ma, M.Y., 213
Macdonald, J.C., 345
Macfarlane, J., 513–14, 587
MacGowan, A.P., 396
MacGregor, J.F., 425
MacKenzie, A.R., 712
MacKenzie, F.M., 105, 117, 379, 380, 383, 460, 479, 706

Macris, M.H., 204
Madaras-Kelly, K.J., 391
Magee, J.T., 493
Magee, W.E., 311
Mahieu, L.M., 283
Maiden, M.C., 437
Maillard, J.-Y., 465–7, 472, 474, 479
Mainous III, A.G., 502–3, 513, 521
Mainous, A.G.I., 493–5, 514
Maira-Litrán, T., 477
Maki, D.G., 199, 201, 203, 209, 217
Maley, M.P., 353
Manangan, L.P., 353
Manavthu, E.K., 315
Mandell, L.A., 706
Manfredi, R., 644
Mangi, R.J., 207
Mangram, A.J., 188
Manie, T., 644
Manzoor, S.E., 470
Marcellin, P., 333
Marcinkeviciene, J., 472
Marinker, M., 540
Marolt, M., 252, 254
Marolt-Gomišček, M., 253
Marr, J.J., 26, 217, 228–9, 231
Marr, K., 321
Marr, K.A., 313
Marrazzo, J., 316
Marrie, T.J., 208
Martel, J.L., 655
Martineau, F., 293
Martinez, J.L., 367
Martins, M.D., 316
Martro, E., 361
Maruyama, S., 492
Mascini, E.M., 270
Mason, P.R., 626, 631
Mathai, D., 270
Mathei, C., 332
Matic, I., 374
Mattoes, H.M., 391
Mayer, R.A., 354
Mayhall, C.G., 186–7
Mbithi, J.N., 353
McBain, A.J., 473
McCaig, L., 572–3, 575–6, 594, 708
McCaig, L.F., 493–4, 594
McCall, J., 332
McConnell, T.S., 507–9
McDermott, P.F., 637

McDonnell, G., 467–8, 471, 479
McEwen, S.A., 617, 622
McFarland, L.V., 359
McGarock, H., 633
McGee, L., 437
McGowan, J.E. Jr., 76, 87, 448
McGowan, J.E., 495, 593, 598
McHugh, T.D., 632
McIsaac, W.J., 494–5, 511–12
McKeegan, K.S., 471, 475
McLaws, M.L., 299
McLean, W., 239
McManus, P., 569, 572, 574, 577
McMullin, S.T., 234, 238, 273
McMurry, L.M., 469, 472–3, 475–6
McNeil, M.M., 312–13
McNeil, M.R., 470
McNulty, C., 76, 92
McNulty, C.A.M., 504, 506
Meberg, A., 263
Méchin, L., 471
Meehan, T.P., 203, 207, 209–10, 212
Mehtar, S., 295, 297
Melander, E., 578, 583
Merrer, J., 297
Metlay, J.P., 208
Metz, R.G., 535
Meyer, J.C., 504–6
Meyer, K.S., 75
Meyer, S.K., 209
Michel-Nguyen, A., 316
Milan, E.P., 315
Miller, P.J., 283–4, 299
Mills, S.D., 689
Milovanovič, M., 252–3
Minooee, A., 231
Mitchell, B.A., 478
Mitema, E.S., 621, 633, 637–8
Moerbeek, M., 499
Mokaddas, E.M., 284
Moken, M.C., 474–5
Molbak, K., 671
Møller, J.K., 263
Mölstad, S., 105, 107, 558, 572, 704
Mongkolrattanothai, K., 354, 360
Monnet, D.L., 71, 90, 111, 115, 117, 133, 151,
    255, 262, 272, 294, 422, 448, 460–1,
    557–8, 661, 667, 692, 695
Montelcalvo, M.A., 354
Montgomery, M.J., 394
Moran, G.P., 315

Moreno, M.A., 655
Morfin, F., 333, 335–6
Moro, M.L., 550
Morosini, M.I., 377
Morrell, R., 215
Morrison, D.F., 426, 435
Moser, S.A., 422
Moss, F., 205, 212
Moszynski, 618
Motola, G., 202
Mouton, J.W., 389, 392
Moxon, R.E., 367
Mozillo, N., 202
Muder, R.R., 282, 287, 298
Mueller-Premru, M., 251
Muñoz, A.Y., 447
Munoz, P., 325
Munton, T.J., 471
Murray, G.E., 478
Murray, S., 494–5
Murray-Leisure, K.A., 291–2
Murthy, R., 638
Murtough, S.M., 479
Muto, C.A., 292, 296, 360

Naaber, P., 262
Nagler, J., 554
Nakavuma, J., 637
Narui, K., 475
Nasrin, D., 492, 583
Nathwani, D., 201–2, 204, 207, 228, 241,
    283–4, 287, 295, 298, 545, 600, 685
Natsch, S., 70, 169, 205–7, 217
Nava, J.M., 492
Neal, T.J., 479
Neely, A.N., 353
Negri, M.C., 373
Nei, M., 438
Nettleman, M.D., 290
Newsom, S.W.B., 477
Ng, M.E.G.-L., 467
Ng, W., 391
Nguyen, D.V., 425
Nichol, K.L., 332
Nicholls, T., 637, 656
Nichols, R.L., 186
Nicolau, D.P., 394
Niederman, M.S., 284–5, 287, 297
Niesters, H.G., 338
Nijhuis, M., 336
Nikaido, H., 471, 475

Nilsson, L., 261, 266, 268
Noel, I., 283
Noguchi, N., 475
Noskin, G.A., 354
Nucci, M., 317
Nyamogoba, H., 618, 632
Nyquist, A.-C., 493

O'Brien, K., 618, 626
O'Brien, T.F., 8, 619
O'Connell, D.L., 502–3
O'Connell, N.H., 354
O'Dempsey, T., 609
O'Dempsey, T.J.D., 627
O'Neill, A.J., 377
Obala, A.A., 618, 632
Odds, F.C., 315, 317, 320
Odhacha, A., 611
Okeke, I., 602–4
Oliver, A., 375–6, 382
Oller, A.R., 374
Olson, M., 283
Olson, M.D., 300
Onyeji, C., 391
Oosterheert, J.J., 707
Oražem, A., 252–3
Örtqvist, A., 578
Osterblad, M., 379
Oxman, A.D., 494, 521

Padungton, P., 620
Pahor, M., 72
Paladino, J.A., 45, 49, 206, 208, 284, 631
Pallares, R., 492
Palmer, D.A., 494
Palmer, D.L., 645
Pan, A., 287
Pape, J.W., 626
Papia, G., 288, 293
Paredes, P., 600
Parikh, S.L., 472–3
Parimi, N., 635
Parise, M.E., 631
Park, J., 379–80
Parret, T., 199, 201–2, 204–5, 210, 212, 216–17
Parry, M.F., 619
Passos, S., 198, 207
Pastel, D.A., 241
Paterson, D.L., 86, 273
Patterson, J.E., 102

Paulsen, I.T., 471, 476, 478
Pawlotsky, J.M., 333
Paya, C.V., 345
Pearce, H., 476
Pechere, J.C., 556, 578
Pedersen, K.B., 651
Peetermans, W.E., 169, 554
Pelissolo, A., 587
Peltola, H., 205, 212
Pelz, R.K., 323
Pena, C., 283
Perez-Cuevas, R., 504, 506
Perz, J.F., 515–19, 522–3, 708
Pescovitz, M.D., 334
Pestotnik, S.L., 173, 215, 271–2
Petersen, I.S., 153, 262
Peterson, G.M., 507–9
Peterson, L.R., 422
Peterson, M.L., 403
Pettersen, P.-G., 131
Pfaller, M.A., 285, 313–4, 316
Phillips, I., 479
Phongsamart, W., 618, 630
Piddock, L.J.V., 702
Pillay, D., 333, 335–6, 338, 343
Pinner, R.W., 283
Piot, P., 645
Pittet, D., 274, 283, 288–90, 293, 295, 297–9
Plowman, R., 282–3, 293–4
Poggeler, S., 316
Polk, R.E., 90, 92–4, 96, 706
Poole, K., 468, 471, 474–5
Porwancher, R., 354
Posada, D., 441
Poulsen, K.B., 283
Poupard, J.A., 421, 423–4, 684
Powers, W.J., 671
Pradier, C., 556, 590
Prado, M.A., 231–2
Prazuck, T., 604
Premji, Z., 623
Preston, S.L., 391
Price, D., 532
Price, M.F., 316
Priest, P., 493
Prince, J., 468
Prochaska, J.O., 535
Projan, S.J., 689–90
Puerto Alonso J.L., 630
Pulimood, T.B., 628
Pullen, W., 469

Punt, N., 392–3
Putman, M., 471

Quilitz, R.E., 321
Quindos, G., 315
Quintero, A.C., 447
Quintiliani, R., 208, 211

Radebaugh, J.F., 493
Radetsky, M.S., 492
Radman, M., 374
Raffaele, B., 314
Rahal, J.J., 76, 92, 94
Ramaekers, D., 169
Ramirez, J.A., 241
Rampling, A., 296–7, 299
Rao, N., 287, 294, 297
Rasheed, J.K., 377–8
Rasmussen, B.A., 689–90
Rautakorpi, U.-M., 559–60, 575
Raveh, D., 650
Ray, W.A., 500–1, 507, 509
Raymond, D.P., 208–9, 211–12, 272
Raynaud, J.P., 651
Raynor, D.K., 239
Raz, R., 203, 207
Reboli, A.C., 290
Reeves, D.S., 680, 682–3
Regnier, B., 261
Reichler, M.R., 492
Reid, R., 283, 307
Reilly, J.S., 283
Reimer, K., 283
Rello, J., 270
Resi, D., 87
Reverdy, M.E., 477
Rex, J.H., 315, 321, 325
Reynolds, R., 683
Rheinbaben, F.V., 359
Rhinehart, E., 274
Rice, L.B., 93, 232, 234–6
Richard, P., 92
Richet, H.M., 409, 411
Richter, S.S., 437
Rickman, L.S., 231
Ridley, 623
Riewpaiboon, A., 594
Rifenburg, R.P., 103
Riley, T.V., 447, 457
Robinson, R., 299
Rock, C.O., 477

Rocke, D.M., 425
Rocourt, J., 663
Røder, B.L., 204, 210, 216, 263
Rodriguez, A.J., 607
Rodriguez-Cerrato, V., 369–70
Rodriguez-Villalobos, H., 265, 269–70
Roger, P.M., 210
Rokstad, K., 502–3
Romero-Vivas, J., 282
Ronning, M., 87, 114
Rose, R., 283
Rosenquist, H., 663
Rostel, B., 650
Rotter, M.L., 466
Rouche, D.A., 471
Roujeinikova, A., 472
Rowe, A.K., 645
Rowe, G., 664
Rowe-Jones, D.C., 189
Rubin, L.G., 473
Rubin, M.A., 617–18
Rubin, R.J., 300
Rubinovitch, B., 288, 293, 295, 298
Ruckdeschel, G., 268
Rucker, T.D., 229
Ruiz, J., 368
Ruiz-Bremon, A., 84
Russell, A.D., 360, 465, 467–8, 470–1,
    474, 478–9
Russell, I.T., 171, 538
Russell, N.J., 467
Russo, T.A., 617, 631
Rutala, W.A., 352–3, 358–61, 466
Rutherford, C., 294
Ryan, J.L., 687, 691, 695
Ryan, K.A., 91
Rybak, M.J., 203, 206

Saag, M.S., 321
Saez, M., 452
Saez-Llorens, X., 645
Safrin, S., 332
Sahm, D.F., 426
Salvatierra-Gonzalez, R., 594
Salzman, M.B., 473
Samaranayake, L.P., 638
Samore, M.H., 90, 617–18, 620, 633, 665, 708
Sample, M.L., 354
Sanchez, D., 249
Sanchez, M.L., 231–2, 242
Sanders, C.C., 391

Sanders, P., 649, 654–5, 667
Sandven, P., 313
Sanford, J.P., 468
Sanford, M.D., 282
Sanglard, D., 315, 320
Santoso, B., 504–5, 594
Sanyal, S.C., 284
Sasatsu, M., 468
Sasse, A., 202
Sattar, S.A., 359
Saurina, G., 361, 645
Sbarbaro, J.A., 176
Scaglione, F., 391
Schaaff, F., 377
Schaberg, D.R., 284–5
Schaffner, W., 500–1, 507–9
Scheifele, D., 492
Schentag, J.J., 92–3, 217, 229, 231–2, 234,
    237–8, 294, 370, 665
Schiff, G.D., 229
Schlaes, D.M., 691
Schlemmer, B., 686, 692, 694
Schrag, S.J., 205
Schroeder, C.M., 637
Schüklenk, U., 602
Schuna, A.A., 199, 201, 203
Schwalbe, R.S., 268
Schwartz, B., 637–8, 708
Schwartz, S., 629
Schweizer, H.P., 465, 472–4
Scolan, V., 300
Scott, G., 273
Segal-Maurer, S., 646
Segarra-Newnham, M., 338
Sehgal, R., 638
Seligman, S.J., 207, 211, 213
Selkon, J.B., 291
Semjen, G., 638
Seppala, H., 76, 93, 492–3, 500–1, 518, 522, 559
Sexton, D.J., 206, 286–7, 619
Sexton, J.D., 646
Seyfarth, A.M., 639
Shackley, P., 710
Shah, P.M., 637
Shaikh, Z.H.A., 353
Shalita, Z., 476
Shanson, D., 291
Shears, P., 634
Sherris, J.C., 395, 405
Shlaes, D.M., 27, 31, 76, 160, 228, 687, 694–5
Shojaee, A.F., 637

Shrestha, N.K., 294
Shrimpton, S.B., 203
Šibanc, B., 255
Siegel, E.C., 374
Silber, J.L., 106
Silver, A., 202, 207
Silver, S., 476
Sims, M.E., 323
Sindelar, G., 370
Sing, C.F., 438–9
Singer, M.E., 284
Singh, N., 202, 208–9, 211–12, 272, 317
Sintchenko, V., 598
Sirinavin, S., 638
Sizun, J., 353
Skurray, R.A., 471, 476
Slater-Radosti, C., 472, 474
Slayter, K., 241
Slots, J., 638
Smith, D.L., 75, 661
Smith, D.W., 620
Smith, J.T., 619
Smith, R.D., 562, 637–8
Smith, T.L., 354
Smithson, R., 332
Smyth, E.G., 712
Smyth, E.T., 283
Snell, N.J., 333
Soll, D.R., 316
Solomon, D.H., 228–9
Sommet, A., 583, 585
Sondossi, M., 473
Sörensen, T.L., 557–8, 692, 695
Sørenson, L.T., 255
Soriano, F., 369–70
Sosa, A., 600, 603–4, 607
Soumerai, J.A., 229
Soumerai, S.B., 173, 215, 500–1, 507, 509
Soumerai, S.N., 541
Soussy, C.J., 396
South, M., 594
Souweine, B., 273
Speller, D.C.E., 468
Spengler, R.F., 283
Spicer, W.J., 299
Spratt, B.G., 437
Srinivasan, A., 274
St Germain, G., 315
Stafford, R.S., 494
Stalsby Lundborg, C., 529, 575
Stamm, A.M., 286

Stanhope, M.J., 438, 441
Stearne, L.E., 404
Stege, H., 657
Steinke, D.T., 493–4, 704
Steinman, M., 576, 594
Stelling, J., 607
Stephen, J., 266, 269–70
Sterling, T.R., 609
Stevens, D.A., 321, 325
Stevenson, M., 447
Stewart, F.M., 523
Stewart, J., 515–16, 522
Stewart, M.J., 472
Stewart, P.S., 477
Stickler, D.J., 469, 477
Stobberingh, E., 703
Stobberingh, E.E., 553
Stockley, J.M., 283
Stott, N.C., 493
Straand, J., 575
Stratchounski, L., 571, 599
Strausbaugh, L.J., 710
Stroup, D.F., 452
Struelens, M.J., 202, 257, 271, 554
Sturm, A.W., 210
Sudha, V., 607
Suller, M.T.E., 474, 479
Sundheim, G., 471, 478
Swann, M., 660
Swindell, P.J., 202, 217, 229
Swofford, D.L., 442

Taboulet, F., 493
Tacconelli, E., 316
Taddei, F., 375
Talon, D., 291, 351, 354
Tan, L., 638
Tan, T.Q., 492
Tanner, D.J., 206
Tapsall, J., 607
Tarlow, M.J., 339
Tarp, B.D., 263
Tarzi, S., 291
Tattawasart, U., 469, 471, 474–5, 477
Tauber, M.G., 647
Tawfik, Y., 610
Taylor, B., 493
Taylor, D.J., 702
Taylor, G.M., 209, 211
Teare, E.L., 282, 290
Tediosi, F., 557

Teele, D.W., 495
Templeton, A.R., 438–9
Tennent, J.M., 471
Teshager, T., 655
Tezzo, R., 645
Thirumoorthy, T., 646
Thomas, C., 447, 457–8
Thomas, J.K., 87, 206, 404–5
Thomas, L., 361, 469, 477
Thomson O'Brien, M.A., 521
Thomson, R., 169
Thornsberry, C., 426
Thouvenot, D., 333, 335–6
Threlfall, E.J., 650, 655, 703
Thuong, M., 202, 213–15, 217–18
Timmers, G.J., 354, 360
Tokars, J.I., 92–3
Tokue, Y., 293
Tollefson, L., 621
Tornieporth, N.G., 93
Torres, H.A., 317
Touati, D., 472
Toutain, P.L., 653, 665
Townsend, E.H.J., 493
Townsend, E.J.J., 493
Traisman, H.S., 493
Trape, J.F., 624
Travers, K., 600
Trick, W.E., 312–13
Trotta, M.P., 340
Tullus, K., 263
Tumbarello, M., 626, 628, 631, 633, 636
Turnidge, J., 392

Urassa, W., 646

Van den Bergh, M.F.Q., 296
van den, B.A., 619, 620, 629
Van der Doef, E., 535
Van der Horst, C., 339
Van der Meer, J.W.M., 169, 200, 204, 228, 257
van Duijn, H., 578
Van Houten, M.A., 206, 212
van Kasteren, M., 185, 188, 194
van Kasteren, M.E., 195, 554, 666
Van Kasteren, M.E.E., 201, 205
Van Klingeren, B., 469
Vatopoulos, A., 550
Venezia, R.A., 294
Veninga, C.C.M., 504–6, 578
Verweij, P.E., 317

Vidaver, A.K., 617, 622
Vincent, J.-L., 210, 261, 263, 266, 268
Vink, A.A., 668
Vinken, A., 295
Visser, L.G., 206
Vižintin, T., 257
Vlahovic-Palcevski, V., 262
Vodolo, K.M., 338
Volger, B.W., 199, 217
Vose, D., 618, 651, 658–9, 663
Voss, A., 268, 274, 351
Vriens, M., 287, 294

Wachter, D., 606
Wai, C.T., 339
Wakefield, D.S., 287, 292–4
Walker, A., 283
Walker, P.P., 316
Walmsley, A.R., 471, 475
Walsh, S.E., 469–70, 474
Walsh, T.J., 314, 325
Walsh, T.R., 282
Walters, S., 339
Walther, S.M., 153, 262
Wang, H., 475, 493
Ward, R.L., 359
Ward, S., 261
Wasfy, 628
Waskerwitz, S., 495
Watkins, W.M., 623
Watling, S.M., 237
Watson, J.D., 688
Watson, M.C., 539, 541
Watson, R.L., 494
Waugh, S.M.L., 331, 338, 340
Wazana, A., 710
Weber, D.J., 353, 358–9, 466
Webster, J., 290, 479
Wegener, H., 702
Wegener, H.C., 621, 629, 659, 663
Weidle, P.J., 626
Weingarten, S.R., 208
Weinstein, M.P., 208
Weinstein, R.A., 638
Weinstein, W.M., 186
Weniger, B.G., 353
Wenzel, R.P., 284, 300
Werry, C., 358
Wessling, A., 586
West, R.R., 493

Westh, H., 117, 119
White, A.C., 208, 211–12
White, A.C., Jr., 135, 258
White, A.R., 655
White, D.G., 620–1
White, R.L., 75, 82–3, 86, 89–90, 98, 117, 263
White, T.C., 315
Whitehouse, J.D., 283
Whitney, C.G., 492
Whittle, G., 368
Widmer, A.F., 274
Wierup, M., 56, 666
Wilkins, E.G., 339
Wilkins, E.G.L., 203–4
Willems, R.J., 661
Willems, R.J.L., 661
Williams, D.N., 241
Williams, R.R., 167
Wilson, P., 291
Wilton, P., 257, 608, 637
Winder, C.L., 469, 471
Winston, D.J., 323
Wise, R., 9, 369–71, 396, 637–8
Wisplinghoff, H., 312
Witte, W., 297
Wolday, D., 623, 634
Woodhead, M., 556
Woodward, R.S., 295
Wray, C., 654
Wyllie, S., 235
Wymenga, A., 189

Yamazaki, T., 312–13
Yates, R.R., 91
Yun, H.J., 624–6

Zafar, A.B., 479
Zaidi, M., 271
Zaidi, M.B., 617, 633, 636
Zaman, S.R., 617
Zambon, M., 335
Zambrowski, J.J., 241
Zar, H.J., 626, 631
Zellweger, C., 647
Zenni, M.K., 492
Zgurskaya, H.I., 475
Zhanel, G.G., 492–3, 579
Zhao, X., 402
Zoutman, D., 283

Subject Index

Abacavir, 333
ABC Calc *see* Antibiotic Consumption
    Calculator
ABC-family *see* ATP-binding cassette
Academic detailing *see* Educational outreach
Academic Pharmacy Unit (Hammersmith
    Hospitals, NHS Trust, London), 244
Accelerated Bacteriology Laboratory Evaluation
    for better quality antibiotic prescribing, 711
Aciclovir, 332–3, 339, 341
    in *Herpes simplex* virus (HSV), 332
Acinetobacter baumannii, 94, 361
Acinetobacter johnsonii, 478
Acinetobacter spp, 64, 265, 358, 469
Acute respiratory infections (ARIs), 632
    antibiotics prescribed in, 600, 612
Adefovir, 333
Adelaide, South Australia (SA), 133
    data analysis
        by antimicrobal agent, 142
        cephalosporins, usage in, 142–8
        of usage by antimicrobial class, 139–41
    data collection and reporting, 136–7
        overall trends in antimicrobial usage
            rates, 137–9
    diversity of computer systems used by
        hospital pharmacy departments, 135
    hospitals, comparative intensive care usage
        rates for, 154

Adelaide, South Australia (SA) *contd.*
    metropolitan public and private hospitals,
        133–5
    see also Australia
Adenovirus, 338, 360
Africa, 504, 599, 602
    antibiotics from unsanctioned
        providers, 603
    data on MRSA, 628
    see also Sub-Saharan Africa
African Comprehensive HIV/AIDS
    Partnership (ACHAP), 638
Aggregate antimicrobial use
    data, 79–80
    and resistance, relationships based on
        usage, 92
AGREE *see* Appraisal of Guidelines, Research
    and Evaluation for Europe
Agriculture, 701
    antibiotic use in, 701–2
    benefits from, 660
AIDS epidemic, 312, 618
    efforts to combat, 638
    patients, use of antimicrobial drugs,
        626, 630
    related mycoses, 314
    see also HIV
Akaike Information Criterion
    (AIC), 454

733

Alcohol, 352
  hand disinfectants, 271, 273, 290
  mixture, high concentration, 360
Aldehydes, 352, 466
Alexander Project, 424, 435, 440
  10-year dataset, 436
  antibiotics analyzed from, 425
  antimicrobial susceptibility databases, 437
  collection, 439
  isolates, 441
Algorithm
  for classification of prescriptions in
      different categories of inappropriate
      use, 199
  for evaluation of antimicrobial therapy, 200
Amantadine, 332–3, 336
American Academy of Pediatrics (APP), 515
American College of Clinical Pharmacists
    (ACCP), 243
American College of Clinical Pharmacy
    (ACCP), 243
American college of emergency physicians
    criteria for administration of thrombolytic
    therapy, 131
American College of Physicians, 39, 60
American Society for Microbiology (ASM),
    410, 678
American Society of Consultant Pharmacists
    (ASCP), 242–3
American Society of Health-System
    Pharmacists (ASHP), 237, 242–3
  website, 244
American Society of Internal Medicine,
    39, 60
American Thoracic Society (ATS), 38
  antibiotic guidance, 49
Amine oxide, 474
Aminoglycosides, 94, 139, 149, 178, 190, 209,
    213, 295
  antibacterials, 109
  underdosing of, 203
  use, ICU and total, 147
Amoxicillin, 138, 253, 578, 594, 689
  in Chile, 595
Amoxicillin/clavulanate, 138, 573–4, 577
  usage, 148
Amoxicillin/clavulanic acid, 253
  in Chile, 595
  dosing regimes, four different, 393
Amphenicols, 109

Amphotericin, 317–18, 322–5
  liposomal, 213, 325
Ampicillin, 371, 600, 689
Ampicillin/sulbactam, 77
Amprenivir, 333
AMT *see* Antimicrobial Management Teams
Anaphylaxis, 43
Anatomical Therapeutical Chemical
    Classification system (ATC), 72, 85,
    108–9, 568, 657
  and DDD, 116
  classification system, 110
  see also DDD
Anidulafungin, 319–21
Animal daily dosage (ADD), 657
Animals
  agriculture, benefits from the use of
      antibiotics in, 660
  and animal derived food, development of
      harmonised antimicrobial resistance
      and antimicrobial resistance, 667
  different impact of antimicrobial treatments
      at the individual level and at the herd
      level, 666
  diseases, resistance in DCs, 629
  drug consumption at population level, 656
  experiments by Burke, 186
  production, development of bio-security
      concept in, 663
  sector in European Union, development of
      an antibiotic policy for containment of
      surveillance programmes in, 654
Animals, antibiotic use in,
  communication, 663–4
  feed additives, 660
  health purposes, worldwide use of, 621
  market authorisation, 651–4
  monitoring and surveillance, 654–7
  nutrition, guidelines for assessment of
      additives in, 651
  perceptions about antimicrobial resistance
      due to, 664
  population, spread of antimicrobial
      resistance bacteria at the level of, 665
  policies and control measures around
      Europe, 649–68
  research, 664–5
  risk analysis, 658–64
  veterinary drugs, 653
  ways of progress, 664–7

Annals of Internal Medicine, 496
Antibacterial discovery through genomics,
    process of, 689
    defined stages in process, 689
    development and use, 697
Antibacterial drugs, 109, 252
    companies involved in development of, 676
    prescribing, 694
    resistance and policies, surveillance of
        resistance and usage, 680–4
Antibiogram data, 90
Antibiotic/s
    additional avoidance of, 191
    concentrations at target sites, 369
    costs for MSSA and MRSA, 294
    cycling, 707
    MICs and resistance in *S. pneumoniae*,
        broad patterns of, 434
    misuse, continued, 713
    not to be used therapeutically, 189
    not readily leading to emergence of
        microbial resistance, 190
    order form, 215
        for drugs on restricted list, 166
    prescribing, 713
    in presence of acute respiratory infection
        (ARI)/pneumonia, 600
    stewardship programme, 651
    substandard, 604
    total outpatient consumption of, 253
    for viral infections, 494
Antibiotic consumption
    alternative units of measurement, 106–8
    and bacterial resistance, problems in
        demonstrating relationship, 447–50
    in Chile, 597
    in DDD/1,000 inhabitants and day (DID)
        in 1997 in 15 EU countries, 569
    in ICUs, 262–4
Antibiotic Consumption Calculator
    (ABC Calc), 71, 111–14
    denominator data, 112–13
    numerator data, 112
    in practice, 113–14
Antibiotic Control Committee, 164, 166,
    176, 181
Antibiotic Control Plan (ACP), 164
Antibiotic policies
    components of, 2–3
    early developments, 2–5

Antibiotic policies *contd.*
    ensuring compliance, 8–10
    extension of, 5–6
    historical perspective, 1–11
    for all hospitals, 6
    origins of, 1–8
    pioneers, 1–2
Antibiotic prescriptions
    Australia, 577
    Canada, 577
    Europe, 574–5
    Finland, 575
    indications for, 574–7
    Norway, 575
    per 1,000 inhabitants and year,
        number of, 568
    Sweden, 575
    United States, 575–7
        National Ambulatory Medical Care
        Survey, 576
Antibiotic prophylaxis
    in (clean-) contaminated procedures, 187
    in surgery
        pilot survey of, 193
        state of the art, 185–90
    timing of, 186
Antibiotic resistance/resistant
    acquisition of, 368–9
    bacteria,
        comparison of impacts of infections, 632
        selective pressure, 371
    development of relationships based on
        patient-specific data, 91–2
    Escherichia coli urinary isolates, 492
    in ICU, 264
    pathogens, 168, 271
    quantitative relationships between
        use and, 77
    in veterinary medicine, 629
Antibiotic Resistance, Prevention and Control
    (ARPAC), 113, 116, 706
Antibiotic susceptibility testing (AST), 607
Antibiotic use
    in agriculture, 702
    in animals, 655
    in community, 567–79
    and cost trends, 594–8
    data, 106
    as feed additives, 655–7
    for growth promotion, 702

Antibiotic use *contd.*
  hospital and outpatient use in Nordic
    countries, 571
  quantity and length of application, v
  and resistance in bacteria
    community-acquired pathogens, 492
  for surgical prophylaxis, 190
  as veterinary drugs, 666
Antibiotypes, 444
  method of data mining, 423
Antibiotyping, assurance system for, 655
Antifungal agents
  available for prophylaxis and treatment of
    invasive mycoses, 317
  drugs, 213
    classes and formulations of, 321
    combinations and therapy changes, 325–6
    limitations of recommending uses, 321–2
    rational use of, 317–26
    resistance and rational use, 311–26
    systemic, main properties of, 317–19
Antifungal prophylaxis, 311
  and treatment guidelines, 321
Antifungal resistance, 317
  in *Candida* species, 314
  problem of, 314–7
  transmitted by extrachromosomal DNA,
    315–7
Antifungal therapy in neutropenic patients,
  empirical, 324
Anti-infectives, 605
Antimicrobial/s
  aggregation, level of of, 72–3
  classes, six, 139
  correlation between mean daily dose and
    days of therapy, 98
  costs, 227
    for HAI, 294
  dispensing records, 80
  donations, reality-check for developing
    countries, 605
  drug appropriate category IV,
    choice of, 202
  management programmes, 229–30
  new, development of, 76, 611
  policies, 604–6
  prescribing program, 215
  prophylaxis, limiting, 235
  purchases, 80
  surveillance, definitions used in, 136

Antimicrobial/s *contd.*
  therapy
    based on Sensitivity index (SI) value in
      relation to resistance, 607
    classification of different types of, 198
Antimicrobial Management (or Review) Team
  (AMT), 230
  functions and responsibilities of, 231–2
  intervention activities, 235–7
  lead role in, 238–9
  methods of identifying targets for
    intervention, 234–5
  models of delivery, 232
  multidisciplinary, antibiotic choices
    influenced by, 233
  operational aspects of, 232–4
  structure of, 230
Antimicrobial prescription
  for respiratory tract infection (RTI), 574
  skin and soft tissue infections, 574
  urinary tract infections (UTI), 574
Antimicrobial resistance, 617
  microorganisms, 27
Antimicrobial resistance surveillance
  data, 606
    for key pathogens, need for access
      to, 607–8
    published in medical literature,
      evaluation of, 410–6
    translating into action, 607
  program, 417
    case definition, 417
    pneumonia, 417
    urinary tract infection, 417
Antimicrobial susceptibility testing (AST)
  methods, 395
  results, 421
Antimicrobial usage
  comparison of various measures of, 94
  data, sources of, 80–1
  effect on resistance, not
    contemporaneous, 452
  in grams, 83
  measures, 81–2
    patient-specific, 81–2
  pattern of use in specific institution, 90
  protocolised alternations, 235
  utilisation surveillance programme, 133
Antipseudomonal penicillin/β-lactamase
  inhibitor combinations, 139, 148

Anti-retroviral (ARV) drugs, 626
  therapy for human immunodeficiency virus
    (HIV), 332
  see also Highly reactive anti-retroviral
    therapy (HAART)
Antistaphylococcal agent, 3
Anti-tuberculosis primary drug resistance, 625
Antituberculous and antihuman
  immunodeficiency virus therapy, 178
Antivirals, 213
  adherence of patient to regime, 342
  adverse effects, 335
  correct timing of treatment, 339
  currently available, 333
  doses, 341
  drug choice and dose, 345
  resistance, 335–6
  strategies for rational use of, 331–46
Antiviral prophylaxis strategies, 343–6
  adherence, 346
  alternatives to, 344
  monitoring for infection, 346
  restriction to necessary situations, 344
Antiviral therapy/treatment
  combination, 342–3
  problems associated with, 334–7
  monitoring, 343
  strategies, 337–43
    correct drug choice, 340–1
    sources of information and advice, 337
    timing and duration of, 345
Antiviral use
  generally and locally, limited experience
    in, 334
  limit where possible, 339
Appraisal of Guidelines Research and
  Evaluation in Europe (AGREE), 25
  collaboration, 169
  development methodology, 27
Appropriate practices for antibiotic
  prescribing, 15
Approval for use of antibiotics on restricted
  list, 166
APUA (Alliance for the Prudent Use of
  Antibiotics), 547, 561
  survey on physician antibiotic prescribing
    practices and knowledge in seven
    countries in Latin America, 600
ARBAO II, 655
Ardacin, 651

Area under the concentration-time curve
  (AUC), 389
  (AUC)/MIC relationship, 98
  and probability of emergence of
    resistance, relationship between, 405
  ratio and effect, relationship
    between, 391
  ratio and frequency of emergence of
    resistance, relationship between, 403
ARIMA (autoregressive integrated moving
  average) model proposed by Box and
  Jenkins, 423, 457
  to predict percentage of ceftazidime-
    resistant/intermediate Gram-negative
    bacilli, 454
ARIs *see* Acute respiratory infections
ARPAC *see* Antibiotic Resistance, Prevention
  and Control
ASHP *see* American Society of Health-System
  Pharmacists
Asia, 599, 602
Aspergillosis, 325
  mortality in invasive, 313
Aspergillus flavus, 313
Aspergillus fumigatus, 313, 315–17
Aspergillus spp., 320, 323
  invasive infections caused by, 312–13, 320
ATC *see* Anatomical Therapeutical Chemical
  Classification system
ATP-binding cassette (ABC) family, 471
  transporters, 315
AUC *see* Area under the concentration-time
  curve
Audit/s
  for adherence to guidelines, 218
  of antimicrobial drug use, 198
  in Canada, 202
  in Dundee, 30
  and feedback, 16, 503
  frequency of, 218
  of policy application, 10
  procedures of, 217–18
Audit Commissions Report, 237
Audits for monitoring the quality of
  antimicrobial prescriptions, 197–219
  antimicrobial drug classes, 213
  automated methods, computer-assisted
    prescribing, 215
  data collection techniques, 214
  manpower, 216–17

Audits for monitoring the quality of
    antimicrobial prescriptions *contd.*
    stratification level, 212–13
    use of surveillance data as quality
        measure, 215–16
Audits with intervention
    (before and after) without a control
        group, 210
    (before and after or simultaneous) with
        nonrandomised control group, 211
Australia, 28, 202, 242, 578, 605, 679
    antibiotic policy changes, 600
    data on sale of oral antibiotics, 572
    Drug Utilization Sub-Committee, 574
    Infection Control Service of South
        Australian Department of Human
        Services, 133, 135
    Joint Expert Technical Advisory Committee
        on Antibiotic Resistance
        (JETACAR), 134
    overall profile to antibiotic use by drug
        class, 594
    study, 51
    see also Adelaide, South Australia
Australian hospitals, 51
    benchmarking between, 151
    ICU usage rates for parenteral antimicrobial
        use, 152
    study of economic outcomes, 51
    treatment of community-acquired
        pneumonia, 155
    usage data, 134
Australian Society for Antimicrobials
    (ASA), 243
Austria, 268, 561, 568
Avilamycin, 651, 667
Avoparcin, 651, 703
Azithromycin, 150, 253, 493
    usage, 156–7
Azole resistance, 316

Bacillus anthracis, Ames strain, 688
Bacillus subtilis, 472
Bacitracin zinc, 651
Bacteraemia, 203
    standards of therapy of, 202
Bacteraemic pneumococcal pneumonia, 203
Bacteria
    adaptation to exposure to antimicrobials
        used in medicine, food animals, crop
        production, 622

Bacteria *contd.*
    from animal origin, antimicrobial resistance
        in, 659
Bacterial bronchitis, 509
Bacterial endocarditis, management of, 27
Bacterial food borne illnesses, 621
Bacterial pneumonia, 212
Bacterial resistance
    to antibiotics, v
    to biocides and antibiotics, evidence of, 473–4
Bacterial sepsis and SIRS, difference
    between, 201
Bacterial spores, 470
Bangladesh, 611
Barriers to change in practice, 494
    identified, 522
    interventions oriented to, 17
Bayer, 215
Bed-days per patient managed (BDPM), 89
    surrogate marker of resource use, 51
Behaviorist theories, 18
Belarus, 571
Belgium, 202, 204, 210, 265, 268, 568,
    578, 679
    campaign to inform the public about
        antibiotic resistance, 555
    Committee for the Coordination of
        Antibiotic Policy (BCCAP), 554
    Hospital Management of Anti-Infectives
        training course, 244
Benalkonium chloride, 467
Benchmarking, 119–31
    different types of, 119–21
    ethics in, 125
    example of definitions of, 120
    implementing improvements, 125
    partners, 121–4
    planning for, 121
    with other antimicrobial utilisation data,
        150–4
    for reducing vancomycin use and
        vancomycin-resistant enterococci in
        US ICUs, 125
Benzylpenicillin, 155
    usage, 156–7
Beta-lactam antibacterials, 109
    penicillins, 109
Biguanide to treat urinary tract infections, 469
    resistant bacteria, 479
    resistance in *staphylococci*, 469
    use and antibiotic resistance, 465–80

Biocide/s in hospital environment, 479
  functions, 466
Biocide mechanisms of bacterial resistance,
  468–73
  to alternation of target sites, 472
  to biofilms, 473
  to degradation, 472
  impermeability, 470
  to multidrug efflux pumps, 471–2
  types of, 466–7
Bioterrorism, 593, 674
  acts, 678
Birmingham Accident Hospital, 5–6
Bisbiguanide, 467
  chlorhexidine, 466
Bisphenols, products containing, 477
BLAST (basic local alignment
  search tool), 441
Blood culture prior to antibiotic therapy, 47
BMJ, 496
Bolivia, study examining impact of
  unregulated antibiotics sales in, 599
Boston City Hospital, 8
Bovine mastitis in Uganda, 629
Box and Jenkins auto regressive integrated
  moving average (ARIMA) models, 451
  see also ARIMA models
Breakpoints
  clinical use of PK/PD to set for quinolones,
    399–401
  and different resistance rates, 396
  harmonization in Europe, 398
  short history and overview, 395–401
  SIR, 423
  systems, international similarities and
    differences in, 398
  two different types of, 399
  for variety of antimicrobials, 399
British HIV Association (BHIVA)
  guidelines, 337
British National Formulary, 340
British Society for Antimicrobial
  Chemotherapy (BSAC), 26, 160, 229,
    396, 551
  Endocarditis Guidelines, 27
  in the United Kingdom, 395
  surveillance projects for respiratory tract
    pathogens, 683
  Working Party, 169, 175
    literature search, 161

British Society for Antimicrobial
  Chemotherapy (BSAC), *contd.*
  Report, 228, 237
  survey, 167
  systematic review, 161
British Society for Medical Microbiology, 31
British Thoracic Society (BTS), 38
  community acquired pneumonia guidelines
    (BTS), 27–8
  guidelines, 38, 54
  Standards of Care Commmittee, 33
Broadness of spectrum, 204
  targets, 690
Bronchitis, 495
Brooklyn Antibiotic Resistance Task
  Force, 284
Brucella, 375
BSE outbreaks, 664
Bulgaria, 561
Burkholderia picketti, 468

C reactive protein (CRP), 202
Campylobacter jejuni, 627
Campylobacter spp, 583, 629, 638,
    653–5, 662
  with quinolone resistance, 703
Canada, 493, 510, 679
  amoxicillin, 573
  antibiotic policy changes, 600
  antibiotic use, 573–4
  antimicrobial resistance, 261
  audit, 202
  community-based parenteral anti-infective
    therapy or CoPAT, 241
  Drug Information Network, 573
  Information Medical Services (IMS), 572–3
  database, 585
Canadian Society of Hospital Pharmacists
  (CSHP), 243
Candida albicans, 311, 314–16, 321
  resistant, 213
Candida dubliniensis, 314, 316
Candida glabrata, 313–16
  infections, 312
Candida krusei, 213, 312, 314–15, 321
Candida parapsilosis, 312–3
Candida spp., 311–12, 320–21, 323
  infections, 312–13, 320
    in HIV-positive patients, 316
  resistance to triazole antifungals, 315

Candida tropicalis, 312–13, 315
Candidaemia, 27, 325
CAP *see* Community-acquired pneumonia
Carbapenems, 91, 139, 676
  against *Enterobacter* spp, 265
  in ICU, 149
  use, 145
Carbapenemases, 712
Carbenicillin, 689
Care pathways: definition, 39
  in management of CAP, 39
Caspofungin, 319–20, 322–3, 325
Catheterization, 469
Cationic biocides, 468
  selective pressure exerted by the use of, 478
Cattle drug usage for mastitis in Sweden and
    Norway, 656
CBA *see* Controlled before and after studies
CD4 count, 345
  in HIV infection, 344
CDC*see* Centres for Disease Control in the US
Cefaclor, 253
Cefazolin, 213
Cefepime, 86, 142
Cefotaxime, 86, 142, 204, 264, 373–4,
    377–8, 380
  breakpoints for *Enterobacteriaceae* in
      Europe and the United States, 398
  distribution of, 397
Ceftazidime, 82, 86, 93, 142, 164, 213, 264–5
  Enterobacter cloacae resistance to, 93
  use in hospital and percent ceftazidime-
      resistant/intermediate Gram-negative
      bacilli,
    monthly, 450
    temporal relationship, 453
    transfer function model, 455
    yearly, 449
Ceftriaxone, 142, 164
  3GC consumption and CDAD
      incidence, 459
  Australia, prescribing patterns for, 155
  policy implementation for usage, 156
Center for Disease Control and Prevention
    (CDC), United States, 99, 188, 274, 337,
    355, 410, 468, 491, 515, 679, 694
  active surveillance for invasive
      S. pneumoniae infections, 492
  guidelines include measures to prevent
      infection originating from
      environmental contamination, 353

Center for Disease Control and Prevention
    (CDC), United States *contd.*
  national campaign against inappropriate
      antibiotic use, 709
  Project ICARE, 126, 262
  Task Force, 679
Center for Research and Documentation on
    Health Economics (CREDES), 585
Centers for Medicare and Medicaid Services
    (CMS), 19
  performance measures, 20
Centre de Resource et de Documentation en
    Economie de la Santé, 585
Cephalexin, 213, 509
Cephaloridine, 7
Cephalosporins, 7, 140, 253, 178, 190, 207,
    256, 295, 579, 594, 600, 676–7, 707
  based antibiotic regimens, 51
  data on usage in South Australia, 142–8
  four "generations" of, 151
  in Greece, 572
  second generation, 706
  third generation, 77, 138–9, 549,
      603, 706
    antibiotics, 457
  use, total hospital, 143
Chemoprophylaxis, 25
Chile
  antibiotic consumption in, 594, 596
  national antibiotic policy, regulatory
      measures towards introduction
      of, 601
Chlamydia pneumoniae as an independent and
    co-infecting pathogen, 37
Chloramphenicol,
  resistance to, 704
    in staphylococci, 4
Chlorhexidine, 467–8, 477
  resistant *Proteus mirabilis*, 469
Chlorine
  based disinfectant, 359
  dioxide, 467
Chloroquine (CQ) resistance, 623
Cholera, 634
Cidofovir, 332–3
Cimulative sum (CUSUM), 458
CINAHL, 496
Ciprofloxacin, 5, 207, 213, 253, 369,
    380, 399
  breakpoints for Enterobacteriaceae in
      Europe and the United States, 398

Ciprofloxacin resistance, 265, 703
  in *Campylobacter*, 702
  in *E. coli*, 264
  in *K. pneumoniae*, 264
  in *P. aeruginosa*, 96
  usage, 98
Citrobacter, 469
Citrobacter freundii, 478
Civil wars and unrest in developing
    countries, 634
Cleaning
  definitions, 351
  or disinfection of environment, 355–60
  equipment, 358
  frequency of routine
    in isolation room, 356
    in nursing department, 356
    in operating department, 357
    in outpatients' clinic, 357
    in various rooms, 357
  methods, dry cleaning and wet
      cleaning, 355
  pre-disinfection, 358
Clinical breakpoints, 398–9
  value dependent on the dosing regimen, 394
Clinical diagnostic microbiology, 710
Clinical effectiveness cycle, 24
Clinical evaluation process, limitations of, 693
Clinical Governance, 539, 553
  importance of, 531
Clinical pharmacists
  with experience in infection
      management, 230
  in infection management, 239
  role in operational aspects of the AMT, 233
  in the United States, 176
Clinical pharmacy specialist, 236
Clinical Resources and Audit Group
    (CRAG), 32
Clinical standards, 29
Clonal spread, 713
  of resistance, 75
  role in worldwide spread of penicillin
      (and multiresistant) pneumococci, 704
Clonality and resistance in
    S. pneumoniae, 439
Clostridia, 653
Clostridium difficile, 359
Clostridium difficile-associated diarrhea
    (CDAD), 457

Cloxacillin, 3–4
Cluster-randomisation, 57
CMV *see* Cytomegalovirus
Coagulase-negative staphylococci (CoNS),
  resistance in ICUs, 265
  surveillance data, 411
Co-amoxiclav, 706
  MRS protective effect, 707
Cochrane Effective Practice and Organisation
    of Care (EPOC), 51, 57, 117, 161, 211,
    219, 229, 495
  Controlled Clinical Trials register, 496
  database of clinical trails, 161
Coercive theories, 19
"Collateral damage" related to antimicrobial
    use, 79
Combination
  antimicrobial therapy febrile
      neutropenia, 609
  of antibacterials, 109
  of interventions, 521
    effectiveness in reducing inappropriate
        antibiotic use, 517
Committee of Veterinary Medicinal Products
    of EMEA, 655
Commonwealth
  Department of Agriculture, Fisheries and
      Forestry, Australia, 134
  Department of Health and
      Aged Care, 134
  supported surveillance of antibiotic
      resistant organisms and antibiotic
      utilisation, 134
Communication
  around antimicrobial resistance in bacteria
      of animal origin, 664
  in risk evaluation process, 663
Community
  acquired bacterial infections, 509
  acquired MRSA, 354
  intervention study in Tennessee, 519
  outbreaks of hepatitis A and acute
      gastroenteritis, 351
  see also Community-acquired pneumonia
Community Strategy against Antimicrobial
    Resistance, 2001, 679
Community-acquired pneumonia (CAP),
    19–20, 37
  administration of appropriate intravenous
      antibiotics for patients with severe, 29

Community-acquired pneumonia (CAP) *contd.*
  antibiotic prescription in management
    of, 43
  audit in Tayside, 30
  BTS guidelines for, 27–8
  critical control points for, 43
  guidelines, 38
  healthcare costs in the United Kingdom and
    the United States, 37
  intervention, designing and
    implementing a, 56
  patients requiring hospitalisation, antibiotic
    prescribing in, 52
  process aspects of care in, 43–7
    impact on clinical and patient-centered
      outcomes, 48
    impact on economic outcomes, 49
    measures in QI, advantages of, 47
    QI studies published in English language
      medical literature, 40–2
Complex relationship between host,
  pathogens, and anti-infective agents, 197
Compliance with policies, measuring, 10
Computer
  information systems, 17
  program "Clinical-decision-support", 215
Computerized physician order entry
  (CPOE), 17
Concentration-time profile of emergence of
  resistance, 403
Consensus strategy, 26–7
Consumer education, 16
Consumption of antibiotics
  in ambulatory care, 252
  of penicillins, 258
  in the UMC Ljubljana, 258
  units of measurements, 567–8
Contaminated environmental surfaces, 353
Continuing Education, 539
Continuing Personal Development (CPD)
  hours of, 539
Continuing Professional Development, 539
Controlled before and after (CBA) studies,
  161, 496, 498, 501
Controlled clinical trials, (CCT), 161
Copenhagen recommendations, 679
Core interventions, 164
  in antibiotics
    automatic stop-order policy, 167–8
    control plan, 164–5
    formulary, 165

Core interventions *contd.*
  educational interventions, 175
  enforcing formulary restrictions, 166
  guidelines for antibiotic prescribing,
    168–74
  laboratory control and role of medial
    microbiologist/infectious diseases
    physician, 174
  role of hospital pharmacist, 176
Costs
  of antimicrobial resistance, 284
  of antimicrobial therapy, 203
  biological, and development of
    resistance, 632
  per capita of healthcare-associated *S. aureus*
    infection in Denmark and the
    Netherlands, 285
  ratios of microbiology tests, 293
  required to develop new antimicrobial
    agents, 284
  of resistance, vi
Co-trimoxazole, resistance to, 704
Counterfeiting of drugs in India and the
  Philippines, 604
Coxsackie B virus, 360
CPD *see* Continuing Personal Development
CREDES, Centre for Research and
  Documentation on Health
  Economics, 589
Critical success factors (CSFs)
  identifying, 121
  use of, 131
Croatia, 561
Cross-correlation function (CCF), 454
Cross-resistance bacterial resistance to
  biocides and antibiotics
  acquisition of genetic determinants, 476
  alteration of target sites, 475
  biofilms, 476
  changes in membrane permeability, 475
  evidence of, 473–7
  induction of multidrug efflux
    systems, 475
  possible mechanisms involved in, 474
Cross-transmission of antibiotic resistant
  clones, 271
Cryptococcus neoformans, 320
  infections, 320
  meningitis, 312
Cumulative sum of squares (CUSUMQ), 458
Cumulative sums (CUSUMs), 423

Cycle of behavioural change and appropriate
  education to manage response to illness
  action, 536
  contemplation, 536
  maintenance, 536
  pre-contemplation, 536
  preparation, 536
  sustained behaviour change, 536
Cycling, data supporting its efficacy, 179
Cystic fibrosis, 609
Cytomegalovirus (CMV), 333
  infections, 332
Czech Republic, 561
Czechoslovakia, 7

Daily review of antibiotic therapy, 707
Danish Integrated Antimicrobial Resistance
  Monitoring and Research Programme
  (DANMAP), 111, 115, 136, 150, 204, 220
  data, benchmarking with, 153
  see also Denmark
Daptomycin, 611
Data
  antimicrobials, 94
    usage, 77–81
  evaluated relationships between four
    measures of antimicrobial use for, 19
  patient-specific, 77
  sufficiency for categorization category, 201
  for surveillance of use and feedback to
    prescribers, 73
Data collection techniques
  abstracts of medical record, 214
  aggregate or "group level", 77
  antibiotic order forms, 214
  denominator, 110
  interviews, 214
  at patient level, 262–3
Data mining
  antibiotype approach to, 424
  to discover antimicrobic resistance
    brief literature review, 422–3
  emerging patterns of antimicrobic
    resistance, 421–44
  evolutionary genetic approaches, 436–40
  of large multinational databases,
    resistance, 422
  surveillance, 423
  multivariate analysis
    methods of, 424–36
    multidimensional scaling, 434–6

Data mining *contd.*
  resistance information, three approaches to,
    423–43
  studies on antibiotic use and drug resistance
    in hospital setting, 422
Data mining surveillance system (DMSS), 422
Davey, general advantages of antibiotic
  policy, 685
Day-care centers
  for child care outbreaks of multiresistant
    pneumococci, 705
  and nursing homes, clonal spread and, 713
Days of therapy, 90
DDD (Defined daily dose), 68–9, 77, 87, 90,
  93, 107, 124, 136, 216, 568, 585
  of antimicrobial per, 1,000 occupied
    bed-days, 452
  as in human medicine, 656
  as an international standard, 85
  method to measure and compare
    antimicrobial use, 84
  numerator, 110
  pseudo, 114–5
  as a unit of measurement, 70
DDD/1,000 inhabitants and day (DID),
  70, 138, 568–72
  Europe, 568–72
  European Union, 568–70
  for β-lactam antibiotics, 578
Deep-tissue mycoses, 312
Defined daily dose *see* DDD
Denmark, 204, 210, 212, 253, 255–6, 268,
  557, 569–70, 621, 661
  emergence of resistant *Staphylococcus
    aureus*, 557
  healthcare system, 558
  hospitals antimicrobials in, 558
  ICUs, 262
  Medical Product Agency, 568
  Medicines Agency, 557, 572
  patient's central personal registration
    (CPR), 557
  see also Danish Integrated Antimicrobial
    Resistance Monitoring and Research
    Programme (DANMAP)
Detergents, 358
  and disinfectant-use, comparing, 358
Developing countries (DCs), 617
  access to essential antibiotics at all levels of
    care, 601–11
  acute respiratory infections (ARI), 627

Developing countries (DCs) *contd.*
  AIDS epidemic, 626
  animal pathogens in veterinary medicine,
    resistance of, 629
  antibiotic policies in, 593–612
  antimicrobial resistance and its containment
    in, 617–39
    effectiveness and cost-effectiveness of
      measures, 607–10
    strategies for, 637–49
  bacterial pathogens, resistance, 623
  consumer awareness, 610
  deaths resulting from major killer
    diseases, 632
  donations of antimicrobials agents, impact
    on country AMR policies, 604
  drug resistance in hospital-acquired
    infections, 628
  economic implications of antimicrobial
    resistance, 631–3
  emergence of new resistant
    pathogens, 611–2
  enteric pathogens, 627
  epidemiology of antimicrobial
    resistance, 618
  excessive use of antimicrobials in, 604
  factors contributing to development and
    spread of resistance, 598, 633–7
  healthcare
    budgets, impact of antibiotic use on,
      598–601
    inadequate systems, 599, 633
    providers, 635
  HIV/AIDS
    implications on the antimicrobial
      resistance problem, 630–1
    resistance in, 626
  laboratory services, 627
  lack of proper diagnostic facilities, 600
  malaria, 622
    resistance in, 622–4
  prophylaxis agents *Pneumocystis carinii*
    pneumonia, 631
  substandard and counterfeit drugs, 598
  training of healthcare workers and other
    individuals, 610–1
  tuberculosis, 624
  use of antimicrobial agents, 620–2
DG Sanco of European Commission, 116
Diagnostic Laboratory, 710
  improving antibiotic prescribing, 710

Diagnostic uncertainty, issue of, 495
Diagnostic virology in clinical practice,
  impact of, 338
Diarrhea, 632–3
  surveillance data, 411
Dicloxacillin/flucloxacillin, 148
Didanosine, 333
Direct-Obsreved-Therapy Short Course
  (DOTS)-Plus, use of, 609
Diseases of poverty, 618
Diseases Pharmacist, 231
Disinfectants, 352, 358
  in farm and household products, 622
  in hospital environment, use of, 466
  susceptibility to, 360–1
  types, actions, and usages, 466–7
  usage and antibiotic resistance, 477–9
Disinfection
  definition, 352
  indications for, 358
  policies in hospitals and the community,
    351–61
Disoproxil, 333
Disposable materials, use of, 358
DNA
  automated sequencing technology, 436
  double helix structure of, 688
  extrachromosomal, 316
  high-density microarrays, 690
  replication and repair, changes in the
    fidelity of, 368
Doha, Qatar, Fourth Ministerial
  Conference
  Declaration on international intellectual
    property laws, 602
Dose/Dosage
  alternations, 235
  of antimicrobial drug, 205
  correct, 205
  interval, 206
Dose-effect relationships, 388–95
  concentration-time curves and the PK/PD
    index, 389
  determination of clinical breakpoints for
    susceptibility testing using PK/PD
    index, 392–3
  Monte Carlo simulations, 393–5
  sigmoid dose-response curve, 390–1
DOTS (Direct-Obsreved-Therapy Short
  Course) strategy, 625
Doxycycline usage, changes in, 157

Drug and Therapeutic Committee (DTCs), 608
  rational use of drugs, 608
Drugs
  expenditure of most hospitals, 227
  retailers in Kathmandu, Nepal, 606
Drug/metabolite transporter (DMT)
  superfamily, 471
Drug-resistant bacteria of animal origin, 621
Drug-resistant *S. pneumoniae* (DRSP), 93
Dual effective therapy (DET), 47–8
Dutch
  hospitals, formularies, 553
  infection control guidelines (WIP), 355
  see also Netherlands, the
Dysentery, 623

EBM *see* Evidence-based medical
Echinocandin antifungal family, 319–20
Ecological issues
  in antibiotic use and required actions, 701–13
  in hospital, 706–8
  use of antibiotics in community, 704–6
Education, 541
  benefits in clinical AMT intervention, 237
  effect on rates of penicillin-resistant
    S. pneumoniae in communities, 515
  example of intravenous vancomycin and
    fluoroquinolone prescribing
    practices, 236
  materials
    alone, effect on patients, 513
    distribution of, 500–1
  meeting, 505–6
    antibiotic prescribing behavior, 521
  outreach, 17
    academic detailing in, 17, 509
  pharmacist, role of, 239
  persuasive vs restrictive/coercive
    interventions, 163–4
  theories, 18
  visits, 507–8
Education of patients and professionals,
  531–42
  example in Scotland, 541–2
Education of patients, 532–6
  behaviour change, 535
  cycle of behavioural change to manage
    response to illness, 536
  other mechanisms, 535
  public expectations, 533

Education of professionals, 536–41
  medical and nursing, 239
  postgraduate, 538–40
  undergraduate, 537–8
Efavirenz, 333
Effective Practice and Organization of Care
  (EPOC) *see* Cochrane Collaboration
Egypt, prevalence of drug resistance of
  Salmonella typhi in, 628
EMBASE, 161, 496
EMEA *see* European Medicine Evaluation
  Agency
Emergence of bacterial resistance
  in clinical practice, 469
  to highly reactive biocides, 469
  multiresistant microorganisms, 460
Empiric antibiotic therapy, reasons for
  inadequate choice of, 198, 712
EMRSA-15, 281
EMRSA-16, 281
EMRSA-17, 281
Enoyl-acyl reductase carrier protein in
  mycobateria, 467
Enrofloxacin, 702
Enteric bacteria, resistance in DCs, 627
Enterobacter aerogenes, 94, 265
Enterobacter cloacae, 94, 164, 264
Enterobacter spp., 77, 204, 264–5, 629, 653
  as common nosocomial pathogen, 354
  resistance in ICUs, 264, 270
  in Belgium, high resistance to
    ciprofloxacin, 264
  surveillance data, 411
Enterococcus faecalis, 270, 619
Enterococcus faecium, 270, 619
  pathogenic, for humans, 661
Enterococcus spp, 76, 270, 354, 621, 629,
  653, 703
  surveillance data, 411
Environment of antibacterial resistance
  usage, discovery, and development,
    increasing and decreasing
    factors in, 697
Environmental contamination, 352–61
EPIC *see* European Prevalence of Infection in
  Intensive Care (EPIC)
Epidemiology, v
  of prescribing practices at a local level, 228
  theories, 18
  herds and food industry, studies in, 660

Epstein-Barr virus, 339
  infections, 332
Erythromycin, 4–5, 494, 704
  early restriction of, 4
  resistance, 93
  use only in combination, 3
ESAC *see* European Surveillance of Antibiotic
  Consumption
  project, collection of DID data in
    the, 579
Escherichia coli, 7, 94, 253, 264, 285, 372,
  374, 472, 619, 621, 654–5, 703
  isolates, high resistance rates to
    fluoroquinolones in, 251
  MICs from various sources,
    distribution of, 401
  resistance to fluoroquinolones, 583
  surveillance data, 411
ESCMID *see* European Society of Clinical
  Microbiology and Infectious Diseases
ESGAP *see* European Study Group on
  Antibiotic Polices
ESGAR *see* European Study Group on
  Antibiotic Resistance
Ethidium bromide in *S. aureus*, gene encoding
  for resistance to, 471
EUCAST *see* European Committee on
  Antimicrobial Susceptibility Testing
Europe, 242, 313
  antibiotic policies, need for uniform, 562
  antimicrobial resistance surveillance
    system, 546
  breakpoint systems used, 396
  development of antimicrobial policies and
    control measures in, 649
  guidelines for good clinical practice, 546
  ICUs, surveillance of antibiotic resistance
    in, 266–7
  regulations
    about feed additives, 651
    and guidelines for antimicrobials
      marketing authorisation, 652
  UTI prescribing studies, 594
  veterinary professional organization FVE,
    666
European Antibiotic Policy, minimum
  standards proposed, 546
European Commission (EC), 702
  Concerted Action project, 113
  DG Research, 116

European Committee on Antimicrobial
  Susceptibility Testing (EUCAST), 388,
  396, 398, 405, 712
  tentative breakpoints for quinolones, 400
European Conference on Antibiotic Use in
  Europe, 693–4
European Federation of Pharmaceutical
  Industries and Associations (EFPIA),
  588, 694
European Food Safety Authority (EFSA),
  659, 667
European Medicine Evaluation Agency
  (EMEA), 653, 679
European Pharmaceutical Market Research
  Association
  classification of medicinal products,
    106, 108
European Prevalence of Infection in Intensive
  Care (EPIC), 263
  study, 268
European Society for Clinical Microbiology
  and Infectious Diseases (ESCMID),
  115, 601
  website, 111
European Society of Clinical Pharmacy
  (ESCP), 243
European Strategy for Antibiotic Prophylaxis
  (ESAP) study, 262
  list of antibiotics subject for restricted use,
    271
European Study Group on Antibiotic Polices
  (ESGAP), 111, 115, 547, 601
European Study Group on Antibiotic
  Resistance (ESGAR), 549, 563
European Surveillance of Antibiotic
  Consumption (ESAC), 88, 150, 563, 570,
  584, 706
  project, 116
European Union (EU), 679
  antibiotics to increase the feed efficacy and
    daily growth of animals, 650
  Conference on "the Microbial Threat", 230
  task forces, 695
Evaluation
  by experts, 216–7
  of infectious diseases service, 210
  of evaluation process itself, 211
  QI initiative, 59
  of quality of prescription without
    intervention, 210

Evidence
  based practice, 531
  as basis for clinical practice guidelines or
    antibiotic prescribing, 32
  systematic and analytic evaluation of, 23
Evidence-based medicine (EBM), 24
  approach, 28
  guidelines, 28
    methodology, 32
  methodology, 26
  working group, 24
Evolution of antibiotic resistance within
    patients, 367–83
  case studies, 377
    infant with aplastic anaemia, 378
    long-standing *E. coli* infection of liver
      cysts, 379–82
Evolutionary genetics, 444
  approaches in data mining, 423
Expenditures, antimicrobial, 83
Experts for quality evaluation of antimicrobial
    therapy, 216–7
Extended-spectrum β-lactamase (ESBL),
    93, 270
  detection of, 712
  producers, 703
    strains amongst *E.coli*, 264

Famciclovir, 333
Farms
  epidemiological studies in, 665
  management and hygiene, good, 638
Farquhar systematic review, 28
Febrile neutropenia, 324
Federal Drug Agency (FDA), 678–9
Feed additives
  antimicrobials used as, 659
    benefit of, 660–1
Filaments
  formation of in Gram-negative bacteria, 467
  in fungi, 313
Finland, 557, 568, 570, 679
  Group A streptococci, resistance of, 93
    reduced use of erythromycin for, 501
  intervention for lowering antimicrobial
    resistance in the community, 559
  macrolide study, 522
  Medical Society Duodecim, 560
  MIKSTRA, optimal antibiotic policy for
    outpatient infections, 559

Finland *contd.*
  Ministry of Social Affairs and Health, 560
  Study Group for Antimicrobial
    Resistance, 560
Flavophospholipol, 651, 667
Flinders Medical Centre, 154
Florida Medical Quality Assurance, Inc., 38
Flucloxacillin, 4
Fluconazole, 207, 214, 311, 315, 318, 322–5
Flucytosine, 318, 320
Fluoroquinolones, 37, 53, 77, 91, 94, 139,
    256, 369, 676, 690
  potency against Mycobacterium, 370
  resistance, 93
    in Gram-negative bacilli in
      gastrointestinal flora, 92
  third generation, 138
  usage, 152
Focus Technologies, 683
Food and Drug Administration (FDA), 599
Food animals,
  antibiotics, 655
  antimicrobial usage in, 620, 638, 655
    impact on public health, 620
    monitoring of, 656
    production in developing
      countries, 637
Food borne agents, 650
  pathogens development of resistance related
    to veterinary usage, 650
Formularies as educational tool and guideline
    or protocol, 539
Foscarnet, 333
France, 202, 204, 210, 212–3, 264, 268, 396,
    432–3, 568, 578, 655–6, 679
  acute otitis media, 586
  AFSSAPS Health Products Safety Agency
    (Agency Française de Sécurité
    Sanitaire et des Produits de Santé),
    584, 588–9
  database, 585
  antibiotic use in community, 583–91
    data sources, 583–5
    units of measurement, 585–6
  breakpoint system used (CASFM), 396
  childhood infection-prevention
    programme, 590
  data
    EPAS, Permanent Sample of Health
      Insured Individuals, 589

France *contd.*
  sources on antibiotic consumption,
    characteristics of different, 589
  determinants of antibiotic consumption,
    586–8
  medical density, 587
  pharmaceutical firms pressure, 588
  sociocultural and historical factors, 587
  drug economy, 557
  EDSSM (Enquête Décennale sur la Senté et
    les Soins Médicaux)
  database, evaluations of antibiotic
    consumption using the, 584
  Decennial Inquiry on Health and Medical
    Care, 583, 589
  and Germany, comparison in terms of
    outpatient antibiotic use, 556
  healthcare centers (Direction des Etudes
    Médico-Economiques et de
    l'Information Scientifique, 588
  INSEE, (National Institute for Statistics
    and Economic Studies), 589
  Institute for Public Health Surveillance (In
    VS, Institut de Veille Sanitaire), 588
  MSA, Mutualité Sociale Agricole, 585
  National Agency for Medical Practice
    Evaluation (Agence Nationale pour
    le Développement de l'Evaluation
    Médicale), 588
  National Employees' Health Insurance,
    Caisse Nationnale d' Assurance
    Maladie des Travailleurs Salariés
    (CNAMTS), 585, 589
  National Independent Workers' Health
    Insurance, Caisse Nationnale
    d'Assurance Maladie des
    Professions Indépendantes
    (CANAM), 585, 589
  national plan of actions for the control of
    antimicrobial resistance, 556
  National Report on the use of antibiotics in
    the community, Observatoire National
    des Prescriptions et Consommations
    des Médicaments, 588
  public health authorities, collective
    awakening and progressive
    mobilization of, 588–90

Fungal diseases/infections, 313–14
  invasive, standards of care for patients
    with, 31

Fungal diseases/infections *contd.*
  in neutropenia, 322
  prevention of, 322
  resistance development and
    transmission, 315
Fusarium spp., 312, 314–5
Fusidic acid, 4–5
*The Future of Antimicrobial Drug Availability,
  An Impending Crisis*, 677

Galen, 23
Ganciclovir, 333
Garrod, three eras of antimicrobial
  chemotherapy, 674
Generalized estimating equation (GEE), 499
Genes
  encoding for resistance determinants, 476
  transfer, v
Genetics
  basis and the phenotypic expression of
    resistance, 619
  of resistance mechanisms, 704
Genital *Chlamydia trachomatis* infection,
  management of, 28
Genomics
  driven approach to antimicrobial
    compound, 690
  and high throughput screening (HTS), 688
  role in antimicrobial discovery, 688
  technologies related to, 689
Genotyping isolates, 434
Gentamicin, 5, 7, 213
  resistant gram-negative bacilli and
    gentamicin use, percentage, 449
Germany, 212, 265, 268, 561, 568,
    605, 655, 709
  antibiotics, 203
  DIN breakpoint system396
  health authorities, 557
  ICU patients, 63
  strict control of veterinary prescription
    and drug delivery in animal
    husbandry, 666
Global coalition agent poverty, 638
Global infectious disease burden, 593
Global mobilization of international
  communities against antimicrobial
  resistance, 650
Global *Salmonella* survey, 655
Global warning systems, 611
Glutaraldehyde, 467

Glycopeptides, 139, 149, 151, 256, 690
  intermediate staphylococci, 712
  resistance
    in CoNS, 268
    in ORSA, 269
  use, 145–6
Glycopeptide-intermediate *S. aureus*
  (GISA), 282
Glycylcycline, 690
GOBSAT *see* Good Old Boys Sat Around a
  Table
"Gold-standard" methodology
  of performance measurement, 29, 31
  for research-based evaluations of
    organisation level interventions, 57
Gonorrhoea, 623
Good Drug Donation Practices, 605
Good Old Boys Sat Around a Table
  (GOBSAT), 26, 32
Gould review of role of antibiotic policies in
  control of antibiotic resistance, 685
Gram-negative aerobic infections, 76
Gram-positive and Gram-negative organisms,
  206, 377, 468, 477
  antimicrobials, 96
  bacteria, multiple, 411
Great Britain *see* United Kingdom
Greece, 268, 568, 578, 713
  National Committee for Nosocomial
    Infections guide for antimicrobial
    chemotherapy and prophylaxis, 549
Group A *streptococci*, 559
  Finland, resistance of, 93, 501–2
  macrolide resistance, 492–3, 518, 559
  susceptible to penicillin, 204
  see also Streptococci
Group educational meetings, results from
  studies of, 506
Guatemala, antimicrobial resistance
  in middle and upper class families due to
    unrestricted access to
    antibiotics, 601
Guidelines
  for antibiotic prescribing
    essential features of, 169–71
    dissemination, 171
    evaluation, 173
    implementation, 171–3
  development
    approaches to, 26–7
    methodology, recommended, 25–6

Guidelines *contd.*
  of the EMEA, 667
  for improving use of antimicrobial agents in
    hospitals, 26
    in NHS Scotland, 32
    on the web, 25
  for Prevention of Antimicrobial Resistance
    in Hospitals, 160
  for prudent use of antibiotics, 271
Guideline implementation, 15–21, 28–9
  generally effective strategies, 17
  ineffective strategies generally, 15
  management of resistant pneumococci in
    day care centres and family day
    care, 559
  theories of facilitating change, 18–21
  variably effective strategies, 16

HAART *see* Highly active anti-retroviral
  therapy
HACCP principles *see* Hazard analysis and
  critical control point
Haematology departments, 212
Haemophilus influenzae, 424, 438,
  472, 609
  antimicrobial resistance surveillance
    efforts, 607
  DNA sequence of, 689
Haemophilus spp., surveillance data, 411
HAI *see* Hospital-acquired infection
Halogenated compounds, 352
Halogen-releasing agents, 467
The Hammersmith Hospital, 3
Hand hygiene, 273
Handwash products, 290
Harvard Medical School, 209
  emergency department quality study,
    128–31
Hazard analysis and critical control point
  (HACCP) principles,
  development of, 664
  risk reduction following, 664
Hazard identification, 658
HBV Hepatitis B virus
HCV *see* Hepatitis C virus
Health
  databases, 493
  education campaigns, evaluation of, 542
  facilities, 634
  institutions in DCs, 636
Health Protection Agency (HPA), 244

Healthcare
  acquired infection standards, evaluation of
    national performance against, 30
  delivery systems, 636
  facility, 212
  provider education, 515
HealthSTAR, 161, 496
Hepatitis A virus, 360
Hepatitis B virus (HBV), 333, 354
Hepatitis C virus (HCV), 332–3, 354
  combination of interferon and ribavirin, 342
  in environment, data on survival of, 355
Herd level, information about drug
    prescription at, 657
Herpes family, 332–4
  Cytomegalovirus (CMV), 333, 336
  Herpes simplex virus (HSV), 332, 335–6,
    340–1, 344, 360
  genital, 337
  Varicella zoster virus (VZV), 332–3, 339
  in pregnancy, 337
Highly active anti-retroviral therapy
    (HAART), 312, 314, 342, 626
  in HIV, 345
  see also Anti-retroviral (ARV)
Histoplasmosis, 312
HIV, 354
  Africans infected, 631
  combination therapy, 342
  in environment, data on survival of, 355
  infections, 626
HIV/AIDS, 593, 609, 623, 632–3
  infected persons, increased use of
    antibiotics in, 631
  related bacterial infection, 678
  see also AIDS
Holland, 713
  see also Netherlands
Homeland security, 593
Hong Kong, 432–3
Hormone administration to calves, 664
Hospital
  accreditation, 10
  antimicrobial consumption data, 71
    collecting, converting, and making sense
      of, 67–74
    frequency of data collection, 71
  drug use
    measurement units, 69–71
    sources of data, 68–9

Hospital *contd.*
  pharmacists pharmacists, 217
    in North America, UK, and
      Australasia, 232
Hospital-acquired infection (HAI), 282
  benefits of handwashing in the
    control of, 289
  cost of, 284
  due to MRSA, 288
  due to resistant organisms, economic
    evaluation, 285
  with multiresistant bacteria, 351
Housekeeping genes, 438, 441
HSV *see* Herpes
Human Genome Project, 688
Human Genome Sequencing Consortium, 441
Human immunodeficiency virus *see* HIV
Human parainfluenza virus, 359
Human population growth in developing
    countries, 635
Hungary, 561, 571
Hydrogen peroxide, 467
Hypochlorite, 359

Iceland, 559–60
  penicillin resistance
    in pneumococci, 560
    related to high co-trimoxazole and
      macrolide use, 705
Icelandic Medical Journal, 560
ICUs (Intensive Care Units), 212, 261–74
  antibiotic cycline and their role in reducing
    resistance, 272
  cephalosporin use, 143
  fluoroquinolone use, 146
  as genesis units for antibiotic resistant
    bacteria, 706
  gentamicin, usage of, 149
  and other hospital wards, studies of
    comparison of antibiotic use in, 263
  impact of antibiotic policies and antibiotic
    consumption on antibiotic resistance,
    270–2
  infections
    acquired in, 271
    control, 273
  interventions aimed at controlling the use of
    antibiotics, 273
  IT and benchmarking to improve antibiotic
    prescribing, 272

ICUs (Intensive Care Units) *contd.*
  macrolide use, 148
  monthly aggregate antimicrobial usage, 139
  problem of resistance in, 261
  surveillance programme, 273
  usage, separate rates for, 135
  use by antimicrobial class, total, 141
  vancomycin, use of, 127, 149
ID *see* Infectious Diseases
IDSA *see* Infectious Diseases Society of
    America
Illegal distribution
  of prescription drugs, 602
  by unsanctioned dealers of antibiotics, 603
Imipenem, 77, 86, 213
  resistant *Acinetobacter baumannii*, 461
Immunisation, 25
Immunocompromised patients, 367
Imperial College Faculty of Medicine
    (London), 244
Implementation of antibiotic management
    programmes, effective, 229
Inanimate environment, critical role in
    transmission of organisms, 353
Inappropriate antibiotic prescriptions, 202, 245
  reducing number of, 708
  for viral respiratory tract infections,
    underlying reasons for, 494
Inappropriate antimicrobial use, 635
Incidence studies, 210
Indinavir, 333
Indonesia, 594, 610
Industrialized countries, 600
  guidelines for hospital and veterinary
    use of antibiotics, 600
Infection
  community worldwide, 26
  control, 25
  practice guidelines, variability of
    quality of, 27–8
  prevention, vi
  with resistant pathogens, 284
Infection control
  measures, 90
    effectiveness of, 273
  programmes, 284, 299
  staff, 292
Infection management for pharmacists
  information networking and specialised
    practice interest groups, 243–4

Infection management for pharmacists *contd.*
  post-graduate training opportunities in, 244
  training and support in, 242
Infectious Disease (ID)
  pharmacist, 238
    Microbiology or Antibiotic Use Review
      (AUR), 237
    practice, 244
  physician or clinical microbiologist, 230–1
Infectious Disease Society of America
    (IDSA), 26, 76, 202, 237, 679, 695
  guidelines, 230
    for managing candidiasis, 28
  Joint Committee on the Prevention of
    Antimicrobial Resistance, 27
  newsletter, 677
Influenza virus infection, 332–3
Information Sharing, Feedback, Contracting,
    and Ongoing Monitoring
    (INFECTOM), 610
Inhabitant-days, 89
INRUD (International Network for the
    Rational Use of Drugs), 496
INSPEAR, 611
Insurance, health, 253–4, 259, 510
  additional, 252
  database, 500
  French national, 585, 589
Intellectual property protection, 604
Intensive Care and Special Care Baby Units, 283
Intensive care unit *see* ICUs
Intercontinental Marketing Services (IMS), 106
Intercontinental Medical Service,
    Melbourne, 572
Interferon, 333
International Federation of Pharmaceutical
    Manufacturers Association (IFPMA), 694
Internet, 535, 560
Interrupted time series (ITS), 161–2, 501
  analysis, 57
  model, 58
  studies, 496, 498, 500
Interscience Conference on Antimicrobial
    Agents and Chemotherapy (ICAAC), 675
Interventions, 163–81
  effects
    on antibiotic resistance, 518–9
    on patient outcomes, 519–20
    on rate of penicillin resistance, effect of, 518
  studies, 194

Interventions to improve antibiotic
  prescribing
 audit and feedback with or without other
   educational materials, 497, 502–4
 barriers are amenable to change, 495
 in the community, 491–524
 computer-assisted decision support, 180
 discussion
  of included studies, 500–17
  of results and implications for practice,
    520–4
 distribution of educational materials, 497
 educational group meetings or seminars,
    497, 504
 educational outreach/academic detailing,
    497, 507–10
 financial interventions, 498
 financial/healthcare system changes,
    510–11
 literature review, methods of, 495–8
 local consensus processes, 497
 local opinion leaders, 497
 marketing, 497
 mass media, 497
 multifaceted interventions, 515, 17
 outcome measures, 180–1
 patient mediated interventions, 497
 patient-based interventions, 513–15
 problems with interpretation of published
    studies, 498–500
 reminders, 497, 511–13
Interventions to optimise antibiotic
  prescribing in hospitals, 177
combination therapy, 178
 "cycling" (rotation), 179
intravenous (iv)-oral switch therapy, 177
 "streamlining", 177
 therapeutic substitution, 178
 UK approach, 159–81
Intravenous (IV) antibiotics, 20
 to oral switching (sequential therapy), 235
Invasive bacterial infections in young
  children, 495
Invasive fungal disease, 311
 epidemiology of, 311–11
 patients at risk of, 322
Inventory, 634
Investment in R&D for new antibacterials, 696
 see also R & D
Iodine, 352, 473

Ireland, 203, 568
 SARI, Strategy for the Control of
   Antimicrobial Resistance in
   Ireland, 550
 see also United Kingdom
Isoniazid (INH), 467, 625
Isoxazolyl penicillins, 282
Israel, 202–4
IT systems adapted to needs of AMT, 231
Italy, 202, 269, 568
Itraconazole, 318, 320, 322–4
ITS *see* interrupted time-series

Japan, 432–4, 710
Joint British Society for Antimicrobial
   Chemotherapy/hospital infection
   society working party on optimising
   antibiotic prescribing in hospitals
   (working party), 160
Joint Commission on Accreditation of
   Healthcare Organizations (JCAHO), 19
 measures, 20
 performance measures, 20
Joint Commission on Accreditation of
   Hospitals, 167
Judicious antimicrobial use for pediatric upper
   respiratory infections, principles of, 515

Kanamycin, 7
Kenya, 621, 625
Ketolides, 676
Killer diseases, major occurrence in world
   and proportion in developing
   countries, 633
Klebsiella pneumoniae, 94, 264
 ceftazidime-resistant, 93
 infection, 5
Kunin, original criteria of, 199

Laboratory
 costs as part of an MRSA screening
   programme, 293
 methods for susceptibility testing,
   standardisation of, 712
 virology provision, good, 338
β-lactamases, 476
 enzymes, small concentration differences
   in, 373
 inhibitor combinations in Denmark, 151
 TEM-1, evolution in vitro, 371–2

β-lactams, 94, 369–70, 425, 706
 and aminoglycoside combination, 178
 fluctuating pressure of, altering pattern of
  evolution, 372
 resistance to, 704
Lamivudine, 333
Lancet, 496
Latin America, 269, 313, 599, 602
 antimicrobial resistance, 261
Latvia, 570
LDS pharmacy database, 215
Levofloxacin, 49, 86, 399
Liberia, 618
Lincomycin, 5
Lincosamides, 109
Linezolid, 296
Lipopeptide, 676, 690
Listeria outbreaks, 664
Livestock producers, 660
Local guideline adherence, barriers to, 194
Local health organisations websites, 535
Local hospital guideline on antibiotic
  prophylaxis, 191
Local opinion leaders and local consensus
  conferences, use of, 16
Local susceptibility patterns, 165
Lopinavir, 333
Lower respiratory tract infection, 593
 management of, 28
Luxembourg, 568–9

Macrolides, 109, 140, 150, 256, 282, 370,
  425, 579, 676, 706
 antibiotics, substitution for
  fluoroquinolones, 511
 in Austria, Finland, and Germany, 572
 and Group A streptococcal infections,
  effect of reduction in consumption, 518
 in Finland, 521
 resistant, 493
 use, 93, 705
 resistance to, 704
 third generation, 138
Major facilitator superfamily (MFS), 471
Maki and Schuna's study on administration of
  antimicrobial drugs, 204
Malaria, 593, 609, 623–4, 632–4
Malnutrition and hunger in developing
  countries, 634
Mannheima haemolytica, 654

Mannheimia, 655
Market authorization for feed additives, 651–3
Marketing theories, 18
Mathematical modeling
 of rates of change of antibiotic resistance
  among bacteria, 523
 of relationships between antimicrobial use
  and resistance, 90
Maximum Residue Levels (MRL) for drugs
 maximum acceptable level of antibiotics in
  animal products, 622
MDR *see* Multidrug-resistance
Measurements of antibiotic consumption
 relate to antibiotic resistance, 75–98
 antimicrobial usage, 77
 resistance, relationships between, 89–97
Mechanisms of bacterial resistance to
 antibiotics, 472, 705
Media, 509
Medical and surgical use of
 immunosuppressive procedures, 312
Medical Letter, 218
Medical Research Council (MRC)
 framework, 51
Medicare, 19
Medicated soaps, 271
Medicinal products categories, 252
Medicines Crossing Borders Project, 605
MEDLINE, 161, 411, 496
Med-Vet-Net, major veterinary and human
 health institute, 667
Meningitis, surveillance data, 411
Mercuric chloride, 466
Methicillin (oxacillin), 3–5, 689
 resistance, 4, 251, 285
Methicillin-resistant *Staphylococcus aureus*
 (MRSA), 126, 202, 251, 261, 270, 281,
  360, 612, 628, 706–7
 acquisition of hospital hygiene and the
  "superbug", 300
 antibiotic consumption in the Intensive Care
  Unit, 295
 bacteriaemia, 712
 communty/long-term care, 298–9
 control measures, debate on the type and
  extent of, 299
 costs
  antibiotic, 294–6
  of eradicating, 287
  general, 285–7

Methicillin-resistant *Staphylococcus aureus*
 (MRSA) *contd.*
  hospital-acquired infection, 282–4
  real, 281–300
  screening/surveillance cultures, 287
   specific, 287–99
  cross-infection, 297
  eradication, 288
  gentamycin, susceptible to, 712
  handwashing, 289–90
  hospital cleaning, control of, 296
  in various Scottish hospitals, 460
  infections, 353, 710
   aggressive containment strategies, 298
   intensive care units, 297–8
   isolates, proportion, 292
   isolating, cohorting, and contact isolation,
    290–2
  laboratory costs, 292–4
  outbreaks, 115, 296–7
  prevalence, 127
  reservoir, early identification of the, 288
  savings from infection control policies
    and programmes in use of
    antibiotics, 294
  screening methods, 293
  strains of, 296
   resistance to triclosan, 474
  surveillance data, 411
Methicillin-sensitive *Staphylococcus aureus*
 (MSSA), 282, 479
  infections, 285–6
Methylglyoxal (MG), 375
MIC *see* Minimum Inhibitory Concentration
Micafungin, 319, 321
  in Japan, 320
Microbial and genetic aspects of resistance,
  research into the, 708
Microbial ecology of hospitals, 706
"The Microbial Threat", conference, 409
Microbiological outcome, emergence of
  resistance, spread of resistance, 208–9
Middle East, 602
Minimum Inhibitory Concentration (MIC) of
 the drug, 205, 388
  data, 427–8
  frequency distributions, 421
  for β-lactams, 377
  as measure of susceptibility of
    microorganism to drug, 388

Minimum Inhibitory Concentration (MIC) of
 the drug *contd.*
  measurement wild-type population
    distribution, 395
  testing, reproducibility and accuracy of, 395
Model
  standards setting, 31
  on theory of behavioural change, six stages
    of subject change, 535
Moldova, Republic of
  antibiotics in, 561
  DRUGS, 561
  training programs for physicians in an
    attempt to change antibiotic
    prescribing, 561
Molecular mechanisms of antibiotic
  resistance, 382
Monitoring
  antibiotic use, 116–17
  of patients prescribed the targeted drugs, 167
  programmes
    of antimicrobial resistance for *E. coli* and
      Enterococcus faecium, 655
    of veterinary pathogens, zoonotic
      bacteria, and indicator bacteria, 654
    and surveillance, antimicrobial resistance,
      654–5
Monnet, ABC Calc *see* Antibiotic
  consumption calculator
Moraxella catarrhalis, surveillance data, 411
Moving average (MA) control charts, 423
Moxifloxacin, 399
MPC *see* Mutant prevention concentration
MRSA *see* Methicillin-resistant
  Staphylococcus aureus
MSSA *see* Methicillin-sensitive
  Staphylococcus aureus
Multidimensional scaling (MDS), 435
Multidisciplinary antimicrobial management
  teams, 230–7
  application working to improve
    antimicrobial prescribing, 227
  and role of pharmacist in management of
    infection, 227–45
Multidisciplinary approach to antimicrobial
  prescribing practice, 237, 244
Multidisciplinary or integrated care
  networks, 227
Multidrug and toxic compound extrusion
  (MATE) family, 471

Multidrug efflux systems, 475
  transporters in fungi, 315
Multidrug resistance in bacteria, 465
  Acinetobacter anitratus species, emergence
    of in neonatal and paediatric intensive
    care units in Durban, South Africa, 608
  P. aeruginosa and *Acinetobacter
    baumannii*, 611
  tuberculosis (MDR-TB), 626
    cost of treating, 632
    in HIV-infected patients, 625
Multifaceted interventions and reductions, 18
  changes in prescribing, 524
  community, trial in Alaska, 519
  in incidence of antibiotic-resistant
    bacteria, 524
  overall antibiotic use in study
    communities, 518
Multilevel or hierarchical regression, 499
Multiple antibiotic-resistant Gram-negative
    bacilli, 360
  as nosocomial pathogens, 354
Multiple locus sequence typing
    (MLST), 437
Multivariate analysis, 444
  in data mining, 423
Mutation prevention concentration (MPC),
    401–2
  for an antibiotic, 370
  definition, 371
  major shortcoming of, 402
  and mutant selection window, 370–1
  one-step, 190
Mutators
  bacteria, 375
  isolates, 376
  persistence of, 376–7
  strains, 374–7
    in development of antibiotic resistance,
      role of, 377
Myanmar, 604
Mycobacterium chelonae, 469
Mycobacterium smegmatis, 469, 472
Mycobacterium tuberculosis, 359, 472, 491,
    623, 630
Mycoplasma genitalium, 689
Mycoses, 312
  diagnosed invasive, rational
    therapy of, 325
  resistant to antimycobaterial drugs, 625

Nafcillin, 86
National Antibiotic Committees
  at ministerial level in Russia, Czech
    Republic, and Slovak Republic, 561
National Antibiotic Formulary
  Czech Republic, Slovak Republic, and
    Russia, 561
National Antibiotic Policies, 546–9
  influence on antibiotic prescribing, 545–62
  practice in Central and Eastern Europe,
    560–2
National Antibiotic Policies, practice in
    Northern Europe, 550–60
  ambulatory care, 252
  antibiotic use in, 553
  Belgium, 554–5
  Denmark, 557–8
  Finland, 559–60
  France, 555–7
  Germany, 556–7
  Greece, 549
  Iceland, 560
    practice in Ireland, National Disease
      Surveillance Centre, 550, 565
  Italy, 550
  practice of, 547–62
  Scandinavia, 557–8
  Southern Europe, 547–50
  Sweden, 558–9
  UK, 550–2
National Committee for Clinical Laboratory
    Standards (NCCLS) in United States,
    388, 396
National Corporation of Swedish
    Pharmacies, 572
National Health Service (NHS) of
    United Kingdom, 160, 237
  Management Executive, 34
  clinical guidelines supported by, 24
  quality improvement Scotland, 30
  reforms of, 553
National Institute of Clinical Excellence of the
    United Kingdom, 25, 337
National Institute of Health (NIH),
    United States, 694
National Institute of Pharmacy, Medical
    and Pharmaceutical University of
    Moldova, 561
National Nosocomial Infections Surveillance
    (NNIS) system, United States, 99, 354

National Pathway Association of the
United Kingdom, 39
National Programme to Preserve
Antimicrobial Efficacy in France, 590
National Report on the Control of Bacterial
Resistance in France, 588
National Staphylococcal Reference
Laboratory of the United Kingdom, 4
Natural catastrophies in developing
countries, 634
Nei and Li's index of nucleotide diversity, 438
Neisseria gonorrheae, 491, 623
antimicrobial resistance surveillance
efforts, 607
Nelfinavir, 333
Neomycin, 4
Nepal, 607
Netherlands, the, 194, 202–4, 206–7, 210,
256, 264, 271, 504, 568, 570, 579, 656
antibiotic use in the, 533
breakpoint systems used, 396
multicenter audit in hospitals of adherence
to guidelines for surgical
chemoprophylaxis, 554
CRG, 396, 398
national guidelines for use of antibiotics in
hospitals, 554
surveys on antimicrobial use in hospital
setting, 228
SWAB (Stichting Werkgroep
Antibioticabeleid–Foundation
Antibiotics Policy Work Group),
189, 554
Nucleotide diversity index, 438
Nevirapine, 333
New drugs
agents and immunomodulators, 676
antibacterials currently in late-stage
development, 690
see also R & D
NHS *see* National Health Service, UK
Non-human use, information on problem of
antimicrobial resistance in, 666
Norway, 268, 557, 570
breakpoint system used (NWGA), 396
general practitioners, 503
study, 522
Norwegian Medical Depot (NMD), 85, 108
Nosocomial infections, 351, 353, 609, 623
resistance in DCs, 628
Nova Scotia, 510

Novobiocin, 3–4
Nuffield Institute for Health, 24, 34
Nystatin, 323

Occupied bed-days (OBDs), 136
Ofloxacin, 399
OIE, 654–5
ad hoc group recommendations and the
precautionary principle, 661
Online prescription writer, 511
Oral antibiotics, 20
administration, 206
Organic acid, 352
Organic food, 663
Organisation of Care checklist, 51
Organizational theories, 18
ORSA *see* Oxacillin-resistant *S. aureus*
Oseltamivir, 333
Otho-phthalaldehyde, 467
Otitis media in children, 491, 507–9, 511–19
computer point-of-care evidence on optimal
duration of antibiotics, 511
delayed prescriptions for, 523
study, 520
Outbreaks
acute gastroenteritis, 352
of hepatitis A, 352
of multiresistant *Salmonella* spp., 703
Outcomes
parameters, 198–209
and process of care measures for clinical
CAP research or QI, 44–5
Outcome Measures for Arthritis Clinical Trials
(OMERACT), 59
Outpatient or home parenteral anti-infective
therapy (OHPAT), UK
national guidelines, 241
Overconsumption of antimicrobial
drugs, 202
Over-the-counter (OTC) antibiotics, 601
OVID, 161, 496
Oxacillin, 603
resistance, 266
Oxacillin-resistant *S. aureus* (ORSA)
strains, 268
prevalence, 274
resistant to multiple drugs, 269
Oxapenems, 676
Oxazolidines, 676
Oxazolidinones, 370, 611
Oxidising agents, 467

Pan-American Health Organization, 610
Pan-European survey demand index for
    antibiotic prescription, 556
Paper-based decision support tool for
    diagnosis and management of
    sore throat, 512
Parainfluenza virus, 353
Passive education, 229
    efforts, 15–16
Pasteurella multocida, 654–5
Paternalistic model of professional
    practice, 531
Patient
    administration records, 81
    with bacteraemia, 212
    centered healthcare, 16
    and community education, 515
    current knowledge, 532–3
    information leaflet regarding antibiotics for
        acute bronchitis, 513
    mediated interventions, 514
    outcomes based on, 208, 522
        variables as indicators of quality-of-use
            in intervention studies, 209
    prescription profiles as a source of data, 68
    and public as partners in healthcare, 531
    quality of life, 50
    specific data, 78–9
Patient-days, 89
PDL modelling *see* Polynomial distributed lag
    modelling
Peer review audits, 219
Penciclovir, 333
Penem, 690
Penicillins, 1, 4–5, 140, 148, 152, 252, 494,
    584, 600, 677
    assay of, 2
    extended-spectrum, 572
    marneffei infection, 312
    and multidrug resistant pneumococcal
        infection, 38
    narrow-spectrum, 578
        in Denmark and Sweden, 572
    resistance, 522, 704
    sensitive microbes, 1
    usage, total hospital and ICU, 142
Penicillin G, 710
    development of, 674
Penicillin-resistant pneumococci, 206, 437,
    523
    Streptococcus pneumoniae, 583, 612

Peracetic acid, 467
Peroxide, 352
Peroxygen compounds, 466
Personnel for data collection, 216
Peru, 611
    Lima study on use of antibiotics to treat
        diarrhoea, 600
Pharmaceutical companies/industry
    ability to demonstrate clinical efficacy
        against emerging resistant
        organisms, 692
    antibacterials
        development, labeling, and benefits, 674,
            690–3
        discovery and genomics, 688–90
        resistance and policies, 678–84
    approach to antibiotic polices, 673–97
    clinical development of antibacterials key
        points of concern, 691
    and the "crisis", 675–8
    discovery, development, and
        commercialization in the face of
        policies, 687–90
    in the future, 693–6
    integrated activities of discovering,
        developing, and marketing
        antibacterials, 673
    involvement and funding in surveillance, 682
    policies
        guidelines, and education on antibacterial
            use, 684
        and strategies to combat resistance,
            potential consequences of, 681
    at present, 675–93
    R&D (research and development)
        of antibacterials, 674; collaboration, 697
        and commercialization of therapies, 678
        role of, 693
    top 12 leaders in the field of
        antibacterials, 677
Pharmacists, 176, 240
    in adult immunisation programmes, 242
    antimicrobial control programmes, led
        by, 234
    audit functions, 240
    in formulary development, 242
    implementation of streamlining and
        sequential therapy, 241
    infection
        control activities, 241
        management, role in, 237–42

Pharmacists *contd.*
  overseeing appropriate drug prescription
    and usage, role in, vi, 245
  and prescribing errors, 240
  resources and support networks, 243
  specialist advice and education, 239–40
Pharmacodynamics
  dosing schedules, 707
  impact on dosing schedules, 387–405
  measures, relationships between antibiotic
    concentration and bactericidal effects
    and post-antibiotic effects, 370
  and emergence of resistance, 401
Pharmacokinetics (PK)
  consultation services, formal, 239
  of drug in animals, 654
  estimates of antimicrobial exposure, 87
  importance of, 2
  and pharmacodynamics
    of antibiotics, 369
  properties through distribution, 690
Pharmacokinetic/pharmacodynamic (PK/PD)
    index, 390, 401–4
  concentration time-curves and, 389–90
  mutation, 401–4
  values and emergence of resistance,
    relationship between, 404
Pharmacy
  databases, 500
  deliveries to wards or units as source of
    data, 68
  drug surveillance networks, formal, 240
  monitoring services, 240
  undergraduates, 538
Pharyngitis, 495
Phenolics, 360, 466–8
Philippines, pharmaceutical market, 604
Physicians
  advice against need for antibiotic for viral
    infections, 514
  education and patient materials in the office
    and sent to homes, 517
  reminders, 511
    effect on altering prescribing
      behavior, 513
Piperacillin/tazobactam, 86, 92, 148,
    213, 265
Plasmodium falciparum, 623
  clinical failure rates to chloroquine (CQ) and
    sulfadoxine-pyrimethamine (SP), 624
  see also Malaria

Pleconaril, 339
PLS (projection to latent structures)
  modeling antibiotic MIC or antibiotype
    patterns, 432
Pneumocystis jiroveci pneumonia, 312, 314
Pneumonia, 623
  guidelines
    for hospital and ventilator-associated, 26
    in practice, 37–61
  severe bacteraemic pneumococcal, 47
Pneumonia Outcomes Research Team (PORT)
    study, 50
Pneumonia Severity Index (PSI), 37, 43, 49
Poland, 561, 571
Policies
  in controlling antibacterial use, 696
  in developing countries, 636
  negatively impacting pharmaceutical
    industries' investment in
    antibacterials, 697
  practical applications of, 8–10
Polyenes, 318
Polymerase chain reaction (PCR), 293, 338,
    436
Polynomial distributed lag (PDL) modeling,
    458
  third-generation cephalosporin (3GC)
    consumption, 459
Population
  increases and urbanization, 635
  studies on antibiotic use in community,
    567
Portugal, 268, 568
Posaconazole, 319, 321
Postal terrorist attacks, 688
Post-graduate training in anti-infective
    therapy, 233
Poverty in developing countries, 599, 618,
    622, 633
Povidone-iodine, 467, 473
Practice guideline programmes, 29
Pre-printed antibiotic prescription/order form,
    167
Prescribed daily doses (PDD), 86, 216
Prescribing practices
  altering, 229
  understanding, 228–9
Prescriptions/1,000 Inhabitants and Year,
    572–4
  European Union, 572
  United States, 572–3

Presumed viral respiratory tract infection
(PVRTI), 586
Prevalence or incidence studies, 209
Prevention of selection of resistant strains by
combination therapy, 707
Primary resistance surveillance data, 423
Principal components analysis (PCA), data
modeled, 426
Process outcomes
prescribing behaviour, 198–207
definition, 198
Professionals
selection of, 536–7
as teachers
for patients, 540
of professionals, 540
Prognostic and clinical-decision support
tools, 37
Programmes
conducting surveillance of antimicrobial
consumption in Europe and the United
States, 150
to manage or control antimicrobial use,
228
Projection to latent structures (PLS) modeling,
432
Prolongation of hospital stay as measurable
index, 287
Prophylaxis, 198
for abdominal surgery, 186
antibiotics given for a short duration, 189
to cover microorganisms causing infection,
189
against fungal disease, rational, 322–3
in neutropenia, choice of agent for, 323
penicillin, use of, 2
in practice, surveying the quality of
antibiotic, 190–1
in surgery, auditing and improving the
quality of antibiotic, 190–4
Protein synthesis inhibitors, 676
Proteus mirabilis, 94
Pseudomonas, 76–7
Pseudomonas aeruginosa, 91, 93–4, 164,
264–5, 358, 361, 375–6, 448, 468–9, 472,
475, 607
amikacin-resistant, 460
study of piperacillin-tazobactam
resistant, 92
surveillance data, 411
and third-generation cephalosporin, 448

Pseudomonas stutzeri, 475
biguanides, resistance to, 469
chlorhexidine resistance in some strains of,
360
Public education
health education leaflets, 533
methods of, 533–5
Public health perspective, risk of spread of
resistant virus, 337
Publication of research findings, 15
PubMed, 161
Pulse field gel electrophoresis (PFGE), 437
PVL gene, 299
Pyrimidine analogue, 318

QACs *see* Quaternary ammonium compounds
QI *see* Quality improvement
Qualitative audit, 218
Quality
assurance and development of standards,
29–30
of empiric therapy and antimicrobial
prophylaxis, 197
evaluation criteria of antimicrobial therapy,
201
measurement, audits of practice, 198
of prescribing of antimicrobial drugs by
audits, criteria to evaluate, 199
of use assessment of antimicrobial drug
prescriptions
flow chart for, 192
of vancomycin, 210
Quality improvement (QI)
activities, monitoring and benchmarking, 73
initiatives for CAP, 37, 54
clinical outcomes, 50
development of, 61
economic outcomes, 51
evidence on improved outcomes,
49–51
evidence on improved process of care,
48–9
long-term surveillance, 60
patient-based outcomes, 50
interventions
impact on patient-centred outcomes, 39
linking process of care to outcomes in,
43–7
studied in CAP, 38–43
Quantification of drug use, 67
Quantitative audit, 218

Quantitative measurement of antibiotic use, 105–17
Quasi-randomized clinical trials (QRCT), 498, 506
Quaternary ammonium, 352, 360
Quaternary ammonium compounds (QACs), 361, 466–8, 474
Quinolones, 178, 425, 579, 677, 706
    antibacterials, 109
    resistance, 265, 704
        among *P. aeruginosa*, 265

R&D
    for antibacterials, 695
    "crisis", reasons for, 678
    cost for single new antimicrobial, 612
    Scare Care, 695
Ramirez criteria, 20
Randomized controlled trials, (RCT), 161, 211, 496, 498, 501
Rational use of antibiotics, 509
    in ICU, 298
    WHO guidelines for viral respiratory tract infection and acute diarrhea, 504
Ravuconazole, 319, 321
Reduction
    in macrolide use in Finland, 518
    use of oral cephalosporin, cephalexin, 501
    use of penicillin with repeated small group meetings, 506
    use of specific classes of antibiotics, 509
Relationships
    between antibiotic usage and resistance, 76, 620
    between countries and isolates, 432
Relative return on investment (ROI) or net present value (NPV) of antibacterials, 687
Reminders
    to healthcare providers, 17–18, 28, 497, 511–13
Report of International Collaborative Study on Antimicrobial Susceptiblity Testing, 395
Research
    activity relating to CAP, 37
    based pharma, 674
    design, 499
    on determinants of antibiotic abuse, 708
Research and development *see* R&D
Resistance Alert Thresholds (RAT), 607

Resistance
    among β-lactams, 434
    consecutive measurements of levels, 452
    of *E. cloacae* to ceftazidime, 264
    Enterobacter strains, 204
    of *Escherichia coli* to triclosan, 469
    to fluoroquinolones in mycobacteria, 371
    to heavy metals, 472
    to hydrogen peroxide and peracetic acid, 470
    of *Mycobacterium tuberculosis* to antibiotics, 624
    to non-quinolone agents, 432
    of *Pseudomonas aeruginosa* to fluoroquinolones, 251
    selection among the microbial community, 369
    of *Streptococcus pyogenes*, 251
    Staphylococcus aureus starins, 395
    strains, growth of, v
    via horizontal gene transfer, 476
    zoonotic organisms, 629
Resistance-nodulation-division (RND) family, 471
Respiratory synscytial virus, (RSV), 333, 353
Respiratory tract infections, 575–7, 579
    resistance in developing countries, 627
    surveillance data, 411
Respiratory viruses, 353
Reviews
    of antimicrobial formulary, 231
    process for antibiotic therapy, 707
Rhinoviruses, 332, 353
Ribavirin, 332–3
    for Lassa fever, 332
Rifampicin, (RMP), 625
Risk analysis, principles, 658–60
Risk assessment, 658
    Process, four components, 659
    avoidance action, 661
Risk management, 658
Ritonavir, 333
RNA viruses, 335
Rotavirus, 360
    transmission of, 359
Routine cleaning of environmental surfaces, 358
Royal College of General Practitioners, UK, 539
Royal College of Nursing, UK, 539

Royal College of Obstetrics and Gynaecology, UK, guidelines, 337
Royal Pharmaceutical Society of Great Britain, 538–9
Rural hospital antimicrobial utilisation surveillance network, 135
Russia, 561, 571
   antimicrobial resistance, 261
   cities, study to inventory antibiotics with non-medical population, 599
   Institute of Antimicrobial Chemotherapy Smolensk, 599

Safety of intravenous (IV) to oral switch therapy, 37
Sales data from wholesalers to hospital pharmacies, 69
Salinomycin, 651, 667
Salmonella spp, 375, 491, 583, 627, 629, 631, 637–8, 653–5, 662
   surveillance data, 411
Salmonella typhi, 623
   in Egypt, 627
Sanford Guide to Antimicrobial Therapy, 218
Saquinavir, 333
SARS (Severe Acute Respiratory Syndrome) virus, genome of, 688
Scandinavia, 204, 702
   antimicrobial resistance, 261
   consumption of antibacterials, 557
Scedosporium
Scedosporium spp., 312, 314–15
   infections, 325
SCOPE, 706
Scotland, 552
   Consensus Statement on Qualitative Research in Primary Health Care, 55
   Deans Medical Curriculum group, 538
   development
      of clinical standards in, 30–2
      methodology, 27
   guidelines on acute sore throat and tonsillectomy, 38
   implementation of, 32
   Medical Schools, 538
   study of public's awareness of issues around antibiotic use, 532
The Scottish Doctor, 538
Scottish Infection Standards and Strategy Group (SISS), 707
   good practice guidance statements, 31

Scottish Intercollegiate Guideline Network (SIGN), 24–6, 35, 169, 539
Scrip, 604, 614
Selection pressure
   determining rate and extent of emergence of AR, 619
   on commensal bacteria, 619
Selective decontamination of the digestive tract (SDD), 262, 707
   use of, 271
Self help leaflets, 534
Self-medication, 635
Semmelweis, 466
SENTRY Antimicrobial Surveillance Program, 285, 313
Sepsis by Gram-negative bacteria, 609
Serratia marcescens, 94, 468
Service providers in developing countries, 635
Severity prediction rules, 43
SF-36, 50
SHEA *see* Society for Healthcare Epidemiology of America
Shigella dysentariae, 623
Shigella spp, 583, 607, 627
Simple audits or intervention audits
   "Case Review" or "Recommendations audit", 210
Simple statistical techniques, 499
Simple, single intervention studies, 521
Single effective therapy (SET), 47
Sinusitis, 495
Sir Charles Gairdner Hospital (SCGH), 457
SIREVA-Vigía project, 610
SISS *see* Scottish Infection Standards and Strategy Group
Slovak Republic (Slovakia) 432–4, 561, 571
Slovenia, 252, 261
   antibiotic policy
      experiences, 251
      hospital, 254–7
      national, 252–7
      in tertiary care centre, 257
   antibiotic resistance, 251
   consumption of antibiotics in, 258
   decrease in total use of antibiotics, 258
   hospital
      antibiotic policy, 254–7
      consumption of antibacterials for systemic use, 255
      therapeutic committee (TC), 254

Slovenia *contd.*
    use of antibacterials in, 256
    in the ESAC (European Surveillance
        of Antimicrobial Consumption)
        project, 255
    pattern of antimicrobial use in 2001, 254
    penicillins, 256
    reimbursement, 251
    restricted antimicrobials, 258
    total consumption of antibacterials in, 253,
        255
Slovenia University Medical Centre Ljubljana
    (UMC), 254–5, 257
    utilization of some problem antibiotics in,
        257
SMAC *see* Standing Medical Advisory
    Committee
Small multidrug resistance (SMR)
    family, 471
Small soluble proteins (SASPs), 470
Social influence theories, 18
Society for Healthcare Epidemiology of
    America (SHEA), 27, 76
    and Infectious Diseases Society of America
        Joint Committee on Prevention of
        Antimicrobial Resistance, 160
Society of Infectious Diseases of America
    see Infectious Disease Society of
    America (IDSA)
Society of Infectious Diseases Pharmacists
    (SIDP), US, 243
Socrates, 23
Sodium dichloroisocyanurate (NaDCC), 467
Sodium hypochlorite, 360, 467
Sodium monensin, 651
Sosa's study of physician prescribing practices
    in seven countries in Latin America and
    the Caribbean, 608
South East Thames Regional Health Authority,
    United Kingdom, 7
Southern European countries, 570
    antimicrobial resistance, 261
Spain, 204, 265, 568, 578, 655, 713
    effect of primary care reform in a region of,
        510
    ICUs, 263
    problem of antimicrobial resistance, 547
    recommendations of task force, 547–9
Specialist advice on aspects of antimicrobial
    use, 239

Spiramycin, 651
Sri Lanka, 504
St Thomas' Hospital, 5, 8
Standard treatment guidelines (STGs), 598
Standing Medical Advisory Committee report
    (SMAC), United Kingdom, 532, 539
    Sub-group on Antimicrobial Resistance,
        160
Staphylococcus aureus, 75, 186, 285, 377,
    472, 623, 661
    antibiotic resistance in ICUs, 268–9
    bacteraemias, treatment of, 218
    infections, 3
        medical costs of, 286
    isolates, 8
    resistance
        genes, 471
        β-lactam, 474
        to methicillin, levels in, 127, 258, 286
    surgical site infection, 286
    surveillance data, 411
    trends in antibiotic resistance of, 5
    see also Methicillin-resistant *S. aureus*
        (MRSA); Methicillin-sensitive
        S. aureus (MSSA); Oxacillin-resistant
        S. aureus (ORSA) *and* Vancomycin-
        resistant *S. aureus* (VRSA)
Staphylococcus spp., 76, 469, 654–5
    infection, diminished use of streptomycin
        and tetracycline for treatment, 4
Statistical parsimony networks, 439–40
Stavudine, 333
Strategies to implement antimicrobial policies,
    229
Streamlining or narrowing of empirical
    therapy, 235
Streptococcus pneumoniae, 38, 251, 253, 424,
    438, 491, 609, 623, 627, 631, 704
    antimicrobial resistance, 610
        surveillance efforts, 607
    development of antibiotic resistance in, 650
    genetic diversity, 438
    horizontal transfer of resistance loci in,
        440–3
    housekeeping genes, 438
        divergence typical of, 442
    isolates, multivariate analysis to Alexander
        Project collection, 425
    and questions of clonal spread, 437–40
    resistance to fluoroquinolones, 93

Streptococcus pneumoniae contd.
  strains by treatment with penicillin, 370
  surveillance
    data, 411
    network, 610
Streptococcus pyogenes, 704
Streptococcus spp., 654–5
Streptogramins, 109, 611
Streptomycin, 4–5
  on fruit trees, 701
Structured drug interruption, 342
Studies
  to compare DDD values to PDD values, 86
  of delayed antibiotics in common cold, 520
  of dynamics of pneumococcal carriage
      during antibiotic treatment, 492
  of outpatient antibiotic use and prevalence
      of antibiotic resistant pneumococci,
      709
  partners, 124
  of prescribing for acute bronchitis in adults,
      513
Sub-Saharan Africa, 638
  HIV/AIDS, 638
  malaria-induced fever, 622, 624
  prevalence of MRSA in, 629
  tuberculosis in, 626
  see also Africa
Substandard antibiotics, 602
Substitution of macrolide antibiotics for
      fluoroquinolones, 511
Subtherapeutic doses, 604
Sulfadoxine-pyrimethamine (SP), 623
Sulfamethoxazole, 369
Sulfonamides, 2, 109, 253
Surfaces contaminated by blood, 359
Surgery and antibiotic prophylaxis
  guidelines, 28
  historical background, 185–6
  improving prescribing in, 185–94
  perioperative, procedures for, 188
  pilot investigation of postoperative patients,
      191
  principles of, 186–90
  quality assessment, 193–4
  state of the art of, 185–90
  see also antibiotic prophylaxis
Surgical wounds
  classification, 187
  infections, occurrence of, 186

Surveillance programs of antimicrobial
      resistance, 419
  acceptability, 416
  of antibiotic usage and resistance, 8
  and benchmarking, experiences with
      antimicrobial utilisation, 133–57
  choice of specimens, 418
  critical evaluation of, 412
  data
    to collect, 418
    types and meaningful indicators for
        reporting antimicrobial resistance,
        409–20
  definition of, 410
  denominator, 419
  design, 416
  expression of data, 418
  goal of, 606
  method, choice of, 418
  microbiological representativeness, 418
  model of, 418
  numerator for, 418
  practical aspects of the implementation of
      the, 417
  predictive positive value, 416
  quality control program, 419
  sensitivity, 416
  for antimicrobial resistant bacteria,
      monitoring, 659
Surveillance system
  flexibility, 413
  integrating within institution, 67
  representativeness, 413
  simplicity of, 412
  timeliness, 414
  usefulness, 414–16
Surveys
  of consultant microbiologists and hospital
      pharmacists in the United Kingdom,
      169
  of hospitals in United States participating in
      Project ICARE, 169
Susceptibility
  of antibiotic-susceptible and antibiotic-
      resistant hospital bacteria, study, 361
  of *Enterobacter*, 93
  of non-fermenting Gram-negative rods, 712
Susceptible-intermediate-resistant (SIR) in
      antimicrobial susceptibility testing,
      421–4

SWAB *see* Netherlands
Swann report on additives in feeding-stuffs, 651
Sweden, 253, 255–6, 265, 268, 504, 558, 569–70, 679
  breakpoint systems used, 396
  ICUs, 263
  study of decision process for antibiotic prescriptions in, 272
  Institute for Medical Statistics (IMS), 568
  National Corporation of Swedish Pharmacies, 568
  STRAMA (Swedish Strategic Program for the Rational Use of Antimicrobial Agents and Surveillance of Resistance), 255, 558
Swedish Reference Group for Antibiotics (SRGA) 395–6, 558
Switzerland, 202, 210, 268
  paediatric ICU, 263
Synercid, 296, 611
Synthesis of proteins, 472

Tanzania, 611
Targets and discovery methods, 676
Tayside Community-Acquired Pneumonia Project (TAYCAPP), 52–4
  improving management of CAP, 51
  model of micro-determinants of delivery and appropriateness of antibiotic therapy, 56
  pathway based on, 46
  pre-implementation audit of process of care measures, 52
TB see Tuberculosis
TCS computer program, 439–40, 442
Team approach to improve antibiotic turnaround times, 237
Teicoplanin use, 149
TEM-1 β-lactamase *see* lactamase
Tenofovir, 333
Tetracyclines, 4–5, 109, 253, 509
  on fruit trees, 701
  resistance to, 704
  third generation, 138
Thailand, drug expenditure in, 594
Therapeutic drug monitoring programmes, 239
Therapeutic prescriptions, 202
Therapeutic substitution, challenges of, 178
Therapy and prophylaxis, unnecessarily prolonged prescriptions for, 167

Third/fourth generation cephalosporins *see* cephalosporin
Ticarcillin/clavulanate, 148
Time series analysis
  antibiotic prescribing and resistance rates, 159
  to antibiotic resistance and consumption data
    application of, 447–61
    from Denmark, 461
    deterministic and stochastic processes, 450–1
    evaluation of interventions to control antibiotic resistance, 457–60
    evolution of resistance
      example to predict short-term, 455–7
      techniques to study, 460
    examples of application of, 453
    modeling stochastic processes using, 451–2
    study of antimicrobial consumption and bacterial resistance relationship, 452–5
Timing, appropriate category, 207
"Tomorrow's Doctors", 537–8
Toronto Declaration, 2000, 679
Toxicity/allergy, 203
Training in antibiotic use in animals, 665
Transmission
  of blood-borne viruses, risks of, 354
  of hepatitis B virus (HBV) in renal dialysis units, 332
  of resistance, 452
Treatment failure, rates of, 284
Triazoles, 318
  antifungal agents, 311
  new, 322
Triclosan, 290, 361, 467, 472, 477
Trimethoprim, 109, 369
Trimethoprim/sulfamethoxasol (TMP/SMX), 94, 252, 258
  for prophylaxis in HIV/AIDS patients, 631
TRIPS (trade-related aspects of intellectual property rights) 602
Tuberculosis (TB), 593, 623, 632
  control, standard international method of, 625
  combined antimicrobial drugs for the treatment of, 609
  and HIV/AIDS, interaction of, 630
  resistance in developing countries, 624–6
  treatment costs, 632
Tufts Center for the Study for Drug Development (CSDD), 612

Turkey, 265
Tylosin phosphate, 651
Typhoid, 623

Uganda, 611
  AIDS patients, 626
  clinical failure rates, 624
  free medical care to population, 636
  resistance patterns of bacteria isolates from
    bovine clinical mastitis in, 630
UK *see* United Kingdom
UK Clinical Pharmacy Association (UKCPA),
  243
UKNEQAS, 712
United Kingdom (UK), 166, 176, 178, 202,
  206, 256, 568, 655–6, 679
  antibiotic policy changes, 600
  Antimicrobial Resistance Strategy, 32
    and Action Plan, 35
  breakpoint systems used, 396
  British Thoracic Society guidelines, 706
  Clinical Pharmacy Association, 242
  community consumption, 105
  Drugs and Therapeutics Committee, 164
  guidelines
    evidence-based, 24
    infection, type and sources of, 24–5
    methodology and standards of care,
      23–33
  HIV drug resistance, monitoring
    transmission of, 337
  Hospital Infection Society, 160, 229
  hospitals
    antibiotic consumption, 105
    Working Party Report, 551
  House of Lords
    recommendations about development of
      resistance, 684
    Select Committee on Science and
      Technology Report, 160, 228, 552
  medicine, human use per capita in, 703
  model for healthcare reform, 227
  overall profile to antibiotic use by drug
    class, 594
  National Institute of Clinical Excellence
    (NICE), 337
  part-time MSc in Infection Management for
    pharmacists, 244
  strategy to tackle problem of resistance, 552
  therapeutic guidelines, 25
  undergraduate medical training, 537

United States, iv, 178, 212–13, 285, 313,
  432–4, 605, 679
  administration, 206
  agriculture, antibiotic use by weight, 701
  antibiotic policy changes, 600
  antimicrobial resistance, 261
  annual cost of healthcare system, 75
  azithromycin/clarithromycin, quinolones, 573
  breakpoint systems used, 396
  Centers for Disease Control *see* Centers for
    Disease Control
  community-based parenteral anti-infective
    therapy or CoPAT, 241
  Federal Interagency Task Force on
    Anti-microbial Resistance, 605
  guidelines on virology issues, 337
  hospital-based AMR studies, 607
  ICARE see Centers for Disease Control
  identification of the MRSA reservoir, 289
  mortality records, 312
  National Ambulatory Medical Care Surveys
    (NAMCS), 493
  ORSA rates in *S. aureus*, 269
  Pharmacopoeia, 561
US Department of Agriculture (USDA), and
  CDC, 694
Upper respiratory tract infections in children,
  treatment of, 650
Urinary tract infection (UTI), 377, 504, 509,
  600
  adult, management of, 28
  surveillance data, 411
  in women, 503
Uropathogenic strains, 377
Uruguay, regulatory measures towards
  introduction of national antibiotic policy,
  601
Usage rates, 135
  for intensive care units (ICUs), 134
Use of antibiotics, 573
  in ambulatory care from data collection by
    ESAC, 570
  data, normalization of, 88–9
  for growth promotion, 701
  possible causes for observed variations in,
    578–9
UTI *see* Urinary tract infection

Vaccination, 713
  campaigns against infectious diseases, 242
  for *S. pneumoniae* and *H. influenzae*, 609

Vaccine preventable disease, 332
Valaciclovir, 333
Valganciclovir, 333–4
Vancomycin, 7, 86, 213, 295, 703
    Continuation Form, 215
    dosing, 162
    usage in this study, 128
Vancomycin-resistant enterococci (VRE), 76,
        91–2, 126, 162, 270, 274, 354, 360,
        611–12, 707
    of animal origin, 703
    and avoparcin use, 661, 703
    selection by avoparcin and streptogramin
        resistance by virginiamycin, 650
Vancomycin-resistant *S. aureus* (VRSA), 269,
        611
Varicella zoster virus (VZV), 332–3
    infection, 339
Venezuela, regulatory measures towards
        introduction of a national antibiotic
        policy, 601
Verbal Case Review (VCR), 610
Veterinarians, 665
Veterinary drugs, 662–3, 666
    antibiotic use for prophylaxis and therapy,
        702
    benefits, 662
    process of harmonisation of risk-benefit
        assessment for, 653
    residue, risk assessment of, 654
    risks, 662–3
VetStat, data on consumption of prescription
        medicines, 657
Vibrio cholerae, 627
    antimicrobial resistance surveillance efforts,
        607
    surveillance data, 411
Viral infections
    and their management, 331
    of respiratory tract, 353
ViResiST
    microbiology data, 456
    pharmacy data, 456
    project, 115, 117
    resistance surveillance using time series
        analysis techniques, 115, 456
Virginiamycin, 650–1
Viruses
    mutation, 335
    transmitted from environmental surfaces, 359

Voluntary restriction of selected antibiotics,
        164
Voriconazole, 319–20, 322–4
VRE see Vancomycin-resistant enterococci
VZV *see* Varicella zoster virus

Ward pharmacy, development of, 9
Ward-based pharmacists, 240
Water-borne infections, 635
West Midlands General Practice Research
        Database, United Kingdom, 574
WHO (World Health Organisation), 50, 70,
        104, 136, 216, 409, 417, 547, 679, 695
    Collaborating Centre for Drug Statistics
        Methodology, 72, 85, 108, 111, 118,
        136, 567, 581, 585
    website, 110
    Defined Daily Doses (DDD) per 1,000
        patient-days, 262
    Drug Utilisation Research Group (DURG),
        108
    Global Strategy for Containment of
        Antimicrobial Resistance (2001),
        679
    Integrated Management of Children
        Illnesses (IMCI), 611
    International Working Group for Drug
        Statistics, 85
    recommendations for drug utilisation
        research, 108–11
Wild-type cut-off and clinical breakpoints,
        396, 398–9, 401
Withdrawal/withholding of antibiotics
    for certain conditions, 522
    scheduled, 179
World Health Organisation *see* WHO, 50,
        70, 104, 136, 216, 409, 417, 547,
        679, 695

Yersinia spp., 621, 637, 662

Zaire, 634
Zalcitabine, 333
Zanamivir, 333
Zidovudine, 333
Zoonosis, development of network of
        excellence for, 667
Zoonotic bacteria, 638, 654
Zygomycota, 312, 314–15
Zyvox, 611b